D1717422

*Edited by*
*Klaus Wandelt*

**Surface and Interface Science**

*Surface and Interface Science*

**Edited by Klaus Wandelt**

Volume 1: Concepts and Methods

Volume 2: Properties of Elemental Surfaces

Print ISBN 978-3-527-41156-6

oBook ISBN 978-3-527-68053-5 (Volume 1)

oBook ISBN 978-3-527-68054-2 (Volume 2)

Volume 3: Properties of Composite Surfaces: Alloys, Compounds, Semiconductors

Volume 4: Solid-Solid Interfaces and Thin Films

Print ISBN 978-3-527-41157-3

oBook ISBN 978-3-527-68055-9 (Volume 3)

oBook ISBN 978-3-527-68056-6 (Volume 4)

Volume 5: Solid-Gas Interfaces I

Volume 6: Solid-Gas Interfaces II

Print ISBN 978-3-527-41158-0

oBook ISBN 978-3-527-68057-3 (Volume 5)

oBook ISBN 978-3-527-68058-0 (Volume 6)

Volume 7: Solid-Liquid and Biological Interfaces

Volume 8: Applications of Surface Science

Print ISBN 978-3-527-41159-7

oBook ISBN 978-3-527-68059-7 (Volume 7)

oBook ISBN 978-3-527-68060-3 (Volume 8)

*Edited by Klaus Wandelt*

# Surface and Interface Science

Volume 5: Solid-Gas Interfaces I

**The Editor**
***Prof. Dr. Klaus Wandelt***
University of Bonn
Institute for Physical and
Theoretical Chemistry
Germany
k.wandelt@uni-bonn.de

**Cover Picture:**
Design by Klaus Wandelt and
Spiesz Design, Neu Ulm

Pictures by
Fraunhofer IFAM Dresden

Florian Mittendorfer, Universität Wien



**Library of Congress Card No.:** applied for

**British Library Cataloguing-in-Publication Data**
A catalogue record for this book is available from the British Library.

**Bibliographic information published by the Deutsche Nationalbibliothek**
The Deutsche Nationalbibliothek lists this publication in the Deutsche Nationalbibliografie; detailed bibliographic data are available on the Internet at <http://dnb.d-nb.de>.



**Print ISBN:** 978-3-527-41158-0
**oBook ISBN:** 978-3-527-68057-3

**Cover Design** Grafik-Design Schulz, Fußgönheim, Germany
**Typesetting** SPi Global, Chennai, India
**Printing and Binding** Markono Print Media Pte Ltd, Singapore

Printed on acid-free paper

# Contents

# Preface

Surfaces and interfaces shape our world in two senses. On the one hand, they structure our world and make it so diverse and beautiful. On the other hand, surfaces and interfaces are locations of gradients. These gradients drive spontaneous and man-controlled processes, which affect our living conditions. Living behind a coastal dike makes you care about its stability all your live, or move. Heterogeneous catalysis of chemical reactions at solid surfaces has contributed to the explosion of the human population. The physics of interfaces in artificial electronic nanostructures is just, in a revolutionary way, changing our communication behavior and by that, our social life. Our body functions by processes at and through interfaces of membranes, which in turn can be influenced by traces of drugs. It is, thus, a great scientific challenge to investigate the properties of surfaces and interfaces, and it even appears to be a necessity of vital importance for our future to understand the processes occurring at them and to make wise use of them.

Although theoretical predictions about properties of surfaces as well as intuitive models of surface processes existed much earlier, modern experimental surface science started about 40 years ago with the commercial availability of ultrahigh vacuum (UHV) technology. Under UHV conditions, it was possible to prepare clean surfaces and to develop and apply a number of methods based on particles beams. Unlike photon beams, as for instance, used in X-ray crystallography, electron, ion, and atom beams interact only with the outermost layers of a solid and therefore provide information that pertains only to the surface. While in the beginning, practical surface investigations were concentrated on the *changes* of surface properties due to exposure to gases or vapors, it soon turned out that the properties of the bare surfaces *themselves* posed a lot of scientific surprises.

Now, 40 years later, the so-called reductionist "surface science approach", that is, the use of well-defined, clean single-crystal surfaces under UHV conditions, enables a microscopic and spectroscopic characterization of these bare surfaces atom by atom. The achievements of this research may ultimately be summarized by the general statement: Surfaces are a different state of matter! Likewise, nowadays, it is possible not only to study the interaction of individual atoms and molecules with a surface but also to manipulate them on the surface according to our will.

The present series of books aims at giving a broad overview of the present state of understanding of the physics and chemistry peculiar to surfaces. This account not only reflects the "success story" of surface science but also becomes more and more important for a number of other disciplines and technologies that increasingly rely on the established knowledge about surfaces. These are the science of composite and low-dimensional materials including nanoscience and nanotechnology, heterogeneous catalysis in gaseous and liquid phases, electrochemistry, and biology, to name only some. The intention of this series of books is not only to give an introduction to those who enter the field of surface research but also to provide an overview for those whose work needs conceptual and analytical input from surface science. Emphasis is placed on the results of the basic physics and chemistry of surfaces and interfaces. The most important experimental and theoretical methods that led to these results are grouped in classes and described to an extent so that the reader may just gain confidence in "what surface scientists are able do": more detailed descriptions of these methods can be found in existing publications.

The vast material is presented in eight volumes and nearly hundred chapters and is structured according to increasing complexity of the systems in question. Each chapter is written by experts of the respective subject and is supposed to start with an introduction of the basic phenomenon, to develop the problem from simple to more specific examples, and to end "wherever appropriate" with the identification of open questions and challenges for future research.

When starting this project, the first volume was planned to describe "Bare surfaces and Methods", that is, all the physical properties of clean surfaces of elemental and composite solids as well as the most relevant analytical methods. It soon turned out that an adequate treatment of all these subjects was far beyond any reasonable size of a single volume, and the material now easily fills the first three of the eight volumes as they stand now: Volume 1: Concepts and Methods, Volume 2: Properties of Elemental Surfaces, Volume 3: Properties of Composite Surfaces: Alloys, Compounds, Semiconductors, Volume 4: Solid/Solid Interfaces and Thin Films, Volumes 5 and 6: Solid/Gas Interfaces, Volume 7: Solid/Liquid and Biological Interfaces, and Volume 8: Applications of Surface Science.

The editor is extraordinarily thankful to all authors who have contributed to this series of books and have accepted the concept how to structure and compose their chapters. The editor is also very grateful to the publisher for his understanding and flexibility when the original concept of the whole project had to be "adapted" to new circumstances, as for example, described above in the case of the original Volume 1. Finally, one important factor that is crucial for the realization of such project is *patience*, not only the patience of the authors and the publisher with the editor but also the patience of the editor with some authors. A result of this mutual patience of all three parties involved is now in the hands of the reader.

*Klaus Wandelt*

# The Editor

**Klaus Wandelt** is currently Professor Emeritus at the University of Bonn, Germany, where he was also Director of the Institute of Physical and Theoretical Chemistry until 2010. He is a Guest Professor at the Universities of Wroclaw, Poland, and Rome, Tor Vergata, Italy. He received his Ph.D. on electron spectroscopy of alloy surfaces in 1975, spent a postdoctoral period at the IBM Research Laboratory in San Jose, California, from 1976 to 1977, and qualified as a professor in 1981. Since then his research focuses on fundamental aspects of the physics and chemistry of metal surfaces under ultrahigh vacuum conditions and in electrolytes, on the atomic structure of amorphous materials, and, more recently, on processes at surfaces of plants. Professor Wandelt has chaired the Surface Physics divisions of the German and the European Physical Societies as well as of the International Union of Vacuum Science, Techniques, and Applications, has organized numerous national and international conferences and workshops, was editor of journals, conference proceedings and books.

# List of Contributors


*Boris V. Andryushechkin*
A.M. Prokhorov General Physics Institute of the Russian Academy of Sciences
Vavilov Str. 38
119991 Moscow
Russia

*Klaus Christmann*
Freie Universität Berlin
Physikalische und Theoretische Chemie
Institut für Chemie und Biochemie
Takustr. 3
14195 Berlin
Germany

*Reinhard Denecke*
Universität Leipzig
Fakultät für Chemie und Mineralogie
Wilhelm-Ostwald-Institut für Physikalische und Theoretische Chemie
Linnéstr. 2
04103 Leipzig
Germany

*J. Michael Gottfried*
Philipps-Universität Marburg
Fachbereich Chemie
Physikalische Chemie
Hans-Meerwein-Straße
35032 Marburg
Germany

*Jan Haubrich*
Universität Bonn
Institut für Physikalische und Theoretische Chemie
Wegelerstreet 12
D-53115 Bonn
Germany

*Sabine Maier*
University of Erlangen-Nürnberg
Department of Physics
Erwin-Rommel-Street
91058 Erlangen
Germany

*Anton.G. Naumovets*
National Academy of Sciences of Ukraine
Institute of Physics
46 Prospect Nauki
UA-03680 Kiev 28
Ukraine

***Günther Rupprechter***
Vienna University of Technology
Institute of Materials Chemistry
Getreidemarkt 9
A-1060 Vienna
Austria

***Miquel Salmeron***
Materials Sciences Division
Lawrence Berkeley National
Laboratory
1 Cyclotron Road
Berkeley
CA 94720
USA

***Rolf Schuster***
Karlsruher Institut für
Technologie
Institut für Physikalische Chemie
Fritz-Haber-Weg 2
76131 Karlsruhe
Germany

***Hans-Peter Steinrück***
Friedrich-Alexander-Universität
Erlangen-Nürnberg
Department Chemie und
Pharmazie
Lehrstuhl für Physikalische
Chemie II
Egerlandstr. 3
91058 Erlangen
Germany

***Klaus Wandelt***
Universität of Bonn
Institut für Physikalische und
Theoretische Chemie
Wegelerstreet 12
D-53115 Bonn
Germany

***Peter Zeppenfeld***
Johannes-Keppler-Universität
Linz
Institut für Experimentalphysik
Altenberger Str. 69
4040 Linz
Austria

# Abbreviations

| | |
|---|---|
| AB | azobenzene |
| AES | Auger electron spectroscopy |
| AFM | atomic force microscopy/microscope |
| AI | adatom island |
| AIMD | *ab initio* molecular dynamics |
| ALD | atomic-layer deposition |
| AR | added-row |
| ARPES | angle-resolved photoelectron spectroscopy |
| ARUPS | angle-resolved UV photoelectron spectroscopy |
| as | asymmetric stretch |
| AvH | Alexander-von-Humboldt |
| β-PVDF | β-polyvinylidene fluoride |
| BAM | Brewster-angle microscopy |
| BBO | bridge-bonded oxygen |
| BD | bulk dislocation |
| BD | 1,3-butadiene |
| BE | binding energy |
| BEP | Brønsted–Evans–Polanyi |
| BET | Brunauer, Emmett, Teller |
| BIS | Bremsstrahlung isochromat spectroscopy |
| BLAG | buffer-layer-assisted growth |
| BOA | Born–Oppenheimer approximation |
| BZ | Belousov–Zhabotinsky |
| CC | capillary condensation |
| CCW | counterclockwise |
| CE | counter electrode |
| CI | configurational interaction |
| CIP | Cahn–Ingold–Prelog |
| CL | crossed leg |
| CPM | close-packed monolayer |
| CuPc | copper phthalocyanine |
| CV | cyclic voltammogram |

| | |
|---|---|
| CW | clockwise |
| 1D-IS | one-dimensional incommensurate structure |
| 2D | two-dimensional |
| 3D | three-dimensional |
| DF | density functional |
| DFG | difference frequency generation |
| DFG | Deutsche Forschungsgemeinschaft |
| DFT | density functional theory |
| DFT–GGA | density functional theory–generalized gradient approximation |
| DIET | desorption induced by electronic transition |
| DIMET | dynamics induced by multiple electronic transition |
| DOFs | degrees of freedom |
| DOS | density of states |
| e–h | electron–hole |
| EAM | embedded atom method |
| EBL | electron beam lithography |
| EBSD | electron backscatter diffraction |
| EDC | energy distribution curve |
| EDX | energy-dispersive X-ray fluorescence |
| ee | enantiomeric excess |
| EELS | electron energy loss spectroscopy |
| EH | extended Hückel |
| EMSI | ellipsomicroscopy for surface imaging |
| EMT | effective medium theory |
| ER | Eley–Rideal |
| ESD | electron-stimulated desorption |
| ESQC | elastic scattering quantum chemistry |
| ETEM | environmental transmission electron microscopy |
| ETS | extended transition state |
| EXAFS | edge X-ray absorption fine structure |
| fcc | face-centered cubic |
| FCI | Fonds der Chemischen Industrie |
| FEF | field emission fluctuation |
| FEM/FIM | field-emission microscopy/field-ionization microscopy |
| FFT | fast Fourier transform |
| FHI | Fritz Haber Institute |
| FIM | field ion microscopy |
| FK | Frenkel–Kontorova |
| FMT | fundamental measure theory |
| GC | gas chromatography |
| GGA | generalized gradient approximation |
| GIXD | grazing incidence synchrotron X-ray diffraction |
| H.c. | Hermitian conjugate |
| HA | hot atom |
| HAS | helium atom scattering |

| | |
|---|---|
| HB-HPB | hexa-*tert*-butyl-hexaphenylbenzene |
| HBC | hexabenzocoronene |
| hcp | hexagonal close-packed |
| HCRL | Hitachi Company Research Laboratory |
| HF | Hartree–Fock |
| HNC | hypernetted chain |
| HOMO | highest occupied molecular orbital |
| HOMO–LUMO | highest occupied molecular orbital–lowest occupied molecular orbital |
| HOPG | highly oriented pyrolytic graphite |
| HREELS | high-resolution electron energy loss spectroscopy |
| HRTEM | high-resolution transmission electron microscopy |
| hs | hard-sphere |
| IC | incommensurate |
| ice Ih | hexagonal ice |
| IES | infrared emission spectroscopy |
| IESH | independent electron surface hopping |
| IET | inelastic electron tunneling |
| IETS | inelastic electron tunneling spectroscopy |
| IMC | intermetallic |
| IPE | “inverse” photoemission |
| IPES | inverse photoemission spectroscopy |
| IR | infrared |
| IRAS | infrared reflection absorption spectroscopy |
| IS | incommensurate structure |
| IUPAC | International Union of Pure and Applied Chemistry |
| IVR | internal vibrational redistribution |
| LAPW | linear augmented plane wave |
| LCEP | lower critical end point |
| LCT | lower critical temperature |
| LDA | local density approximation |
| LDOS | local density-of-states |
| LEED | low-energy electron diffraction |
| LEEM | low-energy electron microscopy |
| LEIS | low-energy ion scattering |
| LH | Langmuir–Hinschelwood |
| LIF | laser-induced fluorescence |
| LJ | Lennard-Jones |
| MAA | methylacetoacetate |
| MAL | mass action law |
| MB | molecular beam |
| MBE | molecular beam epitaxy |
| MBRS | modulated beam relaxation spectroscopy |
| MC | Monte Carlo |
| MCT | mercury–cadmium–telluride |

| | |
|---|---|
| MD | molecular dynamics |
| MDC | methanol decomposition |
| MDEF | molecular dynamics with electronic friction |
| MEP | minimum energy path |
| MGR | Menzel, Gomer, and Redhead |
| ML | monolayer |
| MLE | monolayer equivalent |
| MO | molecular orbital |
| MOF | metal–organic frameworks |
| MR | missing-row |
| MRT | magnetic resonance tomography |
| MS | mass spectrometry |
| MSA | mean spherical approximation |
| MSD | mean square displacement |
| MSR | methanol steam reforming |
| NA | Newns–Anderson |
| NAP–XPS | near–ambient pressure X-ray photoelectron spectroscopy |
| nc-AFM | noncontact atomic force microscopy |
| NEA | negative electron affinity |
| NEXAFS | near-edge X-ray absorption fine structure |
| NOCV | natural orbitals for chemical valence |
| NRA | nuclear-reaction analysis |
| Nd:YAG | neodymium yttrium–aluminum garnet |
| OPA | optical parametric amplification |
| ORD | optical rotatory dispersion |
| OZ | Ornstein–Zernike |
| PAH | polyaromatic hydrocarbon |
| PAW | projector-augmented wave |
| PAX | photoelectron spectroscopy of adsorbed xenon |
| PBE | Perdew, Burke, and Ernzerhof |
| PED | Photoelectron diffraction |
| PEEM | photoemission electron microscopy |
| PEM | photoelastic modulator |
| PES | potential energy surface |
| PL | parallel leg |
| PM-IRAS | polarization-modulation infrared reflection absorption spectroscopy |
| PR | pairing-row |
| PT | periodic table |
| PT-1 | first-order phase transition |
| PTCDA | perylene-3,4,9,10-tetracarboxylic dianhydride |
| PVBA | pyridylvinyl benzoic acid |
| PY | Percus–Yevick |
| PZC | potential of zero charge |
| QC | quasi-classical |

| | |
|---|---|
| QCM | quartz crystal microbalance |
| RAIRS | reflection absorption infrared spectroscopy |
| RAM | reflection anisotropy microscopy |
| RAS | reflection anisotropy spectroscopy |
| RC | reaction coordinate |
| RE | rare-earth |
| RE | reference electrode |
| REMPI | resonantly enhanced multi-photon ionization |
| RFD | reference fluid density |
| RH | relative humidity |
| RKKY | Rudermann–Kittel–Kasuya–Yosida |
| RPH | reaction path Hamiltonian |
| RPM | restricted primitive model |
| SAFT | statistical associating fluid theory |
| SBZ | surface Brillouin zone |
| SCAC | single-crystal adsorption calorimetry |
| SERS | surface-enhanced Raman spectroscopy |
| SF | sealing flange |
| SFG | sum frequency generation |
| SH | sample holder |
| SHG | second-harmonic generation |
| SMM | surface mass model |
| SMSI | strong metal support interaction |
| SNS | scanning noise spectroscopy |
| SOM | surface oscillator model |
| SP | saddle point |
| SP | surface potential |
| SPA | spot profile analysis |
| SPM | scanning photoemission microscope |
| SPT | scaled particle theory |
| SRP | specific reaction parameter |
| SS | surface state |
| ss | symmetric stretch |
| ssd | standard square deviation |
| STM | scanning tunneling microscopy/microscope |
| SU | succinic acid |
| subPC | subphthalocyanine |
| SXRD | surface X-ray diffraction |
| TA | through adsorbate |
| TB | tight binding |
| TBPP | tetra-di-*t*-butylphenyl-porphyrin |
| TDS | thermal desorption spectroscopy/spectra |
| TED | transmission electron diffraction |
| TEM | transmission electron microscopy |
| THG | third harmonic generation |

| | |
|---|---|
| Ti:Sa | titanium sapphire |
| TM | transition metal |
| TMA | trimesic acid |
| TOF | time of flight |
| TOF | turnover frequency |
| TON | turnover numbers |
| TP-XPS | temperature-programmed X-ray photoelectron spectroscopy |
| TPD | temperature programmed desorption |
| TPT | thermodynamic perturbation theory |
| TR | time-resolved |
| *trans*-TBA | *trans*-tetra-tert-butylazobenzene |
| TS | through space |
| TST | transition state theory |
| TWC | three-way catalytic converters |
| UCEP | upper critical end point |
| UCT | upper critical temperature |
| UHV | ultrahigh vacuum |
| UPD | underpotential deposition |
| UPS | ultraviolet photoelectron spectroscopy |
| VASP | Vienna *ab initio* Simulation Package |
| VB | valence band |
| vdW | van der Waals |
| VI | vacancy island |
| WCA | Weeks, Chandler, and Andersen |
| WE | working electrode |
| WKB | Wentzel–Kramers–Brillouin |
| WMSI | weak metal support interaction |
| XAS | X-ray absorption spectroscopy |
| XPD | X-ray photoelectron diffraction |
| XPS | X-ray photoelectron spectroscopy |
| XRD | X-ray diffraction |

# 31
# Basics of Adsorption

*Klaus Wandelt*

## 31.1
## Introduction

Adsorption is a process in which particles, that is, atoms, molecules, ions, from one phase are accumulated at the interface to an adjacent phase: There is a driving force that attracts them to the interface. Atoms and molecules from a gas or vapor phase may stick at a solid surface, ions from an electrolytic solution may be attracted to an electrode, and surfactants are enriched at the surface of their aqueous solution. Obviously, adsorption is the first fundamental step in all surface confined processes, among which are technologically so important ones such as heterogeneous catalysis, corrosion, flotation, thin film growth, and so on. The great interest in these phenomena and their basic mechanisms has found its expression in a correspondingly broad and fundamental literature starting decades ago [1–15]. If the adsorption process may be reversed, we speak about desorption. While adsorption/desorption phenomena at solid/liquid interfaces are subject of Volume 7 of this series of books, Volumes 5 and 6 concentrate on solid/gas interfaces.

This chapter is an attempt to give an introduction to the topic of gas adsorption on solid surfaces by illustrating a number of important phenomena that result from the interaction of gas particles with the substrate surface, such as sticking, structure formation including substrate reconstruction, segregation, and film growth as well as redesorption. These processes are, of course, accompanied by changes in the electronic and vibronic properties of adsorbate and substrate. The "complete story" of these phenomena is laid out in detail in the following chapters of Volumes 5 and 6 for different classes of adsorbates, and was partly subject of chapters in the preceding volumes.

If adsorption and desorption are forward- and backward-reaction of a dynamic equilibrium at the interface between two (three-dimensional, 3D) bulk phases, for instance a gas and a solid

$$A_{gas} + \square \underset{k_{ad}}{\overset{k_{des}}{\rightleftarrows}} A_{ad} \tag{31.1}$$

*Surface and Interface Science: Solid-Gas Interfaces I*, First Edition. Edited by Klaus Wandelt.

the equilibrium constant $K$ is given by

$$K = \frac{k_{\text{ad}}}{k_{\text{des}}} \tag{31.2}$$

where □ is a free adsorption site on the solid surface and $k_{\text{ad}}$, $k_{\text{des}}$ are the rate constants of the adsorption and desorption process, respectively. It is common, but somewhat imprecise to call both $A_{\text{gas}}$ and $A_{\text{ad}}$ the "adsorbate," where $A_{\text{gas}}$ denotes the "adsorbate" in the gas phase (prior to adsorption) and $A_{\text{ad}}$ the "adsorbate" bound to the "substrate" surface.

At given gas pressure, the equilibrium constant $K$ and the rate constants $k$ are temperature dependent:

$$\frac{\delta \ln K}{\delta T} = \frac{\Delta G}{RT_2} \text{ (van't Hoff isobare)} \tag{31.3}$$

and

$$k = k_0 e^{(-E^*/RT)} \text{ (Arrhenius equation)} \tag{31.4}$$

The Gibbs free energy $\Delta G$ is often approximated by the mere energy $\Delta E$, and $E^*$ stands for the activation energy of the adsorption or desorption process, $E^*_{\text{ad}}$, $E^*_{\text{des}}$, respectively.

This temperature dependence enables a decoupling of the adsorption and desorption process. At sufficiently low temperature $T_s$ of the substrate, the particles may adsorb, but not redesorb, because $E^*_{\text{des}} > RT_s$. This permits a *dosewise* gas adsorption on the surface in question and its investigation in the absence of a surrounding gas pressure. Similarly, the subsequent desorption into vacuum, that is, without the presence of a 3D gas phase, may be studied by controlled increase in the sample temperature $T_s$, thereby "activating" the desorption process.

A combination of both approaches is to expose the surface *in vacuum* only strictly *locally* either to a constant or a pulsed influx of particles, namely, to a collimated continuous or pulsed beam of atoms or molecules, and to monitor their adsorption/desorption behavior at a given surface temperature. This beam may even be well prepared in terms of kinetic and inner energy and the orientation of the molecules. Figure 31.1 illustrates these scenarios at a solid–gas interface.

The advantage of separate dosewise adsorption and subsequent desorption by controlling the sample temperature accordingly, or by applying the "molecular beam" technique, that is, the absence of a 3D gas phase, is the possibility to use all electron-, ion-, and atom-beam-based surface-sensitive analytical methods described in Volume 1 to study the properties of $A_{\text{ad}}$ and the substrate.

In the following sections, we unroll the phenomena of adsorption and desorption in the order of the most obvious questions: (i) How do the adsorbing particles reach the surface/interface? (ii) How many of them "stick" on the surface at given external parameters such as pressure, flux, temperature, and so on? (iii) What holds the $A_{\text{ad}}$ particles bound at the surface? (iv) What are the specific properties of the adsorbed particles (compared to those in the gas phase) and of the adsorbate layer as a whole like surface coverage, thermal behavior, structure, and electronic properties? (v) To what extent do the substrate surface properties change upon interaction with $A_{\text{ad}}$?

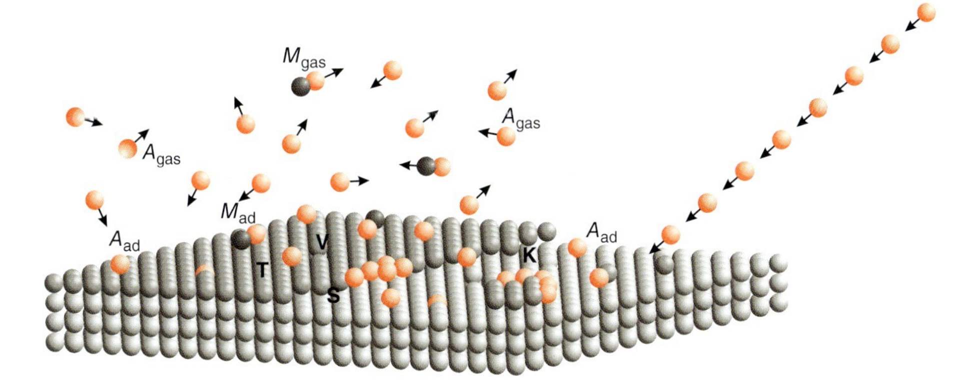


**Figure 31.1** Scenario of adsorption on a solid surface. Atomic ($A_{gas}$) or molecular ($M_{gas}$) particles become bound to the surface ($A_{ad}$, $M_{ad}$). On the atomic scale, the surface offers adsorption sites on terraces (T), at steps (S) and kinks (K), and at vacancies (V). The $A_{gas}$ /$M_{gas}$ exposure can occur via an isotropic background pressure or a directed particle beam.

### 31.1.1
### Adsorbate Delivery

There are various ways how to expose a surface to an adsorbate. The approach of the adsorbing particles to the surface/interface requires their mobility and may happen via diffusion in an isotropic gas phase, which is described by Fick's laws. Restricted to one dimension $z$ (normal to the surface) the first and second Fick's law

$$F = -D\frac{dc}{dz} \tag{31.5}$$

and

$$\frac{dc}{dt} = D\frac{d^2c}{dz^2} \tag{31.6}$$

refer to stationary and nonstationary conditions, respectively, with $F$ = flux (particles $m^{-2}\,s^{-1}$), $D$ = diffusion coefficient ($m^2\,s^{-1}$), and $dc/dz$ = concentration gradient. The required mobility depends on the temperature with $D \sim T^{3/2}$ for gases, $D \sim T$ in liquids, and $D \sim \exp(-E^*_{diff}/RT)$ in solids with $E^*_{diff}$ = activation energy of diffusion. Typical values for $D$ are in the range of $10^{-5}$ ($m^2\,s^{-1}$) in gases, $10^{-9}$ ($m^2\,s^{-1}$) in liquids, and $10^{-13}$ ($m^2\,s^{-1}$) in solids. Assuming, for instance, an ideal gas of pressure $p$ the impact rate of particles of mass $m$ onto a surface of area $A$ per second and unit area is

$$\frac{dN_{gas}}{Adt} = F = \frac{p}{\sqrt{2\pi mkT_{gas}}} \tag{31.7}$$

Ions in an electrolytic cell are additionally exposed to an electrostatic force, which, however, interestingly, has no accelerating but only an orientational effect on the motion of the charged particles (see Volume 7).

The pressure $p$ of a 3D gas phase is, according to the ideal gas law, at given volume and temperature, determined by the number of enclosed gas particles, which can be controlled either "batch-wise" by filling a closed volume (equilibrium pressure) or under "flow conditions" by a given combination of gas inlet and "pumping-speed $S$" of an open vessel (steady state pressure).

Highly anisotropic, directed atom or molecular beams are prepared by effusion cells, for example, a Knudsen cell. At temperature $T$, the substance $M$ in the Knudsen-cell acquires its equilibrium vapor pressure $p_M$ within the cell if the cross section of the exit hole is sufficiently small. Assuming further that the vapor pressure within the cell behaves like an ideal gas, Equation 31.7 can be used to calculate the number of particles traversing the exit hole, that is, the flux $F_{\text{eff}}$ leaving the cell toward the sample. Similarly, this particle flux can be determined by directing the beam into a mass spectrometer (Figure 31.14). The time $t$ the sample surface is actually exposed to the particle beam can be precisely controlled by opening ($t=0$) and closing ($t=t$) a shutter in front of the hole, or by means of a chopper, which cuts the beam in pulses of defined length and frequency (see Figure 31.18).

### 31.1.2
### Accommodation, Sticking, and Coverage

Not all particles that impinge on the surface do stick. Their kinetic (in the case of atoms) or total (in the case of molecules) energy, determined by the gas temperature $T_{\text{gas}}$, may prevent that they stick on the surface (of area $A$) but are rather scattered back. The adsorption rate is thus given by

$$r_{\text{ad}} = \frac{dN_{A_{\text{ad}}}}{Adt} = s(T_s, \theta)\,\frac{dN_{A_{\text{gas}}}}{Adt} = s(T_s, \theta)\frac{p}{\sqrt{2\pi mkT_{\text{gas}}}} = s(T_s, \theta)\alpha p \qquad (31.8)$$

The sticking coefficient $s$ is defined as the ratio of the numbers of sticking and impinging particles

$$s = \frac{N_{A_{\text{ad}}}}{N_{A_{\text{gas}}}} \qquad (31.9)$$

and describes the probability of a particle to stick and, therefore, has a value between 0 and 1. On the one hand, $s$ depends on the availability of adsorption sites $\square$ (see Equation 31.1) or conversely on the already existing coverage of the surface with particles (definition see Equations 31.11 and 31.12). On the other hand, any excess energy of the incoming particles must eventually be dissipated at the surface until the average energy of $A_{\text{ad}}$ and the substrate are the same, otherwise particles may be reflected ($A_{\text{gas},r}$) from the surface with $E_{A_{\text{gas}}} > E_{A_{\text{gas},r}} > E_{A_{\text{ad}}}$.

$$\alpha = \frac{E_{A_{\text{gas}}} - E_{A_{\text{gas},r}}}{E_{A_{\text{gas}}} - E_{A_{\text{ad}}}} \qquad (31.10)$$

is the so-called accommodation coefficient. Low-temperature STM studies could actually show that it takes some time, that is, some hops of adsorbing molecules or hot atoms over the surface, to release their excess energy to the substrate [16]. For a detailed description of the adsorbate–substrate interaction dynamics see Chapter 47 in Volume 6.

Also ubiquitous surface defects such as steps, kinks, point defects, and foreign atoms (see Figure 31.1) have an enormous influence on the sticking probability [17]. This is elegantly shown with molecular beam experiments in which, for instance, a Maxwellian beam of different temperature $T_{gas}$ is directed in up-step and down-step direction onto a stepped surface. This is exemplarily shown in Figure 31.2. At $T_{gas} = 90\,K$ huge differences of the initial sticking coefficient $s_o$ are found between both directions. A much higher sticking coefficient in up-step direction shows the dominant influence of the steps. This difference vanishes with increasing beam temperature. It is, after all, not surprising that published data of sticking coefficients on surfaces of unknown defect density may show a huge scatter and that measurements of trustworthy values are a difficult task.

Continued sticking of particles leads to an increase in the surface coverage until the surface is covered with one complete monolayer (ML). There are two common ways to define the coverage. It may either be expressed by the number of ad-particles per *substrate surface atom*

$$\theta = \frac{N_{A_{ad}}}{N_o} = \frac{\text{number of adsorbed particles}}{\text{number of surface atoms}} \qquad (31.11)$$

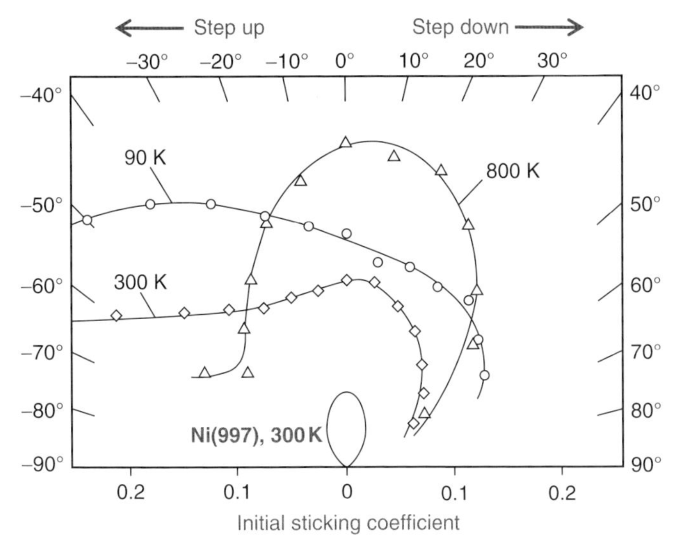


**Figure 31.2** Polar plot of the variation of the initial sticking coefficient for a Maxwellian beam of $H_2$ molecules on a Ni(997) surface. The beam is directed in the direction either step up or step down onto the surface at fixed temperature [17].

or by the number of adsorbed particles relative to the maximal possible number of *adsorbed particles*

$$\delta = \frac{N_{A_{ad}}}{N_{A_{ad,max}}} = \frac{\text{number of adsorbed articles}}{\text{maximal possible number of adsorbed particles}} \tag{31.12}$$

$\theta$ is an "absolute coverage" if besides the known number of substrate surface atoms per cm$^2$ either the absolute number of ad-particles per cm$^2$ can be determined or the lateral distribution of ad-particles with respect to the substrate surface atoms, that is, the ad-layer structure, is known. The "relative coverage" $\delta$ requires only a reliable signal that depends linearly on coverage and clearly indicates when the saturation coverage $N_{A_{ad,max}}$ is reached, for example, XPS, AES, or thermal desorption spectra (TDS) (see Section 31.3).

Of course, at sufficiently low temperatures, the amount of adsorbed particles may exceed 1 ML, in that particles "adsorb" on top of the first, second, and so on saturated layer. But these particles are not in direct contact with the substrate and are more or less "condensed" on itself. It is an interesting subject to study the difference in bonding of particles in the second, third, and so on layer compared to that of the first layer, but here, except in Section 31.2.2 we concentrate on the properties of the first monolayer.

Under equilibrium conditions, the surface coverage depends on – besides the nature of substrate and adsorbate – temperature and pressure of the gas phase. Measurements of the coverage as a function of pressure at constant temperature result in "adsorption isotherms," while such measurements as a function of temperature at constant pressure yield "adsorption isobars" as illustrated in Figure 31.3. The slope at any point along, for instance, an adsorption isotherm represents the sticking coefficient $s$ (relative to $s_o = s(\theta = 0)$ at zero coverage as long as $s_o$ is not absolutely known) at the corresponding coverage. Structural surface defects tend to enhance sticking (see Figure 31.2); therefore, steps on a Ru(0001) surface are populated earlier, that is, already at lower pressures, than terraces (see Figure 31.12).

Under nonequilibrium conditions, the sticking coefficient is measured as the slope of a plot $\theta \sim N_{A_{ad}}$ as a function of dosage:

$$\theta = \frac{1}{N_o}\int_0^{N_{A_{ad}}} dN_{A_{ad}} = \frac{s(T,\theta)}{\sqrt{2\pi mkT_{gas}}}\int_0^t pdt \tag{31.13}$$

with the dosage

$$D = \int_0^t pdt \quad (\text{Torr } s), \tag{31.14}$$

often being given in units of L (Langmuir), with $10^{-6}$ Torr $s = 1$ L. The slope $d\theta/dD$ yields again the relative sticking coefficient at any point along the uptake curve. The initial slope at $\theta = 0$ gives the initial sticking coefficient $s_o$, which in general decreases with increasing coverage $\theta$ (see however Section 31.2.1.4).

It is difficult to measure absolute sticking coefficients. Even though the absolute number of impinging particles may be known from kinetic gas theory (see

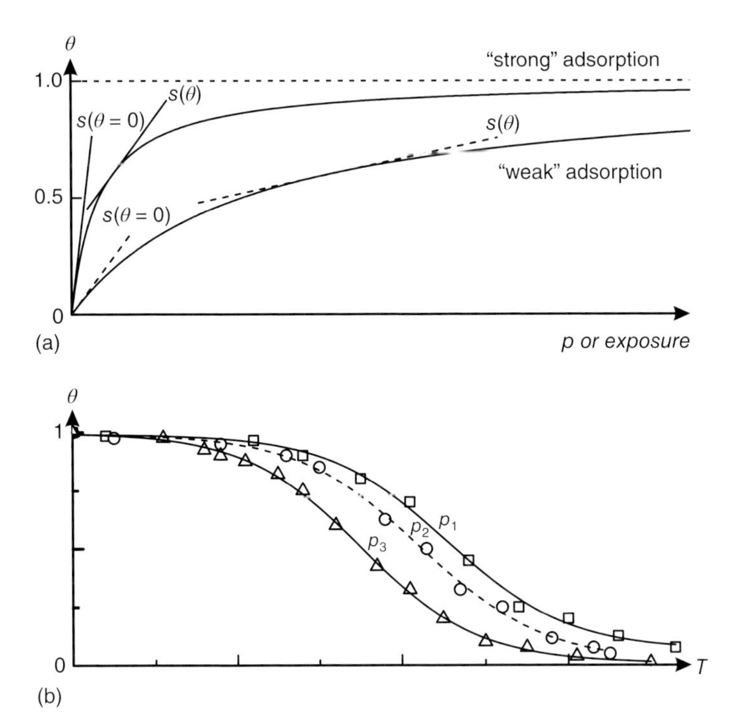


**Figure 31.3** (a) Adsorption isotherms measured at the same temperature on a "strongly" or "weakly" binding substrate, respectively. The slope at any point along such an uptake curve yields the sticking coefficient at the corresponding coverage. (b) Adsorption isobars measured on one substrate at different ambient pressures $p_1 < p_2 < p_3$.

Equation 31.7), it is not trivial to calibrate the coverage $N_{A_{ad}}$ in absolute terms. This problem can be circumvented in an elegant way. Instead of measuring the number of stuck particles, one may detect the number of back-scattered particles in a molecular beam experiment [18]. If $F_o = dN_{A_{gas}}/dt$ is the particle flux of the incoming beam and $F$ that of the back reflected beam (see Figure 31.4) then

$$F_o = \frac{dN_{A_{gas}}}{dt} = F(t) + \frac{dN_{A_{ad}}}{dt} = F(t) + s\,\frac{dN_{A_{gas}}}{dt} = F(t) + s(t)F_o \tag{31.15}$$

or

$$s(t) = \frac{F_o - F(t)}{F_o} = 1 - \frac{F(t)}{F_o} \tag{31.16}$$

$F_o$ and $F(t)$ are monitored with a mass spectrometer. As an example, Figure 31.5 shows the variation of the sticking coefficient as a function of $\theta$ and $T$ [19].

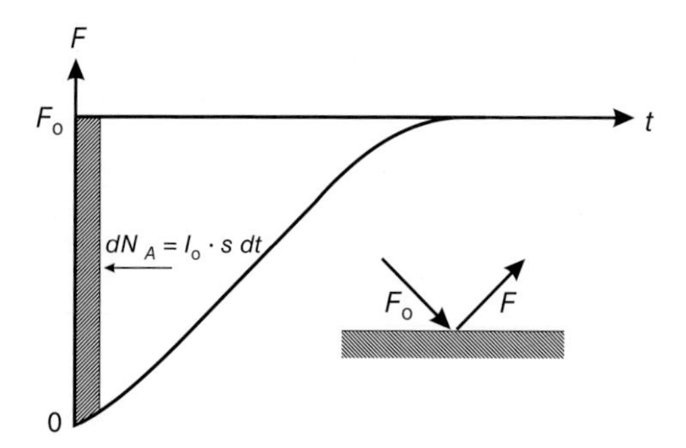


**Figure 31.4** Determination of sticking coefficients according to Equation 31.16 by measuring the flux of an impinging and back-scattered particle beam.

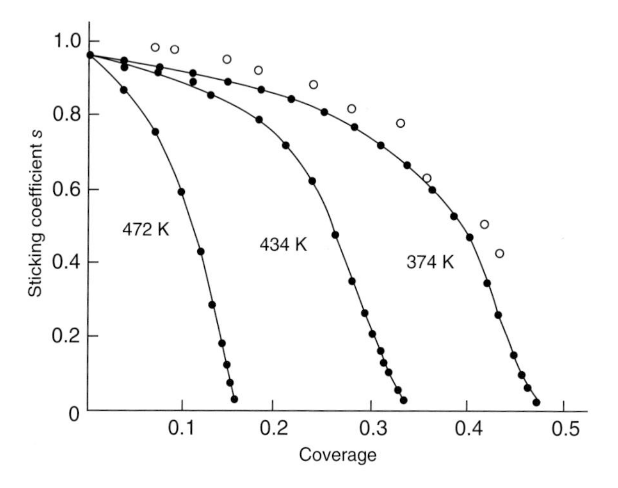


**Figure 31.5** Sticking coefficient of CO on a Pd(111) surface as a function of coverage and temperature. (•) with beam modulation and (o) without beam modulation [19].

## 31.2
## Adsorption Isotherms

Uptake curves as shown in Figure 31.3 can be described by thermodynamic "state equations," which describe the coverage of a surface in equilibrium with a gas (liquid) phase of pressure $p$ (concentration $c$) and temperature $T$:

$$\theta = f(\text{substrate}, p, T)$$

or

$$\theta = f(\text{substrate}, c, T)$$

Since it is easy to control the temperature of the sample, often isotherms are measured. The most common way to derive an adsorption isotherm is via kinetic considerations of the assumed underlying processes. The resultant model isotherms are then compared with experimentally determined ones, which allows conclusions about the validity of the microscopic model assumptions, namely, the detailed microscopic nature of the substrate and the adsorbate and their mutual interactions, in particular, the structural and chemical homogeneity of the substrate (also as a function of the coverage) and the inherent properties of the *adsorbed* particles. As, for instance, mentioned earlier, surface defects affect the sticking probability, and, in a sense, the mere presence of already adsorbed immobile particles may act like “defects” for impinging particles. Moreover, if upon interaction with the substrate, an adsorbed particle assumes, for instance, a strong dipole moment, which it did not have in the gas phase (see Section 31.4), this may result in repulsive interactions with other ad-particles and thereby create “exclusion zones” between them. Both effects, the mere physical presence and, for example, the lateral repulsion of ad-particles, introduce “coverage dependence” in the adsorption process.

## 31.2.1 Isotherms of Monolayer Adsorption

### 31.2.1.1 Langmuir Isotherm

The most commonly applied approach to interpret measured isotherms is the Langmuir isotherm [20]. The basic assumptions for this model isotherm are as follows:

- Already adsorbed particles are statistically distributed across the surface and are immobile.
- Adsorption occurs on free surface sites only (□); an occupied site is a “forbidden site.”
- The bond energy of each particle is the same, the adsorption energy is constant, that is, site and coverage independent. All possible adsorption sites are equivalent, and there is no interaction between the adsorbed particles.

As a consequence, the coverage will saturate at 1 ML with all adsorbed particles in immediate contact with the substrate.

A further input into the derivation of the model isotherm is whether or not the adsorbing particles remain intact or dissociate upon interaction with the surface. Diatomic molecules, for instance, are prone to dissociate.

#### 31.2.1.1.1 Atomic and Associative Adsorption

If the adsorbing particles remain intact, the equilibrium process is described by (see Equation 31.1)

$$A_{gas} + \square \underset{k_{ad}}{\overset{k_{des}}{\rightleftarrows}} A_{ad} \tag{31.17}$$

with the rate of adsorption

$$r_{ad} = k'_{ad} A_{gas} \square = k_{ad} p_A (1 - \theta) \tag{31.18}$$

and the rate of desorption

$$r_{des} = k'_{des} \cdot A_{ad} = k_{des} \cdot \theta \tag{31.19}$$

The number of adsorbed particles $A_{ad}$ is proportional to $\theta$, while the number of still free sites is proportional to $(1 - \theta)$. In equilibrium $r_{ad} = r_{des}$ and thus:

$$\frac{\theta}{1-\theta} = \frac{k_{ad}}{k_{des}} p_A \tag{31.20}$$

or

$$\theta = \frac{k_{ad} p_A}{k_{des} + k_{ad} p_A} = \frac{K_A p_A}{1 + K_A p_A} \quad \text{(Langmuir isotherm)} \tag{31.21}$$

with the equilibrium constant:

$$K_A = \frac{k_{ad}}{k_{des}} = \frac{k_{ad,0} e^{-E^*_{ad}/RT}}{k_{des,0}\, e^{-E^*_{des}/RT}} = \frac{k_{ad,0}}{k_{des,0}} \cdot e^{-(E^*_{ad} - E^*_{des}/RT)} \tag{31.22}$$

or

$$K_A = b_o e^{+E_{ad}/RT} \tag{31.23}$$

with

$$b_o = \frac{k_{ad,0}}{k_{des,0}} \text{ and } E_{ad} = E^*_{des} - E^*_{ad} \tag{31.24}$$

as illustrated in Figure 31.6, so that

$$\frac{\theta}{1-\theta} = b_o p_A \cdot e^{+E_{ad}/RT} \tag{31.25}$$

$E^*_{ad}$ and $E^*_{des}$ are the activation energies of adsorption and desorption, respectively. While for atomic and associative adsorption $E^*_{ad}$ is generally zero, this may be different in the case of dissociative adsorption as discussed in the next section. Obviously, desorption always requires an activation energy to overcome the adsorbate–substrate bond (see Section 31.3).

At low pressures $K_A p_A \ll 1$, according to Equation 31.21, so that the Langmuir isotherm simplifies to

$$\theta \sim K_A p_A = b_o p_A \cdot e^{+E_{ad}/RT} \tag{31.26}$$

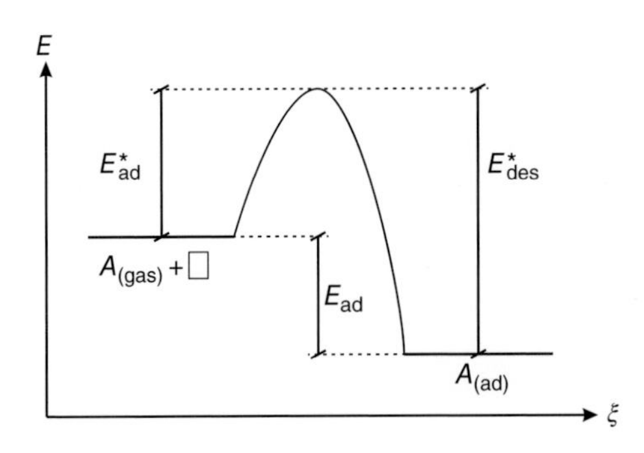


**Figure 31.6** Energy diagram for an activated process, here the adsorption of $A_{gas}$ on a free surface site to form $A_{ad}$. But generally only dissociative adsorption may be activated; $\xi$ = reaction coordinate.

which is also known as the *Henry isotherm* and which applies at low coverages, which are most probable if $E_{ad}$ is small.

At high pressures $K_A p_A \gg 1$ Equation 31.21 simplifies to

$$\theta = \theta_m = 1 \quad \text{(monolayer, 1 ML).}$$

Figure 31.7 displays theoretical Langmuir isotherms for three different temperatures of the same substrate. The higher the temperature, the higher must be the pressure to achieve the same equilibrium coverage.

Writing the Langmuir isotherm in the form

$$\frac{1}{\theta} = \frac{1}{K_A p_A} + 1 \tag{31.27}$$

shows that an experimentally determined isotherm satisfies the Langmuir model if a plot $(1/\theta)$ versus $(1/p_A)$ of the measured data yields a straight line.

**31.2.1.1.2 Dissociative Adsorption** Adsorption of *molecules* often occurs dissociatively, the dissociating molecules require at least as many adsorption sites as there are dissociation fragments, even though many more may be necessary, for example, eight in the case of dissociative oxygen adsorption on a Ni(100) surface [21]. For the sake of simplicity, we assume two fragments and two sites:

$$A_{2,\text{gas}} + 2\square \underset{k_{\text{ad}}}{\overset{k_{\text{des}}}{\rightleftarrows}} 2A_{\text{ad}} \tag{31.28}$$

In this case, the adsorption rate

$$r_{\text{ad}} = k_{\text{ad}} p_{A_2} (1 - \theta)^2 \tag{31.29}$$

and desorption rate

$$r_{\text{des}} = k_{\text{des}} \theta^2 \tag{31.30}$$

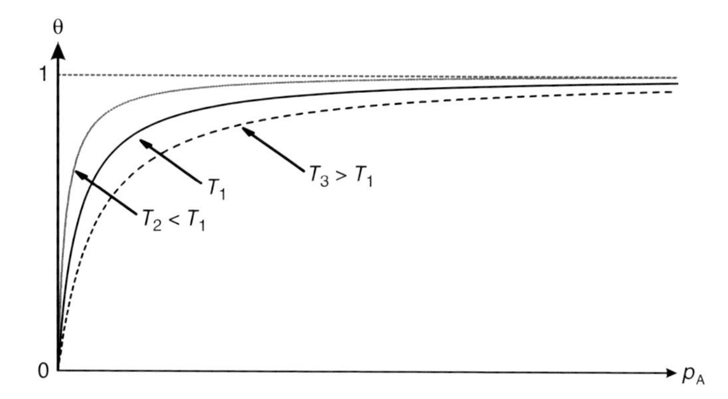


**Figure 31.7** Langmuir isotherms for adsorption at different temperatures on the same surface.

are of second order, which leads to the equilibrium coverage

$$\theta = \frac{\sqrt{K_{A_2} p_{A_2}}}{1 + \sqrt{K_{A_2} p_{A_2}}} \quad (K_{A_2} = \text{equilibrium constant}) \tag{31.31}$$

Thus, comparison of measured with model isotherms allows to draw first conclusions about the adsorption mechanism. The adsorption is nondissociative if $(1/\theta)$ versus $(1/p_A)$ gives a straight line, but dissociative if a plot $(1/\theta)$ versus $(1/\sqrt{p_{A_2}})$ is linear.

**31.2.1.1.3 Competitive Adsorption** According to the Langmuir–Hinshelwood mechanism of heterogeneously catalyzed reactions, both (all) reactants are adsorbed on the catalyst's surface. Thus, two (or more) adsorbates compete for the available adsorption sites:

$$A_{\text{gas}} + \square \underset{k_{A,\text{des}}}{\overset{k_{A,\text{ad}}}{\rightleftarrows}} A_{\text{ad}} \tag{31.32}$$

$$B_{\text{gas}} + \square \underset{k_{B,\text{des}}}{\overset{k_{B,\text{ad}}}{\rightleftarrows}} B_{\text{ad}} \tag{31.33}$$

Under equilibrium conditions, adsorption and desorption rate of each individual component are equal:

$$k_{A,\text{ad}} \cdot p_A (1 - \theta_A - \theta_B) = k_{A,\text{des}} \cdot \theta_A \tag{31.34}$$

$$k_{B,\text{ad}} \cdot p_B (1 - \theta_A - \theta_B) = k_{B,\text{des}} \cdot \theta_B \tag{31.35}$$

whereby adsorption can only take place at sites that are neither covered by $A_{\text{ad}}$ nor $B_{\text{ad}}$. The resultant equilibrium coverages are given as follows:

$$\theta_A = \frac{K_A \cdot p_A}{1 + K_A \cdot p_A + K_B \cdot p_B} \tag{31.36}$$

$$\theta_B = \frac{K_B \cdot p_B}{1 + K_A \cdot p_A + K_B \cdot p_B} \tag{31.37}$$

with the equilibrium constants $K_A = (k_{A_{\text{ad}}}/k_{A_{\text{des}}})$ and $K_B = (k_{B_{\text{ad}}}/k_{B_{\text{des}}})$.

It is obvious that more than two components and additional equilibria in the gas (liquid) phase quickly lead to more complex expressions.

Even though it is also obvious that in reality not all assumptions of the Langmuir isotherm are fulfilled, it is still a very useful model. In the following sections, we stepwise lift the stringent Langmuir assumptions.

#### 31.2.1.2 Temkin Isotherm

Most unrealistic is the Langmuir assumption of a constant, coverage independent adsorption energy. The successive population of different surface sites, for example, step, kink, vacancy, and terrace sites (see Figure 31.1), as well as lateral interactions between ad-particles may very well change $E_{\text{ad}}$ as a function of coverage. Different models start from different assumptions about the influence of site heterogeneity and lateral interactions. While, for example, the Freundlich isotherm [22] assumes an exponential dependence of $E_{ad}$ on coverage $\theta$, the Temkin model [23] lifts the assumption of constant adsorption energy by assuming a linear decrease of $E_{ad}$ with increasing coverage. In the latter case:

$$E_{ad} = E^{o}_{ad}(1 - \alpha\Theta),\ E^{o}_{ad} = E_{ad}(\Theta = 0) \tag{31.38}$$

Inserting this into the Langmuir isotherm (Equation 31.25) yields

$$\frac{\Theta}{1 - \Theta} = b_{o} p_{A} e^{(E^{o}_{ad}(1-\alpha\Theta)/RT)} \tag{31.39}$$

The logarithmic form

$$\ln p_{A} = -\ln(b_{o} e^{(E^{o}_{ad}/RT)}) + \frac{E^{o}_{ad}\,\alpha\Theta}{RT} + \ln\frac{\Theta}{1 - \Theta} \tag{31.40}$$

shows that the term $\ln(\Theta/1 - \Theta)$ becomes small, and even zero for $\Theta = 0.5$. Its neglection leads to

$$\Theta = \frac{RT}{E^{o}_{ad}\,\alpha} \ln(b_{o} p_{A} e^{(E^{o}_{ad}/RT)}), \tag{31.41}$$

that is, a logarithmic correlation between $\Theta$ and $p_{A}$. This is a consequence of a Boltzmann-like population of adsorption sites of different $E_{ad}$.

#### 31.2.1.3 Fowler – Guggenheim Isotherm

This isotherm introduces explicitly lateral interactions between nearest-neighbor ad-particles [24]. If $z$ is the number of possible nearest neighbor sites to a given ad-particle, and $w$ the constant interaction energy between pairs of ad-particles, independent of the actual number and arrangement of nearest-neighbor ad-particles (see inset in Figure 31.8), the occupation of the $z$ sites varies with $\Theta$, and it follows

$$E_{ad} = E^{o}_{ad} + zw\Theta, E^{o}_{ad} = E_{ad}(z = 0) \tag{31.42}$$

Inserting this into the Langmuir isotherm (Equation 31.25) yields

$$\frac{\Theta}{1 - \Theta} = b_{o}\, p_{A_{gas}}\, e^{((E^{o}_{ad} + zw\theta)/RT)} = b_{o} p_{A_{gas}}\, e^{(E^{o}_{ad}/RT)}\, e^{(zw\theta/RT)} = bp_{A_{gas}}\, e^{(zw\theta/RT)} \tag{31.43}$$

A plot of $\Theta$ as a function of $\ln(bp_{A})$ with variable $\beta = zw/RT$ is shown in Figure 31.8. The line for $\beta = 0$ reproduces the Langmuir isotherm of noninteracting ad-particles. Repulsive lateral interactions are represented by $\beta < 0$. The lateral repulsion leads to a less steep increase of the coverage $\Theta$ as a function of the gas pressure, or vice versa the same coverage as predicted by the Langmuir isotherm requires a higher pressure (chemical potential) of the gas. Conversely, if $\beta > 0$ the ad-particles attract each other and an uptake comparable to the Langmuir isotherm requires a lower gas pressure. Interestingly, for the case depicted in Figure 31.8, $\beta > 4$ yields isotherms with S-shape, that is, for one value of bp there are three values of $\Theta$, which is physically unrealistic. In analogy to the behavior of a 3D van der Waals gas, this paradox is explained by the consideration of 2D condensation within the layer of ad-particles, that is, an equilibrium between a 2D condensed phase (2D liquid or 2D solid) and its

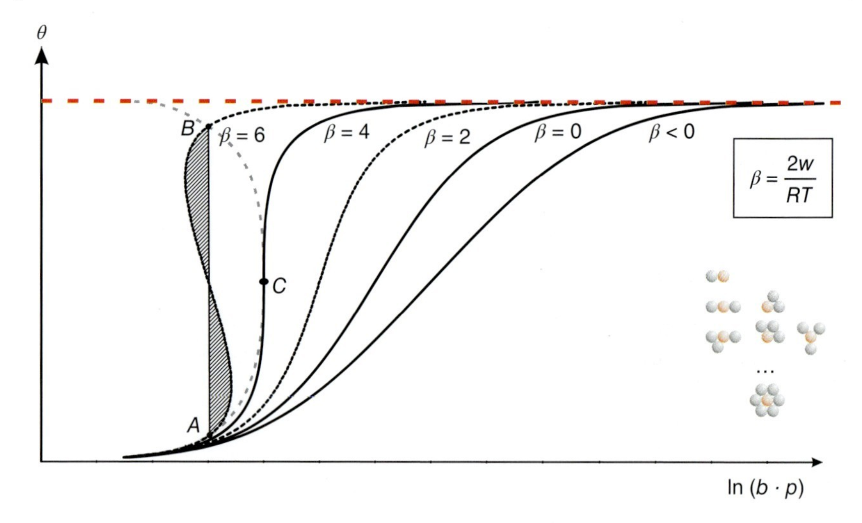


**Figure 31.8** Fowler–Guggenheim isotherms including lateral repulsive ($\beta < 0$) or attractive ($\beta > 0$) interactions w between *pairs* of neighboring ad-particles independent of their further coordination (inset). For the particles set of chosen parameters $w$ and $T$, isotherms with $\beta > 4$ include a 2D gas ↔ 2D condensed + 2D gas phase transition. The dotted line through points $A$, $B$, and $C$ correspond to the phase coexistence line (see text).

2D vapor phase. The physically correct course of the isotherm is along the straight line from $A$ to $B$, such that the hatched areas are equal (see text-books about the van der Waals equation of real gases). For $\beta = 6$, the values of $\theta$ and $\ln(bp)$ at point $A$ define the onset of 2D condensation. The critical point $C$ lies on that isotherm, of which $\beta$, that is, the combination of attraction $w$ and thermal energy $RT$, does first no longer allow 2D condensation at any pressure like for all values $\beta < 4$ in the illustrated case. The dashed line through all points $A$ and $B$, and $C$, corresponds to the coexistence line between the 2D gas (at low $\Theta$) and the 2D condensed (at high $\Theta$) phase (see also Section 31.4.3.2).

Figure 31.9a shows adsorption isotherms of Xe on a substrate that consists of a Pd(100) surface already precovered with one complete monolayer of Xe as derived from Xe(5p) photoemission spectra [25]. Since these spectra (Figure 31.9b) permit a clear distinction of Xe atoms adsorbed in the second layer (and higher layers) from those (preadsorbed) in the first layer [26, 27], the coverage data in Figure 31.9a refer layer-specific only to the population of the second and third layer. Xe-atoms in the *second* layer interact so weakly with the metal that they do no longer acquire such a strong dipole moment as those in the first layer (see Section 31.4.2.2). As a consequence, their weakened mutual repulsion does no longer prevent 2D condensation in contrast to the first Xe layer (see Figure 31.12). The dashed line indicates again the coexistence region between 2D Xe gas (points $A$) and 2D condensed Xe (points $B$) at the different temperatures. $T_{c,2}$ ($T_{c,3}$) marks the critical point of 2D condensation in the second (third) Xe layer.

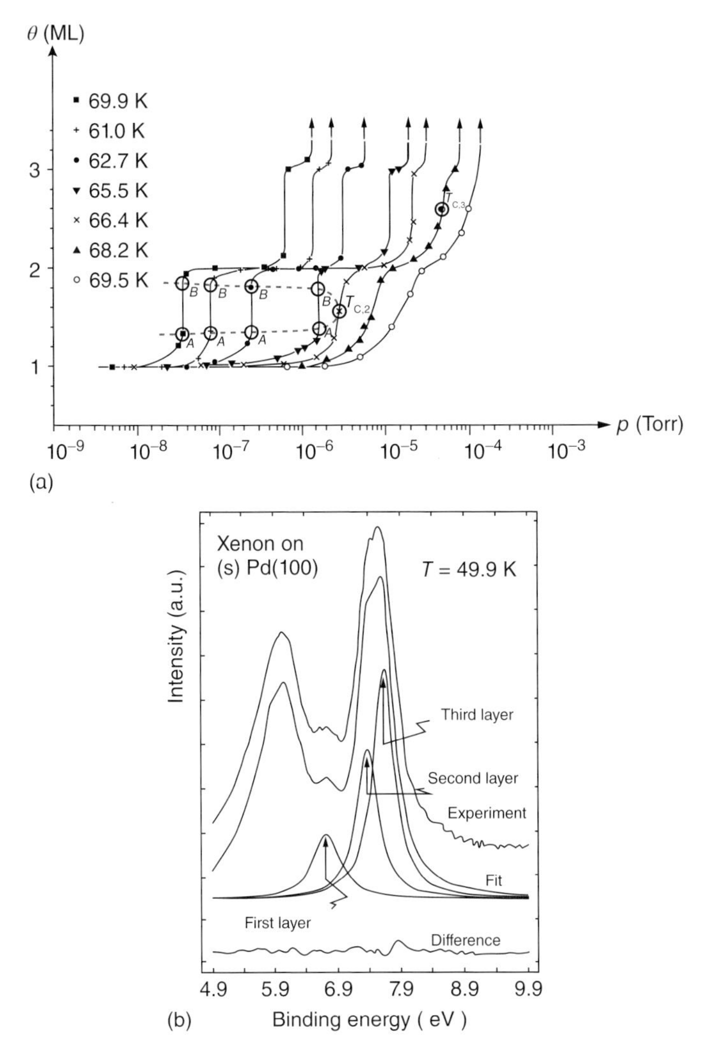


**Figure 31.9** (a) Isotherms for Xe adsorption on a Pd[8(100) × (110)] surface deduced from the integrated layer specific Xe($5p_{1/2}$) UPS intensities from Xe atoms in the first, second, and third layer. The symbols in (a) refer to the different temperatures. The arrows indicate condensation of a thick Xe film. The dashed line through points *A*, *B*, and *C* correspond to the phase coexistence line. (b) Layer-specific ($5p_{3/2,1/2}$) UPS spectra of xenon adsorbed on a Pd[8(100) × (110)] surface enabling a distinction of Xe atoms in the first, second, and third layer [25].

#### 31.2.1.4 **The Kisliuk – Model**

All model isotherms discussed so far, starting with the Langmuir model, are based on the assumption that adsorption of particles can only occur on *empty* adsorption sites [28]. There are, however, many adsorption results where the sticking coefficient does not immediately decrease with increasing coverage, as sketched in Figure 31.10a and exemplified by the system CO/Pd(111) at 374 K in Figure 31.5.

This can be understood if adsorption proceeds through a so-called precursor, a weakly bound "physisorbed" state of the adsorbate is temporarily trapped in a physisorption minimum (see Section 31.4.1). This can occur above the free surface ("intrinsic precursor" $P_i$) or also above an existing island of ad-particles ("extrinsic precursor" $P_e$) as sketched in Figure 31.10b. During their lifetime near the surface both types of precursors can diffuse across the surface or the island (see gray regime in Figures 31.30 and 31.31) before they either redesorb into the gas phase or transform into an adsorbed particle $A_{ad}$. To this end, the extrinsic precursor, during its lifetime, has to reach the edge of the island and drop down on the free surface ($k_3$ in Figure 31.10b).

For the sake of simplicity, we assume "single point," that is, nondissociative adsorption. The total reaction scheme of sequential and parallel processes then reads

$$A_{gas} \underset{k_2^e,\, k_2^i}{\overset{k_1}{\rightleftarrows}} P_i, P_e \xrightarrow{k_3} A_{ad} \xrightarrow{k_4} A_{gas} \tag{31.44}$$

The total adsorption rate is thus

$$r_{ad} = \frac{dN_{A_{ad}}}{dt} = r_1 - r_2 - r_4 \tag{31.45}$$

with $r_1$ = adsorption rate into a precursor state, $r_2$ = desorption rate of a precursor, which may be different for an intrinsic or extrinsic precursor particle, and $r_4$ = desorption rate of a finally adsorbed particle at given temperature. A solution of this rate equation can be achieved by using the so-called "stationary principle,"

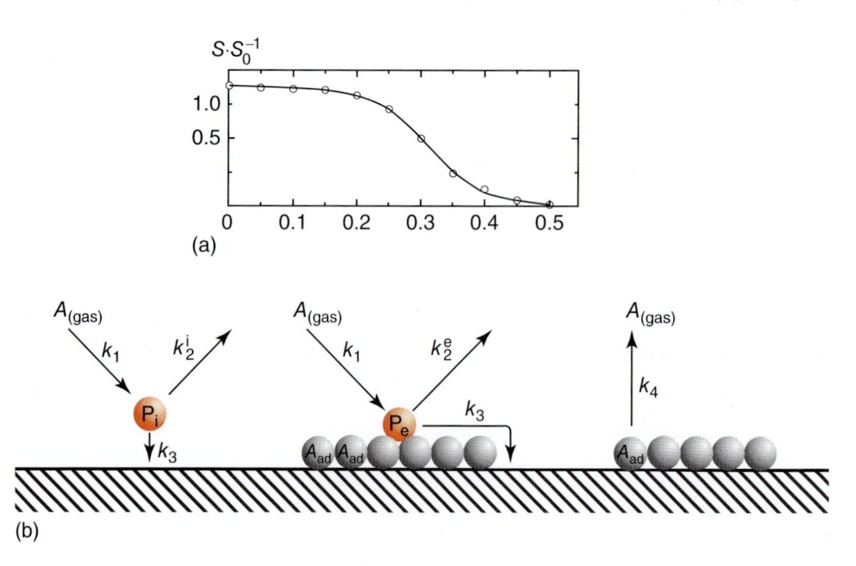


**Figure 31.10** (a) A constant sticking coefficient up to high coverages may be explained by (b) the Kisliuk model: Chemisorption via a physisorbed precursor state.

which assumes that the total precursor concentration (coverage) on the surface is small and constant:

$$\frac{dN_P}{dt} - 0 \tag{31.46}$$

In view of the energetically much more favorable final adsorbate state $A_{ad}$, this condition is certainly true at not too low temperatures, which in turn would lead to an accumulation of particles in the physisorption state (see Section 31.4.1), that is, the precursor state.

From $N_P = \text{const}$ it follows immediately

$$r_1 = r_2 + r_3 \tag{31.47}$$

that is, the number of "sticking" *precursor* particles is equal to the sum of precursor particles disappearing either by redesorption or by transformation into $A_{ad}$ with $r_3$.

Combining Equations 31.45 and 31.47 yields

$$r_{ad} = r_2 + r_3 - r_2 - r_4 = r_3 - r_4 \tag{31.48}$$

Assuming that the sticking coefficient of intrinsic and extrinsic precursor particles over free and precovered sites, respectively, is the same, gives:

$$r_1 = s_P AF \tag{31.49}$$

with $s_P$ = sticking coefficient of the *precursor* particles, $A$ = area, and $F$ = impact rate of gas particles (see Equation 31.7).

The desorption of precursor has two contributions, namely, from free as well as precovered surface areas. Desorption of intrinsic precursor from free surface patches is proportional to $(1 - \Theta)$:

$$r_2^i = k_2^i(1 - \Theta)N_P, \tag{31.50}$$

and desorption of extrinsic precursor from precovered surface areas is proportional to $\Theta$:

$$r_2^e = k_2^e \Theta N_P \tag{31.51}$$

The total desorption rate of precursor particles is therefore

$$r_2 = r_2^i + r_2^e = N_P\,[k_2^i(1 - \Theta) + k_2^e \Theta]. \tag{31.52}$$

The transformation rate of precursor into adsorbed particles $A_{ad}$ depends on the *total* number $N_P$ of precursor particles and the *still free* surface area $(1 - \Theta)$. Assuming the same transformation probability for both types of precursors yields

$$r_3 = k_3 N_P (1 - \Theta) \tag{31.53}$$

Combining Equations 31.49, 31.52, and 31.53 yields:

$$s_P AF = N_P[k_2^i(1 - \Theta) + k_2^e \Theta] + k_3 N_P(1 - \Theta) \tag{31.54}$$

or rearranged

$$N_P = \frac{s_P AF}{[k_s^i(1 - \Theta) + k_2^e \Theta] + k_3(1 - \Theta)} \tag{31.55}$$

Inserting this into Equation 31.53 gives with Equation 31.48

$$r_{ad} = r_3 - r_4 = \frac{k_3 s_P AF(1 - \Theta)}{[k_2^i(1 - \Theta) + k_2^e\Theta] + k_3(1 - \Theta)} - r_4 \tag{31.56}$$

If desorption of $A_{ad}$ is negligible, that is, $r_3 \gg r_4, r_4$ can be neglected in Equation 31.56, and the probability to stick as $A_{ad}$ particle

$$s = \frac{r_{ad}}{AF} \tag{31.57}$$

becomes

$$s \approx s' = \frac{s_P k_3(1 - \Theta)}{k_2^i - k_2^i\Theta + k_2^e\Theta + k_3 - k_3\Theta} = \frac{s_P k_3(1 - \Theta)}{k_2^i + k_3 - \Theta(k_3 + k_2^i - k_2^e)} \tag{31.58}$$

($s'$ accounts for the neglect of $r_4$ in Equation 31.58).

For $\Theta = 0$:

$$s' = \frac{s_P\, k_3}{k_2^i + k_3} = s_P\, s_o \tag{31.59}$$

with

$$s_o = \frac{k_3}{k_2^i + k_3} \tag{31.60}$$

Dividing the left and right sides of Equations 31.58 and 31.60 yields

$$\frac{s'}{s_o} = \frac{s_P(1 - \Theta)(k_2^i + k_3)}{k_2^i + k_3\, \Theta(k_3 + k_2^i) + k_2^e\Theta} \tag{31.61}$$

and rearranged

$$\frac{s'}{s_o} = \frac{s_P(1 - \Theta)}{1 - \Theta + \Theta\frac{k_2^e}{k_3 + k_2^i}} = \frac{s_P}{1 + \frac{\Theta}{1 - \Theta}\frac{k_2^e}{k_3 + k_2^i}} = \frac{s_P}{1 + K\,\frac{\Theta}{1 - \Theta}} \tag{31.62}$$

With

$$K = \frac{k_2^e}{k_3 + k_2^i}. \tag{31.63}$$

If $K$ is very small, that is, the transformation into $A_{ad}$ very efficient, one obtains with $k_3 \gg k_i$, $k_2^e$:

$$\frac{s'}{s_o} = s_P \tag{31.64}$$

that is, the sticking coefficient is equal to $s_P$ and, thus, independent of $\Theta$.

If $K = 1$, that is, $k_3 = k_2^e - k_2^i$, one obtains

$$\frac{s'}{s_o} = \frac{s_P}{1 + \frac{\Theta}{1 - \Theta}} = s_P(1 - \Theta), \tag{31.65}$$

which corresponds to the Langmuir model.

Reconsideration of the term $r_4 = k_4\ \Theta$, which was neglected in Equation 31.58, leads to the true sticking coefficient

$$s = s' - \frac{k_4 \Theta}{A\ F}. \tag{31.66}$$

The correction term $k_4\Theta/AF$ may become significant above $\Theta \sim 0.95$.

In the case of high coverage it is $k_2^{\mathrm{i}} < k_2^{\mathrm{e}}$, so that

$$K \approx \frac{k_2^{\mathrm{e}}}{k_3} = \frac{e^{(-E_2^*/RT)}}{e^{(-E_3^*/RT)}} = e^{(-(E_2^* - E_3^*)/RT)} \tag{31.67}$$

with $E_2^* = -E_{2,\mathrm{p}}$ = physisorption energy (see Section 31.4.1) and $E_3^*$ = activation energy for the transformation of $P$ into $A_{\mathrm{ad}}$, where $E_2^* - E_3^*$ can be determined from the temperature dependence of $K$.

### 31.2.2
### BET Isotherm of Multilayer Adsorption

At sufficiently low temperature and/or high pressure ad-particles may also *permanently* populate the second (or a higher) layer, that is, after saturation of the first layer (see, e.g., Figures 31.9a and 31.12). This physisorption may lead to adsorption of multilayers due to the long-range van-der-Waals forces (see Section 31.4.1).

The most popular model to describe equilibrium multilayer adsorption on a homogeneous surface is the BET-isotherm (Brunauer, Emmett, Teller, 1938) [29], which is based on the following assumptions:

- Population of the first layer ($n = 1$) obeys the Langmuir model with $E_{A_{\mathrm{ad}}}$ = const and no interactions between the adsorbed particles.
- Every particle in the first, second, and so on, layer is an *allowed* adsorption site for an ad-particle in the next higher layer.

This leads to vertical columns of ad-particles on the surface, whereby these columns do not interact with each other (see inset in Figure 31.11). The adsorption energy for a particle in the $(n+1)$th layer, $n = 1,2,3\ldots$, is taken equal to the condensation/evaporation energy of the adsorbate, $E_{\mathrm{ad},n+1} = E_{\mathrm{cond}} = E_{\mathrm{vap}}$.

Kinetic as well as statistical derivation leads to

$$\frac{p_A}{\theta(p_{A_\mathrm{o}} - p_A)} = \frac{1}{\theta_\mathrm{m}\ b} + \frac{b-1}{\theta_\mathrm{m}\ b}\cdot\frac{p_A}{p_{A_\mathrm{o}}} \quad \text{(BET isotherm)} \tag{31.68}$$

with

$\Theta$ = total coverage at the equilibrium pressure $p_A$

$p_{A_\mathrm{o}}$ = saturation vapor pressure of the pure adsorbate ($A$) at the chosen temperature

$\theta_\mathrm{m}$ = saturation coverage of the first layer, and

$b = e^{((E_{\mathrm{ad}} - E_{\mathrm{vap}})/RT)}$

Figure 31.11 shows model BET isotherms for different values of $b$.

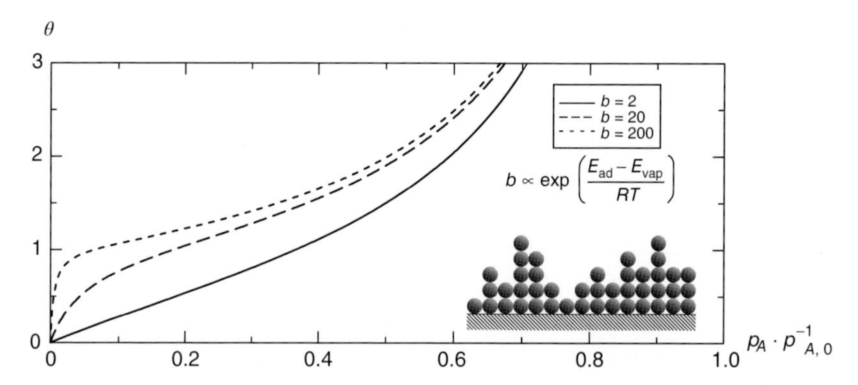


**Figure 31.11** BET adsorption isotherms for different values of $b = \exp[(E_{ad} - E_{vap})/RT]$; see text.

Experimental data that meet the BET model isotherm, plotted in the form $(p_A/\theta(p_{A_0} - p_A))$ versus $(p_A/p_{A_0})$, give a straight line with the slope $((b - 1)/(\theta_m \cdot b))$ and the intersect $(1/(\theta_m \cdot b))$.

With known values for $\theta_m$, $b$, and $E_{vap}$ the adsorption energy $E_{ad}$ of the first layer can be calculated. Vice versa, if $E_{ad}$ and $\theta_m$ from a Langmuir isotherm and $E_{vap}$ and $p_{A_0}$ for the adsorbate are known, $\theta$ can be estimated.

The BET isotherm is used to determine surface areas of powder or porous materials, for example, catalysts, because if $\theta_m$ and the size of one particle of the adsorbed gas are known one can calculate this area. Generally, $N_2$, Ar, or Kr is used to determine such "BET areas." The method is based on the requirement that the experimental isotherm shows a clear indication of monolayer saturation, that is, a plateau (Figure 31.11), and the assumption that the ad-particles in the monolayer are densely packed. In reality, the assumption of a uniform evaporation energy for all layers above the first one is too crude. There are very well distinguishable adsorption energies $E_{ad}(n)$ up to several layers $n$; an example is shown for Xe adsorption on a Pd(100) surface in Figure 31.9 (see also Ref. [30]).

### 31.2.3
### Heat of Adsorption

The process of adsorption, that is, the bond formation between ad-particles and the substrate, is accompanied by the release of heat $Q_{ad,}$ which can be measured calorimetrically as outlined in detail in Chapter 32 of this volume. But also adsorption isotherms or isobars can be analyzed with regard to the heat of adsorption. Since this heat may, and most likely will, be coverage dependent as considered in all models beyond the Langmuir isotherm, an important quantity is the "isosteric heat of adsorption," that is, the heat of adsorption at given coverage.

Figure 31.12 shows selected adsorption isobars of Xe on a stepped Pd(100) surface as determined by recording the intensity of the $M_5M_{4,5}N_{4,5}$ (530 eV) Xe Auger transition as a measure of the equilibrium Xe uptake at different temperatures for given Xe pressure [26]. Curve 1 saturates already at the lowest pressure ($7.4 \times 10^{-11}$ Torr) and the relatively high temperature of ~90 K, and corresponds to the saturation of the step sites. The plateau value corresponds to a coverage of $\theta_s = 0.16$ ML, which is in perfect agreement with one row of Xe atoms per available step edge on this surface as deduced from LEED observations. With increasing Xe pressure (curves 4–9), the temperature, at which these curves indicate the same Xe coverage $\theta_1, \theta_2, \theta_3$, increases as marked by the horizontal dashed lines. Monolayer population and final saturation (second plateau) proceeds – at given pressure – over a wide temperature range, a consequence of a relatively strong dipole moment of the adsorbed Xe atoms and the resultant repulsive Xe–Xe interaction [26].

The isobars can be used to determine the isosteric heat of adsorption on the steps and terrace sites, respectively, by applying the Clausius–Clapeyron equation as follows: First, the equilibrium values of $p$ and $T$ at constant coverage $\theta$, that is,

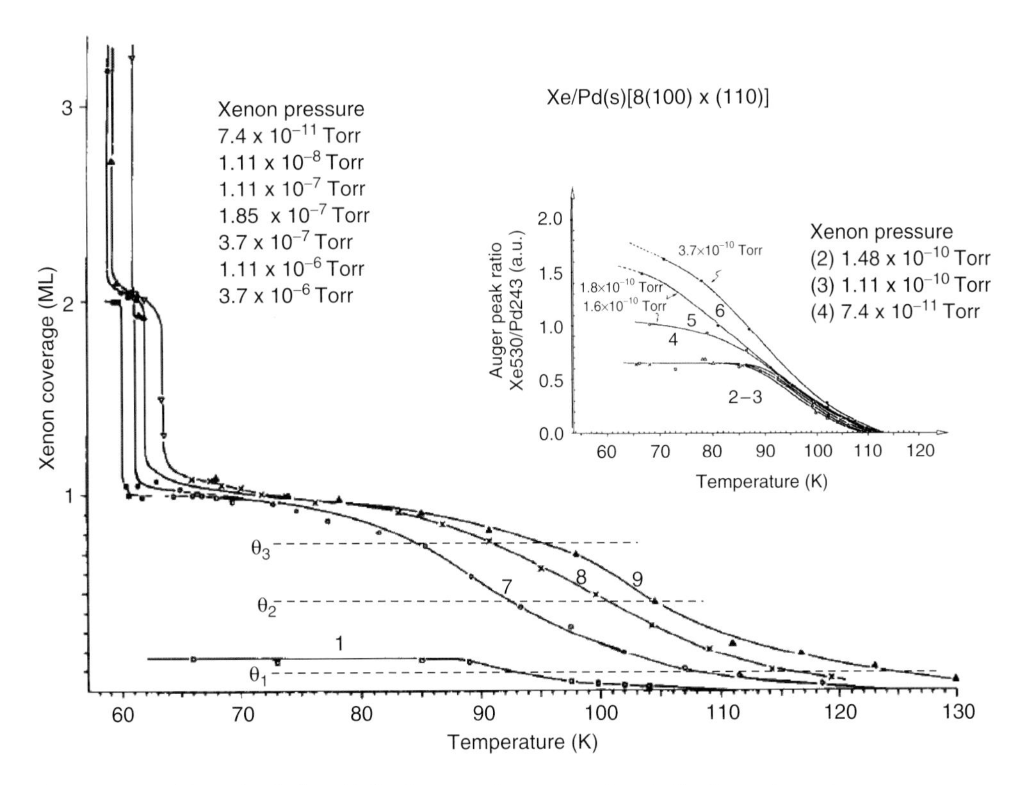


**Figure 31.12** Adsorption isobars (1–9) of Xe on a stepped Pd(s)[(110) × (110)] surface as measured by the Xe $M_5M_{4,5}N_{4,5}$ (530 eV) Auger intensity. Data were taken at all pressures given in the figure, but only selected curves are shown. $\theta_1$, $\theta_2$, $\theta_3$ are selected coverages for isosteric heat determination (see text) [26].

the coordinates of the intersection points of the dashed horizontal lines for $\theta_1$, $\theta_2$, and $\theta_3$ with all isobars are determined. As expressed by the Clausius–Clapeyron equation

$$\left(\frac{d \ln p}{d\left(\frac{1}{T}\right)}\right)_{\theta=\text{const}} = -\frac{Q_{\text{ad}}(\theta)}{R} \tag{31.69}$$

a plot of $\ln p$ versus $1/T$ (at given $\theta$) gives a straight line whose slope yields $Q_{\text{ad}}(\theta)$. The resulting isosteric heats of adsorption first up to saturation of all step sites and then up to monolayer saturation on all terraces are plotted in Figure 31.13. In accord with the stronger dipole moment of Xe atoms adsorbed at step sites [26] than on terrace sites, the adsorption energy decreases more steeply up to saturation of the step sites than for the following population of the terrace sites. It is worth mentioning that the adsorption energy near completion of the monolayer is still higher than the adsorption energy of the second Xe layer, the latter still being bound more strongly than even higher layers as shown in Figure 31.13. This enables the layer-by-layer growth of Xe films on this (and other materials') surface as already conveyed by Figure 31.9.

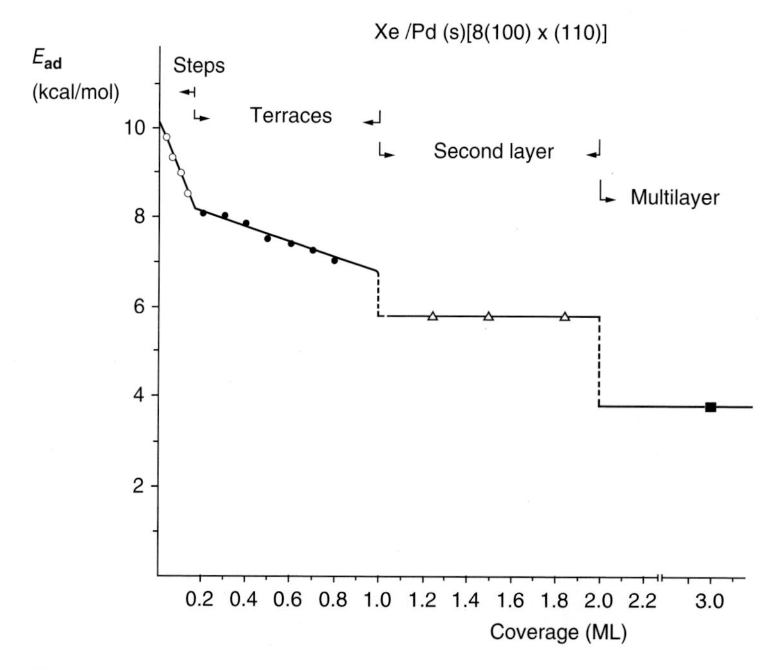


**Figure 31.13** Isosteric adsorption energies of Xe on a stepped Pd(s)[(110) × (110)] surface as obtained from the isobars shown in Figure 31.12, using the Clausius–Clapeyron equation (Equation 31.69) [26].

## 31.3 Desorption

### 31.3.1 Desorption Kinetics

Desorption – as the reverse process of adsorption – may be regarded as a dissociation reaction

$$A_{ad} \xrightarrow{k_{des}} A_{gas} + \square \tag{31.70}$$

and is, thus, always associated with an activation energy to overcome the bond between $A_{ad}$ and the substrate:

$$k_{des} = \nu\, e^{(-E^*_{des}/RT)} \tag{31.71}$$

The activation energy $E^*_{des}$ is, for instance, supplied by heating the substrate. The preexponential factor $\nu$ may be interpreted as an "attempt frequency," that is, the number of vibrations normal to the surface from which at temperature $T$ only the fraction $1/e^{E^*_{des}/RT}$ is successful and leads to desorption/dissociation. In a different picture, the desorption may be regarded as an excitation from the energy level of the adsorbed particle to that of the escaping particle. The difference in population of the two levels is controlled by the Boltzmann term $e^{-E^*_{des}/RT}$.

The desorption rate can be described by the rate equation

$$r_{des} = -\frac{dN_{A_{ad}}}{dt} = k_{des} N^n_{A_{ad}} = \nu N^n_{A_{ad}}\, e^{-\,(E^*_{des}/RT)} \tag{31.72}$$

which in the literature has also received the name "Polanyi–Wigner–equation" [e.g., 9, 10]. A measurement of $r_{des}$ may provide information about the order $n$ of the desorption reaction, the activation energy of desorption $E^*_{des}$ which is often taken as a first indication of the bond strength between $A_{ad}$ and the substrate, and the "frequency factor" $\nu$. In principle, a coverage dependence of $\nu$, $n$, and $E^*_{des}$ must also be considered:

$$r_{des} = \nu(\theta)\, N_{ad}{}^{n(\theta)} e^{((-E^*_{des}(\theta))/RT)}. \tag{31.73}$$

The desorption rate $r_{des}$ is determined by measuring, with a mass-spectrometer, the partial pressure of desorbing particles as a function of time if the heating rate $\beta$, in the easiest case a linear function of the time,

$$T = T_o + \beta t \tag{31.74}$$

is known. Assuming ideal gas behavior a simple mass balance gives the time dependence of the partial pressure:

$$\frac{dp_{A_{gas}}}{dt} = -A\frac{dN_{A_{ad}}}{dt}\frac{kT}{V} - \frac{Sp_{A_{gas}}}{V} = \frac{AkT}{V} r_{des} - \frac{Sp_{A_{gas}}}{V} \tag{31.75}$$

with $r_{des}$ being the actual desorption rate (see Equation 31.72), $A$ the desorbing area, $p_{A_{gas}}$ the partial pressure of the desorbing particles $A_{gas}$, and $S$ the pumping speed on the UHV vessel of volume $V$. The pumping speed has to be high

$$\frac{Sp_{A_{\text{gas}}}}{V} \gg \frac{dp_{A_{\text{gas}}}}{dt} \tag{31.76}$$

so that the desorption is irreversible, that is, readsorption excluded, and from Equation 31.75

$$\frac{Sp_{A_{\text{gas}}}}{V} \approx \frac{AkT}{V} r_{\text{des}}. \tag{31.77}$$

Thus, for sufficiently large $S$ the desorption rate $r_{\text{des}}$ is proportional to the measured partial pressure of released $A_{\text{gas}}$.

"TDS" or "Temperature Programmed Desorption (TPD)" was, due to its simplicity, besides work function measurements, actually one of the first surface science techniques [3]. An experimental setup is sketched in Figure 31.14. The desorbing sample is placed in UHV very near to a small aperture of area $A$ of a differentially pumped mass spectrometer in order to avoid detection of particles desorbing from other locations (e.g., sample edge, sample holder) than the well-prepared surface of interest.

Figure 31.15a shows series of Xe desorption spectra from a Pt(111) surface after (a) high and (b) very low Xe exposure [31]. The information from a TDS spectrum is manifold. First of all, the area under a desorption curve is proportional to the initial surface coverage. A plot of this area as a function of the dosage, which led to the respective coverage $\Theta$, yields an uptake curve. The slope of this curve, at any given value of $\Theta$, yields again the sticking coefficient $s$ at this coverage (see Figure 31.3a), that is, important information about the *adsorption* kinetics. Of course, an uptake curve can also be measured by any other technique that provides a signal proportional to the coverage of the adsorbate $A_{\text{ad}}$, like, for example, AES, XPS, or UPS as demonstrated by Figures 31.9 and 31.12. This has actually the advantage that spurious contributions to the desorption not coming from the sample surface do not affect the measurement.

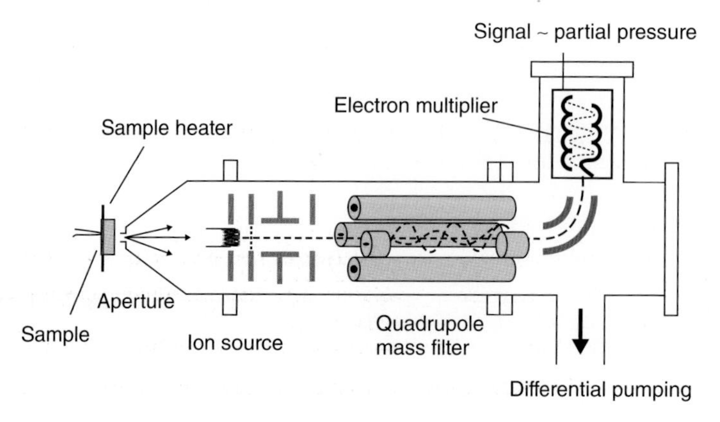


**Figure 31.14** Experimental setup for thermal desorption spectroscopy (TDS).

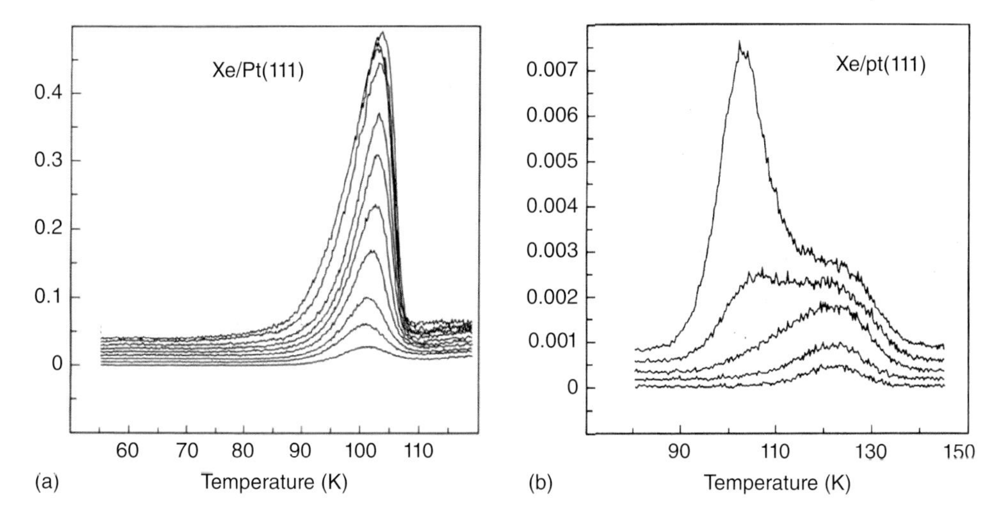


**Figure 31.15** Thermal desorption spectra of Xe from a Pt(111) surface after (a) high exposure and (b) particularly low Xe exposures. The desorption peak at 122 K corresponds to Xe adsorbed at defect sites, while the main peak at 102 K represents desorption from flat (111) regions [31].

In turn, the temperature range and form of a TDS spectrum contain information about the *desorption* kinetics. Each curve is basically a "spectrum of activation energies of desorption." If the *adsorption* process $A_{gas} \rightarrow A_{ad}$ was nonactivated, a desorption curve is even a "spectrum of adsorption energies" of $A_{ad}$. Figure 31.15b displays desorption spectra of very low initial Xe coverages from the same Pt(111) surface as in (a). These desorption curves show clearly two peaks, which suggest the presence of coexisting states of $A_{ad}$ as a function of coverage, for example, at inequivalent adsorption sites. First, the lowest coverages of Xe on this Pt(111) surface desorb at ~20 K higher temperature, suggesting that they are bound more strongly at defect sites (see also Section 31.2.3 and Figure 31.12). Only when the defect sites are saturated, Xe population and, consequently, desorption from flat (111) regions at 102 K as in Figure 31.15a sets in. This may even be exploited to "titrate" the number density of inequivalent surface sites.

Desorption spectra from homogeneous surfaces yield information about the order of the desorption process, that is, information about adsorption- and desorption-mechanisms [e.g., 9, 12].

#### 31.3.1.1 Kinetic Order of Desorption

Assuming, for instance, a first-order desorption kinetics

$$A_{ad} \xrightarrow{k^{(1)}_{des}} A_{gas} \tag{31.78}$$

with $n = 1$, the Polanyi–Wigner–equation simplifies to

$$r^{(1)}_{des} = k^{(1)}_{des} N_{A_{ad}} = \nu^{(1)} N_{A_{ad}} e^{(-E^*_{des}/RT)}. \tag{31.79}$$

In the case of a second-order desorption kinetics, the actual desorption step is preceded by a recombination process:

$$A_{ad} + A_{ad} \xrightarrow{k^{(2)}} A_{2,ad} \xrightarrow{k^{(1)}_{des}} A_{2,gas} \tag{31.80}$$

which is rate limiting if $k^{(1)}_{des} \gg k^{(2)} = k^{(2)}_{des}$; once formed on the surface $A_{2,gas}$ desorbs spontaneously. For $n = 2$ the rate equation reads

$$r^{(2)}_{des} = k^{(2)}_{des} N^2_{A_{ad}} = \nu^{(2)} N^2_{A_{ad}} e^{(-E^*_{des}/RT)} \tag{31.81}$$

In the simplest case second-order desorption is expected for dissociatively adsorbed molecules or for reaction products from two surface species

$$A_{ad} + B_{ad} \rightarrow AB_{ad} \rightarrow AB_{gas} \tag{31.82}$$

with

$$r^{(2)}_{des} = k^{(2)}_{des} N_{A_{ad}} N_{B_{ad}} = \nu^{(2)} N_{A_{ad}} N_{B_{ad}} e^{(-E^*_{des}/RT)} \tag{31.83}$$

An interesting case is desorption of zeroth order: For $n = 0$ the desorption rate becomes

$$r^{(0)}_{des} = k^{(0)}_{des} N^0_{A_{ad}} = k^{(0)}_{des} = \nu^{(0)} e^{(-E^*_{des}/RT)} \tag{31.84}$$

that is, independent of the coverage. This can be explained by desorption from a condensed multilayer; $r^{(0)}_{des}$ is then equivalent to the evaporation/sublimation rate leading to the known exponential temperature dependence of the vapor pressure. If also sub-monolayer coverages exhibit a zeroth-order desorption behavior, this may be roughly explained by desorption from the edges of condensed adsorbate islands. A rough microscopic picture behind this explanation is the equilibrium between $2D_{cond}$ islands and their 2D vapor of $A_{ad}$ as illustrated in Figure 31.41 (see Section 31.4.3.2). Actual desorption occurs out of the $2D_{vapor}$ state whose particle density, that is, 2D pressure $\Pi$, is constant as long as the total edge length (and shape) of all islands does not change significantly.

Model TDS spectra of clear-cut zeroth, first, and second order are displayed in Figure 31.16 for different initial coverages.

For *zeroth* order, the coverage independence manifests itself in the common exponential rise of all spectra for different initial coverages $\Theta_i$ (Figure 31.16). However, for different $\Theta_i$ (different area under the respective curve) the rise of the desorption rate $r_{des}$ rises until all $A_{ad}$ particles are desorbed and then abruptly falls to zero (if the pumping speed is $S = \infty$). A linearization of the flank

$$\ln r^{(0)}_{des} = \ln \nu^{(0)} - \frac{E^*_{des}}{RT} \tag{31.85}$$

and a plot $\ln r^{(0)}_{des}$ versus $1/T$ yields the activation energy of desorption $E^*_{des}$.

At the maximum of the *first*- and *second*-order desorption peak (Figure 31.16), the following condition holds for the desorption rate:

$$\frac{d^2 N_{A_{ad}}}{dt^2} = 0 \tag{31.86}$$

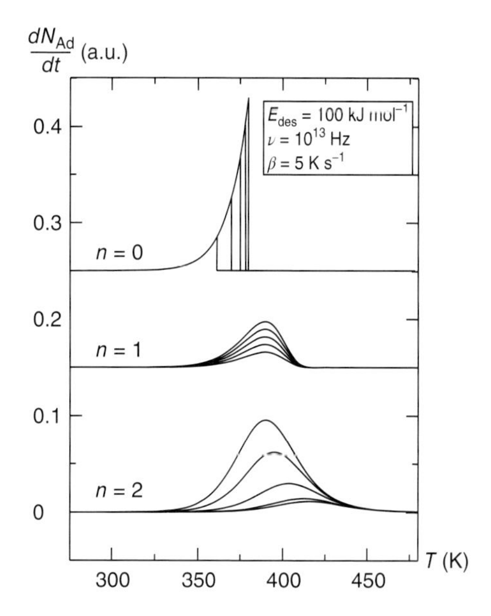


**Figure 31.16** Simulated thermal desorption spectra of desorption order $n = 0, 1, 2$ using the Polanyi–Wigner-equation (Equation 31.72) and $E^*_{des} = 100$ kJ/mol, $\nu = 10^{13}\ s^{-1}$, and $\beta = 5$ K/s.

With Equation 31.73 this leads to

$$\frac{E^*_{\text{des}}}{RT^2_{\text{m}}} = \frac{\nu^{(1)}}{\beta} \exp\left(\frac{-E^*_{\text{des}}}{RT_{\text{m}}}\right) \quad \text{for } n = 1 \tag{31.87}$$

and

$$\frac{E^*_{\text{des}}}{RT^2_{\text{m}}} = \frac{2\,N_{\text{m}}\nu^{(2)}}{\beta} \exp\left(\frac{-E^*_{\text{des}}}{RT_{\text{m}}}\right) \quad \text{for } n = 2 \tag{31.88}$$

With $\beta$ = heating rate, $T_m$ = temperature at maximum position, and $N_m$ = rest coverage at $T_m$.

Since a desorption peak of second order is nearly symmetric (in contrast to a first-order desorption peak, see Figure 31.16) one can write $2N_m = N_o$, so that

$$\frac{E^*_{\text{des}}}{RT^2_{\text{m}}} = \frac{N_{\text{o}}\nu^{(2)}}{\beta} \exp\left(-\frac{E^*_{\text{des}}}{RT_{\text{m}}}\right) \quad \text{for } n = 2 \tag{31.89}$$

This shows that $T_m$ of a first-order desorption peak is independent of the initial coverage $N_o$, while $T_m$ of a second-order desorption peak decreases with increasing initial $N_o$. This allows immediately a distinction between a first- and second-order desorption process (but see Section 31.3.2.3). A plausible explanation of the shift of $T_m$ for $n = 2$ as a function of coverage is the preceding recombination equilibrium (see Equation 31.80) before the actual desorption. The more particles $N_o$ are on the surface, the more often they collide and recombine even at lower temperatures. Vice versa, at lower coverages, the same recombination and, thus, desorption rate is obtained only with higher mobility of the adsorbed particles, that is, at higher temperatures. Note that according to Equations 31.87–31.89, the peak maximum position depends also on the heating rate $\beta$.

#### 31.3.1.2 Analysis of Desorption Data

Several methods are described in the literature to retrieve the characteristic parameters $\nu$, $n$, and $E^*_{\text{des}}$ from measured TDS spectra of a given desorption system.

1) Differentiating the logarithmic form of Equations 31.87 and 31.88 yields for both first- and second-order kinetics,

$$\frac{d\left[\ln\left(\frac{T_{\text{m}}^2}{\beta}\right)\right]}{d\left(\frac{1}{T_{\text{m}}}\right)} = \frac{E^*_{\text{des}}}{R}. \tag{31.90}$$

Thus, varying $\beta$ and plotting $\ln(T_{\text{m}}^2/\beta)$ versus $1/T_{\text{m}}$ yields $E^*_{\text{des}}$; $\beta$, however, needs to be varied by at least a factor of 50–100. Inserting $E^*_{\text{des}}$ into both Equations 31.87 and 31.88 gives $\nu$.

2) The most frequently applied method is the "Redhead analysis" [32]. Equation 31.87 shows that for a first-order desorption process $T_{\text{m}}$ is independent of coverage. Consequently, $E^*_{\text{des}}$ can be obtained directly from a measurement of $T_{\text{m}}$ provided a value of $\nu^{(1)}$ is assumed. The so-called "Redhead approximation"

$$E^*_{\text{des}} = RT_{\text{m}}\left[\ln\frac{\nu^{(1)}T_{\text{m}}}{\beta} - 3.64\right] \tag{31.91}$$

is a "very nearly linear" [32] relationship between $E^*_{\text{des}}$ and $T_{\text{m}}$ in the range $10^{13} > \nu^{(1)} > 10^8$. Assuming the usual "attempt frequency" $\nu^{(1)} = 10^{13}\ \text{s}^{-1}$, and varying $\beta$ by at least 2 orders of magnitude shows that the deviation from linearity is acceptably small.

3) The "leading edge method" [33] assumes that at the very onset of a TDS trace the "rest coverage" is very nearly equal to the initial coverage $\theta_{\text{o}}$. Hence:

$$r_{\text{des}} = -\frac{d\theta}{dt} = \nu\,\theta_{\text{o}}^n \exp\frac{-E^*_{\text{des}}}{RT} \tag{31.92}$$

With the known coverage $\theta_{\text{o}}$ and order $n$, a plot $\ln r_{\text{des}}$ versus $1/T$ yields again $E^*_{\text{des}}$. The method requires very high-quality data within the analyzed interval at the onset of the measured TDS spectra.

4) The so-called "complete analysis" [12, 34] is illustrated in Figure 31.17. Panel (a) shows selected TDS spectra of silver from a Ru(0001) surface (see also Figure 31.20). While at low coverage ($\theta_{\text{o}} = 0.18$ ML), the peak shape suggests a first-order desorption process, with increasing initial coverage the desorption curve develops more and more the form characteristic of a zeroth-order desorption process (see Figure 31.16). This suggests that *at the high desorption temperature* low Ag coverages form a 2D gas phase of silver atoms which desorb in a first order process, while with increasing Ag coverage the critical 2D gas pressure is exceeded leading to a 2D equilibrium between 2D condensed Ag islands and a constant 2D vapor pressure, which results in a zeroth order like desorption

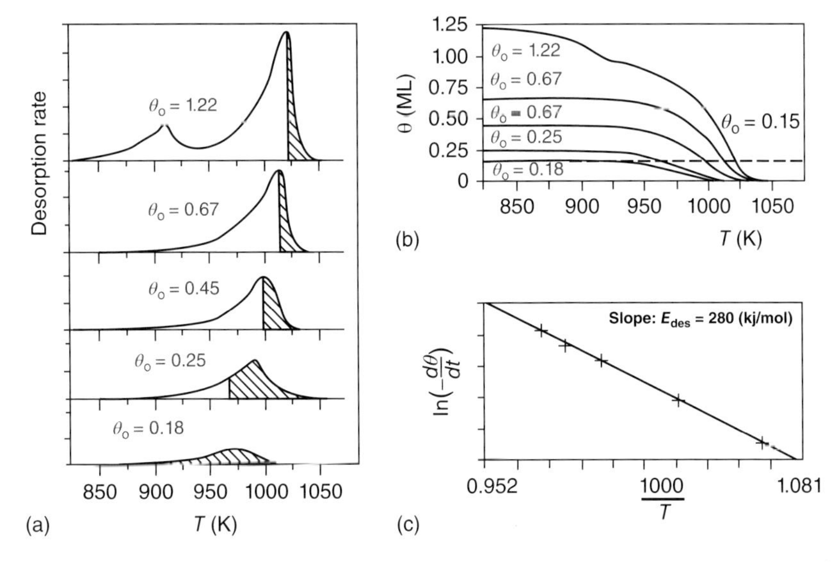


**Figure 31.17** (a–c) Illustration of the "complete analysis" of thermal desorption spectra of Ag from a Ru(0001) surface, see text [12].

process (see also Section 31.4.3.2). Beyond $\theta_0 = 1$ ML a second zeroth-order desorption peak develops at about 100 K lower temperature, which corresponds to desorption/evaporation of Ag atoms from the second layer.

Integration of these spectra from the high temperature side $T(\theta_r = 0)$ to a certain temperature $T(\theta_r) < T(\theta_r = 0)$ yields the rest-coverage $\theta_r$ at $T(\theta_r)$ and eventually at very low temperatures the initial coverage $\theta_0$ as plotted in (b). Depending on $\theta_0$ a certain rest-coverage $\theta_r$ is reached at different temperatures $T(\theta_r)$. As an example $\theta_r = 0.15$ ML is marked in the spectra of Figure 31.17a,b. The height of each original desorption trace *at* $T(\theta_r)$ in (a) represents the desorption rate $r_{des}(\theta_r)$. The slope of an Arrhenius plot $\ln(r_{des}(\theta_r))$ versus $1/T(\theta_r)$ from corresponding pairs of $r_{des}(\theta_r)$ and $T(\theta_r)$ yields the activation energy of desorption $E^*_{des}$ at the particular rest-coverage $\theta_r$, and Arrhenius plots for different values of $\theta_r$ may result in different $E^*_{des}$ values, as well as different intercepts $\rho(\theta_r)$ equal to $\rho(\theta_r) = \ln \nu + n \ln(\theta_r)$. Hence, a plot $\rho(\theta_r)$ versus $\ln(\theta_r)$ yields $\nu$ and $n$.

5) The technique of modulated beam relaxation spectroscopy (MBRS) [35] can also be used very beneficial to study the kinetics of desorption. Figure 31.18 shows the experimental setup. A periodically chopped particle beam strikes the sample surface, and the back-scattered particles are detected with a mass-spectrometer. The principle of this technique for the elucidation of a first-order desorption process is illustrated in Figure 31.19a [35]: The intensity $I_0$ of the primary beam is modulated by a rotating disc chopper producing a square shaped wave form of frequency $\omega$. Due to the interaction with the surface, the wave form of the scattered particle intensity is changed and may be characterized by its amplitude $I_S$ and the phase lag $\tan\varphi$ of the first Fourier component by means

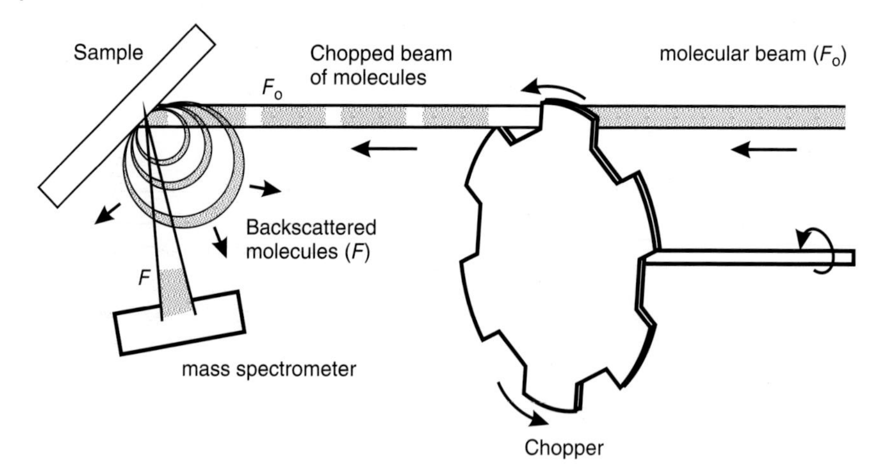


**Figure 31.18** Experimental setup of a chopped molecular beam experiment.

of lock-in techniques. In the simple case of first-order desorption, the variation of the surface concentration is given by

$$\frac{dN_{\mathrm{ad}}}{dt} = sF_{\mathrm{o}} - k_{\mathrm{des}}N_{\mathrm{ad}} \tag{31.93}$$

and the amplitude and phase lag of the signal are, respectively, [36]

$$F_{\mathrm{S}} = \frac{F_{\mathrm{o}}}{\left(1 + \frac{\omega^2}{k_{\mathrm{des}}^2}\right)^{1/2}} \tag{31.94}$$

and

$$\tan\varphi = \frac{\omega}{k_{\mathrm{des}}} \tag{31.95}$$

For a first-order desorption process the inverse of the rate constant, that is, $\tau = 1/k_{\mathrm{des}}^{(1)}$, corresponds to the mean residence time of the adsorbed particles on the surface:

$$\tau = \tau_{\mathrm{o}}\, e^{(E_{\mathrm{des}}^{*}/RT)} \quad \text{(Frenkel equation)} \tag{31.96}$$

which causes the phase lag:

$$\tan\varphi = \omega\tau = \omega\tau_{\mathrm{o}}\, e^{(E_{\mathrm{des}}^{*}/RT)} \tag{31.97}$$

or

$$\ln\frac{\tan\varphi}{\omega} = \ln\tau_{\mathrm{o}} + \frac{E_{\mathrm{des}}^{*}}{RT} \tag{31.98}$$

Thus, measurement of $\varphi$ and a plot $\ln(\tan\varphi/\omega) = \ln\tau$ versus $(1/T)$ yields both the activation energy of desorption $E_{\mathrm{des}}^{*}$ and the preexponential factor $k_{\mathrm{o,des}}^{(1)} = 1/\tau_{\mathrm{o}}$. Figure 31.19b shows such a plot for the system CO/Pt(111), resulting in $E_{\mathrm{des}}^{*} = 32$ kcal/mol and $k_{\mathrm{o,des}}^{(1)} = 1/\tau_{\mathrm{o}} = 5 \times 10^{14}\ \mathrm{s}^{-1}$. Both values are true "initial values" for vanishingly small CO coverages, which are hardly accessible with other methods [35].

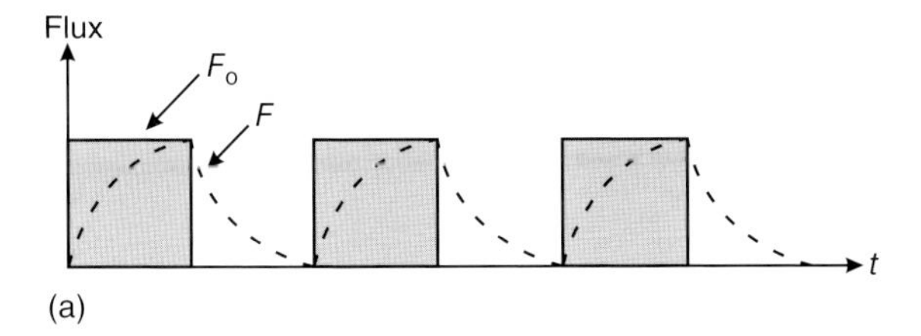


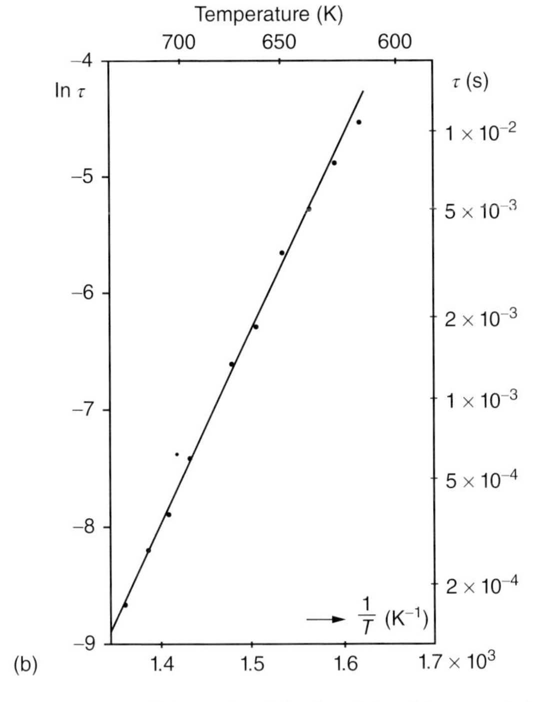


**Figure 31.19** (a) Principle of the "modulated beam technique" for studying the kinetics of desorption and surface reactions [35]. (b) Mean surface residence time $\tau$ of CO molecules on Pt(111) as a function of temperature [35].

### 31.3.2 Complications with TDS Spectra

Four major complications in the interpretation of TDS spectra shall be addressed here. These complications arise from

- the fact that the necessary heating is a drastic disturbance of the system to be measured;
- surface heterogeneities with the consequence of a distribution of adsorption energy values at;
- unequal adsorption sites;

- coverage dependent lateral interactions between the adsorbate particles; and
- a so-called "compensation effect."

and, of course, a combination of all four.

#### 31.3.2.1 **Influence of Heating**

Obviously, both the application of heat and the heating rate may affect the TDS signal. The supply of heat causes surmounting of activation barriers. Assuming that the adsorbate on the surface in question may exist in different states, for example, of different structure and density or on inequivalent sites of a heterogeneous surface, the relative population of different states depends on their energy difference and on activation barriers between them. If the activation barriers are low, the TDS spectrum will represent the equilibrium population of the different states *at the desorption temperature* if the heating *rate* is adjusted to the finite kinetics of exchange processes between the different desorption states. If the activation barriers are very high, the TDS spectrum will be more representative of the population *at the adsorption temperature.*

As an example, Figure 31.20 displays TDS spectra of Cu, Ag, and Au from a Ru(0001) surface [38–40]. In all three cases, the initial coverage is $\theta_0 > 1$ ML, so that desorption from the first metal layer (in direct contact with the Ru(0001) substrate) at high temperature and desorption from the higher layers at lower temperature is detected. All three metals are known to grow via nucleation, island formation, and initial layer-by-layer growth mode on Ru(0001). It may therefore be expected that the desorption, except at very low initial coverages, follows zeroth-order kinetics. As indicated by the spectra in Figure 31.20, this is, indeed, the case for Cu and Ag for both the first and the higher layers. Surprising, however, is the fact that the high-temperature, that is, the first layer, desorption peak of *gold* clearly reflects a first order like desorption kinetics: The maxima of all submonolayer desorption peaks occur at the same temperature, namely, ~1425 K (see Figure 31.16). This can be explained as follows: The desorption of the first Au layer occurs at such a high temperature that intermixing and surface alloy formation between gold and the Ru(0001) substrate is activated, so that Au desorption actually takes place out of an alloy phase. In turn, all other desorption temperatures of the higher Au layers as well as of the first and higher layers of Cu and Ag are too low in order to cause surface alloy formation and therefore exhibit a zeroth-order desorption behavior, except at very low population of the respective deposit due to the prevalence of a 2D vapor phase (at the desorption temperature) addressed in more detail in Section 31.4.3.2. This interpretation is supported by Monte-Carlo simulations of the TDS spectra of all three metal-on-metal systems [38].

#### 31.3.2.2 **Surface Heterogeneity**

Chemically and/or structurally inhomogeneous surfaces offer a variety of adsorption sites of different adsorption energy. This may be reflected in multipeaked TDS spectra as seen in Figure 31.15b for structural defects. As a second example Figure 31.21 shows a desorption spectrum of CO from an ordered $Cu_3Pt(111)$ alloy surface [41].

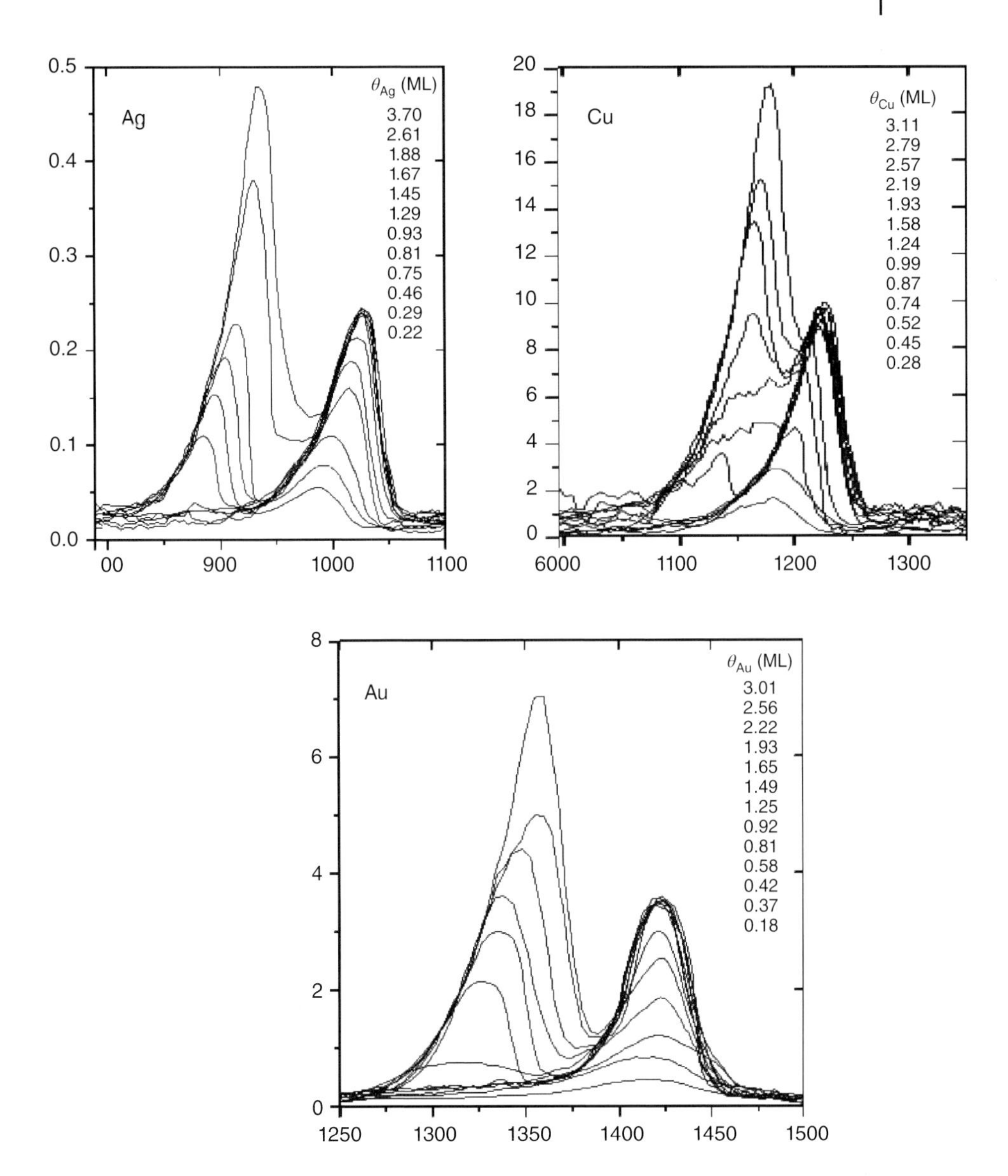


**Figure 31.20** Thermal desorption spectra of Cu, Ag, and Au from a Ru(0001) surface. The high temperature peak in all three panels corresponds to desorption of the first metal layer (in direct contact with the Ru(0001) substrate), while the lower temperature peak corresponds to desorption from the higher layers. Note the different shape of the first layer Au desorption peak at 1425 K compared to all other zeroth-order desorption peaks [37].

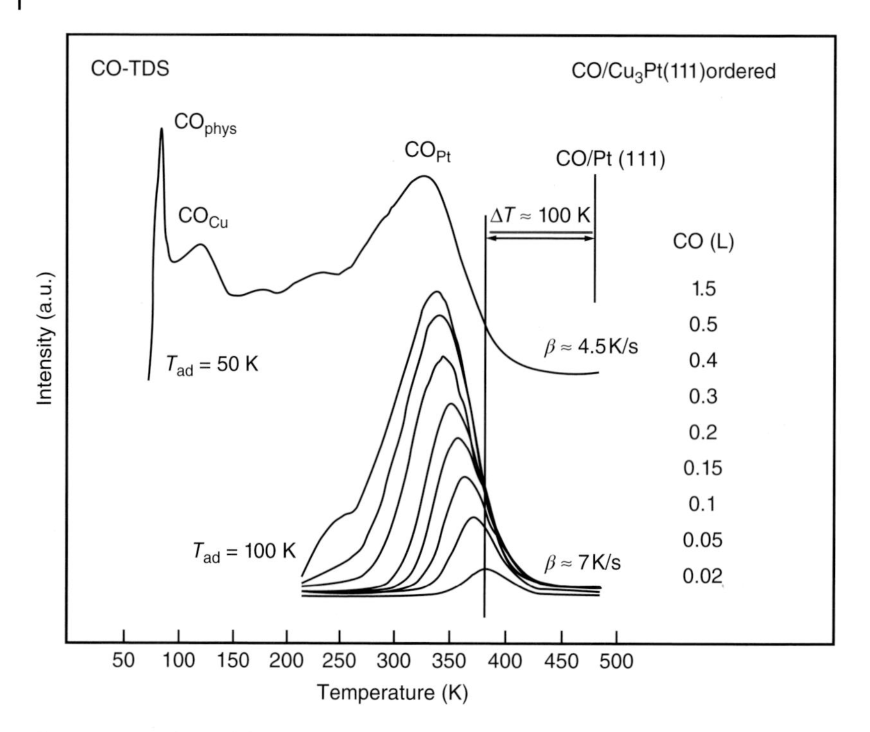


**Figure 31.21** Thermal desorption spectra of CO from an ordered $Cu_3Pt(111)$ surface. Adsorption at 110 K enables only population of the Pt sites, while adsorption at 50 K leads to population of ensemble sites of different local $Cu_xPt_{1-x}$ composition, which leads to the different desorption peaks. Note that desorption of CO from the Pt sites of the alloy occurs at ~100 K lower temperature than from a Pt(111) surface [41].

The various peaks between 100 and 350 K correspond to CO desorption from sites ("ensembles") of different local $Pt_xCu_{1-x}$ composition (see also Section 31.4.3.5, and Chapter 38 in this volume). The analysis of such spectra is undoubtedly complicated by the overlap of the component peaks and/or temperature-induced phase transitions within the adsorbed layer during the temperature ramp as mentioned in the previous section. In this case, an interpretation of the TDS spectra is only possible on the basis of kinetic simulations with model systems assuming the microscopic structure of the substrate and interaction energies between all participating particles on the part of the substrate as well as adsorbate.

#### 31.3.2.3 **Lateral Interactions**

As considered by the Fowler–Guggenheim isotherm, the adsorbed particles may laterally interact, either directly or indirectly through adsorbate-induced modifications of the substrate ('through substrate interactions'). The case of *attractive* lateral interactions was already explicitly demonstrated for Xe atoms on a Pd(100) surface (precovered with 1 ML of Xe) in Figure 31.9.

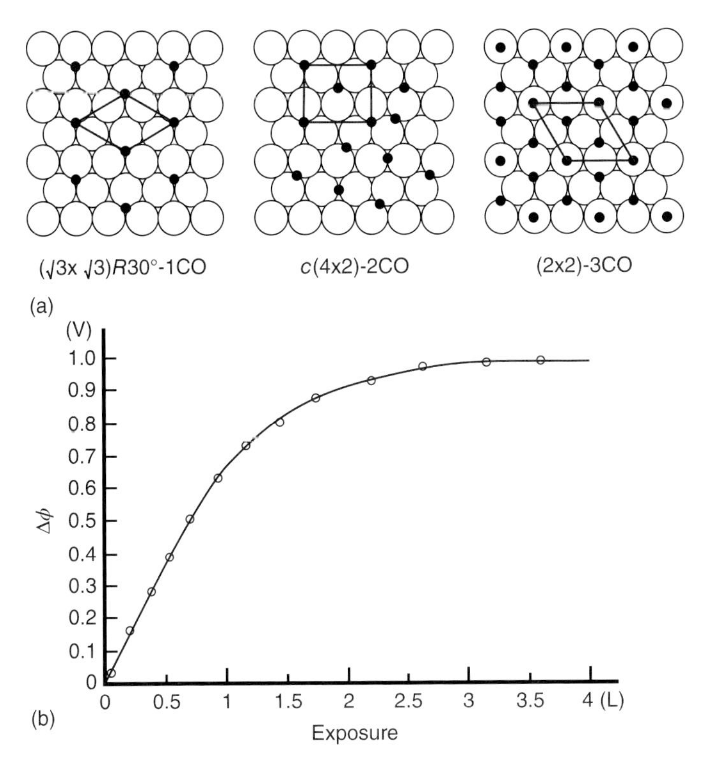


**Figure 31.22** (a) Selected structures of increasing density of CO adsorbed on a Pd(111) surface and (b) CO-induced work function increase on Pd(111) [42].

In turn, repulsive interactions are to be expected if, for instance, the adsorbate particles acquire a strong dipole moment upon interaction with the substrate. A prototypical example is the adsorption of carbon monoxide on Pd(111), which with increasing coverage passes through a series of adsorbate structures (Figure 31.22a) and is accompanied by a significant work function increase of ~1 eV (Figure 31.22b) corresponding to a dipole moment of 0.33 D at maximum coverage [42, 43]. The parallel dipoles of the adsorbed CO molecules with the positive end on the metal-near carbon atom (see Section 31.4.2.3) repel each other. As a consequence, the isosteric heat of adsorption is found to decrease with increasing CO coverage and each CO phase of increased density. TDS, thus, exhibit even a rather multipeaked structure, which with increasing coverage extends to lower and lower temperatures as displayed in Figure 31.23 [44].

#### 31.3.2.4 Compensation Effect

The assumption of a constant, coverage independent preexponential factor ($\nu$ or $k_0$) is per se not justified. In fact, many cases are known, which show a linear correlation [45]

$$\ln \nu = a \cdot E^*_{\text{des}} + b \tag{31.99}$$

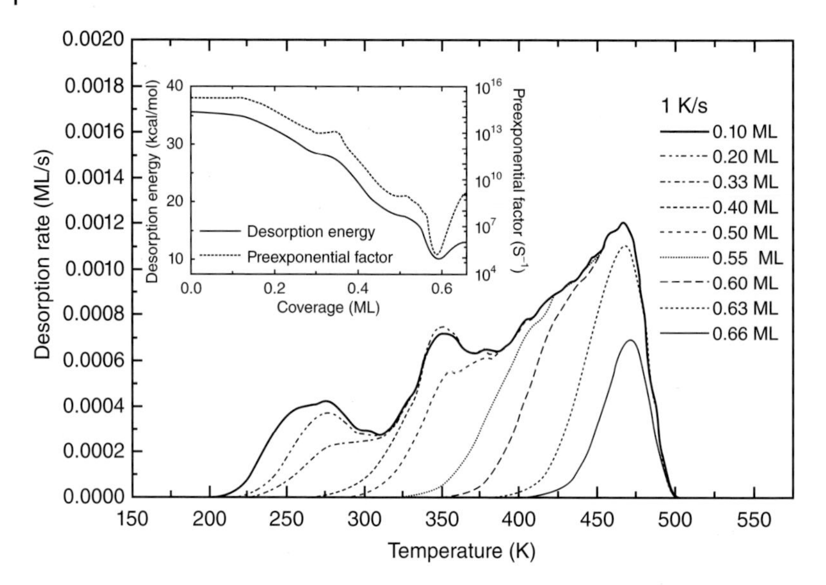


**Figure 31.23** Thermal desorption traces for different CO coverages on Pd(111) as deduced from laser desorption experiments [44].

with $a$ and $b$ being constants; a change of $E^*_{des}$, for example, as a function of coverage, is accompanied by a respective change in $\nu$. Two important cases must be distinguished as displayed in Figure 31.24. The two Arrhenius lines $\ln \nu$ versus $(1/T)$ shown in both panels intersect at a value of $1/T = 1/T_x$ with

$$0 < \frac{1}{T_x} < +\infty$$

or

$$-\infty < \frac{1}{T_x} < 0 \tag{31.100}$$

A comparison of Equation 31.99 with the logarithmic form of the Arrhenius equation

$$\ln k_{des} = \ln \nu - \frac{E^*_{des}}{RT} \tag{31.101}$$

yields for $E^*_{des} = 0$: $b = \ln \nu_o = \ln k_{des,0}$ and $a = 1/RT_x$. In the case $0 < 1/T_x < +\infty$ the point of intersection lies in the regime of "real temperatures" (Figure 31.24a), with $E^*_{des,2} < E^*_{des,1}$ and $\nu_2 < \nu_1$. Consequently, an increase of the preexponential factor $\nu$ is counterbalanced by a decrease in the Boltzmann term $e^{(-E^*_{des}/RT)}$ ("compensation effect"). Vice versa, if $-\infty < (1/T_x) < 0$ the point of intersection lies in the regime of "negative temperatures" (Figure 31.24b), so that an increase in $E^*_{des}$ leads to a decrease of $\nu$; for $E^*_{des,2} < E^*_{des,1}$ Figure 31.24b shows $\nu_2 > \nu_1$ ("anti-compensation effect"). Microscopic interpretations of the "compensation effect" can be found in Refs. [46, 47].

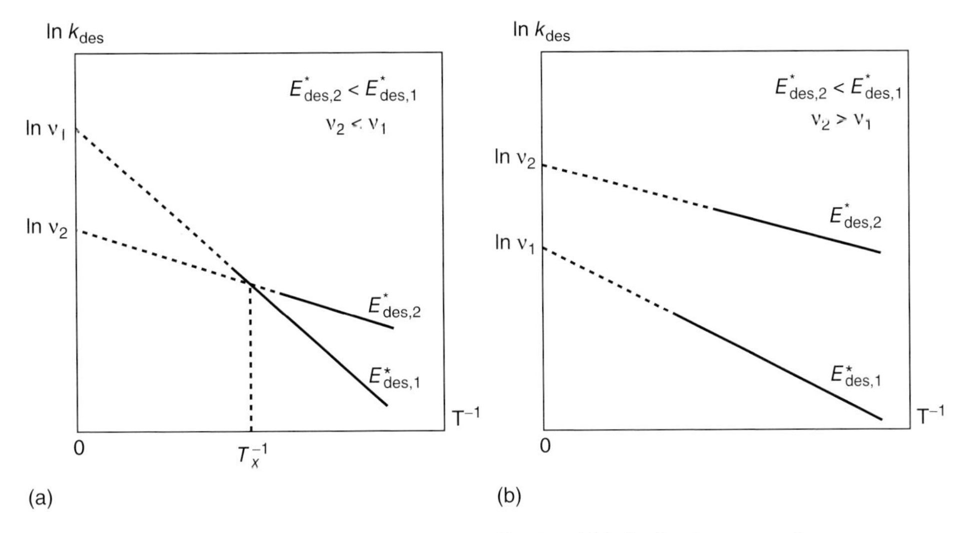


**Figure 31.24** Illustration of (a) the "compensation effect" and (b) the "anti-compensation effect" in thermally activated processes, e.g. thermal desorption (see text).

## 31.4 Properties of the Adsorbate

### 31.4.1 Physisorption, Chemisorption

The driving force for adsorption is the fact that the total energy decreases when an adsorbate particle $A_{gas}$ is brought to the bonding distance of $A_{ad}$ on the surface. This distance dependence of the energy is described by a "potential" curve whose characteristics depend on the nature of the bonding mechanism. The two limiting cases are "physisorption" and "chemisorption".

"Physisorption" arises from mere van der Waals interactions between any kind of matter, for example, two atoms. Ground state fluctuations of the electronic charge of an atom lead to a fluctuating dipole moment of the atom, which in turn induces a fluctuating dipole moment on an adjacent atom. Both dynamic dipoles and, thus, both atoms attract each other until, with decreasing distance $r$, their "electron clouds" start to repel each other (Pauli repulsion). Both contributions are combined in the so-called Lennard-Jones (LJ) Potential sketched in Figure 31.25.

In the case of two interacting atoms, the attractive part of this potential varies with $r^{-6}$ and dominates at greater distances, while the short-range repulsive contribution is proportional to the overlap of the orbitals, which – in principle resulting in an exponential increase of the energy with decreasing distance – is approximated by an $r^{-12}$ dependence:

$$E(r) = 4\varepsilon \left[ \left(\frac{\sigma}{r}\right)^{12} - \left(\frac{\sigma}{r}\right)^{6} \right] \tag{31.102}$$

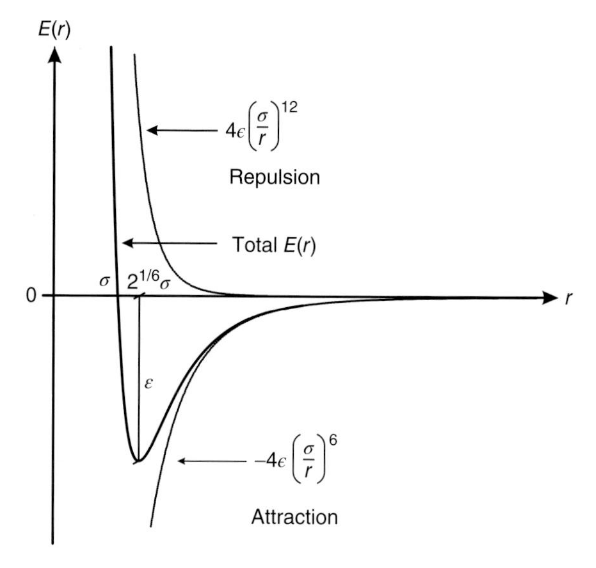


**Figure 31.25** Lennard-Jones (12,6) Potential and decomposition in attractive and repulsive contribution.

This potential has two distinct points, a zero at $r = \sigma$ and a minimum of depth $\varepsilon$ at $r_o = 2^{1/6}\ \sigma$. Both parameters $\sigma$ and $\varepsilon$ can be derived from the virial coefficient of the equation of state of the respective real gas. $E(r)$ can also be determined from a comparison of experimental and theoretical surface – atom/molecule scattering results. For scattered beams of polarized NO molecules, the potential is even found to be different depending on whether the NO molecules hit the surface N- or O-head on [48].

A physisorbed atom on a surface interacts via the LJ-potential with all atoms in the solid. However, while only the nearest-neighbor surface atoms cause repulsion, the attraction results from the summation over all pair interactions with all substrate atoms, which leads to a $z^{-3}$ – dependence of the attractive contribution ($z$ perpendicular to the surface). Since in this summation the nearest neighbors are the major contributors, physisorbed species tend to seek the highest coordinated adsorption sites on the surface, for example, kink, step, or hollow sites with as many nearest neighbors as possible.

In the case of "chemisorption," a true chemical bond is formed between $A_{ad}$ and the substrate due to overlap of electron orbitals/states. This overlap depends on the relative energy and the nature of the involved orbitals and occurs only at very short distances, dominantly with nearest-neighbor atoms. The minimum of the "chemisorption potential $C$" in Figure 31.26 occurs, thus, at much shorter distances and may also be much deeper than the "physisorption minimum $P$" indicating a stronger bond, a real chemical bond involving charge exchange between the participating atoms. Within a simple one-electron picture, the possible orbital interactions between an ad-particle and the substrate are sketched in Figure 31.27. The overlap of two unoccupied orbitals (1) above the Fermi-level $E_F$ of the substrate

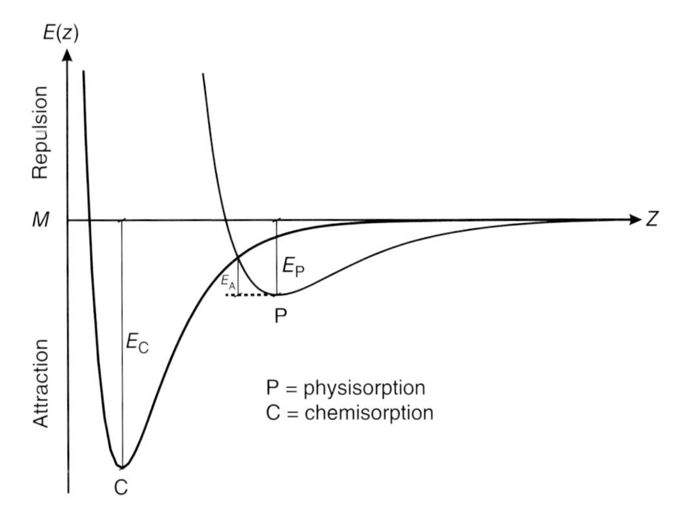


**Figure 31.26** Potential diagram for atomic adsorption with physisorption and chemisorption minimum.

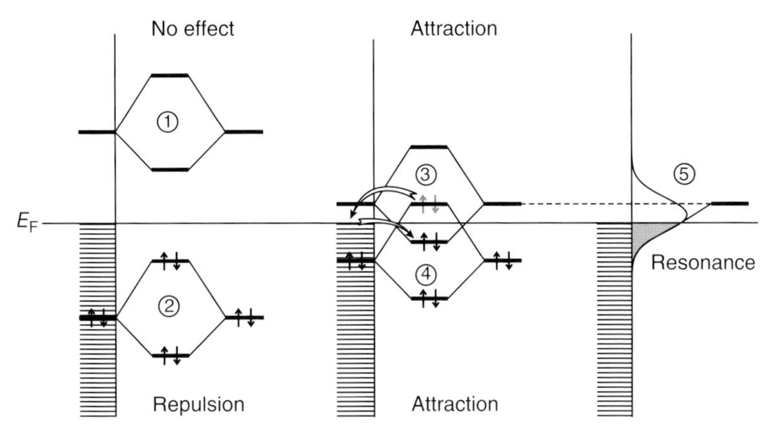


**Figure 31.27** Schematic diagram of hybridization between adsorbate and substrate electronic states. The five possibilities 1–5 are explained in the text [49].

has no bonding effect, and the overlap of two fully occupied orbitals (2) is even repulsive, unless the unoccupied orbitals (3) and the filled orbitals (4) are so close to the Fermi level that their hybridization leads to an empty orbital below $E_F$ (3) or an occupied orbital above $E_F$ (4). Both situations are unstable, and a corresponding charge flow to ((3)) or from ((4)) the Fermi level as indicated by the arrows makes both interactions attractive [49]. This charge flow is possible due to the huge charge reservoir of the substrate and suggests that the dominantly localized bonding interaction has, to a certain extent, also a longer ranging effect on the substrate, in particular, since the involved valence electron states on the substrate side are *not* localized atomic orbitals but delocalized band states (see Chapter 5, Volume 2). A molecular filled or unfilled orbital therefore does not only interact with a singled-out surface atom state but more or less with all valence band states leading to a broadening and the formation of a so-called "resonance" (5) in Figure 31.27. Filling of the resonance tail below $E_F$ may, thus, add to attraction [50].

It is obvious that the hybridization and charge transfer from/to the ad-particle involved in a chemisorption bond leads to changes of the properties of both bonding partners, which manifest themselves in energetic shifts, population, and depopulation of electronic states; in altered structural parameters between atoms of both the ad-particle and the substrate surface; in changed vibrational properties, and so on, and in the extreme case in the dissociation of adsorbed molecules. In the latter case, the charge exchange between the adsorbing molecules and the surface destabilizes the intramolecular bond such that the molecule dissociates and the interaction of all molecular *fragments* with the surface is energetically favored.

For instance, diatomic molecules $A_{2,\text{gas}}$ such as $H_2$ (see Chapter 36 in this volume) or $O_2$ are known to adsorb dissociatively at room temperature on many surfaces. The final product are two adsorbed atoms $A_{\text{ad}}$ per adsorbed molecule. The same product, of course, can also be obtained by first dissociating the $A_2$ molecule in the gas phase and adsorbing the resultant atoms. This, however, requires first the "expenditure" of the full molecular dissociation energy $E^*_{\text{diss}}$, which is then more than recovered by the energy gain of two separately chemisorbed atoms as illustrated in Figure 31.28, namely, $2E_{A_{\text{ad}}} = E_{\text{diss}} + E_{\text{c}}$. Direct molecular dissociative adsorption saves this expenditure, but may proceed stepwise via several precursor states as illustrated in Figure 31.29. The incoming molecule $A_{2_{\text{gas}}}$ may first be trapped in a molecular physisorbed state $A_{2_{\text{phys}}}$, which via a small activation energy may convert into a *still molecular* chemisorbed state $A_{2_{\text{chem}}}$, which ultimately, again via a certain activation barrier, dissociates into two separate chemisorbed atoms $A_{\text{ad,chem}}$. With each step, the respective species comes closer to the surface and becomes bound more strongly (see also Chapter 47 in Volume 6).

Originally an adsorbate system was classified as "physisorbed" or "chemisorbed" by its adsorption energy. A measured adsorption energy (see Chapter 32 in this volume) or activation energy of desorption (see Section 31.3) below 30–40 kJ/mol was taken to be indicative of mere "physisorption," while higher values up to several hundreds of kJ per mol were considered characteristic of "chemisorption" [3]. This definition, however, is too imprecise. Even in the case of the adsorbed rare gas xenon, originally considered to be a prototypical "physisorption system," both experimental results and theoretical calculations favor a weak chemisorption as discussed further in what follows.

A more precise distinction between "physisorption" and "chemisorption," is based on whether the adsorptive bond is or is not accompanied by hybridization and charge redistribution, which for instance, in photoemission spectroscopy manifests itself in relative shifts, depopulation, and population of selected orbitals with respect to others. Instead, photoemission spectra of physisorbed species show, apart from a line broadening, basically the same rigid shift of all orbitals due to screening effects (see Chapter 5 in Volume 1).

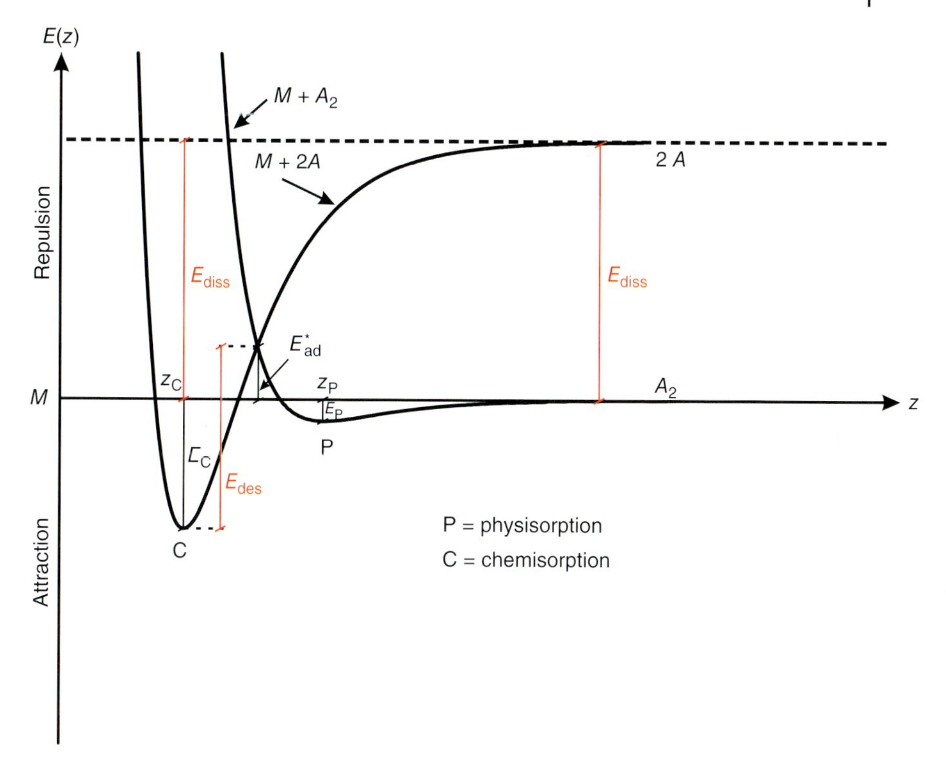


**Figure 31.28** Potential curves for molecular physisorption ($M + A_2$) and dissociative chemisorption ($M + 2A$); $M$ = metal, $z_p$, $z_c$ = equilibrium distances in the physisorbed and chemisorbed state, $E_p$, $E_c$ = physisorption, chemisorption energy. $E^*_{ad}$ = activation barrier between physisorbed and chemisorbed state.

The next section includes a few characteristic examples illustrating the addressed changes of adsorbed molecules and the substrate upon adsorption. Many more related examples for different adsorbates are the content of the following chapters of Volumes 5 and 6.

## 31.4.2
## Properties at Individual Ad-Particles

### 31.4.2.1 Corrugation, Localization, Mobility

The most prominent feature of an "adsorbate–substrate-complex" is the adsorption site. The equilibrium adsorption site of an individual ad-particle may be ontop of a substrate atom ("ontop site"), between two substrate atoms ("bridge site"), in a hollow site between three or four substrate atoms on a (111) or (100) surface, respectively, or on any other less symmetric site on any crystallographic surface plane (e.g., Figure 31.22a). Structure sensitive methods such as IV-LEED, STM, angular- and energy-dependent UPS and XPS, and ISS as well as HREELS enable a clear distinction between these various possibilities as laid out in the chapters

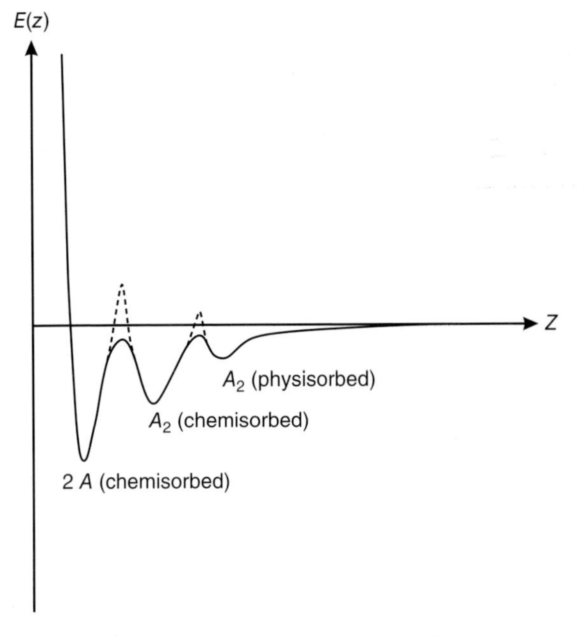


**Figure 31.29** Molecular adsorption may actually proceed via several precursor states, for example, molecularly physisorbed, and *molecularly* chemisorbed until final dissociative chemisorption (see also Figure 31.38).

of Volume 1. The adsorption site may depend on the coverage; with increasing coverage "crowding" may lead to a displacement of ad-particles from their initial equilibrium site at vanishing coverage due to lateral interactions. The change of the interaction energy between an ad-particle and the substrate at varying adsorption sites is called *corrugation function* or simply *corrugation* and is sketched along one dimension in Figure 31.30. The shaded region in Figure 31.30 marks those surface positions, which ad-particles will occupy depending on their thermal energy. At very low temperatures, the particles will be trapped in the valleys of the corrugation function, that is, at fixed surface positions. The movement of an ad-particle from one minimum energy site to an adjacent one obviously requires the activation energy of migration $E^*_{\mathrm{m}}$. The probability of such site exchange ("hopping") is expressed by the rate constant of this process

$$k_{\mathrm{m}} = k_{\mathrm{m,o}} e^{(-E^*_{\mathrm{m}}/RT)} \tag{31.103}$$

or by the residence time of a particle in a particular site

$$\tau = \tau_{\mathrm{o}}\, e^{(+E^*_{\mathrm{m}}/RT)} \quad \text{(Frenkel-equation)} \tag{31.104}$$

Once the thermal energy of the ad-particles is higher than $E^*_{\mathrm{m}}$ but still lower than $E_{\mathrm{gas}}$, the particles are still attracted by the substrate but may freely move across the surface. Depending on the size of $E^*_{\mathrm{m}}$, that is, the amplitude of the corrugation function, ad-particles may therefore at given temperature be "localized" or "mobile".

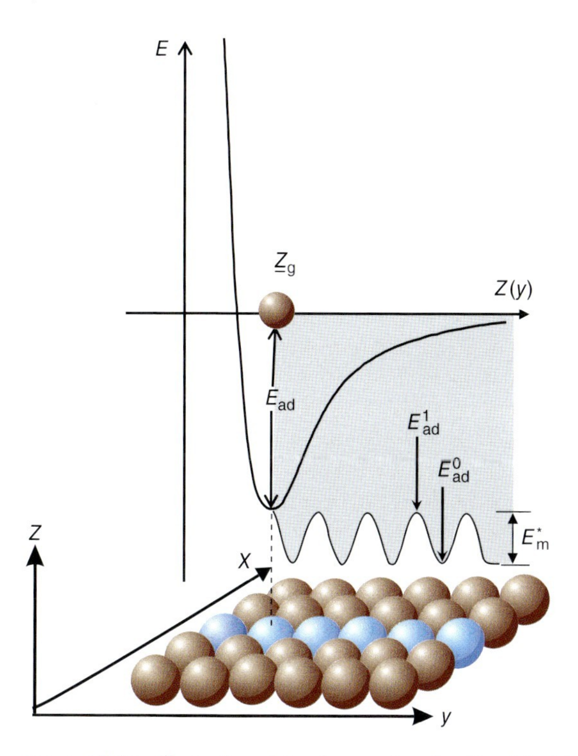


**Figure 31.30** Illustration of the "corrugation function," that is, the site-specific modulation of the adsorption energy $E_{ad}$ and the resultant activation energy of surface migration $E^{*}_{m}$.

Figure 31.31 illustrates the ratio

$$\frac{\text{mobile}}{\text{immobile}} = \frac{e^{(-E^{*}_{m}/kT)} - e^{\left(-E^{\square}_{gas}/kT\right)}}{1 - e^{(-E^{*}_{m}/kT)}} \tag{31.105}$$

of mobile versus localized particles for two different corrugation amplitudes at three different temperatures. This ratio is obviously important for reactions between adsorbed species. Only those particles in the tail of the respective Boltzmann distribution with energies $E_{gas} > E > E^{*}_{m}$ are near the surface and mobile enough to diffuse across the surface and to collide and react with other particles.

The degree of localization or nonlocalization is also a decisive parameter in a statistical description of adsorbate properties. Localized atoms have no translation degree of freedom, while mobile atoms have two degrees of translational freedom (parallel to the surface) and only one vibrational degree of freedom perpendicular to the surface. An exhaustive description of the statistical thermodynamics of adsorbates is given in Chapter 50 of Volume 6.

Obviously, the size of the adsorbed species compared to periodicity lengths of the substrate surface matters in this context. Large flat-lying organic molecules spanning several substrate atoms may "have" a much smother corrugation than atoms and molecules smaller than interatomic distances between the substrate atoms (see Chapter 50 in Volume 6).

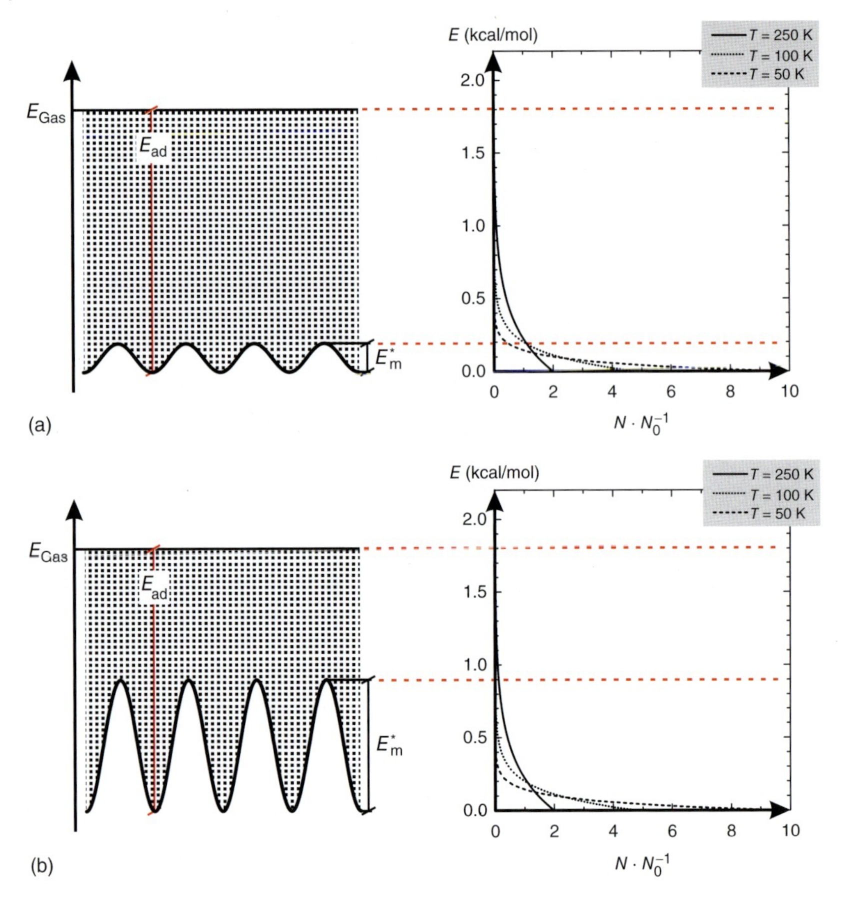


**Figure 31.31** Illustration of the ratio of mobile versus immobile ad-particles (see Equation 31.105) on two surfaces of different corrugation.

The properties of both the adsorbed species and the substrate change depending on the interaction strength between both. Through their bond to the substrate adsorbed *atoms and molecules* may accept or donate charge, which shows up in work function changes (see, e.g., Figure 31.22) and "chemical shifts" of electron states. Adsorbed *molecules*, in addition, exhibit altered vibrational properties and may, in the extreme case, even dissociate.

#### 31.4.2.2 **Adsorbed Xenon**

Adsorbed xenon atoms, originally considered a prototypical example of "physisorption," actually occupy sites at upper step edges on a Pt(111) surface, that is, *not* high-coordination down-step or hollow sites, and donate a minute amount of charge into the substrate [51, 52], as verified by the charge contour plot in Figure 31.32 [53] (see also Chapter 33 in this volume). This charge donation is consistent with a measured Xe-induced work function decrease of 0.6 eV and a dipole moment of 0.26

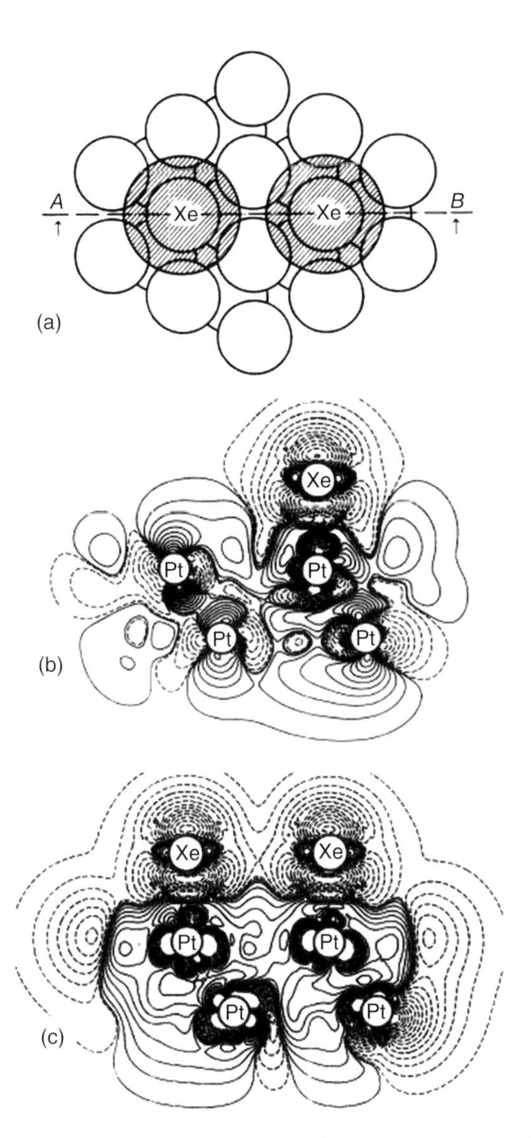


**Figure 31.32** (a) $Pt_{22}Xe_2$ cluster showing the symmetry plane *A*–*B* where the difference charge densities (b) $\Delta\rho = \rho(Pt_{22}Xe) - \rho(Pt_{22}) - \rho(Xe)$, and (c) $\Delta\rho = \rho(Pt_{22}Xe_2) - \rho(Pt_{22}) - \rho(Xe_2)$, induced by the adsorption of one and two Xe atoms, respectively, are being displayed. The adsorption distance of the Xe atoms is 3.07 Å. The continuous ($\Delta\rho > 0$) and the dashed ($\Delta\rho < 0$) contour lines represent densities given by $n^3 \times 10^{-5}$ e Å$^{-3}$, for $n = -15 \ldots 15$ [53].

Debye per Xe atom [52]. Distinctly electronegative atoms such as oxygen and chlorine (see Chapter 35) on the one hand and electropositive atoms such as alkali atoms (see Chapter 34) on the other hand are known to lead to core level shifts to lower and higher binding energies, respectively, and to strong work function increases and decreases, respectively [54].

#### 31.4.2.3 **Adsorbed Carbon Monoxide**

Carbon monoxide as a prototypical diatomic molecule (also termed the *drosophila of surface science*) binds intact to metals such as Pt, Pd, Ru, Ni, Cu, and so on via the carbon atom [55–58]. The bonding mechanism consists by and large of a competition between charge donation from the CO(5σ) orbital into metal d-states and charge-backdonation from the metal d-states into the $2\pi^*$ orbital of the molecule. Both contributions depend on the position of the metal d-states. The closer the center of gravity of the d-states are to the Fermi level, the more are 5σ-donation into empty d-states and d-backdonation into the empty, antibonding $2\pi^*$-states of the CO molecule favored (see (3) and (5) in Figure 31.27). As a consequence, CO adsorbs readily at room temperature on Ni, Pd, Pt, and so on while it adsorbs on Cu and Ag only at low temperatures [58–60], accompanied by a work function increase (decrease) on the former (latter) metals due to a dominant (weak) $d \rightarrow 2\pi^*$ backdonation (Figure 31.22b). A strong involvement of $5\sigma \rightarrow d$ donation manifests itself in a strong selective shift of the CO(5σ) orbital to higher electron binding energies as verified by valence band photoemission spectra using polarized UV light. As an example, the spectra in Figure 31.33 prove that, in contrast to gas phase CO and CO *adsorbed* on copper, the 5σ orbital of CO adsorbed on Pd(111) is shifted to

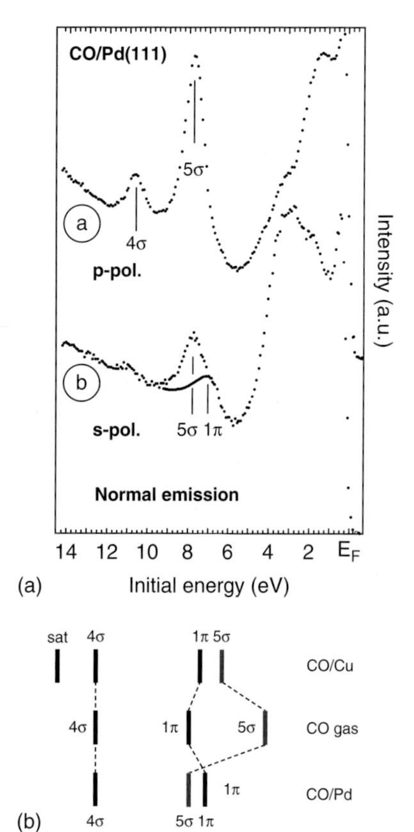


**Figure 31.33** (a) UV photoemission spectra of CO adsorbed on Pd(111) taken at normal emission with polarized light, enabling a discrimination between σ- and π-orbitals, and thereby a separation of 5σ- and 1π-states [61]. (b) Relative 5σ- and 1π-orbital positions for CO adsorbed strongly on palladium compared to gas phase CO, and CO being weakly bound on copper, respectively.

even higher binding energy than the CO(1π) states. A strong d → 2π* backdonation into the antibonding CO(2π*) orbitals causes a weakening of the intramolecular CO bond, which is seen in IR and HREELS spectra and which is the basis for catalytically facilitated CO reactions. High-resolution XPS core level spectra (Figure 31.34) and HREELS spectra (Figure 31.35) permit even a distinction of CO in different adsorption sites, for example, at different coverages [13, 62].

As a further example, Figures 31.21 and 31.36 show TDS and HREELS spectra of CO adsorbed on an ordered $Cu_3Pt(111)$ *alloy* surface, which obviously provides a whole series of well-defined surface sites ("ensembles") of different local composition

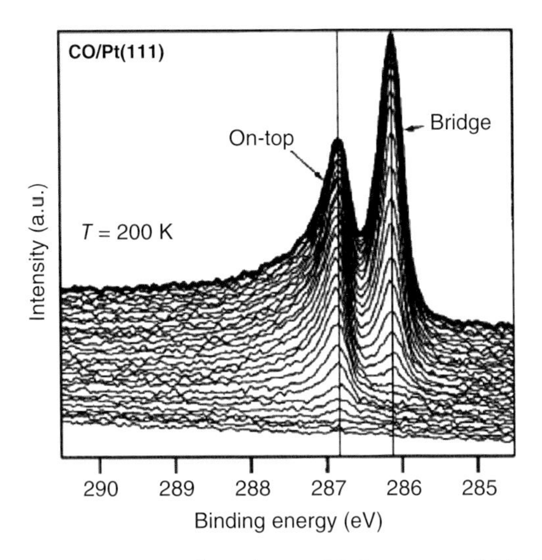


**Figure 31.34** High-resolution XPS C1s spectra of CO adsorbed on Pt(111) showing the successive population of on-top and bridge sites [62].

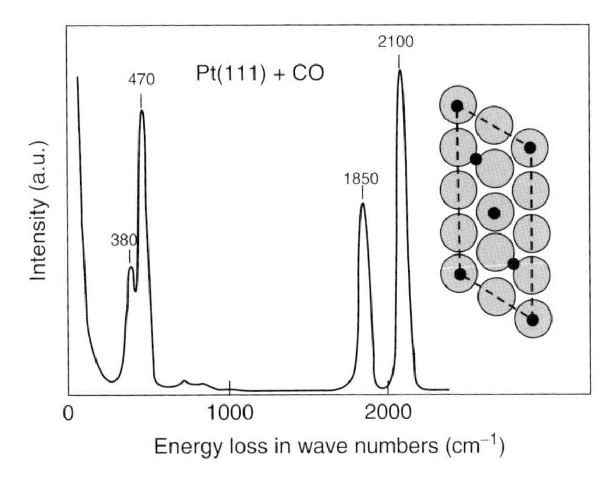


**Figure 31.35** HREELS spectra of CO adsorbed on Pt(111), which enable a distinction between CO molecules bound atop a Pt atom (2100, 470 $cm^{-1}$) and bridge-bonded CO (1859, 380 $cm^{-1}$), respectively [13].

and coordination [41, 63]. A CO molecule could either sit ontop or between two or three atoms, and each of these atoms could be either Pt or Cu. This is reflected in the multipeaked TDS in Figure 31.21. Note that the highest temperature peak, according to what has been said earlier about the CO–metal interaction strength, corresponds to CO molecules being bound ontop of Pt atom. Note also that this peak is shifted to lower temperatures by about 100 K compared to CO on a bare Pt(111) surface due to the so-called "ligand effect," that is, a change of the electronic properties of the Pt atoms being embedded in the Cu surrounding. Only after saturation of all Pt sites, CO adsorbs also on mixed Pt–Cu- and, finally, pure Cu sites at much lower temperatures as indicated by the corresponding desorption peaks below 250 K.

HREELS spectra of CO adsorbed on the $Cu_3Pt(111)$ surface are presented in Figure 31.36. The stronger CO–Pt interaction is reflected in the higher energy of the Pt–CO vibration at 55 meV and the lower PtC–O vibration at 257 meV compared to the corresponding Cu–CO and CuC–O vibrations. Note the very low intensity of

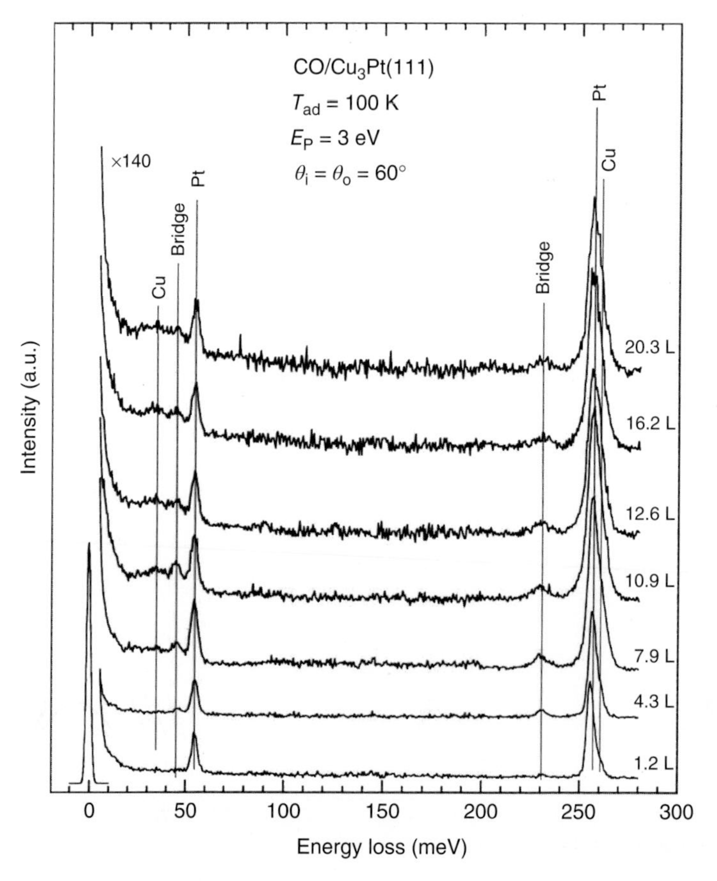


**Figure 31.36** HREELS spectra of a monolayer of CO adsorbed on an ordered $Cu_3Pt(111)$ surface. The position of the signals indicates a successive population of Pt and Cu sites, and is consistent with a stronger (weaker) interaction to Pt (Cu), respectively [63].

the "bridge" signal at 230 meV compared to CO adsorbed on Pt(111) (Figure 31.35), indicative of a few deviations from the perfect chemical order of the surface in the form of paired Pt atoms.

#### 31.4.2.4 Dissociative Oxygen Adsorption

Oxygen molecules are more prone to dissociative adsorption than CO, whereby depending on the nature of the substrate and the adsorption conditions several intermediate reaction steps may be isolated. Unlike CO (or NO), the oxygen molecule is homo-atomic; thus, there is no preference for one or the other oxygen atom to interact with the surface. As a consequence, parallel configurations of the molecular axis to the surface are more likely than with CO and have indeed been verified. Besides truly physisorbed molecular oxygen and chemisorbed *atomic* oxygen also the existence of superoxo- and peroxo-species with differing backdonation of charge into the $O_2(2\pi^*)$ orbital and differing oxygen–oxygen bond order has in particular been concluded from TDS, HREELS, and STM measurements for instance, on Pt(111) [64–67], Pd(111) [68], and Ag(111) [69]. The existence of not only three (as illustrated in Figure 31.29) but up to five different adsorption states of oxygen has been suggested for Pd(111) [68]. TDS spectra distinguish between species of different interaction strength to the substrate. As an example, Figure 31.37 shows a TDS spectrum of oxygen from Pt(111) [70]. Note the first- (second-) order shape of the peak at 140 K (~800 K). UV-photoemission spectra reveal immediately changes of the valence states, namely, the existence and occupation of molecular orbitals. This is illustrated in Figure 31.38 for oxygen adsorption on a gallium film between 10 and 300 K [71]. Warming an initially condensed oxygen film ($O_{2,cond}$) first leaves a monolayer of physisorbed oxygen ($O_{2,phys}$) recognizable by the still complete orbital scheme of *molecular* oxygen. This species transforms into chemisorbed diatomic oxygen ($O_{2,chem}$) as suggested by the shifted and somewhat modified, but still molecular orbitals seen in spectrum

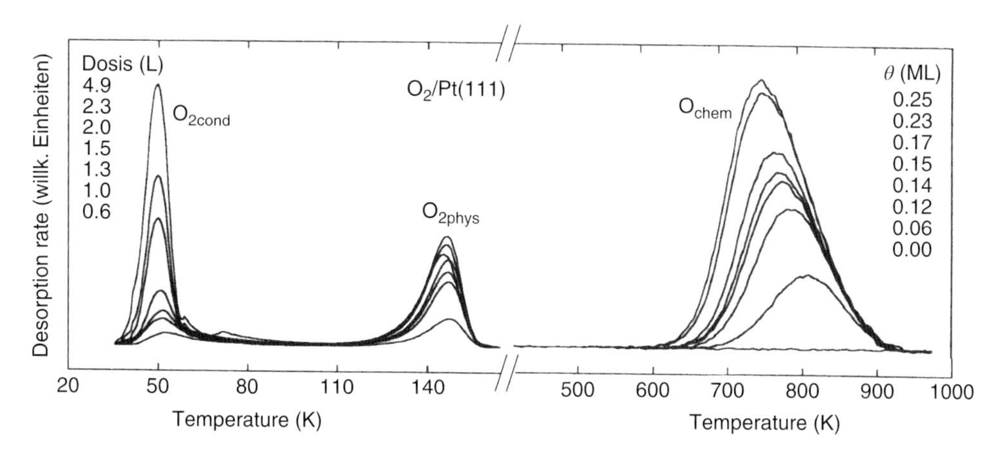


**Figure 31.37** Thermal desorption spectra of oxygen from Pt(111), showing a clear distinction of evaporation of condensed molecular oxygen, physisorbed molecular oxygen, and chemisorbed atomic oxygen. Note the different desorption order of $O_{2,chem}$ and $O_{chem}$ [70].

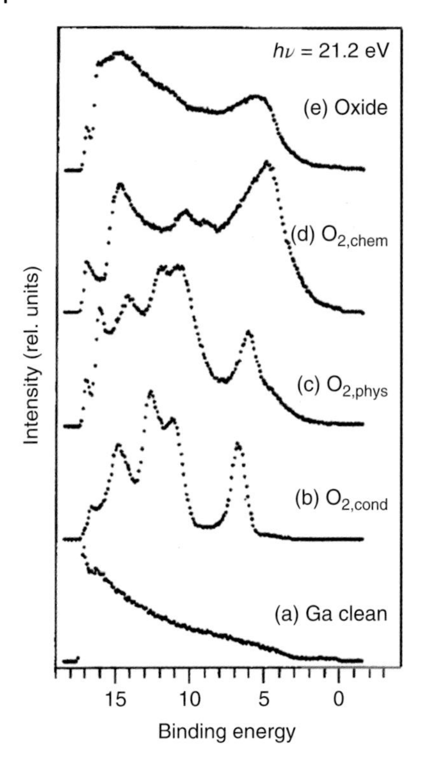


**Figure 31.38** UV (HeI) photoemission spectra of a (a) clean and (b–e) oxygen covered gallium film. Four different states of adsorbed oxygen are distinguishable: (b) condensed molecular oxygen, (c) physisorbed molecules, (d) chemisorbed molecular ("diatomic") oxygen, and (e) atomically adsorbed and oxidic oxygen [71].

denoted $O_{2,chem}$. This chemisorbed molecular species finally transforms into atomic oxygen on, in, or under the surface as chemisorbed or oxidic oxygen, as verified by the characteristic O(2p) signal at about 6 eV electron binding energy.

The strong interaction between electronegative oxygen atoms and a metal substrate, of course, causes also significant changes of the substrate electronic states. In particular, oxygen-adsorption-induced surface core level spectroscopy proved to be a very sensitive probe on the local reactivity and concomitant electronic structure changes. Owing to the reduced coordination of the substrate *surface* atoms of a *bare* surface, their core levels may be measurably shifted from those of the bulk atoms (see Chapter 3, Volume 1). As an example Figure 31.39a shows only the $d_{5/2}$ spin–orbit component of a high-resolution 3d photoemission spectrum of a Rh(100) surface, split into a bulk component at 307.2 eV and a component from the very first layer of Rh atoms at 306.6 eV [72]. It obviously appears that chemisorbed oxygen atoms will, above all, modify the electronic properties of the immediate neighbor Rh atoms. Figure 31.36b shows only the *surface component* of the $Rh(3d_{5/2})$ signal of the bare Rh(100) surface ($Rh_0$) and at oxygen coverages of $\theta_o = 0.185$ ML ($Rh_{1,4}$) and $\theta_o = 0.375$ ML ($Rh_{2,4}$), respectively [73]. At the low coverage ($C_{1,4}$), the oxygen atoms occupy fourfold hollow sites and form a $p(2 \times 2)$ structure (see Figure 31.40), and each Rh atom interacts at most with one oxygen atom, which leads to a new $Rh3d_{5/2}$ component at 220 meV higher binding energy denoted $Rh_{1,4}$ (one oxygen atom per four rhodium atoms). At the higher oxygen

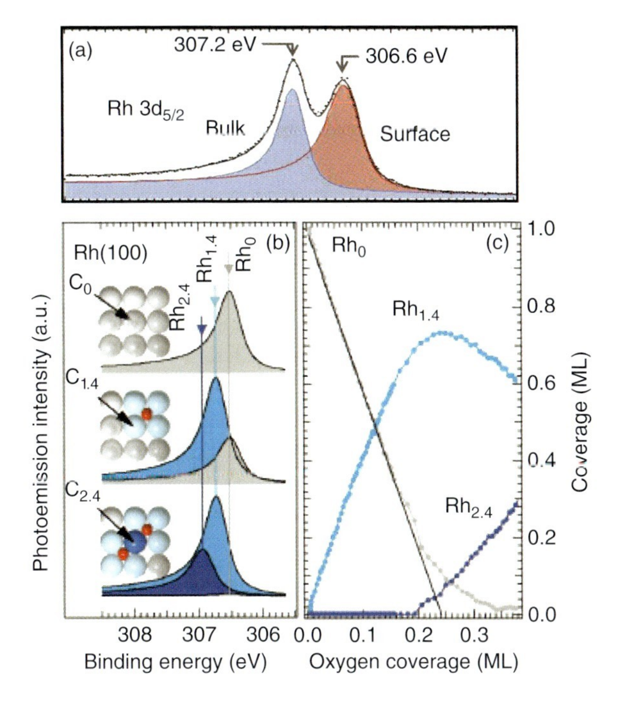


**Figure 31.39** High-resolution core level spectra of the $Ph3d_{5/2}$ spin–orbit component only of (a) a clean [72] and (b) a step-wise oxygen covered Rh(100) surface. Panel (b) shows only the *surface component* of the $Rh3d_{5/2}$ emission for oxygen-free ($C_0$), the $p(2\times2)$-O($C_{1,4}$) and the $c(2\times2)$-O($C_{2,4}$) covered Rh(100) surface. Panel (c) presents the intensity of the $Rh3d_{5/2}$ surface component of Rh atoms in contact with no, one or two adsorbed oxygen atoms ($Rh_0$, $Rh_{1,4}$, $Rh_{2,4}$) [73].

coverage ($C_{2,4}$), a denser $c(2\times2)$-O structure is formed in which inescapably Rh atoms interact with two oxygen atoms, giving rise to a further Rh($3d_{5/2}$) component ($Rh_{2,4}$) at 435 meV below the peak of the bare surface. Thus, a plot of the normalized intensity of the three components $Rh_0$, $Rh_{1,4}$, and $Rh_{2,4}$ as a function of oxygen coverage (Figure 31.39c) illustrates the relative abundance of coexisting Rh surface atoms in contact with no, one, or two oxygen atoms. The low coverage $Rh_{1,4}$ state grows linearly right from the beginning, while the $Rh_{2,4}$ state starts to grow once the $p(2\times2)$ structure is (nearly) saturated. In turn, the $Rh_0$ state decreases linearly with slope $(d\text{Rh}/d\theta) = -4.2$ and vanishes once the $p(2\times2)$-O structure is saturated. The slope of −4.2 is a clear confirmation of the oxygen atoms adsorbing in fourfold hollow sites with no common Rh-atom. Similar adsorption experiments with oxygen on Rh(111) resulted in, besides the $Rh_0(3d_{5/2})$ peak, the three signals $Rh_{1,3}$, $Rh_{2,3}$ and $Rh_{3,3}$ shifted by +345, +780, and +1120 meV, respectively [74]. The linear decrease of the $Rh_0$ component as a function of oxygen coverage was found to be −3 in agreement with oxygen adsorption in threefold hollow sites with no common Rh-atom. Note also that on both the Rh(100) and the Rh(111) surface the incremental peak shift per O-Rh bond is constant; the shift of $Rh_{2,4}$ (+435 meV) is twice as large as that of $Rh_{1,4}$ (220 meV), and the shift of $Rh_{2,3}$ and $Rh_{3,3}$ is two

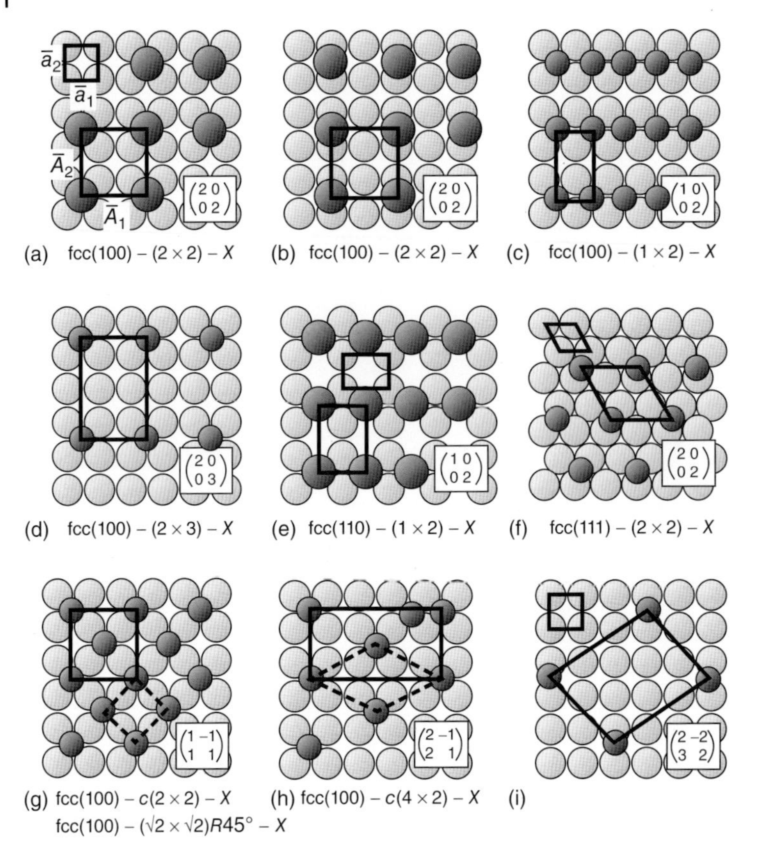


**Figure 31.40** (a–i) Examples of simple adsorbate (*X*) superstructures on fcc(100), fcc(110) and fcc(111) surfaces (figure reproduced from Chapter 4.2, Volume 2).

and three times larger than that of $Rh_{1,3}$, indicating a dominantly local interaction between oxygen atoms and the Rh substrate.

A very useful techniques to verify the adsorption site is also photoelectron forward scattering as nicely demonstrated for oxygen adsorption on Cu(100); see Ref. [75] and Chapter 3.2.2 in Volume 1.

## 31.4.3
## Collective Ad-Layer Properties

### 31.4.3.1 Ad-Layer Structure

With increasing coverage, the properties of $A_{ad}$ are no longer solely determined by their interaction with the local site on the substrate, but lateral interactions come into play. Depending on the relative strength of adsorbate–adsorbate and adsorbate–substrate interactions collective properties within the ad-layer as a whole will emerge at certain coverages. Attractive adsorbate–adsorbate interactions will cause a "self-assembly" of ad-particles leading, for example, to a two-dimensional condensation into 2D $A_{ad}$ islands and filling of the *first* monolayer whose internal structure may be independent of that of the substrate surface (Section 31.2.1.3). If these attractive interactions are very strong, the formation of 3D ad-particle clusters may energetically be even more favorable than the growth of 2D islands. As detailed in Chapter 1 of Volume 1 or Chapter 20 in Volume 3, the relative strength of adsorbate–adsorbate versus adsorbate–substrate interactions is correlated with the values of the surface tension $\gamma$ (surface Gibbs free energy) of the substrate ($\gamma_S$) and the bulk material A ($\gamma_A$), respectively, and plays a decisive role in deposition processes on surfaces (see Chapter 20, Volume 3). As long as

$$\gamma_A(n) + \gamma_I(n) \leq \gamma_S \tag{31.106}$$

with $\gamma_A(n)$ being the surface tension of an $n$-layer thick film of A on the substrate (S), and $\gamma_I(n)$ the interface energy between this film and the substrate, adsorbing ad-particles will preferentially be incorporated into the substrate-nearest $A_{ad}$ layer leading to a layer-by-layer or so-called Frank–van der Merwe growth. The dominant substrate–adsorbate interaction may also enforce structural registry (i.e., epitaxy) between the completed adsorbate film and an unequal substrate surface lattice, thereby causing strain within the ad-layer. With increasing film thickness, that is, with increasing number $n$ of $A_{ad}$ layers, this strain and the surface tension of the layer change, eventually leading to a change from layer-by-layer to 3D cluster growth on the initial 2D film, the so-called Stranski–Krastanov growth with the internal structure of the clusters approaching that of bulk A. If, however,

$$\gamma_A(n) + \gamma_I(n) \geq \gamma_S \tag{31.107}$$

$A_{ad}$ particles will form 3D clusters right away leaving as much as possible low-energy substrate surface uncovered (Volmer–Weber growth).

Depending on the relative adsorbate–adsorbate and adsorbate–substrate interaction strength, and the adsorbate coverage, there may thus be a structural coincidence between the first (and higher) ad-layer and the substrate surface lattice or not. Strong adsorbate–substrate interactions favor the formation of 2D- or at least 1D-commensurate ad-layers, in – at least partial – registry with the substrate surface lattice. The surface structure acts like a 2D or 1D "template" [76]. Dominant adsorbate–adsorbate interactions favor the growth of incommensurate layers. A few examples of typical adsorbate structures are displayed in Figure 31.40 as reproduced from Chapter 4.2, Volume 2, where a much more extensive and systematic discussion of adsorbate structures can be found.

#### 31.4.3.2 Two-Dimensional Phase Transition

The activation energy of migration $E^*_m$ (Figures 31.30 and 31.31) is decisive for the *kinetics* of the formation of equilibrium surface structures. If $E^*_m$ is small, the ad-particles may form a 2D gas phase, which with increasing coverage and depending on the adsorbate–adsorbate interaction may condense into 2D islands of their energetically most favorable structure. Such 2D-phase transitions, that is, the co-existence of 2D gas phase and 2D islands, can immediately be visualized by STM imaging [77], but are also discernible with integrating methods as illustrated with the following two examples.

Due to the Smoluchowski electron smoothing effect (see Chapter 5, Volume 2) at surfaces, the dipole moment of single ad-atoms in a 2D gas phase is significantly larger than that of ad-atoms embedded within a 2D condensed phase. Thus, the same number of ad-particles in 2D gas form versus 2D condensed phase (see Figure 31.41) will cause a different work function change. This effect was exploited to determine 2D gas $\rightarrow$ 2Dgas + 2D condensed phase transitions during the growth of ultrathin metal films using work function change measurements [78]. As an example, Figure 31.42 shows the work function change ($\Delta\Phi$) measured in situ with a Pendulum Kelvin probe [79] during Ag deposition onto a Ru(0001) surface in the sub-monolayer regime for different substrate temperatures. At a substrate temperature of 350 K, the work function $\Phi$ decreases linearly [78] as sketched in Figure 31.42. In turn, all curves registered at the higher substrate temperatures first show a much steeper work function decrease, which then slows down to the same value (slope) as the 350 K line. The "knee" in the respective curve occurs at higher Ag coverage the higher the substrate temperature [78].

The initially steeper work function decrease at $T_S > 645$ K can be explained by the existence of a 2D gas phase of single Ag atoms. While at 350 K adsorbing Ag atoms are mobile enough to be immediately incorporated into 2D Ag islands, the higher thermal energy at $T > 645$ K keeps/drives them apart, until at the distinct change of the slope of all curves at a higher temperature a 2D condensation of the 2D Ag gas sets in, which at a higher substrate temperature requires a higher Ag atom coverage. The "knee" in these curves, thus, occurs at the critical density of isolated Ag atoms, that is, the critical 2D Ag gas pressure $\Pi$ where 2D condensation sets in and 2D islands start to coexist with their 2D equilibrium vapor pressure ("spreading pressure"). Defining the "knee" in each curve as the intersection point of the two straight lines fits to the two slopes of each curve, the thus determined critical coverage $\theta_{crit}$ (i.e. critical 2D vapor pressure) and the related substrate temperature can be used to calculate the 2D heat of condensation/vaporization of Ag on Ru(0001). Assuming

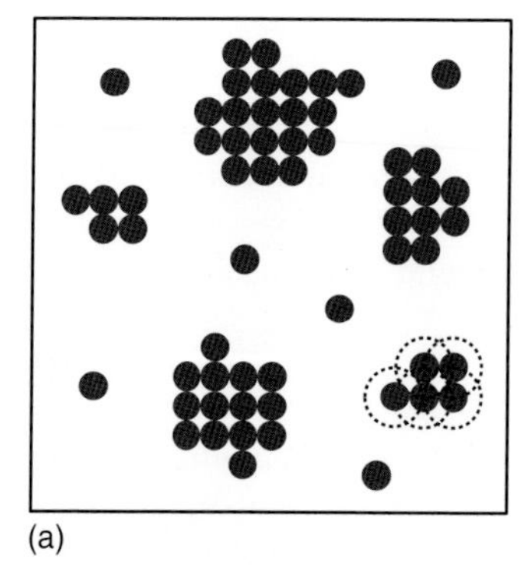


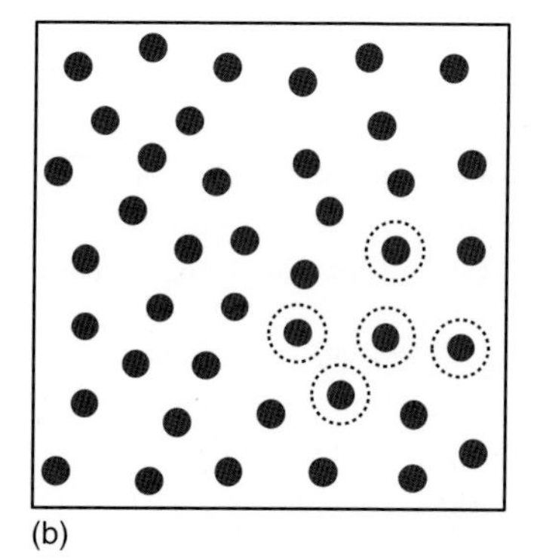


**Figure 31.41** Schematic representation of a two-dimensional (2D) gas (b), and two-dimensionally condensed islands in equilibrium with their 2D vapor pressure $\Pi$ (a). The dotted circles represent the scattering cross-section of adsorbed Xe atoms on Pt(111) for thermal He atoms.

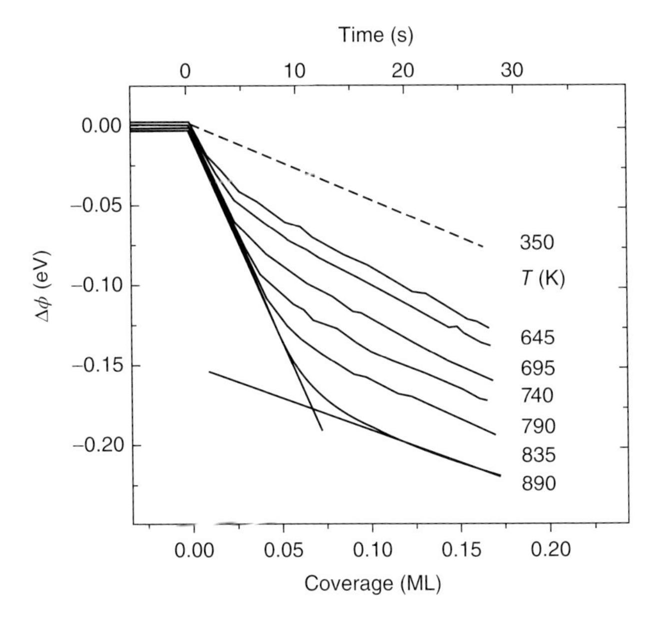


**Figure 31.42** Work function change during Ag deposition at various substrate temperatures and constant deposition rate (036 ML/min). The 2D gas – 2D condensed phase transition is observed in the low coverage regime ($\theta = 0.02$–$0.06$ ML). The intersection of the extrapolated linear $\Delta\Phi$ segments (full straight lines) yields the onset of 2D condensation [78].

an ideal state for the 2D vapor phase, its 2D pressure is given by

$$\Pi = NRT \tag{31.108}$$

The number density $N$ of ad-particles is proportional to $\Delta\Phi$ as indicated by the linear portions of the curves in Figure 31.42, so that

$$\Pi = \text{const}\ \Delta\Phi RT \tag{31.109}$$

In the coverage regime of the phase transition, that is, $\theta < 0.06$ ML, the area covered by the 2D condensate is negligible compared to the area $A = 1/N$ covered by the 2D vapor phase. Under equilibrium conditions, the temperature dependence of $\Pi$ is then determined by a 2D analog of the Clausius–Clapeyron equation of the form

$$\left(\frac{\partial \Pi}{\partial T}\right)_{\text{crit}} = \frac{\Delta H_{\text{2D, vap}}}{TA} \tag{31.110}$$

$\Delta H_{\text{2D,vap}}$ being the heat of 2D vaporization. Combining Equations 31.109 and 31.110 and integrating yields

$$\ln \Delta\Phi + \ln T = -\frac{\Delta H_{\text{2D, vap}}}{TA} + \text{const.} \tag{31.111}$$

A plot of ($\ln \Delta\Phi_{\text{crit+ln}\,T}$) versus $1/T$ yields a straight line, and the slope is proportional to $\Delta H_{\text{2D,vap}}$.

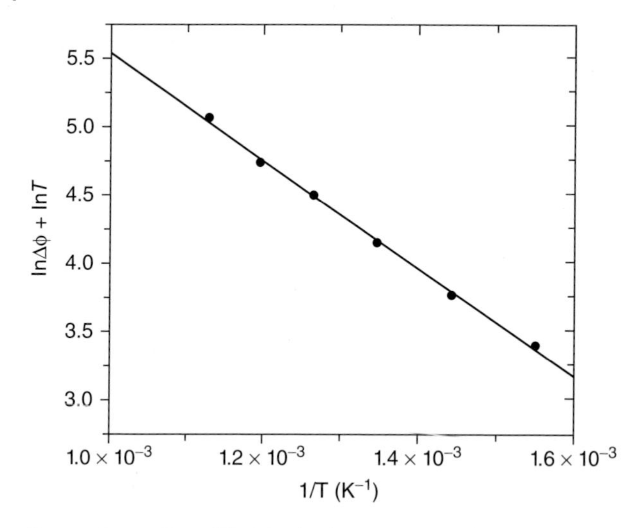


**Figure 31.43** Plot of (ln ΔΦ + ln T) versus 1/$T$ at the 2D gas-condensed transition (knees in Figure 31.42) yields a straight line, the slope of which gives the 2D heat of vaporization (see text) [78].

Such a plot using ($\Delta\Phi_{crit}$,$T$) pairs from the traces in Figure 31.42, which were proved to be equilibrium curves, is shown in Figure 31.43. The slope of the remarkably straight line yields a value of $\Delta H_{2D,vap} = 0.33 \pm 0.02\,\text{eV/Ag}$ atom. This result may be converted into a lateral interaction energy between two sixfold coordinated Ag atoms of $E_6(\text{Ag–Ag}) = 1/3\ \Delta H_{2D,vap} = 0.11 \pm 0.02\,\text{eV}$. This value is more trustworthy than results deduced from a "complete analysis" (see Section 31.3.1.2) of TDS of Ag from Ru(0001) (see Figure 31.20) [38].

Thermal He atom scattering has been applied to study the 2D gas → 2D gas + 2D condensed phase transition, for instance, in an adsorbed Xe layer on Pt(111) [80]. In this case, the very large cross section $\sigma$ of diffusely scattered He atoms from adsorbates is exploited. This large cross section of $\sigma \approx 120\,\text{Å}^2$ for Xe, illustrated by the dashed circles around *isolated* 2D Xe gas atoms in Figure 31.41, makes the method particularly sensitive to small coverages. As long as there are only isolated Xe atoms (gas phase) on the surface, the decay of the scattered intensity as a function of the number density $N$ is (almost) exponential, because of statistical overlap of the "scatterers" [80]

$$-d\left(\frac{I}{I_0}\right) \approx \sigma\, e^{-N\sigma} dN \qquad (31.112)$$

Upon 2D condensation, the cross sections of the Xe atoms incorporated in an island overlap with each other and their "effective" cross section shrinks basically to their physical cross section, namely to $a \approx 20\,\text{Å}^2$ within hexagonally packed Xe islands on Pt(111), and at the limit of large islands ($N > N_{crit}$) the attenuation is

$$-d\left(\frac{I}{I_0}\right) \approx a \frac{e^{-N_{crit}\sigma}}{1 - N_{crit}a} dN \qquad (31.113)$$

As a consequence, the attenuation of the scattered intensity becomes much smaller as soon as $N > N_{crit}$. This is reflected in the measured data shown in Figure 31.44 for different substrate temperatures $T_S$. The measured intensity traces show a distinct discontinuity at the onset of 2D condensation (at $N = N_{crit}$), that is, at the 2D gas → 2D gas + 2D condensed phase transition. Using again the coordinates of the "knees" in the different traces registered at different temperatures $T_S$ for a plot $\ln \Pi_{crit} \sim \ln(N_{crit} k T_S)$ versus $1/T_S$ yields a heat of 2D vaporization of $\Delta H_{2D,vap} = 1.1 \pm 0.1$ kcal/mol.

#### 31.4.3.3 Ad-Layer Band Structures

With increasing density of the ad-layers, the lateral interactions between the ad-particles manifest themselves also in collective electronic and vibrational properties [60]. As an example, Figure 31.45 shows a clear dispersion of the $4\sigma$- and $(5\sigma + 1\pi)$-derived electronic bands of a $c(4 \times 2)$ CO overlayer on Pd(111) along the $\overline{\Gamma}\,\overline{M}$ and $\overline{\Gamma}\,\overline{K}$ high symmetry directions of the Pd(111) surface Brillouin zone (SBZ) (see inset) [61]. The vertical bars indicate the position of the $\overline{K}$ and $\overline{M}$ points of the SBZ of the clean Pd(111) surface, while the arrows point to the maxima of the CO-derived bands. The observation that the overall bandwidth is 0.3 eV for *both* bands is, at first glance, surprising. Considering the spatial extent of the CO orbitals, a much larger dispersion for the $1\pi$ and $5\sigma$ states than for the $4\sigma$ band would be expected, as for instance predicted by simple tight-binding calculations of the band dispersion for an isolated CO layer, namely 0.4 eV. 1.4 and 1.7 eV for the $4\sigma$, $1\pi$, and $5\sigma$-derived bands, respectively [81]. The reason for the same width of both bands lies in the strong shift of the $5\sigma$-states to higher binding energy upon CO interaction with the surface (see Figure 31.33) and a resultant modified hybridization between the $5\sigma$-states with the $1\pi$- and $4\sigma$-levels, respectively. More examples of adsorbate band structures, including adsorbed Xe layers, can be found in Chapter 3.2.2, Volume 1 and Chapter 38 in this volume.

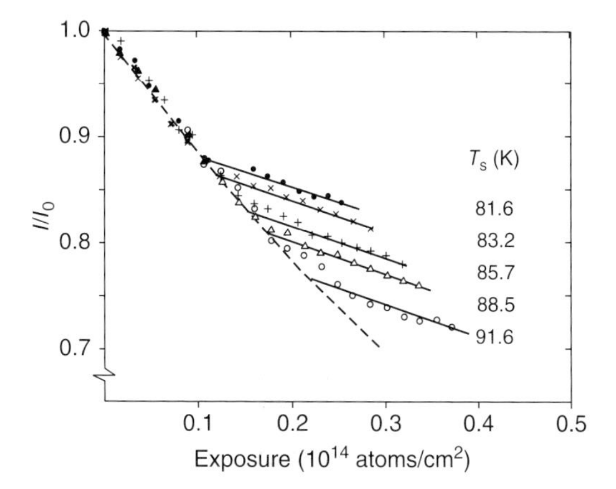


**Figure 31.44** Relative specular He intensity versus Xe exposure. Dashed curve (exponentials) is a plot of Equation 31.112 with $\sigma \approx 120\,\text{Å}^2$ and $s = 1$.

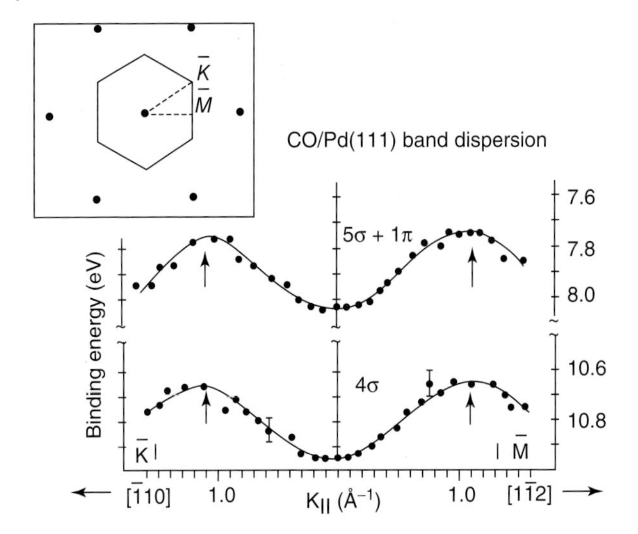


**Figure 31.45** Band dispersion of the 4σ-derived and (5σ + 1π)-derived levels along the $\bar{\Gamma}\bar{M}$ and $\bar{\Gamma}\bar{K}$ high-symmetry directions of the clean Pd(111) surface Brillouin zone (see inset). Bars indicate the position of the $\bar{K}$ and $\bar{M}$ points of the clean SBZ. Arrows denote the experimental maxima of the bands [61].

Surface phonons were extensively treated in Chapter 8, Volume 2. Excellent treatises of adsorbate vibrations and phonons can also be found in [82].

#### 31.4.3.4 **Adsorbate-Induced Substrate Restructuring**

Chemisorption, associated with the observed changes of the electronic properties not only of the adsorbate but also of the substrate, is likely to change also the structure of the substrate, first locally underneath isolated ad-particles, and eventually, with increasing coverage, of the whole surface. Two important cases shall be distinguished. In the first case, the *bare* surface is already reconstructed and adsorption *lifts* this reconstruction. In the second case, an adsorbate *causes* reconstruction of a previously unreconstructed surface.

The equilibrium structure of the bare Ir(100) surface does not exhibit the expected square symmetry like the parallel bulk planes underneath, but forms a so-called $(5 \times 1)$ reconstructed quasi-hexagonal densely packed surface atom layer, within which densely packed atom rows are running *parallel* to densely packed atom rows of the quadratic structure of the layer underneath (see Chapter 4 in Volume 2). This mismatch between the (111) mesh of the surface layer and the square atomic arrangement underneath causes *surface* Ir atoms to sit in inequivalent sites and, thus, at different heights, which leads to two orthogonal *domains* of a long-range $(5 \times 1)$ superstructure as displayed in Figure 31.46 reproduced from Chapter 4, Volume 2 [15]. The atom density of this quasi-hexagonally reconstructed surface layer is 20% higher than that in an unreconstructed (100) surface. Oxygen adsorption at about 470 K followed by annealing to 700 K lifts this reconstruction and leads to an unreconstructed, but oxygen covered Ir(100)$1 \times 1 - O$ surface.

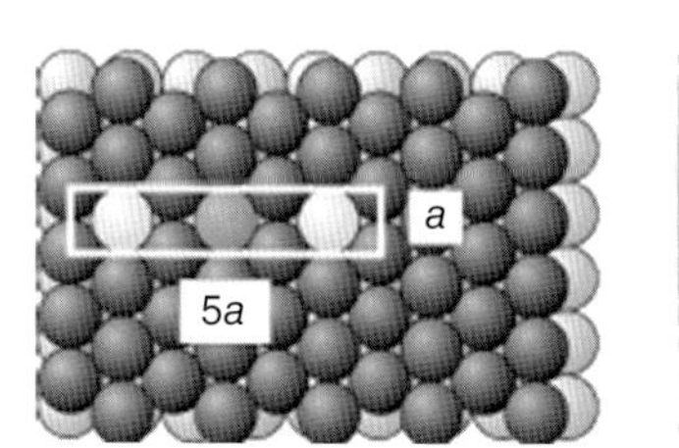

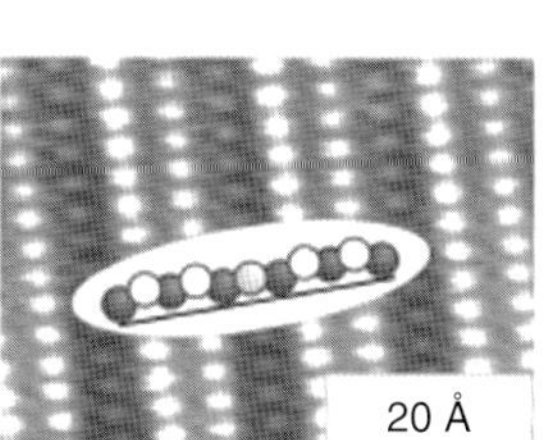

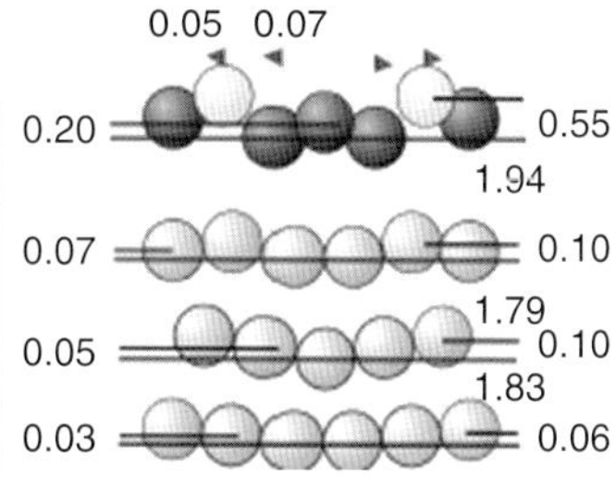


**Figure 31.46** Quasi hexagonal reconstruction of the Ir(100) surface. The top layer atoms are not quadratically but quasi-hexagonally arranged, so that a (5 × 1) coincidence super-lattice results. By bonding to the quadratic substrate layer, the surface top layer buckles so that two surface atoms within the (5 × 1) unit mesh protrude significantly out of the surface (brighter atoms). The top layer buckling induces also buckling in the subsurface layer. The values are given in Å. (Reproduced from Figure 4.343 in Volume 2.)

Exposure of this surface to hydrogen enables even removal of the oxygen (in the form of water), and, eventually, the preparation of a clean but meta-stable Ir(100)1 × 1surface. Annealing this (1 × 1) modification causes this surface to return to the thermodynamically stable bare Ir(100)5 × 1 surface phase by a (5 × 1) nucleation and growth process [83] (see Chapter 4, Volume 2).

The Pt(100) surface exhibits a reconstruction similar to that of the Ir(100) surface, with the exception that the hexagonal surface layer is, in addition, slightly rotated against the atom rows of the quadratic layer underneath (Figure 31.47) [84], resulting in *four* rotational domains of an again long-range wavy superstructure. Cu deposition onto this surface at 340 K leads to elongated Cu island growth

**Figure 31.47** STM image of the reconstructed Pt(100) surface [84] showing one of four possible rotational domains.

preferentially parallel to the wave crests and to a local lifting of the Pt reconstruction underneath. This transition from quasi-hexagonal to the quadratic (100) structure is associated with an expulsion of the 20% of excess Pt atoms, which – in growth direction – are built into the Cu islands in the form of one Pt-atom row per wave width of the previously reconstructed Pt substrate [85].

In turn, adsorption on an unreconstructed (reconstructed) surface may *cause* a reconstruction (change in reconstruction) as demonstrated by the following three selected examples.

Experimentally, it is found that the *clean* Pd(110) surface is not spontaneously reconstructed in contrast to some other fcc(110) surfaces, which exhibit $(1\times r)$, $r=2\ldots 5$, mostly 2, reconstructed surface structures like, for example, the Au(110) or Pt(110) surfaces dealt with below. Theoretical calculations demonstrate, however, that the surface energy difference between an unreconstructed and a $(1\times 2)$ reconstructed Pd(110) surface is only very small, so that the tiniest trigger may induce a reconstruction. In fact, already very small amounts of adsorbed hydrogen lead to a reconstruction of the surface, and a rich set of stable and meta-stable hydrogen adsorption configurations can be associated with a number of experimentally observed reconstruction phases [86, 87]. Theoretical calculations showed that the most stable hydrogen adsorption configuration leads to a $(1\times 2)$ missing-row reconstructed surface, which offers (111) microfacets with the most stable adsorption sites for hydrogen atoms [88].

Figure 31.48 shows an atomically resolved section of a well-prepared Pd(110) surface kept in vacuum with a residual hydrogen partial pressure of $p_{H_2} \sim 5\times 10^{-9}$ Pa for 2–3 h. The rectangular unit cell of the Pd(110) surface lattice is clearly visible. Interesting are some single-atom wide missing rows (dark) and an added row (bright) next to one of them. These structural peculiarities are due to a local beginning of a hydrogen-induced reconstruction of the surface [86]. Figure 31.49 shows the results of a systematic exposure of the surface to a hydrogen pressure of

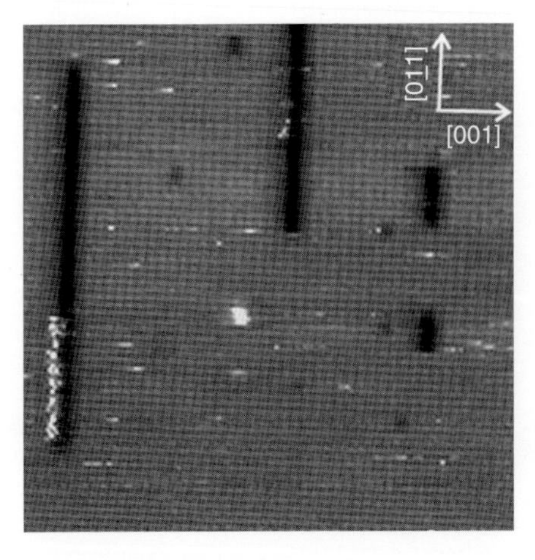


**Figure 31.48** Atomically resolved STM image of the clean Pd(110) surface; tunneling parameters $U_b = -7$ mV, $I_t = 150$ pA [89].

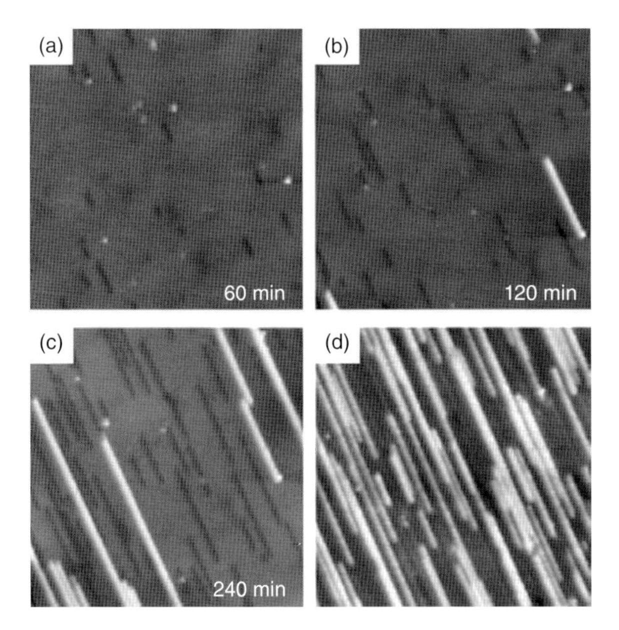


**Figure 31.49** (a–c) STM images obtained during hydrogen exposure ($p(H_2) = 6 \times 10^{-8}$ Pa). (d) STM image taken at the same surface position, after a large addition dosage (see text). Image sizes: $81 \times 81\ nm^2$; tunneling parameters: $U_b = -7$ mV, $I_t = 120$ pA [89].

$p_{H_2} \sim 7 \times 10^{-8}$ Pa for 240 min (a–c), followed by an exposure at $p_{H_2} \sim 5 \times 10^{-5}$ Pa for 10 min (d), both exposures at room temperature. The number and length of missing and added rows increase continuously from (a) to (d), and a detailed statistical analysis supports the notion [89] that the missing rows are one atom wide and that the expelled Pd atoms diffuse in $[1\bar{1}0]$ direction, that is, within the troughs, and are incorporated at the next step edge forming dimer chains extending onto terrace regions (Figure 31.50) or forming dimer chains on terraces (Figure 31.49c,d). The final result is by and large a hydrogen-induced $(1 \times 2)$ reconstruction of the whole surface [89]. Of course, desorption of the hydrogen restores the original unreconstructed Pd(110) surface structure.

In contrast to Pd(110), the bare Au(110) surface does spontaneously reconstruct forming a variety of $(1 \times r)$ reconstructions of nearly equal energy, $r$ ranging from 2 to 5 [90]. Among these, the $(1 \times 2)$ reconstruction is again observed most often. The $(1 \times 3)$ reconstruction, however, is slightly lower in energy, so that it is not surprising that weak perturbations, such as, for example, the adsorption of molecules, can change the reconstruction. This could actually be shown when using the anisotropic Au(110)$1 \times 2$ surface as a template for the growth of one-dimensional organic structures, for instance, metal-free phthalocyanine molecules [91].

Figure 31.51 shows a grating of parallel lines in the $[1\bar{1}0]$ direction, which correspond to the rows of gold atoms ("ridges") on this surface (white arrows). In addition, rows of big bright double spots (more or less resolved) are seen running in the same direction (black dotted arrows). Each double dot corresponds to

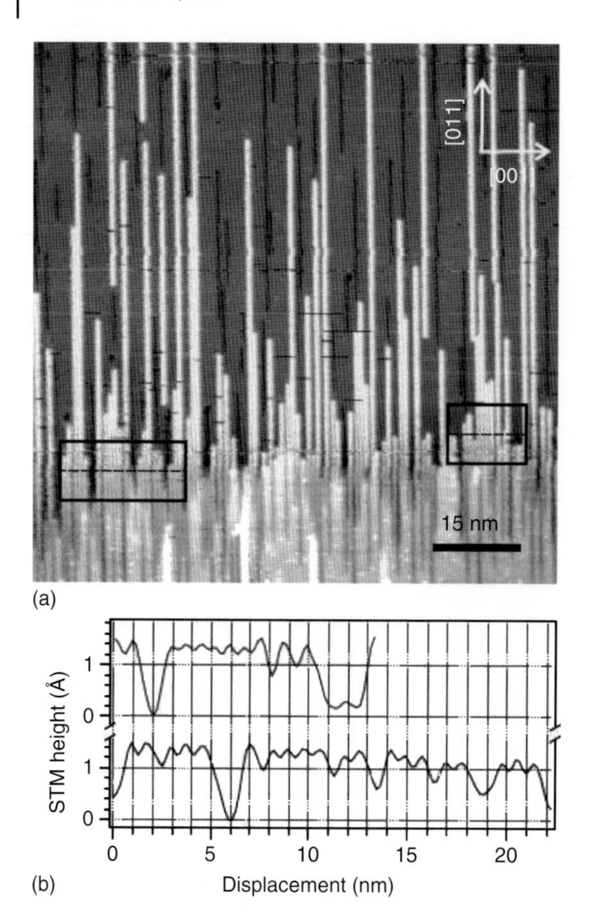


**Figure 31.50** (a) STM image of a rough step obtained after 25 L of hydrogen were dosed at $p(H_2) = 1 \times 10^{-8}$ Pa. Tunneling parameters: $U_b = 1.27$ V, $I_t = 120$ pA. (b) Line profiles taken at the positions indicated in (a) [89].

one phthalocyanine molecule, and the distance between the molecules along the molecular rows is, depending on the coverage, seven, six, or five times the interatomic Au–Au distance in the same direction, that is, $[1\bar{1}0]$, termed d7, d6, and d5 structure. Occasionally, single phthalocyanine molecules are also resolved mainly at dislocations of the substrate structure underneath (green arrows and dotted circles). Close inspection of this image representing a phthalocyanine coverage of 0.5 ML reveals that the separation of all gold atom rows covered with molecules is wider than the separation of uncovered Au atom rows. In fact, as indicated by the height profiles in Figure 31.51b,c the Au–Au separation underneath the molecules corresponds to a $(1 \times 3)$ reconstruction in agreement with LEED observations for the same adsorption system [91].

In total, three dominant adsorbate structures could be identified for different molecular coverages by comparing the experimental data with DFT calculations [91]. At low coverages, the molecules form a d7 row structure which is energetically

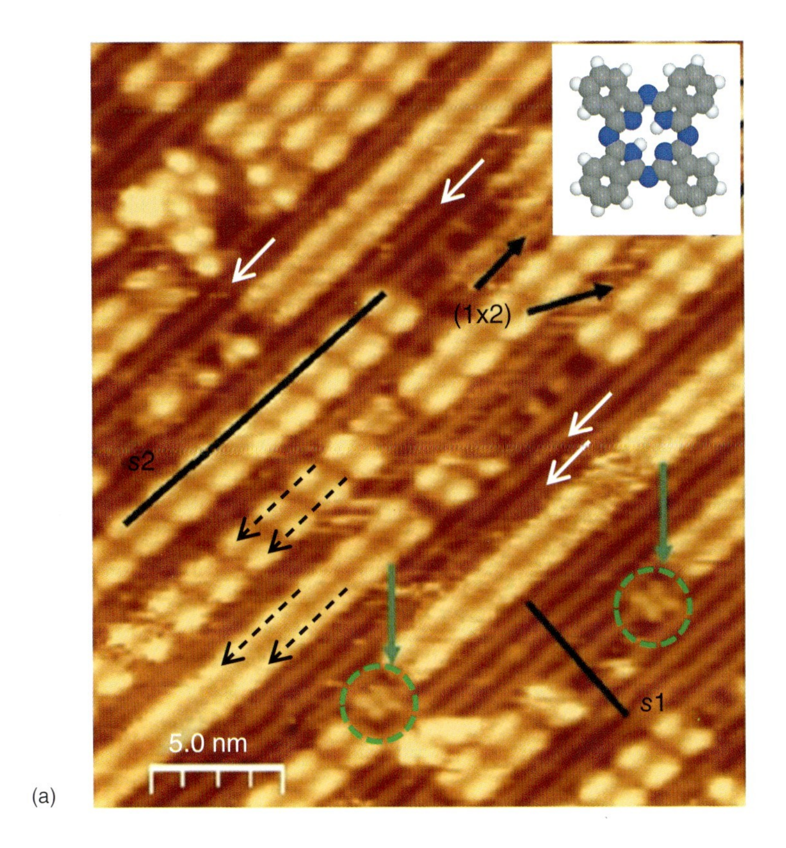


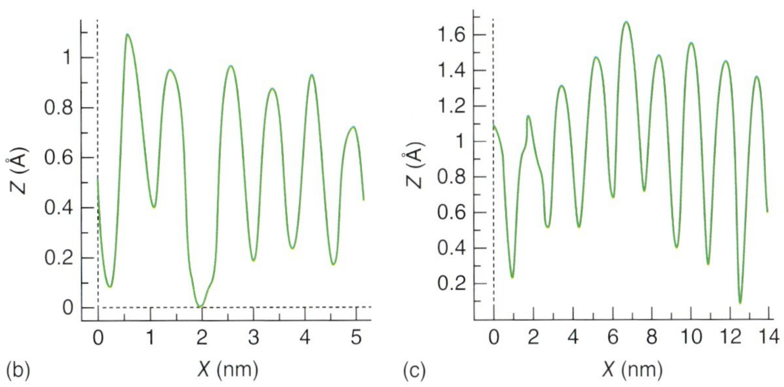


**Figure 31.51** (a) STM image of the Au(110) surface with 0.5 ML coverage of phthalocyanine molecules as shown in the inset; tunneling parameters: $U_b = -2.3$ V, $I_t = 53.4$ pA, and $T = 90$ K; the white arrows point to Au atom rows, the black dashed arrows point along phthalocyanine rows; green arrows and circles mark individual phthalocyanine molecules. (b) Line scan along the black line s1 indicated in the STM image, that is, perpendicular to the Au rows. (c) Line scan along the $(1\bar{1}0)$ direction as indicated by the black line s2 in the STM image [91].

more stable on the (1 × 2) reconstructed than on the (1 × 3) reconstructed substrate (Figure 31.52b), and in which the phthalocyanine molecules bind covalently with two of their phenyl groups to the gold ridges while the two other phenyl groups are pending over the trough. At higher coverages, the d6 structure displayed in Figure 31.52a evolves from this d7 phase by moving the molecules closer together by one Au–Au interatomic distance. The resultant intermolecular repulsion, however, is now compensated by a (1 × 3) gold reconstruction underneath. The bright double spots in Figures 31.51 and 31.52a thus correspond to the two phenyl groups resting on the gold ridges of the (1 × 3) structure. At even higher coverages, the molecules move even closer together on the (1 × 3) troughs, but in the resultant d5 structure, the molecules need to be rotated compared to the d7 and d6 phase in order to reduce the phenyl–phenyl repulsion (Figure 31.52d). Thus, one important ingredient in the stabilization of the final structure is the adsorbate-induced change in gold reconstruction from Au(110)(1 × 2) to Au(110)(1 × 3) [91].

This adsorbate-induced substrate reconstruction can go much further and even lead to a drastic restructuring, that is, faceting, of the substrate surface altogether. Already in his early studies on the platinum catalyzed oxidation of carbon monoxide, Langmuir [92] speculated about "changes in the structure of the surface itself, brought about by the reaction," "changes in the direction of becoming a better catalyst." Later experiments verified this assumption by showing that even macroscopic changes of the catalyst morphology may occur during reaction and, in turn, influence its activity [93, 94]. Following up on the original idea of Langmuir, Ladas *et al.* [95] proved the prediction of Langmuir in well-defined model studies with a Pt(110) surface. The bare surface shows, similar to the Au(110) surface above,

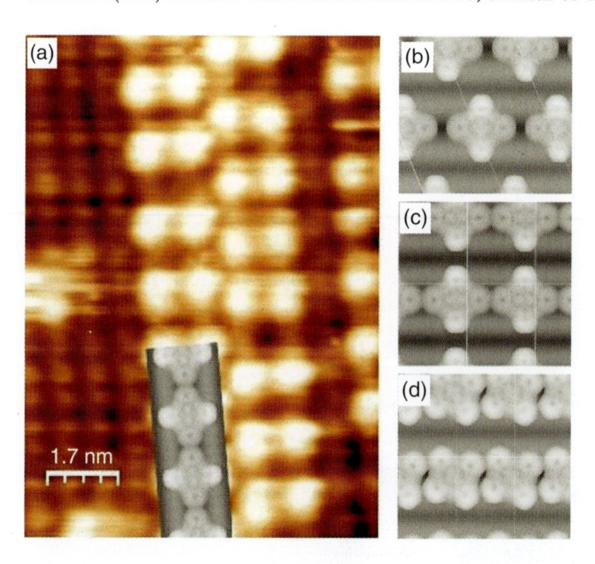


**Figure 31.52** (a) Superposition of the experimental STM image and the STM simulation for the d6 structure (c); constant current STM simulations for $U_b = -2.5$ V of the (b) d7 structure, (c) d6 structure, and (d) d5 structure with phthalocyanine molecules adsorbed on the (1×3) troughs of the Au substrate [91].

a dominant (1×2) reconstruction. Under conditions of catalytic CO oxidation, this surface, however, may evolve into a heavily faceted surface morphology, which depending on pressure and temperature conditions can reach macroscopic dimensions in the micrometer range [93, 96].

Faceting is driven by an increased anisotropy of the surface free energy in the presence of an adsorbate [97, 98] and transforms the originally planar surface into a new one of larger total surface area, but lower *total* surface free energy. Thus, the process is thermodynamically driven, but the involved massive mass transport requires thermal activation. The process is driven just by an adsorbate *on the surface* not to be confused with an in-depth reaction between the substrate and adsorbate.

In particular, open, that is, atomically rough surfaces are prone to faceting. A particularly impressive example in this respect is the oxygen-induced faceting of the Ir(210) surface. Panel (a) in Figure 31.53 shows a hard-sphere model of the Ir(210) surface in which atoms of the first *four* atomic layers a–d are visible [99]. This structure is stable as long as the surface is clean and adsorbate free. Preexposing this surface for 1 min to $1 \times 10^{-7}$ Torr oxygen followed by flash annealing in $1 \times 10^{-7}$ Torr to 1800 K and cooling to room temperature in oxygen leads to a completely new surface morphology as displayed in Figure 31.54a,b [100]. There are no flat surface regions anymore, instead the surface is dotted with pyramids of rather uniform size

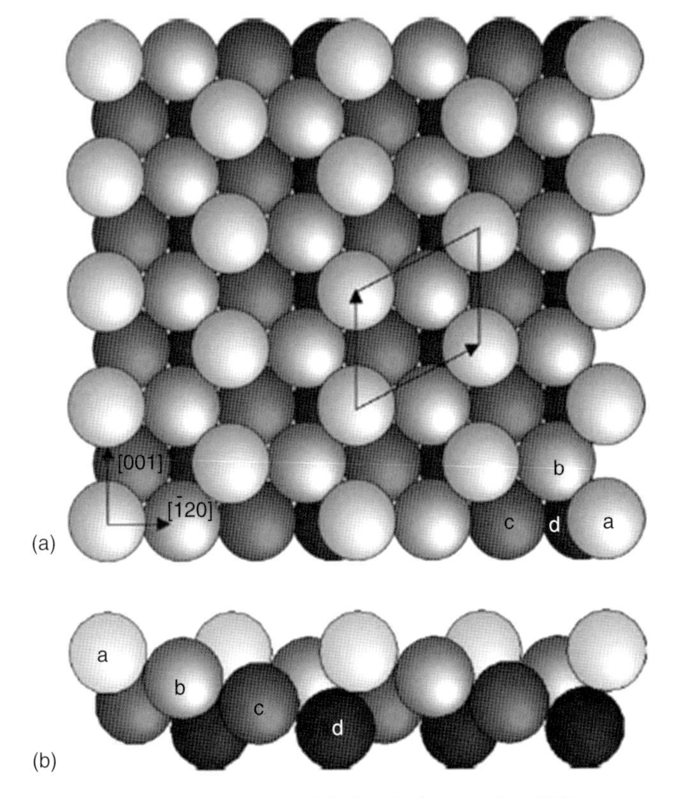


**Figure 31.53** Hard sphere model of an fcc(210) surface [99].

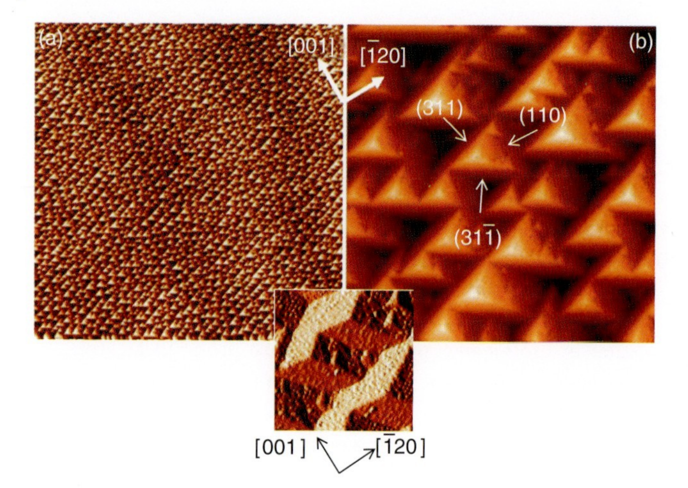


**Figure 31.54** STM scans of the fully faceted Ir(210) surface prepared by flash annealing to 1800 K and cooling to room temperature on oxygen background; facets form as the sample is cooled below ~1150 K. Tunneling parameters: (a) $1000 \times 1000\,\text{nm}^2$, $U_b = 0.1\,\text{V}$, $I_t = 20\,\text{pA}$. (b) $100 \times 100\,\text{nm}^2$, $U_b = 0.1\,\text{V}$, $I_t = 10\,\text{pA}$, the *z*-range is ~3.7 nm. Inset: Differentiated STM scan of a fully faceted surface; scan parameters $U_b = 30\,\text{mV}$, $I_t = 10\,\text{pA}$ [100].

and distribution. The small area scan in (b) shows that all pyramids have similar shape and expose faces of identical crystallographic orientation, namely, (311) and (110) as marked in the image. The facet orientation does not depend on annealing temperature, but the average pyramid size increases with annealing temperature from ~5 nm at 600 K (minimum temperature needed for complete faceting) to ~14 nm after flash annealing to 1800 K and cooling to room temperature in oxygen. The oxygen can be removed reactively by exposing the surface to CO at ~550 K or to hydrogen at ~400 K without lifting the faceted morphology. The clean faceted surface remains stable as long as $T < 600\,\text{K}$, but irreversibly relaxes to its planar state (Figure 31.53) for $T > 600\,\text{K}$.

#### 31.4.3.5 Adsorbate-Induced Segregation

So far we have discussed the influence of an adsorbate on the *structure* of an *elemental* surface, namely, the *lifting* (or change) of an inherent reconstruction, and the restructuring and even faceting of an unreconstructed surface *caused* by adsorption.

A further effect must be considered if adsorption takes place on a composite, for example, alloy, surface (see Volume 3). In this case, adsorption may not only cause a change of surface structure but also a change of surface composition in terms of chemical distribution and concentration at the very surface. This will happen if the adsorbate interacts differently strong with the constituent species in the surface and may lead to a lateral redistribution of the components *within* the surface and/or even to an enrichment of a component due to segregation *from the bulk*, as illustrated by the following example. Since the chemical redistribution parallel and perpendicular to the surface involves transport (diffusion) of the respective component, these processes require thermal activation.

A $Ni_3Al$ alloy has fcc structure and is chemically ordered, and so is its (111) surface. Figure 31.55 illustrates the chemical order of the $Ni_3Al(111)$ surface. Adsorption of oxygen on this highly reactive surface at 800 K results in the formation of triangular islands of same orientation and similar size as seen in Figure 31.56 [101, 102]. Zooming into these islands (c and d) discloses a structure similar to the one found after oxygen adsorption on bare Al(111) at 440 K. This suggests that, as a first step, adsorbing oxygen "assembles" the aluminum atoms in the $Ni_3Al(111)$ surface to form an Al-O bi-layer on the alloy substrate (Figure 31.57) because

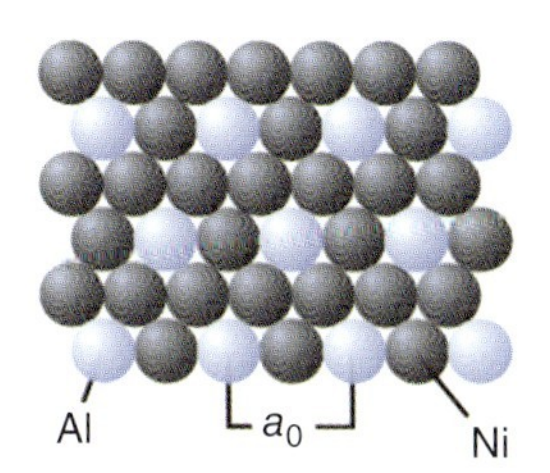


**Figure 31.55** Structure model of the ordered $Ni_3Al(111)$ surface. The Al–Al distance is $a_0 = 5.074$ Å.

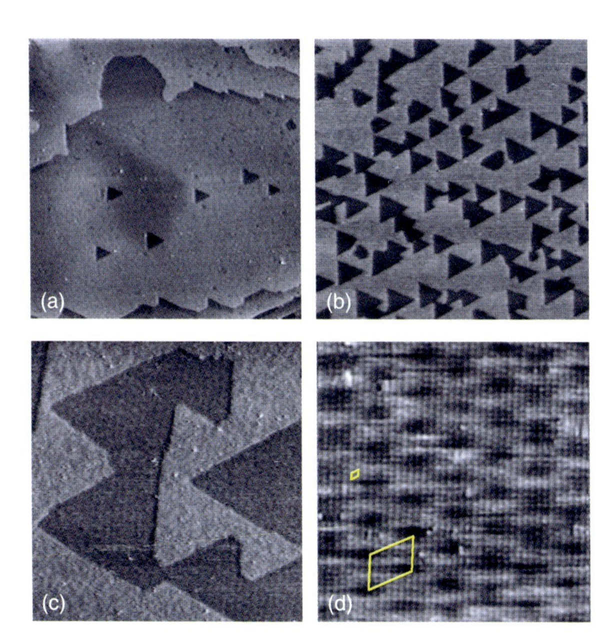


**Figure 31.56** STM images of the $Ni_3Al(111)$ surface after oxidation at 800 K with (a) 10 l and (b) 33 l oxygen. Image size $290 \times 290\ nm^2$, tunneling parameters (a) $U_b = 500$ mV, $I_t = 0.15$ nA, (b) $U_b = 545$ mV, $I_t = 0.16$ nA. (c) and (d) close-ups of the islands in (a) and (b); tunneling parameters (c) $500 \times 500\ nm^2$, $U_b = 1$ V, $I_t = 0.15$ nA; (d) $4 \times 4\ nm^2$, $U_b = 0.16$ mV, $I_t = 4$ nA. Indicated in (d) are the unit cells of the oxygen and the superstructure, respectively [101, 102].

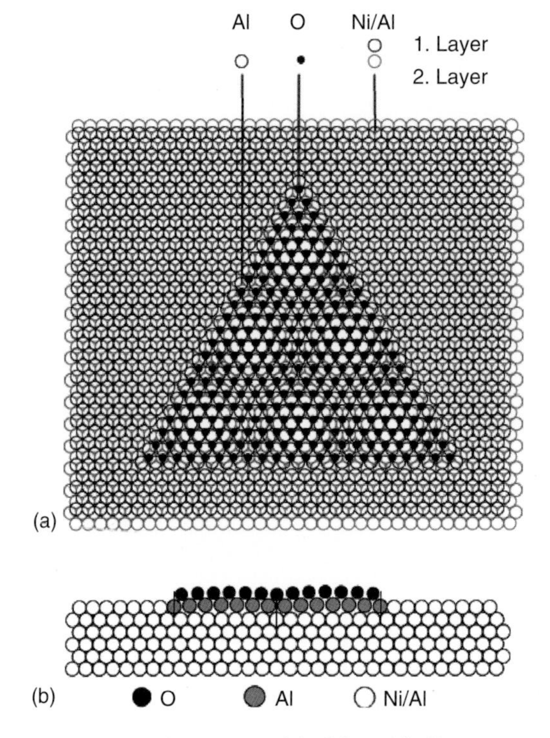


**Figure 31.57** Structure model of the oxide film grown at 800 K on $Ni_3Al(111)$ alloy surface. (a) Top view of an island showing the Moiré structure as a result of the lattice mismatch between oxide film and substrate. (b) side view [101].

the interaction of oxygen with aluminum is much stronger than that with nickel as suggested by the heats of formation of their oxides ($\Delta H_f$ (NiO) = 240 kJ/mol; $\Delta H_f(Al_2O_3)$ = 1675 kJ/mol). The necessary lateral convergence of the Al atoms in the *alloy* surface explains the higher activation temperature of 800 K compared to 440 K on the bare Al(111) surface in order to produce the similar structure.

Exposing the $Ni_3Al(111)$ surface at a higher temperature of 1000 K to oxygen enables even segregation of aluminum from the subsurface region and the formation of a closed oxygen terminated aluminum Al–O–Al–O film on the alloy substrate. This aluminum oxide film has extremely interesting structural and electronic properties, which have exhaustively been studied with a large number of surface techniques [101, 103–105] and which have partly been summarized in Chapter 23, Volume 4 of this book series. In short, the aluminum oxide film possesses a Moiré-type superstructure, which arises from the mismatch between the hexagonal lattices of the underlying (111) structure of the substrate and of the oxide film. On the long range, this superstructure exhibits two hexagonal sublattices, the so-called "net-structure" and the "dot-structure" of 2.4 and 4.1 nm periodicity length, respectively, as displayed in Figure 31.58. While the first arises from a topographic modulation, the second results from a respective modulation

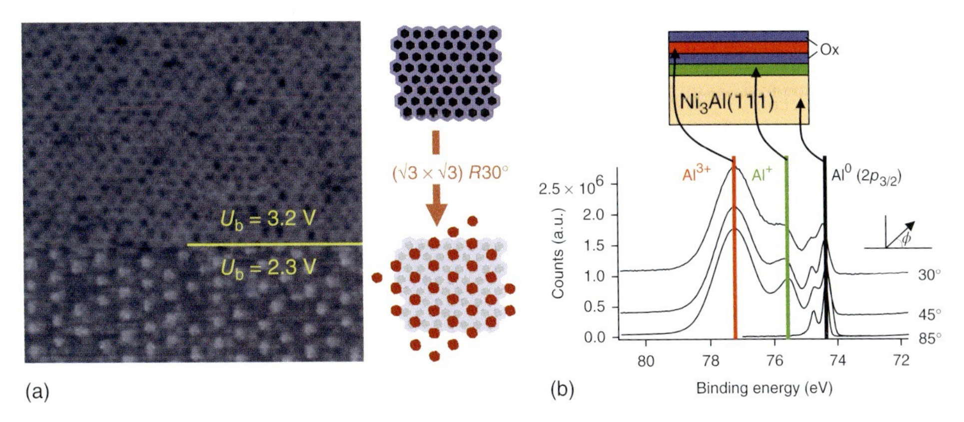


**Figure 31.58** (a) STM image of the Al-oxide film grown at 10 000 K on the $Ni_3Al(111)$ alloy surface. The structure change between the top and bottom part of the image is a result of a change of bias voltage from $U_b = 2.3$ V to $U_b = 3.2$ V. There is a $\sqrt{3}R30°$ relationship between the upper and the lower structure. (b) Angle-resolved XPS spectra of the Al-oxide film grown at 1000 K on the $Ni_3Al(111)$ surface suggesting a distinction and distribution of $Al^0$, $Al^+$, and $Al^{3+}$ ions as indicated in the figure [101] (A. Rosenhahn and C.S. Fadley, unpublished results).

of the electronic properties at every $\sqrt{3}$ site of the net structure as verified by simply changing the bias voltage between surface and STM-tip (see pannel (a) in Figure 31.58). The double bilayer structure is supported by angular resolved XPS measurements (Figure 31.58b) (A. Rosenhahn and C.S. Fadley, unpublished results). These oxide films are excellent templates for the nanostructured growth of metal particles [76, 106].

In the present context, it shall be emphasized that it is the adsorbate-driven segregation of aluminum in and to the surface of the Ni3Al(111) surface, which causes these drastic chemical and structural changes at the alloy surface. Very similar effects have, for instance, also been observed with oxygen adsorption on an ordered $Pt_3Ti(111)$ surface, which leads to Ti segregation and the formation of different titanium oxide films again depending on the oxygen exposure and temperature during the film preparation [107, 108].

## Acknowledgment

I am very grateful for the support of Dr. Stephan Breuer in the preparation of figures.

## References

As an editor of journals and books over many years I realized that citations of publications older than say 15 years are increasingly ignored in recent research papers. Since these publications, however, often laid the basis for the present research, I have deliberately selected such early references here.

1. Ponec, V., Knor, Z., and Cerny, S. (1974) *Adsorption on Solids*, Butterworth.
2. Adamson, A. (1990) *Physical Chemistry of Surfaces*, 5th edn, Wiley-Interscience, New York.
3. Wedler, G. (1971) *Chemisorption: An Experimental Approach*, translated by Klemperer, D.F., Butterworth.
4. Ertl, G. and Küppers, J. (1985) *Low Energy Electrons and Surface Chemistry*, Wiley-VCH Verlag GmbH, Weinheim.
5. Rhodin, T.N. and Ertl, G. (1979) *The Nature of the Surface Chemical Band*, North-Holland.
6. Wissmann, P. (ed) (1987) *Thin Metal Films and Gas Chemisorption*, Studies in Surface Science and Catalysis, vol. 32, Elsevier.
7. Somorjai, G. (1993) *Introduction to Surface Chemistry and Catalysis*, Wiley-Interscience.
8. Zangwill, A. (1988) *Physics at Surfaces*, Cambridge University Press.
9. Christmann, K. (1991) *Introduction to Surface Physical Chemistry*, Springer-Verlag, New York.
10. Kolasinski, K.W. (2001) *Surface Science – Foundations of Catalysis and Nanoscience*, John Wiley & Sons, Ltd.
11. Chung, Y.-W. (2001) *Practical Guide to Surface Science & Spectroscopy*, Academic Press.
12. Niemantsverdriet, J.W. (2007) *Spectroscopy in Catalysis: An Introduction*, 3rd edn, Wiley-VCH Verlag GmbH.
13. Ibach, H. (2006) *Physics of Surfaces and Interfaces*, Springer-Verlag, Heidelberg.
14. Ertl, G. (2009) *Reactions at Solid Surfaces*, John Wiley & Sons, Inc.
15. Wandelt, K. (ed.) (2012) *Surface and Interface Science*, vol. 1 Concepts and Methods, vol. 2 Properties of Elemental Surfaces (2014) vol. 3 Properties of Composite Surfaces: Alloys, Compounds, Semiconductors, vol. 4 Solid-Solid Interfaces and Thin Film Growth, Wiley-VCH Verlag GmbH.
16. Barth, J.V., Zambelli, T., Wintterlin, J., and Ertl, G. (1997) *Chem. Phys. Lett.*, **270**, 152.
17. Rendulic, K.D. (1988) *Appl. Phys. A*, **47**, 55.
18. King, D.A. and Wells, M.G. (1972) *Surf. Sci.*, **29**, 454.
19. Engel, T. (1978) *J. Chem. Phys.*, **69**, 373.
20. Langmuir, I. (1918) *J. Am. Chem. Soc.*, **40**, 1361.
21. Brundle, C.R. and Hopster, H. (1981) *J. Vac. Sci. Technol.*, **18**, 663.
22. Freundlich, H. (1906) *Z. Phys. Chem.*, **57** (A), 385.
23. Tempkin, M.I. and Pyzhev, V. (1940) *Acta Phys. Chem. USSR*, **12**, 327.
24. Guggenheim, E.A. (1949) *Thermodynamics*, North-Holland Publishing Co.
25. Miranda, R., Albano, E.V., Daiser, S., Wandelt, K., and Ertl, G. (1984) *J. Chem. Phys.*, **80**, 2931.
26. Miranda, R., Daiser, S., Wandelt, K., and Ertl, G. (1983) *Surf. Sci.*, **131**, 61.
27. Miranda, R., Albano, E.V., Daiser, S., Ertl, G., and Wandelt, K. (1983) *Phys. Rev. Lett.*, **51**, 782.
28. Kisliuk, P. (1957) *J. Phys. Chem. Solids*, **3**, 95.
29. Brunauer, S., Emmett, P.H., and Teller, E. (1938) *J. Am. Chem. Soc.*, **68**, 309.
30. Thomy, A., Duval, X., and Regnier, J. (1981) *Surf. Sci.*, **1**, 1.
31. Pennemann, B. (1991) Adsorption von Xenon und Wasserstoff auf Platin-Einkistalloberflächen. PhD thesis. University of Bonn, Germany.
32. Redhead, P.A. (1962) *Vacuum*, **12**, 203.
33. Habenschaden, E. and Küppers, J. (1984) *Surf. Sci.*, **138**, L147.
34. King, D.E., Madey, T.E., and Yates, J.T. Jr., (1971) *J. Chem. Phys.*, **55**, 3236.
35. Ertl, G. (1979) *Surf. Sci.*, **89**, 525.
36. Jones, R.H., Orlander, D.R., Siekhaus, W.J., and Schwarz, J.A. (1972) *J. Vac. Sci. Technol.*, **9**, 1429.
37. Parthenopoulos, M. (2009) Elektrochemische und spektroskopische Untersuchungen dünner Edelmetallfilme auf einer Ru(0001)-Oberfläche. PhD thesis. University of Bonn, Germany.
38. Michels, J. (1998) *J. Chem. Phys.*, **108**, 4248; and (1997) Modellierung der Kinetik von Adsorptions- und Desorptionsprozessen dünner Edelmetallfilme. PhD thesis. University of Bonn, Germany.

39. Niemantsverdriet, J.W., Dolle, P., Markert, K., and Wandelt, K. (1987) *J. Vac. Sci. Technol.*, **A5**, 875.
40. Wandelt, K., Markert, K., Dolle, P., Jablonski, A., and Niemantsverdriet, J.W. (1987) *Surf. Sci.*, **189/190**, 114.
41. Schneider, U., Castro, G.R., and Wandelt, K. (1993) *Surf. Sci.*, **287/288**, 146.
42. Ertl, G. and Koch, J. (1970) *Z. Naturforsch.*, **25A**, 1906.
43. Canrad, H., Ertl, G., Koch, J., and Latta, E.E. (1974) *Surf. Sci.*, **43**, 462.
44. Carrez, S., Dragnea, B., Zeng, W.Q., Dobost, H., and Bourguignon, B. (1999) *Surf. Sci.*, **440**, 151.
45. Niemantsverdriet, J.W., Markert, K., and Wandelt, K. (1988) *Appl. Surf. Sci.*, **31**, 211.
46. Palmer, W.G. and Constable, F.H. (1924) *Proc. R. Soc. London, Ser. A*, **106**, 250.
47. Constable, F.H. (1925) *Proc. R. Soc. London, Ser. A*, **108**, 355.
48. Fecher, G.H., Volkmer, M., Prawlitzky, B., Böwering, N., and Heinzmann, U. (1990) *Vacuum*, **41**, 265.
49. Hoffmann, R. (1988) *Solids and Surfaces: A Chemist's View of Bonding in Extended Structures*, VCH-Publishers, New York.
50. Gumhalter, B. and Newns, D.M. (1979) *Phys. Lett.*, **56A**, 423.
51. Lang, N.D. (1981) *Phys. Rev. Lett.*, **46**, 842.
52. Wandelt, K. and Gumhalter, B. (1984) *Surf. Sci.*, **140**, 355.
53. Müller, J.E. (1990) *Phys. Rev. Lett.*, **65**, 3021.
54. Hölzl, J. and Schulte, F.K. (1979) *Work Function of Metals*, Springer Tracts of Modern Physics, vol. 85, Springer-Verlag, Berlin.
55. Blyholder, G. (1964) *J. Phys. Chem.*, **68**, 2772.
56. Sung, S.S. and Hoffmann, R. (1985) *J. Am. Chem. Soc.*, **107**, 578.
57. Norskov, J.K. (1990) *Rep. Prog. Phys.*, **53**, 1253.
58. Gajdos, M. and Hafner, J. (2005) *Surf. Sci.* **509**, 117.
59. Brundle, C.R. and Wandelt, K. (1977) Proceedings of the 3rd International Conference on Solid Surfaces, Vienna, p. 1171.
60. Freund, H.-J. and Neumann, M. (1988) *Appl. Phys.*, **A47**, 3.
61. Miranda, R., Wandelt, K., Rieger, D., and Schnell, R.D. (1984) *Surf. Sci.*, **139**, 430.
62. Bondino, F., Comelli, G., Esch, F., Locatelli, A., Baraldi, A., Lizzit, S., Paolucci, G., and Rosei, R. (2000) *Surf. Sci.*, **459**, L467.
63. Becker, C., Pelster, T., Tanemura, M., Breitbach, J., and Wandelt, K. (1999) *Surf. Sci.*, **427/428**, 403.
64. Artsyukhovich, A.N., Ukraintsev, V.A., and Harrison, I. (1996) *Surf. Sci.*, **347**, 303.
65. Zambelli, T., Barth, J.V., Winterllin, J., and Ertl, G. (1997) *Nature*, **390**, 495.
66. Campbell, C.T., Ertl, G., Kuipers, H., and Segner, J. (1981) *Surf. Sci.*, **107**, 220.
67. Avery, N.R. (1983) *Chem. Phys. Lett.*, **96**, 371.
68. Imbihl, R. and Demuth, J.E. (1986) *Surf. Sci.*, **173**, 395.
69. Campbell, C.T. (1985) *Surf. Sci.*, **157**, 43.
70. Becker, C. (1993) Adsorption von Wasserstoff und Kohlenmonoxid auf $Cu_3Pt(111)$. PhD thesis, University of Bonn, Germany.
71. Schmeisser, D. and Jacobi, K. (1981) *Surf. Sci.*, **108**, 421.
72. Baraldi, A., Bianchettin, L., Vesselli, E., de Gironcoli, S., Lizzit, S., Pelaccia, L., Zampieri, G., Comelli, G., and Rosei, R. (2007) *New J. Phys.*, **9**, 143.
73. Baraldi, A., Lizzit, S., Comelli, G., Kiskinova, M., Rosei, R., Honkala, K., and Norskov, J.K. (2004) *Phys. Rev. Lett.*, **93**, 046101-1.
74. Ganguglia-Pirovano, M.V., Scheffler, M., Baraldi, A., Lizzit, S., Comelli, G., Paolucci, G., and Rosei, R. (2001) *Phys. Rev.*, **B63**, 205415.
75. Kono, S., Fadley, C.S., Hall, N.F.T., and Hussain, Z. (1978) *Phys. Rev. Lett.*, **41**, 117.
76. Becker, C. and Wandelt, K. (2009) *Templates in Chemistry III – Topics in Current Chemistry*, vol. 287, Springer-Verlag, p. 45.

77. Wintterlin, J., Trost, J., Renisch, S., Schuster, R., Zambelli, T., and Ertl, G. (1997) *Surf. Sci.*, **394**, 159.
78. Nohlen, M., Schmidt, M., Wolter, H., and Wandelt, K. (1995) *Surf. Sci.*, **337**, 294.
79. Wilms, M., Schmidt, M., Bermes, G., and Wandelt, K. (1998) *Rev. Sci. Instrum.*, **69**, 2696.
80. Poelsema, B., Verheij, L.K., and Comsa, G. (1983) *Phys. Rev. Lett.*, **51**, 2410.
81. Batra, I., Hermann, K., Bradshaw, A.M., and Horn, K. (1979) *Phys. Rev.*, **B20**, 801.
82. Ibach, H. (1982) *Electron Energy Loss Spectroscopy and Surface Vibrations*, Academic Press.
83. Küppers, J., Michel, H., Nitschke, F., and Wandelt, K. (1979) *Surf. Sci.*, **89**, 361.
84. Ritz, G., Schmid, M., Varga, P., Borg, A., and Ronning, M. (1997) *Phys. Rev*, **B56**, 10518.
85. Schaefer, B., Nohlen, M., and Wandelt, K. (2004) *J. Phys. Chem.*, **108**, 14663.
86. Kampshoff, E., Waelchli, N., Menck, A., and Kern, K. (1996) *Surf. Sci.*, **360**, 55.
87. Yoshinobu, J., Tanaka, H., and Kawai, M. (1995) *Phys. Rev.*, **B51**, 4529.
88. Ledentu, V., Dong, W., Sautet, P., Kresse, G., and Hafner, J. (1998) *Phys. Rev.*, **B57**, 12482.
89. Kralj, M., Becker, C., and Wandelt, K. (2006) *Surf. Sci.*, **600**, 4113.
90. Landmann, M., Rauls, E., and Schmidt, W.G. (2009) *Phys. Rev.*, **B79**, 045412.
91. Rauls, E., Schmidt, W.G., Pertram, T., and Wandelt, K. (2012) *Surf. Sci.*, **606**, 1120.
92. Langmuir, I. (1921) *Trans. Faraday Soc.*, **17**, 607.
93. Flytzani-Stephanopoulos, M. and Schmidt, L.D. (1979) *Prog. Surf. Sci.*, **9**, 83.
94. Inioui, A., Eddovasse, M., Amariglio, A., Ehrhardt, J.J., Lambert, J., Alnot, M., and Amariglio, H. (1985) *Surf. Sci.*, **162**, 368.
95. Ladas, S., Imbihl, R., and Ertl, G. (1988) *Surf. Sci.*, **197**, 153.
96. Calwey, A.K., Gray, P., Griffiths, J., and Hasko, S.M. (1985) *Nature*, **313**, 668.
97. Madey, T.E., Guan, J., Nien, C.-H., Dong, C.-Z., Tao, H.-S., and Campbell, R.A. (1996) *Surf. Rev. Lett.*, **3**, 1315.
98. Che, J.G., Chan, C.-T., Kuo, C.H., and Leung, T.C. (1997) *Phys. Rev. Lett.*, **79**, 4230.
99. Ermanoski, I., Pelhos, K., Chen, W., Quinton, J.S., and Madey, T.E. (2004) *Surf. Sci.*, **549**, 1.
100. Ermanoski, I., Kim, C., Kelty, S.P., and Madey, T.E. (2005) *Surf. Sci.*, **596**, 89.
101. Rosenhahn, A. (2000) Ultradünne Aluminiumoxidfilme auf Ni3Al(111): Template für nanostrukturiertes Metallwachstum. PhD thesis. University of Bonn, Germany.
102. Rosenhahn, A., Schneider, J., Becker, C., and Wandelt, K. (2000) *J. Vac. Sci. Technol.*, **A18**, 1923.
103. Gritschneder, S., Degen, S., Becker, C., Wandelt, K., and Reichling, M. (2007) *Phys. Rev.*, **B76**, 0141213.
104. Hamm, G., Barth, C., Becker, C., Wandelt, K., and Henry, C.R. (2006) *Phys. Rev. Lett.*, **97**, 126106.
105. Degen, S., Krupski, A., Kralj, M., Langner, A., Becker, C., Sokolowski, M., and Wandelt, K. (2005) *Surf. Sci. Lett.*, **576**, L57.
106. Becker, C., Rosenhahn, A., Wiltner, A., von Bergmann, K., Schneider, J., Pervan, P., Milun, M., Kralj, M., and Wandelt, K. (2002) *New J. Phys.*, **4**, 75.1.
107. Le Moal, S., Moors, M., Essen, J.M., Breinlich, C., Becker, C., and Wandelt, K. (2013) *J. Phys. Condens. Matter*, **25**, 045013.
108. Breinlich, C., Buchholz, M., Moors, M., Le Moal, S., Becker, C., and Wandelt, K. (2014) *J. Phys. Chem.*, **C118**, 6186.

# 32
# Surface Microcalorimetry

*J. Michael Gottfried and Rolf Schuster*

## 32.1
## Introduction

*Calorimetry* is the measurement of energy in the form of heat. Heat $q$ results from phase transitions, chemical reactions, or other processes and is measured quantitatively by observing the effects that it produces, such as temperature changes. Since heat is related to the thermodynamic state functions, calorimetry provides access to the thermodynamic properties of a chemical system. At surfaces and interfaces, the most fundamental processes are adsorption, desorption, and surface reactions. The heat released during adsorption at the solid–gas interface is a direct measure of the *adsorption energy* and thus of the strength of the adsorbate–surface interaction [1]. At electrochemical interfaces which are the main subject of Volume 7 of this book series, however, the exchange of electric work causes the heat to be related to the *entropy change* of the electrochemical process, as will be discussed in the following.

Adsorption energies play an important role in heterogeneous catalysis, where the activation of a molecule requires sufficiently strong interaction of the intermediate with the surface of the catalyst, that is, sufficiently high adsorption energy. However, very high adsorption energies can prevent desorption and thus lead to poisoning of the catalyst. Therefore, intermediate adsorption energies are often the most effective in catalysis (principle of Sabatier) [2]. Precise experimental adsorption energies are also required as reference data for the improvement of theoretical methods such as density functional theory (DFT). Since DFT-based methods still struggle with the correct calculation of van der Waals contributions to the adsorbate–substrate bond, such benchmark values are especially important for weakly bound adsorbates such as large organic molecules or biomolecules.

In electrochemical systems, knowledge about the entropy may provide important information on the details of the electrochemical reaction at the electrode–electrolyte interface. Often, the charge transfer at the interface, for example, at the deposition of a metal ion, is accompanied by side processes such as double-layer charging, coadsorption processes of anions, and reorganization of the solvent at the interface. Particularly at small conversions, such as during the deposition of the first layer of a metal at the electrode, such side processes may have strong influence on

*Surface and Interface Science: Solid-Gas Interfaces I*, First Edition. Edited by Klaus Wandelt.

the overall entropy change of the reaction. Because of the side processes, the surface layer may be entropically stabilized, as was recently found for Cu underpotential deposition (UPD) on Au. Furthermore, the entropy change provides complementary information to the net charge that flowed during the reaction. This may help to identify side reactions such as coadsorption or restructuring of the solvent. It should also be mentioned that information on entropy changes in electrochemical systems is technically important, for example, for the heat management in batteries.

Many different types of calorimeters have been devised for measuring heat by observing quantitatively the effects produced when energy in other forms is transformed to heat. In this chapter, the focus is on calorimeters designed to measure heats of adsorption and reaction on surfaces on single crystals and thin molecular or polymeric films, and on electrochemical interfaces with thin metal films or single crystals. Not included is calorimetry on powders [3–5] and other interfaces of systems with a high surface-to-volume ratio, because these surfaces are generally not well defined.

The most rigorous calorimetric technique for studying adsorption on well-defined solid–gas interfaces under ultrahigh-vacuum (UHV) conditions is known as single-crystal adsorption calorimetry (SCAC), which was originally developed by the group of Sir David King [6]. Later, Charles T. Campbell and coworkers introduced a novel detection method that made measurements below room temperature possible [7]. The SCAC approach makes use of molecular beam (MB) techniques and requires, in addition to the heat measurement, precise measurements of the beam flux and sticking probability. Interpretation of the energies of adsorption and reaction is possible only on the basis of detailed insight into the adsorption process and related surface reactions such as dissociation. Adsorption calorimetry is therefore typically performed in combination with surface analytical and microscopic techniques. Prior to these studies on well-defined single-crystal surfaces, adsorption calorimetry was performed on metal wires, ribbons, and thin films [8].

Calorimetry is not the only thermodynamic approach for measuring enthalpies of adsorption at solid–gas interfaces. An alternative method is to observe the change of the equilibrium pressure with temperature at constant coverage. From these data, the isosteric heat, a measure for the adsorption energy, can be derived by means of the Clausius–Clapeyron equation. If the equilibrium pressure can be measured at different temperatures, this method gives direct access to adsorption energies with the advantage of being technically less demanding than adsorption calorimetry. For many systems, however, the equilibrium pressure is too low to be practically measured. (This case is also, somewhat loosely, named "nonreversible adsorption.") Examples include large organic molecules, which would undergo temperature-induced decomposition on the surface before developing a practically measurable equilibrium pressure, but also reactive intermediates of surface reactions or metals on thermally sensitive organic materials. All these systems require the calorimetric approach.

Insight into the strength of the adsorbate–substrate bond can also be obtained by kinetic measurements. A widely used method is temperature-programmed desorption (TPD) [9], in which the temperature of an adsorbate-covered surface is increased until desorption is complete. Arrhenius analysis of the temperature dependence of the desorption rate constant yields a desorption activation energy $E_d$, from which the adsorption energy can be estimated. TPD has similar limitations as isosteric measurements, that is, it cannot be applied when adsorption in nonreversible.

Modern calorimetry at the solid–electrolyte interface employs similar detection methods as calorimetry at the solid–gas interface, but controlled adsorption is achieved differently. Instead of the pulsed MB used in SCAC, a sudden change in the electrode potential is used to induce an electrochemical reaction, which, for example, leads to adsorption or deposition of the material at the interface. Whereas upon adsorption of a gas molecule on a surface in UHV the measured heat essentially corresponds to the heat of adsorption, that is, the change of the internal energy of the gas–surface system, the heat generated at an electrode upon an electrochemical reaction is related to the entropy change of the electrode–electrolyte interface. The reason can be sought in the difference between the electrical work $w_{el}$, which for reversible electrochemical processes is given by the Gibbs free energy of the cell reaction $\Delta_R G$, and the enthalpy of the cell reaction $\Delta_R H$, which would be transformed into heat upon a purely chemical reaction. This difference is given by the entropy term $T\Delta_R S$ of the electrode reaction. Although this holds true for the complete cell reaction, the situation for the heat exchange at the individual electrodes is more complicated. It requires the consideration of the detailed transport processes in the cell within the framework of irreversible thermodynamics. To elucidate the origin of those complications, we present a brief review of the current theory as well as the historical developments in Section 32.2.2.

## 32.2
## Thermodynamic Principles of Surface Microcalorimetry

### 32.2.1
### Solid–Gas Interfaces

#### 32.2.1.1 Approaches for the Measurement of Adsorption Energies

Adsorption energies can be determined by thermodynamic and, indirectly, by kinetic methods (see also Chapter 31 in this Volume). TPD (also known as thermal desorption spectroscopy, TDS) [9, 10] belongs to the kinetic methods. In a TPD experiment, an adsorbate-covered surface is heated such that the temperature $T$ increases with a constant rate $dT/dt$ until the adsorbate is completely desorbed. The temperature-dependent desorption rate $r$ is typically measured by observing the transient change of the partial pressure of the respective substance in a pumped recipient using a mass spectrometer. To determine desorption activation energies $E_d$, a simple $n$th-order rate equation is combined with an Arrhenius expression for the rate constant. The resulting rate equation is sometimes referred to as Polanyi–Wigner equation [9]:

$$r = k_d \cdot \Theta^n = \nu \cdot \exp\left(-\frac{E_d}{RT}\right) \cdot \Theta^n \tag{32.1}$$

where $\Theta$ is the coverage, and $n$ is the desorption order. In the absence of an activation barrier for adsorption, the desorption activation energy approximates the negative adsorption energy. Otherwise, the adsorption activation energy, which can be difficult to measure, must be known to estimate the adsorption energy. A TPD trace, which is a plot of $r$ versus $T$, has at least one maximum, because the desorption rate constant $k_d$ increases with temperature (and thus with time), while the coverage decreases with time (and thus with temperature) until it reaches zero.

Another non-calorimetric approach for the measurement of adsorption energies is based on equilibrium thermodynamics and considers the change of the adsorption/desorption equilibrium pressure $p_{eq}$ with temperature at constant coverage. It can be shown that the variation of $p_{eq}$ with $T$ at constant $\Theta$ is determined by the ratio between the changes of the molar entropy and the molar volume of this process. Assuming an ideal gas phase and taking into account that the reversible entropy equals $q/T$ result in a special form of the Clausius–Clapeyron equation defining the isosteric heat $q_{st}$ of adsorption, which is the heat of adsorption at constant coverage [1]:

$$q_{st} = -R\left(\frac{\partial \ln p}{\partial (1/T)}\right)_{\Theta}. \tag{32.2}$$

The major limitation of both methods is that they require a measurable equilibrium pressure of the adsorbate at the temperature of the surface. This usually requires that molecules desorb faster than they decompose on the surface, a condition that is referred to as “reversible adsorption,” but must not be confused with the concept of thermodynamic reversibility. An example for “reversible” adsorption is carbon monoxide desorbing from a Pt(111) surface (Figure 32.1a). Many molecules, especially large organic molecules or biomolecules, however, dissociate on the surface and only their fragments desorb. An example is benzene on Pt(111) (Figure 32.1b), which adsorbs intact at and below room temperature, but decomposes at elevated temperatures. In this case, neither TPD nor methods based on the Clausius–Clapeyron Equation 32.2 can be used. For such systems, the adsorption energy can be determined only by direct calorimetric measurement, that is, by measuring the heat released during adsorption.

#### 32.2.1.2 Thermodynamic Considerations

In a typical experiment with a UHV-based calorimeter, a pulsed MB strikes the surface of a single-crystal sample. When molecules contained in the MB pulse adsorb, heat is released and measured as a temperature change of the sample. Simultaneously, the sticking probability (i.e., the fraction of incident molecules adsorbing on the sample) is measured for each pulse. Using the sticking probability and the separately measured flux, the number of molecules that is adsorbed on the sample with each pulse is calculated.

The conditions inside a constant-volume UHV apparatus are approximately isochoric. Therefore, the heat measured in the calorimetric experiment, $q_{cal}$, is equivalent to the change of the internal energy of the sample, $\Delta u_S$. (Note that lowercase letters are used for extensive quantities and uppercase letters are used for

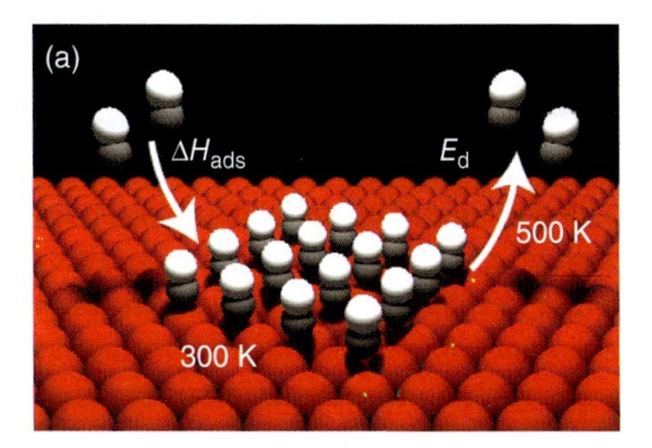


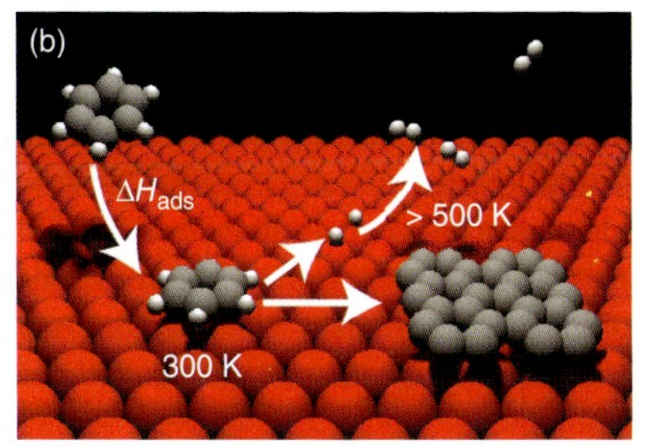


**Figure 32.1** Examples of reversible and irreversible adsorption. (a) Reversible: CO on Pt(111). CO desorbs intact and the molar adsorption enthalpy $\Delta H_{\mathrm{ads}}$ can be estimated from the desorption activation energy $E_{\mathrm{d}}$. (b) Irreversible: benzene on Pt(111). Benzene adsorbs at 300 K as an intact molecule, but dissociates above 500 K forming hydrogen and carbon. In this case, calorimetry is needed to determine $\Delta H_{\mathrm{ads}}$. (Image courtesy E.K. Vestergaard. Adapted with permission from Ref. [11]. Copyright 2011, Bunsengesellschaft für Physikalische Chemie.)

intensive quantities in this chapter.) Two processes contribute to $\Delta u_{\mathrm{S}}$: (i) adsorption of molecules ($\Delta u_{\rightarrow|}$) and (ii) reflection of molecules with the exchange of energy ($\Delta u_{\rightleftarrows|}$) [12]:

$$q_{\mathrm{cal}} = \Delta u_S = \Delta u_{\rightarrow|} + \Delta u_{\rightleftarrows|}. \tag{32.3}$$

The contribution $\Delta u_{\rightarrow|}$ is the sum of three terms: (i) the actual adsorption energy $\Delta u_{\mathrm{ads}}$, which is defined as the change of the internal energy of the gas–surface system during adsorption of gas with the sample temperature $T_{\mathrm{S}}$; (ii) the term $\frac{1}{2}RT_{\mathrm{MB}}$ for the energy difference between a gas *flux* and a gas *volume* at the translational temperature of the MB, $T_{\mathrm{MB}}$ [13]. This term takes into account that the adsorbate is dosed with an MB and not from a gas phase; (iii) a contribution that arises when the gas temperature $T_{\mathrm{MB}}$ in the MB is different from the sample temperature $T_{\mathrm{S}}$. In this case, the molar isochoric heat capacity of the gas, $C_{\mathrm{v}}$, must be taken into account:

$$\Delta u_{\rightarrow|} = -\Delta u_{\mathrm{ads}} + n_{\mathrm{ads}}\left(\frac{1}{2}RT_{\mathrm{MB}} - \int_{T_{\mathrm{MB}}}^{T_{\mathrm{S}}} C_{\mathrm{v}}\, dT\right). \tag{32.4}$$

In Equation 32.4, $n_{\mathrm{ads}}$ is the adsorbed amount of gas. Note that $\Delta u_{\mathrm{ads}}$ contributes with a negative sign, because the gas–surface system releases heat (therefore,

$\Delta u_{ads} < 0$ from the point of view of the gas–surface system), but the sample (which is considered here as the sign-defining thermodynamic system) gains this amount of heat (therefore, $\Delta u_{\rightarrow|} > 0$).

The integral term in Equation 32.4 comes into play when the gas and the surface have different temperatures. If this is the case, also the reflected molecules can contribute to $q_{cal}$ by exchanging energy with the sample, as expressed by the term $\Delta u_{\rightleftarrows|}$ in Equation 32.3. If all reflected molecules adopt the surface temperature $T_S$, $\Delta u_{\rightleftarrows|}$ is given by

$$\Delta u_{\rightleftarrows|} = -n_{refl} \int_{T_{MB}}^{T_S} \left( C_v + \frac{1}{2} R \right) \, dT \tag{32.5}$$

with $n_{refl}$ as the reflected amount of gas. The term $\frac{1}{2}R$ accounts for the fact that a flux of molecules contains more kinetic energy than the corresponding volume of gas at rest. If the volume of gas is in thermal equilibrium, this difference corresponds to $\frac{1}{2}RT$ [13].

The *adsorption enthalpy* $\Delta h_{ads}$ at the sample temperature $T_S$ equals the adsorption energy $\Delta u_{ads}$ plus the volume work, which would under isobaric conditions result from the compressions of the gas phase (assumed to be ideal) to the negligibly small volume of the adsorbed phase:

$$\Delta h_{ads} = \Delta u_{ads} - n_{ads} R T_S. \tag{32.6}$$

The *molar heat of adsorption* is defined as the negative molar adsorption enthalpy, $-\Delta H_{ads}$, and thus is always positive. Combining Equations 32.3–32.6 results in the following relation between $-\Delta H_{ads}$ and the calorimetrically measured heat $q_{cal}$:

$$-\Delta H_{ads} \equiv \frac{-\Delta h_{ads}}{n_{ads}} = \frac{1}{n_{ads}} (q_{cal} - K_{ads} - K_{refl}). \tag{32.7}$$

Equation 32.7 includes the correction terms

$$K_{ads} = n_{ads} \cdot \left[ -\int_{T_{MB}}^{T_S} C_v dT + \frac{1}{2} R T_{MB} - R T_S \right] \tag{32.8}$$

$$K_{refl} = n_{refl} \cdot \left[ -\int_{T_{MB}}^{T_S} \left( C_v + \frac{1}{2} R \right) dT \right] \tag{32.9}$$

where $-\Delta H_{ads}$ is the differential heat of adsorption, which means that it describes the heat per mol during adsorption at a constant coverage. It is suitable for comparison with tabulated standard enthalpies after conversion to standard temperature.

Integration of $-\Delta H_{ads}$ over the coverage $\Theta$ yields the *integral* heat of adsorption $-\Delta H_{ads,\, int}$:

$$-\Delta H_{ads,\, int} = -\int_0^{\Theta} \Delta H_{ads} \, d\Theta'. \tag{32.10}$$

The integral heat of adsorption is the heat per mol that is released when a clean surface is covered by the adsorbate up to a certain coverage (see Figure 32.2).

To compare calorimetric heats of adsorption with adsorbate–substrate bond energies obtained by *ab initio* calculations, it is necessary to take the changes in the

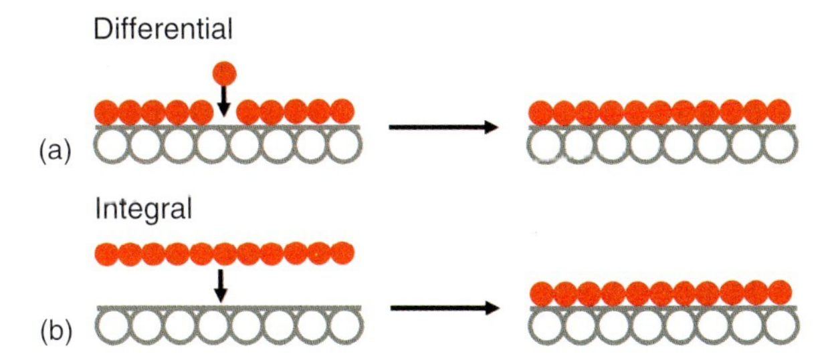


**Figure 32.2** Differential (a) and integral (b) heats of adsorption near saturation coverage. The differential heat of adsorption considers differentially small changes of coverage, while the integral heat of adsorption considers finite changes of coverage starting from the clean surface. Adsorption calorimetry measures differential heats of adsorption, whereas theoretical calculations typically provide integral heats of adsorption. (Adapted with permission from Ref. [11]. Copyright 2011, Bunsengesellschaft für Physikalische Chemie.)

internal and external degrees of freedom (DOFs) of the adsorbing molecules into account. In a first approximation, their influence on the heat of absorption can be estimated by basic equipartition considerations. In the simplest case of an ideal mono-atomic gas with three translational DOFs, the molar enthalpy of the gas is given by

$$H_{gas} = U_{gas} + pV_{gas} = \frac{3}{2}RT + RT. \tag{32.11}$$

The molar enthalpy of the adsorbed phase, $H_{ads}$, depends on the translational and vibrational DOF of the adsorbed atoms. In the case of completely mobile adsorption, the atom has two lateral translational DOFs ($F_{trans} = 2$) and one vibrational DOF ($F_{vib} = 1$) perpendicular to the surface. In the case of full excitation of all DOF, these considerations result in the equation

$$H_{ads} \approx U_{ads} = -|E_0| + 2RT, \tag{32.12}$$

where $|E_0|$ is the adsorbate–substrate bond energy at $T = 0$. This leads to the following relationship between the heat of adsorption and the adsorbate–substrate binding energy:

$$-\Delta H_{ads} = H_{gas} - H_{ads} = |E_0| + \frac{1}{2}RT. \tag{32.13}$$

In contrast, for localized adsorption with $F_{trans} = 0$ and $F_{vib} = 3$, one obtains in the case of full excitation of all DOFs

$$-\Delta H_{ads} = |E_0| - \frac{1}{2}RT. \tag{32.14}$$

$|E_0|$ can be compared to the calculated adsorbate–substrate bond energies.

For the adsorption of large molecules, the relationship between the calorimetric heat and the adsorbate–substrate binding energy may be difficult to calculate, because all internal DOFs and their degrees of excitation in gas and adsorbed phase must be taken into account. For the adsorbed phase, adsorption-induced changes in the excitation of the internal DOFs must be considered. This holds true in particular for chemisorption, which influences the strengths of chemical bonds in the molecule, and thus, via the partition functions, influences the contributions of the vibrations to the total energy. These contributions are in the order of $RT$ (~2.5 kJ/mol at 300 K) per vibrational DOF. For large molecules with many internal DOFs, this can lead to substantial differences between the calorimetric heat of adsorption and the adsorbate–substrate binding energy as calculated by theory.

## 32.2.2 Solid–Liquid Interfaces

### 32.2.2.1 Thermodynamic Aspects of Heat Evolution at Single Electrodes

The heat exchange at an electrode under current flow in an isothermal electrochemical cell close to thermal equilibrium is caused by the transport of charge across the boundary between the electrode and the electrolyte. It is identical to the Peltier heat at metal–metal junctions, with just two complications. First, electrons and ions contribute to charge transport in an electrochemical cell, and, second, the transport processes are accompanied by an electrochemical half-cell reaction at the interface. Although the electrochemical Peltier phenomenon was discovered in 1879 by Bouty, a thorough theoretical explanation became possible only with the advent of nonequilibrium thermodynamics. In the following, we first present a summary of the current theory of the Peltier effect, and then a brief historical review how the conception of the Peltier effect changed between its discovery and its explanation.

### 32.2.2.2 The Electrochemical Peltier Effect

The first theoretical explanation of the Peltier effect implementing the new ideas of Eastman on the transport of entropy upon ion migration [14] was proposed in 1929 by Wagner [15, 16] and simultaneously by Lange, Miščenko, and Monheim [17–19]. Later, the theory was extended and summarized by several authors, for example, Refs. [20–25]. A modern introduction into the topic can be found, for example, in Refs. [26, 27]. Below, we will exemplarily derive formulas for the Peltier heat upon the deposition of a simple metal from a solution of its salt, closely following the argumentation and notation of Agar [22].

Figure 32.3 shows a sketch of an electrode of metal M with a wire of the same material attached to it. The electrolyte is a solution of metal ions $M^+$ and anions $A^-$. For simplicity, no supporting electrolyte is considered. The half-cell reaction is the reduction reaction of the metal ion: $M^+ + e^- \rightarrow M$. Let us first define appropriate boundaries for the electrode–electrolyte interface. On the metal side, this may be a plane P at a fixed position of the wire, through which the electrons pass. In solution, we choose a plane H, which is far enough from the electrode to experience bulk properties of the electrolyte. Furthermore, the plane may slightly move under current flow in order to keep the net amount of solvent constant between the boundaries. Such a boundary is called "Hittorf reference plane." Temperature and pressure of the cell are kept constant, and the current will be infinitesimally small. Then, after progress of the reduction reaction by $d\xi$, the thermodynamic state of the

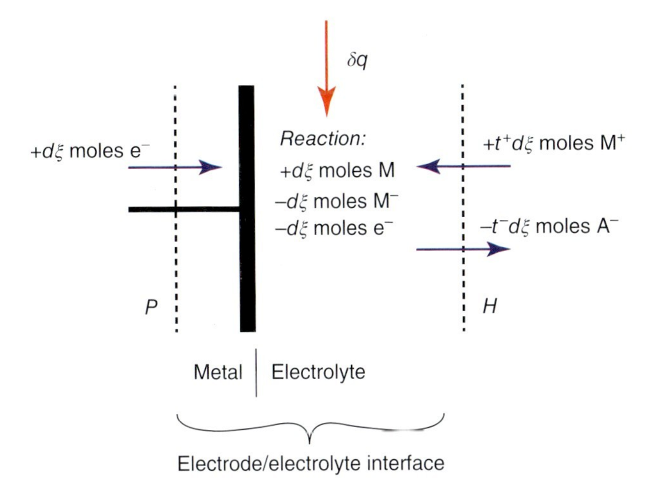


**Figure 32.3** Interface of a single electrode of metal M in an electrolyte of its salt ($M^+$, $A^-$). Upon progress of the metal reduction reaction by $d\xi$, the quantities of the reactants change accordingly, due to the electrochemical reaction as well as due to ion and electron transport. *P* and *H* denote the borders of the interface volume (see the text).

electrode–electrolyte interface changes accordingly: $d\xi$ moles of metal ions $M^+$ are reduced to metal M. By migration of ions through the Hittorf plane, $t^+ d\xi$ moles of $M^+$ are transferred into the interface and $t^- d\xi$ moles of $A^-$ migrate into the bulk of the electrolyte. $t^+$ and $t^-$ are the Hittorf transference numbers, which state how the current in the electrolyte is distributed between the ions. The net number of electrons in the metal part of the interface does not change. Since temperature and pressure are kept constant, the change of the entropy of the electrode–electrolyte interface volume can be readily calculated.

$$dS = (s_{\mathrm{M}} - s_{\mathrm{M^+}} + t^+ s_{\mathrm{M^+}} - t^- s_{\mathrm{A^-}})\, d\xi, \tag{32.15}$$

where $s_{\mathrm{M}}$, $s_{\mathrm{M^+}}$, and $s_{\mathrm{A^-}}$ give the molar entropies of the metal and the ions.

The change in entropy is caused by the transport of electrons and ions across the interface boundaries as well as by the transfer of heat $\delta q$ from the surrounding to the interface volume. The heat transferred to the system during the infinitesimally small, reversible progress of the reaction is called Peltier heat, $\delta q = \Pi d\xi$, where $\Pi$ is the molar Peltier heat or the Peltier coefficient in molar quantities. The corresponding entropy change of the system is given by $\delta q/T = (\Pi/T)d\xi$. However, the entropy transported by the charge carriers upon traversing the boundary planes needs closer inspection. In a seminal paper, Eastman showed (in 1926) that, in general, the transport of a charge carrier $i$ is accompanied by the transport of heat, the so-called (molar) heat of transport $\widehat{Q}_i$. The corresponding entropy change $\widehat{s}_i = \widehat{Q}_i/T$ is called Eastman's entropy of transfer, and adds to the entropy change due to the conventional molar entropy of the ion or electron $s_i$. In total, when 1 mol of charge carriers $i$ is transported across the boundary, the entropy change associated with the transport process is $\overline{\overline{S}}_i = (s_i + \widehat{s}_i)$. $\overline{\overline{S}}_i$ is called transported entropy. Note that the terminology is not generally accepted and that different authors use different names; for example, the heat of transport is sometimes also called heat of transfer. A comparison of conventions in use is given in Table 10 of Ref. [22]. In this reference, the reader can also find a summary of models for a physical reasoning of the heat of transport of ions. In brief, the heat of transport is traced back to local thermal effects, for example, from affecting the structure and hydrogen-bond network of the surrounding water beyond the first solvation shell of the ions. Ions can increase or decrease the entropy of the surrounding water and are accordingly called structure-breaking or structure-making [28, 29]. When an ion moves, the structural effects on the water are released behind and built up ahead. The involved heat (positive or negative) is liberated behind and absorbed ahead, which results in transport of heat together with the movement of the ion.

When metal ions are reduced, such ions and electrons are transported across the boundaries into the interface volume, and the anions migrate outwards. Following the above consideration on the transported entropy, the entropy change due to the transport of ions and electrons across the boundaries and due to transfer of heat is given by

$$\left(t^+\overline{\overline{S}}_{M^+} - t^-\overline{\overline{S}}_{A^-} + \overline{\overline{S}}_{e^-} + \frac{\Pi}{T}\right) d\xi$$
$$= \left(t^+\left(s_{M^+} + \widehat{s}_{M^+}\right) - t^-(s_{A^-} + \widehat{s}_{A^-}) + (s_{e^-} + \widehat{s}_{e^-}) + \frac{\Pi}{T}\right) d\xi. \tag{32.16}$$

Since entropy is a variable of state, the entropy change in Equation 32.16 must be equal to the one derived from the change of the thermodynamic state of the system in Equation 32.15. On equating Equations 32.15 and 32.16 and solving for $\Pi/T$, we get

$$\begin{aligned}\frac{\Pi}{T} &= s_M - s_{M^+} + t^+ s_{M^+} - t^+ s_{M^+} - t^+\widehat{s}_{M^+} - t^- s_{A^-} + t^- s_{A^-} + t^-\widehat{s}_{A^-} - s_{e^-} - \widehat{s}_{e^-} \\ &= (s_M - s_{M^+} - s_{e^-}) - (t^+\widehat{s}_{M^+} - t^-\widehat{s}_{A^-} + \widehat{s}_{e^-}) \\ &= \Delta_R S - \frac{1}{T}(t^+\widehat{Q}_{M^+} - t^-\widehat{Q}_{A^-} + \widehat{Q}_{e^-}).\end{aligned} \tag{32.17}$$

$\Delta_R S$ represents the first bracket in the second line of Equation 32.17 and corresponds to the molar entropy change of the half-cell reaction. The Peltier heat exchanged at the electrode is hence determined by the difference of the entropy change of the chemical half-cell reaction and the entropy change due to the Eastman's entropies of transfer or the heats of transport of the charge carriers, respectively. This result can be straightforwardly extended to any electrochemical reaction and electrolyte composition. It is *a priori* not clear which term dominates, and the total heat of transport can overcompensate the heat effects due to the half-cell reaction entropy. Eastman's entropy of transfer is strongly dependent on the kind of the ion and it is particularly high for protons. Hence, the Peltier heats in acidic solutions may considerably differ from those in neutral electrolytes [30].

#### 32.2.2.3 Historical Conception of the Electrochemical Peltier Effect

Although the electrochemical Peltier effect cannot be understood within the framework of equilibrium thermodynamics, it may be instructive to follow the historical developments starting from classical thermodynamics. The first calorimetric investigations of electrochemical systems are closely connected to the progress of modern equilibrium thermodynamics in the second half of the nineteenth century. Precise calorimetric experiments, amongst others by Raoult [31] and Thomsen [32], pointed to deviations between the "chemical energy" – in modern terms the enthalpy of the chemical cell reaction – and the "electrical energy" or electric work performed by the electrochemical cell. Gibbs [33] and, independently, Helmholtz [34] found that upon delivering electrical work, an electrochemical cell must in general exchange heat with its surrounding and that the difference between the "chemical energy" and the electrical work is given by the change in entropy. It was in this context that Helmholtz coined the expression "free energy" for the amount of energy, that can be converted to electrical work. These findings were concisely summarized, for example, by Ostwald [35]. Translated to modern symbols (and slightly rearranged), he stated for the exchanged molar heat $q_m$ of an reversible, isobaric, isothermal electrochemical system $q_m = zFTdE/dT = T\Delta_R S$, where $F$ is the Faraday constant, $z$ gives the number of involved electrons, $dE/dT$ is the temperature coefficient of the electromotoric force $E$, and $\Delta_R S$ is the molar entropy change of the cell reaction. Helmholtz and Gibbs derived their formulas for complete electrochemical cells, and no further discussion on the microscopic origin of the heat exchange or the detailed role of the half-cell reactions and transport processes in the cell was given by the authors.

In parallel to Gibbs' and Helmholtz's considerations, Bouty [36] observed that individual electrodes in electrochemical cells change their temperatures upon current flow. He immersed two Cu-covered thermometers into a $CuSO_4$ solution and found that, upon current flow, the thermometer where the Cu was deposited cooled down, while the other one where Cu dissolution occurred heated up (Figure 32.4). In analogy to the Peltier effect in metal–metal junctions under electron flow, Bouty named the observed heat effect at the electrode–electrolyte interface Peltier heat. He supported this assignment by comparing careful calorimetric measurements of the Peltier effect at single electrodes with calculations employing Thomson's second relation $\Pi = TdV/dT$, also called second Kelvin relation [37], which relates the Peltier coefficient, that is, the molar Peltier heat $\Pi$ to the Seebeck coefficient $dV/dT$, that is, the thermoelectric power of an appropriate non-isothermal cell. Bouty, for example, employed a $Cu|CuSO_4|Cu$ cell whose electrodes were kept at different temperatures for the determination of the thermopower of a Cu electrode in $CuSO_4$ solution [38, 39]. It should be noted that Mills had observed the electrochemical Peltier effect already in 1877 [40] but attributed it to mechanical deformation of the thermometer by electrostriction forces.

*A priori* it is not clear how the entropy change of the cell reaction is connected to the Peltier heat. While the latter is clearly the result of a transport process of charge carriers at the interface, the Gibbs and Helmholtz formulas involve the reversible entropy change of the complete electrochemical cell. Helmholtz was obviously aware

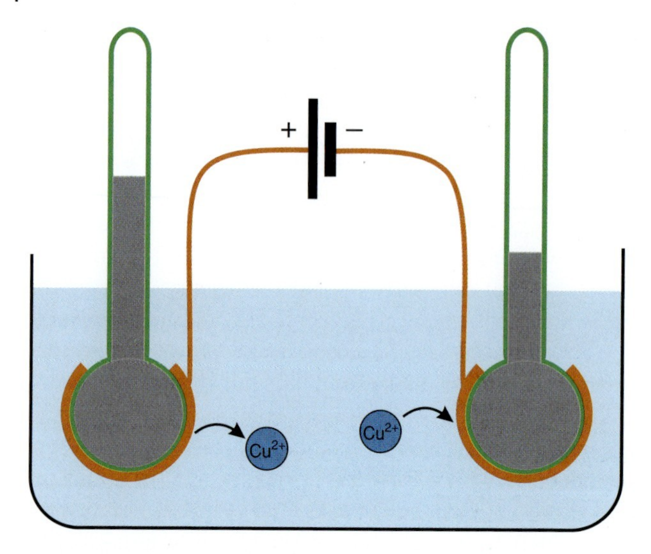


**Figure 32.4** Sketch of Bouty's first measurement of the electrochemical Peltier effect. The temperature differences between the Cu-covered thermometer electrodes and the electrolyte were quantitatively determined.

of those complications and explicitly stated in his 1882 article that Peltier heat might even hinder the experimental observation of the heat effects he was discussing [34, p. 25]. Nonetheless, both concepts were intermixed in the following years by several authors, mostly without detailed discussions. For example, Gockel in 1885 compared the sum of the Peltier heats of the electrodes of electrochemical cells, which he derived analogously to Bouty from the thermopowers of non-isothermal cells, with calorimetrically determined values for the heat effects of complete isothermal electrochemical cells [41]. Since he was considering complete cells, his approach was correct also from current point of view, and Gockel found reasonable agreement between both heat values.

It is worth mentioning that, in the middle of the nineteenth century, the physical reason for the electromotive force and the distribution of the electric potential inside an electrochemical cell was still under strong dispute. This situation got clarified only at the end of that century, when the concept of half-cells was established, strongly influenced by the work of Nernst and Ostwald. Yet, it became clear that heat evolution in isothermal electrochemical cells close to thermal equilibrium (in the absence of electrolyte–electrolyte junctions) occurs in direct vicinity to the electrodes' surfaces, that is, where the elementary electrochemical half-cell reactions occur. Close to equilibrium, Joule heat associated with the current flow through the electrolyte can be disregarded, since the current is infinitesimally small.

In the context of single electrode potentials, Ostwald also discussed the Peltier effect in metal–metal junctions. He derived Kelvin's second relation by ascribing the Peltier heat to the entropy change of the electrons by traversing the junction, multiplied by the temperature [35, p. 920]. Ostwald was probably not aware that he was implicitly talking about the *transported entropy* of the charge carriers. He also extensively employed Kelvin's second relation for the explanation of single electrode potentials and their temperature variation. His interpretation of the Peltier effect in metal–metal junctions strongly suggests transferring this concept also to the solid–liquid junction and identifying the Peltier heat at a single electrode with the entropy change of the half-cell reaction, multiplied by the temperature. From the current point of view, since entropies of transfer would have been disregarded, such an interpretation is clearly wrong. Ostwald himself refrained from this misinterpretation. But, in 1929, Bružs [42–44] measured the Peltier heats of several metal electrodes in the respective salts and presumably calculated the involved entropy change of the half-cell reaction by dividing the Peltier heat by the temperature. Bružs compared his experimental results with the half-cell reaction entropies calculated from data on the ionic entropy of ions in solution, obtained a few years earlier by Latimer and Buffington, and found reasonable agreement of the data, probably due to the small net effect of the Eastman's entropies of transfer in the investigated electrolytes. Fortunately, Bružs understandable misconception did not lead to much confusion. As stated above, Wagner and Lange and coworkers soon clarified the theoretical description by introducing Eastman's concept of entropies of transfer.

#### 32.2.2.4 Irreversible Heat Effects

Up to now, we considered electrochemical half-cells close to thermal equilibrium, that is, under infinitesimally small current flow in isothermal cells, where the process is reversible. Experimental determination of the Peltier coefficient, however, requires measurable heat changes and hence considerable cell currents. In the extreme case, the charge transport in the cell will be accompanied by transport of heat and mass and the complete flux equations have to be solved [27]. In most experimental cases, however, the temperature change at the electrode is kept small compared to the cell temperature and the cell can be considered to be isothermal. The overall conversions are usually small, such that the concentrations of the constituents of the electrolyte stay approximately constant. In this case, the irreversible effects reduce to the evolution of Joule heat, due to current flow through the electrolyte with finite conductivity and heat evolution due to overpotential, that is, due to deviations from the electrochemical equilibrium, in order to drive the reaction with a finite rate. If $\eta$ denotes the overpotential, that is, the difference between the applied half-cell potential and the equilibrium potential, $z$ the charge number of the electroactive species, $F$ the Faraday constant, and $d\xi$ the increment of the reaction variable, then the surplus electric work that is transferred into heat becomes $|\eta|zFd\xi$, irrespective of the sign of the overpotential. The Joule heat $\delta q_{\text{Joule}}$ can be calculated from the electrolyte resistance $R$, the current $I$, and the increment of time $dt$ during which the current is flowing: $\delta q_{\text{Joule}} = I^2Rdt$. Both terms are positive, and heat is transferred from the system to the surroundings under isothermal conditions. With our sign convention, the total heat that is transferred to the surrounding heat reservoir, including the Peltier heat, becomes

$$\delta q = \Pi d\xi - |\eta|zFd\xi - I^2Rdt. \tag{32.18}$$

Evolution of Joule heat will occur along the complete current path inside the electrolyte. Depending on the type of the experiment, it might be only partly measured. See, for example, Bouty's historic experiment, where a temperature gradient will evolve close to the thermometer electrode and only Joule heat generated close to the thermometer electrode is expected to influence the electrode's temperature. Heat generation due to overpotential is more difficult to conceive. It contains contributions attributable to different rate-determining steps of the electrochemical reaction. Those are the hindrance of the charge-transfer reaction itself, mass transport limitations, or rate-limiting chemical reactions accompanying the charge transfer [45, 46]. Usually, all those processes occur in close vicinity to the electrode and are quantitatively measured together with the Peltier heat [47]. However, particularly in experiments without supporting electrolyte and when concentration gradients evolve over considerable distances, transport processes and the accompanying heat fluxes may have to be considered.

## 32.3 Experimental Setup and Method

### 32.3.1 Approaches for Single-Crystal Adsorption Calorimetry at the Solid–Gas Interface

#### 32.3.1.1 Historical Development

Long before the advent of SCAC in the 1990s, heats of adsorption on metal surfaces have been measured using thin metal wires and ribbons. Since the 1930s, J. K. Roberts studied the adsorption energetics of small molecules such as $H_2$, $N_2$, and $O_2$ on wires of refractory metals. Adsorption-induced temperature changes were detected as changes of the wires' resistance, that is, the samples served simultaneously as resistance thermometers [48]. Using metal ribbons instead of wires, P. Kisliuk later achieved better sensitivity due to the higher ratio between the surface area and heat capacity [49]. Until the 1970s, the wire/ribbon approach was pursued by various groups [8]. However, it became increasingly evident that the polycrystalline nature of the samples with their wide spectrum of different adsorption sites led to problems with the reproducibility of results, especially as the wire and ribbon samples were also structurally and chemically insufficiently well defined.

Adsorption calorimetry on thin metal films produced by physical vapor deposition is another approach that has been pursued concomitantly to the studies on wires and ribbons. Around 1940, Beeck [50, 51] reported on a thin-film calorimeter in which metal films were vapor-deposited on the inside of a thin glass bulb, to which a resistance thermometer was attached from the outside. This technique was later refined by Wedler [52, 53] and several other groups [8]. The Wedler calorimeter, as shown in Figure 32.5, consists of a thin-walled spherical glass bulb (1) housing a metal filament (4) from which the metal for the film is evaporated and deposited onto the inner wall of the bulb as a thin polycrystalline film. A spiral-shaped metal wire is attached to the outside of the bulb as a resistance thermometer,

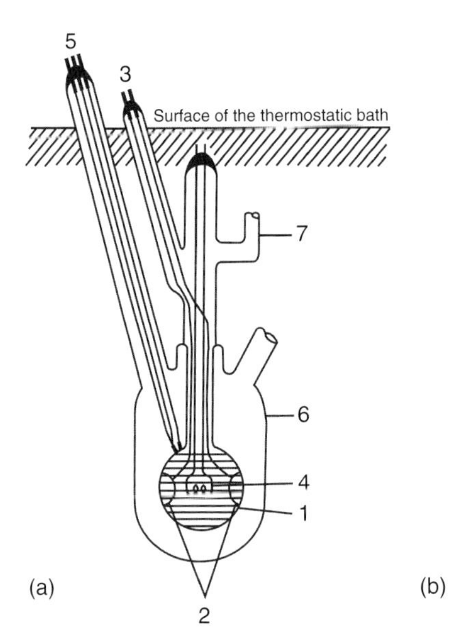


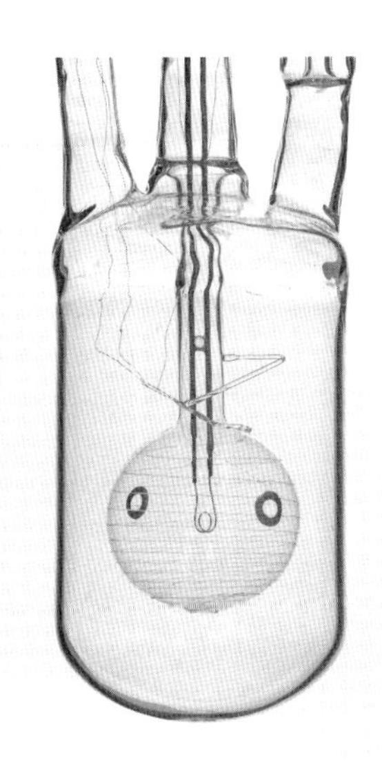


**Figure 32.5** Spherical thin-film adsorption calorimeter by Wedler [54, 55]. (1) Thin-walled glass bulb (Ø 5 cm, wall thickness 0.1 mm) with resistance thermometer on the outside (tungsten, length 2 m, Ø 10 μm). (2) Platinum contact foils and (3) the related electrical feedthroughs for measuring resistance of the film. (4) Evaporant (metal wire). (5) Electrical feedthroughs for the resistance thermometer. (6) Evacuated glass bulb for thermal insulation. (7) Gas inlet. (Left figure reprinted from Ref. [8], Copyright 1996, with permission from Elsevier. Photography adapted with permission from Ref. [11]. Copyright 2011, Bunsengesellschaft für Physikalische Chemie.)

using an induction-free loop arrangement. After deposition of the metal in the evacuated bulb, the gas is introduced in small pulses, while the adsorption-induced temperature change of the bulb is measured. For thermal insulation, the bulb is located inside an evacuated glass cylinder (6), which is immersed in a thermostat. The heat capacity of this calorimeter is in the range of 1 J/K, and temperature changes as low as 1 μK can be detected [54, 55]. With an active sample surface of ~75 cm$^2$, absolute sensitivities are the range of 10 nJ/cm$^2$. For calibration, the resistance thermometer is used as an electrical resistance heater to produce a precisely known amount of heat. This calorimeter type was used for adsorption studies of various small molecules on transition metal films [8], although it has similar disadvantages as the wire/ribbon calorimeters: The polycrystalline nature of the metal films and the dependence of the film morphology on the deposition parameters led to reduced reproducibility of the results. Since the films cannot be accessed inside the bulb, structural and chemical characterization of the surface is even more difficult than in the case of wires and ribbons.

An UHV adsorption calorimeter much more compatible with contemporary surface science techniques was introduced by Černý and coworkers [56] by using a single-crystalline $LiTaO_3$ pyroelectric detector following a previously suggested approach [57]. The substrate was a thin metal film, which was directly deposited onto a $LiTaO_3$ wafer. The adsorbate was dosed using a molecular beam, which was chopped into short pulses, because a pyroelectric detector is sensitive to temperature changes rather than absolute temperature (see Section 32.3.2.1). For calibration, a laser beam was sent along the MB path to deposit precisely known amounts of heat in the sample (Figure 32.6). The sticking probability was determined with a mass spectrometer using the King–Wells technique [59, 60]. The Černý calorimeter shares most major features with the calorimeters by King and coworkers [6] and Campbell and coworkers [7] described below. Its drawbacks were the simple construction of glass and quartz, which made it difficult to obtain a well-defined MB and, again, the polycrystalline nature of the metal film.

Adsorption calorimetry on well-defined single crystal surfaces was first attempted in the 1980s by Kyser and Masel [61], who equipped a Pt(111) single crystal (diameter 10 mm, thickness approximately 1 mm) with two small thermistors. This sample was mounted in a UHV recipient immersed in a thermostated water bath. One thermistor was used to measure adsorption-induced temperature changes with sensitivity in the $10^{-5}$ K range, while the second thermistor was used to deposit heat for calibration. After a cleaning procedure, which included thermal treatment, the Pt(111) sample needed up to 12 h to reach thermal equilibrium with the environment before an adsorption calorimetry experiment could be started. During this time, the sample accumulated typically 0.5 monolayers of carbon monoxide (CO), making high-coverage CO adsorption and co-adsorption studies the only feasible type of experiments. Gas adsorption on the Pt(111) sample caused temperature rises of typically 0.1–2.0 μK. The smallness of these temperature changes explains why thorough thermal equilibration was so important to achieve

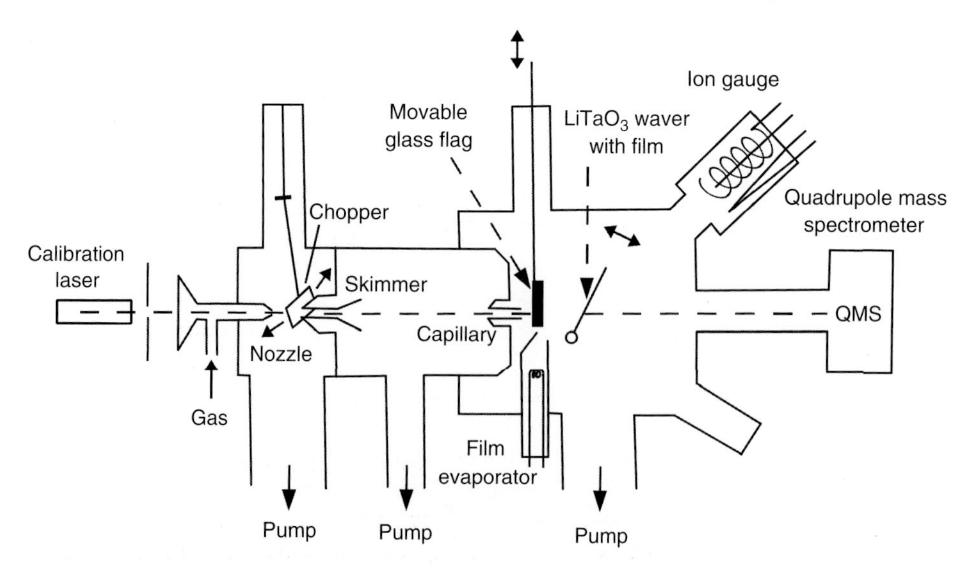


**Figure 32.6** UHV adsorption calorimeter by S. Černý and coworkers for polycrystalline films with a pyroelectric temperature sensor, a pulsed molecular beam source, and laser calibration [56, 58]. (Reprinted from Ref. [8]. Copyright 1996, with permission from Elsevier.)

a stable baseline. The major drawback of this approach was clearly the large heat capacity of the conventional single crystal sample: small adsorption-induced temperature changes, slow equilibration, and thus high sample contamination were the results.

A rather unique approach is the micromechanical calorimeter [62–65], which uses the temperature-induced bending of a bimetallic cantilever for the detection of heats of adsorption and reaction. A typical Si/Al cantilever has a length of ~400 μm, a width of 35 μm, and a thickness of ~1.5 μm. The bending is measured using the deflection of a laser beam by the cantilever, similar as in an atomic force microscope. The calorimeter can reach an absolute sensitivity limit of $\sim 10^{-12}$ J, which corresponds to $\sim 10\,nJ/cm^2$. Thus, the sensitivity per area is similar to that of the Wedler calorimeter, and both are limited to polycrystalline substrates.

For a more detailed description of the history of adsorption calorimetry on solid–gas interfaces, we refer to the review by Černý [8].

#### 32.3.1.2 **Present Approaches to Single-Crystal Adsorption Calorimetry**

**32.3.1.2.1 Single-Crystal Adsorption Calorimeter with Infrared Detector** A major breakthrough in adsorption calorimetry on well-defined surfaces was achieved in the early 1990s by D. A. King and coworkers, who realized that single crystals with very low heat capacity are needed to obtain sufficient sensitivity [66]. The single crystals used in their experiments had a thickness of 0.2 μm and a surface area on the order of $1\,cm^2$ [6, 67, 68] and were produced by epitaxial growth on single-crystalline substrates with suitable lattice constants such as sodium chloride [69, 70]. After dissolving NaCl in water, the remaining ultrathin crystal was mounted on a support ring. Because of the very low heat capacity of these samples, even adsorption of a fraction of a monolayer leads to temperature changes in the range of 1 K [71]. For the detection of adsorption-induced temperature changes, the intensity changes of the emitted infrared radiation were used. To enhance the emissivity of the sample, its back was covered with a thin layer of carbon. The infrared light was focused onto a broadband photoconductive mercury-cadmium-telluride (MCT) infrared detector located outside the UHV apparatus. The gas was dosed in small portions using a pulsed supersonic MB with a diameter of 2 mm. The pulses had a length of 50 ms, a repeat frequency of $0.5\,s^{-1}$, and contained approximately $10^{12}$ molecules per pulse. A pulsed beam was chosen to maximize the temperature change of the sample and thus to improve signal detectability. Integrated over the area exposed to the beam, the heat capacity of the sample was of the order of only $10^{-6}$ J/K. The temperature rise of the crystal for increments of 1% of a monolayer amounted to typically 0.01–0.1 K [66].

For calibration of the calorimeter, heat was deposited in the sample with a pulsed laser beam of known intensity and with temporal and spatial characteristics closely matched to the MB. This method requires precise measurement of the sample reflectivity at the laser wavelength. Alternatively, a known heat of reaction or phase transition, such as the multilayer adsorption energy of the adsorbate, may be used for calibration. To determine the heat of adsorption per molecule (or per mole), the number of adsorbed molecules in each pulse was determined by separate measurements of flux and sticking probabilities. Flux measurements were performed with a stagnation gauge, as shown in Figure 32.7. The sticking probability was determined by measuring the amount of reflected molecules with a mass spectrometer, employing the King–Wells techniques [59, 60]. The reference signal corresponding to complete reflection was measured with an inert gold flag placed in front of the sample. Because the sticking probability is thus always referenced to complete reflection, its relative uncertainty increases when the sticking probability becomes smaller. As a result, reliable adsorption energies cannot be determined for sticking probabilities below 10%. From the measured amounts of heat and the corresponding numbers of molecules, the molar heat of adsorption was calculated for each pulse. In the limit of very small pulses ($\Delta\Theta \rightarrow 0$), this is a differential heat of adsorption (Figure 32.2). The corresponding coverage was determined by integration of the product of dosage and sticking probability.

When saturation coverage is reached, which is possible only for adsorbates that do not form multilayers, there is still a measurable heat of adsorption from molecules that adsorb during the pulse-on time and desorb between the pulses. Summation

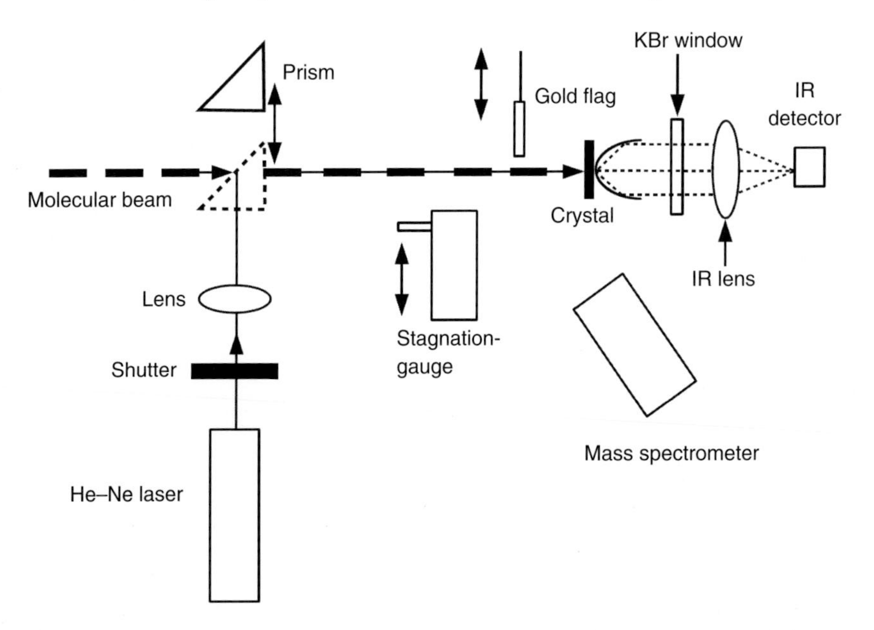


**Figure 32.7** Single-crystal adsorption calorimeter with infrared detector as introduced by D. A. King. Adsorption-induced temperature changes of the thin single-crystal sample (thickness 0.2 μm) are measured with an infrared detector outside the vacuum recipient (right). For delivery of the gas, a pulsed supersonic molecular beam is used (left). Further components include a pulsed laser beam for calibration (left) and a stagnation gauge for flux measurements (center). A mass spectrometer and a gold flag are used for sticking probability measurements after King and Wells [59, 60]. (Reprinted from Ref. [68], Copyright 1996, with permission from Elsevier.)

of the product of dosage and sticking probability in this steady-state regime [72] results in an apparent coverage which increases indefinitely, while the true coverage oscillates around the saturation value.

The sensitivity of the calorimeter depends on the absolute sample temperature and will be strongly reduced at low temperatures, because the radiant power $\Delta P_{rad}$ follows the Stefan–Boltzmann law: $\Delta P_{rad} \propto T^3 \cdot \Delta T$. Measurements below or above ambient temperature are further complicated by the fact that cooling or heating of the sample via the support ring is impossible, since lateral heat transfer is extremely ineffective. Thus, the sample temperature is dominated by radiative thermalization with the environment. To date, no attempts have been made with this approach to measure at sample temperatures above or below room temperature, for example, by means of a cryoshield surrounding the sample. Calorimetry at low temperatures was, instead, performed with a pyroelectric detector in direct contact with the sample [73], similar as in the Černý calorimeter [56]. In particular, the ultrathin metal single crystal was cold-welded to one face of a 0.3-mm-thick $LiTaO_3$ wafer coated with metal electrodes. With this detector, measurements in the temperature range 90–410 K could be performed and temperature changes in the $10^{-5}$ K range could be detected. However, the pyroelectric detector dramatically increases the heat capacity of the sample–detector assembly, which conflicts with the original aim pursued with thin single crystals. In addition, the permanent contact between the sample and the detector limits the temperature range that can be used during sample preparation and annealing. A more versatile solution was introduced later by C. T. Campbell and coworkers, who employed a retractable pyroelectric polymer detector with low heat capacity. This technique will be described below.

A problem immanent to King's approach and its successors is related to the effective area of the sample exposed to the MB. Because of surface roughness, the real surface area will always be larger than the ideal surface area of an atomically flat surface. Already, Kyser and Masel [61] remarked that the greatest uncertainty in their data resulted from the measurement of the total amount of gas adsorbed, because the effective surface area was not precisely known. This is especially critical for thin single crystals, which are not entirely flat and may contain more surface defects than conventionally fabricated single crystals. Another problem is related to the beam diameter. Similar to the case of an optical shadow, the central region with a laterally uniform flux (umbra) is surrounded by a penumbra ring in which the flux is lower. In the penumbra region of the sample, the coverage is lower than in the central umbra region, and thus the heat of adsorption per area will be different if the heat of adsorption depends on coverage. Therefore, care must be taken to minimize the penumbra region, for example, by placing the beam-defining orifice of the MB source close to the sample.

###### 32.3.1.2.2 Single-Crystal Adsorption Calorimeter with Pyroelectric Contact Detector

The major drawback of the single-crystal adsorption calorimeter with an IR detector is its limitation to room-temperature measurements. Attempts to use pyroelectric detectors in direct mechanical contact with the sample were hampered by the large heat capacity of the detector and the permanent contact between the sample and the detector (King's calorimeter [73]), or by the polycrystalline nature of the metal film that was deposited directly onto the pyroelectric detector (Černý's calorimeter [56, 58]).

Both problems were elegantly addressed with the retractable pyroelectric contact detector introduced by Campbell and coworkers [7]: a thin β-polyvinylidene fluoride (β-PVDF) foil coated with metal electrodes on both faces was shaped as a bent ribbon and brought into mechanical and thermal contact to the backside of a thin single-crystal sample during the measurement, as shown in Figure 32.8. Retraction of the detector between the measurements made sample preparation at high temperature possible.

Because of the direct mechanical contact between the sample and the detector, the single-crystal samples must be mechanically more robust than the crystals used with contact-free IR detectors. For this reason, the thickness of the samples was increased to typically 1 μm (and even up to 100 μm in some cases [7, 74–76]). The higher heat capacities and the resulting smaller temperature changes are counterbalanced by the high sensitivity of the pyroelectric material in combination with the direct contact between sample and detector, such that sample temperature changes in the microkelvin (μK) range can be measured. Samples of 100 μm thickness have the additional advantage that they can be produced by mechanical cutting and polishing of macroscopic single crystals, rather than by epitaxial growth, which works only for certain elements and surface orientations. Compared to the typically slightly buckled surfaces of the thin 0.2–1 μm single-crystal foils, the smooth and mirror-like surfaces of the 100 μm samples also facilitate the optical reflectivity measurements necessary during calibration of the calorimeter and allow a more accurate determination of the effective surface area hit by the MB.



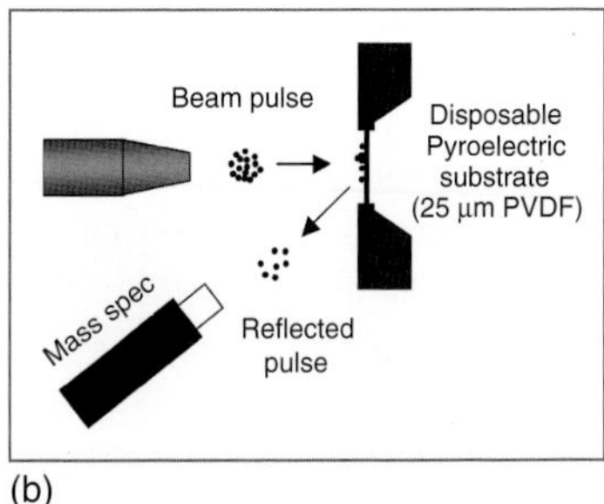


**Figure 32.8** Single-crystal adsorption calorimeter with pyroelectric contact detector. (a) Setup for use with single-crystalline samples. The pyroelectric polymer detector contacts the backside of the sample. As a pulse of molecules adsorbs on the single-crystal sample, heat is deposited and is detected as a temperature change. Simultaneously, the sticking probability is measured by a mass spectrometer detecting the reflected fraction of the pulse. (b) The setup for polymer samples, organic films, and other polycrystalline samples, which are deposited as thin films directly onto the detector. (Adapted with permission from Ref. [11]. Copyright 2011, Bunsengesellschaft für Physikalische Chemie.)

Because the heat transfer between the sample and the detector is efficient even at low temperatures, measurements below room temperature are possible. Calorimetric measurements under cooling with liquid nitrogen have been successfully performed [12]. The upper temperature limit for the Campbell design is around 400 K, where degradation of the detector sets in. During sample preparation, which often includes annealing, the detector is retracted from the sample and thus does not limit the maximum annealing temperature.

As in previous designs, pulsed molecular beams, either supersonic or effusive, are used for dosing the adsorbates. The pulses with a typical length of 100 ms and a repeat rate of 0.5 Hz deposit heat in the sample and cause a detector signal, which is proportional to the temperature increase and, thus, to the amount of heat (Figure 32.9). Note that a pyroelectric detector is a current source and that the current is proportional to the time derivative of the temperature, $I_{PE} \sim dT/dt$. Calibration is achieved with a laser, a procedure that requires, as in King's setup, precisely measured sample reflectivities.

In the Campbell design, the flux of the molecular beam is measured with a quartz crystal microbalance (QCM) positioned in front of the sample. If a substance is too volatile to form multilayers at room temperature in UHV, the QCM is cooled with liquid nitrogen to allow condensation. As in King's setup, the sticking probability is determined by the King–Wells technique [59, 60] using a mass spectrometer, as shown in Figure 32.8. For metals and other low vapor pressure compounds, the inert flag, which is used to provide the reference signal for a zero sticking coefficient, must be heated to ensure complete reflection. From the shape of the reflected molecular

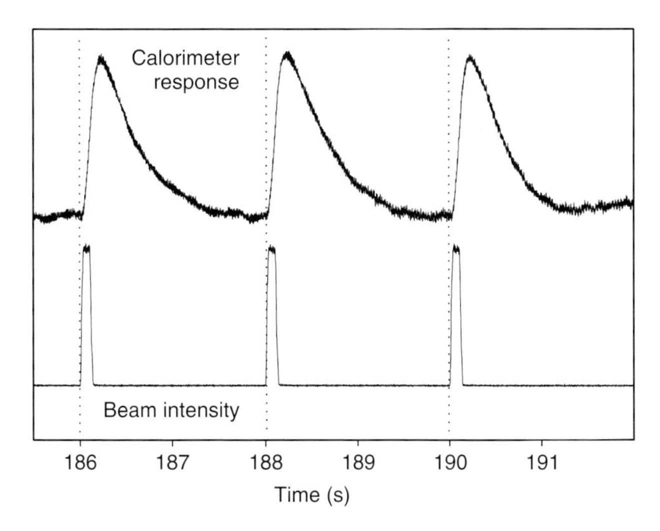


**Figure 32.9** Calorimeter signal (top curve) and molecular beam intensity (bottom curve) for the adsorption of cyclohexene on Pt(111) at 100 K. Each pulse had a length of 102 ms, contained $2.5 \times 10^{12}$ molecules (0.011 ML), and led to a heat input of ~250 nJ. (Adapted with permission from Ref. [12]. Copyright 2008, American Chemical Society.)

beam pulse, kinetic data about the transient adsorbed molecules or atoms can be obtained [77–79].

Infrared radiation emitted by the molecular beam source is absorbed by the sample and thus also contributes to the calorimeter signal. These contributions are measured separately after blocking the molecular beam with an IR transmissive window ($BaF_2$, KBr). This is especially critical for adsorption studies of catalytically relevant transition metals, which require high source temperatures resulting in high IR emissions. Calorimetry has already been used to study the adsorption of Ag on oxide films [80], but attempts to use metals with even lower vapor pressures require refining the techniques for measuring radiative contributions. Radiation also represents a challenge at lower source temperatures if both the evaporant and the sample have high emissivity in the far-infrared region, were the transmission of infrared window materials such as $BaF_2$ or KBr gets small. This is, for example, the case for large organic molecules. Possible solutions include Au or Ag mirrors or rotating velocity filters, which are only transmissive for the molecules and completely eliminate the radiative contribution to the calorimeter signal [81].

The single-crystal adsorption calorimeter with a pyroelectric contact detector exists in several variants, which are all derived from the Campbell calorimeter. R. Schäfer and coworkers use this detection technique, but introduced a novel pulsed MB, which creates pulses by opening and closing a vessel containing a gas of known volume using a piezoelectric plunger [82, 83]. The average gas dose per pulse is determined by measuring the pressure drop in the gas-containing volume during the experiment and dividing by the total number of pulses.

Another advancement of the Campbell approach was achieved by Schauermann, Freund, and coworkers [84, 85] with an adsorption calorimeter specifically designed to study the interaction of gases with planar model catalysts consisting of metal nanoparticles supported on single-crystal surfaces or epitaxial thin oxide films on single crystals. Technical improvements include the reduction of systematic calibration errors by *in situ* measurements of the reflectivity and an *in situ* photodiode for measuring the power of the calibration laser. In addition, the setup addresses the problem of mechanical vibrations, which contribute to the calorimetry signal due to the piezoelectric properties of the pyroelectric polymer. To reduce the influence of vibrations, the calorimeter is placed on a vibration-isolating, polymer-based support similar to those used in scanning probe microscopy (Figure 32.10). The calorimeter is mounted on a rotating platform, which also carries a photodiode and a stagnation gauge as a flux monitor. By rotating the platform, these auxiliary devices can be positioned in exactly the same place as the sample. This represents a great advantage compared to previous designs, in which the positioning of these devices in the space between the MB source and sample was an additional source of error.

In the future, we are likely to see calorimeters that are increasingly adapted to the investigated systems, resulting in a further differentiation of the calorimetric technique. This also holds true for the molecular beam sources, which will be adapted to the properties of the deposited materials. Regarding the thin single-crystal samples, very thin single-crystal foils of 1 μm will be further used when high sensitivity

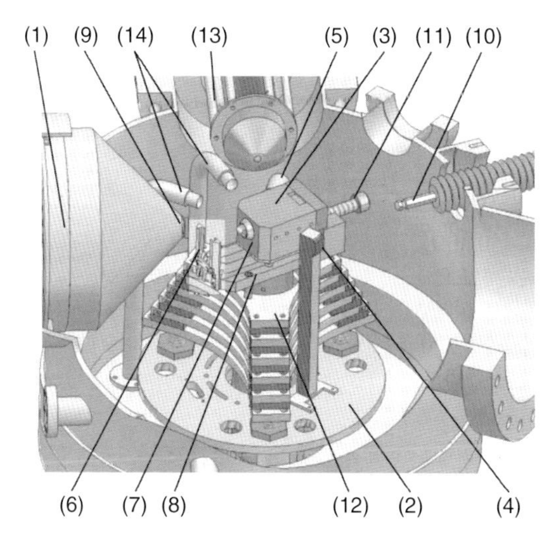


**Figure 32.10** Adsorption calorimeter for the investigation of gas adsorption on planar model catalysts. Main components: (1) molecular beam source, (2) rotatable platform, (3) microcalorimeter, (4) *in situ* photodiode, (5) beam monitor, (6) sample holder mounting, (7) calorimeter detector head, (8) Cu platform carrying sample holder mounting and detector, (9) outer molecular beam aperture, (10) Allen wrench mounted on wobble stick, (11) translation screw, (12) vibration damping stack, (13) QMS, and (14) gas dosers. (Reprinted with permission from Ref. [84]. Copyright 2011, AIP Publishing LLC.)

is required. Thicker crystals of typically 100 μm, however, may be preferred for maximum accuracy, because reflectivity measurements are more reliable for thicker crystals. The overall accuracy may be increased by performing the laser calibration at wavelengths at which the samples have low reflectivities.

### 32.3.2
### Experimental Setup and Methods for Electrochemical Calorimetry

Microcalorimetric investigations of surface electrochemical reactions at single electrodes rely mainly on the methods available for the study of bulk electrochemical reactions. Historically, electrochemical calorimetric studies aimed at the determination of Peltier heats and heats of transport in macroscopic electrochemical cells, for example, upon bulk copper or silver deposition. Since detailed surface processes were not within the focus of those investigations, comfortably high conversions could be used. Nevertheless, since the techniques for the study of surface electrochemical reactions are based on those experimental developments, we briefly summarize them in the following. More details can be found in a review by Kuhn and Shams El Din [86]. A concise summary is also given by Boudeville [87]. Early experiments are listed in a review by Lange and Monheim [19].

#### 32.3.2.1 **Traditional Experimental Approaches**

There are, in principle, two experimental approaches for the measurement of electrochemical heat changes at single electrodes: (i) measurement of the temperature difference evolving between the electrode and its surrounding during the progress of the electrochemical reaction, and (ii) measurement of the heat evolved in the half-cells in an adiabatic differential calorimeter.

The measurement of the temperature difference arising between the electrode and its surrounding dates back to the very early work of Bouty on the electrochemical Peltier effect, where he observed the temperature changes of Cu-plated thermometer electrodes [36, 39] (Figure 32.4). Various methods for the temperature measurement have been introduced since then. Gill used a resistance thermometer [88], and Bružs employed thermocouples in the form of a thermopile [43] with which he measured the difference of the temperature between the two electrodes arranged back to back in a symmetrical cell. Later, the use of thermistors was established for the electrodes' temperature measurement, for example, by the work of Holmes and Jonich [89] and Graves [90]. In most of the following work on Peltier heats associated with bulk electrochemical reactions, thermistor electrodes were used. Among the other prominent examples are the studies by Tamamushi [91, 92], Ozeki *et al.* [30, 93, 94], Boudeville [87], and Donepudi and Conway [95]. The temperature change was also measured indirectly via the mechanical deformation of the electrode due to the temperature change. For this purpose, the cell current was modulated with a small AC current, and the cyclic mechanical deformation was detected either by a piezo element attached to the electrode [96] or acoustically by a microphone [97].

The temperature difference evolving between the electrode and its surrounding is determined by the balance between the heat flux due to the electrochemical reaction (Peltier heat and irreversible heat, due to overpotential and Joule heat), the heat exchanged with the surrounding (by heat conduction and convection), and the heat associated with the heat capacity of the system upon warming or cooling. Since most variables that determine the heat flux balance are poorly known, in the majority of the experiments the steady temperature difference, which evolves with a constant electrochemical heat flux, that is, constant electrochemical current, is measured. In common experimental setups, a glass-encapsulated thermistor is attached to the electrode, and typical times before reaching the steady state amount to several tens of seconds. Mostly it was assumed that the heat flux into the surrounding is proportional to the steady-state temperature difference [98], but this does not necessarily need to be true [87]. Complications arise from changes of the heat transport mechanism from conduction to convection with increasing temperature difference or from Joule heat, which evolves along the complete current path, that is, also across the temperature gradient. From the principle of the measurement, there arise two problems: First, since the concentrations of the reactants in front of the electrode change during the electrochemical reaction, under galvanostatic operation also the overpotential of the electrochemical reactions will vary with time. This leads to time-dependent contributions from irreversible heat. Second, the calorimeter has to be calibrated. For this purpose, Graves incorporated an electric Peltier heater directly at the electrode [90]. More commonly, the Peltier heat of a well-studied reaction such as Ag or Cu deposition or the $[Fe(CN)_6]^{3-/4-}$ electron transfer reaction is used for calibration. Alternatively, Kuz'minskii and Gorodyskii and also Boudeville used the heat due to overpotential for an intrinsic calibration by comparing how the temperature difference changes with varying overpotential, that is, cell current [98, 99]. Boudeville critically discussed the different methods for calibration [87].

Most problems with calibration are circumvented in the second experimental approach, namely the measurement of the heat changes of the half-cells in an adiabatic differential calorimeter (Figure 32.11). This approach was first demonstrated by Lange and Monheim in 1928. They inserted the two half-cells of a symmetrical electrochemical cell, for example, a $Cu|CuSO_4||CuSO_4|Cu$ cell, into an adiabatic calorimeter. Upon current flow, Peltier heat was produced in one compartment and absorbed in the other one, which would have led to a temperature difference between the two compartments. The evolving temperature difference was measured by a thermopile and adjusted to zero by an electrical heater in the colder compartment. By this procedure, the difference of the Peltier heats and the accompanying irreversible heat contributions in the two compartments was measured on an absolute scale by the electrical energy that was supplied to the heater. Since the complete cell system was kept isothermal, heat fluxes between the two cell compartments were avoided. No further calibration was necessary, besides knowing the calorimeter constant, which could be determined with very high precision. This allowed very accurate absolute measurements of the heat changes. A similar approach was used by Sherfey and Brenner [100]. Also, Calvet-type

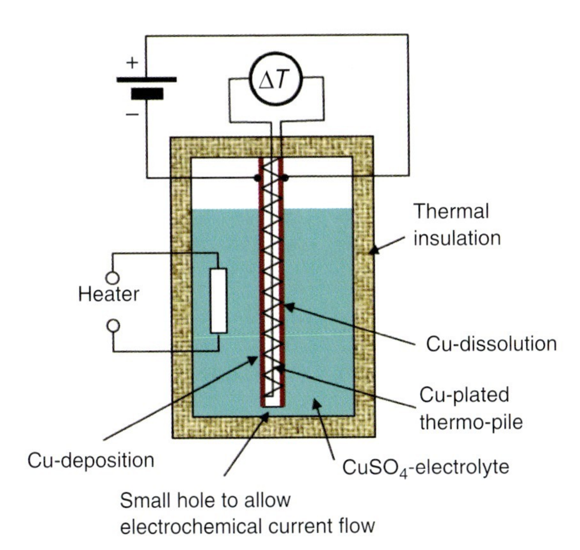


**Figure 32.11** Sketch of Lange's adiabatic differential calorimeter for the determination of the electrochemical Peltier heat. The temperature difference between the two compartments that evolves upon Cu dissolution or Cu deposition on either side of the Cu-plated thermopile is compensated by electric heating.

calorimeters were used for direct quantification of the heat flux caused by the electrochemical reaction [24, 101].

An essential problem of all electrochemical calorimetric experiments is the distinction between irreversible heat effects and reversible electrochemical Peltier heat. In principle, irreversible heat effects could be kept infinitely small by reducing the cell current and hence overpotential and Ohmic potential drop in solution. Particularly in experiments employing adiabatic calorimeters, due to the large electrode size, the current density could be kept small [18, 102]. However, to yield measurable temperature changes or heat fluxes, typical current densities often exceed $1\,mA/cm^2$ [30, 99], which for many electrochemical reactions will cause considerable irreversible heat effects. Sherfey and Brenner corrected for the irreversible heat by measuring the potential drop across the complete half-cell, including overpotential and potential drop in the electrolyte [100]. Multiplication by the cell current and integration over the time of the experiment yielded the electric energy fed into the irreversible processes. Alternatively, multiple experiments were conducted at different electrochemical current densities or overpotentials. Since the dependence of the irreversible heat on the current density and the overpotential is known (Equation 32.18), the experimental data of the heat at varying current densities could be fitted and extrapolated to zero overpotential. There exist various realizations of this method; see, for example, [30, 98, 99]. Several authors [30, 87, 95] have utilized the fact that the Peltier heat changes its sign under current reversal, while the irreversible part of the heat is always deposited in the system. Hence, the sum of the irreversible heat of the cathodic and anodic processes corresponds to the sum of the total heat effects for both processes. Although for reversible electrochemical reactions anodic and cathodic irreversible heat may be about the same, separation of both parts becomes difficult for less reversible systems.

It should be noted that calorimetric techniques and local temperature measurements are intensively applied also for the study of battery systems. Most of those studies concentrated on battery systems under operating conditions, where strong temperature gradients exist across the cells and irreversible heat effects become prominent. But, also the contribution of the reversibly exchanged heat was discussed in those instances. Sensitive heat measurement was often not required in such systems. Since the current review is devoted to the study of individual electrode reactions with small conversions, the reader is referred, for example, to an article by Kjelstrup *et al.*, where more references and some reviews on heat measurements in battery systems can be found [103]. Measurements on reversible heat effects on battery-related single electrodes mostly employ the methods summarized above. Examples for such systems can be found in Section 32.4.2.

#### 32.3.2.2 **Methods for Studying Electrochemical Surface Reactions**

All of the above-mentioned experiments required rather high electrochemical conversions. For example, in Lange's calorimetric experiments on Hg deposition, the typical charge that flowed during the reaction was between 2 and $60\,mC/cm^2$, which corresponded to the deposition of 20–400 monolayers of a typical monovalent metal. Similar conversions were also applied in the thermistor experiments cited above. Hence, for studying surface electrochemical systems, the sensitivity of the heat measurement has to be considerably enhanced. In temperature-difference experiments, the sensitivity was limited by two factors. First, the rather high heat capacity of the electrode sensor assembly, mostly a glass-encapsulated thermistor, led to long response times of the sensor. Second, heat conduction into or from the electrolyte lowered the temperature difference, that evolved between the electrode and its surrounding. Similarly, in the experiments employing an adiabatic calorimeter, the complete half-cell had to be at thermal equilibrium, and hence the heat effects were "damped" by the heat capacity of the half-cell, including the electrodes as well as the electrolyte.

In 1988, Shibata *et al.* considerably improved the sensitivity of the heat measurement [104, 105] by significantly reducing the heat capacity of sensor and electrode. They used a thick-film thermistor as temperature sensor onto whose surface a thin Pt foil was glued. To increase the effective surface area, the electrode was platinized with Pt black. With this sensor, Shibata and coworkers could measure heat effects upon reversible adsorption and desorption of oxygen and the adsorption of hydrogen on the Pt surface. To our knowledge, those were the first experiments in which surface electrochemical reactions with submonolayer coverage changes became accessible to electrochemical calorimetry. Shibata and coworkers performed pulse experiments in which they galvanostatically switched back and forth between anodic and cathodic reactions at a fixed average potential as well as potential scan experiments, where they recorded the heat effects in parallel with recording a cyclic voltammogram (CV). Since the heat capacity of the sensor and its mount were still relatively high, they did not reach steady-state temperature differences on the timescale of their experiments (the typical scan rate for the CVs was 50 mV/s), and they therefore derived the magnitude of the heat effects from the rate of the temperature change. However, quantitative Peltier heats were not given. Jiang and coworkers [106–108] employed a 50-μm-thick pyroelectric PVDF foil mounted on a stainless steel support, on which a 50-μm-thick platinized Pt foil was mounted as the working electrode (WE). They modulated the potential by a small sine voltage with frequencies between 1 and 20 Hz and detected the AC heat signal accompanying the reversible electrochemical reaction via a lock-in amplifier. Since the irreversible heat is always evolved in the system, irrespective of the sign of the cell current, heat due to irreversible processes will vary with double the modulation frequency, while reversible heat will follow the sine perturbation of the potential and can be easily separated by the lock-in technique. The method was sensitive enough to follow hydrogen adsorption on platinized platinum as well as several other bulk electrochemical reactions. Similarly, also Hai and Scherson used a WE directly attached to the temperature sensor, in the case a ZnSe crystal. They measured the temperature change of the electrode by a laser beam, penetrating the ZnSe crystal parallel to the WE. The electrochemically produced heat led to minute thermal gradients accompanied by changes of the refractive index of the crystal adjacent to its surface, which resulted in the deflection of the laser beam (the so-called "mirage effect"). This allowed following the heat evolution upon the formation of a layer of surface oxide on a Au film in sulfuric acid [109]. The mirage effect was employed previously for the investigation of electrochemical heat effects, for example, upon Cu bulk deposition or the $[Fe(CN)_6]^{3-/4-}$ redox reaction by Decker and coworkers [110, 111]. However, in those experiments the laser beam was deflected inside the electrolyte, being susceptible to temperature as well as concentration changes.

An essential drawback of all electrochemical experiments so far cited was that the heat capacity of the electrode and the sensor was still considerably high, thus leading to small temperature changes at the electrode. With the knowledge about the techniques for the measurements of heats of adsorption in vacuum, it becomes straightforward to adopt those approaches for the use in electrochemical systems. Schuster *et al.* introduced an electrochemical setup in which a thin WE, typically between 10 and 100 μm thickness, was in contact with a 9-μm-thick PVDF foil as a pyroelectric sensor [112, 113]. To ensure intimate thermal contact between the electrode and the sensor, both were pressed together by removing the air in between the sensor and the foil. A sketch of the sensor is shown in Figure 32.12. In addition, the authors conducted the electrochemical reactions only for a short time, typically only 10 ms, which was long enough to achieve thermal equilibration between the sensor and the electrode and short enough to lose only negligible amounts of heat into the electrolyte. Vice versa, also the Joule heat produced in the electrolyte was not transferred to the electrode and could hence be ignored in those measurements. The heat effects were determined by the temperature difference between the electrode and its surrounding at the end of the electrochemical potential pulse. Calibration of the calorimeter was performed mostly by comparison of the heat changes with those of well-studied electrochemical bulk reactions, such as the $[Fe(CN)_6]^{3-/4-}$ redox reaction, for which the Peltier heats were known from the literature. The latter reaction is an electron-transfer reaction and is expected to proceed without side reactions such as anion adsorption, and so on, and can hence be used for calibration also at submonolayer conversions. With this apparatus, heat effects upon the underpotential deposition and dissolution of a few percent of a monolayer of Cu on Au became accessible [114]. It is noteworthy that a freely suspended PVDF temperature sensor was already used by Tasaki and Iwasa in 1981 for the detection of temperature changes associated with nerve excitation, however, without an electrochemical cell [115].

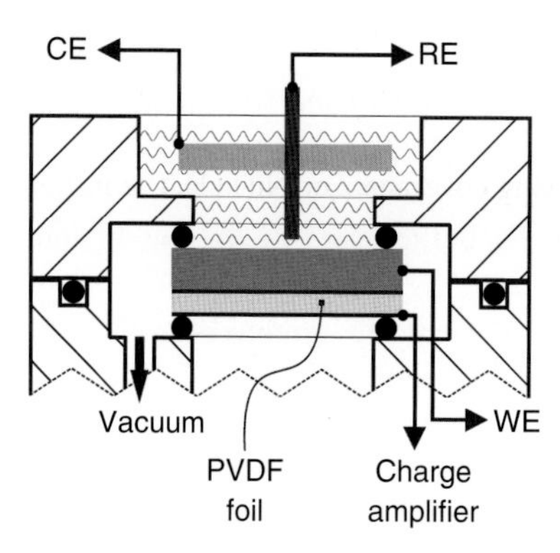


**Figure 32.12** Sketch of an electrochemical microcalorimeter for the study of surface electrochemical reactions. (From Ref. [112].) The temperature of the working electrode (WE) is sensed by a pyroelectric PVDF foil. By evacuating the space between the WE and the PVDF foil, both are brought into intimate contact by the outside air pressure. On the upper side of the WE, the electrochemical cell with counter electrodes (CEs) and reference electrodes (REs) is mounted. (Reproduced with permission from Ref. [112]. Copyright 2010, American Institute of Physics.)

Although with the recent calorimetric developments heat changes of electrochemical reactions with minute conversions of a few percent of a monolayer have become accessible, still some experimental challenges exist. Utmost sensitivity was achieved for relatively fast electrochemical reactions, where heat was measured on timescales of about 10 ms in order to avoid heat losses into the electrolyte and the surrounding. On the other hand, it might become desirable to study slow reactions like corrosion processes or electrochemical reactions with slow consecutive steps such as ordering processes in surface layers, and so on. For such reactions, adiabatic differential calorimetry as introduced by Lange and coworkers might become attractive. Miniaturization of the cell and particularly of the electrolyte volume might reduce the overall heat capacity of the electrochemical system, and sensitivity toward submonolayer electrochemical conversions might become possible with adiabatic calorimeters, too. Another drawback of the current experiments often is the poorly defined crystallographic structure of the investigated electrodes. Most experiments were performed with polycrystalline metal foils as electrodes. In few cases gold films with (111)-texture were used [47, 109]. However, in principle, single-crystalline foils could be produced sufficiently thin from bulk single crystals, as described in Section 32.3.1.2. In conclusion, with the current experimental tools for electrochemical calorimetry, many electrochemical systems can already be addressed. With future improvements, slow time-resolved measurements as well as studies on well-defined single-crystal surfaces may come within our reach.

## 32.4 Applications

### 32.4.1 Solid/Gas Interfaces

#### 32.4.1.1 Adsorption of Molecules

As mentioned previously, adsorption calorimetry is especially useful if desorption of the intact adsorbate is not possible (nonreversible adsorption) and therefore TPD or isosteric measurements cannot be applied. An example is the adsorption of $O_2$ on metal surfaces, which often leads to dissociation, diffusion into the metal bulk, and formation of oxide films. Adsorption of $O_2$ on Ni surfaces was among the first single-crystal calorimetric measurements performed by King and coworkers [66, 67]. The heat of adsorption of $O_2$ on Ni(100) at 300 K was shown to start at 550 kJ/mol on the clean surface and to drop rapidly with increasing coverage. For Ni(111), the initial heat of adsorption was lower, 420 kJ/mol, indicating a pronounced dependence on the surface orientation [116]. With a thin Ni(100) crystal permanently mounted on a 0.3-mm-thick $LiTaO_3$ pyroelectric detector, measurements at different sample temperatures of 90, 300, and 410 K were performed. The initial heat of $O_2$ adsorption on Ni(100) was higher at 410 K than at 300 and 90 K, which temperatures were too low for the formation of the most stable equilibrium structure. Formation of an oxide film (four layers) on Ni yielded the highest integral heat of formation for Ni(111),

320 kJ/mol, and considerably less for Ni(100) and Ni(110) (220 and 290 kJ/mol); all these heats were lower than the bulk heat of NiO formation, 479 kJ/mol. This result quantitatively illustrates that the stability of thin oxide films depends on the structure of the substrate, although defects may also contribute to the lower heats of formation of the films.

Large organic molecules represent another example of species that are likely to undergo nonreversible adsorption, because even relatively weak van der Waals interactions with the surface add up to large binding energies per molecule if only the number of atoms in the molecules is large enough. An example is benzene, which adsorbs molecularly on Pt(111) at 300 K. However, only at coverages above ~0.6 ML (1 ML = saturation coverage at 300 K) the molecules can be desorbed intact by heating. At coverages below ~0.6 ML, the benzene molecules dissociate completely to $H_2$ and graphitic carbon. Therefore, heats of adsorption can be measured only with calorimetric techniques. This system was the first example for adsorption of a large organic molecule studied with single-crystal adsorption calorimetry and was performed with the Campbell calorimeter [78]. The resulting heat of adsorption as a function of coverage is shown in Figure 32.13. The data fit well to the indicated second-order polynomial. Extrapolation to zero coverage results in an initial heat of adsorption of 197 kJ/mol, while extrapolation to saturation coverage yields only 66 kJ/mol. From the coverage dependence of the heat of adsorption, quantitative information about the strength of intermolecular repulsive

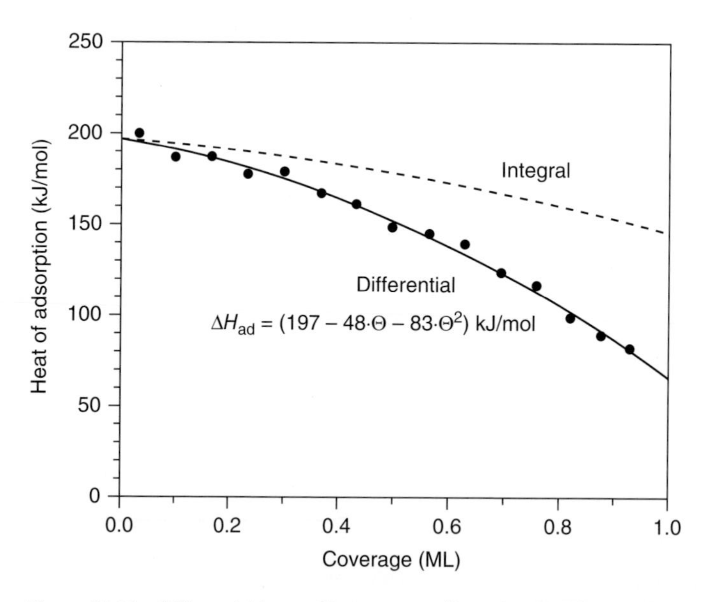


**Figure 32.13** Differential heat of benzene adsorption on Pt(111) as a function of coverage at 300 K (full circles). Also shown is the best fit to the data using a second-order polynomial dependence on coverage (solid line). The integral heat of adsorption (Equation 32.10) was computed using this polynomial and is shown as the dashed curve. (Adapted with permission from Ref. [78]. Copyright 2004, American Chemical Society.)

interactions was extracted. Adsorption of naphthalene on Pt(111) was studied with the same approach [77]. The initial heat of naphthalene adsorption, 317 kJ/mol, indicates that the heat of adsorption per C atom is almost identical for benzene and naphthalene (32–33 kJ/mol). As in the case of benzene, repulsive interactions lead to a decrease of the heat of adsorption with increasing coverage.

#### 32.4.1.2 Energetics of Reaction Intermediates

Many organic molecules dissociate upon adsorption on reactive metal surfaces. Hydrocarbons have found particular attention because of their importance in catalytic fuel conversion. Experimental insight into the energetics of these reactions can be obtained only by calorimetric techniques. As typical examples, the adsorption and reaction of acetylene ($C_2H_2$) and ethylene ($C_2H_4$) on various single-crystal surfaces of Ni, Pd, and Pt were studied by King and coworkers. Figure 32.14 shows the differential heat of adsorption and reaction of $C_2H_4$ on Pt(110)-(1 × 2).

The coverage dependence of the heat of adsorption is associated with the formation of different reaction intermediates, as depicted in Figure 32.14. The chemical nature of these species was determined by complementary measurements. The initial heat of 235 kJ/mol is related to the formation of ethylidyne (≡C–$CH_2$–). At higher coverages, ethylidyne (≡C–$CH_3$–) is formed, followed by di-$\sigma$-bonded ethylene, which leads to a plateau at 160 kJ/mol. The heat of adsorption at saturation coverage, 140 kJ/mol, is attributed to $\pi$-bonded molecular ethylene. By means of a thermodynamic cycle, these values can be used to calculate average values for C–Pt bond energies. A selection of such carbon–metal bond energies is shown in Table 32.1.

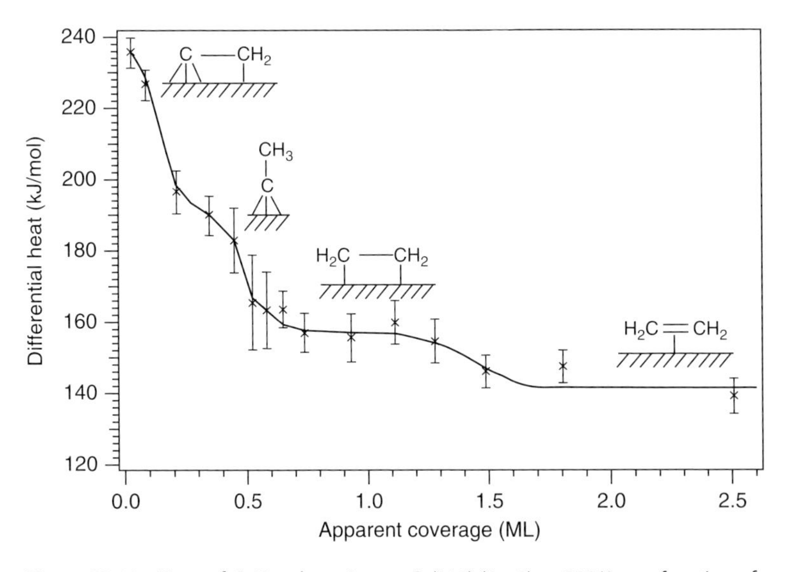


**Figure 32.14** Heat of $C_2H_4$ adsorption on Pt(110)-(1 × 2) at 300 K as a function of coverage. (Adapted with permission from Ref. [6]. Copyright 1998, American Chemical Society.)

**Table 32.1** Average metal–carbon bond energies on Ni, Pd, and Pt single-crystal surfaces [6].

| Surface | Average M–C single bond energy (kJ/mol) |
|---|---|
| Pt(110) | 242 |
| Pt(111) | 244 |
| Pt(100) | 240 |
| Pd(100) | 171 |
| Ni(100) | 205 |

Formation of different intermediates from the same precursor molecule can also be controlled by temperature. An example that was studied by Campbell and coworkers is cyclohexene, which forms a di-$\sigma$-bonded species when adsorbed on Pt(111) at 100 K [12]. In contrast, adsorption above 180 K leads to the scission of one C–H bond per molecule and to the formation of a 2-cyclohexenyl (or $\pi$-allyl) species. At 300 K, abstraction of four hydrogen atoms per molecule results in the formation of benzene. Above room temperature, the adsorbed hydrogen desorbs, and around 400 K the benzene further dehydrogenates to form graphitic carbon. Figure 32.15 shows the heat of adsorption of cyclohexene on Pt(111) at 100 K.

In the monolayer range (here: 1 ML = 1 molecule per surface Pt atom; 0.24 ML corresponds to a saturated monolayer), the heat decreases from the initial value of 130 to ~50 kJ/mol at monolayer saturation coverage. In the multilayer range, the heat

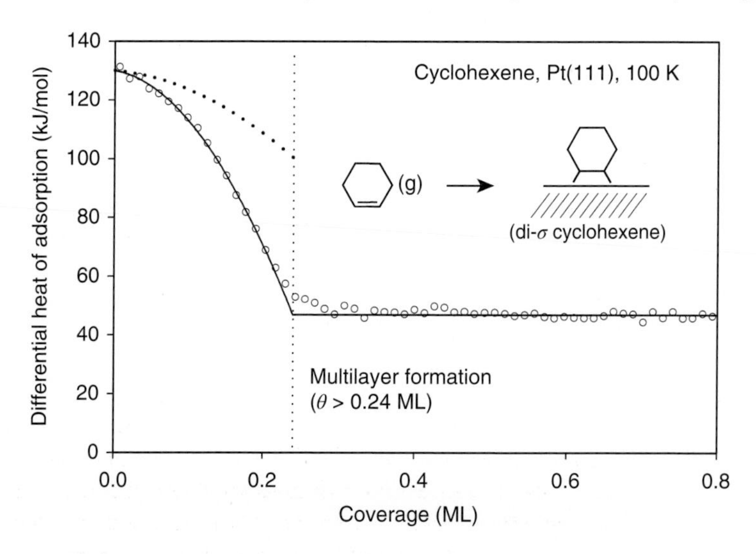


**Figure 32.15** Differential heat of adsorption of cyclohexene on Pt(111) at 100 K (open circles) and the integral heat of adsorption (dotted line in the range $0 < \Theta < 0.24$ ML), calculated using Equation 32.10. (Adapted with permission from Ref. [12]. Copyright 2008, American Chemical Society.)

of adsorption (47 kJ/mol) is identical to the heat of sublimation under the conditions of the experiment. The pronounced coverage dependence in the monolayer range reveals repulsive interactions between the di-$\sigma$ cyclohexene moieties. The standard deviation between the data points in the multilayer range indicates that a heat of adsorption of ~5 kJ/mol represents the detection limit for the calorimeter used in this experiment. This corresponds to a sensitivity limit of ~100 nJ/cm$^2$ if the active sample area is taken into account.

Compared to the initial heats of cyclohexene adsorption at 100 K, higher initial heats were measured at 263 K (174 kJ/mol, formation of 2-cyclohexenyl) and 293 K (202 kJ/mol, formation of benzene). From the initial heats of adsorption, standard enthalpies of formation of di-$\sigma$-bonded cyclohexene (−135 kJ/mol) and 2-cyclohexenyl (−143 kJ/mol) were derived. The average C–Pt bond energy can be calculated by a thermodynamic cycle, as shown in Figure 32.16. The resulting value of 205 kJ/mol is lower than that obtained for $C_2H_4$ (244 kJ/mol, Table 32.1), which shows the influence of ring strain.

Based on these measurements and values for the other above-mentioned temperatures, a complex energy landscape for the hydrogenation of benzene to cyclohexane on Pt(111) can be derived (Figure 32.17) [117]. Such data are of great importance for the quantitative understanding of heterogeneous catalytic reactions.

The energetics of adsorbed $CH_3$ on Pt(111) were studied by measuring the heat of adsorption of $CH_3I$, which dissociates into $CH_3$ and iodine atoms upon adsorption. The standard heat of formation of adsorbed $CH_3$ (−53 kJ/mol) and the Pt(111)–$CH_3$ bond energy (197 kJ/mol) were calculated using thermodynamic cycles [118, 119]. Low-temperature adsorption of $H_2O$ or $CH_3OH$ on oxygen-precovered Pt(111) leads to the formation of deutroxyl (–OD) [120] or methoxy (–$OCH_3$) [121], respectively. These reactions were employed to determine the standard heats of formation of –OD (−226 kJ/mol) and –$OCH_3$ (−170 kJ/mol), as well as the bond energies for DO–Pt(111) (263–274 kJ/mol) and $H_3CO$–Pt(111) (187 kJ/mol).

#### 32.4.1.3 Adsorption of Metals on Oxides and Organic Materials

Interface energies of metal/oxide or metal/organic interfaces can be determined by calorimetry through measurement of the energy released when metal atoms adsorb on oxides or organic materials. Desorption-based methods cannot be used: organic materials would decompose before the metal layer desorbs. Desorption of metals from oxides is possible in many cases; however, at the temperatures of the TPD experiment, the atoms would desorb from a two-dimensional gas phase, which is usually not the state of the system one is interested in.

Metal–oxide interfaces occur in heterogeneous catalysts with oxide-supported metal particles (see also Chapter 28 in Volume 4). The metal–support interaction, for which the interface energy is a quantitative measure, influences the catalytic activity and is thus subject to extensive investigation. Calorimetric measurements require a well-defined beam of metal atoms with stable flux, which represents a considerable experimental challenge. Therefore, most studies were focused on metals with relatively high vapor pressures, such as Pb, Li, and Ca. Even though these metals are not typically used in catalysis, their investigation provides important insight into

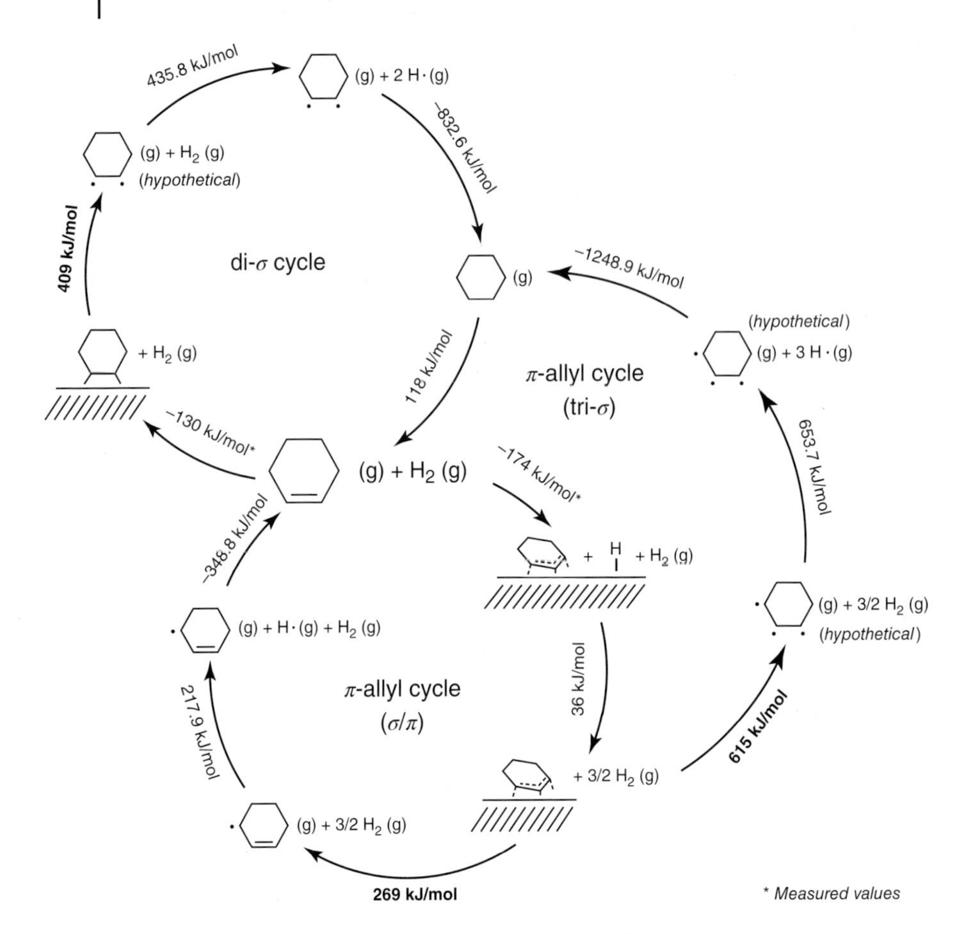


**Figure 32.16** Thermodynamic cycle for the calculation of the average C–Pt bond enthalpy for di-σ-bonded cyclohexene and 2-cyclohexenyl on Pt(111). Two different cycles are shown for 2-cyclohexenyl: one assuming a tri-σ-bonded species and one assuming a σ/π-bonded species. Values marked with an asterisk (*) are the standard enthalpies of adsorption measured by calorimetry, and the values in bold are adsorption energies relative to the corresponding gas-phase radicals, calculated by setting the sum of the enthalpies for that cycle equal to zero. The di- and triradicals are hypothetical species wherein the energy costs for removing the second and third hydrogen were assumed to be identical to that for removing the first hydrogen, so that any radical–radical interactions between neighboring carbon atoms are neglected. (Adapted with permission from Ref. [12]. Copyright 2008, American Chemical Society.)

phenomena such as particle sintering. An example is the adsorption of Pb on MgO, which gives insight into the dependence of the surface energy of the metal particles on their size [122]. It was shown that the stability of the Pb atoms in a particle (i.e., their heat of adsorption relative to gaseous Pb) decreases rapidly with decreasing particle size: for particles with radii larger than 4 nm, the heat of adsorption equals the bulk heat of sublimation, 195 kJ/mol, in contrast to only 100 kJ/mol for

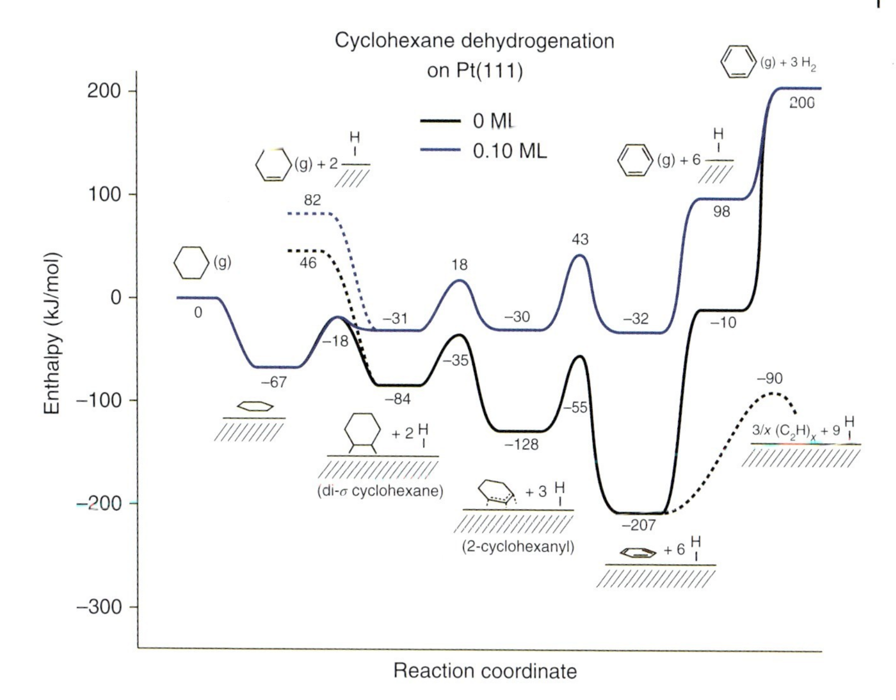


**Figure 32.17** Reaction enthalpy diagram for cyclohexane dehydrogenation to benzene (and benzene hydrogenation to cyclohexane) over a Pt(111) model catalyst, showing the energetics of the key adsorbed intermediates as determined by adsorption calorimetry, at 0.1 ML and in the limit of zero coverage. Also shown are the activation energy barriers for their interconversions determined by a variety of other surface reaction kinetic techniques. (Adapted from Ref. [74]. Copyright 2009, with permission from Elsevier.)

particles of 0.5 nm radius. This size dependence is stronger than predicted by the Gibbs–Thompson relation. The low stability of small particles makes sintering more likely, which would lead to reduced activity in the case of catalytically active metal particles.

For Ag, which is more relevant for catalysis, comparison was made between the heats of adsorption on different oxide supports: Mg(100), $Fe_3O_4$(111), and $CeO_x$(111) with $x = 1.8$ and 1.9 [80]. The data in Figure 32.18 show that the heat of adsorption and its change with particle size depend on the nature of the oxide surface. The adsorption energy (and thus, the particle stability) is lowest on MgO(100) and highest on $CeO_{1.8}$(111). From these data, the adhesion energies of the metal particles can be extracted (Table 32.2). As a result of the lower stability of Ag particles on MgO, the driving force for sintering of the particles is higher.

Surface microcalorimetry has also been applied to the vapor deposition of metals on organic substrates [124–127]. Such interfaces occur in organic electronic or optoelectronic devices [128, 129] and in metal-coated polymer materials. An important parameter of metal–organic interfaces is their energy of formation, because it determines the stability of the interface and is related to its electronic properties.

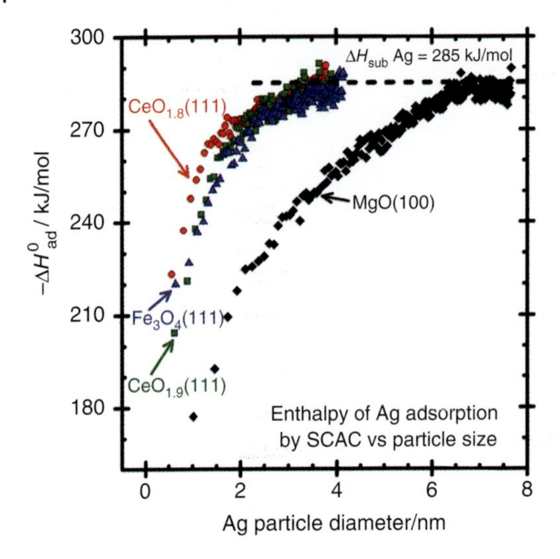


**Figure 32.18** Heat of Ag atom adsorption on oxide surfaces at 300 K versus the diameter of the Ag particles to which the atoms add. (Reproduced with permission from Ref. [123]. Copyright 2013, American Chemical Society.)

**Table 32.2** Calorimetrically measured adhesion energies of Ag nanoparticles to oxide surfaces and bulk Ag for the indicated Ag particle sizes [123].

| Surface | Ag adhesion energy (J/m$^2$) | Ag particle size (nm) |
|---|---|---|
| MgO(100) | $0.3 \pm 0.3$ | 6.6 |
| $CeO_{1.9}$(111) | $2.3 \pm 0.3$ | 3.6 |
| $CeO_{1.8}$(111) | $2.5 \pm 0.3$ | 3.6 |
| $Fe_3O_4$(111) | $2.5 \pm 0.3$ | 3.6 |
| Ag(solid) | 2.44 | ∞ |

For example, the charge injection rates at the interface depend on the overlap of wave functions and thus on the character of the interfacial chemical bond [130]. In addition, energies of interface formation provide insight into (often unwanted) chemical processes at the interface. Desorption-based methods are not applicable to such systems because, in general, the organic substrate decomposes before the metal desorbs. In calorimetric experiments, the organic material can be deposited directly onto the detector by vacuum sublimation or spin coating, which improves the sensitivity (Figure 32.8).

An example is the adsorption of Ca on poly(3-hexylthiophene), an organic semiconductor. The differential heat of Ca adsorption as a function of coverage is shown in Figure 32.19. From the initial value of 625 kJ/mol (adsorption on defect sites), the curve drops rapidly to 405 kJ/mol, which is still larger than the sublimation enthalpy of Ca, 178 kJ/mol. This indicates a strong metal–polymer interaction or a chemical

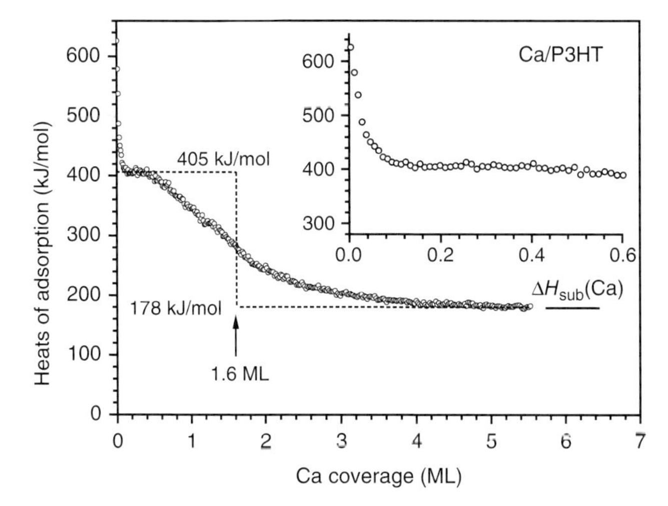


**Figure 32.19** Differential heat of adsorption of Ca on poly(3-hexylthiophene) (P3HT) at 300 K. A coverage of 1 ML corresponds to a closed-packed Ca(111) layer. (Adapted with permission from Ref. [124]. Copyright 2012, American Chemical Society.)

reaction at the interface. Complementary spectroscopic investigations show that Ca reacts with the sulfur of the polymer, forming calcium sulfide (CaS) up to a maximum depth of 3 nm at 300 K. Deposition at lower temperatures leads to smaller reaction depths [124, 131]. This reaction prevails during the early stages of deposition, whereas Ca particles and finally a closed Ca film are formed at higher Ca coverages (>0.5 ML). As a result, the heat of adsorption approaches the heat of sublimation of Ca around 5 ML, where newly adsorbing Ca atoms add exclusively to the Ca layer. The third competing process, the reflection of impinging Ca atoms, is predominant at small coverages (initial sticking probability $S_0 = 0.35$), but becomes less important at higher coverages, when the sticking probability approaches unity.

#### 32.4.1.4 **Energy Differences Between Two Solid Surface Phases**

SCAC can also be used to determine the energy differences between different surface reconstructions of a solid [132, 133]. Such energies are important as references for theoretical calculations and give insight into the energetics of adsorbate-induced surface restructuring.

For example, Pt(100) can be prepared at 300 K as a metastable (1 × 1) structure, which reconstructs to the more stable hexagonal Pt(100)-hex structure when heated above 500 K. Adsorption of CO or NO on Pt(100)-hex lifts this reconstruction and leads to the (1 × 1) phase. The differences of the heats of adsorption on the two surface phases can be used to determine the energy difference between the two phases.

The initial differential heats of CO adsorption on the two surfaces are 220 kJ/mol for Pt(100)-(1 × 1) and 193 kJ/mol for Pt(100)-hex [6]. The differential heat versus coverage curves decrease for both surfaces with increasing coverage and overlap above 0.5 ML, because Pt(100)-hex reconstructs to Pt(100)-(1 × 1) at this coverage, making both samples identical. The difference in the *integral* heats of adsorption

(Equation 32.10) at 0.5 ML can therefore be attributed to the different surface energies of the initial Pt(100)-hex and Pt(100)-(1 × 1) phases. This energy difference, which is identical to the energy of the hex ↔ (1 × 1) phase transition, amounts to 12.5 kJ per mol of surface Pt atoms. An energy diagram for the clean and CO covered Pt(100)-hex and Pt(100)-(1 × 1) phases is shown in Figure 32.20. As can be seen, the very different heats of adsorption of CO on both surfaces result in their reversed relative stability in the presence of adsorbed CO.

#### 32.4.1.5 **Adsorption on Model Catalysts: Particle-Size-Dependent Adsorption Energies**

Heats of adsorption on catalytically active supported metal nanoparticles and the dependence of this heat on the particle size have been studied with the calorimeter setup shown in Figure 32.10. Desorption-based methods are unsuitable here because the high temperatures needed for desorption often lead to particle restructuring, particle sintering, or other unwanted processes.

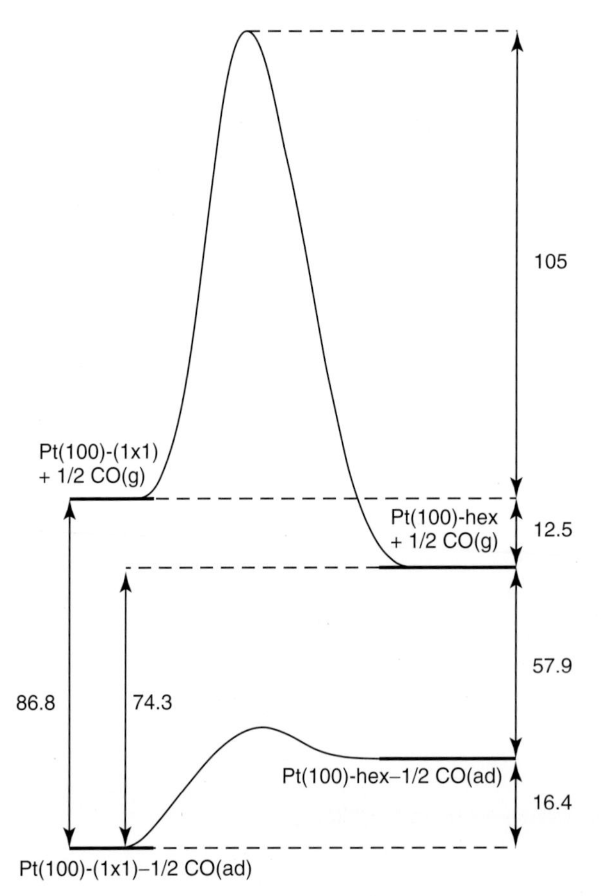


**Figure 32.20** Schematic energy diagram illustrating the phase transition between clean and CO covered Pt(100)-hex and Pt(100)-(1 × 1). (Adapted with permission from Ref. [6]. Copyright 1998, American Chemical Society.)

An example of high relevance for heterogeneous catalysis is the adsorption of CO and $O_2$ on Pd nanoparticles grown on a $Fe_3O_4$ thin film. Both molecules show pronounced but different dependencies of their initial heats of adsorption on the particle size (Figure 32.21): the initial heat of oxygen adsorption increases from Pt(111), where oxygen occupies threefold hollow sites, to large Pd nanoclusters, where oxygen first occupies particle edges. When the particles get smaller, the heat of oxygen adsorption decreases again and reaches a value similar to that for Pd(111) for the 2-nm Pd particles (Figure 32.21a). The fact that the heat of adsorption has its maximum for large particles was explained by the interplay of two factors that determine the binding energy of O atoms: the local configuration of the adsorption site, and the particle size. The change of the adsorption site from threefold hollow on Pd(111) to an edge site of a Pd particle causes an increase of the heat of adsorption. Reduction

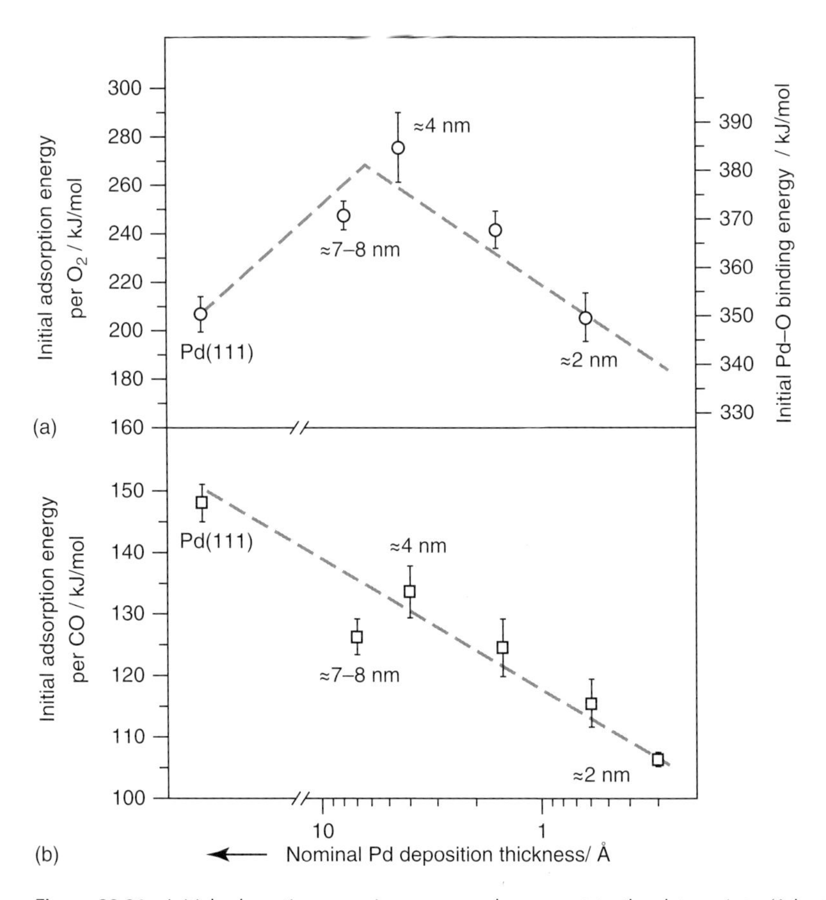


**Figure 32.21** Initial adsorption energies for (a) $O_2$ and (b) CO molecules versus the nominal Pd coverage on Pd/$Fe_3O_4$/Pt(111) and on the Pd(111) single-crystal surface. The average diameters of the Pd particles are shown next to the data points. (Adapted with permission from Ref. [134]. Copyright 2013, Wiley-VCH Verlag GmbH & Co. KGaA, Weinheim.)

of the particle size with the oxygen atom in the same adsorption site (edge site), however, results in a decrease in the heat of adsorption.

The latter trend – the heat of adsorption decreasing with decreasing particle size – was also observed for CO (Figure 32.21b). However, the highest value here is found for Pd(111), which means that the change of the adsorption site from a threefold hollow site in Pd(111) to the favorite binding site on the Pd nanoparticles has much less influence on the heat of adsorption than in the case of oxygen.

### 32.4.2
### Applications of Electrochemical Single-Electrode Calorimetry

Whereas UHV surface calorimetry was particularly designed for the study of molecule–surface interactions at coverages far below one monolayer, electrochemical calorimetry at single electrodes was traditionally devoted to the investigation of bulk electrochemical reactions (see also references in Section 32.3.2). Measuring the Peltier heat of metal deposition reactions and electron transfer reactions helped to clarify fundamental aspects of electrochemical transport processes. Heats of transfer and Eastman's entropies of transport of various anions and cations were studied this way [30, 135, 136]. In another example, Zhang and co-authors [137] attempted to derive the absolute entropy of the hydrogen electrode from measurements of the Peltier heat of $[Fe(CN)_6]^{3-/4-}$, which was, however, critically discussed by Rockwood [137]. Additionally, in a few experiments, information on the reaction mechanism of electrochemical reactions was derived from the heat changes. Franklin and McCrea investigated the temporal evolution of the electrode temperature during and after 1 min of Ag deposition or dissolution from cyanide-containing solutions and deduced hints on the involved chemical reactions, such as complexation [138]. Graves studied the oxidation processes of formaldehyde in perchloric acid [90]. Scholz *et al.* showed that the entropy changes upon oxidation or reduction of solid metal hexacyanoferrates can be attributed to the insertion of $K^+$ or $Na^+$ ions into the lattice of the solid electrode [139].

Also, the Peltier heat of battery-related half-cells was studied. Starting with the work of Donepudi and Conway on zinc and bromine electrodes [95], calorimetry was applied, for example, to the hydrogen electrode in acidic and alkaline electrolytes [104, 108, 140]. Recently, also the Peltier heat of a hydrogen electrode with a Nafion polymer electrolyte was reported [103]. Li bulk deposition was studied from a commercial carbonate-based Li-ion battery electrolyte. From the Peltier heat, the coordination number of $Li^+$ in the carbonate-based solvent was derived [141]. Also, $Li^+$ and $K^+$ intercalation in graphite electrodes were addressed by calorimetry, but no quantitative values for the Peltier heat could be given [142]. It should be briefly mentioned that reversible heat changes of complete batteries were intensively studied calorimetrically. Separation into half-cell reaction entropies might be difficult; however, by appropriate choice of the cell under study, information on entropy changes of half-cell processes can be derived. An example is the determination of the entropy of Li intercalation into $Li_xMo_6Se_8$ by studying the heat evolution of a $Li_xMo_6Se_8|Li^+|Li$

cell close to equilibrium conditions [143, 144]. Similarly, entropy changes were also determined for other cell types [145, 146].

Although electrochemical calorimetry was frequently used for the study of bulk electrochemical reactions, calorimetric experiments on surface electrochemical reactions are rather scarce. As already discussed in Section 32.3.2, this may be attributed to the sensitivity that is required to detect temperature changes from such small conversions as, for example, the adsorption of fractions of a monolayer of an electroactive species. Shibata and coworkers were the first to measure the temperature changes upon submonolayer hydrogen adsorption on a platinized Pt electrode. In contrast to the evolution of gaseous hydrogen, adsorption of hydrogen leads to heating of the electrode, which signals the reduction of entropy of the system. Shibata *et al.* could identify the relative entropy changes for the different hydrogen species and discussed their result together with IR data on the water and $H_3O^+$ structure at the interface [104]. Their results are in agreement with recent results of Etzel [147], who determined the Peltier heat of a polycrystalline Pt foil in perchloric and sulfuric acid with an AC modulation method similar to that of Jiang *et al.* [108]. Also, measurement of the heat changes during the formation of platinum and gold surface oxides became possible [105, 109]. In those cases, the surface oxide formation was governed by irreversible processes and information on the entropy could not be derived. Recently, Cu and Ag UPD on Au surfaces was studied calorimetrically [114, 147, 148]. Distinct differences between the entropy of the metal bulk deposition and the UPD processes point to charge-compensating coadsorption of anions. It was also found that entropy plays an important role for the stabilization of the Cu-UPD layer: about one-third of the energy difference between the Cu bulk phase and the first Cu-UPD layer has to be attributed to the entropic contributions to the Gibbs free energy [114]. The achieved sensitivity of electrochemical calorimetry was sufficient to detect heat effects upon adsorption or desorption of a few percent of a monolayer of the electroactive species. This allowed studying electrochemically controlled surface aggregation of amphiphilic dodecyl sulfate molecules and determination of their aggregation entropy, a quantity that is often difficult to determine. It was demonstrated that the aggregation process was entropy-driven, which pointed toward the importance of the hydrophobic interactions of the hydrocarbon tail of the dodecylsulfate molecules with the surrounding water [149, 150]. Similarly, entropy changes upon electrochemically induced swelling of polyelectrolyte multilayers could be determined [151]. Also, here entropic effects were found to be essential for the formation of the layer structures. As demonstrated by these examples, entropies of electrochemical surface systems can be reliably determined by electrochemical calorimetry.

#### 32.4.2.1 Alternatives for the Determination of Entropy Changes upon Electrochemical Reactions

Beside electrochemical calorimetry, there exist several approaches to determine the entropy changes of electrochemical reactions. The most straightforward method is based on the variation of the electromotoric force of a complete cell with temperature. If the cell reaction is at equilibrium, that is, in the absence of mixed potentials and side reactions, the derivative of the electromotoric force with respect to the temperature corresponds to the derivative of the Gibbs free energy change of the cell reaction with respect to the temperature, that is, to the reaction entropy [152]. Textbook examples are reaction entropies of a $H_2$–$O_2$ fuel cell or combinations of other gas electrodes such as $H_2$ and $Cl_2$ [153]. Temperature coefficients of the cell voltage were also determined for Li-battery related systems (see e.g., [154–157]). Those measurements yield the reaction entropy of the complete cell reaction; however, when measuring, for example, the entropy of a Li-graphite battery, the intercalation entropy of Li into graphite can be directly determined [156].

In principle, for the determination of the electromotoric force and its variation with temperature, the existence of a well-defined equilibrium potential is required. However, in the case of surface electrochemical processes, such as adsorption of an electroactive species, the equilibrium state of the system usually changes with the adsorbate coverage and is hence strongly dependent on the applied potential. This may complicate the determination of temperature coefficients of the respective equilibrium potentials. However, Breiter *et al.* demonstrated that a variation of the above method is also applicable to the study of surface electrochemical systems: in their case, H-adsorption on polycrystalline Pt. They studied the temperature dependence of the current waves in CVs and derived the heat of adsorption of hydrogen and the involved entropy changes [158, 159]. To overcome the strong assumptions made by Breiter and Böld on the equilibrium of gaseous hydrogen and the adsorbed species, Conway *et al.* introduced an analysis of the temperature-dependent data on the basis of a generalized adsorption isotherm [160]. This analysis was also applied for the determination of the entropy and enthalpy of hydrogen adsorption on other Pt metals, on Pt-catalyst nanoparticles, and on single-crystalline surfaces [161–167].

Information on the entropy of the electrode–electrolyte interface can also be inferred from the basic electrocapillary equation [168–170]. Harrison and coworkers derived a generally valid equation that related the derivative of the entropy of formation of the interface with respect the surface charge to the temperature coefficient of the cell potential at fixed surface charge density [170]. The entropy of formation of the interface is the difference in entropy of the components in the electrode–electrolyte interface, when they are forming part of it, compared to their entropies in the bulk phases of the electrolyte or metal, respectively, and is hence closely related to the entropy change during the electrochemical surface reaction. In their experiments, Harrison *et al.* derived the entropy of formation of the double layer of a mercury electrode in aqueous electrolytes from temperature-dependent capacitance measurements. Recently, Garcia-Araez *et al.* applied the electrocapillary equation to the evaluation of temperature-dependent cyclic voltammetry data of hydrogen adsorption on Pt single-crystal surfaces [171, 172]. A critical discussion of the evaluation of temperature-dependent CV data can be found in Ref. [173]. It should be noted that temperature-dependent cyclic voltammetry was also performed for other systems such as UPD of Cu and Ag on Pt [174, 175]. However, in those cases information on the entropy changes was not derived, although data evaluation based on the electrocapillary equation may in principle be applicable.

Similar to the temperature-dependent CV measurements, the entropy of formation of the double layer, or more precisely its derivative with respect to the surface charge, was determined by measuring the change of the open-cell potential, that is, the potential of the cell at fixed electrode charge density, following a laser-induced temperature jump [176–178]. From those data, for example, the reorientation of the water dipoles at potentials positive to the point of zero charge on Pt single-crystalline surfaces was inferred [178]. For liquid surfaces, the Gibbs adsorption isotherm can be more directly applied by measuring the surface tension. Its temperature coefficient then corresponds directly to the excess entropy of the interface (see e.g., [179]).

As an alternative to the temperature variation of the complete cell under isothermal conditions, also nonisothermal cells were studied. The variation of the cell voltage with the temperature difference between the cell compartments, that is, the thermopower of the cell is related to the Peltier coefficient of the cell system via Kelvin's second relation. This approach was already applied by Bouty in the very first determinations of the Peltier coefficients of electrochemical cells (see Section 32.2.2) and was later used for various systems (see also Agar [22] for a detailed discussion of the theoretical analysis and experimental examples). Some recent examples are investigations of solid metal hexacyanoferrates [139], the hydrogen electrode [140], and of reference electrodes (REs) in nonaqueous media [140].

In conclusion, there exist several alternatives for the determination of entropy changes involved in electrochemical reactions. Particularly, for the study of bulk electrochemical reactions as in batteries, temperature-dependent measurements of the open cell voltage or of the current waves in CVs may be a practicable alternative to calorimetric studies. On the other hand, when it comes to surface electrochemical reactions, where the electrochemical processes cause considerable changes of the interface itself, methods based on temperature-dependent measurements have hitherto been applied mainly to the adsorption of hydrogen or the study of double-layer formation. This may find its reason in the relatively complicated data analysis as well as in the necessity for additional information on the system. For example, the analysis in terms of Gibbs adsorption isotherm requires knowledge of the temperature coefficients of the RE and of the point of zero charge of the surface system, in order to extract the entropy. On the contrary, with sufficiently reversible reactions, electrochemical calorimetry provides direct access to the entropy changes at single electrodes by just measuring the evolved heat. With recent experimental improvements, the sensitivity has become sufficient to directly address processes such as the UPD of metals or the adsorption of hydrogen or organic molecules. Also, very fundamental processes such as double-layer charging may come within our reach for electrochemical calorimetry.

## 32.5 Perspectives

Surface microcalorimetry on single crystals and other low-surface-area systems is a versatile approach for the measurement of heat effects upon adsorption processes and surface reactions under well-defined conditions. It has been applied to many gas–solid interfaces, including the deposition of metal vapors on solids, where it provides a direct measure for the adsorbate–substrate binding energy, as well as to liquid–solid interfaces in electrochemical systems, for which information about the entropy changes due to the electrochemical reactions and their side processes are obtained.

Technical developments that greatly profited from the use of thin single-crystal samples in combination with very sensitive pyroelectric polymer detectors are expected to continue. In the future, we may see calorimetric measurements involving adsorption of even larger molecules, intercalation of metals in organic films, or measurements of the heat of formation of metal-organic or covalent networks on surfaces. Such complex surface reactions often require precise temperature control, which makes it necessary to perform calorimetric measurements below or above room temperature.

Research on electrochemical interfaces may be extended to other liquid–solid interfaces with the adsorption of neutral species, for example, large organic or biomolecules, which cannot be deposited from the gas phase via thermal evaporation. A technical challenge for such measurements would be the dosing of the molecules in the liquid. Another future option for studying large molecules is adsorption through electrospray and ion beam techniques.

Further improvements are to be expected in the field of data analysis. In some instances, the original adsorption or surface reaction may be followed by a slower secondary reaction, such as dissociation, diffusion of species into the substrate, or desorption, which will lead to additional delayed heat effects. These may strongly influence the temporal evolution of the temperature. Careful analysis of the temperature transients may hence elucidate the kinetics of the secondary reaction. Surface microcalorimetry profited immensely from its long history, during which its thermodynamic principles were thoroughly investigated and experimental approaches were intensively explored and improved. Together with the advent of modern surface science techniques, it became a very versatile method for the study of the thermodynamic properties of adsorption processes and reactions at well-defined surfaces, which were hitherto often difficult or even impossible to be obtained with other methods.

## References


1. Freund, H.-J. (1997) in *Handbook of Heterogeneous Catalysis* (eds G. Ertl, H. Knözinger, and J. Weitkamp), Wiley-VCH Verlag GmbH, Weinheim, pp. 911–942.
2. Boudart, M. (1997) in *Handbook of Heterogeneous Catalysis* (eds G. Ertl, H. Knözinger, and J. Weitkamp), Wiley-VCH Verlag GmbH, Weinheim, pp. 1–13.
3. Damjanovic, L. and Auroux, A. (2008) in *Handbook of Thermal Analysis and Calorimetry*, Recent Advances, Techniques and Applications, vol. 5 (eds M.E. Brown and P.K. Gallagher), Elsevier B.V., pp. 387–438.
4. Naumann d'Alnoncourt, R., Bergmann, M., Strunk, J., Löffler, E., Hinrichsen, O., and Muhler, M. (2005) The coverage-dependent adsorption of carbon monoxide on hydrogen-reduced copper catalysts: the combined application of microcalorimetry, temperature-programmed desorption

and FTIR spectroscopy. *Thermochim. Acta*, **434**, 132–139 (CALORIMETRY AND THERMAL EFFECTS IN CATALYSIS A Collection of Papers from the Third International Symposium on Calorimetry and Thermal Effects in Catalysis).

5. Spiewak, B.E., Cortright, R.D., and Dumesic, J.A. (1997) in *Handbook of Heterogeneous Catalysis* (eds G. Ertl, H. Knözinger, and J. Weitkamp), Wiley-VCH Verlag GmbH, Weinheim, p. 698.
6. Brown, W.A., Kose, R., and King, D.A. (1998) Femtomole adsorption calorimetry on single-crystal surfaces. *Chem. Rev.*, **98**, 797–831.
7. Stuckless, J.T., Frei, N.A., and Campbell, C.T. (1998) A novel single-crystal adsorption calorimeter and additions for determining metal adsorption and adhesion energies. *Rev. Sci. Instrum.*, **69**, 2427–2438.
8. Cerný, S. (1996) Adsorption microcalorimetry in surface science studies sixty years of its development into a modern powerful method. *Surf. Sci. Rep.*, **26**, 1–59.
9. Christmann, K. (1991) *Introduction to Surface Physical Chemistry*, Steinkopff Verlag, Springer-Verlag, Darmstadt, New York.
10. King, D.A. (1975) Thermal desorption from metal surfaces: a review. *Surf. Sci.*, **47**, 384–402.
11. Lytken, O. and Gottfried, J.M. (2011) Nanojoule-Adsorptionskalorimetrie: Messung von Adsorptionsenergien auf wohldefinierten Oberflächen. *Bunsenmagazin*, **13**, 17–25.
12. Lytken, O., Lew, W., Harris, J.J.W., Vestergaard, E.K., Gottfried, J.M., and Campbell, C.T. (2008) Energetics of cyclohexene adsorption and reaction on Pt(111) by low-temperature microcalorimetry. *J. Am. Chem. Soc.*, **130**, 10247–10257.
13. Comsa, G. and David, R. (1985) Dynamical parameters of desorbing molecules. *Surf. Sci. Rep.*, **5**, 145–198.
14. Eastman, E.D. (1926) Thermodynamics of non-isothermal systems. *J. Am. Chem. Soc.*, **48**, 1482–1493.
15. Wagner, C. (1929) Über die thermodynamische behandlung stationärer Zustände in nicht isothermen Systemen. *Ann. Phys.*, **395**, 629–687.
16. Wagner, C. (1930) Über die thermodynamische behandlung stationärer Zustände in nicht isothermen Systemen II. *Ann. Phys.*, **398**, 370–390.
17. Lange, E. and Miščenko, K.P. (1930) Zur Thermodynamik der Ionensolvatation. *Z. Phys. Chem. A*, **149**, 1–41.
18. Lange, E. and Monheim, J. (1930) Über elektolytische Peltier-Wärmen und ihre Messung mittels isotherm-adiabatischer Differentialkalorimetrie. *Z. Phys. Chem. A*, **150**, 177.
19. Lange, E. and Monheim, J. (1933) Peltierwärmen an elektrochemischen Zweiphasengrenzen, in *Handbuch der Experimental Physik* (eds W. Wien, F. Harms, and H. Lenz), Akademische Verlagsgesellschaft m. b. H., Leipzig.
20. Haase, R. (1953) Heats of transfer in electrolyte solutions. *Trans. Faraday Soc.*, **49**, 724.
21. Rysselberghe, P.V. (1955) *Electrochemical Affinity*, Hermann, Paris.
22. Agar, J.N. (1963) in *Advances in Electrochemistry and Electrochemical Engineering*, vol. 3 (ed. P. Delahay), Interscience Publishers, London, pp. 31–121.
23. deBethune, A.J. (1960) Irreversible thermodynamics in electrochemistry. *J. Electrochem. Soc.*, **107**, 829–842.
24. Thouvenin, Y. (1963) Contribution á l'étude théorique et experimentale de l'effet Peltier éléctrolytique. *Electrochim. Acta*, **8**, 529–541.
25. Tyrrell, H.J.V. (1961) *Diffusion and Heat Flow in Liquids*, Butterworth, London.
26. Førland, K.S., Førland, T., and Ratkje, S.K. (1988) *Irreversible Thermodynamics*, John Wiley & Sons, Ltd, Chichester.
27. Kjelstrup, S. and Bedaux, D. (2008) *Non-Equilibrium Thermodynamics of Heterogeneous Systems*, Series on Advances in Statistical Mechanics, World Scientific, Singapore.
28. Frank, H.S. and Evans, M.W. (1945) Free volume and entropy in condensed systems. *J. Chem. Phys.*, **13**, 507–532.
29. Marcus, Y. (1985) *Ion Solvation*, John Wiley & Sons, Ltd, Chichester.

30. Ozeki, T., Ogawa, N., Aikawa, K., Watanabe, I., and Ikeda, S. (1983) Thermal analysis of electrochemical reactions influence of electrolytes on Peltier heat for Cu(0)/Cu(II) and Ag(0)/Ag(I) redox systems. *J. Electroanal. Chem.*, **145**, 53–65.
31. Raoult, M.F.-M. (1865) Recherches sur les forces électromotrices. *Ann. Chim. Phys.*, **4**, 392–426.
32. Thomsen, J. (1880) Chemische Energie und electromotorische Kraft verschiedener galvanischer Combinationen. *Ann. Phys.*, **247**, 246–269.
33. Gibbs, J.W. (1878) On the equilibrium of heterogeneous substances. *Trans. Conn. Acad.*, **3**, 343–524.
34. Helmholtz, H. (1882) Die Thermodynamik chemischer Vorgänge. *Sitzungsber. Königl. Preuss. Akad. Wissenschaften, Berlin*, **1**, 22–39.
35. Ostwald, W. (1903) *Lehrbuch der Allgemeinen Chemie*, Wilhelm Engelmann, Leipzig.
36. Bouty, E. (1879) Sur un phénomène analogue au phénomène de Peltier. *Compt. Rend.*, **89**, 146–148.
37. Callen, H.C. (1985) *Thermodynamics and an Introduction to Thermostatistics*, John Wiley & Sons, Inc., New York.
38. Bouty, E. (1880) Mesure absolue du phénomène de Peltier au contact d'un métal et de sa dissolution. *Compt. Rend.*, **90**, 987–990.
39. Bouty, E. (1880) Mesure des forces électromotrices thermo-electriques au contact d'un métal et d'un liquide. *Compt. Rend.*, **90**, 917–920.
40. Mills, E.J. (1877) On electrostriction. *Proc. R. Soc. London*, **26**, 504–512.
41. Gockel, A. (1885) Ueber die Beziehungen der Peltier'schen Wärme zum Nutzeffect galvanischer Elemente. *Ann. Phys.*, **260**, 618–642.
42. Bružs, B. (1929) Temperaturmessungen an arbeitenden Elektroden. II. *Z. Phys. Chem.*, **145**, 470–476.
43. Bružs, B. (1929) Temperaturmessungen an arbeitenden Elektroden. *Z. Phys. Chem.*, **145**, 283–288.
44. Bružs, B. (1930) Temperaturmessungen an arbeitenden Elektroden. III. *Z. Phys. Chem.*, **146**, 356–362.
45. Bockris, J.O.M., Reddy, A.K.N., and Gamboa-Aldeco, M. (2000) *Modern Electrochemistry 2A: Fundamentals of Electrodics*, Kluwer Academic Publishers, New York.
46. Vetter, K.J. (1968) *Electrochemical Kinetics*, Academic Press, New York.
47. Bickel, K.R., Etzel, K.D., Halka, V., and Schuster, R. (2013) Microcalorimetric determination of heat changes caused by overpotential upon electrochemical Ag bulk deposition. *Electrochim. Acta*, **112**, 801–812.
48. Roberts, J.K. (1934) The heat of adsorption of hydrogen on tungsten. *Proc. Cambridge Philos. Soc.*, **30**, 376–379.
49. Kisliuk, P. (1959) Calorimetric heat of adsorption – nitrogen on tungsten. *J. Chem. Phys.*, **31**, 1605–1611.
50. Beeck, O. (1945) Catalysis – a challenge to the physicist (as exemplified by the hydrogenation of ethylene over metal catalysts). *Rev. Mod. Phys.*, **17**, 61–71.
51. Beeck, O., Cole, W.A., and Wheeler, A. (1950) Determination of heats of adsorption using metal films. *Discuss. Faraday Soc.*, **8**, 314–321.
52. Wedler, G. (1960) Elektronische Wechselwirkung und Adsorptionswärme bei der Chemisorption von Gasen an aufgedampften Metallfilmen. I. Adiabatischer Kalorimeter zu gleichzeitigen Messung von Adsorptionswärme und Filmwiderstand. *Z. Phys. Chem.*, **24**, 73–86.
53. Wedler, G. (1961) Elektronische Wechselwirkung und Adsorptionswärme bei der Chemisorption von Gasen an aufgedampften Metallfilmen. II. Einwirkung von Sauerstoff auf Eisenfilme bei 273°K. *Z. Phys. Chem.*, **27**, 388–401.
54. Wedler, G., Ganzmann, I., and Borgmann, D. (1993) Calorimetric investigations of processes at the solid gas interphase. *Ber. Bunsen-Ges.-Phys. Chem. Chem. Phys.*, **97**, 293–297.
55. Wedler, G. and Strothenk, H. (1966) Kalorimetrische Messung der differentiellen Chemisorptionswärmen von Wasserstoff an Titanfilmen bei 77°K. *Ber. Bunsen-Ges. Phys. Chem.*, **70**, 214–220.

56. Kovar, M., Dvorak, L., and Černý, S. (1994) Application of pyroelectric properties of $LiTaO_3$ single crystal to microcalorimetric measurement of the heat of adsorption. *Appl. Surf. Sci.*, **74**, 51–59.
57. Coufal, H.J., Grygier, R.K., Horne, D.E., and Fromm, J.E. (1987) Pyroelectric calorimeter for photothermal studies of thin films and adsorbates. *J. Vac. Sci. Technol., A*, **5**, 2875–2889.
58. Dvorak, L., Kovar, M., and Cerny, S. (1994) A new approach to adsorption microcalorimetry based on a $LiTaO_3$ pyroelectric temperature sensor and a pulsed molecular beam. *Thermochim. Acta*, **245**, 163–171.
59. King, D.A. and Wells, M.G. (1972) Molecular-beam investigation of adsorption kinetics on bulk metal targets – nitrogen on tungsten. *Surf. Sci.*, **29**, 454–482.
60. King, D.A. and Wells, M.G. (1974) Reaction mechanism in chemisorption kinetics – nitrogen on (100) plane of tungsten. *Proc. R. Soc. Lond. A-Math. Phys. Sci.*, **339**, 245–269.
61. Kyser, D.A. and Masel, R.I. (1987) Design of a calorimeter capable of measuring heats of adsorption on single crystals surfaces. *Rev. Sci. Instrum.*, **58**, 2141–2144.
62. Antonietti, J.M., Gong, J., Habibpour, V., Rottgen, M.A., Abbet, S., Harding, C.J., Arenz, M., Heiz, U., and Gerber, C. (2007) Micromechanical sensor for studying heats of surface reactions, adsorption, and cluster deposition processes. *Rev. Sci. Instrum.*, **78**, 054101.
63. Barnes, J.R., Stephenson, R.J., Welland, M.E., Gerber, C., and Gimzewski, J.K. (1994) Photothermal spectroscopy with femtojoule sensitivity using a micromechanical device. *Nature*, **372**, 79–81.
64. Barnes, J.R., Stephenson, R.J., Woodburn, C.N., Oshea, S.J., Welland, M.E., Rayment, T., Gimzewski, J.K., and Gerber, C. (1994) A femtojoule calorimeter using micromechanical sensors. *Rev. Sci. Instrum.*, **65**, 3793–3798.
65. Gimzewski, J.K., Gerber, C., Meyer, E., and Schlittler, R.R. (1994) Observation of a chemical reaction using a micromechanical sensor. *Chem. Phys. Lett.*, **217**, 589–594.
66. Borroni-Bird, C.E. and King, D.A. (1991) An ultrahigh vacuum single crystal adsorption microcalorimeter. *Rev. Sci. Instrum.*, **62**, 2177–2185.
67. Borroni-Bird, C.E., Al-Sarraf, N., Andersoon, S., and King, D.A. (1991) Single crystal adsorption microcalorimetry. *Chem. Phys. Lett.*, **183**, 516–520.
68. Stuck, A., Wartnaby, C.E., Yeo, Y.Y., Stuckless, J.T., AlSarraf, N., and King, D.A. (1996) An improved single crystal adsorption calorimeter. *Surf. Sci.*, **349**, 229–240.
69. Besenbacher, F., Stensgaard, I., and Mortensen, K. (1987) Adsorption position of deuterium on the Pd(100) surface determined with transmission channeling. *Surf. Sci.*, **191**, 288–301.
70. Stensgaard, I. and Jakobsen, F. (1985) Adsorption site location by transmission channeling – deuterium on Ni(100). *Phys. Rev. Lett.*, **54**, 711–713.
71. Kose, R. (1998) New frontiers in single crystal adsorption calorimetry. Dissertation, Downing College. University of Cambridge, Cambridge.
72. Stuckless, J.T., Alsarraf, N., Wartnaby, C., and King, D.A. (1993) Calorimetric heats of adsorption for CO on nickel single crystal surfaces. *J. Chem. Phys.*, **99**, 2202–2212.
73. Dixon-Warren, S.J., Kovar, M., Wartnaby, C.E., and King, D.A. (1994) Pyroelectric single crystal adsorption microcalorimeter at low temperatures – oxygen on Ni(100). *Surf. Sci.*, **307**, 16–22.
74. Campbell, C.T. and Lytken, O. (2009) Experimental measurements of the energetics of surface reactions. *Surf. Sci.*, **603**, 1365–1372.
75. Diaz, S.F., Zhu, J.F., Shamir, N., and Campbell, C.T. (2005) Pyroelectric heat detector for measuring adsorption energies on thicker single crystals. *Sens. Actuators, B: Chem.*, **107**, 454–460.
76. Lew, W., Lytken, O., Farmer, J.A., Crowe, M.C., and Campbell, C.T. (2010) Improved pyroelectric detectors for single crystal adsorption

calorimetry from 100 to 350 K. *Rev. Sci. Instrum.*, **81**, 024102.
77. Gottfried, J.M., Vestergaard, E.K., Bera, P., and Campbell, C.T. (2006) Heat of adsorption of naphthalene on Pt(111) measured by adsorption calorimetry. *J. Phys. Chem. B*, **110**, 17539–17545.
78. Ihm, H., Ajo, H.M., Gottfried, J.M., Bera, P., and Campbell, C.T. (2004) Calorimetric measurement of the heat of adsorption of benzene on Pt(111). *J. Phys. Chem. B*, **108**, 14627–14633.
79. Starr, D.E. and Campbell, C.T. (2008) Large entropy difference between terrace and step sites on surfaces. *J. Am. Chem. Soc.*, **130**, 7321–7327.
80. Farmer, J.A. and Campbell, C.T. (2010) Ceria maintains smaller metal catalyst particles by strong metal-support bonding. *Science*, **329**, 933–936.
81. Murdey, R., Liang, S.J.S., and Stuckless, J.T. (2005) An atom-transparent photon block for metal-atom deposition from high-temperature ovens. *Rev. Sci. Instrum.*, **76**, 023911.
82. Schießer, A., Hörtz, P., and Schäfer, R. (2010) Thermodynamics and kinetics of CO and benzene adsorption on Pt(111) studied with pulsed molecular beams and microcalorimetry. *Surf. Sci.*, **604**, 2098–2105.
83. Schießer, A. and Schäfer, R. (2009) Versatile piezoelectric pulsed molecular beam source for gaseous compounds and organic molecules with femtomole accuracy for UHV and surface science applications. *Rev. Sci. Instrum.*, **80**, 086103.
84. Fischer-Wolfarth, J.H., Hartmann, J., Farmer, J.A., Flores-Camacho, J.M., Campbell, C.T., Schauermann, S., and Freund, H.J. (2011) An improved single crystal adsorption calorimeter for determining gas adsorption and reaction energies on complex model catalysts. *Rev. Sci. Instrum.*, **82**, 024102.
85. Fischer-Wolfarth, J.-H., Farmer, J.A., Flores-Camacho, J.M., Genest, A., Yudanov, I.V., Rösch, N., Campbell, C.T., Schauermann, S., and Freund, H.-J. (2010) Particle-size dependent heats of adsorption of CO on supported Pd nanoparticles as measured with a single-crystal microcalorimeter. *Phys. Rev. B*, **81**, 241416.
86. Kuhn, A.T. and Shams El Din, A.M. (1983) Thermometric and calorimetric methods in electrochemical and corrosion studies. *Surf. Technol.*, **20**, 55–69.
87. Boudeville, P. (1994) Thermometric determination of electrochemical Peltier heat (thermal effect associated with electron transfer) of some redox couples. *Inorg. Chim. Acta*, **226**, 69–78.
88. Gill, J. (1890) Ueber die Wärmewirkungen des electrischen Stromes an der Grenze von Metallen und Flüssigkeiten. *Ann. Phys.*, **276**, 115–138.
89. Holmes, H.H. and Jonich, M.J. (1959) Thermal electroanalysis. *Anal. Chem.*, **31**, 28–32.
90. Graves, B.B. (1972) Differential voltammetric scanning thermometry of tenth formal formaledyde solution in formal perchloric acid. *Anal. Chem.*, **44**, 993–1002.
91. Tamamushi, R. (1973) An experimental study of the electrochemical Peltier heat. *J. Electroanal. Chem. Interfacial Electrochem.*, **45**, 500–503.
92. Tamamushi, R. (1975) The electrochemical Peltier effect observed with electrode reactions of Fe(II)/Fe(III) redox couples at a gold electrode. *J. Electroanal. Chem.*, **65**, 263–273.
93. Ozeki, T., Watanabe, I., and Ikeda, S. (1979) The application of the thermistor-electrode to Peltier heat measurement $Cu/Cu^{2+}$ system in aqueous perchlorate solution. *J. Electroanal. Chem.*, **96**, 117–121.
94. Ozeki, T., Watanabe, I., and Ikeda, S. (1983) Analysis of copper(I) ion in chloride solution with cyclic-voltammothermometry. *J. Electroanal. Chem.*, **152**, 41–54.
95. Donepudi, V.S. and Conway, B.E. (1984) Electrochemical calorimetry of the Zinc and Bromine electrodes in Zinc-Bromine and Zinc-Air batteries. *J. Electrochem. Soc.*, **131**, 1477–1485.
96. Gokhstein, A.Y. (1968) *Elektrokhimiya*, **4**, 886.
97. Decker, F., Fracastoro-Decker, M., Cella, N., and Vargas, H. (1990) Acoustic detection of the electrochemical

Peltier effect. *Electrochim. Acta*, **35**, 25–26.
98. Kuz'minskii, Y.V. and Gorodyskii, A.V. (1988) Thermal analysis of electrochemical reactions Part I. Kinetic method of determining Peltier heats. *J. Electroanal. Chem.*, **252**, 21–37.
99. Boudeville, P. and Tallec, A. (1988) Electrochemistry and calorimetry coupling. IV. Determination of electrochemical Peltier heat. *Thermochim. Acta*, **126**, 221–234.
100. Sherfey, J.M. and Brenner, A. (1958) Electrochemical calorimetry. *J. Electrochem. Soc.*, **105**, 665–672.
101. Zhang, H., Zhang, P., and Fang, Z. (1997) Coupling microcalorimeter with electrochemical instruments for thermoelectrochemical research. *Thermochim. Acta*, **303**, 11–15.
102. Lange, E. and Hesse, T. (1932) Elektrolytische Peltierwärmen am System Ag/$AgNO_3$, aq. *Z. Elektrochem.*, **38**, 428–441.
103. Kjelstrup, S., Vie, P.J.S., Akyalcin, L., Zefaniya, R., Pharoah, J.G., and Burheim, O.S. (2013) The Seebeck coefficient and the Peltier effect in a polymer electrolyte membrane cell with two hydrogen electrodes. *Electrochim. Acta*, **99**, 166–175.
104. Shibata, S. and Sumino, M.P. (1985) The electrochemical Peltier heat for the adsorption and desorption of hydrogen on a platinized platinum electrode in sulfuric acid solution. *J. Electroanal. Chem.*, **193**, 135–143.
105. Shibata, S., Sumino, M.P., and Yamada, A. (1985) An improved heat-responsive electrode for the measurement of electrochemical Peltier heat. *J. Electroanal. Chem.*, **193**, 123–134.
106. Jiang, Z., Xiang, Y., and Wang, J. (1991) Study of the oxidation layer on the nickel surface in 1 M NaOH solution using in-situ photothermal spectroscopy method. *J. Electroanal. Chem.*, **316**, 199–209.
107. Jiang, Z., Zhang, W., and Huang, X. (1994) AC electrochemical-thermal method for investigating hydrogen adsorption and evolution on a platinised platinum electrode. *J. Electroanal. Chem.*, **367**, 293–296.
108. Jiang, Z., Zhang, J., Dong, L., and Zhuang, J. (1999) Determination of the entropy change of the electrode reaction by an ac electrochemical-thermal method. *J. Electroanal. Chem.*, **469**, 1–10.
109. Hai, B. and Scherson, D. (2009) In situ calorimetry at metal-electrode liquid electrolyte interfaces as monitored by probe beam deflection techniques. *J. Phys. Chem. C*, **113**, 18244–18250.
110. Fracastoro-Decker, M. and Decker, F. (1989) The mirage effect under controlled current conditions. *J. Electroanal. Chem.*, **266**, 215–225.
111. Rosolen, L.M., Fracastoro Decker, M., and Decker, M. (1993) The mirage effect: a sensitive probe for electrochemical cell calorimetry. *J. Electroanal. Chem.*, **346**, 119–133.
112. Etzel, K.D., Bickel, K.R., and Schuster, R. (2010) A microcalorimeter for measuring heat effects of electrochemical reactions with submonolayer conversions. *Rev. Sci. Instrum.*, **81**, 034101.
113. Schuster, R., Rösch, R., and Timm, A.E. (2007) Microcalorimetry of electrochemical reactions at submonolayer conversions. *Z. Phys. Chem.*, **221**, 1479–1491.
114. Etzel, K.D., Bickel, K.R., and Schuster, R. (2010) Heat effects upon electrochemical copper deposition on polycrystalline gold. *ChemPhysChem*, **11**, 1416–1424.
115. Tasaki, I. and Iwasa, K. (1981) Temperature changes associated with nerve excitation: detection by using polyvinylidene fluoride film. *Biochem. Biophys. Res. Commun.*, **101**, 172–176.
116. Stuckless, J.T., Wartnaby, C.E., AlSarraf, N., DixonWarren, S.J.B., Kovar, M., and King, D.A. (1997) Oxygen chemisorption and oxide film growth on Ni{100}, {110}, and {111}: sticking probabilities and microcalorimetric adsorption heats. *J. Chem. Phys.*, **106**, 2012–2030.
117. Lytken, O., Lew, W., and Campbell, C.T. (2008) Catalytic reaction energetics by single crystal adsorption calorimetry: hydrocarbons on Pt(111). *Chem. Soc. Rev.*, **37**, 2172–2179.
118. Karp, E.M., Silbaugh, T.L., and Campbell, C.T. (2013) Energetics of

adsorbed CH3 on Pt(111) by calorimetry. *J. Am. Chem. Soc.*, **135**, 5208–5211.

119. Karp, E.M., Silbaugh, T.L., and Campbell, C.T. (2013) Energetics of adsorbed CH3 and CH on Pt(111) by calorimetry: dissociative adsorption of CH3I. *J. Phys. Chem. C*, **117**, 6325–6336.
120. Lew, W., Crowe, M.C., Campbell, C.T., Carrasco, J., and Michaelides, A. (2011) The energy of hydroxyl coadsorbed with water on Pt(111). *J. Phys. Chem. C*, **115**, 23008–23012.
121. Karp, E.M., Silbaugh, T.L., Crowe, M.C., and Campbell, C.T. (2012) Energetics of adsorbed methanol and methoxy on Pt(111) by microcalorimetry. *J. Am. Chem. Soc.*, **134**, 20388–20395.
122. Campbell, C.T., Parker, S.C., and Starr, D.E. (2002) The effect of size-dependent nanoparticle energetics on catalyst sintering. *Science*, **298**, 811–814.
123. Campbell, C.T. (2013) The energetics of supported metal nanoparticles: relationships to sintering rates and catalytic activity. *Acc. Chem. Res.*, **46**, 1712–1719.
124. Bebensee, F., Zhu, J.F., Baricuatro, J.H., Farmer, J.A., Bai, Y., Steinruck, H.P., Campbell, C.T., and Gottfried, J.M. (2010) Interface formation between calcium and electron-irradiated poly(3-hexylthiophene). *Langmuir*, **26**, 9632–9639.
125. Murdey, R. and Stuckless, J.T. (2003) Calorimetry of polymer metallization: copper, calcium, and chromium on PMDA-ODA polyimide. *J. Am. Chem. Soc.*, **125**, 3995–3998.
126. Zhu, J., Bebensee, F., Hieringer, W., Zhao, W., Baricuatro, J.H., Farmer, J.A., Bai, Y., Steinrück, H.-P., Gottfried, J.M., and Campbell, C.T. (2009) Formation of the calcium/poly(3-hexylthiophene) interface: structure and energetics. *J. Am. Chem. Soc.*, **131**, 13498–13507.
127. Zhu, J., Goetsch, P., Ruzycki, N., and Campbell, C.T. (2007) Adsorption energy, growth mode, and sticking probability of Ca on poly(methyl methacrylate) surfaces with and without electron damage. *J. Am. Chem. Soc.*, **129**, 6432–6441.
128. Faupel, F., Zaporojtchenko, V., Strunskus, T., Erichsen, J., Dolgner, K., Thran, A., and Kiene, M. (2002) Fundamental aspects of polymer metallization, in *Metallization of Polymers 2* (ed. E. Sacher), Kluwer Academic Publishers, New York.
129. Friend, R.H. (2001) Conjugated polymers. New materials for optoelectronic devices. *Pure Appl. Chem.*, **73**, 425–430.
130. Schwalb, C.H., Sachs, S., Marks, M., Schöll, A., Reinert, F., Umbach, E., and Höfer, U. (2008) Electron lifetime in a Shockley-type metal-organic interface state. *Phys. Rev. Lett.*, **101**, 146801.
131. Bebensee, F., Schmid, M., Steinruck, H.P., Campbell, C.T., and Gottfried, J.M. (2010) Toward well-defined metal-polymer interfaces: temperature-controlled suppression of subsurface diffusion and reaction at the calcium/poly(3-hexylthiophene) interface. *J. Am. Chem. Soc.*, **132**, 12163–12165.
132. Yeo, Y.Y., Vattuone, L., and King, D.A. (1996) Energetics and kinetics of CO and NO adsorption on Pt{100}: restructuring and lateral interactions. *J. Chem. Phys.*, **104**, 3810–3821.
133. Yeo, Y.Y., Wartnaby, C.E., and King, D.A. (1995) Calorimetric measurement of the energy difference between 2 solid-surface phases. *Science*, **268**, 1731–1732.
134. Peter, M., Camacho, J.M.F., Adamovski, S., Ono, L.K., Dostert, K.H., O'Brien, C.P., Cuenya, B.R., Schauermann, S., and Freund, H.J. (2013) Trends in the binding strength of surface species on nanoparticles: how does the adsorption energy scale with the particle size? *Angew. Chem. Int. Ed.*, **52**, 5175–5179.
135. Lange, E. and Hesse, T. (1933) Experimenteller Nachweis von Überführungswärmen in elektrolytischen Peltier-Wärmen. *Z. Elektrochem.*, **39**, 374–384.
136. Lange, E. and Hesse, T. (1933) Concerning the existence of the so-called

heats of transfer (Q* values) in Peltier heats. *J. Am. Chem. Soc.*, **55**, 853–855.
137. Rockwood, A.L. (2009) The electrochemical Peltier heat of the standard hydrogen electrode reaction. *Thermochim. Acta*, **490**, 82–84.
138. Franklin, T.C. and McCrea, R. (1973/74) Heat effects, another method of studying electrodeposition processes. *Electrodeposition Surf. Treat.*, **2**, 191–203.
139. Soto, M.B. and Scholz, F. (2002) Cyclic voltammetry of immobilized microparticles with in situ calorimetry; part II: application of a thermistor electrode for in situ calorimetric studies of the electrochemistry of solid metal hexacyanoferrates. *J. Electroanal. Chem.*, **528**, 27–32.
140. Kamata, M., Ito, Y., and Oishi, J. (1987) Single electrode Peltier heat of a hydrogen electrode in $H_2SO_4$ and NaOH solutions. *Electrochim. Acta*, **32**, 1377–1381.
141. Schmid, M.J., Bickel, K.R., Novák, P., and Schuster, R. (2013) Microcalorimetric measurements of the solvent contribution to the entropy changes upon electrochemical lithium bulk deposition. *Angew. Chem. Int. Ed.*, n/a-n/a.
142. Maeda, Y. (1990) Thermal behavior on graphite due to electrochemical intercalation. *J. Electrochem. Soc.*, **137**, 3047–3052.
143. Dahn, J.R., McKinnon, W.R., Murray, J.J., Haering, R.R., McMillan, R.S., and Rivers-Bowerman, A.H. (1985) Entropy of the intercalation compound Li_{x}Mo_{6}Se_{8} from calorimetry of electrochemical cells. *Phys. Rev. B*, **32**, 3316–3318.
144. McKinnon, W.R., Dahn, J.R., Murray, J.J., Haering, R.R., McMillan, R.S., and Rivers-Bowerman, A.H. (1986) Entropy of intercalation compounds: II. Calorimetry of electrochemical cells of the Chevrel compound $Li_xMo_6Se_8$ for $0 < x < 4$. *J. Phys. C: Solid State Phys.*, **19**, 5135–5148.
145. Kobayashi, Y., Miyashiro, H., Kumai, K., Takei, K., Iwahori, T., and Uchida, I. (2002) Precise electrochemical calorimetry of $LiCoO_2$/graphite lithium-ion cell understanding thermal behavior and estimation of degradation mechanism. *J. Electrochem. Soc.*, **149**, A978–A982.
146. Lu, W., Belharouak, I., Liu, J., and Amine, K. (2007) Thermal properties of $Li_4$/3 T5/3O4/$LiMn_2O_4$ cell. *J. Power Sources*, **174**, 673–677.
147. Etzel, K. (2012) *Untersuchungen der Metall-Unterpotentialabscheidung und der Wasserstoffadsorption mittels elektrochemischer Mikrokalorimetrie*, Karlsruhe Institute of Technology, Karlsruhe.
148. Schuster, R. (2007) Electrochemical microstructuring with short voltage pulses. *ChemPhysChem*, **8**, 34–39.
149. Bickel, K.R. (2012) *Mikrokalorimetrische Untersuchungen elektrochemisch induzierter Adsorptions- und Abscheidungsprozesse von Ionen, Komplexen und Amphiphilen*, Karlsruhe Institute of Technology, Karlsruhe.
150. Bickel, K.R., Timm, A.-E., Nattland, D., and Schuster, R. (2014) Microcalorimetric determination of the entropy change upon the electrchemcially driven surface aggregation of dodecyl sulfate. *Langmuir*, **30**, 9085–9090.
151. Zahn, R., Bickel, K.R., Zambelli, T., Reichenbach, J., Kuhn, F.M., Voros, J., and Schuster, R. (2014) The entropy of water in swelling PGA/PAH polyelectrolyte multilayers. *Soft Matter*, **10**, 688–693.
152. Newman, J. and Thomas-Alyea, K.E. (2004) *Electrochemical Systems*, Wiley-Interscience, Hoboken, NJ.
153. Hamann, C.H., Hamnett, A., and Vielstich, W. (2007) *Electrochemistry*, Wiley-VCH Verlag GmbH, Weinheim.
154. Dahn, J.R. and Haering, R.R. (1983) Entropy measurements on $Li_xTiS_2$. *Can. J. Phys.*, **61**, 1093–1098.
155. Wen, C.J. and Huggins, R.A. (1981) Thermodynamic study of the lithium-tin system. *J. Electrochem. Soc.*, **128**, 1181–1187.
156. Reynier, Y., Yazami, R., and Fultz, B. (2003) The entropy and enthalpy of lithium intercalatation into graphite. *J. Power Sources*, **119-121**, 850–855.

157. Reynier, Y., Graetz, J., Swan-Wood, T., Rez, P., Yazami, R., and Fultz, B. (2004) Entropy of Li intercalation in $Li_xCoO_2$. *Phys. Rev. B*, **70**, 174304.
158. Böld, W. and Breiter, M.W. (1960) Bestimmung der Adssorptionswärme von Wassrstoff an aktiven Platinmetallelektroden in schwefelsaurer Lösung. *Z. Elektrochem.*, **64**, 897–902.
159. Breiter, M. (1962) Über die Art der Wasserstoffadsorption an Platinmetallelektroden. *Electrochim. Acta*, **7**, 25–38.
160. Conway, B.E. and Currie, J.C. (1978) Temperature and pressure effects on surface processes at noble metal electrodes. Part 2.-Volume of adsorbed H and oxygen species at Pt and Au. *J. Chem. Soc., Faraday Trans. 1 F*, **74**, 1390–1402.
161. Jerkiewicz, G. and Zolfaghari, A. (1996) Determination of the energy of the metal – underpotential-deposited hydrogen bond for rhodium electrodes. *J. Phys. Chem.*, **100**, 8454–8461.
162. Zolfaghari, A., Chayer, M., and Jerkiewicz, G. (1997) Energetics of the underpotential deposition of hydrogen on platinum electrodes. *J. Electrochem. Soc.*, **144**, 3034–3041.
163. Zolfaghari, A. and Jerkiewicz, G. (1999) Temperature-dependent research on Pt(111) and Pt(100) electrodes in aqueous $H_2SO_4$. *J. Electroanal. Chem.*, **467**, 177–185.
164. Marković, N.M., Schmidt, T.J., Grgur, B.N., Gasteiger, H.A., Behm, R.J., and Ross, P.N. (1999) Effect of temperature on surface processes at the Pt(111)-liquid interface: hydrogen adsorption, oxide formation, and CO oxidation. *J. Phys. Chem. B*, **103**, 8568–8577.
165. Elezović, N.R., Babić, B.M., Krstajić, N.V., Gajić-Krstajić, L.M., and Vračar, L.M. (2007) Specifity of the UPD of H to the structure of highly dispersed Pt on carbon support. *Int. J. Hydrogen Energy*, **32**, 1991–1998.
166. Gómez, R., Orts, J.M., Álvarez-Ruiz, B., and Feliu, J.M. (2004) Effect of temperature on hydrogen adsorption on Pt(111), Pt(110), and Pt(100) electrodes in 0.1 M HClO4. *J. Phys. Chem. B*, **108**, 228–238.
167. Zolfaghari, A., Jerkiewicz, G., Chrzanowski, W., and Wieckowski, A. (1999) Energetics of the underpotential deposition of hydrogen on platinum electrodes. *J. Electrochem. Soc.*, **146**, 4158–4165.
168. Hills, G. (1969) The compact double layer as a function of temperature and pressure. *J. Phys. Chem.*, **73**, 3591–3597.
169. Hills, G.J. and Payne, R. (1965) Temperature and pressure dependence of the double layer capacity at the mercury-solution interface. *Trans. Faraday Soc.*, **61**, 326–349.
170. Harrison, J.A., Randles, J.E.B., and Schiffrin, D.J. (1973) The entropy of formation of the mercury-aqueous solution interface and the structure of the inner layer. *J. Electroanal. Chem. Interfacial Electrochem.*, **48**, 359–381.
171. Garcia-Araez, N., Climent, V., and Feliu, J.M. (2008) Determination of the entropy of formation of the Pt(111) | perchloric acid solution interface. Estimation of the entropy of adsorbed hydrogen and OH species. *J. Solid State Electrochem.*, **12**, 387–398.
172. Garcia-Araez, N., Climent, V., and Feliu, J.M. (2009) Separation of temperature effects on double-layer and charge-transfer processes for platinum|solution interphases. Entropy of formation of the double layer and absolute molar entropy of adsorbed hydrogen and OH on Pt(111). *J. Phys. Chem. C*, **113**, 199913–199925.
173. Garcia-Araez, N., Climent, V., and Feliu, J.M. (2010) Analysis of temperature effects on hydrogen and OH adsorption on Pt(111), Pt(100), and Pt(110) by means of Gibbs thermodynamics. *J. Electroanal. Chem.*, **649**, 69–82.
174. Jerkiewicz, G., Perreault, F., and Radovic-Hrapovic, Z. (2009) Effect of temperature variation on the underpotential deposition of copper on Pt(111) in aqueous $H_2SO_4$. *J. Phys. Chem. C*, **113**, 12309–12316.
175. Radovic-Hrapovic, Z. and Jerkiewicz, G. (2002) Temperature-dependence of the under-potential deposition of Ag on Pt(111), in *Thin Films: Preparation,*

*Characterization, Applications* (eds M.P. Soriaga *et al.*), Kluwer Academic Publishers, New York.

176. Benderskii, V.A. and Velichko, G.J. (1982) Temperature jump in electric double-layer study, part I. Method of measurement. *J. Electroanal. Chem.*, **140**, 1–22.

177. Climent, V., Coles, B.A., and Compton, R.G. (2002) Laser-induced potential transients on a Au(111) single-crystal electrode. Determination of the potential of maximum entropy of double layer formation. *J. Phys. Chem. B*, **106**, 5258–5265.

178. Garcia-Aráez, N., Climent, V., and Feliu, J.M. (2008) Evidence of water reorientation on model electrocatalytic surfaces from nanosecond-laser-pulsed experiments. *J. Am. Chem. Soc.*, **130**, 3824–3833.

179. Habib, M.A. (1977) in *Modern Aspects of Electrochemistry* (eds J.O.M. Bockris and B.E. Conway), Plenum Press, New York, pp. 131–182.

# 33
# Rare Gas Adsorption

*Peter Zeppenfeld*

## 33.1
## Introduction

After 50 years of research, the adsorption behavior of the rare gases on solid surfaces has recently attracted renewed interest. On the one hand, some fundamental aspects have come within the reach of modern experimental and theoretical techniques, such as the very nature of physisorption and the rare gas–substrate interactions [1, 2] or the possibility to study the growth, adlayer dynamics, and electronic structure at the atomic scale [3–5]. Moreover, new topics have emerged, for example, the adsorption on low-dimensional systems such as carbon nanotubes [6–8] or nanoscale surface friction and related issues in tribology [9]. On the other hand, rare gas adsorption is being used as a nondestructive and quantitative surface analytical tool in "photoemission of adsorbed xenon" (PAX) [10, 11], for titration analysis of heterogeneous surfaces based on the site specificity of the interaction strength [12–14], and in "buffer-layer-assisted growth" (BLAG) of metal and semiconductor clusters [15, 16].

The adsorption of rare gases on surfaces has been reviewed extensively in the past. The most recent report by Bruch, Diehl, and Venables was published in 2007 [1] and is highly recommended for further reading and details. A compilation and critical discussion of data as of 2001 can be found in the Landolt–Börnstein database for the adsorption of rare gases on metal substrates and semiconductors [17] as well as on graphite, lamellar halides, MgO, and NaCl [18]. Other reviews on selected topics were already mentioned in the previous paragraph or will be indicated in the relevant sections below.

The aim of this chapter is not to review the extensive research and the wealth of data dealing with the adsorption of rare gases on surfaces, nor to provide an update of Ref. [1]. We rather want to focus on the fundamental characteristics of rare gases on surfaces and the lessons that can be learned from rare gas adsorption studies. Many of the concepts described below are valid and useful far beyond their relevance to a specific rare gas/substrate system. The focus will thus lie on describing these concepts and illustrating them with instructive examples. The following "tour d'horizon" gives a brief historical overview of the developments and main

*Surface and Interface Science: Solid-Gas Interfaces I*, First Edition. Edited by Klaus Wandelt.

achievements over the last 50 years of research on rare gas adsorption. Thereafter, the main scientific topics relevant to the field will be discussed.

## 33.2 A Brief Overview of 50 Years of Research

Historically, the investigation of rare gas adsorption on surfaces ranges among the earliest endeavors in surface science, aiming at a microscopic understanding of the adsorption and ordering of atoms and molecules on well-defined substrates. The first report entitled "Déscription d'un appareil à adsorption fonctionnant à basse pression" by Larher appeared in 1960 [19], long before ultrahigh vacuum (UHV) chambers and, nowadays, "standard" surface analytical tools became routinely available. The instrument described by Larher and similar ones developed by Thomy and Duval [20, 21] or computer-controlled refinements [22] were used to determine adsorption isotherms $p(V)$ at constant temperature by measuring the pressure rise $\Delta p$ within a small container upon incremental doses of small gas volumes $\Delta V$ of the adsorbate (see Figure 33.1a). The container or "cell" was filled with a powder of exfoliated layer compounds such as graphite, boron nitride, or lamellar halides, and later microscopic MgO(100) cubes or carbon nanotubes. These samples exhibit a large specific surface area of up to 100 $m^2/g$ but the individual grains are terminated by microscopically extended surfaces of a single crystallographic orientation. In a sense, these early adsorption experiments resemble the breathing of humans, where the lung provides a large respiratory surface of $\sim$100 $m^2$, allowing macroscopic quantities of air to reach its surface and thus promote the oxygenation of the blood. Adsorption isotherms using $N_2$, Kr, or Ar as probing gases have actually become a standard to determine the specific surface area and morphological details of porous materials following the BET method, as already described in 1938 by Brunauer *et al.* [23] (see also Chapter 31 in this Volume).

Coming back to the pioneering work of Larher, Thomy, a Duval and others in the 1960s and early 1970s, it was soon realized that the adsorption isotherms $p(V)$ recorded from homogeneous samples may exhibit sharp steps and kinks at well-defined pressure values $p_n$ corresponding to the sudden condensation of the $n$th monolayer via a first-order phase transition. At higher temperatures and/or coverages, these monolayer steps could become rounded, indicating supercritical behavior ($T > T_c^{(n)}$) or changes in the wetting behavior, that is, transitions from layer-by-layer to 3D island growth. On comparing the isotherms recorded at different temperatures, the position and shape of the steps are seen to vary in a systematic way, which allows the reconstruction of the two-dimensional (2D) phase diagram (see Figure 33.1b). These measurements also provide quantitative thermodynamic numbers, such as the isosteric heat of adsorption $q_{st}$, which can be extracted from an Arrhenius plot of the characteristic pressure $p$ as a function of temperature $T$ at constant adsorbed amount (layer number $n$ or surface coverage $\theta$):

$$q_{st}(\theta) = -k_B \left.\frac{\partial \ln p}{\partial(1/T)}\right|_\theta . \qquad (33.1)$$

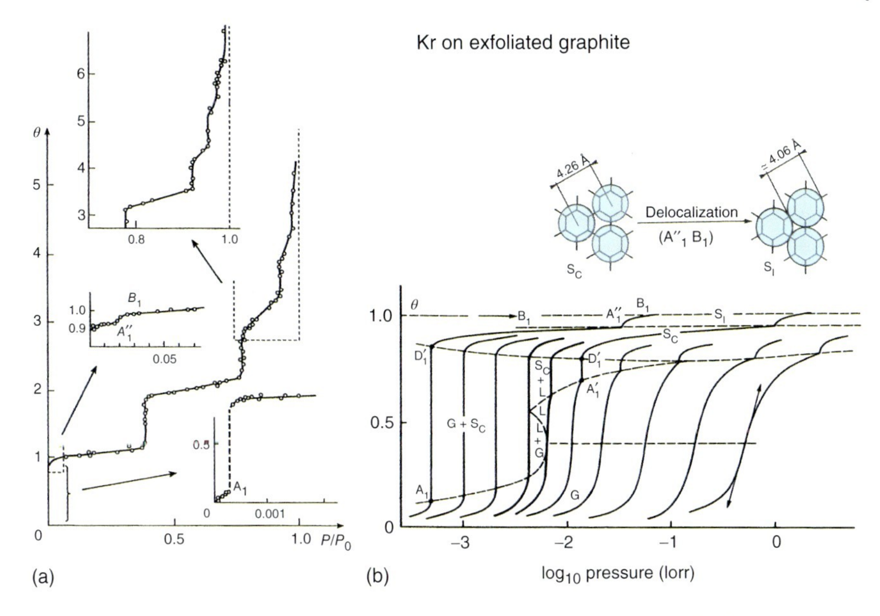


**Figure 33.1** Volumetric analysis of large-area samples. (a) Adsorption isotherm for Kr on exfoliated graphite at 77.3 K (from Refs [24] and [20]). Coverage $\theta$ (in monolayers) versus gas pressure $p$ normalized to the 3D vapor pressure $p_0 = 1.75$ Torr of bulk Kr (at 77.3 K). The "sub-step" $A_1$ corresponds to the phase transition between a 2D gas (G) and a 2D liquid (L); the sub-step $A_1''B_1$ marks the transition from a commensurate solid ($S_C$) to an incommensurate solid ($S_I$), as illustrated in the structure model above panel (b) on the right. (b) 2D phase diagram constructed from adsorption isotherms like those in (a), recorded at various temperatures between 77.3 and 109.5 K (from Ref. [21]). The different phases are denoted by G: 2D gas; L: 2D liquid; $S_C$: 2D commensurate solid; $S_I$: 2D incommensurate solid. Note that, upon counter-clockwise rotation of (b) by 90°, the 2D phase diagram closely resembles the more familiar $p$–$V$ diagrams of 3D systems.

Besides the steps at integer layer completion, "sub-steps" at low coverage and sometimes close to layer completion were found and correctly interpreted in terms of 2D phase transitions. At low coverage, a transition from an initial 2D gas phase occurs to a condensed (2D liquid or solid) phase, whereas at higher coverages transitions between liquid and solid or between different solid phases may occur. The latter were later attributed to different commensurate and incommensurate structures.

By the early 1970s, a large amount of data from adsorption isotherms of the rare gases on graphite [25–27] and on various lamellar halides [28] were available. These results (later reviewed in Refs [20, 21] and [29], respectively) set the stage for further investigations of the adsorption of the rare gases and other small molecules, both by experiments and theory.[1)]

1) The "trilogy" by Thomy and Duval [25–27] is cited more than 800 times in total (as of July 2015). Their review in 1981 appeared as the very first article of the *Surface Science Reports* – quite appropriate in view of the authors' pioneering contributions to surface science.

On the experimental side, large-area samples continued to be used extensively and their microcrystalline quality was improved. The volumetric measurements were supplemented by calorimetry, and the data were subjected to refined thermodynamic and critical point analyses [30, 31]. In addition, neutron and X-ray diffraction experiments were carried out on the same type of samples, yielding complementary information not only on the 2D crystalline structures [32–34] but also on the phonon dynamics, surface mobility, and melting by means of inelastic and quasielastic neutron scattering [32, 35]. At the same time, studies using single crystals of graphite, metals, and semiconductors became more and more popular. Different analytical tools such as Auger electron spectroscopy (AES), low-energy electron diffraction (LEED) [36–38], ellipsometry [39, 40], X-ray diffraction [41, 42], and helium-atom scattering [43, 44] were applied to investigate the adsorption of rare gases on various substrates and to explore the nature of the different 2D phases and phase transitions. He-atom scattering was not only employed for thermodynamic and structural analysis: like neutrons for bulk-like materials, He atoms can also be used to study inelastic processes on surfaces. Since 1984, the phonons of rare gas adlayers were studied systematically [45–48] by means of inelastic He-atom scattering. In addition, quasielastic experiments became feasible [49] and were used to investigate the diffusion and friction of atoms and molecules on surfaces [49, 50]. Recently, the energy resolution was considerably improved ($\Delta E \simeq \mu$eV) by using the $^3$He spin echo technique [51, 52] instead of time-of-flight measurements.

Field emission [53] and photoelectron spectroscopy [54–57] were used early to investigate the origin and systematics of the large work function changes induced by the rare gases on metal surfaces [10] and to investigate their electronic structure. By means of angle-resolved photoelectron spectroscopy (ARPES) it was also possible to explore the electronic band structure of adsorbed rare gas layers [58], Xe atomic chains [59], and the quantum well states of Xe multilayers [60].

With the advent of scanning probe techniques (STM and AFM), local studies came finally within reach. But low temperature (LT) is required to immobilize rare gas atoms on surfaces, and thus an LT-STM had first to be developed. Nevertheless, the first experimental investigations of Xe on Pt(111) and Ni(110) at 4 K were already performed in 1989 and 1990 in Don Eigler's lab at the IBM-Almaden research center [61–63]. A little later, Xe atoms were the first atoms to be ever moved across a surface in a controlled fashion by means of an STM tip, as shown in Figure 33.2.[2)]

It was actually surprising that Xe atoms could be imaged in the STM despite the insulating properties of the rare gases in the bulk. This was the beginning of another fascinating and still ongoing story in the history of rare gas adsorption. In fact, the electronic interaction of the rare gases, especially on metal surfaces, has puzzled experimentalists and theorists a lot. The "visibility" of Xe in the STM experiments was first explained by Lang and coworkers [64] in terms of a small but finite density of states at the Fermi level, which stems from the Xe 6s resonance state and its modification upon the interaction of the Xe atom with the metal substrate. In this context, there came another surprise: the Xe atoms appeared to be preferentially adsorbed on

2) As a side remark only: Two of Don Eigler's dogs at that time were named "Xenon" and "Argon." You may guess what the next one was named after – right: "Neon."

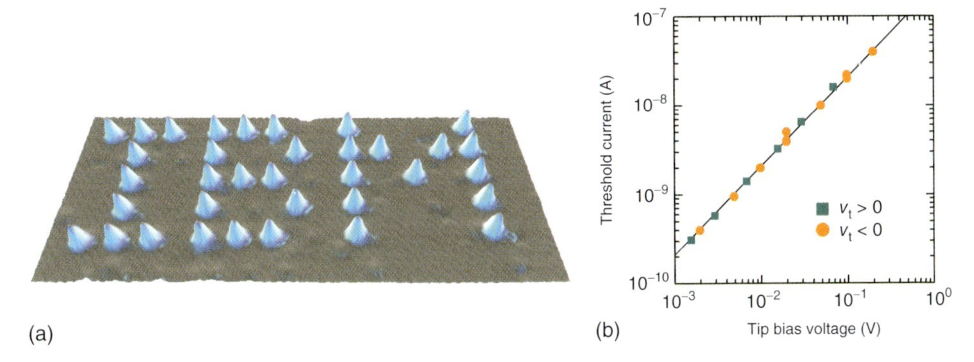


**Figure 33.2** Manipulation of individual Xe atoms on a Ni(110) surface. (a) 35 Xe atoms arranged into a nanoscopic "IBM" logo by sliding the atoms individually along the close-packed $[1\bar{1}0]$ direction with an STM tip [62]. Each letter is only 5 nm high (image originally created by IBM Corporation). (b) Characterization of the "sliding process". Xe atoms can be moved reproducibly along the close-packed rows if the tunneling resistance $V_t/I_t$ is below a threshold value of 4.8 MΩ (corresponding to a tip–sample distance of a few angstroms), independent of the sign of the bias voltage $V_t$ and the tunneling current $I_t$.

the low coordination sites! What the early STM experiments [61, 64, 65] had vaguely suggested became a certainty after systematic dynamical LEED $I - V$ investigations of rare gas adlayers on various metal substrates [66–68]. In fact, the on-top adsorption site on metal surfaces appears to be the rule rather than the exception [69].

This was bad and good news for the theorists working in the field at that time: traditional potentials had to be critically revised, but there was a great challenge to develop new ideas and models. In fact, the preference for low coordination (on-top) sites is opposite to what would be expected on the basis of pairwise sums of two-body potentials.

Until then, models based on interatomic pair potentials known from the gas phase were employed to describe the interaction between adsorbed rare gas atoms on surfaces. Likewise, the interaction with the substrate was often constructed from pairwise sums between rare gas atoms and substrate atoms as well. Although it was realized that simple pairwise potentials might not be adequate to describe the interactions of rare gases on metals, many of these simple models performed surprisingly well and were able to correctly reproduce the adsorption energies and monolayer structures [24, 70] as well as the vibrational properties and phonon dispersion curves [48, 71]. Even much simpler, generic models were able to capture the essence of many of the phenomena discovered in the 1970s and 1980s.

As an example, the famous Frenkel–Kontorova (FK) model [72], also referred to as the Frank–van-der-Merwe (FvdM) model [73, 74], is depicted in Figure 33.3a. The model reduces the interaction with the substrate to a simple sinusoidal corrugation potential with amplitude $V_c$ and period $b$ and mimics the adlayer as an arrangement of particles interconnected by springs with a given force constant $k$ and equilibrium spacing $a_0$. Despite its simplicity, the model is extremely rich and able to explain the existence of commensurate and incommensurate phases,

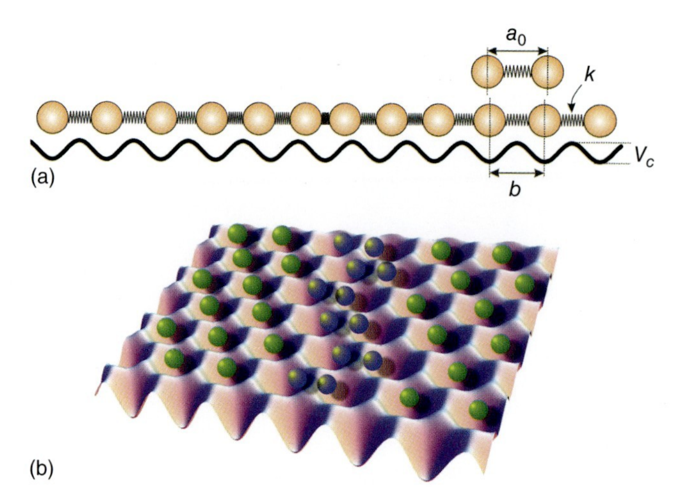


**Figure 33.3** The Frenkel–Kontorova (Franck–van-der-Merwe) model. (a) Schematic of the model, where particles interacting via a simple harmonic potential (spring constant $k$, equilibrium spacing $a_0$) are placed in a sinusoidal corrugation potential with period $b$ and amplitude $V_c$. A misfit dislocation is shown at the center of the image. Here, the chain of particles is locally compressed, yielding a "heavy" domain wall or "kink." (b) Formation of misfit dislocations (kinks and antikinks) in colloidal monolayers of charged polystyrene spheres driven across a periodically corrugated energy surface created by interfering laser beams [84]. Like rare gas adsorption on crystalline surfaces and atomic-scale friction, the phenomena occurring at a micrometer length scale can be described within the framework of the FK model. (Reproduced with permission of Th. Bohlein and C. Bechinger.)

depending on the natural misfit $\delta = (a_0 - b)/b$ and on the relative strength of the interatomic forces and the corrugation energy via a dimensionless parameter $\propto V_c/(kb^2)$. Moreover, the transition from a commensurate (C) to an incommensurate (IC) phase as a function of coverage can be captured by the model. The 1D FK model has been extended to two dimensions and finite temperatures [75] and theories on the nature and critical behavior of C–IC phase transitions [76], the epitaxial rotation of adlayers [77], and 2D melting [78, 79] were built upon it [80]. The FK-based models predict that the C–IC transition proceeds via the formation of misfit dislocations or "domain walls": atoms in unfavorable adsorption positions are restricted to regions with a certain width, lateral separation, and symmetry [73, 76]. Indeed, incommensurate phases with striped (SI) or hexagonal (HI) domain wall arrangements do actually occur in many rare gas adlayers. These long-range modulations were first identified experimentally by the appearance of satellite spots in the LEED, neutron, or X-ray diffraction patterns. Yet, the picture of domain wall structures that people had in mind stemmed from illustrations inspired by the FK model and later from snapshots of large-scale molecular dynamics simulations [81]. To "see" such long-range modulations in the experiment took another decade: in the 1990s, STM images unveiled the full beauty of domain wall phases and their dynamics [3, 82, 83].

It should be emphasized that the charm and wide applicability of the FK model (and similar "generic" models) lies in its reduction to a few most relevant potential parameters. In the case of the FK model, this is the equilibrium distance $a_0$ between atoms of the adsorbate, the effective nearest-neighbour force constant $k$ (i.e., the second derivative of the interatomic potential at the equilibrium distance), as well as the period $b$ and amplitude $V_c$ of the first Fourier component of the lateral variation (corrugation) of the adlayer-substrate potential. Tuning the values of these parameters within a reasonable range leads to fundamental insights (and maybe to an agreement with experimental observations) even though supposedly "simple" questions like "what is the actual adsorption site" cannot be answered at all. In Figure 33.3b we present an image that could well be an artist's view of a rare gas adlayer with domain walls adsorbed on a particular surface, but actually it demonstrates the formation of "kinks" and "antikinks" in colloidal monolayers of charged polystyrene spheres driven across a periodically corrugated potential energy surface, which was created by optical interference of laser beams. Despite the different physical nature and length scale of this artificial "adsorption system," it can also be adequately described within the framework of the FK model [84]. The paper discusses the importance of (heavy and light) domain walls ("kinks" and "antikinks") for the tribolgical properties of the adlayer and also addresses the case of quasicrystalline substrates. Note that both issues, namely tribology and adsorption on quasicrystalline surfaces, are also topical issues in rare gas adsorption research [9, 85]. The comparison with Ref. [84] also illustrates the model character of the rare gas adsorption systems, which were studied many years before.

Coming back to the unexpected on-top adsorption site, there were a few indications from theory that the heavier rare gases might prefer low coordination sites on metal surfaces [86, 87] and there was a single DFT-LDA based cluster calculation by Jorge Müller that actually predicted the on-top site for Xe on a (111)-oriented cluster of 25 Pt atoms [88]. It took another 10 years and considerable effort until density functional schemes were developed to a point where they could systematically reproduce on-top site adsorption of the rare gases on many metal substrates [89, 90]. Nowadays, the inclusion of the nonlocal van der Waals interactions in DFT [91–93] is a promising development with impact in many other topical fields, such as the adsorption and (opto)electronic properties of organic molecules and thin films [94, 95].

## 33.3 The Interaction of Rare Gases on Surfaces

The total potential energy $V$ experienced by an individual rare gas atom adsorbed on a surface can be decomposed into the contribution of the substrate holding potential $V_{AS}$ (i.e., the interaction of the atom with the surface) and the effective lateral interaction $V_{AA}$ shared with all its neighbors. The holding potential is a function of the lateral position $(x, y)$ of the adsorbate and of its height $z$ above the surface. The interaction between the rare gases and metal surfaces is attractive at larger distances (van der Waals interaction) and thus proportional to $-z^{-3}$ (or $-z^{-4}$ at very large distances where retardation effects have to be taken into account). Close to the surface, the interaction becomes strongly repulsive (Pauli repulsion) and can be described by an exponential or algebraic decay. The combination of both the repulsive and the attractive part gives rise to a shallow physisorption well in the holding potential whose depth ranges from a few meV for He up to about 300 meV for Xe. The potential minimum determines the binding distance $z_0(x, y)$ as well as the adsorption site $(x_0, y_0)$. The lateral variation of the well depth $V_{AS}(x, y, z_0)$ reflects the atomic and electronic structure of the substrate lattice; its peak-to-peak amplitude $\Delta V_{AS}$ is called the surface corrugation. Various theoretical models have been proposed to describe the atom–surface interactions for the rare gases [96]. In the simplest approach, the holding potential is obtained as the pairwise sum of atomic pair potentials, for example, of the Lennard–Jones or Morse type. Alternatively, the potential may be expanded into a Fourier series with the first few terms ($V_0$ and $V_{\pm G}$) being taken into account. For instance, in the FK model depicted in Figure 33.3a, only the first Fourier component of this corrugation potential with amplitude $V_c = V_{\pm G}$ is retained. A more sophisticated type of potential has been constructed by Barker and Rettner [87] for the case of Xe/Pt(111) in which noncentral terms are added to the usual pairwise contributions in order to mimic the repulsive interaction with the delocalized metal electrons. In this case, the potential minimum and hence the Xe adsorption site are located on top of the Pt surface atom in contrast to the high coordination (hollow sites) obtained with pair potentials.

The lateral interaction between the adsorbed rare gas atoms is, to a large extent, determined by the dispersion–repulsion usually described by well-known gas-phase potentials [97]. However, significant contributions may also arise from surface-induced dipole repulsions as well as many-body and substrate-mediated interactions, as discussed in detail in Ref. [24]. These contributions tend to weaken the attractive interactions from the expected gas-phase value. For instance, according to the gas-phase Xe–Xe Lennard–Jones pair potential, the lateral interaction in a hexagonal close-packed 2D solid phase would be 72 meV per Xe atom, whereas a value of 43 meV has been reported for Xe/Pt(111) [98]. In a few but notable cases (Ni, W, Pd), the repulsive terms may even outweigh the attractive van der Waals potential, resulting in an overall repulsive (or weakly attractive) lateral interaction [99, 100]. Such repulsive contributions were attributed to surface-induced dipole–dipole interactions and have initiated further investigations on the precise nature of the rare gas–metal interaction. Modifications of the electron density distribution within the metal were already revealed in the early LDA-DFT calculation for Xe on a $Pt_{25}$ cluster [88] and have been interpreted in terms of hybridization of atomic orbitals (see also Chapter 31 in this Volume). This has led to an ongoing (partly semantic) debate concerning the role of "chemical contributions" to the rare gas–metal interaction [89, 90, 101, 102] (which, by the way, were already suggested in the early work of Xe adsorption on W and Ni in the 1970s). Further support for the role of hybridization of the rare gas atomic orbitals with the substrate electronic states has been obtained from the experimental evidence of "anticorrugation" effects in He versus Ne scattering from surfaces [103] and related density functional calculations [104, 105]. Furthermore, intrinsic surface electronic states have been suggested to affect the binding (both $V_{AS}$ and $V_{AA}$) of the rare gases on metals [106]. Meanwhile, van der Waals interactions can be incorporated in DFT codes via nonlocal functionals [91, 92, 102], which promises significant improvements of the theoretical description of the rare gases on surfaces in the near future [93].

## 33.4 Sticking and Accommodation

A large number of studies have dealt with the scattering (adsorption–desorption) of rare gas atoms from surfaces as a function of kinetic energy, angle of incidence, and surface temperature. From a fit to these data, rare gas–surface potentials have been derived, such as the aforementioned Xe–Pt(111) potential [87]. In the present context, we will limit ourselves to those experiments that are relevant to the adsorption properties; that is, we will address the sticking probabilities for rare gas atoms incident with thermal energies, but not the angular and energetic distributions of the backscattered atoms. For an overview of the latter, the reader may consult Refs [107, 108]. At thermal kinetic energies and low enough surface temperature, the heavier rare gases (Xe, Kr, Ar) usually have a sticking probability close to unity, although on some surfaces such as Ru(0001) [109] and Cu(110) [110], values as low as 0.2 have been reported. Such small values are due to the high elastic reflection probability, which is a consequence of the inefficient energy transfer (phonon mismatch) and phonon quantization [109]. The situation is even more dramatic for the lighter rare gases, which exhibit pronounced quantum behavior. For instance, Ne on Ru(0001) has an initial sticking probability below 0.01 at low incident kinetic energy [111]. The sticking coefficient is generally dependent on the adlayer coverage $\theta$. Especially in those cases where the initial sticking coefficient $s_0 = s(\theta \rightarrow 0)$ is significantly smaller than 1, surface areas already covered with rare gas species yield a higher trapping probability for the incident rare gas atoms due to the more efficient energy transfer on these areas [112]. The trapping efficiency of the rare-gas-covered areas strongly depends on the surface temperature as well as on the adlayer morphology. If, for instance, 2D rare gas islands are formed on the surface during the adsorption process, rare gas atoms impinging on these islands may be either trapped on top of an island and become incorporated at its edge or they may desorb from the island before reaching the edge if the surface temperature is too high for the bilayer to be stable and if the islands have grown sufficiently large. In this case, a more complex variation of the sticking coefficient is observed, as depicted in Figure 33.4 [110]. In addition, the presence of surface defects or impurities as well as the "hyperthermal" or "transient" mobility of a rare gas atom upon adsorption [110, 113] may strongly affect the sticking coefficient.

## 33.5 Adsorption Energies

Experimentally, adsorption energies can be inferred from adsorption isotherms in which the surface coverage is recorded as a function of the equilibrium pressure

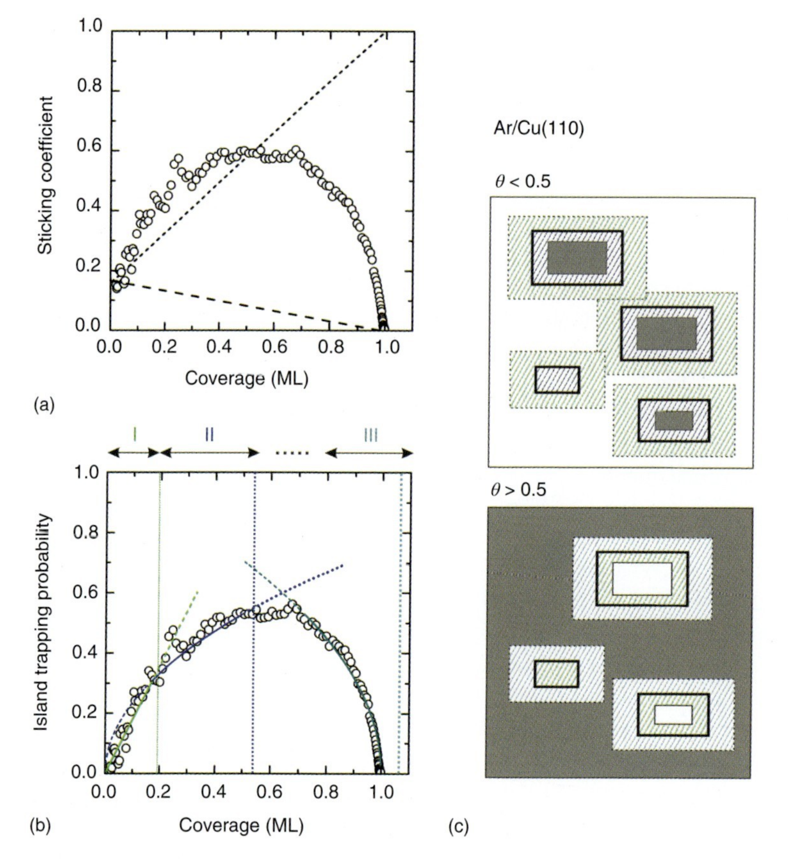


**Figure 33.4** Sticking and island-mediated trapping of Ar on Cu(110) (adapted from Ref. [110]). (a) Sticking coefficient $s(\theta)$ for Ar on Cu(110) obtained by monitoring the Ar surface coverage with specular He-atom scattering while the surface was exposed to Ar gas at constant pressure $p = 8 \times 10^{-8}$ mbar and surface temperature $T = 30$ K. Under these conditions, the initial sticking coefficient is (only) $s_0 = s(\theta = 0) \sim 0.2$. The dotted line shows the "best case" scenario assuming that all the Ar atoms impinging on (2D) Ar islands are trapped with a probability of 1. Note that for low coverages, the experimental data points actually lie *above* this curve! (b) Island trapping probability obtained from (a) by subtracting the substrate contribution $s(\theta) = s_0(1 - \theta)$ (dashed line in (a)). The coverage dependence can be explained with the model depicted in (c): Ar atoms impinging on the bare Cu surface have a sticking probability larger than $s_0$ if they land in the vicinity of an island, since their initial "transient mobility" allows them to reach the nearby island edge where their hyperthermal kinetic energy is efficiently dissipated. These capture zones in front of the island edges (thick black lines in (c)) are illustrated by the hatched green areas in (c). Likewise, atoms impinging on top of Ar islands and close to an island edge (hatched blue zones in (c)) have a chance to reach the edge before desorbing from the bilayer (which is not stable at 30 K), drop down the island edge, and bind to the substrate. The total area of the green and blue capture zones around the island edges scale as $\theta$ for low coverages (regime I), $\sqrt{\theta}$ at intermediate coverages (regime II), and $\sqrt{1-\theta}$ for coverages above the percolation threshold $\theta \gtrsim 0.5$ (regime III). The corresponding curves are shown in (b) and are in good agreement with the experimental observations [110].

and temperature. The surface coverage $\theta$ can be determined in many different ways, such as:

- volumetric analysis on large area samples (see Section 33.2 and Figure 33.1);
- monitoring the variation of the specularly reflected or diffracted intensity of electrons, X-rays, or thermal He atoms scattered from the surface;
- changes of the optical reflectivity or the ellipsometric angles $\Psi$ or $\Delta$;
- the evolution of the photoemission signal; or
- the work function change $\Delta\phi$ upon adsorption.

For a quantitative evaluation, the isosteric heat $q_{\mathrm{st}}$, which is the difference $\Delta H$ between the enthalpies of the adsorbed and the 3D gas phase, can be obtained from the August equation (Eq. 33.1) as an approximation to the Clausius–Clapeyron equation $dp/dT = \Delta H/(\Delta V{\cdot}T)$. It should be noted that in most experiments the 3D gas and the physisorbed adlayer are not in true thermodynamic equilibrium, since the temperature of the 3D gas (usually at room temperature) and the surface temperature (typically between 20 and 100 K) are not the same. The corresponding corrections, however, are small (of the order of $k_B T$) and are usually neglected.

In many experiments, the isosteric heat is only determined for adsorption into the monolayer phase (denoted by $q_1$), which can be assumed to be constant over the coverage range of coexistence between a 2D solid and a 2D gas phase, and before any phase transition or compression of the monolayer solid phase takes place. At low coverage, where only a dilute phase (2D gas) is formed, the isosteric heat may increase or decrease depending on whether the lateral interactions between the adsorbed rare gas atoms are attractive or repulsive. In this context, the isosteric heat of adsorption approaching zero coverage $q_0 = q_{\mathrm{st}}(\theta \to 0)$ is another important quantity. Its experimental determination is more delicate and can be obscured by the fact that the initial adsorption may preferentially occur at surface defects (steps) or impurities due to a stronger binding of the rare gas atoms to such defects. In fact, the increase of $q_{\mathrm{st}}$ at very low coverage observed, for example, on stepped or high index surfaces, allows the determination of the adsorption energy at the defect sites and can even be used as a quantitative tool to evaluate their concentration.

The isosteric heat of adsorption $q_{\mathrm{st}}(\theta)$ is related to the (differential) potential energy $V(\theta)$ of the adsorbate on the surface and the (temperature-dependent) kinetic, rotational, and vibrational degrees of freedom in the adsorbed ($f_{\mathrm{ad}}$) and the 3D gas phase $f_{3D}$, respectively:

$$q_{\mathrm{st}}(\theta) = h_{\mathrm{ad}}(\theta) - h_{\mathrm{3D}} \approx u_{\mathrm{ad}} - h_{\mathrm{3D}} = V(\theta) + \left[\frac{f_{\mathrm{ad}} - (f_{\mathrm{3D}} + 2)}{2}\right] k_B T \qquad (33.2)$$

Assuming an ideal mono-atomic gas ($f_{\mathrm{3D}} = 3$) and high enough surface temperature where the equipartition theorem holds, the kinetic correction becomes $-k_B T/2$ for adsorption into an (ideal) 2D gas phase ($f_{\mathrm{ad}} = 4$ including two degrees of freedom for the translational motion and another two for the perpendicular vibration) and $+k_B T/2$ for adsorption into a 2D solid ($f_{\mathrm{ad}} = 6$ degrees of freedom for the three vibrational modes). Therefore, the kinetic correction for atoms is of the order of $k_B T$, which is often smaller than the experimental error.

The individual contributions to the binding energy $V$ are not all independently accessible to the experiment. Nevertheless, the lateral interaction energy $V_{AA}$ can be inferred from the variation of the isosteric heat with coverage or simply from $q_1 - q_0$. It can also be measured directly by determining the coverage $\theta_c$ at which the transition from the 2D dilute into the 2D solid phase takes place at different surface temperatures. The 2D condensation can be detected, for example, in a helium scattering experiment because of different effective cross sections for diffuse scattering from the 2D dilute and the 2D solid phase, respectively [98, 114]. The 2D heat of condensation $q_{2D}$ is then determined from the 2D analog of the Clausius–Clapeyron and August equations, where the 2D spreading pressure $p_{2D}$ is related to the condensation coverage $\theta_c$ via $p_{2D} = N_0\theta_c k_B T$ (assuming an ideal 2D gas). Taking into account the kinetic contributions, the lateral interaction is obtained as $V_{AA} = q_{2D} - k_B T$.

## 33.6
## Desorption Energies and Desorption Kinetics

Aside from isothermal adsorption experiments, the binding energy of the rare gases is often inferred from thermal desorption measurements. The desorption rate is recorded as a function of time during a linear temperature ramp (temperature-programmed desorption, TPD) or at constant temperature (isothermal desorption). If the adlayer remains in quasi-2D equilibrium during the desorption process, the rate can be expressed by the Polanyi–Wigner equation:

$$r_d(\theta, T) = \nu_d(T)\theta^n N_0 e^{-\frac{E_d}{k_B T}}. \tag{33.3}$$

The exponent $n$ denotes the order of desorption, which is related to the nature and the statistics of the desorption pathway, and $E_d$ is the desorption energy. In accordance with the shape of the holding potential (see Section 33.3), the desorption of the rare gases is not activated and $E_d$ is a direct measure of the binding energy. More precisely, it is the difference between the differential enthalpy of an atom in the adsorbed phase and in the 3D gas phase (generally not in thermodynamic equilibrium with the adsorbed phase). Up to kinetic corrections of the order of $k_B T$ the desorption energy can, therefore, be directly compared to the isosteric heat of adsorption defined above. The prefactor $\nu_d$ in the Polanyi–Wigner equation is related to the entropy term in the relevant free enthalpy balance $\Delta G = \Delta H - T\Delta S$, that is, $\nu_d = e^{-\Delta S/k_B}$. For adsorbed rare gases, the prefactor is usually of the order of $10^{12}$–$10^{13}$ s$^{-1}$ and can be interpreted as an “attempt frequency” for desorption associated with the (perpendicular) atom-surface vibration frequency. As far as the desorption kinetics is concerned, desorption orders $n = 0$ and $n = 1$ have been observed for the rare gases on different surfaces and in different coverage regimes. In the range of phase coexistence (2D equilibrium) between a 2D dilute (gas) and a 2D solid phase, as well as for rare gas multilayers, zero-order desorption is expected [115, 116]; that is, the desorption rate (Eq. 33.3) is independent of coverage and follows a simple exponential increase with temperature. This implies that, in a series of TPD curves recorded for different initial coverages, the leading edges of these curves will coincide. The TPD curves are thus conveniently plotted in an Arrhenius type diagram (see Figure 33.5) from which the desorption energy and the prefactor are readily obtained as the slope and offset of the common leading edge, respectively [117]. First-order desorption kinetics is observed if the rare gas adlayer is in a single

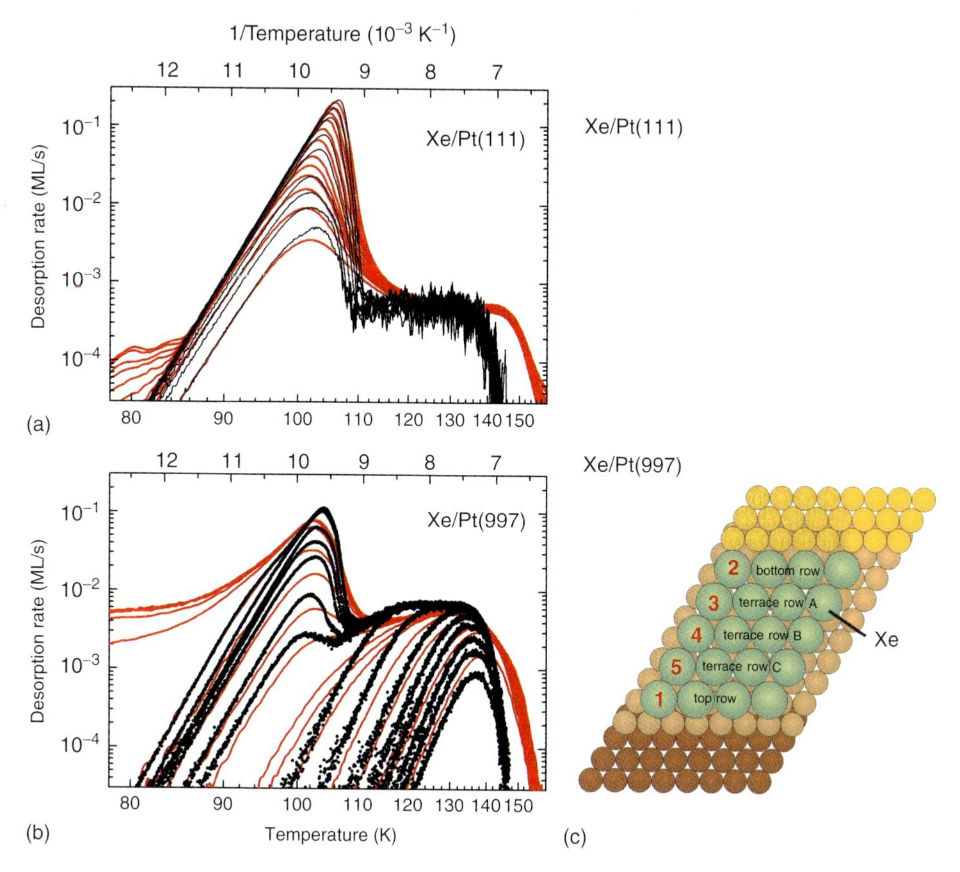


**Figure 33.5** Thermal desorption of Xe from Pt(111) and Pt(997) (adapted from Refs [119] and [13]). (a) TPD spectra in an Arrhenius representation (Menzel–Schlichting plot) of Xe on Pt(111) for different initial coverages ranging between 0.03 and 1 monolayer. Thin solid lines: experimental curves from Ref. [13]; black dots: simulated TPD spectra using the Monte Carlo scheme described in Ref. [120]. (b) Same as (a) but for Xe adsorbed on the vicinal Pt(997) surface. The simulation takes into account the different binding strengths and lateral interactions of Xe atoms adsorbed in the vicinity of a step edge or on a large terrace, respectively. The best match between the simulation and the experimental data is obtained for an adsorption/desorption scenario in which the Xe atoms are most strongly bound to the upper step edges (correlated with an effectively repulsive lateral interaction along the upper step edges), followed by the sites at the lower step edge and the adjacent rows on the lower terrace. (c) Model of the Xe structure on Pt(997) with the hierarchy of the adsorption sites indicated by (row) numbers 1–5 (see Tables I and II (model No. 4) in Ref. [119] for details).

phase. In particular, this situation occurs at low coverage (or high desorption temperature) where a pure 2D gas phase is stable. An extended coverage regime of first-order desorption (even up to monolayer completion) has been observed in those cases where effectively repulsive lateral interactions between the adsorbed rare gas atoms prevent the condensation of the dilute phase into a 2D solid phase, such as for Xe on W, Ni, and Pd surfaces [99, 100]. In most other cases, the regime of first-order desorption and, hence, the stability of a 2D gas phase are limited to a small fraction of a monolayer [118]. Finally, the presence of surface defects (steps or impurities) may significantly alter the desorption kinetics, affecting the desorption order as well as the desorption energy and prefactor, as shown in Figure 33.5 for Xe/Pt(111) versus Xe/Pt(997) [13, 119] (see also Chapter 31 in this Volume).

## 33.7
## Structure and Phase Diagram

Rare gas monolayers are often considered as model systems in the study of 2D phases and phase transitions. As in bulk matter, the analogs of a gas, fluid, and solid phase also exist in purely 2D "floating" adlayers, although the translational ordering of a floating solid is not of long-range order – with important consequences on the nature of the 2D solid–fluid (melting) transition [79, 80]. The adsorption on corrugated crystalline substrates leads to important modifications, such as the enhanced stability of a 2D lattice gas due to the discrete spacing of the adatoms on preferred lattice sites located at the potential minima of the holding potential [121]. In addition, a 2D liquid and possibly several ordered solid phases may exist besides a floating 2D incommensurate phase. These phases can be commensurate or high-order commensurate, that is, in full or partial registry with the 2D substrate lattice. The type of solid phase that is formed depends on the relative size of the lateral interaction $V_{AA}$ between the adsorbed atoms and the corrugation $V_c = \Delta V_{AS}$ of the holding potential as well as on the natural misfit between the 2D floating adlayer and the substrate lattice. Perfectly commensurate (C) monolayer solids are expected for strongly corrugated substrates, weak lateral interactions ($V_c/V_{AA} \geq 1$), and a small lattice misfit $m$. On the other hand, incommensurate (I) structures are favored for $V_c/V_{AA} \ll 1$ or large lattice mismatch. For intermediate cases, high-order commensurate (HOC) phases, characterized by a commensurate superstructure whose unit cell contains several atoms on inequivalent lattice sites, can be obtained. Experimentally, all these cases have been observed [17, 18]. On graphite, various incommensurate phases (hexagonal, striped, and rotated) and a $(\sqrt{3} \times \sqrt{3})$R30°commensurate structure are found for Kr and Xe (see Figures 33.1 and 33.6). For Ne, an HOC $(\sqrt{7} \times \sqrt{7})$R19.1° phase can be stabilized at low temperatures, while Ar on graphite forms a hexagonal incommensurate 2D solid only. Further details and numbers on the structures and phase diagrams of the rare gases on graphite, MgO, NaCl, and lamellar halides can be found in Ref. [18].

On a few close-packed metal substrates, namely Pt(111), Pd(111), and Cu(111), Xe may form perfectly commensurate phases with a $(\sqrt{3} \times \sqrt{3})$R30°structure, and

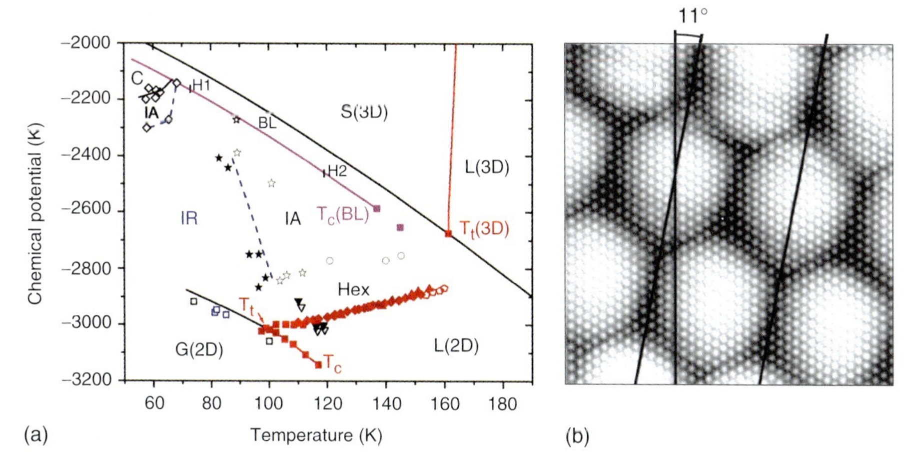


**Figure 33.6** Structures and phase diagram for Xe on graphite (from Refs [1] and [3]). (a) Chemical potential versus temperature ($\mu - T$) phase diagram, spanning the 2D monolayer, bilayer, and 3D bulk phases as well as the monolayer melting transitions. The different symbols are the supporting experimental data from many groups, assembled into a coherent picture by Bruch, Diehl, and Venables in Ref. [1]. Within the monolayer range, the phase boundaries between the incommensurate aligned (IA) and rotated (IR) monolayers (dashed lines) and between the IA and commensurate (C) solid phase (solid line in the upper left corner) can be distinguished. (b) STM image (16×16 nm$^2$) of Xe on graphite, showing atomically resolved hexagonal domains separated by domain walls. The domain walls form a honeycomb-like pattern with a mesh size of about 5 nm. The walls are rotated by ~11°with respect to the close-packed Xe rows. This is consistent with a Novaco–McTague rotation of the Xe lattice by ~0.5° with respect to the high-symmetry direction of the graphite substrate [3].

for Xe on W(100) a c(2×2) structure is observed. On most other surfaces with hexagonal or square lattice symmetry, such as Ag(111), Al(111), Ni(100), Pd(100), etc., the rare gases form purely incommensurate hexagonal close-packed structures with an interatomic spacing similar to that in bulk Xe (4.4 Å). Nevertheless, these adlayers are usually orientationally aligned along a high symmetry direction of the substrate. This has led to the misconception in the early literature that "the substrate may influence the orientation but not the structure and internal spacing of physisorbed adlayers" [122]. The fcc(110) faces of various metals can stabilize uniaxially commensurate (UC) phases due to the high corrugation along the [001] direction; that is, the rare gas atoms may be localized within the potential troughs parallel to the [$1\bar{1}0$] direction. The rare gas atoms are then arranged in a quasi-hexagonal c($\alpha$×2) packing with a more or less significant distortion between the interatomic spacings along and across the troughs, respectively. In some cases, such as Xe on Cu(110), $\alpha$ can be a rational number giving rise to a sequence of ($n \times 2$) HOC phases with $n \gg 2$ [70]. In fact, this type of sequence of 1D HOC phases was also predicted theoretically [123] on the basis of the FK model introduced in Section 33.2 and is termed a "devil's staircase." An example of a devil's staircase in two dimensions was found for Ar/Pt(111) [124, 125].

Another frequent experimental observation is the occurrence of azimuthally broadened "sickle-like" diffraction spots or, in some cases, a finite azimuthal rotation angle. Such a "Novaco–McTague" rotation out of the high-symmetry direction can occur as a result of strain energy minimization. The rotation seems to be quite susceptible to the presence of structural imperfections (steps) or impurities (such as K in the case of Ag(111) [126]). As another example, Kr and Ar adlayers on Pt(111) can be prepared in both R0°and R30°orientations depending on whether the substrate steps involved in the 2D island nucleation and growth are clean or precovered with H or CO [98, 127].

Transitions between the various 2D phases can occur as a function of surface temperature and coverage, giving rise to rather complex structural phase diagrams. For instance, the variation of the lattice misfit with temperature (thermal expansion) or coverage (steric compression) can induce a change of the equilibrium structure among the various I, C, UC, and HOC phases. Examples are shown in Figures 33.6 and 33.7 for Xe on graphite and Pt(111), respectively. Both phase diagrams are very rich and include a $(\sqrt{3}\times\sqrt{3})$R30° commensurate (C), a hexagonal aligned, and a hexagonal rotated incommensurate phase (IA, IR in

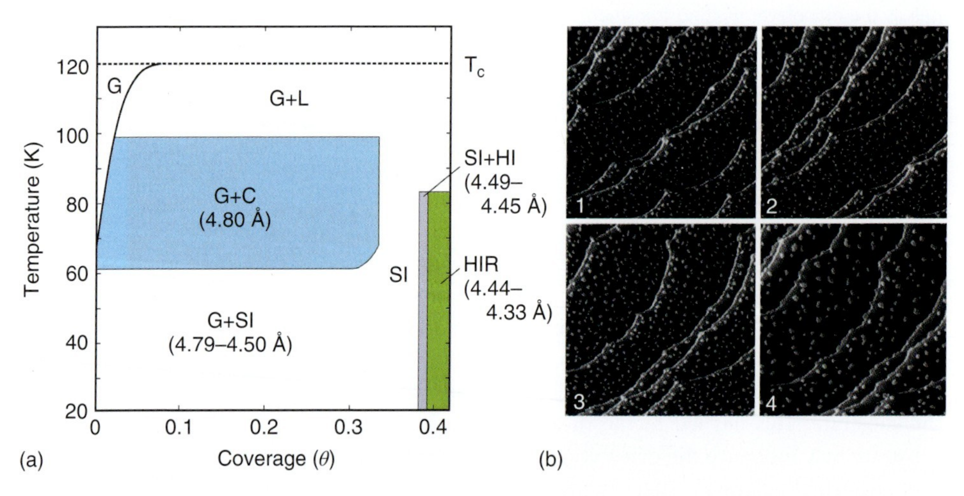


**Figure 33.7** Structures and schematic phase diagram for Xe on Pt(111). (a) Temperature versus coverage ($T$-$\theta$) phase diagram. The $(\sqrt{3}\times\sqrt{3})$R30° commensurate phase (C) has a saturation coverage of $\theta = 1/3$. The other phases are 2D gas (G), 2D liquid (L), uniaxial incommensurate solid (striped phase, SI), hexagonal incommensurate aligned solid (HI), and hexagonal incommensurate rotated solid (HIR). (b) Series of STM images showing the structural evolution with increasing temperature of a sub-monolayer of Xe adsorbed at 17 K onto Pt(111). All images were recorded at 17 K using the same parameters: scan width of 280 nm (tip bias −0.4 V, tunneling current 1 nA). (1) structure after adsorption at 17 K. Before recording (2–4), the sample temperature was raised shortly to 19, 26, and 29 K, respectively. Changes in the morphology of the Xe layer start to become significant at about 27 K [82]. (Adapted from Refs [82, 130].)

Figure 33.6 and HI, HIR in Figure 33.7, respectively). In addition, Xe/Pt(111) also features a uniaxially compressed (striped incommensurate, SI) phase. In this case, the compression of the $(\sqrt{3}\times\sqrt{3})$R30° commensurate phase (C) proceeds in the sequence C → SI → HI → HIR, where first a striped domain wall phase is formed (via a second-order phase transition), which then evolves via a hexagonal phase into a Novaco–McTague R(30 ± 3.3)° rotated phase [128, 129]. This phase diagram as well as the sequence and characteristics of the observed 2D phase transitions are in full agreement with theoretical predictions [76].

## 33.8
## Adsorption Site

In the early literature, the preferential adsorption sites for rare gas atoms on surfaces were generally assumed to be those with highest coordination (i.e., hollow sites on terraces and bottom sites at the substrate steps). More recently, theory [86, 88] and experiment [61, 65, 66] have demonstrated the possibility of an on-top adsorption site. Quantitative results on the adsorption geometry have been obtained from LEED I–V curves for a large number of rare gas–substrate combinations. The best agreement in almost all these cases is found for an on-top adsorption geometry [69]. In particular, Xe adsorbed on the fcc(111) surfaces of Pt, Pd, and Cu [69] and on Ru(0001) [131] were shown to be most strongly bound on top of the substrate atoms and at the upper edge of the substrate steps [65]. In contrast to the results on metal surfaces, Xe atoms in the $(\sqrt{3}\times\sqrt{3})$R30° commensurate phase on graphite were observed to reside in the hollow sites [132].

The preference for low coordination sites on metal surfaces has been ascribed to the peculiar interaction between the rare gas atoms and the metal substrate. This interaction can obviously not be described in terms of a simple sum of pair potentials that would predict high coordination (hollow) sites (see Section 33.3). Both the on top adsorption site and the increased repulsion between the adsorbed rare gas atoms have been correlated with the strength of the holding potential $V_{AS}$ [133]. Strongly binding substrates such as Pd or adsorption sites with enhanced binding energy (such as the step edges on the Pt(111) surface) seem to promote on-top adsorption and repulsive interactions between neighboring adatoms. More recent DFT studies [89, 90, 105] have come to the conclusion that the nature of the interaction of the rare gases is determined by the subtle interplay between electronic polarization and the site-dependent Pauli repulsion. In addition, the contribution of intrinsic surface states to the binding of the rare gases on metals has been proposed to explain the stronger binding of Xe on Pd(111) than on Pt(111), the different lateral interactions of Xe on the two surfaces, and the preferential decoration of the upper step edge on the Pt(111) surface [106]. It should, however, be noted that only recently the nonlocal van der Waals interactions were seamlessly incorporated into DFT schemes [91–93, 102]. Some of the earlier interpretations may thus be revised in the future.

### 33.9
### Electronic Structure

Early work has shown that, upon rare gas adsorption, the surface potential (work function) of metal substrates is lowered by as much as 1 eV [17].[3] These large values can be ascribed to the large polarizability $\alpha$ of the rare gas atoms (in particular Xe) and possible "chemical contributions" to the surface bond. The work function change $\Delta\phi$ has also been used to measure the surface coverage, for example, in isothermal adsorption experiments (see Section 33.5). An analytic relation, accounting for dipole–dipole interactions and mutual depolarization effects as a function of the 2D adsorbate density $n_{\text{ad}} = n_s\theta$, is provided by the Topping formula (in cgs units):

$$\Delta\phi = 4\pi\mu_0 n_{ad} \approx \frac{4\pi\mu_0 n_{\text{ad}}}{1 + 9\alpha n_{\text{ad}}^{3/2}}, \tag{33.4}$$

where $\mu_0$ is the initial adsorption-induced atomic dipole moment. According to Eq. 33.4 , $\mu_0$ can be deduced experimentally from the initial linear slope of the work function change $\partial(\Delta\phi)/\partial\theta$ for low enough adsorbate coverages $\theta$. For rare gas adsorption, the Topping formula has been checked to hold in several (but not all) cases. The initial Xe dipole moment on the low-index Pd (fcc) planes has been found to increase in the order $\mu_0(110) < \mu_0(100) < \mu_0(111)$ and, hence, to exhibit the reverse face specificity as the initial desorption energies [99]. This observation was explained in the framework of the "s-resonance model" [135].

The electronic structure of the adsorbed rare gases is determined by the occupied $n\text{p}_{1/2}$ and $n\text{p}_{3/2}$ atomic orbitals ($n = 3, 4, 5$ for Ar, Kr, and Xe, respectively) and by the unoccupied $(n + 1)$s orbitals which are usually located far above the Fermi level. Because of the coupling with the substrate and the Xe–Xe lateral interactions, the original atomic states are broadened and the $n\text{p}_{3/2}$ photoemission peak is split into two peaks due to the lifting of the degeneracy of the $m_j = \pm 1/2$ and $\pm 3/2$ states.

It has been argued that the binding energy for a given bound electronic state of an adsorbed rare gas atom is "pinned" to the vacuum level [136, 137]. In the case of Xe adsorption, this would imply that $E^V(5\text{p}_{1/2}) = E^F(5\text{p}_{1/2}) + \phi$, where the superscripts $V$ and $F$ denote the vacuum and the Fermi reference levels, respectively (see also Chapter 5, Volume 2). However, the binding energies derived from photoemission are smaller (by about 1.1 eV) than those of the corresponding atomic states, which has been attributed to the final state relaxation (electron hole screening). For a large number of substrates (mainly transition metals), the relaxation effect was found to be essentially substrate independent [137], whereas on the sp-metal Pb(111), for instance, layer-dependent shifts of the photoemission peaks attributed to changes in the final state relaxation have been reported [138]. It has further been argued that the binding energies derived from photoemission depend on the local

3) Table 9 in Ref. [17]) provides a list of the work function changes of the rare gases available at that time. Since then, new Kelvin-probe measurements have been published for monolayers of Ar, Kr, and Xe on the (111) face of the noble metals Cu, Ag, and Au [134].

surface potential of the immediate adsorption site $i = (x_i, y_i)$ [10]. Hence, the electron energy difference between two inequivalent sites $i$ and $j$ reflects the difference in the local work function as for Xe: $\Delta E^F(5p_{1/2})(i,j) = \Delta\phi(i,j)$. This "local work function concept" together with the site specificity of the rare gas adsorption is the basis of "photoemission spectroscopy of adsorbed Xe" (PAX), which has become a powerful tool for the characterization of heterogeneous surfaces, such as stepped surfaces or surface alloys [10, 11].

Upon 2D island formation, the rare-gas-derived $n$p states develop a 2D band structure [17] with typical dispersion amplitudes (band widths) of the order of 0.1–0.5 eV, depending on the strength of the lateral interaction [139]. At step edges or on anisotropic surfaces, such as Pt(110)–(1×2), the rare gas atoms aggregate into chain-like structures exhibiting a 1D band structure [59, 140, 141], as depicted in Figure 33.8. Finally, rare gas multilayers have been shown to exhibit quantized electronic states depending on the layer thickness [142, 143]. Their number and energies can be explained within a simple potential well model [60].

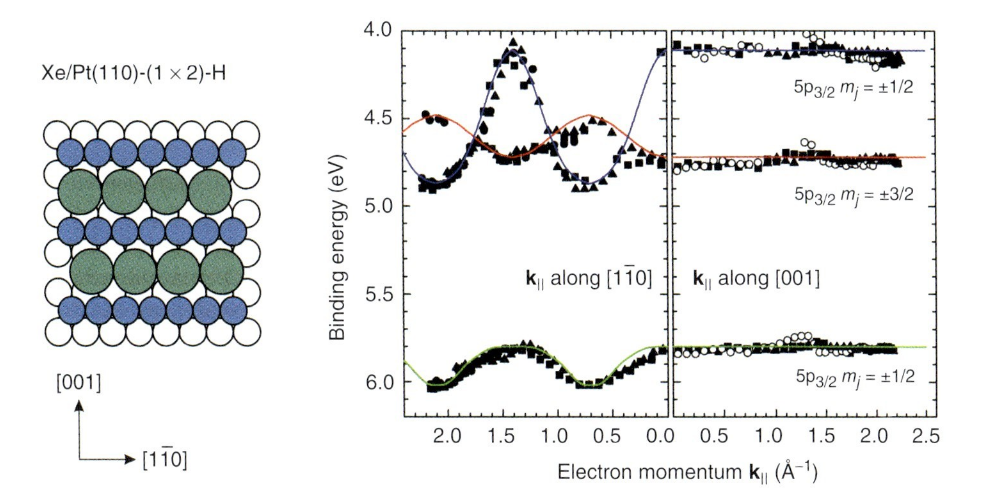


**Figure 33.8** Electronic structure of Xe atomic chains on Pt(110)-(1×2)-H (adapted from Ref. [141]). (a) Ball model of the quasi-1D Xe atomic chains formed upon adsorption of Xe on the hydrogen-modified Pt(110)-(1×2)-H surface. (b) Band structure for dense Xe rows adsorbed on a Pt(110)-(1×2)-H surface along the [1$\bar{1}$0] and [001] directions, determined from ARUPS. Different symbols correspond to datasets with different photon energies (26 and 30 eV) and different angles of light incidence. The result of a fully relativistic Korringa–Kohn–Rostocker calculation for a free-standing Xe row is shown with solid lines. Large dispersion amplitudes are found along the [1$\bar{1}$0] direction (i.e., parallel to the atomic chains) as a result of the electronic coupling between neighboring Xe atoms along each chain. In contrast, no dispersion is discernible along the [001] direction, which indicates the absence of inter-row coupling of the valence states and demonstrates the 1D character of the electronic structure. The amplitude of the dispersion along the [1$\bar{1}$0] direction decreases exponentially with the average Xe–Xe spacing along the chains (i.e., decreasing Xe coverage) [141].

## 33.10 Diffusion Kinetics

Surface diffusion data for the rare gases on surfaces are scarce, and microscopic studies (using FEM and STM) on the diffusion processes, if they exist, are not always quantitative. The experimental determination of the surface mobility (see also Chapter 8.1, Volume 2) is complicated by several effects: (i) Rare gas atoms may exhibit an initial hyperthermal mobility related to the binding energy gained by the atoms upon adsorption and the inefficient energy dissipation via lateral adsorbate–substrate phonon coupling [61, 113]. This "transient mobility" must be separated from the "true" thermal mobility of the thermalized rare gas atoms. (ii) Surface defects and impurities act as efficient traps or nucleation sites for the diffusing rare gas atoms. They have to be included for a correct determination of the diffusion constants from macroscopic measurements or from an analysis of the island density or size distribution in adsorption experiments. (iii) The diffusion process may be complicated by the discrete nature of the vibrational levels within the shallow diffusion potential, especially for the lighter rare gases.

Laser-induced desorption experiments for Xe and Kr on Pt(111), which probe the diffusion on a length scale of a few hundred micrometers, reveal a significant coverage dependence (adatom diffusion vs. chemical diffusion). As a result, the diffusion constants may vary by an order of magnitude in the range between 5% and 25% of the monolayer saturation coverage [144]. The diffusion barrier extracted from these experiments ($E_{dif} = 52 \pm 9$ meV for Xe/Pt(111) in the intermediate coverage range) was assigned to the detachment of Xe atoms from island edges and, therefore, involves the corrugation energy of the holding potential as well as the lateral interaction between the rare gas atoms. This value is consistent with thermodynamic data of adsorption [130]. A similar value ($48 \pm 12$ meV) was found for Xe diffusion on W(110) using the field-emission current fluctuation method [145].

STM data for Xe/Pt(111) at 4 K [61] show a preferential decoration of the step edges during adsorption and, at the same time, very small thermal mobility of isolated thermalized Xe adatoms. This observation has been interpreted in terms of a large transient mobility and a sizeable surface corrugation amplitude. From STM experiments at variable temperature (see Figure 33.7b), the corrugation amplitude (diffusion barrier of a Xe adatom) and the activation energy for detachment from a 2D island edge can be estimated to be of the order of 31 and 64 meV, respectively, at low temperatures [82]. In contrast, the first quasi-elastic He scattering experiments performed on the Xe 2D gas phase on Pt(111) at surface temperatures around 105 K [49] seem to indicate an almost vanishing barrier for adatom diffusion. It should be noted, however, that the effective or "dynamic" barrier probed at such high thermal energies is smaller than the "static" diffusion barrier or surface corrugation [146, 147], which was estimated to be $< 9.6$ meV.

## 33.11
## Lattice Dynamics

The lattice dynamics (phonon spectrum) of a physisorbed rare gas adlayer is characterized by three vibrational modes: one perpendicular to the surface with no or very weak dispersion (Einstein mode), and two other modes polarized within the surface plane. The Einstein mode is determined by the second derivative of the surface holding potential (perpendicular force constant). In the harmonic approximation, the only lateral coupling is due to the stress within the rare gas monolayer (i.e., the first derivative of $V_{AA}$ with respect to the lateral separation), which may be nonzero but is usually quite small. This explains the weak energy variation of this mode with wavevector parallel to the surface. As a consequence, sharp multiple phonon losses can be observed for this mode [24]. Experimentally, the surface phonon spectrum of rare gas monolayers has been investigated on various substrates using high-resolution inelastic He atom scattering (see Figure 33.9). Although the cross section is largest for the perpendicular vibration mode, longitudinally and transversely polarized modes have also been detected, namely on the Cu(100), (110), and (111) surfaces and on Pt(111) [48, 71] (see Figure 33.9b). The vibration energy for the perpendicular mode ranges between 2.5 meV for the heaviest species (Xe) and 5 meV for Ar and is expected to increase with the strength of the holding potential. The vibration energies can be used to fit the holding potential $V_{AS}$ (perpendicular mode), the interatomic potential $V_{AA}$ (longitudinal mode), and the surface stress [148].

As illustrated in Figure 33.9a, the phonon modes may couple to the substrate phonons where they overlap with the substrate bulk phonons and where the perpendicular adlayer mode crosses the edges of the projected bulk phonon bands or the surface Rayleigh mode of the substrate [150]. A linewidth broadening in the overlap region could first be detected for the rare gases on Ag(111) [151] and later on Pt(111). In the latter case, the hybridization between the substrate Rayleigh wave and the rare gas Einstein mode [152, 153] as well as the coupling to the longitudinal bulk band edge (Van Hove anomaly) [149] has also been evidenced (see Figure 33.9a).

With increasing thickness, the phonon spectrum evolves toward the one expected for a semi-infinite rare gas crystal in a layer-by-layer manner [153, 154]. The phonon signal can thus be used to monitor the adsorption and growth of rare gas multilayers.

## 33.12
## Conclusions

The research on rare gas adsorption has come of age. But to translate a German saying: “50 years and still going strong” describes the present state and future. The rare gases can truly be considered model systems in surface science and beyond. They have helped to establish the fundamental concepts of the field, and I have no doubt that this is what the rare gases will also stand for in the future.

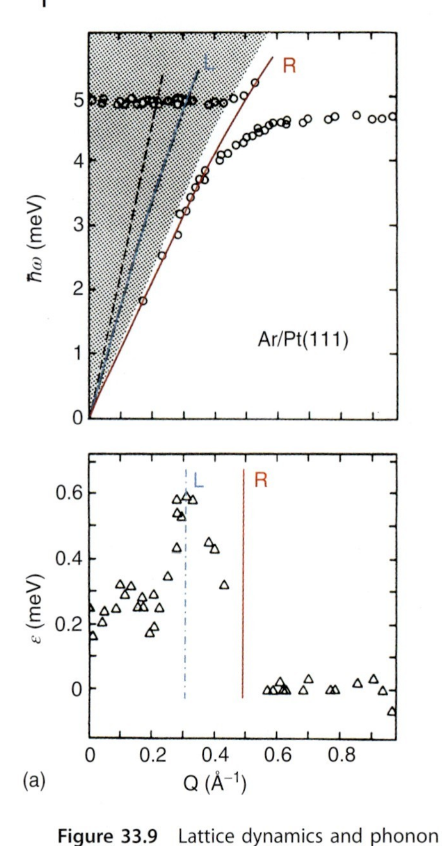


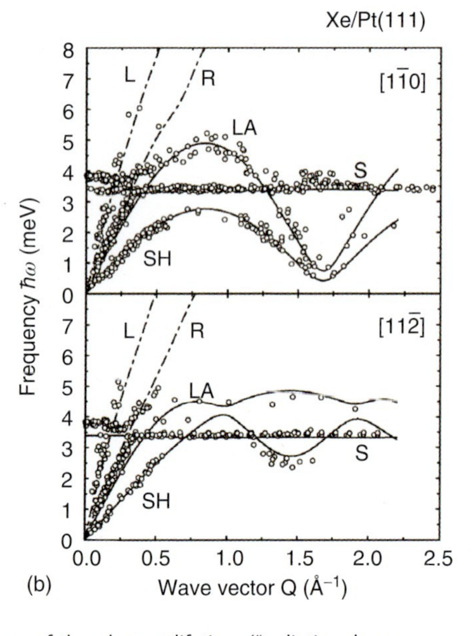


**Figure 33.9** Lattice dynamics and phonon dispersion curves (from Refs [1, 149]). (a) Experimental dispersion curves from a hexagonal aligned monolayer of Ar on Pt(111) and reduced linewidths $\epsilon$ of the adlayer phonon as a function of the parallel wavevector $Q$ along the [11$\bar{2}$] substrate direction ($\overline{\Gamma M}$) obtained with inelastic He scattering (beam energy: 18.3 meV) at a surface temperature of 25 K. The shaded area depicts the projected Pt bulk phonon bands. Also shown are the substrate Rayleigh mode (red solid line, R) and the longitudinal and transverse bulk band edges of the Pt(111) substrate (dash-dotted blue line L and dashed black line, respectively). The data prove the dynamic coupling of the Xe adlayer with the underlying metal substrate leading to (i) a hybridization of the perpendicular Ar vibration mode and the Pt(111) Raileigh mode (avoided crossing), and (ii) a significant increase of the phonon linewidth where the Ar mode overlaps with the projected bulk phonon band, that is, a reduction of the phonon lifetime ("radiative damping"). (iii) This effect is even enhanced in the wavevector region where the adlayer mode crosses the longitudinal bulk band edge of the substrate due to the high phonon density of states (van Hove anomaly). (b) Phonon dispersion curves for a monolayer of Xe on Pt(111) along the substrate [1$\bar{1}$0] and [11$\bar{2}$] directions, respectively. The Xe adlayer is in the HIR phase, that is, rotated by $\sim 2.6°$with respect to the substrate lattice. Besides the dynamical coupling of the perpendicular mode (S) with the substrate Rayleigh mode (R) and the longitudinal bulk band edge (L), the longitudinal acoustic (LA) and the shear horizontal mode (SH) can be distinguished. The experimental data (open circles) were obtained with inelastic He scattering (beam energies from 6 to 17 meV) at a surface temperature of 50 K. The solid lines are the result of calculations with realistic Xe pair potentials and the McLachlan dispersion energy [48].

### Acknowledgments

I would like to take the opportunity to thank all those remarkable scientists and personalities with whom I had the pleasure to collaborate during my 20 years of research in the field of rare gas adsorption on surfaces.

In chronological order: the group at the former IGV, Forschungszentrum Jülich (George Comsa, Klaus Kern and Rudolf David); the people at the former CRMC2 in Marseille (Michel Bienfait, Jean Suzanne and Jean-Marc Gay); Georges Armand (Saclay), Oscar Vilches (University of Washington, Seattle); Don Eigler and John Barker at the IBM-Almaden Research Center; Burl Hall and Doug Mills (UC Irvine); and the theory group at the Université de Franche-Comté in Besançon (Christophe Ramseyer and Claude Girardet). I would also like to acknowledge the many fruitful discussions with experts in the field throughout the world, especially Lou Bruch, Renée Diehl, Bene Poelsema, Dick Manson, and Tom Seyller.

Most of all, I would like to acknowledge the close scientific cooperation and the deep personal friendship with Michel Bienfait, who sadly passed away on January 21, 2014. I would like to dedicate this chapter to Michel, who has been a pioneer and driving force in the field, a mentor and guide at the beginning of my career and a great scientific partner and friend during the last 15 years.

### References

1. Bruch, L.W., Diehl, R.D., and Venables, J.A. (2007) Progress in the measurement and modeling of physisorbed layers. *Rev. Mod. Phys.*, **79** (4), 1381–1454.
2. Lazić, P., Brako, R., and Gumhalter, B. (2007) Structure and dynamics of Xe monolayers adsorbed on Cu(111) and Pt(111) surfaces studied in the density functional approach. *J. Phys. Condens. Matter*, **19** (30), 305 004.
3. Grimm, B., Hövel, H., Pollmann, M., and Reihl, B. (1999) Physisorbed rare-gas monolayers: evidence for domain-wall tilting. *Phys. Rev. Lett.*, **83** (5), 991–994.
4. Brunet, F., Schaub, R., Fédrigo, S., Monot, R., Buttet, J., and Harbich, W. (2002) Rare gases physisorbed on Pt(111): a low-temperature STM investigation. *Surf. Sci.*, **512** (3), 201–220.
5. Andreev, T., Barke, I., and Hövel, H. (2004) Adsorbed rare-gas layers on Au(111): shift of the Shockley surface state studied with ultraviolet photoelectron spectroscopy and scanning tunneling spectroscopy. *Phys. Rev. B*, **70** (20), 205 426.
6. Bienfait, M., Zeppenfeld, P., Dupont-Pavlovsky, N., Muris, M., Johnson, M., Wilson, T., DePies, M., and Vilches, O. (2004) Thermodynamics and structure of hydrogen, methane, argon, oxygen, and carbon dioxide adsorbed on single-wall carbon nanotube bundles. *Phys. Rev. B*, **70** (3), 035 410.
7. Calbi, M.M., Cole, M.W., Gatica, S.M., Bojan, M.J., and Johnson, J.K. (2008) Adsorbed gases in bundles of carbon nanotubes: theory and simulation, in *Adsorption by Carbons*, Chapter 9 (eds E.J. Bottani and J.M.D. Tascon), Elsevier, Amsterdam, pp. 187–210.
8. Migone, A.D. (2008) Adsorption on carbon nanotubes: experimental results, in *Adsorption by Carbons*, Chapter 16 (eds E.J. Bottani and J.M.D. Tascon), Elsevier, Amsterdam.
9. Krim, J. (2012) Friction and energy dissipation mechanisms in adsorbed molecules and molecularly thin films. *Adv. Phys.*, **61** (3), 155–323.

10. Wandelt, K. (1997) The local work function: concept and implications. *Appl. Surf. Sci.*, **111**, 1–10.
11. Baker, L., Holsclaw, B., Baber, A.E., Tierney, H.L., Sykes, E.C.H., and Gellman, A.J. (2010) Adsorption site distributions on Cu(111), Cu(221), and Cu(643) as determined by Xe adsorption. *J. Phys. Chem. C*, **114** (43), 18 566–18 575.
12. Stichler, M., Weinelt, M., Zebisch, P., and Steinrück, H.P. (1996) Argon desorption as a tool to study the growth of molecular layers. *Surf. Sci.*, **348**, 370–378.
13. Widdra, W., Trischberger, P., Frieß, W., Menzel, D., Payne, S., and Kreuzer, H. (1998) Rare-gas thermal desorption from flat and stepped platinum surfaces: lateral interactions and the influence of dimensionality. *Phys. Rev. B*, **57** (7), 4111–4126.
14. Bottani, E.J. and Tascón, J.M.D. (eds) (2008) *Adsorption by Carbons*, Elsevier, Amsterdam.
15. Huang, L., Chey, S., and Weaver, J.H. (1998) Buffer-layer-assisted growth of nanocrystals: Ag-Xe-Si(111). *Phys. Rev. Lett.*, **80** (18), 4095–4098.
16. Antonov, V.N., Swaminathan, P., Soares, J.A.N.T., Palmer, J.S., and Weaver, J.H. (2006) Photoluminescence of CdSe quantum dots and rods from buffer-layer-assisted growth. *Appl. Phys. Lett.*, **88** (12), 121 906.
17. Zeppenfeld, P. (2001) Noble gases on metals and semiconductors, in *Landolt-Börnstein New Series III/42A1*, Landolt-Börnstein - Group III Condensed Matter, vol. 42A1, Chapter 3.1.1 (ed. A.P. Bonzel), Springer-Verlag, Heidelberg, pp. 67–115.
18. Bienfait, M. (2001) Noble gases on graphite, lamellar halides, MgO and NaCl, in *Landolt-Börnstein New Series III/42A1*, Landolt-Börnstein - Group III Condensed Matter, vol. 42A1, Chapter 3.1.2 (ed. A.P. Bonzel), Springer-Verlag, Heidelberg, pp. 117–128.
19. Larher, Y. (1960) Déscription d'un appareil à adsorption fonctionnant à basse pression. *J. Chim. Phys. Phys.-Chim. Biol.*, **57** (11-2), 1107–1108.
20. Thomy, A., Duval, X., and Regnier, J. (1981) Two-dimensional phase transitions as displayed by adsorption isotherms on graphite and other lamellar solids. *Surf. Sci. Rep.*, **1** (1), 1–38.
21. Thomy, A. and Duval, X. (1994) Stepwise isotherms and phase transitions in physisorbed films. *Surf. Sci.*, **299-300**, 415–425.
22. Mursic, Z., Lee, M.Y.M., Johnson, D.E., and Larese, J.Z. (1996) A computer-controlled apparatus for performing high-resolution adsorption isotherms. *Rev. Sci. Instrum.*, **67** (5), 1886–1890.
23. Brunauer, S., Emmett, P.H., and Teller, E.E. (1938) Adsorption of gases in multimolecular layers. *J. Am. Chem. Soc.*, **60** (2), 309–319.
24. Bruch, L.W., Cole, M.W., and Zaremba, E. (1997) *Physical Adsorption: Forces and Phenomena*, Clarendon Press, Oxford.
25. Thomy, A. and Duval, X. (1969) Adsorption de molécules simples sur graphite. I. Homogénéité de la surface du graphite exfolié. Originalité et complexité des isothermes d'adsorption. *J. Chim. Phys. Phys.-Chim. Biol.*, **66** (11-1), 1966.
26. Thomy, A. and Duval, X. (1970) Adsorption de molécules simples sur graphite. II. Variation du potentiel d'adsorption en fonction du nombre de couches adsorbées. *J. Chim. Phys. Phys.-Chim. Biol.*, **67** (2), 286.
27. Thomy, A. and Duval, X. (1970) Adsorption de molécules simples sur graphite. III. Passage de la première couche par trois états successifs. *J. Chim. Phys. Phys.-Chim. Biol.*, **67** (6), 1101.
28. Larher, Y. (1971) Formation of the first layer of argon, krypton, and xenon on a number of layer-like halides by two-dimensional condensation. *J. Colloid Interface Sci.*, **37** (4), 836–848.
29. Larher, Y. (1992) Monolayer adsorption of Ar, Kr, Xe, and $CH_4$ on layered halides, in *Surface Properties Layered Structures* (ed. G. Benedek), Kluwer Academic Publishers, Dordrecht, pp. 261–315.
30. Dash, J.G. (1975) *Films on Solid Surfaces*, Academic Press, New York.

31. Bretz, M. (1977) Ordered helium films on highly uniform graphite – finite-size effects, critical parameters, and the three-state Potts model. *Phys. Rev. Lett.*, **38** (9), 501–505.
32. Taub, H., Passell, L., Kjems, J.K., Carneiro, K., McTague, J.P., and Dash, J.G. (1975) Neutron-scattering studies of the structure and dynamics of $^{36}$Ar monolayer films adsorbed on basal-plane-oriented-graphite. *Phys. Rev. Lett.*, **34** (11), 654–657.
33. Wiechert, H., Tiby, C., and Lauter, H. (1981) Structure and phase transitions of neon submonolayers adsorbed on basal plane graphite. *Physica B+C*, **108** (1-3), 785–786.
34. Stephens, P.W., Heiney, P., Birgeneau, R.J., and Horn, P.M. (1979) X-ray scattering study of the commensurate-incommensurate transition of monolayer krypton on graphite. *Phys. Rev. Lett.*, **4** (1), 47–51.
35. Coulomb, J.P., Kahn, R., and Bienfait, M. (1976) Study of the mobility of a two-dimensional hypercritical fluid by cold neutron quasi-elastic scattering. *Surf. Sci.*, **61** (1), 291–293.
36. Suzanne, J., Coulomb, J.P., and Bienfait, M. (1973) Auger electron spectroscopy and leed studies of adsorption isotherms: xenon on (0001) graphite. *Surf. Sci.*, **40** (2), 414–418.
37. Chinn, M. and Fain, S. (1977) Structural phase transition in epitaxial solid krypton monolayers on graphite. *Phys. Rev. Lett.*, **39** (3), 146–149.
38. Chesters, M., Hussain, M., and Pritchard, J. (1973) Xenon monolayer structures on copper and silver. *Surf. Sci.*, **35**, 161–171.
39. Quentel, G., Rickard, J., and Kern, R. (1975) Étude de l'adsorption par ellipsométrie: adsorption du Xe/(0001) graphite. *Surf. Sci.*, **50** (2), 343–359.
40. Volkmann, U.G. and Knorr, K. (1989) Ellipsometric study of krypton physisorbed on graphite. *Surf. Sci.*, **221** (1-2), 379–393.
41. D'Amico, K.L., Moncton, D.E., Specht, E.D., Birgeneau, R.J., Nagler, S.E., and Horn, P.M. (1984) Rotational transition of incommensurate Kr monolayers on graphite. *Phys. Rev. Lett.*, **53** (23), 2250–2253.
42. Specht, E.D., Mak, A., Peters, C., Sutton, M., Birgeneau, R.J., D'Amico, K.L., Moncton, D.E., Nagler, S.E., and Horn, P.M. (1987) Phase diagram and phase transitions of Krypton on graphite in the extended monolayer regime. *Z. Phys. B: Condens. Matter*, **69** (2-3), 347–377.
43. Kern, K., David, R., Palmer, R.L., and Comsa, G. (1986) Commensurate, incommensurate and rotated Xe monolayers on Pt(111): a He diffraction study. *Phys. Rev. Lett.*, **56** (6), 620–623.
44. Kern, K., Zeppenfeld, P., David, R., and Comsa, G. (1987) Incommensurate to high-order commensurate phase transition of Kr on Pt (111). *Phys. Rev. Lett.*, **59** (1), 79–82.
45. Mason, B.F. and Williams, B.R. (1984) The scattering of He from adsorbed layers of gases on Ag(110). *Surf. Sci.*, **139** (1), 173–184.
46. Gibson, K.D. and Sibener, S.J. (1985) Determination of the surface phonon dispersion relations for monolayer, bilayer, trilayer, and thick Kr(111) films physisorbed on Ag(111) by inelastic He scattering. *Phys. Rev. Lett.*, **55** (14), 1514–1517.
47. Kern, K., Zeppenfeld, P., David, R., and Comsa, G. (1987) Adsorbate-substrate vibrational coupling in physisorbed Kr films on Pt(111). *Phys. Rev. B*, **35** (2), 886–889.
48. Bruch, L.W., Graham, A.P., and Toennies, J.P. (2000) The dispersion curves of the three phonon modes of xenon, krypton, and argon monolayers on the Pt(111) surface. *J. Chem. Phys.*, **112** (7), 3314–3332.
49. Ellis, J., Graham, A.P., and Toennies, J.P. (1999) Quasielastic helium atom scattering from a two-dimensional gas of Xe atoms on Pt(111). *Phys. Rev. Lett.*, **82** (25), 5072–5075.
50. Jardine, A.P., Ellis, J., and Allison, W. (2002) Quasi-elastic helium-atom scattering from surfaces: experiment and interpretation. *J. Phys. Condens. Matter*, **14** (24), 6173–6190.
51. Jardine, A.P., Dworski, S., Fouquet, P., Alexandrowicz, G., Riley, D.J.,

Lee, G.Y.H., Ellis, J., and Allison, W. (2004) Ultrahigh-resolution spin-echo measurement of surface potential energy landscapes. *Science*, **304** (5678), 1790–1793.
52. Alexandrowicz, G. and Jardine, A.P. (2007) Helium spin-echo spectroscopy: studying surface dynamics with ultra-high-energy resolution. *J. Phys. Condens. Matter*, **19** (30), 305 001.
53. Engel, T. and Gomer, R. (1970) Adsorption of inert gases on tungsten: measurements on single crystal planes. *J. Chem. Phys.*, **52** (11), 5572–5580.
54. Waclawski, B. and Herbst, J. (1975) Photoemission for Xe physisorbed on W(100): evidence for surface crystal-field effects. *Phys. Rev. Lett.*, **35** (23), 1594–1596.
55. Küppers, J., Nitschké, F., Wandelt, K., and Ertl, G. (1979) The adsorption of Xe on Pd(110). *Surf. Sci.*, **87** (2), 295–314.
56. Kaindl, G., Chiang, T., Eastman, D., and Himpsel, F. (1980) Distance-dependent relaxation shifts of photoemission and auger energies for Xe on Pd(001). *Phys. Rev. Lett.*, **45** (22), 1808–1811.
57. Jacobi, K. and Rotermund, H. (1982) UV photoemission from physisorbed atoms and molecules: electronic binding energies of valence levels in mono- and multilayers. *Surf. Sci.*, **116** (3), 435–455.
58. Horn, K., Scheffler, M., and Bradshaw, A. (1978) Photoemission from physisorbed xenon: evidence for lateral interactions. *Phys. Rev. Lett.*, **41** (12), 822–825.
59. Weinelt, M., Trischberger, P., Widdra, W., Eberle, K., Zebisch, P., Gokhale, S., Menzel, D., Henk, J., Feder, R., Dröge, H., and Steinrück, H.P. (1995) One-dimensional band structures: rare gases on Pt(110)1 × 2. *Phys. Rev. B*, **52** (24), R17 048–R17 051.
60. Grüne, M., Pelzer, T., Wandelt, K., and Steinberger, I.T. (1999) Quantum-size effects in thin solid xenon films. *J. Electron. Spectrosc. Relat. Phenom.*, **98-99**, 121–131.
61. Weiss, P.S. and Eigler, D.M. (1992) Adsorption and accommodation of Xe on Pt{111}. *Phys. Rev. Lett.*, **69** (15), 2240–2243.
62. Eigler, D.M. and Schweizer, E.K. (1990) Positioning single atoms with a scanning tunnelling microscope. *Nature*, **344** (6266), 524–526.
63. Zeppenfeld, P., Lutz, C.P., and Eigler, D.M. (1992) Manipulating atoms and molecules with a scanning tunneling microscope. *Ultramicroscopy*, **42-44**, 128–133.
64. Eigler, D.M., Weiss, P.S., Schweizer, E.K., and Lang, N. (1991) Imaging Xe with a low-temperature scanning tunneling microscope. *Phys. Rev. Lett.*, **66** (9), 1189–1192.
65. Zeppenfeld, P., Horch, S., and Comsa, G. (1994) Interaction of xenon at surface steps. *Phys. Rev. Lett.*, **73** (9), 1259–1262.
66. Seyller, T., Caragiu, M., Diehl, R.D., Kaukasoina, P., and Lindroos, M. (1998) Observation of top-site adsorption for Xe on Cu(111). *Chem. Phys. Lett.*, **291** (5-6), 567–572.
67. Seyller, T., Caragiu, M., Diehl, R.D., Kaukasoina, P., and Lindroos, M. (1999) Dynamical LEED study of Pt(111)-($\sqrt{3} \times \sqrt{3}$)R30°-Xe. *Phys. Rev. B*, **60** (15), 11084–11088.
68. Seyller, T., Caragiu, M., and Diehl, R.D. (2000) Low-energy electron diffraction study of krypton on Cu(110). *Surf. Sci.*, **454-456**, 55–59.
69. Diehl, R.D., Seyller, T., Caragiu, M., Leatherman, G.S., Ferralis, N., Pussi, K., Kaukasoina, P., and Lindroos, M. (2004) The adsorption sites of rare gases on metallic surfaces: a review. *J. Phys. Condens. Matter*, **16** (29), S2839–S2862.
70. Zeppenfeld, P., Büchel, M., Goerge, J., David, R., Comsa, G., Ramseyer, C., and Girardet, C. (1996) Structure and phase transitions of xenon monolayers on Cu(110). *Surf. Sci.*, **366** (1), 1–18.
71. Ramseyer, C., Pouthier, V., Girardet, C., Zeppenfeld, P., Büchel, M., Diercks, V., and Comsa, G. (1997) Influence of mode polarizations on the inelastic He-scattering spectrum: high-order commensurate Xe monolayer adsorbed

on Cu(110). *Phys. Rev. B*, **55** (19), 13 203–13 212.

72. Braun, O.M. and Kivshar, Y.S. (2004) *The Frenkel-Kontorova Model: Concepts, Methods, and Applications*, Springer Science & Business Media, Berlin.
73. Bak, P. (1982) Commensurate phases, incommensurate phases and the devil's staircase. *Rep. Prog. Phys.*, **45** (6), 587–629.
74. Villain, J. (1980) Two-dimensional solids and their interaction with substrates, in *Ordering in Strongly Fluctuating Condensed Matter Systems*, NATO Advanced Study Institutes Series: Series B, vol. 50 (ed. T. Riste), Plenum Publishing Corporation, pp. 221–260.
75. Pokrovskii, V.L. and Talanov, A.L. (1980) The theory of two-dimensional incommensurate crystals. *Sov. Phys. JETP*, **51**, 134–148.
76. Bak, P., Mukamel, D., Villain, J., and Wentowska, K. (1979) Commensurate-incommensurate transitions in rare-gas monolayers adsorbed on graphite and in layered charge-density-wave systems. *Phys. Rev. B*, **19** (3), 1610–1613.
77. McTague, J.P. and Novaco, A.D. (1979) Substrate-induced strain and orientational ordering in adsorbed monolayers. *Phys. Rev. B*, **19** (10), 5299–5306.
78. Kosterlitz, J.M. and Thouless, D.J. (1973) Ordering, metastability and phase transitions in two-dimensional systems. *J. Phys. C: Solid State Phys.*, **6** (7), 1181–1203.
79. Nelson, D. and Halperin, B. (1979) Dislocation-mediated melting in two dimensions. *Phys. Rev. B*, **19** (5), 2457–2484.
80. Bak, P. (1984) Phase transitions on surfaces, in *The Chemical Physics of Solid Surfaces V*, Chapter 14 (eds R. Vanselow and R. Howe), Springer-Verlag, Berlin, pp. 317–337.
81. Abraham, F.F., Rudge, W., Auerbach, D.J., and Koch, S.W. (1984) Molecular-dynamics simulations of the incommensurate phase of krypton on graphite using more than 100 000 atoms. *Phys. Rev. Lett.*, **52** (6), 445–448.
82. Horch, S., Zeppenfeld, P., and Comsa, G. (1995) Temperature dependence of the xenon-layer morphology on platinum (111) studied with scanning tunneling microscopy. *Surf. Sci.*, **331-333**, 908–912.
83. Grimm, B., Ho, H., Bo, M., Fieger, K., and Reihl, B. (2000) Observation of domain-wall dynamics in rare-gas monolayers at T=5 K. *Surf. Sci.*, **454-456**, 618–622.
84. Bohlein, T., Mikhael, J., and Bechinger, C. (2012) Observation of kinks and antikinks in colloidal monolayers driven across ordered surfaces. *Nat. Mater.*, **11** (2), 126–130.
85. Diehl, R.D., Setyawan, W., and Curtarolo, S. (2008) Gas adsorption on quasicrystalline surfaces. *J. Phys. Condens. Matter*, **20** (31), 314 007.
86. Gottlieb, J.M. (1990) Energy and structure of uniaxial incommensurate monolayer solids: application to Xe/Pt(111). *Phys. Rev. B*, **42** (8), 5377–5380.
87. Barker, J.A. and Rettner, C.T. (1992) Accurate potential energy surface for Xe/Pt(111): a benchmark gas/surface interaction potential. *J. Chem. Phys.*, **97** (8), 5844–5850.
88. Müller, J.E. (1990) Interaction of the Pt (111) surface with adsorbed Xe atoms. *Phys. Rev. Lett.*, **65** (24), 3021–3024.
89. Da Silva, J.L.F., Stampfl, C., and Scheffler, M. (2003) Adsorption of Xe atoms on metal surfaces: new insights from first-principles calculations. *Phys. Rev. Lett.*, **90** (6), 066104.
90. Da Silva, J.L.F., Stampfl, C., and Scheffler, M. (2005) Xe adsorption on metal surfaces: first-principles investigations. *Phys. Rev. B*, **72** (7), 075 424.
91. Dion, M., Rydberg, H., Schröder, E., Langreth, D.C., and Lundqvist, B.I. (2004) van der Waals density functional for general geometries. *Phys. Rev. Lett.*, **92** (24), 246 401.
92. Román-Pérez, G. and Soler, J. (2009) Efficient implementation of a van der Waals density functional: application to double-wall carbon nanotubes. *Phys. Rev. Lett.*, **103** (9), 096 102.
93. Lazić, P., Atodiresei, N., Caciuc, V., Brako, R., Gumhalter, B., and Blügel, S.

(2012) Rationale for switching to non-local functionals in density functional theory. *J. Phys. Condens. Matter*, **24** (42), 424 215.

94. Sony, P., Puschnig, P., Nabok, D., and Ambrosch-Draxl, C. (2007) Importance of Van Der Waals interaction for organic molecule-metal junctions: adsorption of thiophene on Cu(110) as a prototype. *Phys. Rev. Lett.*, **99** (17), 176 401.
95. Puschnig, P., Nabok, D., and Ambrosch-Draxl, C. (2009) Toward an Ab-initio description of organic thin film growth, in *Interface Controlled Organic Thin Films*, Springer Proceedings in Physics, vol. 129 (eds K. Al-Shamery, G. Horowitz, H. Sitter, and H.G. Rubahn), Springer-Verlag, Berlin, Heidelberg, pp. 3–10.
96. Vidali, G., Ihm, G., Kim, H.Y., and Cole, M.W. (1991) Potentials of physical adsorption. *Surf. Sci. Rep.*, **12** (4), 135–181.
97. Tang, K.T. and Toennies, J.P. (2003) The van der Waals potentials between all the rare gas atoms from He to Rn. *J. Chem. Phys.*, **118** (11), 4976–4983.
98. Kern, K., Zeppenfeld, P., David, R., and Comsa, G. (1988) Two-dimensional phase transitions studied by thermal He scattering. *J. Vac. Sci. Technol., A*, **6** (3), 639–645.
99. Wandelt, K. and Hulse, J.E. (1984) Xenon adsorption on palladium. I. The homogeneous (110), (100), and (111) surfaces. *J. Chem. Phys.*, **80** (3), 1340–1352.
100. Zhu, J.F., Ellmer, H., Malissa, H., Brandstetter, T., Semrad, D., and Zeppenfeld, P. (2003) Low-temperature phases of Xe on Pd(111). *Phys. Rev. B*, **68** (4), 045406.
101. Bagus, P., Staemmler, V., and Wöll, C. (2002) Exchangelike effects for closed-shell adsorbates: interface dipole and work function. *Phys. Rev. Lett.*, **89** (9), 096104.
102. Lazić, P., Crljen, V., Brako, R., and Gumhalter, B. (2005) Role of van der Waals interactions in adsorption of Xe on Cu(111) and Pt(111). *Phys. Rev. B*, **72** (24), 245 407.
103. Rieder, K.H., Parschau, G., and Burg, B. (1993) Experimental evidence for anticorrugating effects in He-metal interactions at surfaces. *Phys. Rev. Lett.*, **71** (7), 1059–1062.
104. Annett, J.F. and Haydock, R. (1984) Anticorrugating effect of hybridization on the helium diffraction potential for metal surfaces. *Phys. Rev. Lett.*, **53** (8), 838–841.
105. Petersen, M., Wilke, S., Ruggerone, P., Kohler, B., and Scheffler, M. (1996) Scattering of rare-gas atoms at a metal surface: evidence of anticorrugation of the helium-atom potential energy surface and the surface electron density. *Phys. Rev. Lett.*, **76** (6), 995–998.
106. Bertel, E. (1996) The interaction of rare gases with transition metal surfaces. *Surf. Sci.*, **367** (2), L61–L65.
107. Rettner, C.T. and Ashfold, M.N.R. (eds) (1991) *Dynamics of Gas-Surface Interactions: Advances in Gas-Phase Photochemistry and Kinetics*, Royal Society of Chemistry, Cambridge.
108. Rettner, C.T., Auerbach, D.J., Tully, J.C., and Kleyn, A.W. (1996) Chemical dynamics at the gas-surface interface. *J. Phys. Chem.*, **100** (31), 13 021–13 033.
109. Schlichting, H., Menzel, D., Brunner, T., and Brenig, W. (1992) Sticking of rare gas atoms on the clean Ru(001) surface. *J. Chem. Phys.*, **97** (6), 4453–4467.
110. Zeppenfeld, P., Goerge, J., Büchel, M., David, R., and Comsa, G. (1994) Island mediated sticking of the rare gases on Cu(110). *Surf. Sci.*, **318** (3), L1187–L1192.
111. Schlichting, H., Menzel, D., Brunner, T., Brenig, W., and Tully, J.C. (1988) Quantum effects in the sticking of Ne on a flat metal surface. *Phys. Rev. Lett.*, **60** (24), 2515–2518.
112. Head-Gordon, M., Tully, J.C., Schlichting, H., and Menzel, D. (1991) The coverage dependence of the sticking probability of Ar on Ru(001). *J. Chem. Phys.*, **95** (12), 9266–9276.
113. Tully, J.C. (1985) Stochastic-trajectory simulations of gas-surface interactions: Xe on Pt(111). *Faraday Discuss. Chem. Soc.*, **80**, 291–298.

114. Poelsema, B., Verheij, L., and Comsa, G. (1983) Direct evidence for two-dimensional Xe gas solid phase transition on Pt(111) by means of thermal He scattering. *Phys. Rev. Lett.*, **51** (26), 2410–2413.
115. Bienfait, M. and Venables, J.A. (1977) Kinetics of adsorption and desorption using Auger electron spectroscopy: application to xenon covered (0001) graphite. *Surf. Sci.*, **64** (2), 425–436.
116. Lehner, B., Hohage, M., and Zeppenfeld, P. (2000) Kinetic Monte Carlo simulation scheme for studying desorption processes. *Surf. Sci.*, **454-456** (1-2), 251–255.
117. Schlichting, H. and Menzel, D. (1992) High resolution, wide range, thermal desorption spectrometry of rare gas layers: sticking, desorption kinetics, layer growth, phase transitions, and exchange processes. *Surf. Sci.*, **272** (1-3), 27–33.
118. Lehner, B., Hohage, M., and Zeppenfeld, P. (2003) The influence of weak adsorbate-adsorbate interactions on desorption. *Chem. Phys. Lett.*, **369** (3-4), 275–280.
119. Lehner, B., Hohage, M., and Zeppenfeld, P. (2002) Kinetic Monte Carlo investigation of Xe adsorption and desorption on Pt(111) and Pt(997). *Phys. Rev. B*, **65** (16), 165407.
120. Lehner, B., Hohage, M., and Zeppenfeld, P. (2001) Novel Monte Carlo scheme for the simulation of adsorption and desorption processes. *Chem. Phys. Lett.*, **336** (1-2), 123–128.
121. Doll, J.J. and Steele, W.A. (1974) The physical interaction of gases with crystalline solids. *Surf. Sci.*, **44** (2), 449–462.
122. Papp, H. and Pritchard, J. (1975) The adsorption of Xe and {CO} on a Cu (311) single crystal surface. *Surf. Sci.*, **53** (1), 371–382.
123. Aubry, S. (1983) Exact models with a complete devil's staircase. *J. Phys. C: Solid State Phys.*, **16** (13), 2497–2508.
124. Zeppenfeld, P., Becher, U., Kern, K., and Comsa, G. (1992) Structure of monolayer Ar on Pt(111): Possible realization of a devil's staircase in two dimensions. *Phys. Rev. B*, **45** (10), 5179–5186.
125. Ramseyer, C., Hoang, P.N.M., and Girardet, C. (1994) Interpretation of high-order commensurate phases for an argon monolayer adsorbed on Pt(111). *Phys. Rev. B*, **49** (4), 2861–2868.
126. Leatherman, G.S., Diehl, R.D., Karimi, M., and Vidali, G. (1997) Epitaxial rotation of two-dimensional rare-gas lattices on Ag(111). *Phys. Rev. B*, **56** (11), 6970–6974.
127. Kern, K., Zeppenfeld, P., David, R., Palmer, R.L., and Comsa, G. (1986) Impurity-quenched orientational epitaxy of Kr layers on Pt (111). *Phys. Rev. Lett.*, **57** (25), 3187–3190.
128. Kern, K. (1987) Symmetry and rotational epitaxy of incommensurate Xe layers on Pt(111). *Phys. Rev. B*, **35** (15), 8265–8268.
129. Kern, K., David, R., Zeppenfeld, P., Palmer, R.L., and Comsa, G. (1987) Symmetry breaking commensurate-incommensurate transition of monolayer Xe physisorbed on Pt(111). *Solid State Commun.*, **62** (6), 391–394.
130. Kern, K., David, R., Zeppenfeld, P., and Comsa, G. (1988) Registry effects in the thermodynamic quantities of Xe adsorption on Pt(111). *Surf. Sci.*, **195** (3), 353–370.
131. Narloch, B. and Menzel, D. (1997) Structural evidence for chemical contributions in the bonding of the heavy rare gases on a close-packed transition metal surface: Xe and Kr on Ru(001). *Chem. Phys. Lett.*, **270** (1-2), 163–168.
132. Pussi, K., Smerdon, J., Ferralis, N., Lindroos, M., McGrath, R., and Diehl, R.D. (2004) Dynamical low-energy electron diffraction study of graphite (0001)-($\sqrt{3} \times \sqrt{3}$)R30°-Xe. *Surf. Sci.*, **548** (1-3), 157–162.
133. Jablonski, A., Eder, S., Markert, K., and Wandelt, K. (1986) Two-dimensional gas-solid phase transition of xenon adsorbed on different metal substrates. *J. Vac. Sci. Technol., A*, **4** (3), 1510–1513.
134. Hückstädt, C., Schmidt, S., Hüfner, S., Forster, F., Reinert, F., and Springborg, M. (2006) Work function studies of

rare-gas/noble metal adsorption systems using a Kelvin probe. *Phys. Rev. B*, **73** (7), 075 409.

135. Wandelt, K. and Gumhalter, B. (1984) Face specificity of the Xe/Pd bond and the S-resonance model. *Surf. Sci.*, **140** (2), 355–376.
136. Wandelt, K. (1984) Surface characterization by photoemission of adsorbed xenon (PAX). *J. Vac. Sci. Technol., A*, **2** (2), 802–807.
137. Wandelt, K. (1990) Atomic scale surface characterization with photoemission of adsorbed xenon (PAX), in *The Chemical Physics of Solid Surfaces VIII*, Springer Series in Surface Sciences, vol. 22 (eds R. Vanselow and R. Howe), Springer-Verlag, Berlin, pp. 289–334.
138. Jacobi, K. (1988) Photoemission from Ar, Kr, and Xe on Pb(111). *Phys. Rev. B*, **38** (9), 5869–5877.
139. Hermann, K., Noffke, J., and Horn, K. (1980) Lateral interactions in rare gas monolayers: band-structure models and photoemission experiments. *Phys. Rev. B*, **22** (2), 1022–1031.
140. Trischberger, P., Dröge, H., Gokhale, S., Henk, J., Steinrück, H.P., Widdra, W., and Menzel, D. (1997) One-dimensional xenon band structures on hydrogen modified and stepped platinum surfaces. *Surf. Sci.*, **377-379**, 155–159.
141. Widdra, W. (2001) Electronic band structures of low-dimensional adsorbate systems: rare-gas adsorption on transition metals. *Appl. Phys. A: Mater. Sci. Process.*, **72** (4), 395–404.
142. Schmitz-Hübsch, T., Oster, K., Radnik, J., and Wandelt, K. (1995) Photoemission from quantum-well states in ultrathin Xe crystals. *Phys. Rev. Lett.*, **74** (13), 2595–2598.
143. Paniago, R., Matzdorf, R., Meister, G., and Goldmann, A. (1995) Quantization of electron states in ultrathin xenon layers. *Surf. Sci.*, **325** (3), 336–342.
144. Meixner, D.L. and George, S.M. (1993) Coverage dependent surface diffusion of noble gases and methane on Pt(111). *Surf. Sci.*, **297** (1), 27–39.
145. Chen, J.R. and Gomer, R. (1980) Mobility and two-dimensional compressibility of Xe on the (110) plane of tungsten. *Surf. Sci.*, **94** (2-3), 456–468.
146. Graham, A.P. and Toennies, J.P. (1999) Determination of the lateral potential energy surface of single adsorbed atoms and molecules on single crystal surfaces using helium atom scattering. *Surf. Sci.*, **427-428**, 1–10.
147. Graham, A.P. (2003) The low energy dynamics of adsorbates on metal surfaces investigated with helium atom scattering. *Surf. Sci. Rep.*, **49** (4-5), 115–168.
148. Zeppenfeld, P., Büchel, M., David, R., Comsa, G., Ramseyer, C., and Girardet, C. (1994) Effect of the structural anisotropy and lateral strain on the surface phonons of monolayer xenon on Cu(110). *Phys. Rev. B*, **50** (19), 14 667–14 670.
149. Zeppenfeld, P., Becher, U., Kern, K., David, R., and Comsa, G. (1990) Van Hove anomaly in the phonon dispersion of monolayer Ar/Pt (111). *Phys. Rev. B*, **41** (12), 8549–8552.
150. Hall, B., Mills, D.L., and Black, J.E. (1985) Lattice dynamics of rare-gas overlayers on smooth surfaces. *Phys. Rev. B*, **32** (8), 4932–4945.
151. Gibson, K.D. and Sibener, S.J. (1988) Inelastic helium scattering studies of ordered Ar, Kr, and Xe monolayers physisorbed on Ag(111): dispersion curves, scattering cross sections, and excitation line shapes. *J. Chem. Phys.*, **88** (12), 7862–7892.
152. Hall, B., Mills, D.L., Zeppenfeld, P., Kern, K., Becher, U., and Comsa, G. (1989) Anharmonic damping in rare-gas multilayers. *Phys. Rev. B*, **40** (9), 6326–6338.
153. Zeppenfeld, P., Becher, U., Kern, K., and Comsa, G. (1990) Vibrational spectroscopy of rare gas adlayers. *J. Electron. Spectrosc. Relat. Phenom.*, **54-55**, 265–280.
154. Gibson, K.D. and Sibener, S.J. (1988) Inelastic helium scattering studies of the vibrational spectroscopy and dynamics of ordered Ar, Kr, and Xe multilayers physisorbed on Ag(111). *J. Chem. Phys.*, **88** (12), 7893–7910.

# 34
# Adsorption of Alkali and Other Electropositive Metals

*Anton.G. Naumovets*

## 34.1
## Nature of Chemisorption Bond of Electropositive Adsorbates

Alkali metals belong to the first group of elements in the periodic table and thus have the simplest (s-type) valence electronic shells, which contain just one electron. The alkali atoms have the lowest ionization energies, and the alkali metals have the lowest work functions (Table 34.1).

Investigations of adsorption of alkali metals on refractory metals, particularly of Cs on W, played a remarkable role in the development of basic concepts of surface physics and chemistry, emission electronics, and plasma physics [2–6].

The first studies in this area were made in the years as quantum mechanics was still in its infancy, so all effects found in adsorption were then interpreted in terms of classical physics. For example, it was inferred that, due to the fact that the ionization energy of cesium ($I = 3.89$ eV) is lower than the work function of tungsten ($\varphi = 4.62$ eV, as determined in Ref. [5], p. 444), the valence electron of the Cs adsorbed atom (adatom) is completely transferred to the W substrate. Thus the adsorption bond Cs–W was attributed to the Coulomb attraction between the $Cs^+$ adsorbed ion and the induced negative charge on substrate surface that cancels the electric field within the substrate volume (Figure 34.1).

This model gave an explanation of the fact that practically all Cs atoms impinging upon a hot W surface evaporate from it as positive ions (this phenomenon is called *surface ionization*). The adsorbed $Cs^+$ ions, together with the negative screening charge at the surface, form an electrical double layer, which accelerates the electrons leaving the surface and thus reduces the substrate work function (Figure 34.2).

However, this classical picture, which seemed quite reasonable in the case of the Cs–W system, cannot explain the results obtained with alkaline-earth and other electropositive adsorbates whose ionization energies exceed the work functions of transition and noble metal substrates. Figure 34.3 presents an example of the concentration dependence of the work function recorded for barium on the tungsten(112) surface [7].

*Surface and Interface Science: Solid-Gas Interfaces I*, First Edition. Edited by Klaus Wandelt.

**Table 34.1** Electron characteristics of alkali and alkaline-earth elements.

| Atomic number | Symbol | Electronic structure | Ionization energy | Work function |
|---|---|---|---|---|
| 3 | Li | $1s^2 2s^1$ | 5.41 eV | 2.38 eV |
| 4 | Be | $1s^2 2s^2$ | 9.32 eV | 3.92 eV |
| 11 | Na | $2p^6 3s^1$ | 5.14 eV | 2.35 eV |
| 12 | Mg | $2p^6 3s^2$ | 7.64 eV | 3.64 eV |
| 19 | K | $3p^6 4s^1$ | 4.34 eV | 2.22 eV |
| 20 | Ca | $3p^6 4s^2$ | 6.11 eV | 2.80 eV |
| 37 | Rb | $4p^6 5s^1$ | 4.18 eV | 2.16 eV |
| 38 | Sr | $4p^6 5s^2$ | 5.69 eV | 2.35 eV |
| 55 | Cs | $5p^6 6s^1$ | 3.89 eV | 1.81 eV |
| 56 | Ba | $5p^6 6s^2$ | 5.21 eV | 2.49 eV |
| 87 | Fr | $6p^6 7s^1$ | | 1.5 eV |
| 88 | Ra | $6p^6 7s^2$ | | 3.2 eV |

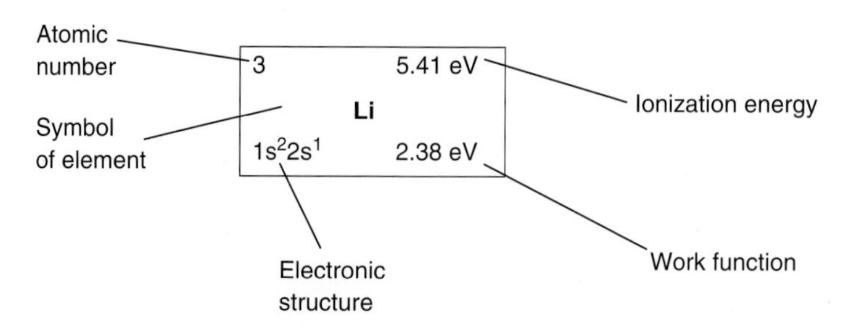


Work function values are averaged data for polycrystalline surfaces.
Fomenko [1].

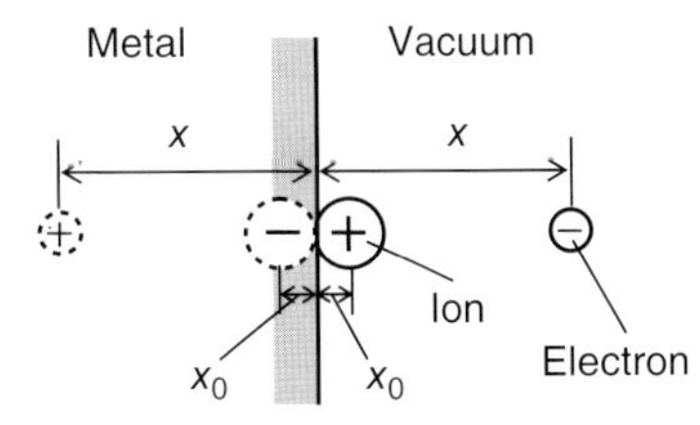


**Figure 34.1** Model of electrical images of an electron and adsorbed ion.

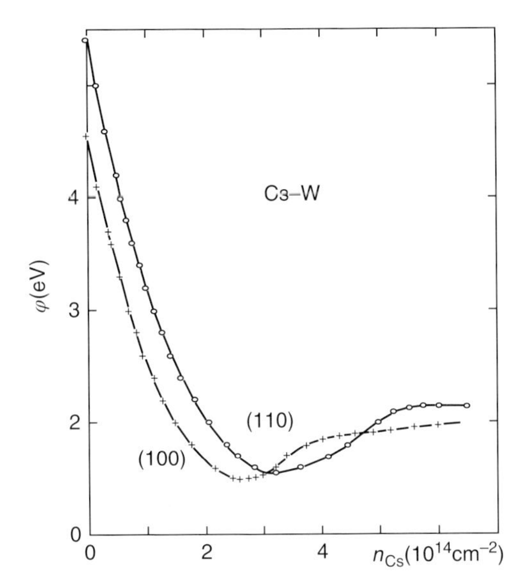


**Figure 34.2** Work function versus concentration of Cs adatoms on the (110) and (100) surfaces of W.

The work function of the W(112) surface is 4.8 eV, and the ionization energy of Ba atom is 5.21 eV (Table 34.1), that is, the parameters of this system obey the inequality $I > \varphi$. Nonetheless, the work function of W(112) is strongly lowered by barium, and the curve $\varphi(n)$ is very similar to the curves measured for Cs on W crystal surfaces when the opposite inequality $I < \varphi$ holds (Figure 34.2). The same behavior is characteristic of other alkaline-earth elements. Moreover, similar $\varphi(n)$ curves were obtained with rare-earth (RE) adsorbates [8, 9], whose ionization energies amount to 5.4–6.3 eV (Table 34.2).

These paradoxical results are explainable in terms of the model proposed by Gurney [12], which takes into account the evolution of the valence energy level when an atom is approaching the surface (Figure 34.4).

There are two major effects in this evolution. The first is an upward shift of the level, that is, reduction of the ionization energy of an atom in the adsorbed state. The origins of this effect can be qualitatively understood even in an electrostatic approach. The ionization energy $I$ of an atom is defined as the minimum energy required to remove an electron from the valence shell (the ground state). For a free

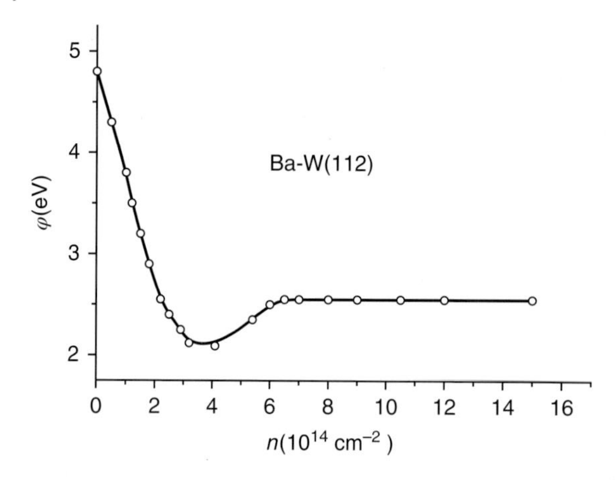


**Figure 34.3** Work function versus concentration of Ba adatoms on the (112) surface of W [7].

**Table 34.2** Electron characteristics of rare-earth elements.

| | | | | | | | |
|---|---|---|---|---|---|---|---|
| 57 5.61eV<br>L<br>a<br>$5d^16s^2$ | 58 5.60eV<br>Ce<br>$4f^26s^2$ | 59 5.42eV<br>Pr<br>$4f^36s^2$ | 60 5.51eV<br>Nd<br>$4f^46s^2$ | 61 5.55eV<br>Pm<br>$4f^56s^2$ | 62 5.63eV<br>Sm<br>$4f^66s^2$ | 63 5.68eV<br>Eu<br>$4f^76s^2$ | 64 5.98eV<br>G<br>d<br>$4f^75d^16s^2$ |
| 65 5.85eV<br>T<br>b<br>$4f^85d^16s^2$ | 66 5.80eV<br>Dy<br>$4f^{10}6s^2$ | 67 6.02eV<br>Ho<br>$4f^{11}6s^2$ | 68 6.08eV<br>Er<br>$4f^{12}6s^2$ | 69 6.14eV<br>T<br>m<br>$4f^{13}6s^2$ | 70 6.25eV<br>Yb<br>$4f^{14}6s^2$ | 71 5.43eV<br>Lu<br>$4f^{14}5d^16s^2$ | |

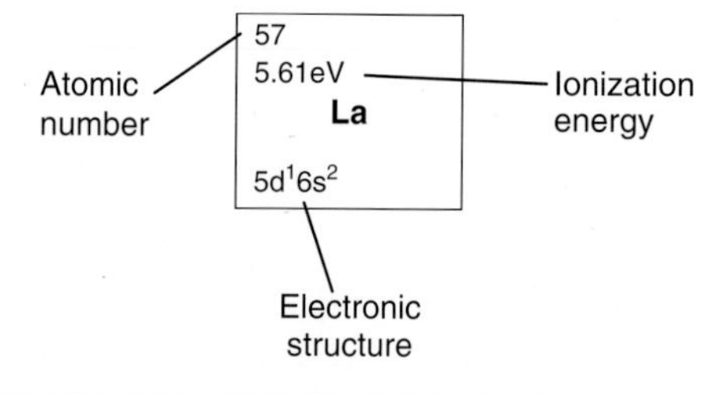


Prokhorov [10] and Zandberg and Ionov [11].

atom, $I$ is determined by the interaction of the valence electron, which is moved away to infinity, with the nucleus and the remaining electrons (the ion core). In the known electrostatic charge image model, two additional forces should be taken into account when an electron is detached from an *adsorbed* atom (adatom) (Figure 34.1): the *attraction* of the electron to its own positive charge image in the metal, $-e^2/4x^2$;

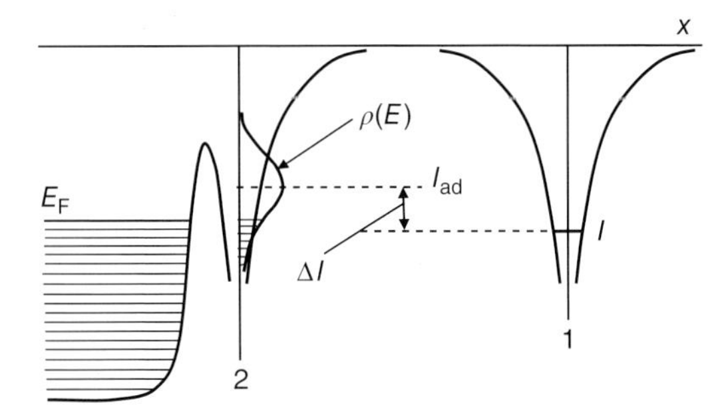


**Figure 34.4** The Gurney model. (1) The potential energy diagram for a free atom and (2) the same for an adsorbed atom.

and its *repulsion* from the negative charge image of the ion core, $e^2/(x+x_0)^2$. Here, $x$ denotes the distance of the electron from the surface, and $x_0$ is the equilibrium distance of the center of adatom from the surface (for a rigorous treatment of the image potential and other characteristics of electrons at surfaces, see, for example, Refs. [13, 14]). The change in the ionization energy, $\Delta I$, caused by these forces can be calculated by their integration over distance in the limits from $x = x_0$ to $x = \infty$. As a result, the ionization energy of the adsorbed atom, $I_{ad}$, is found to be

$$I_{ad} = I - \frac{e^2}{4x_0}. \tag{34.1}$$

Thus, in this approximation, the valence energy level in the adatom is shifted *upwards*, with respect to its position in the free atom, by a value of $\Delta I = e^2/4x_0$ ($= 3.6/x_0$ if the energy is measured in electronvolts and distance in angströms). Actually, this result arises from the fact that the *negative* charge image of the positive ion is permanently located at the distance $-x_0$ beneath the surface, whereas the *positive* charge image of the electron itself is shifting more deeply into the substrate while the electron is moving away from the surface. The net effect is the domination of the *repulsion* of the electron from the negatively charged image of the ion, that is, the decrease in the ionization energy.

The above-presented estimate of the energy level shift due to adsorption is approximate, since it is based on purely electrostatic considerations and on the assumption that the substrate is polarized as an ideal conductor, which is not strictly true at small (atomic-scale) distances. Field desorption experiments [15] and quantum mechanical calculations [16–21] show that the charge image approach is satisfactory down to distances of ≈2–3 Å. At smaller distances, it is necessary to take into account the short-range exchange-correlation interaction and the finite value of the screening length in real metals. Nonetheless, the electrostatic model provides a reasonable semiquantitative assessment of the energy level shift upon adsorption, which, according to stricter calculations incorporating the quantum mechanical effects, amounts to ~1–1.5 eV for alkali metal adatoms [16, 18, 21].

The second important change in the valence energy level of the atom approaching the surface is its broadening. This occurs because the potential barrier between the adsorbing atom and the surface becomes transparent for electron tunneling.

According to the uncertainty principle, the energy spread $\Delta E$ is related to the average time of electron exchange, $\tau$, as

$$\Delta E \sim \hbar/\tau, \tag{34.2}$$

($\hbar = h/2\pi$, where $h$ is the Planck constant). The quantum mechanical estimations for electrons give a value of $\tau \sim 10^{-15}$ s when the potential barrier has an atomic width ($\sim 10^{-8}$ cm). Thus the width of the valence level in an electropositive adatom amounts to

$$\Delta E \sim \frac{6.6 \cdot 10^{-16}\ \text{eV} \cdot \text{s}}{10^{-15}\ \text{s}} \sim 1\,\text{eV}. \tag{34.3}$$

The joint action of level broadening and shift results in an energy diagram given in Figure 34.4. Actually, the adsorbed atom and the substrate represent a unified quantum mechanical system, and the diagram shows a broadened ("virtual") energy level of the adatom filled by electrons up to the Fermi level $E_F$. The negative electric charge $Q$ of the valence electronic shell equals

$$Q = e \int_{-\infty}^{E_F} dE \int_V \rho(r, E) d^3 r, \tag{34.4}$$

where $\rho(r,E)$ is the space and energy function of the charge density in the adatom valence shell. The integration is performed over the volume of adatom, $V$, and up to the energy $E_F$.

Thus, the electropositive adatoms can possess a considerable positive charge $q^*$, which is equal to $e - Q$ for alkali adatoms). This charge is generally fractional in contrast to the integer values that were assumed on the basis of classical views in early works on alkali adsorption. For positive charges of single alkali adatoms, numerical estimations give the values of $q^* \sim 10^{-1}\ e$.

It is understood that the evaluated charge of the adatom depends on the assumed position of the boundary between the adatom and the substrate. Self-consistent quantum mechanical calculations give the coordinates of the nuclei and maps of the spatial distribution of electron density (see, e.g., Refs. [19, 21–23]). However, there exists some ambiguity in tracing the adatom–substrate interface [21, 24]. As a result, some authors attribute the negative screening charge to the adatom, and thus state that adatoms are polarized rather than partially ionized. Even though this point is actually a matter of convention, the electron density distribution is an objective reality. It determines the strength of the electrical double layer at the surface and, according to the Helmholtz formula, the work function change $\Delta\varphi$ caused by adatoms:

$$\Delta\varphi = -4\pi npe, \tag{34.5}$$

where $p$ is the dipole moment of the adsorption bond, $n$ is the surface concentration of adatoms, and $e$ is the elementary electric charge. In this formula, it is assumed that the positive end of the dipole is oriented outward with respect to the substrate, so the work function is reduced ($\Delta\varphi < 0$).

It should be noted that Gurney's model is also useful to understand the nature of the adsorption bond in the case of electronegative adsorbates such as O, S, F, Cl, and other elements of the VI and VII groups of the periodic system. The ionization energies for these elements (we mean here the energies necessary to obtain their *positive* ions) are substantially higher than the work functions of all known solids. Thus the ground-state level in their adatoms, even with allowance made for its possible broadening and upward shift, is completely filled with electrons. However, each of these elements has a considerable electron affinity, which means that considerable energy is needed to detach an electron from their *negative* ions. For example, the electron affinities are (in electronvolts) 1.46 for O, 2.08 for S, 3.40 for F, and 3.62 for Cl.

Similar to the ground-state level, the electron affinity level in the adatom is broadened and shifted. However, its shift occurs *downward*, that is, in the opposite direction to the shift of the ground-state level. The reason can be understood from considerations analogous to those given above in the derivation of Equation 34.1. The shifted and broadened affinity level is filled with electrons up to the Fermi level, which endows the adatom with a fractional *negative* charge.

The evolution of the atom energy levels during adsorption discussed above is the key information for understanding the nature of the adsorption bond, adatom interactions, and other effects in electropositive adlayers.

## 34.2 Charge and Potential Distribution at the Surface

As discussed above, the electric field created by a positively charged adatom induces an excess negative charge distribution at the substrate surface, which cancels the field within the substrate. In the general case, with regard for a concrete geometry of the adsorption site and possible presence of coadsorbed neighbors of various nature, the charge distribution within and around the adatom is rather complex. It can be represented as a sum of multipoles, of which the most important is the electric dipole because it determines the long-range interaction of adatoms and the work function change induced by the adsorbed layer. The dipole moment $\mathbf{p} = q\mathbf{l}$ ($q$ is the value of the positive and negative charges in the dipole and $\mathbf{l}$ the distance separating them) is measured in the debye units: $1\,\text{D} = 3.34 \times 10^{-30}\,\text{C} \times \text{m} \approx 0.21\,\text{e}\cdot\text{Å}$. Typical values of the dipole moments of alkali and alkaline-earth metal adatoms on transition and noble metals and graphite, all of which have work functions of ~4–5 eV, range from a few debyes to about 10 D. It is necessary to underline that, in the strict sense, one should speak about the dipole moment of the *adsorption bond* rather than of the adsorbed atom.

Although the macroscopic data on the work function changes are essential and useful from the basic and practical points of view, they do not provide information about surface electronic characteristics on an atomic scale. Meanwhile, surface chemical reactions, surface properties of semiconductors, and the processes of crystal and thin film growth are extremely sensitive to surface heterogeneities of such a size. It is therefore important to investigate local electronic characteristics in the close vicinity of various surface defects – both intrinsic (surface steps and kinks, dislocations and grain boundaries, single atoms on terraces, vacancies) and extrinsic (adsorbed heteroatoms and clusters, 2D islands, etc.).

For electropositive adatoms, the most detailed information about the local perturbations that they induce in surface potential was obtained by photoelectron spectroscopy of adsorbed xenon (PAX) [25–28]. Xenon and other noble gas atoms proved to be very informative probes of the local surface potential immediately in various adsorption sites located at different distances from an alkali metal adatom. The binding energy of Xe adatom with metal substrates is rather small ($\approx$0.2–0.35 eV). This value is typical of physisorption, which is affected mainly due to the van der Waals forces plus an electrostatic contribution caused by adatom polarization. Thus Xe adatoms can serve as rather gentle sensors that do not substantially perturb the systems under investigation.

Wandelt *et al.* [27–29] showed that the position of the $5p_{1/2}$ valence level of Xe adatom and its core levels with respect to the Fermi level of a metal substrate is a sensitive indicator of the local electrostatic potential. It is important that values of this potential relate to the points located at a distance of $\approx$2.5 Å from the surface and that they can be determined in different adsorption sites, for example, over the atomically flat areas (terraces) of the surface, close to atomic steps, and coadsorbed heteroatoms. The binding energy of the valence electron in Xe adatom, $E_B^V$, derived from the photoelectron energy spectra and counted from the vacuum level $E_V$ (actually, the ionization energy of Xe in the adsorbed state, $I_{ad}$) was found to be *independent* of the adsorption site. However, it is reduced by $\approx$1.1 eV with respect to the ionization energy of the free Xe atom (the reasons for the reduction of the ionization energy of atoms in the adsorbed state have been discussed in Section 34.1). At the same time, the energy gap between the metal Fermi level and the Xe valence level, $E_B^F$, *does depend* on the site type (Figure 34.5).

The difference $E_B^V - E_B^F$, which is determined by the PAX method with a high spatial resolution, actually with an atomic-scale locality, gives a value $\varphi$, which was termed the *local work function* [27–29]. Correspondingly, the difference of $E_B^F$ values

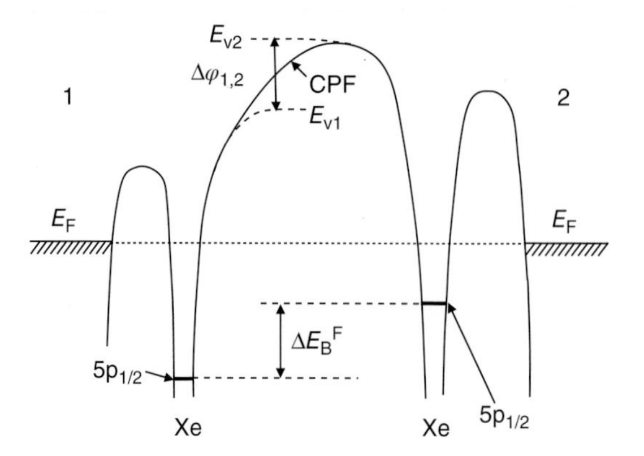


**Figure 34.5** Energy diagram for two Xe atoms adsorbed in sites (1,2) with different local work functions [27, 29]. See text for explanations.

for sites $i$ and $j$, $\Delta E_B^F(i,j)$, determines the local work function difference between these sites:

$$\Delta E_B^F(i,j) = -\Delta\varphi_{i,j}. \tag{34.6}$$

The advantage of the PAX technique is just its ability to probe the potential distribution with a high lateral resolution and in close vicinity to the surface. The nucleus of Xe adatom is located at a distance of $\approx$3 Å from the surface, where the contribution to the work function stemming from the short-range electron–surface interactions (exchange-correlation interaction and the potential drop in electrical double layer) practically levels off [30]. Evidently, this part of the surface potential is strongly site-specific. Contrary to this, the long-range Coulomb (image) potential $-e^2/4x$ gives at $x > 3$ Å a contribution of $\sim$1 eV, which is actually the same for all adsorption sites.

Not only Xe but also other noble gas atoms (Ar, Kr) can be used as surface potential probes. Since their atomic radii differ by 10–15%, one can investigate the variation of the surface potential is some range of distances [28, 29]. The most detailed experiments applying this technique were carried out with K adsorbed on Rh(111), a model system widely used to elucidate the promoting effect of alkali metals in catalytic reactions, for example, in the synthesis of ammonia and the hydrogenation of CO. Close to the K adatom (at spacings <4 Å), the surface potential was found to be reduced by 1–2 eV, this value being dependent on the actual distance between the K and probing adatom. At larger distances from K adatom (>5 Å), the surface potential is nearly constant. Its value depends on the surface concentration of K, $c_K$, and, in the interval $2.7 < c_K < 5.0$ at%, is lower by 0.4–1.0 eV than on the bare (unpromoted) Rh(111). This potential is a cumulative characteristic whose magnitude results from integrated effect produced by all adatoms. Its nature is mostly electrostatic, originating from superposition of the fields created by surface dipoles (charged adatoms + screening electron clouds).

The substantially lowered electrostatic potential at adsorption sites adjacent to an electropositive atom facilitates the transfer of the electron cloud from the metal substrate to electronegative admolecules such as CO or $N_2$. The accepted electrons fill the antibonding orbitals in the admolecules. This entails an easy dissociation of the reactant molecules and the formation of their fragments, which then combine to give the desired chemical products. Such a scenario of the promotive action of electropositive adsorbates is also confirmed by other experimental techniques [26, 31–33] as well as by theoretical calculations [19, 34].

## 34.3
## Work Function versus Coverage

Let us return to experimental data illustrating the effect of electropositive adlayers on the work function. Figure 34.1 shows such data for Cs on the (110) and (100) crystal surfaces of tungsten. It is seen that the curves are very similar to each other, although the Cs adatom concentrations corresponding to work function minima differ by about 30% for these substrates. We recall that cesium ionization energy is lower than the work functions of the tungsten crystal surfaces, which are compared in Figure 34.1 (5.4 eV for the (110)W and 4.5 eV for the (100)W, respectively). Experimental data on the work function changes and the adlayer atomic structures showed that minima in the $\varphi$ versus $n$ dependences appear at a coverage that usually amounts to about 0.5–0.7 of the close-packed monolayer (CPM) [35–39] (see also Section 34.4). Its particular value depends on the chemical composition of the adsorption system and on the atomic structure of the substrate surface. Here and below, we shall use the notion *coverage* (or *degree of coverage*), which is defined as the ratio $\theta = n/n_{\mathrm{m}}$, where $n$ is the actual concentration of adatoms and $n_{\mathrm{m}}$ is their concentration in a monolayer. In some works, $n_{\mathrm{m}}$ is taken as the surface concentration of the substrate atoms (a "geometrical" monolayer), while in other works $n_{\mathrm{m}}$ is attributed to a close-packed ("physical") monolayer. We shall specify this definition in cases where this is essential due to difference of atomic radii of adatoms and substrate atoms.

The existence of the work function minimum results from the variation of the characteristics of adsorption bond with coverage, which, in its turn, is caused by the lateral interaction of the adatoms (see Section 34.4). Each surface dipole creates an electric field around itself, which tends to depolarize the neighboring dipoles. In terms of the Gurney model, this is effected by a downward shift of the broadened adatom valence level, which entails the growth of its occupancy by electrons according to Equation 34.4. Thus the value of the dipole moment $p$ is a decreasing function of the adatom concentration $n$, and the product $pn$ in Equation 34.5 passes through a *maximum* (i.e., the work function should pass through a *minimum*) at some optimum adatom concentration $n_{\mathrm{opt}}$. Although this electrostatic model provides a qualitatively transparent explanation of the nonmonotonic variation of the work function with $n$, it fails to give correct quantitative predictions. The distances between the adjacent adatoms at such coverages are close to the atomic diameter, so the electron shells of the adatoms start to overlap, and the direct electron exchange interaction of the adatoms comes into play. Obviously, the purely electrostatic description of the interaction and mutual depolarization of adatoms becomes unjustified in such a situation.

In a more rigorous approach, the $\varphi(n)$ dependence is treated as a reflection of the transition from the surface covered with partially ionized (or strongly polarized) adatoms to a thick film having properties of a bulk adsorbate metal. Using this approach and a jellium model, Lang [17, 20, 40] calculated the work function variation induced by adsorbed Cs and other alkali metals. The jellium model is a good approximation to evaluate electronic effects in surface phenomena. In this model, the positive charge of ion cores of a metal substrate is uniformly smeared out over the semi-infinite space to obtain a positive jellium. In a similar way, the ion cores of adatoms are replaced by a uniform positive jellium filling a slab whose thickness is equal to the atomic (or ionic at low coverages) diameter of adatoms. It is also assumed that the system contains a corresponding amount of electron gas that is just sufficient to make the whole system electrically neutral. The contours of the positive jellium are fixed, while one calculates the spatial distribution of the electron density that minimizes the total energy of the system. The calculations for such a model [17, 40], as well as for single adatoms on the metal surface represented in the jellium approximation, were performed using the density functional formalism [41]. For single adatoms, the calculations gave the maps of electron density distribution around the adatom ion core and close to it at the substrate surface [22]. In particular, the map obtained for Li adatom on a jellium substrate with $\varphi \approx 4$ eV (corresponding to Al) clearly shows the displacement of the electron cloud from the vacuum side of the adatom toward the substrate, which leads to a substantial depletion of the 2s electron shell of Li adatom. Bormet *et al.* [23] applied the surface Green function method to calculate the charge distributions in Na adlayers of various densities on Al(111), and additionally took into account the substrate atomic structure. Although the structure effects in the charge and in the density-of-states distribution were found to be noticeable, both the works are qualitatively consistent in the statement that the adsorption bond of alkali atoms at low coverages is basically ionic rather than covalent with some degree of polarity [40]. As expected, the adatom ion core was found to be polarized and possess a dipole moment that is much smaller and directed opposite to the moment produced by the charge transfer from the adatom to the substrate. It should be noted that the transferred charge is localized in the close vicinity to the adatom, which is the consequence of the short field screening length in metals. The perturbation of the electron density induced by the charged adatom decays in the metal in a non-monotonic way due to Friedel's oscillations [42]. This is a result of interference of electron waves scattered by the charged adatom. The decay rate of the Friedel oscillations with distance depends on the shape of the Fermi surface and can be rather slow, which gives rise to an oscillating long-range lateral interaction of adatoms (see Section 34.4).

The variation of the work function in a broad range of coverages can be calculated by representing the adlayer as a jellium slab whose charge density and thickness are gradually increased to simulate the progressive buildup of the adsorbed film [17, 20, 40]. The calculated $\varphi(n)$ curves are qualitatively consistent with the experimental ones: they show a minimum at submonolayer coverage and a small maximum at the completion of the first adsorbate monolayer. The quantitative discrepancy between the theory and experiment amounts to $\approx$20%, but, nonetheless, the calculated data give a correct coverage dependence of the work function and provide a pictorial representation of the variation of electron density at the surface during the growth of coverage. At low coverages, the fast decrease of the work function is due to formation of a powerful electrical double layer at the surface (the *charge-transfer region*). On the other hand, at coverages approaching the complete monolayer, the charge transfer reduces, and the film acquires properties that reveal its proximity to the bulk metal (the *metallic region*). In particular, the CPMs of electropositive metals have work functions that are nearly the same (to within $\sim 10^{-1}$ eV) as those of the bulk adsorbates [8, 9, 35, 39, 43–45]. The metallicity of the adlayers past the work function minimum is also confirmed by the emergence of collective electronic oscillations (plasmons) at such coverages (see Section 34.8). The work function minimum appears in the state of the adlayer that is intermediate between the charge transfer state and metallic state.

It is instructive to give here, for comparison with alkali and alkaline-earth metals, some data on adsorption of RE metals. These adsorbates are of much interest because of their peculiar electronic, magnetic, and structural properties caused by the overlap of localized 4f and delocalized 5d6s states, which gives rise to their mixed (intermediate) valence stemming from the relative ease of rearrangement of their electron shells [9, 46, 47]. In some aspects, the adsorption characteristics of RE elements resemble those of alkaline-earth elements. For example, the concentration dependences of the work function for barium and ytterbium on the Mo(110) surface are quite similar: starting from the initial work function of this surface of 5.0 eV, Ba has a reduced value of 2.1 eV [38], and Yb of 2.5 eV [9]. Recall that the ionization energies of Ba and Yb differ by ≈1.0 eV (see Tables 34.1 and 34.2), but the similarity of their adsorption characteristics is easily perceived in terms of Gurney's model. Let us also recall that the $LaB_6$ cathode, whose surface is enriched by La due to its segregation, is one of the best thermal cathodes [48]. Owing to the lower polarity of the adsorption bond of RE metals, their adsorption energy at low coverages is significantly smaller than that of alkaline-earth metals. On the other hand, it decreases more slowly with growing coverage, so that at $\theta \sim 1$ the RE adsorbates are bound more strongly with the substrate. The RE metals reveal also a more pronounced tendency to reconstructive adsorption and to the formation of volume and surface alloys with various substrates [46, 49]. The structure of their adsorption phases is generally more complex than in the case of alkali and alkaline-earth metals [8].

As found in many systems, the work function may pass through a maximum at the coverage corresponding to the CPM. Two possible factors can contribute to this effect. First, the density of the CPM, where adatoms have immediate contact with the substrate and experience a strong attraction to it, can be substantially higher (by 10–15%) than the density of the subsequent monolayers. This should result in a higher work function of the first monolayer, since both experiment and theory show that the denser atom packing entails a higher value of the surface potential barrier [17]. The second factor that may be responsible for the work function maximum at the CPM coverage ($\theta = 1$) is the Smoluchowski effect [50]. Its origin lies in the fact that, at the initial stage of building up the second monolayer over the close-packed first monolayer, the surface becomes atomically rough. However, the electron density at the surface tends to smooth out its contours in order to minimize the energy, since sharp gradients of the density imply the presence of short-wave harmonics in the electron density distribution having a higher energy. This smoothing effect leads to some electron depletion of the adatom apexes, and thus to the rise of a positive dipole moment, which reduces the work function (Figure 34.6).

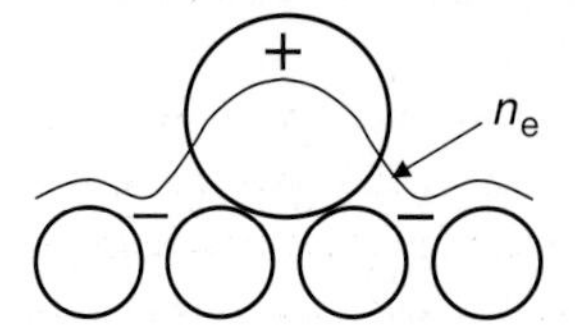


**Figure 34.6** Smoothing of the contour of the electron cloud in a protruding adatom (R. Smoluchowski's effect).

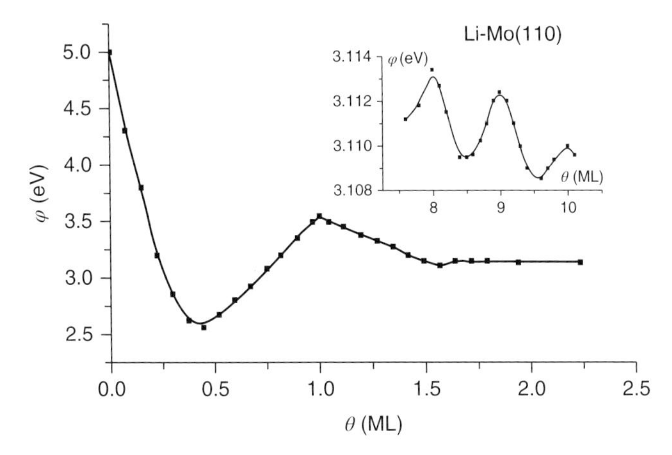


**Figure 34.7** Work function versus coverage for Li on Mo(110) in the coverage intervals $0 < \theta < 2$ (bottom) and $8 < \theta < 10$ (top) [52].

This effect is especially clearly observed in the layer-by-layer (Frank–van der Merwe) growth mode when the next monolayer starts to grow only after the previous monolayer has been completely filled (concerning the epitaxial growth mechanisms, see, e.g., Refs. [8, 51]). In such a case, the periodic variations of the work function allow one to record the formation of successive monolayers in the process of the film growth [52] (Figure 34.7).

These and many other experimental results demonstrate the important role of the atomic structure of both the substrate and the adlayer itself in the work function variation. Frequently, the $\varphi$ versus $n$ dependences are much more similar for the surfaces of different metals having similar structure than for different crystal surfaces of the same metal [43, 44, 53, 54]. As was already noticed in early quantitative studies, various singularities (extrema, breaks, inflection points) in the $\varphi(n)$ curves are often observed at $n$ values that relate with the surface concentrations of substrate atoms as small integers (1/4, 1/3, 1/2, 3/4, and so on) [37, 44, 55, 56]. Such findings suggested that these singularities may correlate with certain structures of adlayers that are coherent (commensurate) with the substrate structure. This idea was confirmed in parallel investigations of the work function and atomic structure of the adlayers as well as of phase transitions in them, which depend on lateral interactions of adatoms. Let us consider the mechanisms of such interactions for electropositive adatoms and their impact on the adlayer structure.

## 34.4
## Lateral Interactions of Adatoms and Structure of Electropositive Adlayers

As indicated in Section 34.1, the salient feature of adsorption of alkalis and other electropositive elements is the strong polarity of their adsorption bond. Now we shall discuss the manifestations of this property in the lateral interaction of adatoms and, as a consequence, in their arrangement on the substrate.

Since the dipole moments of alkali adatoms at low coverages ($\sim 10^{-1}$ of a monolayer) amount typically to a few debyes, the most important contribution to lateral forces is due to the dipole–dipole interaction (Figure 34.8).

As follows from an elementary electrostatic consideration, the energy of interaction of two parallel surface dipoles is given by

$$U_d(r) = 2p^2/r^3, \tag{34.7}$$

where $p$ is the dipole moment, which can be determined from the work function change (Equation 34.5), and $r$ is the distance between the dipoles [39, 57]. Equation 34.7 is valid only at large distances when the dipoles may be considered as point-like. The quantitative criterion for this is $r \gg (a + l_s)$, where $a$ is the adatom radius and $l_s$ the field screening length in the metal substrate. At short distances ($r \approx a$), the interaction energy is determined by the direct exchange interaction, which arises due to the overlap of valence electronic shells of adatoms and decays exponentially with increasing distance. We see that the dipole–dipole interaction has a long-range character ($U_d \propto r^{-3}$), is repulsive, and is isotropic. Its important role in physics and chemistry of adsorbed layers with strong polarity of the adsorption bond was discovered by Langmuir. The work function variation due to the electrical dipole layer was already discussed in Section 34.3.

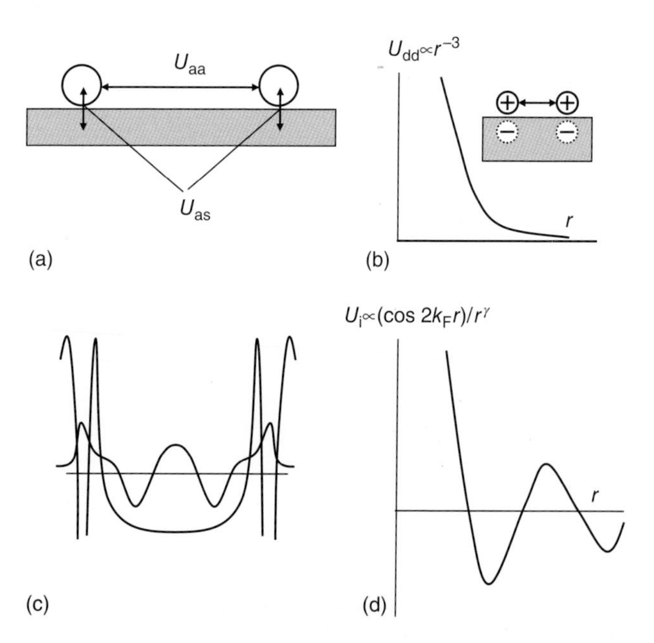


**Figure 34.8** Interactions of adsorbed atoms. (a) Interactions with the surface ($U_{as}$) and lateral interaction ($U_{aa}$). (b) Energy of the dipole–dipole interaction as a function of distance. (c) Superposition of the wave functions via substrate, effecting the indirect interaction of adatoms. (d) Energy of the indirect lateral interaction as a function of distance.

Consider now the manifestations of the dipole–dipole forces in the adsorption energy and atomic structure of electropositive adlayers. Figure 34.9 shows the adsorption energy of cesium atoms as a function of their concentration on the (110) surface of tungsten [36].

It is seen that the energy decreases by a factor of $\approx 2$ as $n$ grows from $n \approx 0$ to $n = n_{opt} = 3.2 \times 10^{14}\ cm^{-2}$, that is, in the charge-transfer region. This decrease is reasonably attributed to mutual dipole–dipole repulsion between adatoms. Further reduction in the adsorption energy occurs in the range $n_{opt}$ to the CPM at $n_m = 5.2 \times 10^{14}\ cm^{-2}$. Here, the metallized adlayer gradually becomes denser until a CPM is formed in which the adsorption energy is close (to within $\approx 0.1$ eV) to the sublimation energy of bulk cesium. On the whole, the adsorption energies of alkali and alkaline-earth elements fall several-fold as one passes from individual adatoms to CPMs [8, 35, 39, 54].

The dipole–dipole repulsion manifests itself most graphically in the atomic structure of electropositive adlayers. It is obvious that repulsive forces will try to keep adatoms as far apart as possible. This means that structures with long lattice periods should form at low coverages. Such structures were actually observed using the method of low-energy electron diffraction (LEED). Since the interaction between adatoms at large distances is in any case rather weak, the ordered long-period structures can exist only at rather low temperatures, so the experimental equipment must provide cooling of the substrates – for example, with liquid nitrogen or helium [43, 56, 58, 59].

As an example, Figure 34.10a shows the structures formed by Sr adatoms on the (110) surfaces of tungsten and molybdenum. It can be seen that at low coverages, very rarefied structures c($7 \times 3$) and c($6 \times 2$) are formed, in which the shortest inter-adatom distances amount to $\approx 10$ Å. As the layers grow denser, the inter-adatom spacings become shorter. However, this process of adlayer compaction is not always smooth.

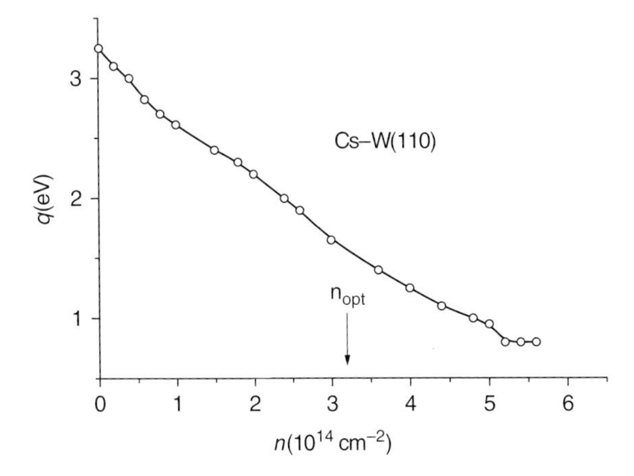


**Figure 34.9** Adsorption energy of Cs on the W(110) surface versus concentration of adatoms.

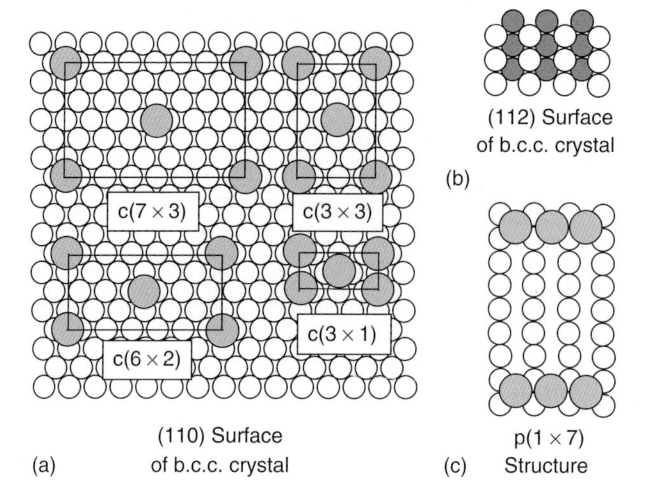


**Figure 34.10** Long-period structures of Sr adatoms on the (110) and (112) surfaces of W and Mo. (a) Structures c(7 × 3) and c(3 × 1) on W(110), structures c(6 × 2) and c(3 × 3) on Mo(110) [42]. (b) Clean (112) surface of bcc crystals (light balls correspond to surface atoms). (c) p(1 × 7) structure on the (112) surface of W.

In the early 1970s, it was found experimentally that the homogeneous distribution of alkali and alkaline-earth metal adatoms on the surface can become unstable in some coverage regions. This effect is revealed as a first-order phase transition (PT-1), in which two adlayer phases of different densities and structures coexist in equilibrium. This was first detected in adlayers of Na on W(110) [37] and later on also for Ba on Mo(110) [38], Li on W(112) [55], Sr on W(110) [57], and many other systems, including alkalis on graphite (see reviews [39, 43, 44, 54, 59, 60]). It should be mentioned that, at the same time (1960–1970s), such transitions were also found in adlayers of Zr and Hf on W and Mo [61, 62]. Although these transition metals, whose electron shell configurations are $4d^2 5s^2$ and $5d^2 6s^2$, are not as electropositive as alkali, alkaline-earth, and RE metals, they provide the work function reduction to ~2.5–3.0 eV.

As the coverage is increased, the two-dimensional (2D) islands of the denser phase grow on the background of the rarefied phase until the whole surface is covered with the denser phase. This is actually the phenomenon of 2D condensation, the occurrence of which in systems with repulsive interaction was rather unexpected. This paradox was first interpreted by Bol'shov in terms of a depolarization model [43, 63]. The strong mutual depolarization of adatoms can lead to an S-shaped dependence of the lateral interaction energy on the distance between adatoms *within the layer* (Figure 34.11).

It can be seen that the interaction energy remains positive at all $r$ values so that in this sense the interaction is repulsive. However, in some interval of distances the depolarization results in a progressive decrease in the repulsion with decreasing interatomic distance ($dU/dr > 0$), which means the occurrence of an effective attraction between adatoms. This attraction operates as a driving force for the 2D

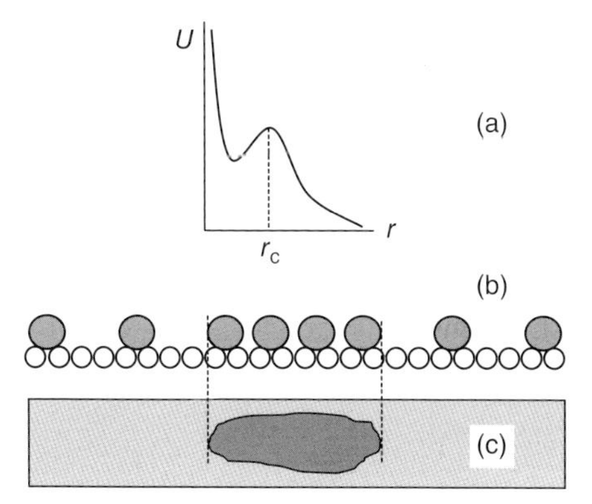


**Figure 34.11** Two-dimensional condensation in an adlayer with repulsive lateral interaction. (a) Variation of the energy of repulsive interaction of adatoms due to their mutual depolarization ($r$ is distance between the adjacent atoms in the adlayer, $r_c$ is the critical distance at which the lateral interaction sets in). (b,c) Models of a heterogeneous adlayer.

condensation. It should be noted, however, that 2D condensation in electropositive adlayers is often observed at such adlayer densities ($\theta \sim 0.3$–$0.5$), where the simple dipole model cannot be expected to give quantitatively correct predictions.

Starting from the mid-1980s, 2D condensation was also detected for alkali metals on Cu, Al, and Ag surfaces (see, e.g., [64–66]). Neugebauer, Scheffler, and Bormet elaborated a more rigorous theory of this phenomenon based on the density functional approach (see Refs [23, 67] and references therein). They calculated the coverage dependence of the adsorption energy for Na and K on Al(111) and also stated its non-monotonic character in some (submonolayer) coverage interval. This singularity implies that in such a region it is energetically profitable for adatoms to arrange into denser islands. Of course, this leads to a substantial change in the dipole moments of adatoms and can result in adlayer metallization (see Section 34.8), accompanied in some cases with a change of the adsorption sites and even with formation of a surface alloy [67]. It is understood that the 2D condensation is also reflected in the coverage dependence of the work function. In particular, the work function minimum, which marks the start of the transition from the charge transfer to metallic state of the adlayer, is tied in many systems to the onset of 2D condensation [37–39, 43, 53].

The 2D condensates of electropositive adsorbates formed in the PT-1s from less dense phases are in most cases commensurate with the substrate structure. However, such structures are usually not close-packed (see Figure 34.11c) and can be further compacted at the sacrifice of the commensuration between the adlayer and the substrate lattices. We will consider the incommensurate adatom structures in more detail in Section 34.5.

Let us again address the dilute, long-period structures and mechanisms of the lateral interaction between adatoms. Previously, we have adduced arguments in favor of the existence of an intense dipole–dipole interaction between electropositive adatoms. This is, however, not the only kind of lateral interaction that determines the structure and other properties of such adlayers. Theory predicts and experimental observations corroborate that an essential contribution to lateral interaction belongs also to the substrate-mediated indirection [68]. The term *indirect* stresses the fact that this kind of lateral forces contrasts with the van der Waals, dipole–dipole, and direct exchange interactions, which do not need substrate mediation to operate (generally, the van der Waals interaction ([69], Chapter 22) plays only a minor role in chemisorbed metal layers). The indirect interaction of adatoms via the substrate is effected through exchange with various quasiparticles excited in the substrate. The most significant contribution originates from exchange with electrons (see, e.g., reviews [39, 57, 68, 70, 71]) (Figure 34.8c). One of the mechanisms of substrate-mediated lateral interactions of adatoms is electron scattering on adatoms and has a long-range character and a few peculiar characteristics. First, its energy $U_i$ is an oscillating function of distance:

$$U_i \cdot \propto \frac{\cos(2k_F r + \beta)}{r^\gamma}, \tag{34.8}$$

where $k_F$ is the Fermi wave vector of electrons, $\beta$ is a phase shift, and $\gamma$ is an exponent depending on the shape of the Fermi surface (Figure 34.8d). Second, the indirect interaction of the electron has a long-range character since, according to theory, the exponent $\gamma$ can vary in the range 1–5. In particular, $\gamma = 1$ is predicted for the direction that is perpendicular to flat areas of the Fermi surface (of course, if such areas do exist). Third, as it is clear from the aforesaid, the indirect interaction must be anisotropic, since Fermi surfaces of real metals always differ from a sphere.

Experimental indications that the indirect interaction may substantially contribute to lateral interaction of electropositive adatoms were found in structural investigations. Using the LEED technique, structures formed by alkaline-earth adatoms were compared on substrates having almost identical atomic structure but different chemical nature [39, 43, 53]. The (110) and (112) surfaces of tungsten and molybdenum were chosen as such substrates. Both tungsten and molybdenum are base-centered cubic (bcc) metals, and their lattice constants differ by no more than 0.6%. Let us recall that, in contrast to the electronic indirect interaction, the dipole–dipole interaction is isotropic and decays with distance monotonically as $r^{-3}$ (see Equation 34.7). If it were the sole mechanism of the lateral interaction, the adatoms should form identical series of 2D lattices on both the substrates. This is, however, not the case: although the structures are similar, they are not identical. For example, strontium adatoms form the structures $c(7 \times 3)$ and $c(3 \times 1)$ only on tungsten, whereas they form the structures $c(6 \times 2)$ and $c(3 \times 3)$ only on molybdenum (Figure 34.10a). These findings conform with the idea that there exists a contribution to lateral forces that depends on the electronic structure of substrate.

Even more spectacular manifestations of the lateral interaction that are sensitive to the substrate electronic structure were observed on surfaces with anisotropic atomic structure, such as (112) surfaces of bcc crystals and (110) surfaces of face-centered cubic (fcc) crystals. An example of such structures is depicted in Figure 34.10c [72]. Adlayer structures on such substrates are highly anisotropic (consist of adatom chains), which is evidently incompatible with pure (isotropic) dipole–dipole interaction [7, 35, 39, 56, 72–74]. On the other hand, the formation, by PT-1s, of dilute chain structures indicates the existence of very long range lateral interaction, giving potential energy minima at distances of the order of 10–20 Å [39, 72, 74]. Again, such lattices formed on surfaces with almost identical atomic structure but with different chemical nature are similar but not the same. These data can be rationalized by assuming a superposition of two dominant interactions – the isotropic dipole–dipole interaction, and the anisotropic oscillating indirect interaction. In particular, the indirect interaction is predicted to decay especially slowly ($\propto r^{-1}$) along the atomic channels on the (112) surfaces of bcc crystals. The dipole–dipole interaction shifts the potential minima in the curve $U_i(r)$ to large distances, which can explain the formation of the chain structures similar to those shown in Figure 34.10c.

The adatom structures discussed above are observed at $n < n_{opt}$, that is, in the charge-transfer region. At $n \geq n_{opt}$, that is, in the metallization region, the character of lateral interaction substantially changes and the structures of electropositive adatoms on anisotropic surfaces become more isotropic. As a rule, the concentration interval $n_{opt} < n < n_m$, where $n_{opt}$ corresponds to work function minimum and $n_m$ to a close-packed ("physical") monolayer, is the region of incommensurate adlayer structures. This is caused by the strong attraction of adatoms to the substrate, which causes some adatoms to leave the substrate potential wells in order to accommodate more adatoms within the first monolayer.

In the region of incommensurate structures (ISs) the adlayer is gradually compressed, which is accompanied by a further strong decrease of the adsorption energy, showing that here the lateral interaction in the adlayer is also repulsive (Figure 34.9). This process continues until the filling of the second monolayer becomes energetically more profitable. Such a moment corresponds to the completion of the formation of the first CPM.

The close-packed ("physical") monolayers of electropositive elements are in most cases incommensurate, except for the rare situations when the favorable (e.g., close to unity) ratio of radii of the adsorbed and substrate atoms allows the formation of a commensurate CPM. An example is the system Li-W(112) [56]. The strong attraction to the substrate results in substantial compression of electropositive adatoms within the CPMs, where their radii can be smaller by 10–20% than the atomic radii in the bulk adsorbates [45, 56]. The growth of electropositive films at *supermonolayer* coverages ($\theta > 1$) occurs in most cases by the Stranski–Krastanov mechanism [8, 51], that is, by building up three-dimensional crystallites over the continuous ("wetting") 2D layer. The initial structure of this layer may rearrange as the film grows thicker to attain a minimum of the free energy in the whole changing system (see, e.g., Refs. [56, 75]).

New possibilities in the study of indirect (substrate-mediated) lateral interactions of adatoms were opened thanks to invention of scanning tunneling microscopy (STM), which allows visualization of electron density distribution around adsorbed particles [71]. It was confirmed very graphically that this distribution and substrate-mediated lateral interaction oscillate with distance, in accordance with Friedel's prediction based on the consideration of scattering (diffraction) of substrate conduction electrons on adatoms. Theoretical works showed that the dominating role in this effect belongs to surface electronic states [39, 71]. Of course, the substrate affects also the interaction of adatoms on short (atomic) distances. This is revealed, for example, in the easy dissociation of adsorbed molecules on some surfaces, which is widely used in heterogeneous catalysis.

## 34.5
## Phase Transitions in Electropositive Adlayers

Now we will focus on some regularities of how 2D phases of electropositive adatoms replace each other when the adatom concentration (coverage) or the temperature is varied. Figure 34.12 shows a schematic diagram of structure states typical of adlayers with repulsive lateral interaction.

The diagram relates to structures that are formed within the first monolayer and represents a simplified section of the phase diagram at a constant (not too high) temperature. General regularities of phase transitions in the adsorbed layers are comprehensively presented in several reviews [53, 76–80]. The experimental examples of phase diagrams for electropositive adlayers on metals and graphite can be found in many reviews [8, 43, 45, 54, 59, 60, 70, 81] and in the original works cited therein. For a number of systems not only the symmetry and periods of 2D lattices were determined but also the exact positions of adatoms on adsorption sites, the length of the adsorption bonds, and the shifts (relaxation) of substrate atoms induced by adsorption. The most comprehensive survey of these results was given by Diehl and McGrath [45, 81].

Let us first discuss the commensurate structures. The abundance of such phases in adlayers where repulsive lateral interaction is dominant offers wide possibilities for experimental investigations of phase transitions in 2D systems having different symmetry, density, and chemical nature. It is known that PT-1s can proceed between phases of arbitrary symmetry, which are separated by an interface and coexist in equilibrium. Such transitions start at proper conditions (in our particular case the coverage and temperature) with spontaneous formation of nuclei of the new phase, usually near some favorable sites such as surface defects of various kinds. The long-range order sets in or changes in this case in jumps.

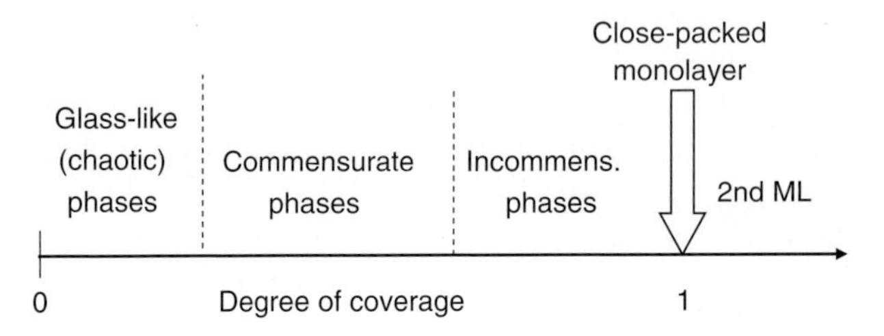


**Figure 34.12** General scheme of structure transitions within the first monolayer of adatoms with repulsive lateral interaction.

The continuous character of second-order transitions imposes some strict constraints on the symmetry of adjacent phases [53, 80]. The reason is that in this case one phase replaces another simultaneously over the entire area of the system, and no interface is formed. The most interesting feature of second-order transitions is their universality. This means that a number of physical characteristics near the transition point vary with temperature and other parameters as power-like functions with exponents depending solely on the lattice symmetry. This statement relates to the order parameter, specific heat, lattice compressibility, correlation length, and correlation function. There are a number of 2D lattices for which the critical exponents were calculated exactly. Experimental results correlate well with the theoretical predictions [53, 80].

Consider now more closely the transitions between commensurate and incommensurate phases (C–I transitions) [51, 53, 76, 82]. At the initial stage, the commensuration between the adlayer and the substrate is broken only locally (obviously, its simultaneous destruction throughout the whole surface would be energetically unprofitable). As a result, the adlayer in this state consists of commensurate domains separated by incommensurate *domain walls*, which are often termed *misfit dislocations* or *solitons* (Figure 34.13).

An important characteristic of the soliton (domain wall) is its width $l_0$. Intuitively, it is clear that $l_0$ should depend on the ratio of the repulsive energy, which seeks to keep adatoms as far apart as possible, to the depth of the substrate potential corrugation, which tends to bind adatoms to substrate's most favorable sites. The repulsive lateral interaction can be characterized by an elastic constant of the adlayer, $\lambda$. It is reasonable to expect that the higher the $\lambda$, the larger the soliton width $l_0$. On the other hand, a deep potential corrugation of the substrate should cause the wall to become narrower. A detailed assessment of the wall width at low temperatures gives

$$l_0 \sim b\sqrt{\lambda/V}, \tag{34.9}$$

where $V$ is the amplitude and $b$ is the period of the substrate potential [52].

Actually, the soliton (domain wall) represents a linear defect of packing in the 2D commensurate lattice. It contains, in the simplest case, one additional row of adatoms when the coverage exceeds the stoichiometric value $\theta_c$ corresponding to the commensurate phase (or a row of vacancies at a substoichiometric coverage $\theta < \theta_c$, in which case one has an antisymmetric situation corresponding to emergence of *antisolitons*). Such defects have a topological character: at $\theta =$ const., the soliton and

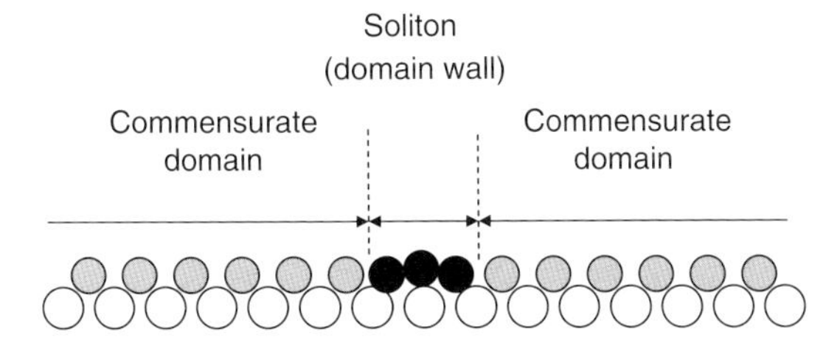


**Figure 34.13** Model of a soliton (incommensurate domain wall) between the commensurate domains at a superstoichiometric adatom concentration.

antisoliton, once created, can disappear either by annihilation (in the case of thermal excitation they can be generated only in pairs) or, if the adlayer has a free boundary, through migration to the boundary and shifting it respectively forward or backward by one lattice period [53, 82–84] (see Section 34.9).

The solitons repel each other. The main contributions to the repulsion originate from the thermally induced collisions of the solitons and from the dipole–dipole interaction. On the whole, the repulsion asymptotically decays with distance $l$ between the solitons as $1/l^2$ and leads to formation of soliton lattices [53, 82]. Their symmetry depends on the symmetry of the substrate lattice and the angular properties of lateral interaction. The simplest soliton lattices, having a structure of parallel stripes, are formed in the situation when the C–I transition occurs by a uniaxial compression of the adlayer. The period of such soliton lattice varies gradually with varying coverage, so the transition is continuous (second order). On isotropic surfaces, an isotropic compression of the adlayer can result, depending on the energy parameters of the system, either in the striped soliton lattice (by second-order transition) or in a honeycomb soliton lattice (by first-order transition).

As temperature is raised, the solitons start to fluctuate by bending and shifting, and their width increases because the growing energy of thermal excitations makes the impact of the substrate potential relief less essential. In the final score, there occurs a transition to a completely incommensurate phase. However, as long as the distance between the solitons is considerably larger than their width, they may be treated as separate objects.

With growing coverage, the surface concentration of solitons increases and the distance between them decreases. As this distance becomes comparable and eventually equal to the soliton width, the commensuration between adlayer and substrate breaks throughout the surface. Such an incommensurate phase has a gapless sliding mode, that is, the adlayer might be shifted without activation if the surface structure were perfect. However, real surfaces always contain some defects, so the nonactivated sliding of the adlayer is unlikely to be observed, at least on macroscopic samples.

As the coverage varies, the orientation of the incommensurate adlayer lattice with respect to the substrate lattice can change either abruptly (by first-order transition) [36] or continuously (by second-order transition) [85]. Such transitions, termed *rotational*, are widely observed in electropositive adlayers (see Refs. [45, 54, 59] for review). They stem from the minimization of the free surface energy by rotating the adlayer lattice. Obviously, the optimum matching between the surface lattice and incommensurate adlayers having different periodicity is attained at different orientation angles.

Incommensurate adlayers, being structures where adatom interaction dominates over the substrate potential corrugation, represent some models of adlayers on a perfectly smooth substrate. This allowed the verification of the theoretical prediction made in the 1930s by Landau and Peierls that strict long-range order is impossible in purely 2D systems [53, 86]. The term "strict long-range order" implies that the binary correlation function, which determines the positions of atoms, does not decay with distance. This is the property of crystals, both three dimensional and also such two-dimensional (actually, quasi-2D) ones as are located on a crystalline substrate and form a commensurate lattice with it. A "purely two-dimensional system" might be modeled by a layer adsorbed on an ideally smooth (structureless) substrate, which exerts no effect on the adlayer structure. Although any real solid surface tries to impose its potential relief on the adlayer structure, the ISs, to some extent, approximate the layers on smooth substrates.

Figure 34.14 shows the temperature dependences of the intensities of diffraction beams recorded in LEED experiments for the commensurate c(2 × 2) and an incommensurate (hexagonal) structure of Na on the W(110) surface [86].

The curves reflect a substantial distinction in the process of thermal disordering of these structures: the long-range order existing in the commensurate structure is destroyed in a rather narrow temperature range, while the disordering of the IS proceeds gradually in a broad temperature range. This result can be understood if one takes into account that, as said above, ISs resemble adlayers on a structureless surface. According to theory, such adlayers may possess only the so-called quasi-long-range order [53]. In this case, at $T > 0$, the binary correlation function does decay with distance, but this decay is power-like ($\propto r^{-\eta}$, where exponent $\eta$ is temperature dependent) and not exponential as in usual liquids. The vibration spectra of incommensurate adlayers, in contrast to spectra of three-dimensional crystals and 2D commensurate adlayers, contain a soft (low-frequency) mode. This gives rise to a peculiar (continuous) character of melting of 2D ISs. It is noteworthy that, despite the absence of the strict *positional* order in the incommensurate adlayers (see above), the long-range *orientational* order does exist in them. For this reason, the IS in the course of its melting, that is, its transformation to an isotropic 2D liquid, should first pass through a 2D liquid crystal phase [87].

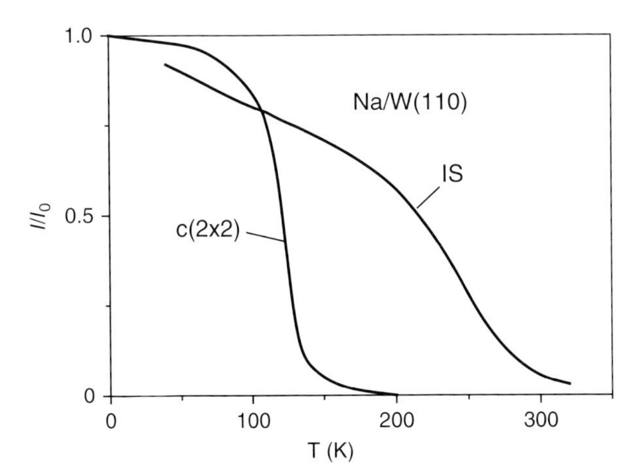


**Figure 34.14** Temperature dependences of the intensity of LEED spots for the commensurate c(2 × 2) structure and an incommensurate hexagonal structure (IS) of Na on W(110). $I_0$ is the intensity at $T = 77$ K [86].

In general, there are numerous observations of the ISs and C–I transitions in electropositive adlayers [35, 39, 45, 53, 59, 88]. In particular, it was found that motion of the solitons can provide fast mass transport within adlayers (see Section 34.9). The mechanism of soliton diffusion was elaborated by Lyuksyutov and Pokrovsky [53, 89, 90]. A soliton senses the periodic potential of the substrate, that is, it is pinned to the substrate lattice. This is analogous to the existence of the Peierls barrier for dislocation motion in solids. Nonetheless, the activation energy for a soliton displacement is substantially lower than that for an individual adatom.

To conclude the discussion of the structures of electropositive adlayers, we shall consider metastable disordered ("chaotic") structures, which take the left side of the structure diagram (Figure 34.12). Earlier (see Section 34.4) we pointed out that the observation of long-period 2D lattices, which reveal the existence of long-range lateral interactions, necessitates experiments at low temperatures. It is understood, however, that at too low temperatures the formation of equilibrium long-period structures is hindered kinetically. Let us dwell on this point in more detail. To observe an ordered adlayer, it is necessary to cool the sample below the temperature of the order–disorder phase transition, which is estimated as

$$T_o = CE_o/k, \tag{34.10}$$

where $E_o$ is the energy gain due to ordering, $k$ is the Boltzmann constant, and $C$ is a coefficient $\sim 1$ determined by the symmetry and dimensionality of the system [91, 92]. Thus the temperature $T_o$ sets the upper limit to the observation temperature. On the other hand, its lower limit is imposed by adatom mobility, which determines the rate of adlayer equilibration. Suppose the adatoms must make, on the average, $N$ elementary jumps to equilibrate the layer. Computer simulations show that $N \approx 10$ is usually a sufficient number for not very complex systems (see, e.g., [93, 94]). $N$ can be written in the form

$$N = \nu e^{-E_d/kT_d} \cdot t, \tag{34.11}$$

where $\nu$ is the adatom vibration frequency, $E_d$ is the activation energy of surface diffusion, $T_d$ is the sample temperature at which the diffusion occurs, and $t$ is the time of experiment. Hence one can evaluate the temperature that enables adatoms to perform $N$ jumps during time $t$:

$$T_d = \frac{E_d}{k \ln(\nu t/N)}. \tag{34.12}$$

A rough estimation with $\nu = 10^{13}\,\mathrm{s}^{-1}$, $t = 10\,\mathrm{s}$, and $N = 10$ gives

$$T_d \approx E_d/30\,\mathrm{K}. \tag{34.13}$$

Note that $\nu$, $t$, and $N$ enter into Equation 34.12 as logarithms, so their variation by orders of magnitude would not change essentially the estimate of $T_d$. Obviously, the observation of an ordered structure is possible only when the inequality $T_d < T_o$ is obeyed. Suppose that the $E_d$ value in Equation 34.13 remains approximately constant for different long-period structures. At the same time, the value of $T_o$ (Equation 34.10) should evidently be lower if the period of the structure is longer.

Therefore, the inequality $T_d < T_0$ will no longer be satisfied at some period length, which means that such a structure cannot be ordered during a reasonable time of the experiment. It is clear that in this case the annealing of the adlayer at high temperatures cannot bring about its ordering, and the disordered (melted) layer will freeze to a 2D glass. For alkali and alkaline-earth adsorbates, this is observed only at very low coverages (~0.1 of a monolayer). For RE electropositive adsorbates, the formation of 2D glasses was found to occur at medium coverages ($\theta \approx 0.2$–$0.6$) after annealing at high temperatures [49]. Such annealing seems to produce a 2D surface alloy that consists of adatoms and substrate atoms and is able to form a 2D metallic glass even at low cooling rates (~1–10°/s). In the important area of amorphous and nanocrystalline materials [95], the 2D metallic glasses may be of both basic and practical interest as low-dimensional solids with extremely high concentration of defects.

## 34.6
## Surface Reconstruction Induced by Alkali Adsorbates

The interaction of electropositive adatoms with metal and semiconductor substrates is rather strong (the adsorption energy amounts to 2–5 eV) and can induce a substantial surface reconstruction in some systems (see Refs [26, 96] and, in particular [97–99] published in them, as well as reviews [100–102]). The best studied case is the reconstruction of the {110} surfaces of fcc metals stimulated by alkali adsorbates. These surfaces have the largest number of broken nearest neighbor bonds in comparison with other singular surfaces of fcc crystals, and for this reason they are most disposed toward reconstruction. In particular, the {110} planes of the 5d metals Au, Pt, and Ir reconstruct even being clean, showing a doubling periodicity in the direction normal to the $[1\bar{1}0]$ close-packed rows of surface atoms. The experimental results are interpreted on the basis of the so-called missing-row (MR) model, in which every second row of atoms is absent on the reconstructed surface. This rearrangement is accompanied by some accommodating displacements of atoms in the second and third subsurface layers.

At the same time, the {110} planes of 3d and 4d fcc metals Ni, Cu, Rh, Pd, and Ag do not reconstruct when clean, but undergo an induced reconstruction to $(1 \times 2)$, $(1 \times 3)$, and sometimes other similar lattices under the effect of alkali adsorbates. This reconstruction is thermally activated, and the lattice $(1 \times 2)$ was found to be the most stable. Calculations based on the local density functional theory showed that a mere 0.05 extra electrons per surface atom is sufficient to induce the $(1 \times 1) \rightarrow (1 \times 2)$ transition [103]. These extra electrons are believed to be donated by the alkali adatoms. On the atomic scale, the $(1 \times 2)$ missing row structure is rougher than the $(1 \times 1)$ structure. This gives more area for electrons and allows them to reduce their charge density and kinetic energy, which actually is the driving force of the $(1 \times 1) \rightarrow (1 \times 2)$ transition on the clean surfaces. In the presence of an adsorbate, another contribution to this driving force is the larger adsorption energy of alkali atoms on the $(1 \times 2)$ surface, where the deeper atomic channels provide more neighbors (a higher coordination) to adatoms.

Since the differences in surface energies of various surface phases are rather small ($\sim 10^{-2}$ eV per surface unit cell), the equilibrium structure is very sensitive to the state of the adlayer. For example, at low K coverages on Cu(110), that is, in the charge-transfer region ($\theta < 0.25$), the missed row structure $(1 \times 2)$ is energetically favorable. However, in the metallic region ($\theta > 0.35$), the donation of electrons from adatoms to the substrate is reduced and the bulk truncated $(1 \times 1)$ surface structure may be restored [104]. A deep reconstruction of the Cu(001) surface induced by Mg adsorption has been found in Ref. [105].

STM studies allowed direct observations of the adsorbate-induced reconstruction on an atomic scale [98, 100, 102, 106]. It has been found that nuclei of the reconstructed phase arise even around single alkali adatoms, and a rich diversity of local surface structures has been discovered with varying coverage.

There have also been a number of experimental observations and theoretical calculations which testify that alkali adatoms on aluminum surfaces and alkaline-earth adatoms on molybdenum mix with the substrate atoms to form surface-limited bimetallic alloys (see [21, 102, 107, 108] and references therein). The surface alloy of RE atoms with transition-metal atoms, formed in the system Dy-Mo(112) at high annealing temperatures, was found to freeze into a glass-like state on cooling [49]. It can thus be stated that reconstruction driven by electropositive adsorbates is an interesting and practically important phenomenon, which presents the possibility of targeted tailoring the surface properties.

## 34.7 Electropositive Adlayers on Semiconductors

In this section we will compare adsorption of electropositive adsorbates on metals and semiconductors. As an example, we will take the system K/Si(100) $2 \times 1$, which is relatively simple and investigated in much detail. The surface Si(100)$2 \times 1$ consists of Si dimers separated by atomically deep channels. The dimers arise as a result of surface reconstruction that reduces the surface energy by partial saturation of the dangling bonds of Si atoms. We shall not go into the details of the surface structure; instead, let us focus on the surface electronic properties. For semiconductors, the work function consists of three contributions (Figure 34.15):

$$\varphi = \Delta + \chi + V_{BB}, \tag{34.14}$$

where $\Delta$ is the energy spacing between the Fermi level and the bottom of the conduction band, $\chi$ is the electron affinity (the energy difference between the vacuum level and the bottom of the conduction band), and $V_{BB}$ is the band bending near the surface. An important role in the surface electronic structure of semiconductors is played by the surface states.

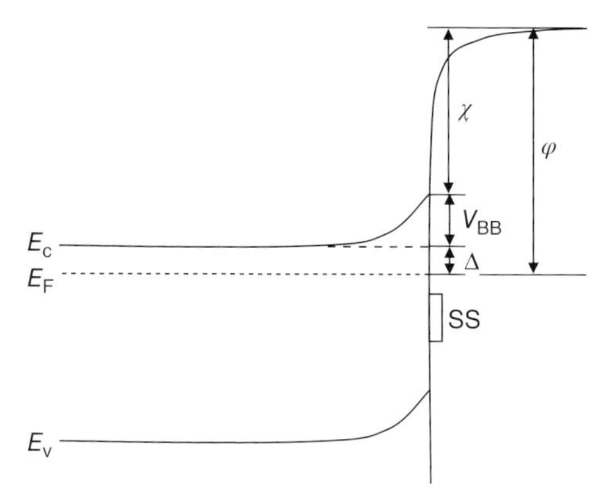


**Figure 34.15** Energy diagram of a semiconductor surface. See text for explanations.

In spite of the basic distinction in the electronic structure of metal and semiconductor substrates, the coverage dependences of the work function and adsorption energy for alkali metals on these substrates are very similar. For example, the adsorption energy of K on Si(100)$2 \times 1$ decreases by $\approx 1.3$ eV within the range of submonolayer coverages [109]. For the densely packed K monolayer, it is close to the K sublimation energy. The work function of this substrate first decreases by more than 3.0 eV to reach a minimum at $\theta = 0.5$, and then grows to a saturation value in the dense monolayer. A distinctive feature of the $\varphi(\theta)$ curves on semiconductor surfaces is their noticeable sensitivity to the substrate temperature. This is connected with the temperature variation of the terms $\Delta$ and $V_{BB}$ in Equation 34.15 (the sum $\Delta + V_{BB}$ is usually called the Schottky barrier). Let us recall that the band bending value, $V_{BB}$, depends on the existence, nature, and population of surface states, which, in turn, depend on the temperature, nature of adatoms, degree of coverage, illumination, external electric field, and other factors [110].

Depending on the position of the adatom valence level with respect to the energy bands and surface states of the semiconductor substrate, electropositive adatoms can interact in a complicated way with its electronic system to form delocalized bonds (like in the Gurney model) or localized "surface molecules," which may possess considerable dipole moments. Photoemission spectroscopy in its various versions is an efficient method to study the surface electronic structure of such systems.

For example, the perturbation of electrostatic potential at the Si(100)$2 \times 1$ surface induced by K adatoms was probed with the PAX method (see Section 34.2). Next to the K atoms on Si, the potential is reduced by 0.5 eV with respect to the overall (long-range) reduction of $\approx 0.23$ eV [109]. The former value is very close to that found for the K/Ru(100) system. These results show that, generally, the effect of K adsorption on the surface potential distribution is similar for metal and semiconductor surfaces. Some distinctions noticeable on these substrates at large distances from alkali metal adatoms may be due to smaller field screening length in metals than in semiconductors.

On the whole, the range of low coverages (before the work function minimum) is "the charge-transfer region" on many semiconductors, as in the case of metal substrates. Past the work function minimum, various experimental techniques detect the metallization of the adlayers on semiconductors, again like on metals.

Systematic studies of the surface electronic properties of Si(100) and (111), GaAs(100), and GaN(0001) covered with Cs, K, and Ba adlayers have been performed by Benemanskaya *et al.* using threshold photoemission spectroscopy [111–113]. Their results exemplify the specificity of the interaction of electropositive metals with semiconductor surfaces. For instance, the electronic properties of Si(100)1 $\times$ 2 covered with Cs and Ba are very similar (semiconductor-like) at low $\theta$'s and determined mainly by the substrate. On Si(111)7 $\times$ 7, the Cs, K, and Ba adlayers at low coverages ($\theta < 0.2$) were found to suppress the initial metallicity (surface conductivity) of this surface. However, at somewhat higher coverages ($\theta \approx 0.35$ for Cs and $\theta \approx 0.5$ for K), there occurs a reentrant nonmetal-to-metal transition, this time immediately within the adlayers. Contrary to this, the metallization of Ba adlayers on Si(111)7 $\times$ 7 was not observed up to $\theta \approx 2$. The complex character of the metallization of Cs and Na adlayers on the gallium-rich GaAs(001) surface was investigated in Ref. [114] (see also [113] and references therein).

Many similarities exist between the structures formed by electropositive adsorbates on metal and semiconductor surfaces. The geometry of these structures on semiconductors suggests that, in addition to the dipole–dipole interaction, an important part is played by substrate-mediated lateral interactions, which can have a long range and strongly anisotropic character. For example, K adatoms on the InAs(110) surface, which has a channeled atomic corrugation, assemble into chain-like structures well resolved in STM [115]. The chains are oriented along the atomic channels, that is, along the $[1\bar{1}0]$ direction. Their formation occurs even at coverages as low as $\sim 10^{-2}$ of a monolayer, and the maximum distance between the chains recorded in Ref. [115] was 73 Å. The chain structures of alkali metals were also found on the (110) planes of other III–V semiconductors (see references in [115]) as well as, much earlier, on the Si(100)2 $\times$ 1 surface, which has a channeled atomic corrugation due to reconstruction [116]. In the latter case, the adatom chains are also oriented along the channels, and at a saturated monolayer coverage, the distance between the chains is 7.68 Å whereas the period within the chains is much shorter. For example, the distance between the nearest K adatoms in the close-packed chain is 3.84 Å, which is shorter by 17% compared to the atomic diameter in bulk potassium. As a result, the energy loss spectra of electrons scattered by such chains reveal plasmon excitations characteristic of a quasi-one-dimensional metal. Let us also recall that many chain structures of electropositive adatoms on metal surfaces with anisotropic corrugation were discovered and studied since the 1970s (see reviews [39, 117, 118]).

The presence of electropositive adatoms on semiconductor surfaces can promote catalytic reactions, in particular surface oxidation and nitridation, which are of technological interest [119, 120]. Besides, cesium as the most electropositive adsorbate is used to attain the regime of negative electron affinity (NEA) at the surface of some semiconductors. In this regime, the vacuum energy level is positioned below the bottom of the conduction band, and thus most of electrons excited to this band can unobstructedly leave the solid and pass to vacuum. The most popular substrate for NEA photoemitters is GaAs(100). To provide a strong downward shift of the vacuum level, it is necessary to create a powerful dipole layer at the semiconductor surface. This is achieved by the deposition of a mixed Cs–O film. The presence of the strongly electropositive (Cs) and the strongly electronegative (O) components in such a film ensures the creation of chemical bonds with a large dipole moment and, as a result, gives the affinity $\chi < 0$. The technology of preparation of Cs-O films is rather involved and we shall not go into its details here. Notice only that the best performance of NEA photoemitters is usually realized when the thickness of the Cs-O coating is larger than a monolayer, and it is also important to optimize its composition [121, 122]. Such photoemitters are widely used in infrared devices and as sources of polarized electrons.

On the whole, considering such complex and interrelated phenomena on semiconductor surfaces as deep and often intricate reconstruction, filling, or depopulation of surface states, inducing new states by adatoms, band bending, size effects, strong impact of temperature and irradiation, and so on, it should be realized that the physical picture of adsorption of electropositive elements on semiconductors is much more involved and specific than on metals. There are rich possibilities of modifying semiconductor surfaces with electropositive metals.

## 34.8 Nonmetal–Metal (NM) Transitions and Plasma Excitations in Electropositive Adlayers

Metallization of electropositive adlayers is a phase transition that entails strong changes in surface properties. Its manifestation in the work function was discussed in Section 34.3, and here we will give some other examples.

The onset of metallization can be revealed from the energy spectrum of scattered electrons, namely from the emergence of characteristic energy losses corresponding to excitation of plasmons in the adlayer [123–132]. Figure 34.16 illustrates the correlation between the work function variation, a C–I structure transition, and the onset of plasma energy losses experienced by a probing electron beam scattered from the Mg adlayer on the Mo(112) surface [128, 131].

Besides, the NM transitions, which represent a crossover from a localized bond-like electronic state to a delocalized band-like state, reveal themselves in the disappearance of the energy gap at the Fermi level, which is observed in the energy spectrum of photoelectrons [128, 133]. The metallization usually occurs in parallel with the commensurate–incommensurate transition, in which the adlayer becomes, in a sense, less dependent on the substrate due to increasing direct overlap of adatom valence shells. The scenarios of metallization are sensitive to the atomic structure and chemical nature of the adlayers. For example, the NM transition in the course of compression of adlayers is predicted to occur gradually for chain-like (linear) structures and sharply for hexagonal structures [133–135]. The evolution of the electronic

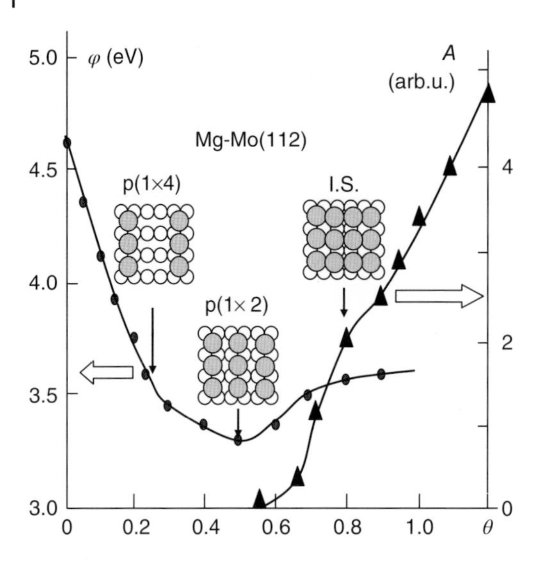


**Figure 34.16** Coverage dependences of the work function ($\varphi$) and intensity of the electron energy loss by surface plasmon excitation ($A$) for Mg adlayer on Mo(112). Structures of Mg correspond to different coverages indicated by arrows. $\theta$ is defined as $n/n_{Mo(112)}$. IS is an incommensurate structure [128, 131].

structure is found to include the hybridization of s orbitals not only with p orbitals but also, for heavier alkaline-earth metals (Ca, Sr, and Ba), with d orbitals [135–137].

The NM transitions that occur in the adlayers with growing coverage are also clearly reflected in the yield of photoemission. The early quantum mechanical models of photoemission considered the photoexcitation of electrons in solids via one-electron transitions described in terms of the band theory. The excited electrons can then pass to vacuum if their energy exceeds the work function. The quantum yield of photoelectrons can be enhanced by increasing the light absorption, minimizing the energy losses of electrons on their way to the surface, and facilitating the exit of electrons to vacuum by reduction of the work function. Traditionally, this combination of requirements could be satisfied in an optimum way for semiconductor photoemitters covered with electropositive adlayers. However, in the 1950s, Borziak detected a peak in the energy spectrum of photoelectrons correlating with the exciton absorption of light (see Ref. [138]). These results suggested the possibility of a photoemission mechanism whose first stage is the generation of collective excitations by the incident light. In the second step, the energy of these excitations is transferred to single electrons that are emitted to vacuum.

Later experimental and theoretical works showed that, under some specific relation between the electron gas density at the metal surface and the frequency of the incident light, the electromagnetic field can be concentrated in a very thin ($\sim 10^{-8}$ cm) layer at the surface [139]. This actually means that only a surface monolayer of atoms of a bare metal or an adsorbed monolayer of foreign atoms will be activated by light [140, 141]. The effect is explained by resonance excitation of the so-called multipole plasmons, whose energy, in the process of their decay, is expended very efficiently to emit electrons [130, 139]. The existence of multipole surface plasmons was predicted by Bennett [142], who considered various modes of plasma oscillations with regard to a smooth decrease of charge density at the surface. The multipole mode of plasma oscillations is well expressed when the charge density distribution is sufficiently wide, which is just the case for electropositive adlayers.

In surface plasmons originally predicted by Ritchie [143], the charge distribution normal to the surface contains only one peak, and for this reason they can be termed *monopole* (Figure 34.17).

Among different plasmon modes, monopole surface plasmons have the lowest energy $\approx 0.7E_\mathrm{p}$ ($E_\mathrm{p}$ is the volume plasmon energy). In contrast, the charge distribution in multipole surface plasmons perpendicular to the surface contains two peaks of opposite signs, that is, it has a dipolar character [130]. In other words, the charge density distribution in the multipole plasmons integrates to zero in the direction orthogonal to the surface plane. Both modes propagate along the surface as plane waves, so at each point the negative and positive charge alternate periodically in time. As a result, the charge distribution *parallel to the surface* in both types of surface plasmons is dipolar. However, taking into account the normal component, the global symmetry of the charge distribution in the multipole plasmons is quadrupolar. According to theoretical predictions, the energy of the multipole plasmon amounts to $\approx 0.8$ of the energy of the volume plasmon:

$$E_\mathrm{mp} \approx 0.8\hbar(4\pi n_\mathrm{e} e^2/m)^{1/2}, \tag{34.15}$$

where $n_\mathrm{e}$ is the electron concentration, and $e$ and $m$ are the charge and mass of the electron, respectively.

The narrow localization of the electromagnetic field at the surface, which occurs in the case of excitation of multipole plasmons, is reflected in the resonance enhancement of the photoelectron yield [124, 144]. Figure 34.18 exemplifies the experimental

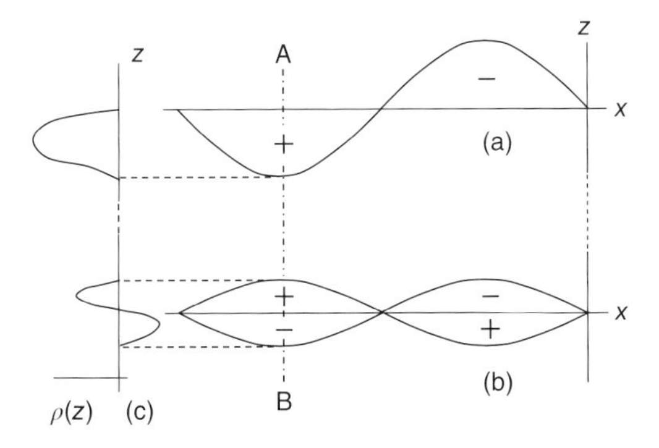


**Figure 34.17** Scheme of charge distributions in a monopole (a) and multipole (b) surface plasmons. Panel (c) shows charge density variation in the plasmons normal to the surface along section AB. The *x*-axis is directed parallel and the *z*-axis normal to the surface. (Adapted from [130].)

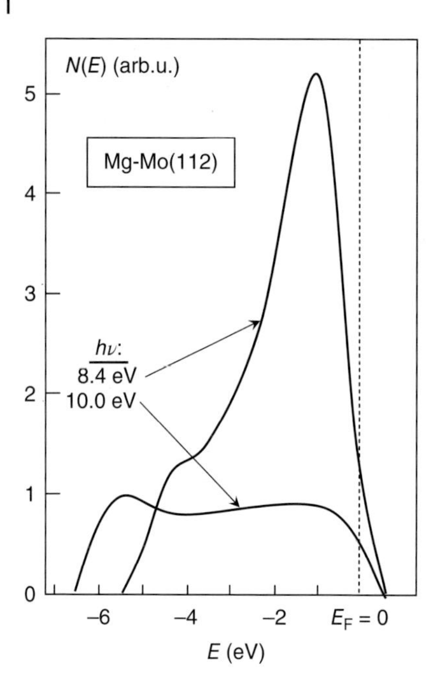


**Figure 34.18** Energy spectra of photoelectrons emitted from the Mg-Mo(112) system illuminated by ultraviolet light with quanta $h\nu = 8.4$ and 10.0 eV. The photoelectron yield (proportional to the area under the curves) is strongly enhanced as $h\nu$ is decreased from 10 eV (energy of volume plasmons) to 8.4 eV (energy of multipole surface plasmons) [131, 138].

results obtained for the system Mg-Mo(112), which demonstrate that the photoelectron current strongly increases when the energy of photons equals $E_{mp}$ [131, 138].

The most probable reason for this effect is that the photoelectrons originate in such a case from the outmost atomic layer of the solid. Multipole plasmons that are generated there transfer their energy to electrons, which can be emitted to vacuum. The pronounced surface character of this emission mechanism is also corroborated by the extremely high sensitivity of the photoelectron yield to the state of the surface [145, 146]. For example, the addition of only 0.1 monolayer of barium or oxygen to the Mg adlayer on Mo(112) suppresses almost completely the resonance photoemission peak, although the work function is decreased by these additives [138]. The coadsorbed species seem to change the energy of plasmons in the Mg adlayer and thereby shift the photoemission resonance to another spectral region or, if the adlayer is changed from a metal to a nonmetal state, suppress the plasmon excitations completely.

As noted above, the plasmon excitations in electropositive adlayers become possible on achieving a critical degree of coverage, which corresponds to the NM transition. This usually occurs in the region of ISs, which allows a gradual variation of the adlayer density and, consequently, of surface electron density as well as of plasmon energy. As a result, it is possible to tune the frequency of the plasmon oscillations to the frequency of incident light and realize a resonance (plasmon-assisted) photoemission. Another way of tuning is to vary the frequency of light while the adlayer remains unchanged.

It is understood that electropositive activating coatings on cathode surfaces can be formed not only by vacuum evaporation but also by means of surface segregation of a metal dissolved in the substrate volume. This method is advantageous in some applications, since in the case of desorption the adlayer can be replenished on the surface by diffusion from the substrate. For example, good performance characteristics were obtained with Mg-Ba alloy photoemitters, which generated strong pulse currents under laser irradiation and showed good longevity even in a technical vacuum of $10^{-4}$–$10^{-3}$ Pa [138]. By the way, just such an activation technology is used in the W-Th cathode, which was invented by Langmuir and Rogers in 1914 and provided a long operation life.

Earlier, we considered the metallization of electropositive adlayers on metal substrates. On dielectrics, this transition was found to occur in a different way. For example, Mg film on the $Cr_2O_3(0001)$ surface becomes metallic only on completing the second monolayer [147]. This is attributed to the fact that the first Mg monolayer forms strong covalent bonds with the substrate.

## 34.9 Surface Diffusion of Electropositive Adsorbates

The rate of surface diffusion is highly sensitive to the atomic structure and the potential corrugation of the surface. This has been known for decades, mainly from experiments carried out in field-emission and field-ion microscopes [148–152]. These devices give strongly (about million-fold) magnified images of the apexes of tips representing mosaics of various crystal planes. It should be noted that surface potential corrugation is *the characteristic of the system adatom + substrate*. This corrugation is probed while tracking random walks of a single adatom. Such a possibility exists in field-ion and scanning tunneling microscopes. However, if one observes diffusion in an ensemble of adatoms, their interaction can dramatically affect the diffusion rate. It is easy to understand that, in the case when diffusion occurs in a concentration gradient, the repulsive lateral interaction should enhance the diffusion rate whereas the attractive interaction will impede it.

As we have seen previously (Sections 34.4 and 34.5), the lateral interactions characteristic of electropositive adlayers manifest themselves in manifold adatom structures and in complex phase diagrams. Suppose the diffusion rate is sufficient to establish a local quasi-equilibrium within small adlayer areas whose size is comparable to the spatial resolution of the probe. Then the region of the concentration gradient (the diffusion zone) represents a series of 2D adsorbate phases corresponding to different coverages at the temperature of diffusion. Each of the phases has its specific diffusion parameters (the activation energy $E_d$ and pre-exponential factor $D_0$), which, as experiments show, may be strongly distinct for different phases [92, 149, 153–157]. This results in a dynamical self-organization of the diffusion zone, which mirrors the phase effects in the diffusion kinetics (actually, the process of 2D reaction diffusion).

Electropositive adlayers, because of the richness of their phase diagrams in which many phases are presented, are convenient objects for studying the regularities of surface diffusion in the systems of different chemical nature and symmetry.

Up to now, there are only a few systematic experimental works in which quantitative data on surface diffusion kinetics were obtained in broad intervals of coverages and temperatures and for systems with documented atomic structures. As an example, let us consider the dependences of the diffusion coefficient versus coverage obtained for Li, Sr, Dy, and Cu on the Mo(112) surface (Figure 34.19) [150, 152]. They were derived by the Boltzmann–Matano method (see, e.g., Refs. [90, 156, 158]) from the adatom concentration profiles recorded experimentally in the process of diffusion of adsorbates out of initial step-function deposits.

Recall that the (112) surfaces of bcc crystals have a highly anisotropic (channeled) structure (Figure 34.10b,c). For this reason, at not too high temperatures, the diffusion of electropositive adsorbates was shown to have a quasi-one-dimensional character, that is, to proceed mainly along the channels. Let us first compare the diffusion rates of Li, Sr, and Dy, which represent alkali, alkaline-earth, and RE elements, respectively. It can be seen that comparable values of their diffusion coefficients are achieved, at least for some coverages, at the temperatures that differ several-fold. Since the dependence of the diffusion coefficient on temperature is exponential, this means that the $D$ values for Li, Sr, and Dy at the same temperature differ by many orders of magnitude. Also, an extremely strong (and generally non-monotonic) variation of $D$ with coverage is observed for each system. It amounts to six orders of magnitude for Li and to more than two orders of magnitude for Sr and Dy. The coverage dependence of $D$ is caused by lateral interactions and structure

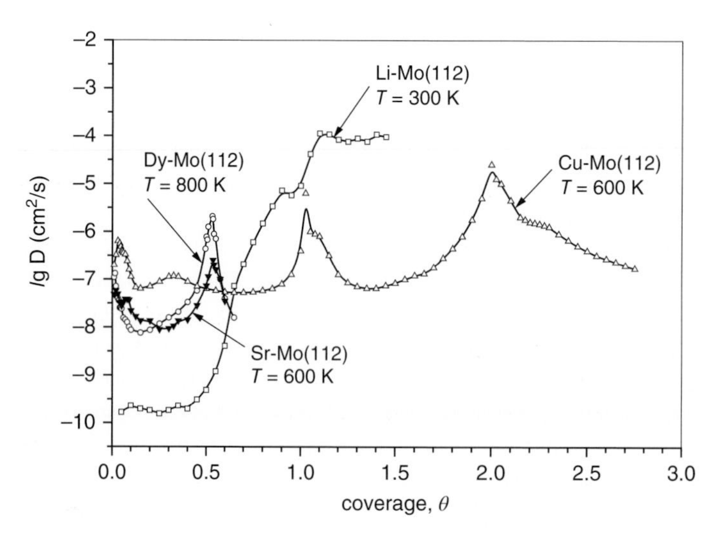


**Figure 34.19** Coverage dependences of the surface diffusion coefficient for Li, Sr, Dy, and Cu on Mo(112). Coverage $\theta$ is defined as $n/n_{Mo(112)}$. The diffusion was recorded along the atomic channels [155].

transitions in the adlayers, which can affect not only the diffusion parameters ($E_d$, $D_0$) but also the diffusion mechanisms.

In the linear approximation, the diffusion flux $J_D$ is driven by the gradient of the chemical potential:

$$\vec{J}_D = -L\nabla\mu, \tag{34.16}$$

where $L$ is a transport coefficient [153, 156]. This equation can be reshaped to the first Fick's law:

$$\vec{J}_D = -D(n)\nabla n, \tag{34.17}$$

where $\nabla n$ is the concentration gradient and $D(n)$ is *the concentration-dependent* diffusion coefficient, sometimes termed collective (or chemical) diffusion coefficient:

$$D(n) = L(n)\frac{\partial\mu}{\partial n} = D_j(n)\frac{\partial(\mu/kT)}{\partial \ln n}. \tag{34.18}$$

The first factor in this expression, $D_j(n)$, is named the *kinetic factor* or the *jump diffusion coefficient*. It is defined as

$$D_j(n) = L(n)kT/n \tag{34.19}$$

and can be cast in a form showing explicitly its dependence on the frequency $\Gamma$ and the length $\Lambda$ of the elementary jumps of the diffusing particles [159]:

$$D_j(n) = \Gamma(n)\Lambda^2(n). \tag{34.20}$$

The dependence of $\Gamma$ and $\Lambda$ on $n$ originates from the structure of the adlayer and lateral interaction, which influence the arrangement of adatoms in the layer, the shape of the potential relief, the adlayer vibration spectrum, and the energy exchange (friction) between the diffusing particle and the substrate. The second factor in Equation 34.18 is termed the *thermodynamic factor* and incorporates the contributions from the coverage-dependent energy and entropy terms in micro.

These considerations reflect the close relation between the kinetics of surface diffusion and surface phase transitions, which is the basis for the interpretation of the coverage dependences of diffusion characteristics. Figure 34.19 shows that, in the region of low coverages, the rate of diffusion usually reveals a more or less expressed decrease with coverage. This may be attributed to transition from a 2D gas, where one observes predominantly the diffusion of individual adatoms, to an also rarefied, but denser phase in which some part of adatoms are integrated into clusters. The mechanisms of diffusion of clusters were found to be remarkably diverse (see, e.g., Refs. [150, 151, 154, 157]). The regions of the PT-1s for all investigated systems are marked with minima of the diffusion rate. This seems to be a natural consequence of the fact that the process of diffusion in heterogeneous adlayers includes a stage of detachment of adatoms from the islands of the denser phase. It must be noted, however, that heterogeneous systems cannot be correctly characterized by a single diffusion coefficient and, therefore, the macroscopic experiments give actually only its effective (averaged) value.

The commensurate–incommensurate transitions in all studied adlayers are marked by a substantial growth of the diffusion rate (sometimes by orders of magnitude). In Section 34.5, we have discussed the possible reasons for this effect. Actually, both the factors in Equation 34.18 can contribute to the rate enhancement. First, the formation of solitons (incommensurate domain walls) leads to lowering of the potential barriers for diffusion and to the growth of the mobility of the mass carriers (solitons). Recall that the diffusion mechanism in this regime has a pronounced collective character. In addition to this, the compression of the adlayer in the incommensurate phase results in a sharp reduction of the adsorption energy and in the corresponding rise of the chemical potential. This also contributes to the increase of the thermodynamic factor and, in turn, of the diffusion rate.

Note also the sharp changes in $D$ that occur in narrow $\theta$ regions where the successive monolayers are completed and the filling of the next monolayers is started (Figure 34.19). A strong effect of this type was recorded for lithium and copper (the data for Cu were obtained up to the third monolayer and are shown here for comparison).

The strong distinctions of the diffusion rate in different 2D phases lead to a dynamical self-organization of the diffusion zone. Indeed, in order to maintain the continuity of the particle flux in the conditions of a quasi-stationary diffusion process, the concentration gradient should adapt to the value of the diffusion coefficient. When $D(n)$ is higher, the same flux can be provided at a lower concentration gradient $\nabla n$. This means that flattened segments of the concentration profiles will be observed in the regions of existence of the phases with the enhanced diffusion rate. Correspondingly, such phases rapidly extend over the surface and take up the major area in the diffusion zone (Figure 34.20) [160].

The collective nature of the soliton diffusion mechanism is manifested, in particular, in its "relay race" character: the displacement of a soliton over large distances is achieved by successive small (atomic-scale) displacements of adjacent adatoms.

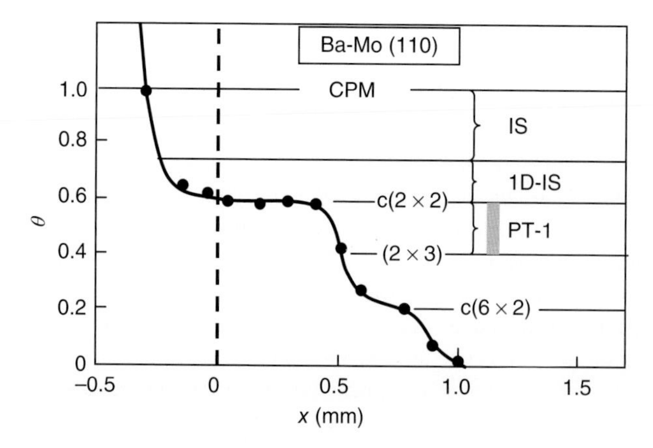


**Figure 34.20** Coverage profile recorded for Ba diffusing on Mo(110). The initial profile was step-like with $\theta > 1$ and the boundary marked by dashed line at $x = 0$. Horizontal lines mark the coverages corresponding to different Ba structures. CPM: close-packed monolayer; PT-1: the region of the first-order phase transition; 1D-ISs: one-dimensional incommensurate structures; and IS: hexagonal incommensurate structure [160].

This process somewhat resembles the mechanisms of martensitic transformations in metals. Manifestations of the high mobility of electropositive adatoms in the C–I transition region were also observed in the measurements of adatom density fluctuations in adlayers, which were found to be compatible with the soliton model [161, 162]. Since the solitons in adlayers are linear objects, their displacements can be substantially affected by pinning on immobile defects (e.g., on impurity atoms that are strongly bound to the surface and can play the role of stoppers) [53, 83, 90, 153]. This statement relates also to the chain structures similar to that shown in Figure 34.10c. The chains can be displaced by the formation and jump-wise motion of kinks in them, so the diffusion mechanism is here pronouncedly collective as well. The strong pinning effect is visible in this case not only in the ordered chain phase but also in the disordered phase well above the melting temperature [163]. A structural modification of the Sr chain lattices on Mo(112) by oxygen doping was investigated in Ref. [164].

On the whole, the alkali and alkaline-earth adatoms on close-packed metal surfaces show a considerable mobility at liquid nitrogen temperature (77 K) and even well below it. This allows observation of ordered long-period lattices of these adatoms. However, the processes of surface reconstruction and surface alloying, which are activated at elevated temperatures, can in some cases inhibit surface diffusion. This can even end up in the formation of 2D (surface) glasses, as was observed for RE adatoms (see Section 34.5).

It should be noted that surface diffusion is an interesting and practically important phenomenon, which calls for further investigation of many unresolved issues. Most of them relate to the collective (many-body) character of diffusion and nonlinear effects that are mostly neglected in existing theories [156, 157]. Electropositive adlayers are excellent model objects for such studies.

## 34.10 Drift of Electropositive Adatoms in Nonhomogeneous Electric Fields

An interesting phenomenon is the surface drift of electropositive adatoms in nonuniform electric fields [90, 165]. It was investigated in detail using field emission microscopy. A dose of adsorbate is deposited onto a tip serving as the field emitter of the microscope and, simultaneously, as an object for study. The tip is first annealed *in the absence of the electric field* at a temperature providing sufficient adatom mobility. By this means, the adlayer is equilibrated, that is, the adatoms are distributed over the tip surface in such a way that their chemical potential is the same everywhere. Then the tip is cooled to suppress the adatom mobility, and the adatom concentration is determined on a chosen crystal plane of the tip apex by recording the field emission current versus voltage characteristic and calculating the work function from it (the curve $\varphi$ vs $n$ is calibrated in separate experiments). In the following steps of the experiment, the annealing (i.e., adlayer equilibration) is carried out *in the presence of the electric field*, when the tip is charged positively or negatively (the field in these cases will be considered hereafter as positive and negative, respectively). The field at the tip apex is the highest, while at the tip shank it is lower by orders of magnitude. In experiments with electropositive adsorbates, it was found that adatoms drift either to the apex or from the apex of the tip, depending on the sign of electric field and the concentration of adatoms (Figure 34.21) [166, 167].

The explanation of this effect is as follows: The field dependence of the adsorption energy of a single adatom can be approximated as [168]

$$E_{aF} = E_a + \mathbf{pF} + \alpha F^2/2, \tag{34.21}$$

where $E_{aF}$ and $E_a$ are the adsorption energies in the presence and absence of the electric field, respectively; $\mathbf{F}$ is the field strength, $\mathbf{p}$ is the dipole moment of the adatom, and $\alpha$ is its polarizability. The experiments showed that the linear (dipole) term, responsible for the drift depending on the mutual orientation of the dipole moment $\mathbf{p}$ and electric field $\mathbf{F}$, is dominant at the submonolayer coverages of electropositive adatoms and at the fields $F \sim 10^7$ V/cm typical of the field emission microscope. At the same time, the third term in Equation 34.22 does not play any considerable role in these conditions [90, 165–167].

Equation 34.21 is valid for a single adatom. Neglecting the third term in Equation 34.21, one can write for *an ensemble of n adatoms*

$$W_F(n) = W(n) - n\mathbf{p}(n)\mathbf{F}, \tag{34.22}$$

where $W_F(n)$ and $W(n)$ are, respectively, the total energy of the ensemble in the presence and absence of the field, and $n$ is the surface concentration of adatoms [86, 159]. Note that here $W_F(n)$ and $W(n)$ are negative values:

$$W_F(n) = -\int_0^n E_{aF}(n)dn, \tag{34.23}$$

$$W(n) = -\int_0^n E_a(n)dn. \tag{34.24}$$

Because of mutual depolarization of adatoms, $\mathbf{p}(n)$ is a decreasing function and thus the product $n\mathbf{p}(n)$, representing the power of the double electrical layer, has a maximum at $n = n_{opt}$, which corresponds to the minimum of work function (Section 34.3). We recall that positive dipoles (having positive ends oriented to vacuum) *reduce* the work function. If the field $\mathbf{F}$ is parallel to $\mathbf{p}$ (the tip is charged positively),

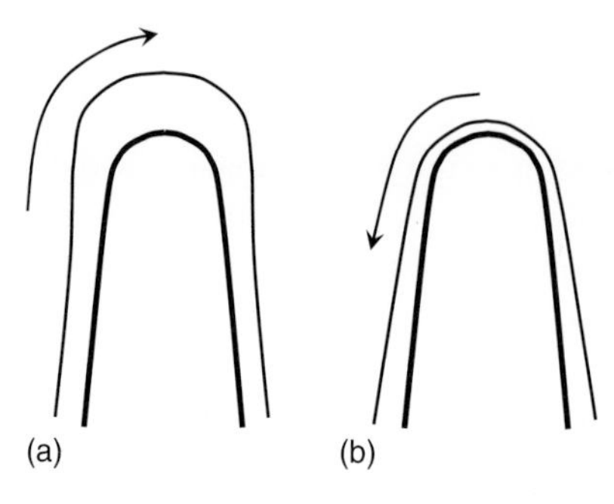


**Figure 34.21** Scheme of the equilibrium distribution of adatoms on the surface of a tip in the case of their drift to the apex (a) and from the apex of the tip.

the product $n\mathbf{p}(n)\mathbf{F}$ will be positive. The system tends to reach a *minimum* of the free energy in the applied field, which means that the product $n\mathbf{p}(n)\mathbf{F}$ in Equation 34.22 should be *maximized* through redistribution of the adatoms.

In the particular case $\mathbf{p}||\mathbf{F}$, which we consider now, this means that the adatom density in the region of the high field (i.e., on the tip apex) will grow at $n_0 < n_{opt}$ and decrease at $n_0 > n_{opt}$ (Figure 34.22).

Here, $n_0$ denotes the equilibrium adatom density at $F = 0$. If $\mathbf{p}$ and $\mathbf{F}$ are antiparallel (the tip is charged negatively), the field-induced changes of $n$ will be opposite to those considered above. The equilibrium state of the adlayer in the presence of the electric field is attained as the drift flux of adatoms to (or from) the tip apex is compensated by the oppositely directed diffusion flux. The latter arises as a result of the buildup of the concentration gradient between the tip apex and the tip shank where the field is close to zero and $n$ remains practically constant (the area of the shank is much larger than the apex area).

The dependence of the direction of the drift of adatoms on field direction is a graphic manifestation of the polarity of the adsorption bond. The evaluation of these experimental results allows an independent assessment of the dipole moment as a function of adatom density, which can be compared with the data gained from the work function measurements. These values show a fair agreement, which confirms that the electrical double layer created at the surface by adatoms does play a dominant part in the work function change at submonolayer coverages [90, 165].

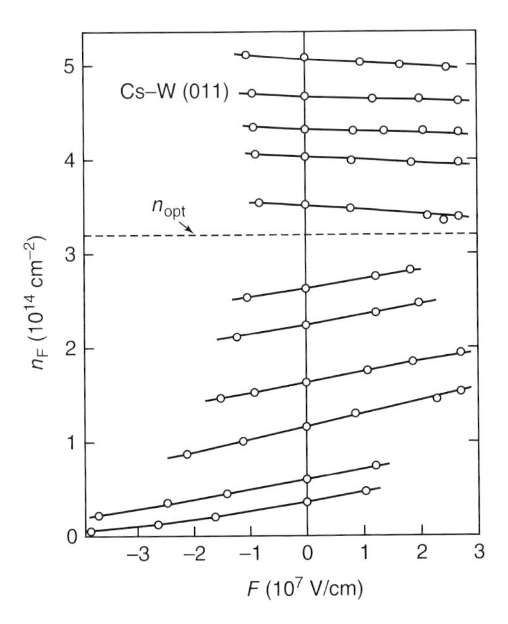


**Figure 34.22** Equilibrium concentration of Cs adatoms on the (110) surface located at the apex of a W tip as a function of electric field and initial concentration set by the deposited Cs dose [165].

Let us now discuss the contribution of the third (polarization) term in Equation 34.21. This term is quadratic in **F**, so the drift of adatoms caused by the field-induced polarization does not depend on the field sign and is *always* directed toward an area where $F$ is maximum and the adsorption energy is increased (i.e., to the tip apex). The analysis of experimental data showed that, in the fields $F \leq 3 \times 10^7$ V/cm and at $\theta \leq 1$, the contribution of the polarization term to $q_F$ for alkali adatoms is negligible, while for alkaline-earth adatoms it is perceptible even though nondecisive. The polarizability $\alpha$ evaluated from these results was found to be $<3 \times 10^{-24}$ cm$^3$ for alkali adatoms and $\sim(3–4) \times 10^{-23}$ cm$^3$ for alkaline-earth adatoms. In the free state, both alkali and alkaline-earth atoms have comparable polarizabilities $\sim(3–5) \times 10^{-23}$ cm$^3$. The difference of their polarizabilities that arises in the adsorbed state may be due to electron transfer from the adatoms to substrate. Judging from the value of the dipole moments, the transferred charge does not exceed the elementary charge $e$. As a result, the partially depleted valence shell of the alkali adatom becomes similar to the shell of the alkali positive ion whose polarizability is known to be very low ($\sim 10^{-24}$ cm$^3$). In contrast, the partially depleted valence shell of the alkaline-earth adatom resembles the shell of the alkali atom with its high polarizability ($\sim 10^{-23}$ cm$^3$) [90, 165].

As the coverage exceeds the CPM ($\theta > 1$), the drift of electropositive adsorbates is directed to the tip apex regardless of the orientation of **F**, which means that in such a case the contribution of the permanent dipole **p** is insignificant and the drift is determined by the third term in Equation 34.21. This effect can be used to controllably grow epitaxial microcrystals of alkali metals on the tip emitters in field emission microscopes [55, 90, 169].

At the surfaces, nonhomogeneous electrostatic fields exist near atomic steps, dislocations, phase and grain boundaries, adatoms and admolecules, and so on. The drift of adatoms in these fields can impact both the kinetics of various surface processes and the equilibrium distribution of adatoms on the surface. This phenomenon is also used when individual adatoms and admolecules are controllably arranged on the surface in an STM [170]. It should be added that electropositive adlayers on semiconductor tips reveal basically the same regularities of the drift as we considered above for metal tips [171, 172].

## 34.11
## Tailoring Surface Properties by Electropositive Adlayers

Chemisorption of electropositive metals represents a limiting case of adsorpion that is distinguished by the strong polarity of the adsorption bond. Thanks to the large transfer of the electronic density from the adatom to the substrate and the existence of long-range lateral interaction, the electropositive adatoms belong to the most efficient surface modifiers.

As we saw in Section 34.3, alkali, alkaline-earth, and RE adlayers can reduce the work function of solids several fold, whose work functions in the clean state are $\sim$4–5 eV. This property is applied in emission electronics to construct various types of cathodes as well as in semiconductor technologies to tune the characteristics of interfaces. The coadsorption of cesium and oxygen is utilized to achieve the NEA and thus to obtain effective photoemitters for infrared devices.

The low ionization energies of electropositive metals are exploited to create sources of positive ions based on the phenomenon of surface ionization. The same method is applicable to generate positive ions of many organic compounds that also have low ionization energies [173, 174]). Such sources are used to produce ion beams in various ion technologies.

When an adlayer is deposited onto a metal tip, the adatoms residing on tip's apex can be desorbed by an electric field $\sim 10^7-10^8$ V/cm as positive ions. The continuous surface transport of the adsorbate to the apex can be provided by surface diffusion from the shank of the tip. Such ion sources combine two important advantages: the point-like geometry, and a narrow energy spread [175]. This principle of operation is also utilized in the construction of the lithium field desorption microscope, in which $Li^+$ ions visualize the tip apex with a spatial resolution of $\sim 10^{-8}$ cm [176, 177].

The formation of positive ions proceeds most efficiently on surfaces with high work function, which facilitates the transfer of electrons from adatoms to the substrate. In contrast, the formation of negative ions occurs easily on surfaces with *low* work function, and Cs adlayers are used to prepare such surfaces in sources of negative ions [178].

The enrichment of solid surfaces with electrons as a consequence of alkali adsorption is the basis of a widely used method of promotion of catalytic reactions [26, 31, 179]. In semiconductor technologies, alkali promoters are applied to accelerate the surface oxidation [119, 120].

In Section 34.8, we considered some examples of interaction of electromagnetic radiation with surfaces covered with electropositive adlayers. The excitation and subsequent nonradiative decay of multipole surface plasmons strongly enhances the yield of photoelectrons. The surface can be tuned to the incident light, which generates the multipole surface plasmons, by choosing the appropriate density of the adlayer.

Electropositive adlayers reveal also strong nonlinear optical effects. For instance, the intensity of second-harmonic generation (SHG) grows by several orders of magnitude when a metal or semiconductor surface is covered with a monolayer of electropositive adsorbate. The resonance in SHG is observed when the condition $2\omega = \omega_{mp}$ is satisfied, where $\omega$ is the frequency of the incident light and $\omega_{mp}$ is the multipole plasma frequency [180, 181]. The SHG signal is extremely sensitive to the presence of electropositive adsorbates on the surface, so their nonlinear optical properties can be used both for studying the adlayers themselves and for optical applications.

Recent decades have seen growing interest in carbon materials – graphite, fullerenes, nanotubes, and graphene. This is motivated by the outstanding (in many cases unique) properties of these materials and also by the abundance of carbon in nature. In particular, they are excellent absorbers of hydrogen, and because of this quality they attract much attention as potential storage media for hydrogen energetics.

Alkali and other electropositive elements adsorbed on surfaces or intercalated into atomic-sized pores within carbon materials can be used to modify their absorbing properties. In any case, adsorption is the first stage of the interaction of alkali modifiers with carbon absorbers. Alkali metal adsorption on graphite shows many similarities to their adsorption on refractory metals [60]. At low coverages ($\theta \leq 0.3$), alkali atoms transfer a substantial part of the electron charge contained in their valence s shell to the graphite substrate and thus become positive ions with a partial charge estimated at $(0.1-0.9)e$. This results in a strong dipole–dipole repulsion of the adatoms and in the formation of dispersed (rarefied) 2D phases where the shortest interadatom distances are tens of angstroms. These phases are liquid-like and show only a short-range ordering even at liquid nitrogen temperatures. At higher coverages ($\theta \sim 0.1-0.3$), there occurs a PT-1 to a condensed 2D phase, which exhibits a long-range order and metallic properties. Such behavior is quite similar to what occurs in alkali adlayers on refractory metals. However, in contrast to the situation on metal substrates, the adsorbed alkalis that have small atomic radii (Li and Na) can intercalate into graphite and produce there many specific phases [182].

Another carbon substrate that has attracted much interest is graphene. A detailed study of lithium, sodium, potassium, and calcium on graphene was carried out using first-principles density functional theory with the generalized gradient approximation [183]. All these adsorbates were predicted to form essentially ionic bonding with graphene. At a coverage where one adatom corresponds to 32 carbon atoms and the distance between the adatoms amounts to $\approx$10 Å, the charge transfer from alkalis to graphene substrate was estimated at $0.90e$ for Li, $0.73e$ for Na, and $0.75e$ for K. The calculated dipole moments in this row of elements are, respectively, 3.46, 2.90, and 4.48 D. Thus alkalis can be used to chemically dope graphene, which is a zero-gap semiconductor with a point-like Fermi surface. Adsorption and thin-film growth of a broad class of metals, including electropositive ones, was also investigated on graphene by *ab initio* calculations [184].

A new carbon-based material is graphane [185]. It represents a hexagonal carbon lattice in which every C atom keeps a hydrogen atom positioned in a chair configuration. A part of hydrogen atoms in graphane can be replaced by metal atoms to form organometallic complexes. It has been predicted theoretically [186] that such complexes formed with alkali atoms can serve as hydrogen storage materials. In particular, lithium seems to be the most efficient modifier for this purpose: the theoretical limit for H storage amount by Li-graphane is estimated at 9.44 wt% at 300 K and pressure in the range 5–250 atm. This corresponds to three adsorbed hydrogen molecules per Li atom on graphane.

Moreover, the alkali metal–graphane complexes may be arranged into a bulk material to form a porous framework structure that would be an effective H storage material. Metal–organic frameworks (MOFs) are synthetic materials that consist of metal oxide groups linked together by organic struts. MOFs have a crystalline structure and contain nanopores where hydrogen molecules can be stored [187, 188]. The hydrogen uptake properties were substantially enhanced when an MOF was modified with lithium [187].

Recently, graphene oxide was shown to be an efficient selective absorber of adverse radioactive waste. Romanchuk *et al.* [189] have investigated the interaction of graphene oxide with simulated nuclear waste solutions that represented water contaminated with actinides Am, Th, Pu, Np, U, and typical fission products such as Sr and Eu. All these elements are rather electropositive (their ionization energies range from 5.7 to 6.2 eV). Thus they form positively charged ions (cations) in a water solution. It was found that flakes of graphite oxide added to such a solution quickly coagulate with the cations and then form aggregates of graphene oxide sheets covered with adsorbed radionuclides. These aggregates can be easily filtered out from the solution. Graphene oxide has shown much better cleaning-up properties than bentonite clays and activated carbon, which are commonly used in nuclear industry. This discovery can also bring about an important breakthrough in the oil and gas extraction to decontaminate the fracking liquids used in drilling operations and also in mining RE metals. There exist also some environmental problems concerned with targeted decontamination of liquid waste from alkali elements (e.g., cesium). In such a case, effective adsorbents of alkalis are demanded (see, e.g., [190]).

Zhu *et al.* [191] suggested an idea to use diamond nanoparticles covered by adsorbed $Na^+$ ions in tumor therapy. This idea is based on the known effect when excess sodium ions inside the cells lead to severe cellular damage induced by osmotic stresses, which increase the intracellular levels of calcium and reactive oxygen species and eventually result in the cell swelling and destruction. The excess $Na^+$ ions can be selectively delivered to malignant cells by diamond nanoparticles.

Above, we have given only a few examples that demonstrate the great application potential of electropositive elements in various technologies.

## 34.12 Summary

The electropositive (alkali, alkaline-earth, and RE) metal atoms have ionization energies lower than or comparable to the work functions of many metals and semiconductors, which lie in the range of 4–5.5 eV. The valence electronic levels of adsorbing electropositive atoms shift upwards and broaden due to adatom–substrate interaction, which results in the formation of their unified quantum mechanical system. The partial transfer of the valence electron cloud from the adatom to the substrate creates a strongly polar adsorption bond effected by delocalized electrons and having an energy of several electronvolts. This nature of the adsorption bond is the origin of all the peculiar properties of electropositive adlayers. These adlayers stand out by the existence of long-range lateral interaction which, at low coverages, is predominantly repulsive (of dipole–dipole nature) but superimposed by indirect electron exchange interaction having an oscillating and, generally, anisotropic character. With increasing coverage within the limits of the first monolayer, there occur interrelated changes in the nature of the adsorption bond and energetics of adsorption, atomic structure, electronic state (transition from nonmetallicity to metallicity), kinetics of surface diffusion, and other properties of the adlayer. The phase diagrams of electropositive adlayers are manifold and present good possibilities for studying physical regularities of phase transitions in 2D systems and, using the ability of such adlayers to strongly change surface characteristics, to reduce the work function, enhance the catalytic activity, chemical reactivity, and electron emission, impact the near-surface electronic structure of semiconductors, intensify the nonlinear optical effects on surfaces, and so on.

The knowledge gained by studies of electropositive adlayers has proven helpful to understand the phenomena that occur in other systems. For instance, many organic molecules have rather low ionization energies and significant dipole moments, so the processes that occur with their participation on various (among them, biological [192, 193]) interfaces bear some similarities to the behavior of inorganic electropositive adlayers.

## Acknowledgments

The support of this work by the National Academy of Sciences of Ukraine within projects VTs-156 and V-158 is gratefully acknowledged. I thank I.N. Yakovkin for helpful comments and O.L. Fedorovich and A.A. Mitryaev for their assistance in preparation of the typescript.

## References

1. Fomenko, V.S. (1981) *Emissionnye Svoistva Materialov (Emission Properties of Materials, in Russian)*, Naukova Dumka, Kiev.
2. Langmuir, I. and Kingdon, K.H. (1923) *Science*, **57**, 58; *Proc. Roy. Soc.* (1925), **A107**, 61.
3. Suits, C.G. (ed.) (1962) *The Collected Works of Irving Langmuir*, vol. 12, Pergamon, New York.
4. Becker, J.A. (1926) *Phys. Rev.*, **28**, 341.
5. Taylor, J.B. and Langmuir, I. (1933) *Phys. Rev.*, **44**, 423–458.
6. Wise, G. (1994) *Vacuum Science and Technology; Pioneers of the 20th Century: History of Vacuum Science and Technology 2*, American Vacuum Society, AIP Press, New York, p. 79.
7. Medvedev, V.K. and Smereka, T.P. (1973) *Sov. Phys. Solid State*, **15**, 507.
8. E. Bauer, in: (eds D.A. King and D.P. Woodruff), *The Chemical Physics of Solid Surfaces and Heterogeneous Catalysis*, vol. 3B, Elsevier, Amsterdam, 1984, p. 1.
9. Stenborg, A. and Bauer, E. (1987) *Phys. Rev. B*, **36**, 5840.
10. Prokhorov, A.M. (ed.) (1988-1998) *Fizicheskaya Encyklopedia*, vol. 1-5 (*Physical Encyclopedia*), Bol'shaya Rossiyskaya Encyklopedia, Moscow (in Russian).
11. Zandberg, E.Y. and Ionov, N.I. (1969) *Poverkhnostnaya Ionizatsiya Surface Ionization*, Nauka, Moscow, pp. 70 and 346 (in Russian).
12. Gurney, R.W. (1935) *Phys. Rev.*, **47**, 479–482.
13. J.E. Inglesfield, B.W. Holland, in: (eds D.A. King and D.P. Woodruff), *The Chemical Physics of Solid Surfaces and Heterogeneous Catalysis*, vol. 1, Elsevier, 1981, Amsterdam, p. 183.
14. Kiejna, A. and Wojciechowski, K. (1996) *Metal Surface Electron Physics*, Pergamon, Kidlington.
15. Naumovets, A.G. (1964) *Sov. Phys. Solid State*, **5**, 1668.
16. Gadzuk, G.W. (1967) *Surf. Sci.*, **6**, 133.
17. Lang, N.D. (1971) *Phys. Rev.*, **B4**, 4234.

18. Gadzuk, G.W. (1975) *J. Vac. Sci. Technol.*, **12**, 289.
19. Nørskov, J.K. (1990) *Rep. Prog. Phys.*, **53**, 1253.
20. Lang, N.D. (1994) *Surf. Sci.*, **299/300**, 284.
21. M. Scheffler, C. Stampfl, in: (eds K. Horn and M. Scheffler), *Handbook of Surface Science*, vol. 2, Elsevier, Amsterdam, 2000, p. 285.
22. Lang, N.D. and Williams, A.R. (1978) *Phys. Rev.*, **B18**, 616.
23. Bormet, J., Neugebauer, J., and Scheffler, M. (1994) *Phys. Rev. B*, **49**, 17242.
24. Aruga, T. and Murata, Y. (1989) *Prog. Surf. Sci.*, **31**, 61.
25. Küppers, J., Wandelt, K., and Ertl, G. (1979) *Phys. Rev. Lett.*, **43**, 928.
26. Bonzel, H.P., Bradshaw, A.M., and Ertl, G. (eds) (1989) *Physics and Chemistry of Alkali Metal Adsorption*, Elsevier, Amsterdam.
27. Wandelt, K., In: (eds R. Vanselow and R. Howe) (1991) in *Chemistry and Physics of Solid Surfaces VIII*, Springer, Berlin, p. 289.
28. Janssens, T.V.W., Castro, G.R., Wandelt, K., and Niemantsverdriet, J.W. (1994) *Phys. Rev.*, **B49**, 14599.
29. Wandelt, K. (1997) *Appl. Surf. Sci.*, **111**, 1.
30. Lang, N.D. and Williams, A.R. (1982) *Phys. Rev.*, **B25**, 2940.
31. Spencer, N.D. and Somorjai, G.A. (1983) *Rep. Prog. Phys.*, **46**, 1.
32. Bonzel, H.P. (1988) *Surf. Sci. Rep.*, **8**, 43.
33. Kiskinova, M. (1987) *J. Vac. Sci. Technol., A*, **5**, 852.
34. Nørskov, J.K. (1989) *Physics and Chemistry of Alkali Metal Adsorption*, Elsevier, Amsterdam, p. 253.
35. Gerlach, R.L. and Rhodin, T.N. (1969) *Surf. Sci.*, **17**, 32; (1970), **19**, 403.
36. Fedorus, A.G. and Naumovets, A.G. (1970) *Surf. Sci.*, **21**, 426.
37. Medvedev, V.K., Naumovets, A.G., and Fedorus, A.G. (1970) *Sov. Phys. Solid State*, **12**, 301.
38. Fedorus, A.G., Naumovets, A.G., and Vedula, Y.S. (1972) *Phys. Status Solidi A*, **13**, 445.
39. Braun, O.M. and Medvedev, V.K. (1989) *Sov. Phys. Usp.*, **32**, 328.
40. Lang, N.D. (1989) *Physics and Chemistry of Alkali Metal Adsorption*, Elsevier, Amsterdam, p. 11.
41. Kohn, W. and Sham, L.J. (1965) *Phys. Rev.*, **140**, A1133.
42. Friedel, J. (1958) *Nuovo Cimento*, **7**, 287.
43. Bol'shov, L.A., Napartovich, A.P., Naumovets, A.G., and Fedorus, A.G. (1977) *Sov. Phys. Usp.*, **20**, 432.
44. Naumovets, A.G. (1984) *Sov. Sci. Rev. A, Phys. Rev.*, **5**, 443.
45. Diehl, R.D. and McGrath, R. (1996) *Surf. Sci. Rep.*, **23**, 43.
46. Kolaczkiewicz, J. and Bauer, E. (1986) *Surf. Sci.*, **175**, 487.
47. P.A. Dowben, D.N. McIlroy, D. Li, in: (eds K.A. Gschneider, Jr., and L. Eyring), *Hadbook on the Physics and Chemistry of Rare Earths*, vol. 24, Elsevier, Amsterdam, 1997, p. 1.
48. Lafferty, J.M. (1951) *J. Appl. Phys.*, **22**, 299.
49. Fedorus, A.G., Mitryaev, A.A., Mukhtarov, M.A., Pfnür, H., Vedula, Y.S., and Naumovets, A.G. (2006) *Surf. Sci.*, **600**, 1566.
50. Smoluchowski, R. (1941) *Phys. Rev.*, **60**, 661.
51. Villain, J. and Pimpinelly, A. (1998) *Physique de la Croissance Cristalline*, Éditions Eyrolles, Paris.
52. Vedula, Y.S. and Poplavsky, V.V. (1987) *Sov. Phys. JETP Lett.*, **46**, 230.
53. Lyuksyutov, I., Naumovets, A.G., and Pokrovsky, V. (1992) *Two-Dimensional Crystals*, Academic Press, Boston, MA.
54. A.G. Naumovets, in: (eds D.A. King and D.P. Woodruff), *The Chemical Physics of Solid Surfaces*, 1994, vol. 7, Elsevier, Amsterdam, p. 163.
55. Gavrilyuk, V.M., Naumovets, A.G., and Fedorus, A.G. (1967) *Sov. Phys.- JETP*, **24**, 899.
56. Medvedev, V.K., Naumovets, A.G., and Smereka, T.P. (1973) *Surf. Sci.*, **34**, 368.
57. T.L. Einstein, in: (ed. W.N. Unertl), *Handbook of Surface Science*, vol. 1, Elsevier, Amsterdam, 1996, p. 577.
58. Kanash, O.V., Naumovets, A.G., and Fedorus, A.G. (1975) *Sov. Phys. JETP*, **40**, 903.
59. Müller, K., Besold, G., and Heinz, K. (1989) *Physics and Chemistry of Alkali*

*Metal Adsorption*, Elsevier, Amsterdam, p. 65.
60. Caragiu, M. and Finberg, S. (2005) *J. Phys. Condens. Matter*, **17**, R995–R1024.
61. Shrednik, V.N. and Odisharia, G.A. (1970) *Sov. Phys. Solid State*, **11**, 1487.
62. Golubev, O.L., Odisharia, G.A., and Shrednik, V.N. (1971) *Izv. Akad. Nauk SSSR, Ser. Fiz.*, **35**, 345.
63. Bolshov, L.A. (1971) *Sov. Phys. Solid State*, **13**, 1404.
64. Aruga, T., Tochihara, H., and Murata, Y. (1986) *Surf. Sci.*, **175**, L725.
65. Aruga, T., Tochihara, H., and Murata, Y. (1986) *Phys. Rev. B*, **34**, 8237.
66. Hohlfeld, F. and Horn, K. (1989) *Surf. Sci.*, **211/212**, 844.
67. Neugebauer, J. and Scheffler, M. (1993) *Phys. Rev. Lett.*, **71**, 577.
68. Grimley, T.B. (1976) *Crit. Rev. Solid State*, **6**, 239.
69. Atkins, P.W. (1998) *Physical Chemistry*, Oxford University Press, Oxford.
70. Zangwill, A. (1988) *Physics at Surfaces*, Cambridge University, Cambridge.
71. Han, P. and Weiss, P.S. (2012) *Surf. Sci. Rep.*, **67**, 19–81.
72. Medvedev, V.K. and Yakivchuk, A.I. (1975) *Ukr. Fiz. Zh.*, **20**, 1900.
73. Chen, J.M. and Papageorgopoulos, C.A. (1970) *Surf. Sci.*, **21**, 377; *J. Vac. Sci. Technol.* (1972), **9**, 570.
74. Kolthoff, D. and Pfnür, H. (2000) *Surf. Sci.*, **459**, 265.
75. Loburets, A.T., Naumovets, A.G., and Vedula, Y.S. (1982) *Surf. Sci.*, **120**, 347.
76. Bak, P. (1982) *Rep. Prog. Phys.*, **45**, 587.
77. E. Bauer, in: (eds W. Schommers and P.V. Blanckenhagen), *Structure and Dynamics of Surfaces II*, Springer, Berlin, 1987, p. 115.
78. Persson, B.N.J. (1996) *Surf. Sci. Rep.*, **23**, 43.
79. L.D. Roelofs, in: (ed. W.H. Unertl), *Handbook of Surface Science*, vol. 1, Elsevier, Amsterdam, 1996, p. 713.
80. Schick, M. (1981) *Prog. Surf. Sci.*, **11**, 245.
81. Diehl, R.D. and Mc Grath, R. (1997) *J. Phys. Condens. Matter*, **9**, 951.
82. V.L. Pokrovsky, A.L. Talapov, P. Bak, in: (eds S.E. Trullinger, V.E. Zakharov and V.L. Pokrovsky), *Solitons*, Elsevier, Amsterdam, 1986, p. 71.
83. Lyuksyutov, I.F., Naumovets, A.G., and Vedula, Y.S. (1986) *Solitons*, Elsevier, Amsterdam, p. 605.
84. Braun, O.M. and Kivshar, Y.S. (2004) *The Frenkel-Kontorova Model*, Springer, Berlin.
85. Novaco, A.D. and McTague, J.P. (1979) *Phys. Rev. B*, **19**, 5299.
86. Naumovets, A.G. and Fedorus, A.G. (1977) *Sov. Phys. JETP*, **46**, 575.
87. Halperin, B.I. and Nelson, D. (1978) *Phys. Rev. Lett.*, **41**, 121.
88. Matsuda, T., Barnes, C.J., Hu, P., and King, D.A. (1992) *Surf. Sci.*, **276**, 122.
89. Lyuksyutov, I.F. and Pokrovsky, V.L. (1981) *Sov. Phys.-JETP Lett.*, **33**, 326.
90. Naumovets, A.G. and Vedula, Y.S. (1985) *Surf. Sci. Rep.*, **4**, 365.
91. Krivoglaz, M.A. and Smirnov, A.A. (1964) *Theory of Order-Disorder in Alloys*, Macdonald, London.
92. Girifalko, L.A. (1973) *Statistical Physics of Materials*, John Wiley & Sons, Inc., New York.
93. Ertl, G. and Küppers, J. (1970) *Surf. Sci.*, **21**, 61.
94. Sadiq, A. and Binder, R. (1984) *J. Stat. Phys.*, **35**, 517.
95. Inoue, A. and Hashimoto, K. (eds) (2001) *Amorphous and Nanocrystalline Materials*, Springer, Berlin.
96. King, D.A. (1994) in *The Chemical Physics of Solid Surfaces*, vol. 7 (ed. D.P. Woodruff), Elsevier, Amsterdam.
97. Behm, R.J. (1989) *Physics and Chemistry of Alkali Metal Adsorption*, Elsevier, Amsterdam, p. 111.
98. Barnes, C.J. (1973) *Statistical Physics of Materials*, John Wiley & Sons, Inc., New York, p. 501.
99. Besenbacher, F. and Stensgaard, I. (1973) *Statistical Physics of Materials*, John Wiley & Sons, Inc., New York, p. 537.
100. Besenbacher, F. (1996) *Rep. Prog. Phys.*, **59**, 1737.
101. Ibach, H. (1997) *Surf. Sci. Rep.*, **29**, 193.
102. Tochihara, H. and Mizuno, S. (1998) *Prog. Surf. Sci.*, **58**, 1.
103. Fu, C.L. and Ho, K.M. (1989) *Phys. Rev. Lett.*, **63**, 1617.

104. Jacobsen, K.W. and Nørskov, J.K. (1988) *Phys. Rev. Lett.*, **60**, 2496.
105. Chen, M.-S., Terasaki, D., Mizuno, S., Tochihara, H., Ohsaki, I., and Oguchi, T. (2000) *Surf. Sci.*, **470**, 53.
106. Schuster, R., Barth, J.V., Ertl, G., and Behm, R.J. (1991) *Surf. Sci.*, **247**, L229; *Phys. Rev. B* (1991), **44**, 13689.
107. Ishida, H. and Liebsch, A. (1998) *Phys. Rev. B*, **57**, 12558.
108. Fedorus, A.G., Godzik, G., Naumovets, A.G., and Pfnür, H. (2004) *Surf. Sci.*, **565**, 180.
109. Michel, E.G., Pervan, P., Castro, G.R., Miranda, R., and Wandelt, K. (1992) *Phys. Rev.*, **B45**, 11811.
110. Mönch, W. (2001) *Semiconductor Surfaces and Interfaces*, Springer, Berlin.
111. Benemanskaya, G.V., Daineka, D.V., and Frank-Kamenetskaya, G.E. (1999) *Phys. Low-Dim. Struct.*, **1/2**, 97.
112. Benemanskaya, G.V. and Vikhnin, V.S. (2001) *Phys. Low-Dim. Struct.*, **1/2**, 9.
113. Benemanskaya, G.V., Frank-Kamenetskaya, G.E., and Evtikhiev, V.P. (2004) *Phys. Low-Dim. Struct.*, **5/6**, 1.
114. Chiaradia, P., Paget, D., Tereshchenko, O.E., Bonnet, J.E., Taleb-Ibrahimi, A., Belkhou, R., and Wiame, F. (2006) *Surf. Sci.*, **600**, 287.
115. Gavioli, L., Padovani, M., Spiller, E., Sancrotti, M., and Betti, M.G. (2003) *Surf. Sci.*, **532-535**, 660.
116. Aruga, T., Tochihara, H., and Murata, Y. (1984) *Phys. Rev. Lett.*, **53**, 372.
117. Yakovkin, I.N. (2001) *J. Nanosci. Nanotechnol.*, **1**, 357.
118. Yakovkin, I.N. (2004) in *Encyclopedia of Nanoscience and Nanotechnology*, vol. 1 (ed. H.S. Nalwa), American Scientific Publisher, New York, p. 169.
119. Miranda, R. (1989) *Physics and Chemistry of Alkali Metal Adsorption*, Elsevier, Amsterdam, p. 425.
120. Soukiassian, P. and Starnberg, H.I. (1989) *Physics and Chemistry of Alkali Metal Adsorption*, Elsevier, Amsterdam, p. 449.
121. Bell, R.L. (1973) *Negative Electron Affinity Devices*, Clarendon, Oxford.
122. Moré, S., Tanaka, S., Tanaka, S., Fujii, Y., and Kamada, M. (2003) *Surf. Sci.*, **527**, 41.
123. Plummer, E.W., Tsuei, K.-D., Song, K.-J., and Murphy, R. (1989) *Physics and Chemistry of Alkali Metal Adsorption*, Elsevier, Amsterdam, p. 141.
124. Tsuei, K.D., Plummer, E.W., Liebsch, A., Kempa, K., and Bakshi, P. (1990) *Phys. Rev. Lett.*, **64**, 44.
125. Gorodetsky, D.A. and Gorchinsky, A.D. (1979) *Izv. Akad. Nauk SSSR*, **43**, 511 (in Russian).
126. Tsuei, K.D., Plummer, E.W., Liebsch, A., Pehlke, E., Kempa, K., and Bakshi, P. (1991) *Surf. Sci.*, **247**, 302.
127. Plummer, E.W., Carpinelli, J.M., Weitering, H.H., and Dowben, P.A. (1993) *Prog. Surf. Sci.*, **42**, 201; *Phys. Low-Dim. Str.* (1994), **4/5**, 99.
128. Katrich, G., Klimov, V.V., and Yakovkin, I.N. (1994) *J. Electron. Spectrosc. Relat. Phenom.*, **68**, 369.
129. Rocca, M. (1995) *Surf. Sci. Rep.*, **22**, 1.
130. Liebsch, A. (1997) *Electronic Excitations at Metal Surfaces*, Plenum Press, New York.
131. Yakovkin, I.N., Katrich, G.A., Loburets, A.T., Vedula, Y.S., and Naumovets, A.G. (1998) *Prog. Surf. Sci.*, **59**, 355.
132. Dowben, P.A. (2000) *Surf. Sci. Rep.*, **40**, 151.
133. Zhang, J., McIlroy, D.N., and Dowben, P.A. (1994) *Phys. Rev. B*, **49**, 13780; *Phys. Rev. B* (1995), **52**, 11380.
134. Yakovkin, I.N. (1998) *Surf. Sci.*, **406**, 57.
135. Yakovkin, I.N. (1999) *Surf. Sci.*, **442**, 431.
136. Wimmer, E. (1984) *J. Phys. F*, **14**, 681.
137. Boettger, J.C. and Trickey, S.B. (1989) *J. Phys. Condens. Matter*, **1**, 4323.
138. Borziak, P.G., Katrich, G.A., and Naumovets, A.G. (1999) *Phys. Low-Dim. Struct.*, **1/2**, 167.
139. Feibelman, P.J. (1982) *Prog. Surf. Sci.*, **12**, 287.
140. Gesell, T.F., Arakawa, E.T., Williams, M.W., and Hamm, R. (1973) *Phys. Rev.*, **B7**, 5141.
141. Monin, J. and Boutry, S.G.A. (1974) *Phys. Rev.*, **B9**, 1309.
142. Bennett, A.J. (1970) *Phys. Rev.*, **B1**, 203.
143. Ritchie, R.H. (1957) *Phys. Rev.*, **106**, 874.
144. Liebsch, A., Benemanskaya, G., and Lapushkin, M. (1994) *Surf. Sci.*, **302**, 303.

145. Lindgren, S.Å. and Walldén, L. (1980) *Phys. Rev.*, **B22**, 5967.
146. Walldén, L. (1985) *Phys. Rev. Lett.*, **54**, 943.
147. Bender, M., Yakovkin, I.N., and Freund, H.-J. (1996) *Surf. Sci.*, **365**, 394.
148. Müller, E.W. and Tsong, T.T. (1969) *Field Ion Microscopy, Principles and Applications*, Elsevier, Amsterdam.
149. Gomer, R. (1990) *Rep. Prog. Phys.*, **53**, 917; *Surf. Sci.* (1994), **299/300**, 129.
150. Tsong, T.T. (1994) *Surf. Sci.*, **299/300**, 153.
151. Ehrlich, G. (1994) *Surf. Sci.*, **299/300**, 628.
152. Antczak, G. and Ehrlich, G. (2010) *Surface Diffusion:Metals, Metal Atoms, and Clusters*, Cambridge University Press, Cambridge.
153. A.T. Loburets, A.G. Naumovets, Yu.S. Vedula, in: (ed. M.C. Tringides), *Surface Diffusion: Atomistic and Collective Processes*, Plenum Press, New York, 1997, p. 509.
154. Naumovets, A.G. and Zhang, Z. (2002) *Surf. Sci.*, **500**, 414.
155. A.T. Loburets, N.B. Senenko, A.G. Naumovets, Yu.S. Vedula, in: (eds M. Kotrla, N.I. Papanicolaou, D.D. Vvedensky, L.T. Will), *Atomistic Aspects of Epitaxial Growth*, Kluwer Academic Publishers, Dordrecht, 2002, p. 1.
156. Ala-Nissila, T., Ferrando, R., and Ying, S.C. (2002) *Adv. Phys.*, **51**, 949.
157. Naumovets, A.G. (2005) *Physica*, **A357**, 189.
158. Mehrer, H. (2007) *Diffusion in Solids*, Springer-Verlag, Berlin.
159. Reed, D.A. and Ehrlich, G. (1981) *Surf. Sci.*, **102**, 588.
160. Vedula, Y.S., Loburets, A.T., and Naumovets, A.G. (1979) *Sov. Phys. JETP*, **50**, 391.
161. Bęben, J., Kleint, C., and Męclewski, R. (1989) *Surf. Sci.*, **213**, 224.
162. Bęben, J., Kleint, C., and Pawełek, A. (1989) *Surf. Sci.*, **213**, 451.
163. Lyuksyutov, I.F., Everts, H.-U., and Pfnür, H. (2001) *Surf. Sci.*, **481**, 124.
164. Godzik, G., Block, T., and Pfnür, H. (2004) *Phys. Rev. B*, **69**, 235414.
165. Klimenko, E.V. and Naumovets, A.G. (1971) *Sov. Phys. Solid State*, **13**, 25; (1974), **15**, 2181.
166. Gavrilyuk, V.M. and Naumovets, A.G. (1964) *Sov. Phys. Solid State*, **5**, 2043.
167. Swanson, L.W., Strayer, R.W., and Charbonnier, F.M. (1964) *Surf. Sci.*, **2**, 177.
168. Drechsler, M. (1957) *Z. Elektrochem.*, **61**, 48.
169. Naumovets, A.G. (1965) *Sov. Phys. Solid State*, **6**, 1647.
170. Barth, J.V., Constantini, G., and Kern, K. (2005) *Nature*, **437**, 671.
171. But, Z.P., Miroshnichenko, L.S., and Yatsenko, A.F. (1978) *Ukr. Fiz. Zh.*, **23**, 978.
172. Suchorski, Y. (1990) *Surf. Sci.*, **231**, 130; 1991, **247**, 346.
173. Rasulev, U.K. and Zandberg, E.Y. (1988) *Prog. Surf. Sci.*, **28**, 181.
174. Rasulev, U.K., Khasanov, U., and Palitsin, V.V. (2000) *J. Chromatogr.*, **A 896**, 3.
175. Medvedev, V.K., Suchorski, Y., and Block, J.-H. (1995) *J. Vac. Sci. Technol.*, **B 13**, 621.
176. Gavrilyuk, V.M. and Medvedev, V.K. (1966) *Sov. Phys. -Tech. Phys.*, **11**, 1282.
177. Medvedev, V.K., Suchorski, Y., and Block, J.-H. (1994) *Ultramicroscopy*, **53**, 27.
178. Sherman, J.D. and Belchenko, Yu.I. (eds) (2005) Production and neutralization of negative ions and beams. AIP Conference Proceedings, Melville, New York, 2005, vol. 769.
179. Schlögl, R. (1989) *Physics and Chemistry of Alkali Metal Adsorption*, Elsevier, Amsterdam, p. 347.
180. Liebsch, A. (1989) *Phys. Rev. B*, **49**, 3421.
181. Barman, S.R., Horn, K., Häberle, P., Ishida, H., and Liebsch, A. (1998) *Phys. Rev. B*, **57**, 6662.
182. Dresselhaus, M.S. and Dresselhaus, G. (2002) *Adv. Phys.*, **51**, 1–186.
183. Chan, K.T., Neaton, J.B., and Cohen, M.L. (2008) *Phys. Rev. B*, **77**, 235430.
184. Liu, X., Wang, C.Z., Hupalo, M., Lu, W.C., Tringides, M.C., Yao, Y.X., and Ho, K.M. (2012) *Phys. Chem. Chem. Phys.*, **14** (25), 9157.
185. Elias, D., Nair, R., Mohiuddin, T., Morozov, S., Blake, P., Hallsal, M., Ferrari, A., Boukhvalov, D., Katznelson, M., Geim, A., and

Novoselov, K. (2009) *Science*, **323**, 610.

186. Antipina, L.Y., Avramov, P.V., Sakai, S., Naramoto, H., Ohtomo, M., Entani, S., Matsumoto, Y., and Sorokin, P.B. (2012) *Phys. Rev. B*, **86**, 085435.
187. Blomqvist, A., Moyses Araujo, C., Srepusharawoot, P., and Ahuja, R. (2007) *Proc. Natl. Acad. Sci. U.S.A.*, **104**, 20173–20176.
188. Murray, L.J., Dinca, M., and Long, J.R. (2009) *Chem. Soc. Rev.*, **38**, 1294.
189. Romanchuk, A.Y., Slesarev, A.S., Kalmykov, S.N., Kosynkin, D.V., and Tour, J.M. (2013) *Phys. Chem. Chem. Phys.*, **15**, 2321.
190. Han, F., Zhang, G.H., and Gu, P. (2012) *J. Hazard. Mater.*, **225-226**, 107.
191. Zhu, Y., Li, W., Zhang, Y., Li, J., Liang, L., Zhang, X., Chen, N., San, Y., Chen, W., Tai, R., Fan, C., and Huang, Q. (2012) *Small*, **8**, 1771.
192. Kasemo, B. (2002) *Surf. Sci.*, **500**, 656.
193. Rod, T.H. and Nørskov, J.K. (2002) *Surf. Sci.*, **500**, 678.

# 35
# Halogen Adsorption on Metals

*Boris V. Andryushechkin*

This chapter presents a brief review of experimental and theoretical investigations of halogen interaction with metal surfaces. The interaction of halogens with metal surfaces, which is a rather complicated process, gives rise to the formation of the chemisorbed monolayer (ML) of halogen atoms at the first stage of reaction and to the subsequent growth of the metal halide film at the second stage [1, 2]. As a sequence, the thermodesorption spectrum of the halogen/metal system contains two peaks corresponding to desorption of multilayers (halide phase) and a monolayer (chemisorbed phase) (see Figure 35.1).

Since halide is a different chemical compound, the chemical state of the halogen atom within halide is strongly different from that in chemisorbed phase. This difference results in a large chemical shift (up to several electronvolts) in the electronic Auger spectra (see Figure 35.2a), which, in turn, allows the identification of chemical states on a surface during halogenation. Indeed, in a number of papers [4–8] it has been shown that application of factor analysis treatment [9, 10] to an array of Auger spectra acquired during halogenation enables following the individual behavior of each chemical compound. In particular, Figure 35.2b presents the concentrations of chemisorbed chlorine and chloride phases shown as functions of molecular chlorine exposure to the Cu(100) surface at 160 K [3]. According to Figure 35.2b, chlorine adsorption leads to the fast growth of chemisorbed chlorine phase. The growth of chloride starts only after saturation of the chemisorbed coverage. Further chlorine dosing results in a continuous decrease of chemisorbed state and growth of chloride. Finally, the signal of chloride saturates when the thickness of the chloride film becomes larger than escape depth of Cl $L_{2,3}VV$ Auger electrons.

Thus, the process of metal halogenation takes place in three stages:

1) Formation of chemisorbed monolayer
2) Nucleation and growth of thin halide film
3) Thick film growth.

In the following sections, we discuss the recent achievements in the identification of atomic structures at each stage.

*Surface and Interface Science: Solid-Gas Interfaces I*, First Edition. Edited by Klaus Wandelt.

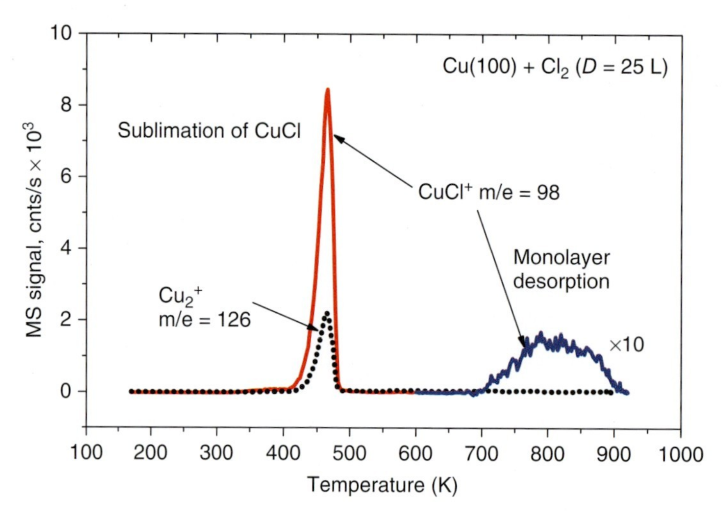


**Figure 35.1** Thermodesorption spectrum obtained after adsorption of molecular chlorine (25 L) on the Cu(100) surface at 160 K. (From Ref. [3].)

## 35.1 Formation of Chemisorbed Layer

Halogens chemisorbed on a metal surface usually form a large variety of ordered two-dimensional (2D) phases [1, 2]. This diversity of structures arises from the competition between the lateral halogen–halogen interaction and the halogen–substrate interaction that tends to impose upon the adsorbate layer the symmetry and periodicity of the substrate.

The interactions between adsorbed atoms on the surface can be either direct or indirect [11, 12]. In the former case, atoms interact directly without participation of the substrate. At short distances, a chemical bond can appear [13]. This kind of interaction weakens exponentially with distance. In the case of charge transfer, the dipole–dipole repulsion between adsorbed atoms appears ($1/r^3$) [14]. In addition, the weak van der Waals attraction is also possible ($1/r^6$) [13, 14].

In the latter case, adsorbed particles interact via the substrate. Such substrate-mediated interactions can be either elastic or electronic. Elastic interactions arise from the local distortion of the substrate lattice caused by the adsorbed atom [15]. Electronic interactions are determined by electrons localized near the surface [12, 14–16]. Scattering of electrons by the adsorbed atom results in oscillations in the electronic density and potential, which can influence the adsorption of other atoms.

### 35.1.1 Submonolayer Coverage

For such electronegative adsorbates as halogens, repulsive dipole–dipole interactions are expected to play a major role in the surface structure formation at any

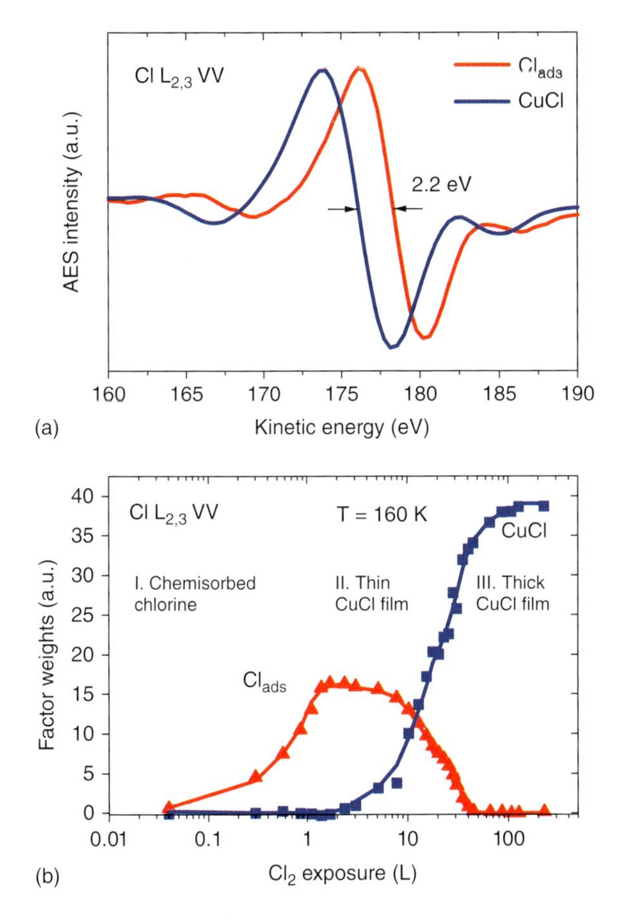


**Figure 35.2** (a) Profiles of Cl $L_{2,3}$VV Auger line acquired for copper chloride film (CuCl) and chemisorbed chlorine ($Cl_{ads}$). (b) Results of AES factor analysis performed for chlorine $L_{2,3}$VV Auger line. The behaviors of CuCl and $Cl_{ads}$ components are indicated by squares and triangles, respectively. (Adapted from Ref. [3].)

coverage. This repulsion has a natural tendency to lead to adatoms forming a hexagonal lattice, with *the period decreasing as the coverage grows.* However, measurements performed with low-temperature scanning tunneling microscopy (LT-STM) demonstrate that indirect interactions between adsorbed halogen atoms can affect the formation of submonolayer structures.

Recently, a case of a structural paradox taking place in the Cl/Au(111) system has been reported [17]: at submonolayer coverage (<0.02 ML), chlorine forms a chain-like structure with abnormally short interatomic distances 3.8 Å (when compared to lattice constant –5.0 Å in a $(\sqrt{3} \times \sqrt{3})R30°$ structure) – see Figure 35.3 [17]. The nature of this anomaly, which was resolved with DFT calculations, was found to be due to elastic interactions through the substrate lattice distortion (up to 19%). Figure 35.4 shows the pair interaction energies between chlorine atoms on Au(111).

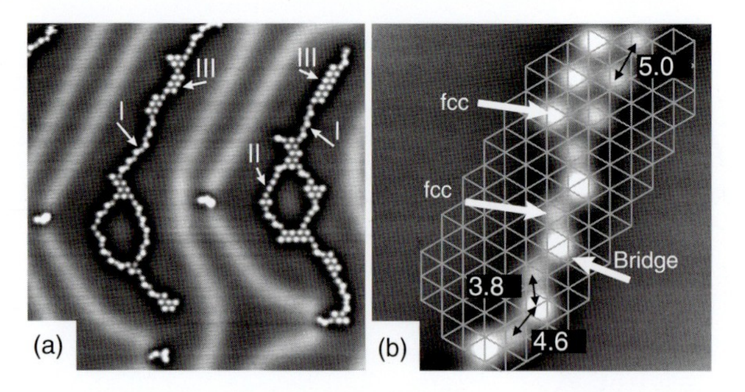


**Figure 35.3** (a) LT-STM image (230×230 Å$^2$, $U_s = -1$ V, $I_t = 0.5$ nA, 5 K) of submonolayer (0.02 ML) chain-like structures formed by chlorine atoms on Au(111). Different local configurations are marked as I (zig-zag chain), II (linear chain), and III (compact ($\sqrt{3} \times \sqrt{3}$)R30° configuration ). (b) A close-up image of a single chain, demonstrating that configuration I corresponds to the alternating fcc bridge-stacking of Cl atoms. A hexagonal grid of the Au(111)-(1×1) lattice was placed in a such way that atoms from configuration III appear to be in fcc threefold hollow positions. (From Ref. [17].)

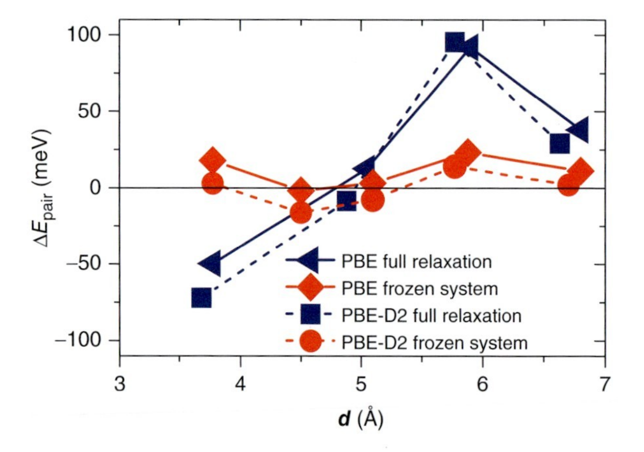


**Figure 35.4** Total interaction energies (blue curves) and their electronic parts (red curves) shown as functions of the distance between two chlorine atoms adsorbed on Au(111)-(1×1). Solid and dashed curves correspond to calculations performed using PBE and PBE-D2 functionals, respectively. The electronic contribution was obtained by freezing the substrate. (From Ref. [17].)

According to the plots, strong attraction at 3.8 Å exists only when the substrate lattice is allowed to relax. In case of the "frozen" substrate, the attraction disappears.

Chain-like structures were also obtained for chlorine adsorbed on Cu(111) and Ag(111) surfaces at submonolayer coverages (<0.05 ML). The interatomic distances within the chains were found to correspond to the fcc–hcp distances (3.8 Å for Cl/Cu(111) and 4.4 Å for Cl/Ag(111)) that were certainly smaller than the ($\sqrt{3} \times \sqrt{3}$)R30° distances (4.43 Å for Cl/Cu(111) and 5.0 Å for Ag(111)) [18]. Recently, short interatomic distances and formation of the chains in the Cl/Cu(111) system have been explained in Ref. [19] by indirect electronic interactions. To date, however, no proper explanation of these structures in the Cl/Ag(111) system has been suggested.

Electronic interactions mediated by surface states can also play a role in the halogen structures formation at low coverage. Nanayakkara *et al.* [20] demonstrated that taking into account this kind of indirect interactions makes it possible to explain the arrangement of small 2D $(\sqrt{3} \times \sqrt{3})$R30° islands of bromine on Cu(111) at coverage lower than 0.33 ML.

In fact, very little is known about submonolayer halogen coverage on metals owing to the lack of structural studies because they have to be performed with low-temperature STM.

### 35.1.2 Commensurate–Incommensurate Phase Transitions

Halogens adsorbed on metal surfaces appear to be good model systems to study the commensurate–incommensurate phase transitions in two dimensions [1]. Indeed, according to the literature published between 1970 and 1980s, halogens often form a commensurate phase at the beginning of adsorption, which can be replaced by a more complex compressed incommensurate phase at further dosing. Two principal models can be suggested to explain the numerous diffraction data of compressed phases: uniform compression, or formation of antiphase domains separated by domain walls (DWs). However, in the case of halogen adsorption on metals, most authors who published their papers in the 1970s–1990s conclude that the model of antiphase domains could give rise to unrealistically small interatomic distances, and preferred to simply postulate that the compression was uniform [1].

At the same time, evidence of DW formation was supported by many diffraction experiments performed for the adsorption of noble gases on graphite [21–24] and on metals (see, e.g., [24, 25]). The peculiarities of the commensurate–incommensurate (C–I) phase transitions in two dimensions were studied in a large number of theoretical papers (for reviews, see [24, 26–28]). A little later, DWs were found in many other physisorbed and chemisorbed systems. In the following sections, we will demonstrate that the DW mechanism dominates over uniform compression for many studied halogen/metal systems.

#### 35.1.2.1 Face-Centered Cubic (111) Surfaces

Almost all halogens form on all fcc (111) surfaces a commensurate structure $(\sqrt{3} \times \sqrt{3})$R30° at 0.33 ML (for the case of Cl/Cu(111) see Figure 35.5b). Further increase of coverage leads to the appearance of a characteristic diffraction pattern containing triangles of six spots with centers in $\sqrt{3}$-positions (Figure 35.5c and d) [29–31]. Similar LEED patterns were observed for many other halogen/metal systems: Br/Cu(111) [32], I/Ag(111) [33], I/Au(111) [34], Cl/Ag(111) [35, 36], I/Cu(111) [37], I/Ni(111), Cl/Ni(111) [38], and Br/Ag(111) [39]. These LEED patterns were explained in the literature by the formation of incommensurate, uniformly compressed, uniaxial c$(p \times \sqrt{3})$ or hexagonal $(n\sqrt{3} \times n\sqrt{3})$R30° lattices.

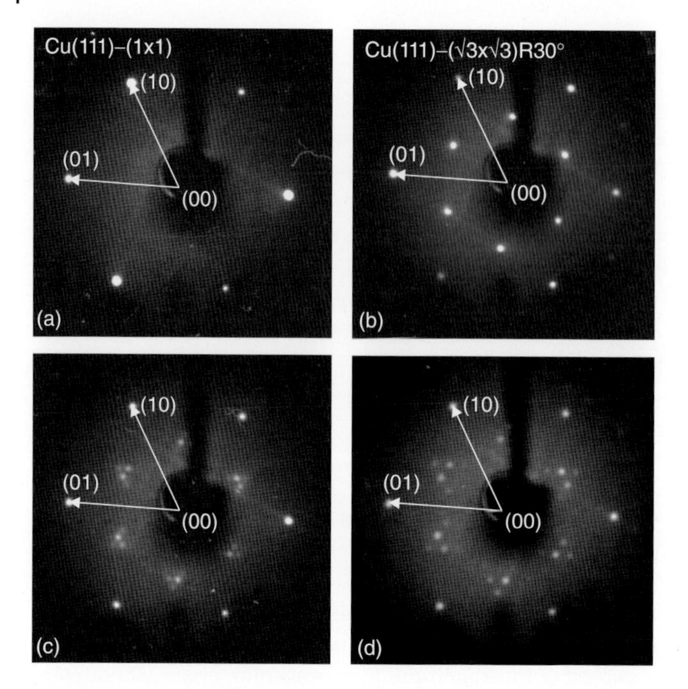


**Figure 35.5** Evolution of LEED patterns ($E_0$ = 75 eV) in the course of continuous chlorine adsorption on the Cu(111) surface at 300 K. (a) Clean Cu(111)-(1×1), (b) Cu(111)-($\sqrt{3} \times \sqrt{3}$)R30°-Cl, (c) and (d) Splitting of the $\sqrt{3}$ spots.

Usually, the structural models were constructed using van der Waals diameter as a natural limit for the packing of halogen atoms on metal surfaces [1].

The characteristic LEED patterns of halogens on (111) fcc metal surfaces (Figure 35.5c and d) are known to be similar to those obtained in the case of noble gas physisorption on graphite [21–23] and Pt(111) surfaces [25] and evidently explained by the DW model. However, this similarity was not taken into account for years, and all authors were in favor of the uniform compression model.

In the first real spaced study performed with STM by Motai *et al.* [40] for chlorine adsorption on Cu(111), the model of uniformly compressed hexagonal lattice was confirmed. However, later Huang *et al.* [41] in their STM study of I/Au(111) system clearly demonstrated the formation of DWs in a iodine layer.

In 2000, Andryushechkin *et al.* [31] investigated the Cl/Cu(111) system again with UHV-STM and concluded that chlorine compression is uniaxial and nonuniform. Figure 35.6a shows an STM image of saturated chlorine monolayer on Cu(111). Three domains with different directions of modulation are clearly seen in the STM image. Strong surface modulation is produced by an alternation of bright and dark stripes. The interatomic distances along the stripes measured in the STM image agreed well with the value in ($\sqrt{3} \times \sqrt{3}$)R30°-Cl lattice (4.4 Å), while the average distance along two other atomic rows in the domain appeared to be much smaller (3.7 Å). The direction of compression indicated in Figure 35.6a for each domain

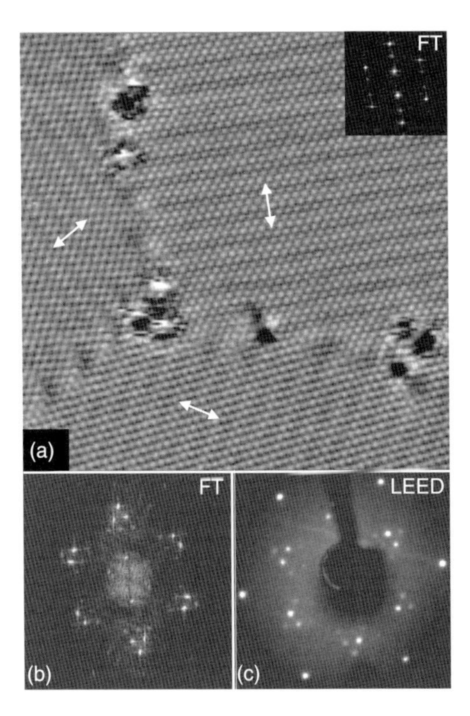


**Figure 35.6** (a) STM image (201 × 201 Å$^2$, $I_t$ = 0.4 nA, $U_t$ = −750 mV) of saturated chlorine monolayer on Cu(111) ($\theta \approx 0.41$ ML), (b) Fourier transform, (c) LEED pattern. Directions of compression are shown by arrows. One domain FT is shown in the inset to (a). The black areas in domain boundaries correspond to CuCl nuclei. (From Ref. [31].)

is parallel to one of the basic directions of the substrate. The one-domain Fourier transform (FT) shown in the inset to Figure 35.6a demonstrates the splitting in the direction perpendicular to the stripes. The three-domain FT (Figure 35.6b), being a simple sum of one-domain FTs, is identical to the LEED pattern (Figure 35.6c). Note that the STM images obtained for lower chlorine coverages also demonstrate striped structures but with a larger period of modulation.

To identify the detailed structure of the striped phases, Andryushechkin *et al.* [31] superimposed an atomic grid of the (111) plane of copper on two STM images with low and high degrees of compression (Figure 35.7). The superposition of the chlorine and copper lattices was done assuming that the brightest spots in STM images correspond to chlorine atoms preserving their positions in $(\sqrt{3} \times \sqrt{3})$R30°-Cl lattice and occupying fcc sites [42]. For low coverage (Figure 35.7a), it is clearly seen that the bright atoms are placed in threefold positions (assigned with fcc sites), so the perfect hexagons of the $(\sqrt{3} \times \sqrt{3})$R30° lattice can be easily constructed. More close-packed darker atoms occupy less symmetrical sites and form DWs (Figure 35.7a). As may be seen in Figure 35.7, every third domain is in phase, while the neighboring domains appear to be antiphase to each other. Note that for the most compressed phase (Figure 35.7b) degeneration of the $(\sqrt{3} \times \sqrt{3})$R30° domain into one atomic row occurs. It was found that at lower coverage (0.33 ML$\leq \theta \leq$ 0.37 ML)

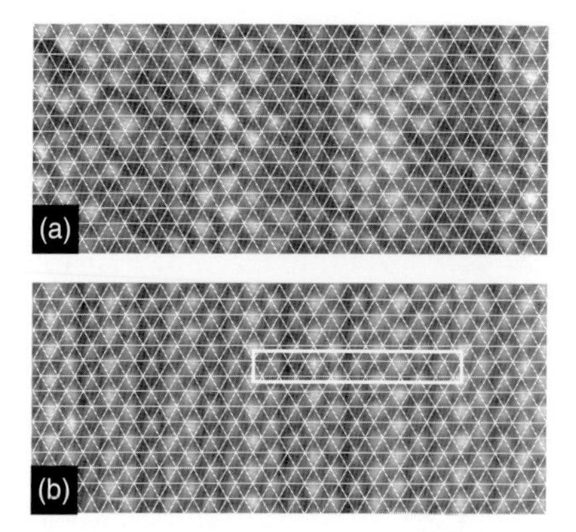


**Figure 35.7** STM images of chlorine monolayer on Cu(111) corresponding to low ($\theta \approx 0.35$ ML) (a) and high ($\theta \approx 0.41$ ML) (b) degrees of compression fitted with the atomic grid of the substrate. For (b), unit mesh ($12\times\sqrt{3}$-R30°) is indicated. (From Ref. [31].)

DWs consist of 3–4 atomic rows (Figure 35.7a), while at higher $\theta$ ($\geq 0.39$ ML) 2- and 3-row DWs have been observed (Figure 35.7b). Such wide walls could be understood as due to the strong relaxation of *superheavy* DWs [43, 44].

The unit mesh of the most compressed phase is denoted as ($12\times\sqrt{3}$-R30°) (Figure 35.7b). It consists of four fcc atomic rows separated by DWs with variable widths. The distances between fcc rows are described by a sequence 3.5*a*, 5*a*, and 3.5*a* (where *a* is the copper (111) interatomic distance). The first and fourth fcc rows seem to be in phase. Several interatomic distances lying in the range 3.5–4.0 Å were derived from STM images for this structure, with the shortest Cl–Cl distance of 3.5 Å in a DW being comparable with the van der Waals diameter of chlorine (3.6 Å).

STM images obtained at lower coverage show the coexistence of domains with different numbers of unit meshes of $(\sqrt{3}\times\sqrt{3})$R30°. As the coverage increases, the average domain size diminishes gradually, producing continuous splitting of the spots in the LEED and STM-FT patterns. Thus, the formation of the incommensurate lattice can be considered to be a phase transition of second order. The displacement of chlorine atoms occurs along atomic rows of the substrate in a channel formed by fcc, hcp, and bridge sites. Calculations of chlorine adsorption energy on Cu(111) [45] shows the fcc hollow to be the deepest adsorption site. Adsorption energies for hcp and bridge sites were found to be higher by ~8.7 and ~75 meV, respectively [45]. The energy corrugation appears to be of the order of $kT$ at room temperature (~24 meV), whereas the value for the top site lies much higher (~450 meV [45]). The calculations support the STM results presented and do not favor the uniform isotropic compression of the chlorine layer at room temperature.

To date, the existence of DWs at room temperature has been proved with STM for several systems: I/Cu(111) [37], I/Ni(111) [46], I/Ag(111) [47], and I/Au(111) [41].

In cases of Cl/Ag(111) [35, 36], and Br/Ag(111) [39], however, triangles in LEED were reported only at low temperatures. In order to look into the transition in the Cl/Ag(111) system, Andryushechkin *et al.* [36] used a low-temperature STM operating at 5 K. They hoped to obtain new data at the initial stage of the C–I transition when the density of DWs is low and diffraction techniques do not work.

Figure 35.8a presents an atomic-resolution LT-STM (5 K) image obtained for chlorine coverage of 0.34 ML on Ag(111). Although $\sqrt{3}$-spots both in the diffraction

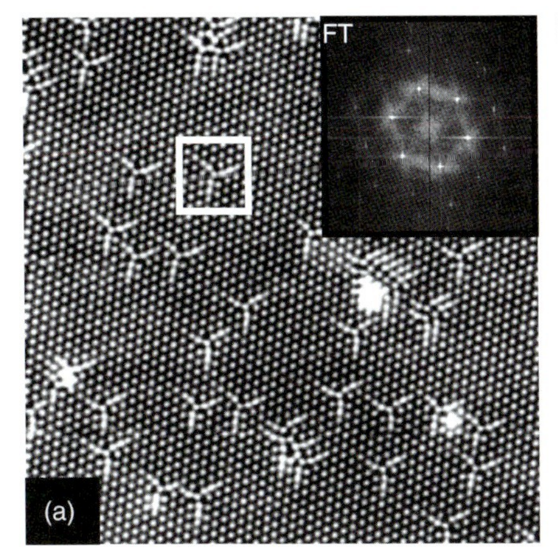


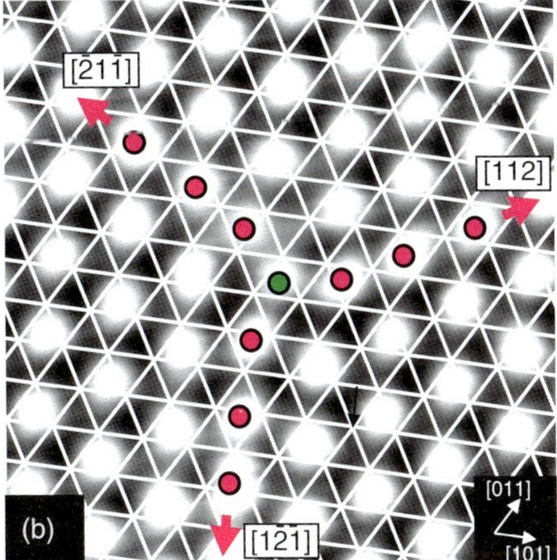


**Figure 35.8** (a) STM Image (330 × 330 Å$^2$, $I_t$ = 2.8 nA, $U_s$ = −60 mV, $T$ = 5 K) of Ag(111) surface covered with $\theta \approx 0.34$ ML of chlorine. In the upper right-hand corner FT, the STM image is shown. (b) Zoom from image (a) superimposed with a grid, where the knots of the grid coincide with position of substrate atoms. A self-interstitial defect (additional Cl atom) in the $(\sqrt{3} \times \sqrt{3})$R30°-Cl structure is highlighted with green circle. The purple circles indicate positions of displaced chlorine atoms. (From Ref. [36].)

pattern and in the Fourier transformation image remain sharp, the real-space STM image changes dramatically, with numerous three-arm star-like objects parallel to the ⟨112⟩ directions. In the reciprocal space, the array of similar star-like objects gives rise to the specific diffuse hexagon seen in the inset to Figure 35.8a.

Figure 35.8b shows a magnified STM image of the star with a superimposed hexagonal grid corresponding to silver (1×1) lattice. The superimposition was made assuming that atoms around the star occupy threefold hollow sites. According to Figure 35.8, the origin of the star is explained by the presence of an additional chlorine atom in the center. This atom can be considered to be a self-interstitial defect in the commensurate $(\sqrt{3} \times \sqrt{3})$R30° lattice, indicating local compression of the chlorine layer. The chlorine lattice remains commensurate with the substrate and does not split into separate domains. The visual "star" effect in the STM images is explained by the narrowing of the interatomic distances in the directions of the stars' arms.

This new object has been assigned as a 2D surface crowdion. Previously, crowdions had been studied in bulk materials since the late 1950s [48, 49]. Recently, the formation of 1D surface crowdions has been suggested as a mechanism for surface adatom transport on strained Cu(100) and Pt(100) surfaces [50, 51].

The additional chlorine atom in the center of a crowdion in Figure 35.8b, occupying a threefold position similar to the surrounding atoms, however, belongs to another $(\sqrt{3} \times \sqrt{3})$R30° sublattice (marked as "2" in Figure 5). There are three equivalent $(\sqrt{3} \times \sqrt{3})$R30° sublattices on the (111) surface of fcc metals. Therefore, another type of crowdion should exist, with a central atom belonging to sublattice "3." Models of both types of crowdions exhibiting different chirality are shown in

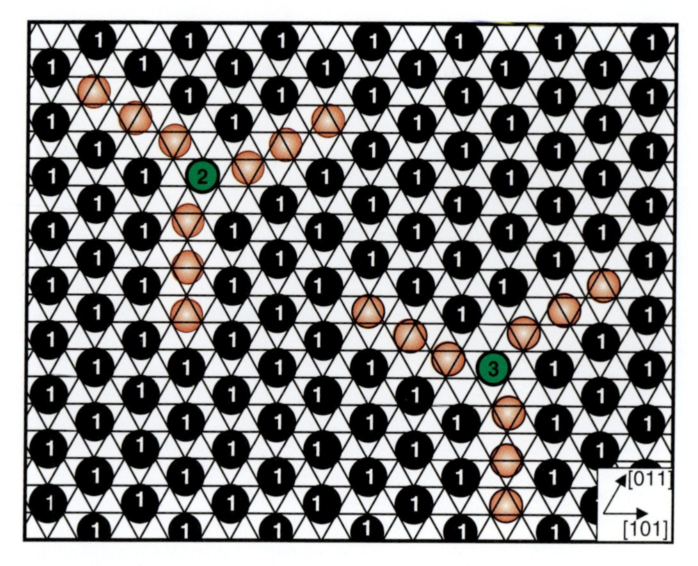


**Figure 35.9** Schematic representation of two types of crowdions (with left- and right-side chirality) in the $(\sqrt{3} \times \sqrt{3})R30°$-Cl structure. Chlorine atoms in different $(\sqrt{3} \times \sqrt{3})R30°$ sublattices are marked with numbers 1, 2, and 3. Disturbed chlorine atoms forming crowdions are shown in pink. Additional chlorine atoms in the centers of crowdions (interstitials) are shown in green. (From Ref. [36].)

Figure 35.9. The size of the crowdions could be estimated by the perturbation of atom positions caused by the interstitial atom. As follows from Figure 35.8b, the perturbation attenuates at 3–4 interatomic distances, which correspond to the crowdion diameter, about 30 Å.

At the intermediate coverage range $\theta<0.37$ ML, both crowdions and DWs coexist on the surface, as shown in Figure 35.10a. At certain scanning parameters ($I_t = 2$ nA, $U_s \approx +2$V), the contrast is such that crowdions and DWs become

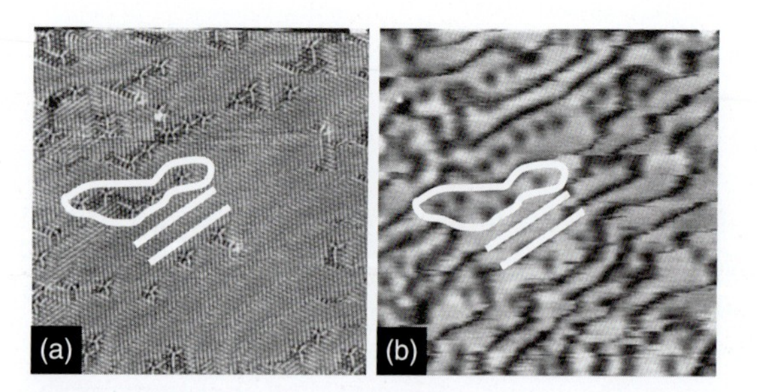


**Figure 35.10** STM images (330×330 Å$^2$, $T = 5$ K) of the same area of Ag(111) surface covered with chlorine. The image (a) was taken with "usual" scanning parameters $I_t = 2.9$ nA, $U_s = -48$ mV, whereas image (b) was taken with a higher bias voltage: $I_t = 2.9$ nA, $U_s = +1980$ mV. Additional contrast in the areas of increased chlorine density can be clearly seen on the image. (b) Crowdions appear as dark spots and DWs as dark lines. (From Ref. [36].)

visible in STM images as dark spots and lines, respectively (see Figure 35.10b). This finding makes it possible to follow the C–I phase transition at a larger scale with higher scanning speed.

Figure 35.11 shows a series of 1000×1000 Å$^2$ STM images corresponding to the gradual increase of the coverage from 0.34 to 0.40 ML. At the first step of $(\sqrt{3} \times \sqrt{3})$R30° compression, only crowdions exist on the surface. Such a situation is shown in Figure 35.11a and corresponds to a coverage of 0.34 ML (this coverage can be obtained directly from the STM image simply by counting the number of crowdions). At some critical density of crowdions (approximately corresponding to the average nearest-neighbor distances about 20 Å), new objects–loop lines (DWs) appeared. The increase of the coverage resulted in an increase of the length of the DW loops. As the coverage increases, the density of DWs grows, while the total number of crowdions decreases. According to Figure 35.11b and c, DW-to-crowdion, crowdion-to-crowdion, and DW-to DW interactions are repulsive. For all three pairs of objects, the separations between them appear to be very close to a distance of ≈20 Å. In Figure 35.11c, the total length of the DWs is high enough, but their ordering just starts at this coverage. Ordering of DWs starts in the areas

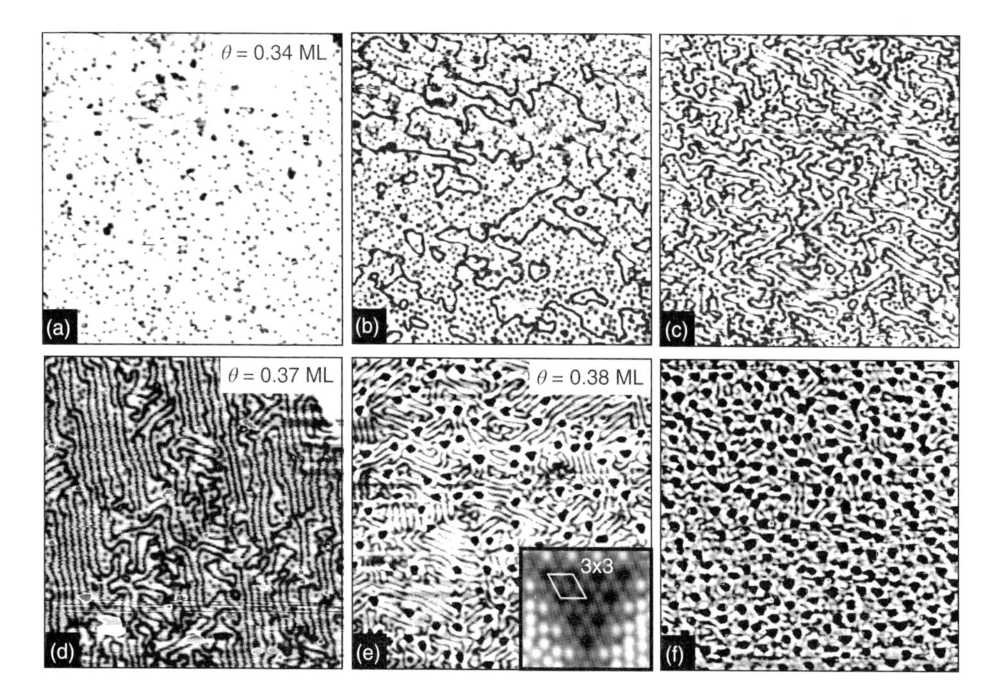


**Figure 35.11** Large-scale STM images (1000×1000 Å$^2$, $I_t$ = 2.9 nA, $U_s$ = +1980 mV, $T$ = 5 K) of Ag(111) surface during continuous increase of surface coverage in the range $0.34 < \theta < 0.40$ ML. (a) 2D gas of crowdions. (b and c) Condensation of crowdion gas into DWs. (d) A striped DW structure (DW crystal) formed at $\theta \approx 0.37$ ML. (e and f) Nucleation and growth of the new (3×3) phase (small dark triangles) in parallel with continuous decrease of the DW distance. The atomic-resolution STM image of the (3×3) island is shown in the inset to (e). (From Ref. [36].)

where all crowdions disappear. The distances between parallel DWs also appear to be equal to 20–25 Å. At coverage 0.37 ML (estimated by the period of the DW lattice), almost all crowdions disappear and large areas with parallel striped DWs can be seen in the STM image, as in Figure 35.11d.

At $\theta > 0.38$ ML (Figure 35.11d), the STM image shows the formation of triangular islands of the 3×3 reconstruction described in detail in our previous work [52]. The 3×3 islands destroy the regular DW lattice. Indeed, now DW lines prefer to end at the 3×3 islands, as seen from Figure 35.11e. As the coverage increases further, the DW lattice breaks up into separate segments between areas with the 3×3 reconstruction (Figure 35.11f). Finally, the DW system degenerates into a compressed quasi-hexagonal lattice. At saturation point, the nonreconstructed areas disappear [52, 53].

Thus, it has been shown that a 2D gas of crowdions condenses into DWs. Since at the beginning of transition the DWs are not ordered, they can be considered to be a DW fluid. An increase in the number of atoms on the surface leads to the solidification of the DW fluid and formation of an equally spaced DW lattice (DW crystal).

Understanding the microscopic mechanism of the crowdion-to-DW transition is of great scientific interest. Figure 35.12 shows an STM image of an interesting object formed as a result of the agglomeration of crowdions (marked as "1"). The superimposition of the atomic grid of the substrate shows that the core of the object contains three chlorine atoms with nearest neighbor distance of 5 Å (the distance in a $(\sqrt{3} \times \sqrt{3})$R30° lattice) occupying threefold hollow sites. However, atoms in the core belong to a different $(\sqrt{3} \times \sqrt{3})$R30° sublattice than the surrounding chlorine atoms. In this connection, such kind of object can be interpreted as a nucleus of the new domain belonging to a different $\sqrt{3}$-sublattice that is surrounded by the DW loop. Atoms from the "arms" of the object occupying nonsymmetrical positions form segments of DWs. As the additional chlorine atoms adsorb, the size of the core domain increases. The STM image in Figure 35.12 shows another object with linear geometry (marked as "2") formed near a surface defect. This object is considered to be formed by several crowdions. In fact, the "arms" in this case form a segment of a DW.

In their theoretical study, Lyuksyutov *et al.* [54] have shown that a 2D gas of the interstitial defects significantly influences the C–I phase transition. In particular,

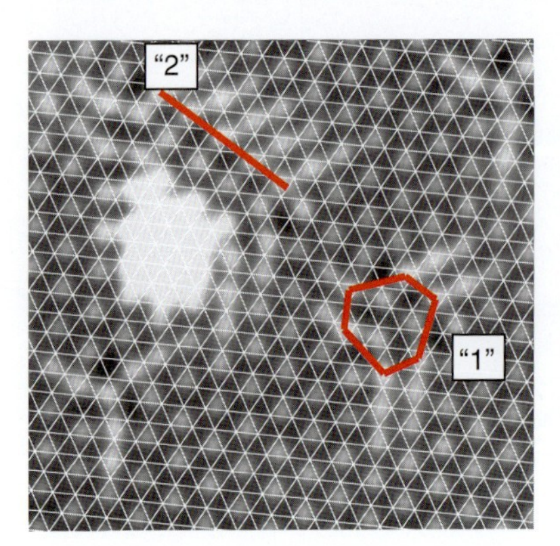


**Figure 35.12** STM images (66×62 Å$^2$, $I_t$ = 2.8 nA, $U_s$ = −60 mV, $T$ = 5 K) showing the process of crowdion condensation. An object "1" is considered to be a nucleus of the DW loop. Three atoms in the core of the object "1" formed a small domain out of phase with surrounding chlorine lattice. A linear object "2" formed by several crowdions is assigned to the small segment of the growing DW. For clarity, DWs are shown by red lines. (From Ref. [36].)

they considered the possibility of the atom exchange between 2D gas of the interstitials and DWs. According to Ref. [54], heating of the DW system leads to an incommensurate–commensurate phase transition via DW evaporation. As a result, all excess atoms ($N$) can be transferred to the gas of self-interstitials, and DWs disappear. Another consequence of the theory by Lyuksyutov *et al.* [54] concerns the stability of the DW lattices at low $N$. The authors have shown that, at $N \rightarrow 0$, the DWs are thermodynamically unfavorable in comparison with a gas of point-like defects. Such a conclusion is in excellent agreement with our data. Indeed, we detected that first DW lines appear at a coverage $\theta > 0.34$ ML.

An older theory by Vilain [55] predicts that at elevated temperatures the striped DW structure with a large period is always unstable against first-order transition to a hexagonal DW lattice. Lyuksyutov *et al.* [54] have shown that, at $N \rightarrow 0$, the phase with point-like defects is always thermodynamically preferred over a hexagonal DW lattice. Such a conclusion is also in line with our experimental data, since the hexagonal DWs were never detected in our STM images.

It is worth noting that in the experiment by Andryushechkin *et al.* [36] the phase separation between crowdions and DW corresponds to a temperature of 5 K. At elevated temperatures, according to Lyuksyutov *et al.* [54], the condensation of crowdions should occur at higher coverages.

**To summarize the results**, the existence of DWs has been clearly proved for Cl,I/Cu(111) [31, 37], I/Au(111) [41], I/Ag(111) [47], Cl/Ag(111) [36], and I/Ni(111) [46]. For the systems Br/Cu(111) [32], Br/Ag(111) [39], and Cl/Ni(111) [38], the DW mechanism is likely due to the similarity of evolution of LEED patterns. For Cl/Au(111), there is no C–I transition. Adsorption of additional chlorine leads to the formation of molecular structures on the surface [56, 57]. For halogens adsorbed on Pt(111) [58–66], Pd(111) [58, 67–71] and Rh(111) [72–74] surfaces, there are no experimental results demonstrating compression of $(\sqrt{3} \times \sqrt{3})$R30° lattice at coverages exceeding 0.33 ML. However, taking into account that DWs were detected even for the I/Ni(111) system, in which a large halogen is adsorbed on the (111) plane with minimal lattice constant, we cannot exclude the possibility of a slight compression of the $(\sqrt{3} \times \sqrt{3})$R30° lattice in all these cases. Therefore, further STM studies are required to clarify the situation.

#### 35.1.2.2 Face-Centered Cubic (100) Surfaces

This section presents a description of halogen phases formed on the (100) faces of fcc metals. Distinction of the Hal/Me(100) systems can be made on the basis of the adsorption position of single halogen atom (see Table 35.1). According to Table 35.1, most of halogens prefer to adsorb into fourfold hollow sites. The adsorption on Au(100) and Pt(100) seems to be different. At least, for the case of Cl/Au(100) [75] and Br/Au(100) [76], bridge sites were found to be preferred to hollow and top positions.

For the case of hollow adsorption, the formation of a single c(2×2) structure was reported by LEED: Cl,Br/Cu(100) [79, 87], Cl,Br,I/Ag(100) [33, 79], Cl,Br/Ni(100) [88, 89], and Cl,I/Pd(100) [90, 91]. STM measurements performed for

**Table 35.1** Preferential adsorption sites for halogens on (100) plane of fcc metals. H, hollow; B, bridge.

| | Cl | Br | I |
|---|---|---|---|
| Ag | H [77, 78] | H [76] | H [8] |
| Au | B [75] | B [76] | – |
| Cu | H [79] | H [80] | H [81] |
| Ni | H | H [82] | H [83] |
| Pd | H [84] | H | H |
| Pt | – | B ? [85] | B ? [86] |

Cl/Cu(100) [92], Br/Cu(100) [87], and I/Ag(100) [8] systems also confirmed the absence of further compression of the halogen layer.

A specific case is realized when large halogen atoms are adsorbed on a metal surface with a small lattice constant. This is the case of iodine adsorption on Cu(100) and Ni(100). Indeed, only for Cu(100) and Ni(100) the distance in the c(2×2) structure (3.61 Å for Cu; 3.52 Å for Ni) appears to be less than the shortest possible I–I distance estimated either by van der Waals diameter (4.3 Å) [93] or STM observations (3.8 Å) [37]. Therefore, it would seem impossible to form stable c(2×2)-I phases on Cu(100) and Ni(100). As a consequence, one would expect the formation of a series of complex incommensurate structures.

Figure 35.13 shows a series of STM images obtained in the course of the step-by-step adsorption of molecular iodine on a Cu(100) surface at 300 K [94].

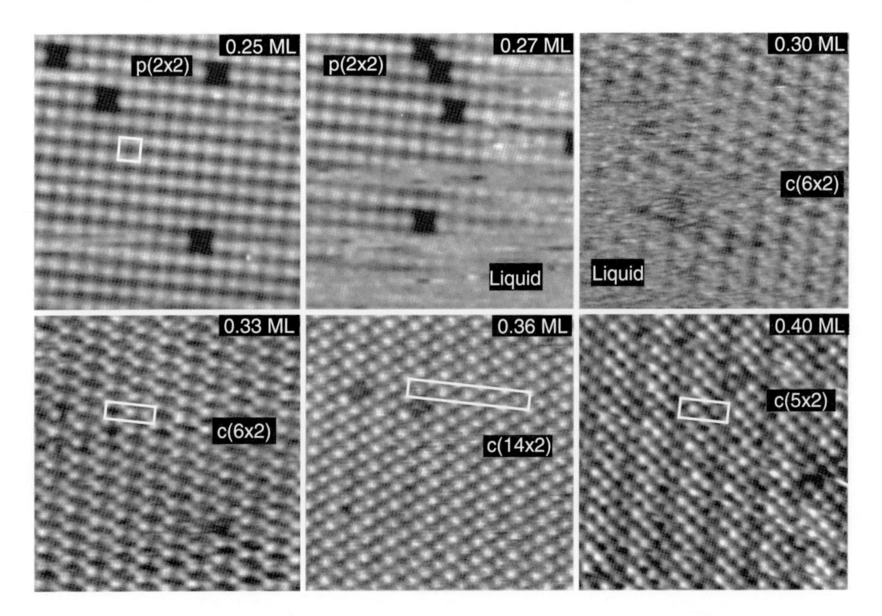


**Figure 35.13** STM images (77×77 Å$^2$, $T$ = 300 K) showing the process of iodine lattice compression on Cu(100). From Ref. [94].

Iodine adsorbed on the Cu(100) surface at room temperature forms five different phases with increasing coverage. These phases have periodicities of p(2×2), liquid (disordered) phase, c(6×2), c(14×2), and c(5×2). In the only "in registry" phase, p(2×2), all iodine atoms occupy equivalent adsorption sites. The other phases can be described as uniaxially compressed structures with atoms in different adsorption sites. The phase transformations p(2×2)⇒ liquid, liquid⇒c(6×2), and c(6×2)⇒c(14×2) can be explained in terms of first-order phase transitions, while c(14×2)⇒c(5×2) appears to correspond to a second-order phase transition.

The I/Ni(100) system was studied in detail using LEED by Jones and Woodruff [95]. A series of different phases were observed, and a model was suggested for the gradual uniaxial compression of the iodine layer along the atomic grooves of Ni(100) [95]. However, these models need to be tested in real space with STM.

Adsorption of halogens onto Au(100) and Pt(100) surfaces leads first to the lift ing of the initial reconstruction [1, 2]. The formation and evolution of the phases on these surfaces is likely determined by a specific energy landscape (Table 35.1). Indeed, since the bridge sites are more energetically favorable than top and hollow sites, the movement of halogen atoms is much easier in the ⟨100⟩ direction than in ⟨110⟩ one. This property makes it possible to form a series of $c(p\sqrt{2} \times \sqrt{2})R45°$ quasi-hexagonal structures [1, 2]. Recent Monte Carlo simulations have confirmed that these types of structures are a sequence of bridge adsorption [96].

#### 35.1.2.3 Face-Centered Cubic (110) Surfaces

Unfortunately, it is not easy to draw a general picture of halogen interaction/adsorption with the (110) planes of fcc metals. In Table 35.2, we present preferred adsorption sites determined for chlorine, bromine, and iodine atoms adsorbed on the (110) planes of fcc metals. For Cl/Cu(110) [97], Br/Cu(110) [98], Cl, Br/Pt(110) [99, 100], and Cl, Br/Ag(110) [78, 101] systems, the twofold short-bridge position is energetically favorable. Experimental evidence for halogen adsorption in short-bridge sites was reported for Cl/Ni(110) [102], but more recent calculations [103] did not confirm it. There are only three systems with halogen atoms adsorbed in high-symmetry hollow sites: I/Cu(110) [106], I/Pd(110) [107], and I/Ag(110) [108]. The adsorption site of halogen atom on the (110) plane of fcc metals seems to be a key point in the formation of surface structures.

**Table 35.2** Preferential adsorption sites for halogens on the (110) plane of fcc metals: H, hollow; SB, short bridge; LB, long bridge.

| | Cl | Br | I |
|---|---|---|---|
| Ni | SB(H) [102, 103] | – | – |
| Pd | SB [104] | LB [104] | H [104, 107] |
| Pt | SB [99] | SB [100] | SB [105] |
| Cu | SB [97] | SB [98] | H [106] |
| Ag | SB [78] | SB [101] | H [108] |
| Au | – | – | – |

Indeed, for iodine on Cu(110), Pd(110), and Ag(110) adsorbed in hollow sites, the continuous compression of the commensurate c(2×2) lattice (second-order phase transition) takes place. The mechanism of this transition can be explained by the DW model. STM data obtained for the I/Cu(110) system [109] evidently demonstrate that at coverage exceeding 0.5 ML the formation of c(2×2) antiphase domains takes place (see Figure 35.14). Domains of the c(2×2) structure are separated by DWs. The density of atoms within DWs exceeds the density in the c(2×2) structure, which means heavily striped DWs are forming [44]. The model of an ideal DW structure with DW separations *l*–20 Å is shown in Figure 35.14d. This model presents sharp DWs formed by atoms occupying long bridge positions. In the present case, the DWs contain at least three atoms situated neither exactly in the long bridge nor in the hollow sites. Such relaxation of the DWs leads to the smoothing of kinks that obviously exist in the ideal model (see Figure 35.14d). Similar STM images have been obtained for iodine adsorption onto Ag(110) [110] and Pd(110) [107] surfaces.

When halogen adsorbs in a short bridge site, the scenario of surface phase transformation is different. In this case, the uniaxial, continuous compression along ⟨110⟩ direction is not possible. Therefore, the transition from the first commensurate structure to a denser phase should take place via a first-order phase transition or

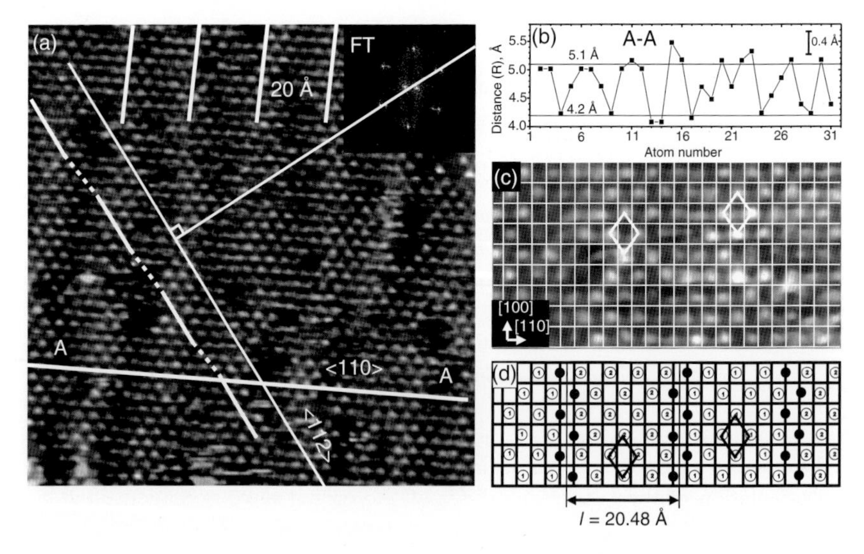


**Figure 35.14** (a) STM image (149×149 Å$^2$, $I_t$ = 0.65 nA, $U_s$ = −149 mV) of the compressed iodine layer on Cu(110). The inset shows the Fourier transformed (FT) image. The ⟨112⟩ direction was determined as perpendicular to the straight line passing through the centers of the doublets (c(2×2) spots positions) in the FT image. Dashed lines show the relaxed kinks between atomic rows in antiphase domains. (b) Plot of iodine–iodine distances measured along the straight line marked in (a) as A–A. Lines at 5.1 and 4.2 Å show the average maximum and minimum distances, respectively. (c) Superposition of the STM image frame and Cu(110) grid, with knots indicating the positions of copper atoms. (d) Model of an ideal DW structure with a DW distance l = 20.48 Å corresponding to separation of the stripes in (a). From Ref. [109].

surface reconstruction. In early works, the surface structure was investigated only by diffraction techniques [111]. Even application of STM (e.g., for Cl/Ni(110) [112], Br/Ni(110) [113]) did not result in any unambiguous determination of the surface structure because of the complexity in the interpretation of STM images. Some experiments were performed in the electrochemical environment (Br/Au(110) [114], I/Au(110) [115], Br/Cu(110) [116]. However, we cannot consider the surface structures for these systems to be completely solved, first of all, due to the lack of DFT data.

Currently, the best investigated system in vacuum is Br/Pt(110) [100, 117–131]. The authors were able to decode the atomic structures of numerous bromine phases on Pt(110) and first-order phase transitions using a combination of different experimental techniques supported by DFT calculations.

#### 35.1.2.4 **Base-Centered Cubic Metals**

Numerous structural models have been suggested in the literature to describe structures formed by halogens on bcc metal surfaces (see reviews of Jones [1] and Altman [2]). In particular, on the (100) plane of bcc metals, halogens form a c(2×2) structure(Cl,Br/Cr(100) [132, 133], Cl,Br,I/Fe(100) [134, 135], I/Ta(100) [136], Cl,Br/V(100) [137, 138], Cl,Br,I/W(100) [139–141]), as in the case of fcc metals. However, in the present case the continuous compression of halogen layer occurs.

On the close-packed (110) face of bcc metals, halogens often (Br/Cr(110) [142], Cl,Br,I/Fe(110) [143, 144]) form a commensurate structure (3×1), which to some extent is analogous to the $(\sqrt{3}\times\sqrt{3})$R30°. Increasing the coverage results in continuous compression of the adsorbate layer, accompanied by rotation.

Unfortunately, all these works date back to the early 1980s. Since no STM studies exist for these systems, a detailed description of the commensurate–incommensurate transitions is impossible.

#### 35.1.2.5 **hcp Metals**

Adsorption of halogens on hcp metal surfaces has not been extensively studied [1, 2]. From the point of view of commensurate–incommensurate phase transition, the Cl/Ru(0001) system is of great interest. Indeed, according to diffraction data, the compression of the $(\sqrt{3}\times\sqrt{3})$R30° lattice is accompanied by splitting of the $\sqrt{3}$ spots into six-spot triangles, as in the case of fcc metals [145, 146]. However, the orientation of the triangles is different. According to classification done by Zeppenfeld *et al.* [44], such type of the diffraction pattern corresponds to heavy rather than superheavy DWs.

### 35.1.3 **Reconstructive Phase Transitions**

For years, halogen-induced reconstruction of metal surfaces was not considered a possible scenario of adsorbate layer compression. However, recent STM measurements have clearly demonstrated the possibility of substrate restructuring under halogen action.

#### 35.1.3.1 Face-Centered Cubic (100) Surfaces

An fcc metal surface as a rule does not reconstruct under halogen gas action. There is only one work by Fishlock *et al.* [147] on Br/Cu(100), in which the authors reported the formation of a chess-board structure involving the reconstruction of the upper copper layer. However, the proposed structural model has not been verified by DFT calculation and therefore the surface reconstruction cannot be considered as rigorously proven.

#### 35.1.3.2 Face-Centered Cubic (111) Surfaces

**n×n structures** For several systems (Cl,Br/Ag(111) [29, 35, 39, 148–153], Cl/Pd(111) [69], Cl/Pt(111) [58–60], Br/Pt(111) [62], I/Pt(111) [64, 66], and Cl/Rh(111) [72]), diffraction patterns labeled as "$n \times n$" were reported in the literature at coverage close to saturation. These LEED patterns were initially interpreted as due to the formation of a compressed hexagonal halogen lattice not rotated with respect to the substrate lattice. Andryushechkin *et al.* [52] proved recently that chlorine-induced reconstruction of the Ag(111) surface was responsible for a "complex" surface structure observed and previously described as (10×10) [29] or (17×17) [152].

The LT-STM image in Figure 35.15 shows an array of triangle-shaped islands of 15–30 Å in size on the surface, surrounded by a slightly disordered structure with atoms approximately arranged in a $(\sqrt{3} \times \sqrt{3})R30°$ lattice. The local structure of each island is well described by a (3×3) lattice with a unit mesh of $8.67 \times 8.67 Å^2$ (see Figure 35.15d). The neighboring (3×3) islands appear to be out of phase with each other, as follows from Figure 35.15d. Therefore, the array of islands can be treated as a system of (3×3) antiphase domains. The atomic structure of the (3×3) phase has been identified using DFT calculations. According to the model, six Ag atoms are left in the topmost substrate layer within the (3×3) unit cell. Three of them keep their original fcc positions, and the other three have moved to the hcp positions. As a result, nearly fourfold hollow adsorption sites for chlorine atoms were created. Additional chlorine atoms can be placed in the corners of the unit cell. Note also that the chlorine atoms in the holes occupy nonsymmetric positions (see Figure 35.15e).

For the Br/Ag(111) system, the observed sequence of LEED patterns is similar to that for Cl/Ag(111). At saturation, the (3×3) superstructure is reported [39]. Unfortunately, the quality of existing STM measurements [153] is quite poor, which makes a direct comparison of the systems difficult. We believe that the same kind of reconstruction can occur for Br/Ag(111). The only difference is in the ability of bromine to form large islands of the (3×3) structure.

The (3×3) superstructure in STM images has also been observed for halogens adsorbed on Pt(111). For the Br/Pt(111) system, the existing LT-STM data shown in Figure 35.16 demonstrate triangular (3×3) islands [62], very similar to Cl/Ag(111) shown in Figure 35.15.

Although no reconstructive structural model (similar to Cl/Ag(111)) has been suggested in Ref. [62], we believe that it should be at least considered for this system. From our point of view, the coexistence of separate (3×3) islands with the $\sqrt{3} \times \sqrt{3})R30°$ structure points at surface reconstruction rather than uniform compression of the iodine layer. The same arguments also apply to iodine and chlorine

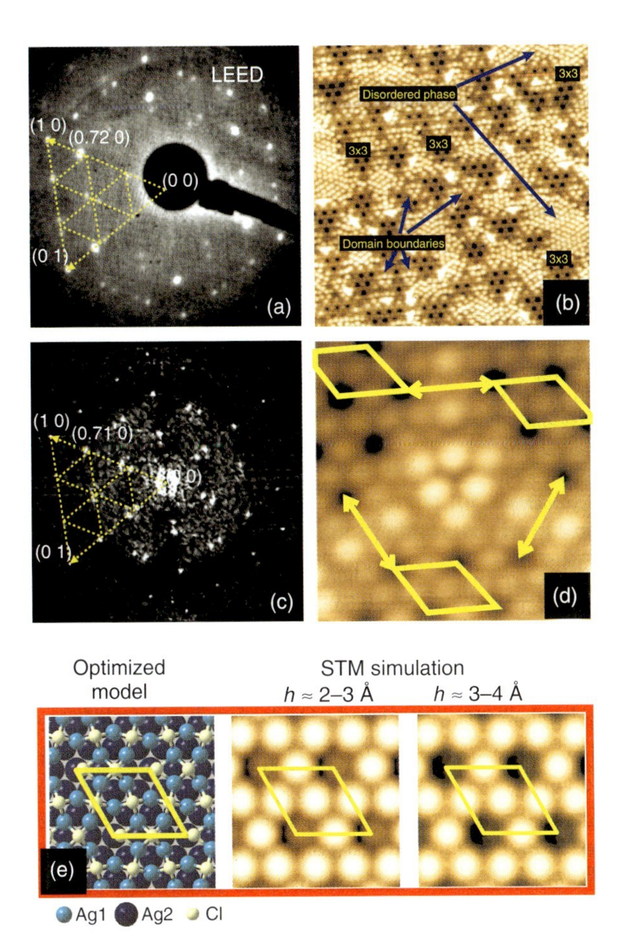


**Figure 35.15** (a) LEED pattern ($E_0 = 76$ eV, 300 K) obtained for chlorine on Ag(111) at coverage close to the saturation. (b) An atomic-resolution STM image (200×200 Å$^2$, $U_s = -1$ V, $I_t = 1.6$ nA, 5 K) recorded at the same coverage as the diffraction pattern from (a). (c) Fourier transformation of the STM image from (b). (d) Fragment of the STM image showing the coexistence of the antiphase (3×3) domains. The positions of the (3×3) spots in the reciprocal space in (a) and (c) correspond to knots of the grid shown by the dashed lines. (e) Structural model and Tersoff–Hamann simulated STM image for the (3×3) phase calculated for two different tip heights: 2–3 Å and 3–4 Å. In the equilibrium structure, one Cl atom in the hole is shifted off-center by 0.7 Å in the direction of the close-packed atomic rows of Ag(111) (coverage 0.44 ML). From Ref. [52].

adsorbed on Pt(111) [61, 154]. In particular, for iodine, several structural models have been considered and analyzed with DFT [155]. However, the possibility of the reconstruction has not been considered.

**35.1.3.2.1 Moiré-Like Superstructures** For iodine adsorbed on Au(111) and Ag(111) surfaces, rosette-like superstructures were observed in LEED at coverages close to saturation [33, 34] (see Figure 35.17a). Such a pattern was interpreted

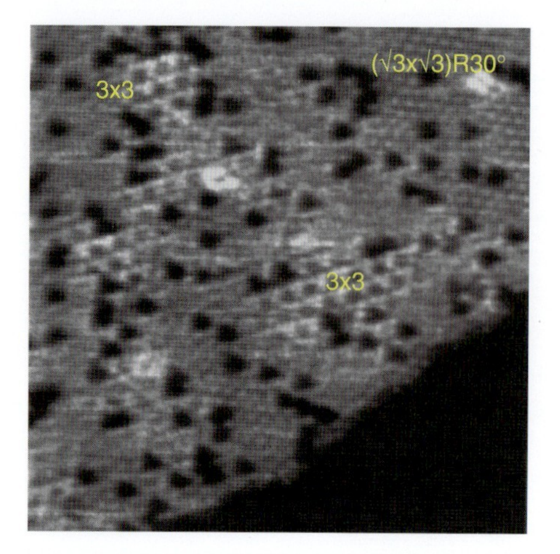


**Figure 35.16** STM image showing coexisting $(\sqrt{3}\times\sqrt{3})$R30° and (3×3) Br island domains on Pt(111) at coverage of 0.15 ML and $T_s = 130$ K. The ordered islands with smaller periodicity have the $(\sqrt{3}\times\sqrt{3})$R30° Br structure, and those with larger periodicity have the (3×3) Br structure. The image size is $166\times166$ Å$^2$. Adapted from Ref. [62].

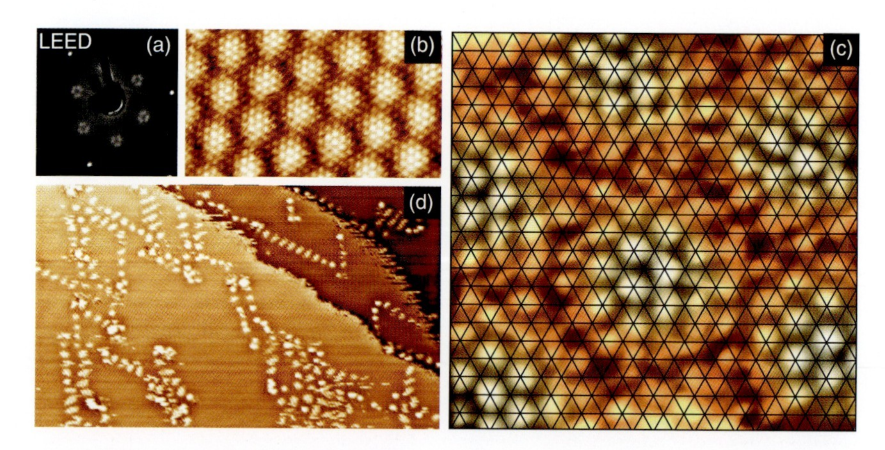


**Figure 35.17** (a) Rosette-like LEED ($E_0$ = 84 eV) pattern obtained near saturation of iodine coverage on Ag(111). (b) STM image ($154\times86$ Å$^2$) corresponding to (a). (c) Superimposition of the Ag(111) lattice on an atomic-resolution STM image. (d) STM image ($930\times530$ Å$^2$) showing nucleation of the individual objects of the Moiré-like reconstruction. Adapted from Ref. [47].

as a compressed iodine layer rotated slightly with respect to the substrate. This model has been supported by the observation of the Moiré-like superstructure with STM in air [41]. However, recent UHV-STM measurements have shown that the situation is much more complicated. Indeed, the corrugation of the Moiré-like superstructure (Figure 35.17b) measured in STM images is equal to ≈0.8–1.0 Å, which is too large for a simple monolayer. Indeed, DFT calculations performed for the I/Ag(111) system have shown that the height difference for an iodine atom adsorbed in fcc hollow and on top positions is equal to 0.28 Å [108]. Figure 35.17c contains a fragment of a Moiré-like structure with the superimposed grid of the Ag(111) substrate. As one can see, the bright atoms from the Moiré-like pattern occupy similar adsorption sites and are arranged in a $(\sqrt{3} \times \sqrt{3})R30°$ lattice. Therefore, we have to conclude that most iodine atoms are adsorbed into unfavorable on-top positions, which is not in line with DFT data [108].

In addition, the nucleation of the Moiré-like phase has been examined. STM image in Figure 35.17d shows that this phase nucleates in the form of individual objects (with a height about 1 Å). The ordering of the Moiré-like phase occurs as a result of their packing. This observation points at the possible surface reconstruction rather than formation of a uniformly compressed, rotated hexagonal lattice.

We believe that the I/Au(111) structures can also be explained by the same model.

#### 35.1.3.3 **Face-Centered Cubic (110) Surfaces**

According to Table 35.2, most halogens prefer to adsorb into short bridge sites on the (110) face of fcc metals. Such adsorption geometry excludes simple uniaxial compression along the ⟨110⟩ direction of the substrate. The likely scenario of the lattice compression in these cases is adsorbate-induced reconstruction.

A good example is the adsorption of chlorine on the Cu(110) surface [156]. At an early stage of chlorine adsorption, the reconstruction of the Cu(110) surface takes place (see Figure 35.18). At coverages $0.4 < \theta < 0.5$ ML, an array of striped light DWs separating the c(2×2) domains forms on the surface. According to DFT calculations, each light DW contains an additional row of copper atoms in the middle. As the coverage increases to 0.5 ML, all the light DWs disappear from the adlayer and a simple c(2×2) structure is formed.

At coverage above 0.5 ML, atomic resolution in STM is completely lost. This result is in line with DFT calculations showing that the direct uniform compression of chlorine layer is unfavorable. According to DFT results, chlorine atoms tend to shift into long bridge, hollow adsorption sites, forming a high-order commensurate structure with weak modulation (Figure 35.19a). The corrugation of the additional modulation does not exceed 0.2–0.3 Å (see Figure 35.19b), which indicates the nonequivalent adsorption sites along the $[1\bar{1}0]$ direction rather than surface reconstruction.

At chlorine coverage $\theta > 0.6$ ML, a significant change in the morphology of the surface is observed. The STM image in Figure 35.20 shows the emergence of ridge-like features running along the [001] direction. The typical width and depth of the ridges are 30–50 Å and 3–15 Å respectively. The ridges are quite long up to several thousands angstroms.

Analysis of the STM images shows that each ridge consists of two atomic {210} planes. Therefore, we can conclude that the adsorption of chlorine induces the facetting of the Cu(110) surface. In the inset to Figure 35.20b, the atomic-resolution image of one side of the facette is shown. For the unit cell marked with white color, we measure parameters 3.7 and 4.4 Å. These values correspond well to the structure Cu(210)-c(1×1)-Cl with parameters, equal to 3.61 and 4.43 Å.

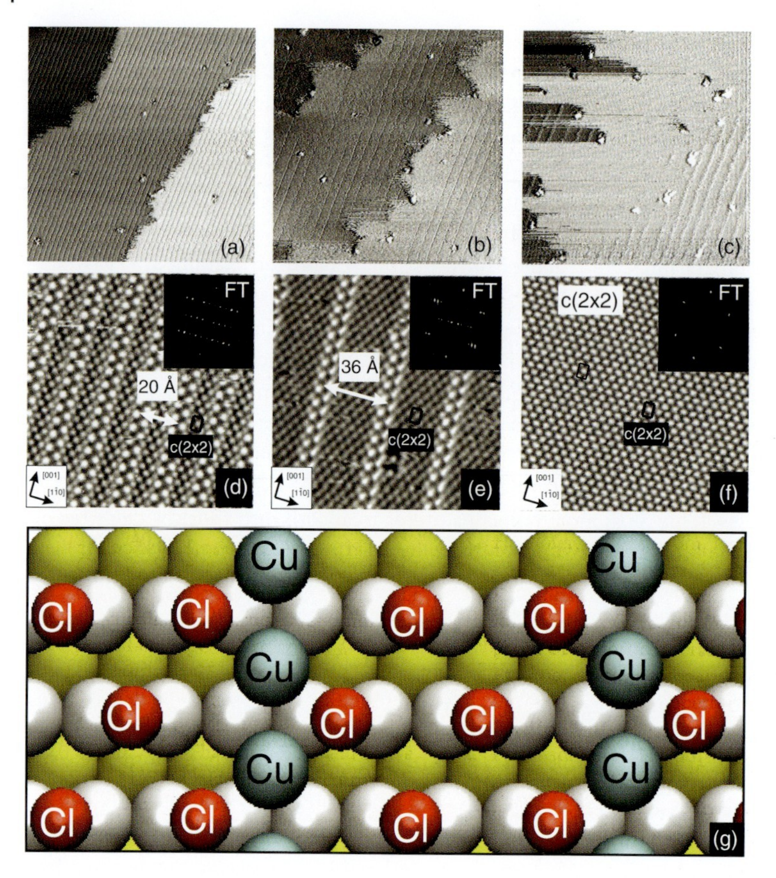


**Figure 35.18** (a)–(c) Panoramic STM images (1041×1020 Å$^2$, $U_s = -1.0$ V, $I_t = 0.3$ nA), acquired during the step-by-step increase of the coverage in the range II. (d)–(f). Atomic-resolution images corresponding to the STM images from (a)–(c) (121×121 Å$^2$, $U_s = -0.8$ V, $I_t = 0.2$ nA). The STM image in (f) corresponds to the c(2×2) structure formed after complete disappearance of the stripes (see (c)). The insets to (d)–(f) show Fourier transforms of the corresponding STM images. (g) Optimized model of the light DW obtained as a result of the DFT calculations. From Ref. [156].

Figure 35.21 shows a model of the atomic structure of the facette. According to DFT calculations, all chlorine atoms occupy positions of the BH-type. The Cl–Cl interatomic distances measured in the [001] direction are equal to 3.61 Å, while in the other direction the separation between chlorine atoms appears to be larger (4.43 Å). Thus, the chlorine atoms form chains parallel to the [001] direction occupying all the adsorption sites of the BH-type. The top of the facette contains a double layer of chlorine atoms, which is in good agreement with STM data from Figure 35.20b.

Facetting has been observed also for other systems. In particular, for the Br/Ni(110) system [113], the adsorption of bromine at room temperature leads to the formation of disordered "butterfly"-like structures. However, heating the

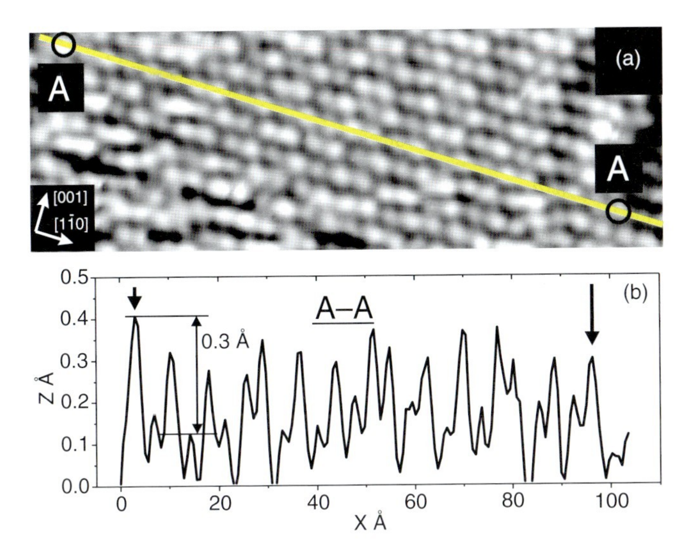


**Figure 35.19** STM image ($34 \times 99$ Å$^2$) of a chlorinated Cu(110)surface ($\theta > 0.5$ ML). The yellow line in (a) indicates the profile of the STM image presented below in (b). (From Ref. [156].)

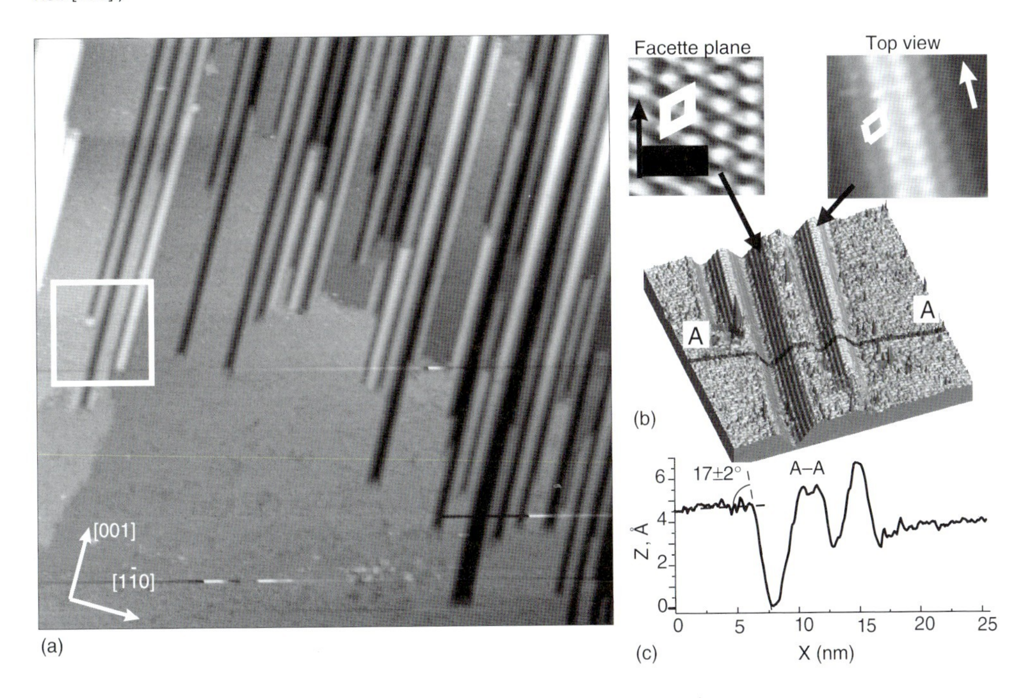


**Figure 35.20** (a) STM image($1500 \times 1500$ Å$^2$, $U_s = -1.5$ V, $I_t = 0.2$ nA) of the initial stage of the Cu(110) surface facetting induced by adsorbed chlorine. (b) The fragment of the STM image ($250 \times 250$ Å$^2$) of the facette shown with a higher magnification. (c) A cross-section of the STM image along the A–A direction. (From Ref. [156].)

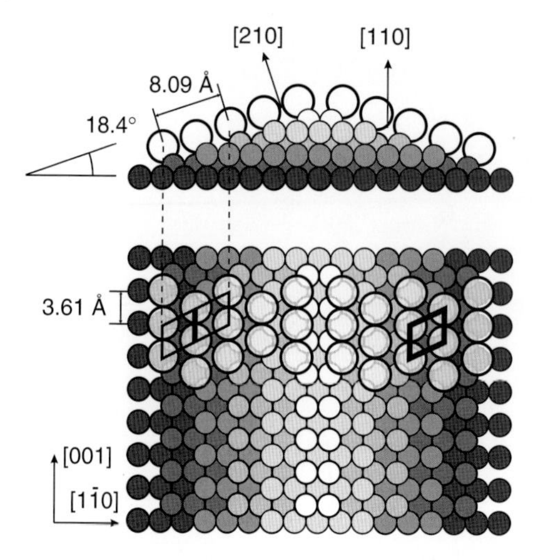


**Figure 35.21** Model of the facette formed by two adjacent {210} planes. Positions of chlorine atoms on {210} planes are shown in accordance with results of DFT calculations. (From Ref. [156].)

system to 200° C resulted in the formation of stripes very similar to the facets in the Cl/Cu(110) system. Moreover, on the terraces between the facets, the STM images demonstrate a high-order commensurate structure similar to the structure shown in Figure 35.19. For bromine adsorption on Cu(110) in the electrochemical cell, the formation of the facets was detected after action of the electrical field of the tip [116].

## 35.1.4 Electronically Driven Phase Transitions

Electronically driven phase transitions are of great importance in condensed matter physics. In halogen/metal systems, this class of phase transition has been studied by Bertel's group [100, 105, 117, 120, 122–131] on the example of Br,I/Pt(110). In particular, the formation of a (3×1) charge density wave (CDW) phase was observed in the detailed investigation of the c(2×2) ⇒ (3×1) transition (see Figure 35.22). Although the presence of CDW was clearly demonstrated, DFT calculations show the major role of local repulsive interactions in the formation of the (3×1) phase. Thus, CDW is then a facilitating but not a driving force of this phase transition. Authors in Ref. [123] called it *CDW-assisted phase transition*.

In the same Br/Pt(110) system, another interesting phase transition has been observed [129]. At room temperature, bromine at 0.5 ML forms a c(2×2) structure characterized by a bright LEED pattern. Cooling of the system to 50 K completely destroys the long-range order. STM image (see Figure 35.23) shows the coexistence of c(2×2) and (2×1) phases. This transition was reported to be fully reversible. Although the (2×1) phase is a real ground state of the system according to DFT calculations, its stability is somewhat unexpected, since it should correspond to a higher inter-adsorbate repulsive energy as compared with the (2×2) phase. To solve

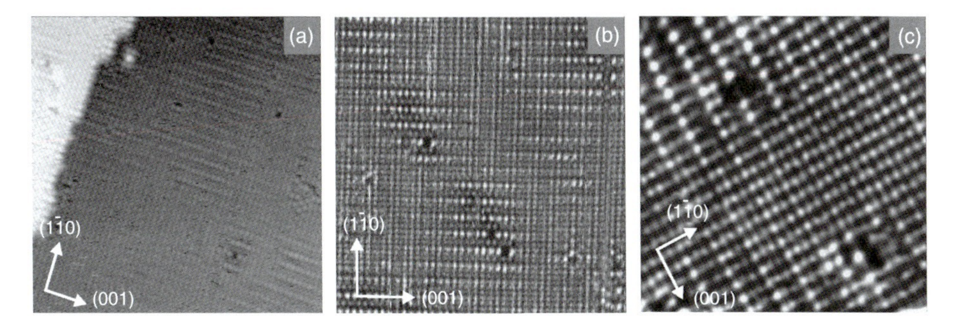


**Figure 35.22** (a) Constant-current STM image (55 mV, 0.64 nA; 210×210 Å$^2$) of the c(2×2) phase with a Br coverage of 0.01 ML additional to 1/2 ML ($\theta$ = 0.51 ML). The c(2×2) structure is preserved only close to steps and defects. On the terraces, extended (3×1) domains surrounded by (1×1) disordered areas are visible. (b) Constant-current image (233 mV, 0.74 nA; 160×160 Å$^2$) as in (a) but with an additional coverage of 0.04 ML ($\theta$ =0.54 ML) dosed onto the c(2×2)-Br/Pt(110) phase. There are three (3×1) antiphase domains. (c) Constant-current image (217 mV, 0.63 nA; 76×76 Å$^2$) as in (b) ($\theta$ = 0.54 ML) prepared by thermal desorption from a higher coverage showing a closeup of the area between adjacent (3×1) domains, clearly demonstrating the continuous change in strength of the (3×1) modulation. (Adapted from Ref. [123].)

**Figure 35.23** Phase transition c(2×2) ⇒ (2×1) occurring upon cooling. (a) The long-range ordered c(2×2) structure observed at room temperature decays into a striped pattern of bright and dark domains. (b) A closeup view reveals the bright domains to be formed by c(2×2) and the dark domains by (2×1) order. (c) Ball model of the (2×1) structure: gray balls are substrate Pt atoms, yellow balls Br atoms. (d) Cross section through the surface showing the buckling of the substrate. (From Ref. [129].)

this problem, one should take into account the periodic lattice distortion associated with a charge density modulation (CDW) [129]. The bromine lines in the model decorate the charge density maxima of the CDW.

The observed phase transition can be explained as follows: At low temperature, additional energy is required to overcompensate the inter-adsorbate repulsion and substrate distortion in the course of the (2×1) phase formation. This extra energy comes from the gain due to the CDW formation. Heating the system to room temperature leads to the thermal suppression of CDW, and the energy balance shifts in favor of the c(2×2), in which the inter-adsorbate repulsion is minimized.

## 35.2 Halide Growth

The growth of metal halides in the second stage of reaction is a good example of the heteroepitaxial growth. Indeed, the result of halogen–metal reaction is the formation of a new two-component crystalline phase–the metal halide. Depending on the system and the temperature, halide can form either a 2D crystalline film (Cl/Cu(111)-160 K [4, 157]; I/Cu(111)-300 K [158], Br/Cu(100)-300 K [159]) or 3D islands (Cl/Ag(111) 300 K [5]); Cl/Cu(111)-300 K [4, 157]); that is, different types of the growth mode are realized. Generally, the layer-by-layer (2D) or Frank–van der Merwe growth mode is distinguished from the 3D or Volmer–Weber model. There is also an intermediate case (Stranski–Krastanov model) in which the film starts growing in 2D mode, and after filling of one or several layers, switches to three dimensions.

The driving force for a particular growth mode is closely related to the necessity to compensate the strain arising at the contact of two noncoincident lattices. One possible mechanism is the formation of an interface layer that separates the growing film and the substrate and has a specific geometrical structure. In case of layer-by-layer growth of thin films, the formation of Moiré structures (Ag/Cu (111), Ag/Ni (111) [160], $Al_2S$/Al(111) [161], iron oxide/Pt (111) [162]), and also regular misfit dislocations (Au/Ni (111) [163], InAs/GaAs (111) [164], InAs/GaAs(110) [165]) is frequently observed. For semiconductor systems, which follow the Stranski–Krastanov scenario (InAs/GaAs(100) [166], Ge/Si (100) [167]), a flat wetting layer forms just prior to the growth of the 3D islands. Note that the atomic structure of the wetting layer formed is not well studied experimentally. Indeed, for InAs/GaAs(001) and Ge/Si(001) [166, 167], the structures of the clean substrate surface and wetting layer are rather complicated, so STM images cannot be analyzed directly and require interpretation.

In this connection, it is difficult to find an answer to an important question: whether the structure of the interface between the growing island film and the substrate really corresponds to the structure of the wetting layer. To our knowledge, there is no direct experimental evidence for this. Our idea is that halogens on metals can be considered a model system where the chemisorbed iodine layer plays the role of the wetting layer.

First investigations of the structure of a halide film grown on a metal surface under halogen gas action were performed with LEED. However, diffraction techniques do not seem to be informative enough in this case. Indeed, the growth of halide phase, as a rule, leads to a diffuse background and blurring of the monolayer spots. Such behavior can be related to the formation of small and imperfectly ordered crystallites of the metal halide.

There are a few papers in which the authors have reported LEED measurements on the halide phase. In particular, Bardi and Rovida [33] reported that iodine adsorption on (100), (110), and (111) silver surfaces leads to additional spots with an angular period of 30°, which was explained by authors as due to the formation of a thin epitaxial AgI layer. As more silver iodide was formed on the surface, the background grew while the spots faded. LEED data by DiCenzo *et al.* [158] suggest the formation of an epitaxial layer of CuI(111) on the Cu(111) surface. Kitson and Lambert [168] studied the structure of the AgCl layer evaporated on Ag(100) and reported a diffuse LEED pattern with 12-fold symmetry associated with a contribution of two AgCl(111) domains rotated by 90° with respect to each other.

The local atomic structure of the halide film has been studied by X-ray photoelectron diffraction (XPD) [157, 169] and surface extended energy loss fine structure spectroscopy (SEELFS) [170, 171]. In particular, for a CuCl film with a thickness ≈20–30 Å, the Cl–Cu nearest-neighbor distance was found to be in agreement with the corresponding bulk CuCl value [170]. A chloride film on Cu(100) and Ag(100) prefers to grow with its (111) plane parallel to the substrate plane [157, 169]. Moreover, XPD data clearly pointed to Cl termination of the both CuCl(111) and AgCl(111) films grown on the Cu(100) [157] and Ag(100) [169] surfaces, respectively.

To recognize the processes of the nucleation and initial stages of halide growth, real-space STM measurements are required. In the next sections, we review atomic structures obtained on thin halide films using STM.

### 35.2.1 Surface Halides

In this section, we focus on surface structures that can be classified as "surface halides." Surface halide phases are intermediate between chemisorbed and real halide phases.

#### 35.2.1.1 Cl/Ag(111)

Indeed, for the Cl/Ag(111) system, the (3×3) reconstruction is accompanied by the extraction of some silver atoms on the surface [52, 53, 172]. These silver atoms can accumulate on the boundaries between (3×3) islands and form small clusters identified as $Ag_3Cl_7$ (see Figure 35.24). Figure 35.25a shows a DFT-optimized model of the $Ag_3Cl_7$ cluster. The sequence of atomic layers in $Ag_3Cl_7$ corresponds well to the sequence of layers in a silver chloride bulk crystal in the direction ⟨111⟩. It is of interest to analyze the atomic structure of the $Ag_3Cl_7$ pyramid in comparison with AgCl lattice (see Figure 35.25b). In the case of bulk AgCl, the interplane distance

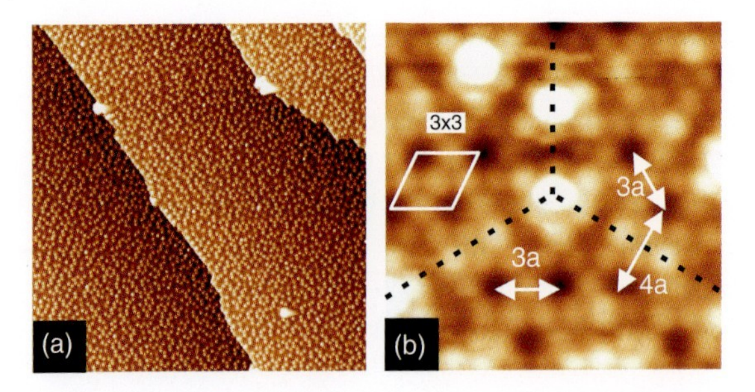


**Figure 35.24** (a) Panoramic STM image ($1000 \times 1000$ Å$^2$, $U_s = -856$ *mV*, $I_t = 0.8$ nA, 5 K) of saturated chlorine coverage on Ag(111) formed after adsorption of $Cl_2$ at 300 K. (b) Atomic-resolution STM image ($68 \times 68$ Å$^2$, $U_s = -1.4$ V, $I_t = 0.5$ nA, 5 K) showing the atomic structure of bright clusters [172].

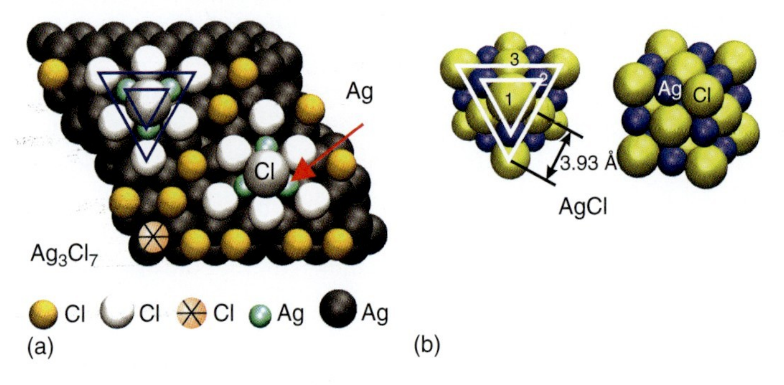


**Figure 35.25** Structural models of the DFT-optimized cluster $Ag_3Cl_7$ (left side). The crystal structure of AgCl (NaCl-type). (a) 3D view. (b) Perspective projection of AgCl structure on the {111} plane. Three upper planes are indicated as "1," "2," and "3" (right side). (Adapted from Ref. [53].)

in the <111> direction is equal to 1.60 Å [173]. In the case of $Ag_3Cl_7$, the distance between the upper chlorine atom and the plane consisting of three silver atoms is equal to 1.468 Å which roughly corresponds to the bulk value. In the model, pairs of atoms Ag1–Ag2, Ag2–Ag3, and Ag3–Ag1 are separated by a distance of 3.636 Å which is slightly less than 3.93 Å, the nearest-neighbor distance in the (111) plane of bulk AgCl crystal [173]. DFT calculations [53] show that silver atoms (Ag1, Ag2, Ag3) occupy approximately threefold hollow adsorption sites above the upper silver layer. The same threefold hollow positions are also occupied by six chlorine atoms from the base of the cluster. That is why the short distance (≈0.39 Å) between the silver plane in $Ag_3Cl_7$ and an average basal chlorine plane is not surprising. DFT data point to a small contraction of the nearest-neighbor chlorine distances in the basal chlorine plane (≈4.13 Å) in comparison with Cl–Cl distances in a six-atom domain boundary from Figure 2b (≈4.4 Å).

Thus, according to the analysis, the upper chlorine atom and three silver atoms occupy positions in a slightly distorted AgCl lattice. The base of the $Ag_3Cl_7$ cluster could be considered in some sense as an interface between the Ag(111) substrate and the $Ag_3Cl$ cluster.

#### 35.2.1.2 I/Cu(111)

Similar to Cl/Ag(111), halide-like clusters ($\approx$10 Å in size) have been observed in the I/Cu(111) system [37] just after the completion of the chemisorbed iodine monolayer and before iodide film growth (see Figure 35.26). The islands are surrounded by a monolayer of chemisorbed iodine with DW structure. Analysis of the STM images shows that these small islands are formed in the places of intersection of domains (domain boundaries) with a different compression direction of the iodine lattice.

#### 35.2.1.3 Cl/Au(111)

Adsorption of $Cl_2$ on Au(111) at 130 K leads to the lifting of the herringbone reconstruction and to the formation of the $(\sqrt{3} \times \sqrt{3})$R30° adsorbate structure ($\theta = 0.33$ ML) with one Cl atom per unit mesh. A small increase in chlorine coverage above 0.33 ML (keeping the sample temperature at 130 K) leads to the appearance of numerous bright oval objects (see Figure 35.27a) [56, 57]. The length of such an object measured in the STM image was 5.5–6.0 Å, while its height was in the range 0.8–1.0 Å irrespective of the bias voltage (at least in the range −1 to +1 V). The distance between the centers of neighboring objects in the line corresponds to the nearest-neighbor distance in the $(\sqrt{3} \times \sqrt{3})$R30° structure (5 Å). Linear structures are usually coupled with 10 Å between the couples, as is clearly seen in Figure 35.27a.

DFT calculations have shown that the oval-shaped objects correspond to quasi-molecules–$AuCl_2$ [57]. The optimized model of $AuCl_2$ on Au(111) is shown in Figure 35.27b. In the final configuration, the gold adatom occupies a bridge position with respect to substrate gold atoms, and two chlorine atoms appear to be in on-top positions. As the coverage increases, the number of $AuCl_2$ quasi-molecules increases, leading to their self-organization into a "honeycomb" structure $\begin{pmatrix} 4 & 0 \\ 1 & 5 \end{pmatrix}$. It was found that each ring of the "honeycomb" structure consists of six $(AuCl_2)_2$

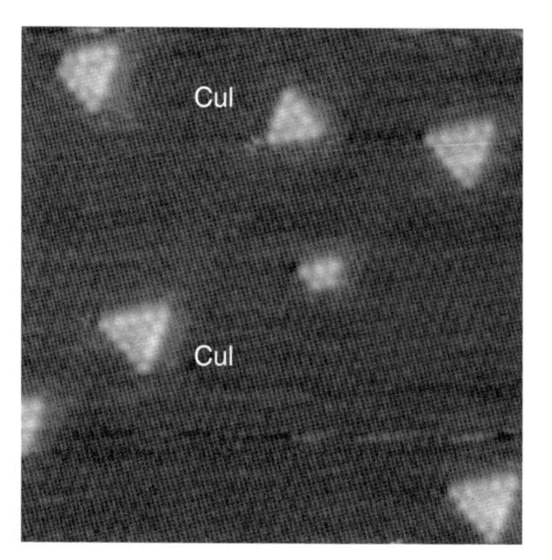


**Figure 35.26** Small islands of 10 Å observed on the atomic plane in the "sea" of chemisorbed iodine layer of DW structure.

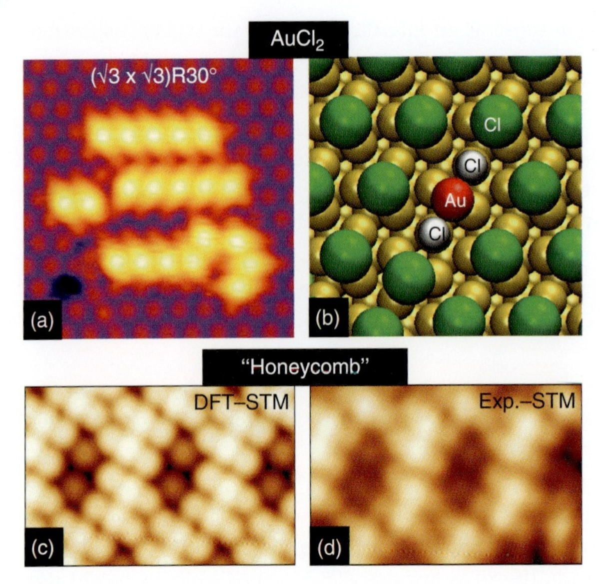


**Figure 35.27** (a) LT-STM image ($T = 5$ K) of the Au(111) surface chlorinated at 130 K demonstrating a simple commensurate $(\sqrt{3} \times \sqrt{3})$R30° structure and appearance of numerous oval-shaped objects. (b) DFT-optimized model of the object–quasi-molecule $AuCl_2$. Theoretical (c) and experimental (d) STM images of the "honeycomb" structure consisting of the $AuCl_2$ structural elements. (Adapted from Ref. [57].)

dimers. Two chlorine atoms are placed inside the ring and occupy fcc and hcp threefold positions. Theoretical and experimental STM images are shown in Figure 35.27c and d, respectively.

#### 35.2.1.4 **Cl/Pt(110)**

In the work by Dona *et al.* [99], the formation of the $PtCl_4$ clusters on Pt(110) surface was reported (see Figure 35.28). Each cluster consists of one platinum

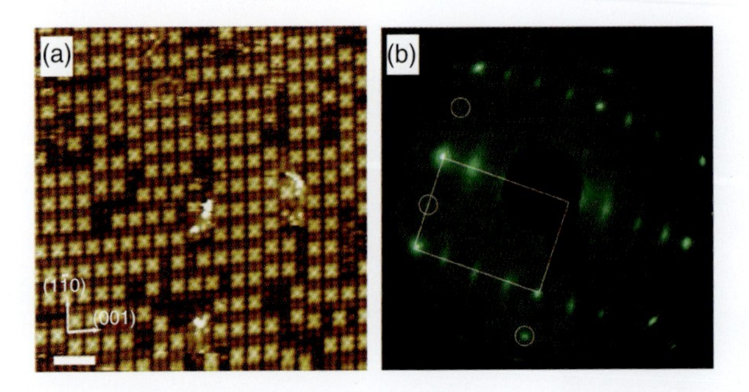


**Figure 35.28** (a) STM image of Cl/Pt(110) with a Cl coverage of ~0.75 ML after annealing at 670 K. The scale bar corresponds to 2 nm. (b) LEED image of the structure shown in (a). The white rectangle denotes the (1× 1) unit cell of Pt(110). The white circles mark half-order spots indicating a preference for (4×2) ordering. (From Ref. [99].)

atom adsorbed in a hollow site and four chlorine atoms occupying positions close to the top. The authors also reported long-range ordering in the Cl/Pt(110) system and a (4×2) structure in LEED. The (4×2) phase was formed by alternative lines of $PtCl_4$ clusters and chlorine atoms adsorbed in short bridge positions (see Figure 35.29).

Similar to the "honeycomb" structure in the Cl/Au(111) system, the (4×2) can be considered to be a 2D quasi-molecular crystal.

## 35.2.2 "Bulk" Halide Growth

### 35.2.2.1 Nucleation of Halide

Figure 35.30 shows the STM images of the Cu(100) surface chlorinated at 160 K after annealing to room temperature. We think the numerous bright objects (50 Å in height) near the step edges or other surface defects are associated with 3D islands of copper chloride. The same surface could be obtained by direct adsorption of molecular chlorine at room temperature. Since the islands of chloride are not flat, one can only report the growth, but an atomic resolution STM study of halide structure is hardly possible.

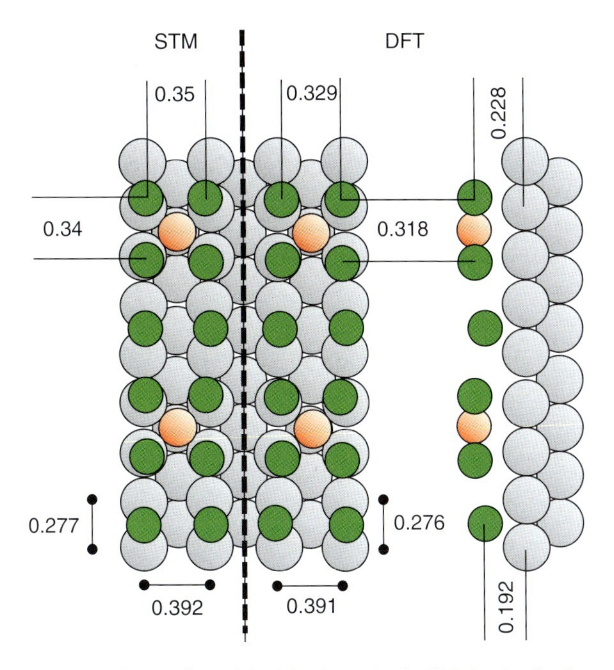


**Figure 35.29** Ball model of the (4×2) $PtCl_4$/Cl/Pt(110) adsorbate system. Gray balls are Pt substrate atoms, green balls Cl atoms, and red balls the central Pt atoms of the $PtCl_4$ pentamers. All distances are given in nanometers. (From Ref. [99].)

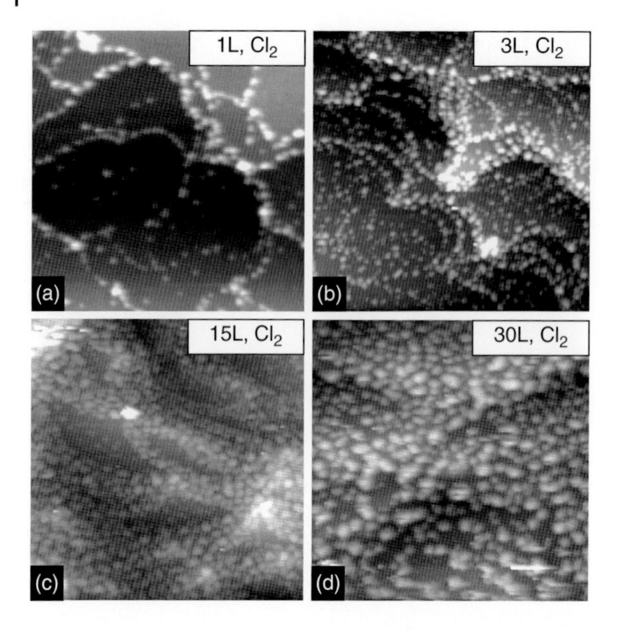


**Figure 35.30** STM images (7500 × 7500 Å$^2$, $U_s = -1.9$ V; $I_t = 0.2$ nA) of the Cu(100) surface after chlorination at 160 K and subsequent annealing to 300 K. Results are shown for different chlorine exposures: (a) 1 L. (b) 3L. (c) 15 L. (d) 30 L.

Detailed structural studies of thin halide films on metal surfaces have been performed for systems exhibiting layer-by-layer (2D) growth at room temperature: Br/Cu(100) [159], I/Cu(111) [174, 175], I/Cu(100) [174, 176], I/Cu(110) [177], I/Ag(100) [8], I/Ag(110) [110], and I/Ni(111) [46].

According to STM measurements, the lattice symmetry and interatomic distances on the surface of halide film in all cases appear to be close to those for the hexagonal close-packed planes of the corresponding bulk halides. For iodide on copper surfaces, it was found that, depending on the substrate symmetry, STM images of the CuI surface demonstrate different Moiré-like patterns and striped misfit dislocation networks [7, 174, 177] (see Figure 35.31).

In the same works, the existence of a specific monatomic layer, playing a role of the interface between copper iodide and copper lattices, was clearly demonstrated. The structure of the interface appeared to correspond to that of a saturated iodine monolayer forming at the first stage of reaction. It is also clear that the detailed atomic structure of halide film cannot be unambiguously determined from the STM images. In particular, due to the proximity of the lattice parameters in the close-packed planes, it is not possible to distinguish between wurtzite and zinc blende types of the crystalline lattice characteristic of the bulk CuBr and CuI [173]. Moreover, the structure of the inner layers of halide and their possible distortions due to the interaction with the substrate are also beyond of the direct interpretation of the STM images.

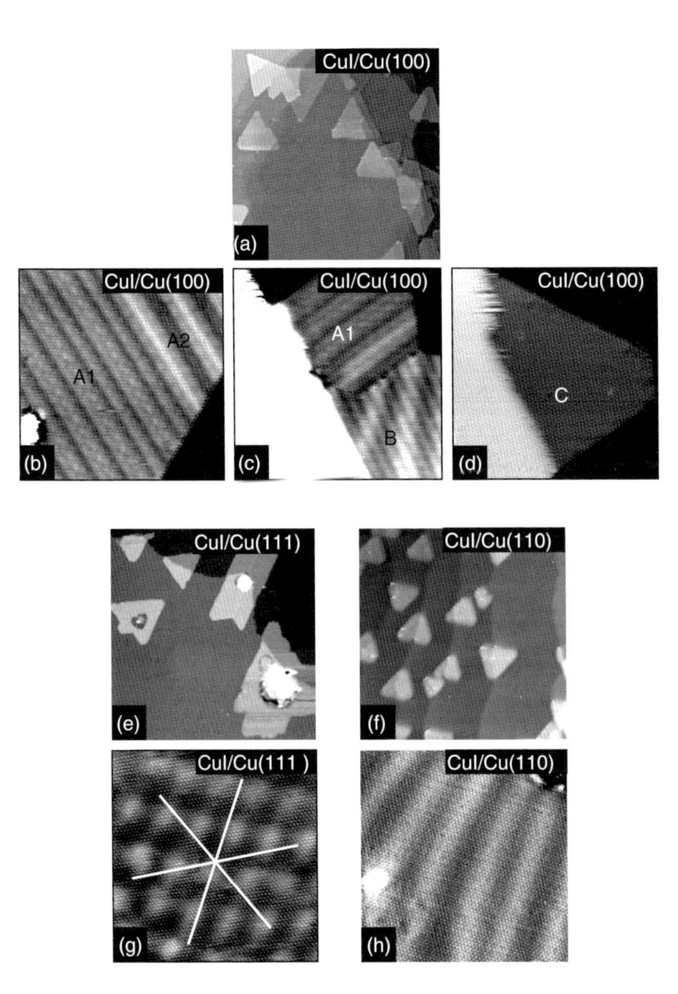


**Figure 35.31** STM images (1961×1961 Å$^2$) showing the formation of CuI islands on (a) Cu(100), (e) Cu(111), and (f) Cu(110) surfaces. (b)–(d) Atomic-resolution STM images (242 × 242 Å$^2$) demonstrating three types of copper iodide islands on Cu(100) characterized by different types of the superstructure: Striped structures ("A1," "A2"), "screw"-like structure "B," and flat structure "C." (g) and (h) Atomic-resolution STM images (242 × 242 Å$^2$) demonstrating quasi-hexagonal and double-striped Moiré-like superstructures corresponding to the CuI growth on Cu(111) and Cu(110), respectively. (Adapted from Refs. [7, 174, 177].)

Therefore, the experimental STM information should be complemented by theoretical structural calculations. The efficiency of such an approach had been already shown on the example of surface oxides on transition metals, usually, as well as halides [174–177], demonstrating complex superstructures with large unit cells [178]. It turned out that DFT calculations were capable of describing the structure of surface oxides in good correspondence with the experimental data obtained with STM and electron diffraction [178].

#### 35.2.2.2 **Atomic Structure of Thin Halide Films**

**35.2.2.2.1 I/Ag(100)** The choice of the appropriate unit cell is a key point in the success of DFT calculations for complex multilayer systems. Since halogens often tend to form quasi-hexagonal structures with large periods, the total unit cell of the halide/monolayer/metal system could turn out to be beyond DFT capabilities. In this connection, one should find a system in which the size of the monolayer unit cell is minimal and 2D growth is realized. A good candidate is iodine adsorbed on the Ag(100) surface. In particular, the formation of the c(2×2) monolayer at the initial stage of adsorption [33, 179] and the hexagonal orientation of the growing silver iodide film [33] were detected by LEED.

Figure 35.32 shows the panoramic STM images of iodine on Ag(100) for different AgI coverages [8]. In addition to the flat terraces separated by monatomic steps, the STM images in Figure 35.32a demonstrate 2D islands (visible as bright objects) situated mainly near the step edges. If iodine is exposed further, the islands enlarge to fill the terraces, while their height remains (≈4.5 Å) unchanged. At the I/Ag AES peak ratio of 0.19, approximately half the surface area is occupied by the AgI film, as shown in the STM frame in Figure 35.32b. Therefore, 2D growth is realized at the initial stages of AgI formation.

Figure 35.33a shows an atomic-resolution STM image of a silver iodide island. The atomic structure roughly corresponds to the hexagonal lattice, as shown in the inset. The STM image of AgI, in addition to atomic modulation, contains the clearly seen superstructure with the periods of 52 Å and 8 Å, as measured along the [001] and [010] directions, respectively.

The STM image shown in Figure 35.33b contains areas covered with both silver iodide and chemisorbed iodine monolayer. Knowing the lattice parameters for the iodine monolayer, it is possible to determine the parameters of silver iodide lattice from this image. A saturated monolayer of chemisorbed iodine is described by a simple square lattice with I–I nearest-neighbor distance of 4.09 Å. Analysis of the STM image shows that one of the directions of the close-packed AgI atomic rows is parallel to the close-packed atomic row of the chemisorbed

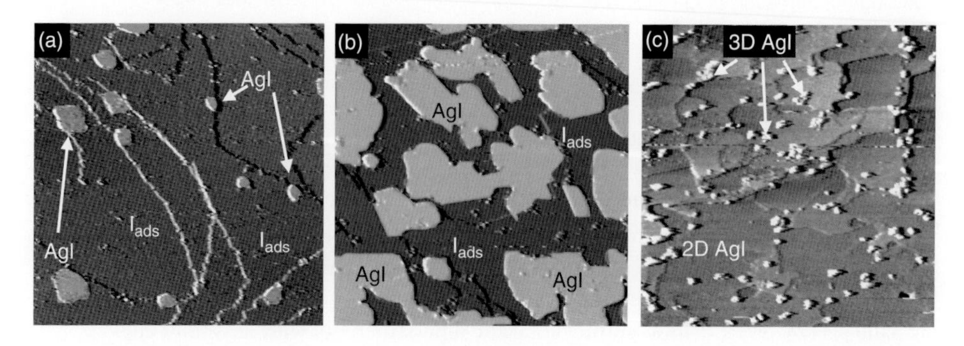


**Figure 35.32** (a)–(c) STM images (6000 × 6000 Å$^2$) showing the formation of AgI islands on Ag(100) for three different coverages characterized by I/Ag AES peak ratio 0.12, 0.19, and 0.59. (From Ref. [8].)

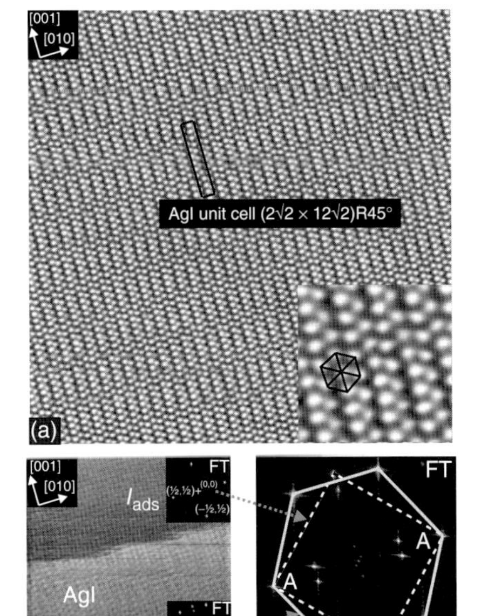


**Figure 35.33** (a) Atomic-resolution STM image (277×277 Å$^2$, $I_t$= 0.23 nA, $U_s$ = −250 mV) of the AgI surface. In the inset, a nearly hexagonal lattice of AgI is indicated by the hexagon. (b) The STM image (159×159 Å$^2$, $I_t$ = 0.2 nA, $U_s$ = −250 mV) demonstrating the edge of the AgI island and the atomic terrace with the c(2×2) iodine monolayer. The inset in the upper right corner shows the FT image of the monolayer region, while that in the lower right corner shows the FT image of AgI. (c) The superimposition of the Fourier transformed images for the c(2×2) monolayer and AgI. (From Ref. [8].)

monolayer and, accordingly, to the [001] direction of the substrate. An accurate reconstruction of the silver iodide lattice was done on the basis of analysis of the Fourier transformed (FT) image. The FT image of the silver iodide lattice (see the bottom inset in Figure 35.33b) contains six main bright spots arranged in a slightly distorted hexagon. Additional four pairs of spots in the FT image are associated with the visible superstructure in the STM image. The FT image of the chemisorbed monolayer, according to the upper inset in Figure 35.33b, contains a square of four spots corresponding to the known c(2×2) structure. In Figure 35.33c, a superposition of the FT images of AgI and chemisorbed iodine is presented. The hexagon and the square intersect in two points marked as “A”. This means that the distance between the atomic rows of AgI is the same as that of chemisorbed iodine in the [010] direction (4.09 Å). Consequently, the upper plane of silver iodide is described by a slightly compressed quasi-hexagonal lattice with average parameters 4.33, 4.62, and 4.62 Å. Additionally, from the FT image in Figure 35.33c, the unit cell of the superstructure was determined as $(2\sqrt{2} \times p\sqrt{2})\text{R}45°, p \approx 12.0 \pm 0.5$.

To explain the superstructure in the STM images, the substrate and the possible interface layer have to be included in the models in addition to the silver iodide layer. To describe the substrate, a three-layer silver slab has been used. As the interface, we have chosen the c(2×2) iodine monolayer, which, according to factor analysis data, remains between the silver iodide layer and the substrate. In the starting models, the silver iodide lattice has been slightly compressed to correspond to the $(2\sqrt{2} \times 12\sqrt{2})\text{R}45°$ unit cell, as determined from the STM in Figure 35.33.

As a starting configuration, a model consisting of two bilayers in wurtzite structure on top of the c(2×2) monolayer was used. Optimization of coordinates significantly modifies the initial configuration; that is, the lower bilayer flips, resulting in a sandwich-like structure. The "sandwich" is formed by two coupled silver layers in the middle and two layers of iodine on the borders (Figure 35.34). All four atomic layers of the sandwich are very close to the ideal planes, since the deviations of the atomic coordinates along the *z*-axis ([100] direction of the substrate) for both iodine and silver layers do not exceed 0.1 Å. The atomic structure of each layer of the "sandwich" is described by the quasi-hexagonal lattice. The close-packed atomic rows parallel to the [001] direction of the substrate also remain straight within 0.1 Å.

Thus, DFT calculations have shown that the saturated iodine monolayer does remain under the silver iodide film as the atomically sharp interface.

This result is of general interest for other halide/metal systems. In particular, according to previous studies [174, 177], the only possible explanation for superstructures detected in the STM images of CuI films grown on copper single-crystal surfaces is that there is an interface preserving the structure of a saturated monolayer (see Figure 35.31).

According to the results presented above, the halide layer appears to be separated from the substrate in the AgI/Ag(100) system. Indeed, the distances between iodine atoms from the interface and iodine atoms from the "sandwich" are close to ≈4 Å, which corresponds well to the iodine van der Waals diameter (4.0–4.3 Å [93]). It is likely that this "independence" of halide layer from the underlying substrate explains why during the halide growth on copper and silver surfaces, independently of the substrate symmetry, the upper halide planes correspond to the hexagonal close-packed lattice [33, 157, 159, 169, 174–177, 180].

**35.2.2.2.2 I/Ni(111)** In this section, we present an example of thin nickel iodide film growth on Ni(111). After saturation of a monolayer exhibiting striped DW structure, the nucleation and growth of nickel iodide islands take place. Figure 35.35 shows a series of STM images obtained after continuous $I_2$ exposure onto Ni(111) at 300 K. As the iodine coverage grows, the surface area covered by 2D islands increases (see Figure 35.35a–c).

The atomic resolution image of the island surface shows not just atomic modulation (with a lattice constant of ≈3.9 Å) but also a Moiré-like superstructure with a period of 26–28 Å. The nearest-neighbor distances correspond well to the

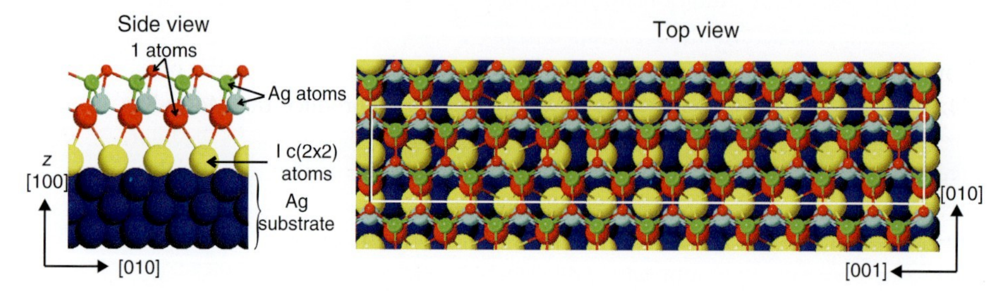


**Figure 35.34** Sandwich-like model of the AgI film obtained as a result of the DFT calculations. (From Ref. [8].)

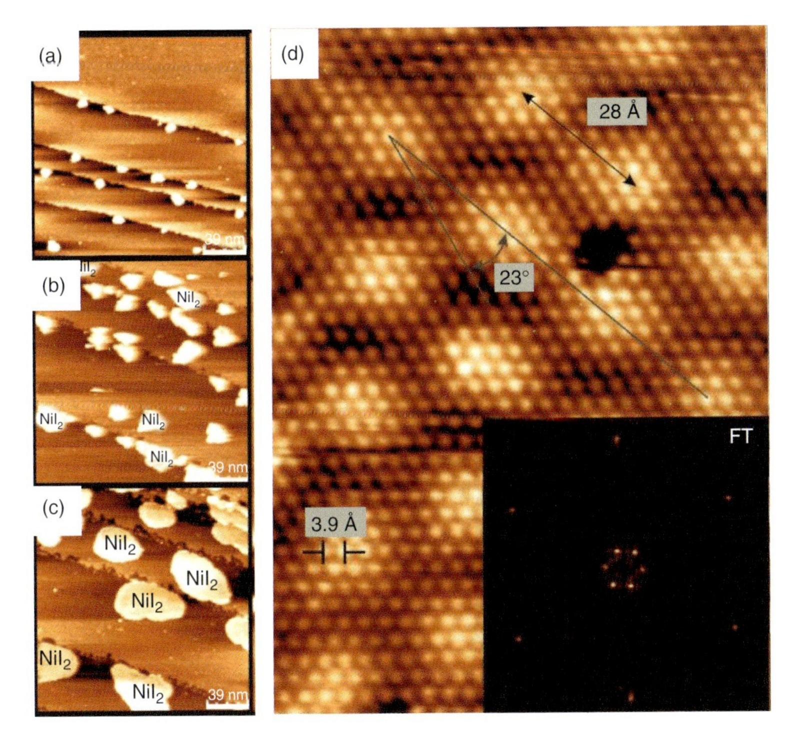


**Figure 35.35** (a)–(c) STM images (2038×2038 Å$^2$) acquired in the course of continuous $I_2$ adsorption on Ni(111) at 300 K after saturation of chemisorbed monolayer. Flat islands formed near the step edges are associated with nickel iodide. (d) Atomic-resolution STM image (154×98 Å$^2$) of the surface of the island from (c). Moiré-like superstructure with a period of 28 Å is clearly seen. The inset shows the Fourier transform of the image from (d). (Adapted from Ref. [62].)

lattice constant of 3.96 Åin the hexagonal plane of $NiI_2$ ($CdCl_2$ bulk lattice type, see Figure 35.36). In addition, the atomic lattice of iodide appears to be rotated with respect to Moiré-lattice by 23°. Therefore, the iodide lattice is rotated with respect to the underlying structure.

To explain the Moiré-like pattern in Figure 35.35d, the existence of an interface layer between Ni(111) and nickel iodide has been suggested. By analogy with other halide/metal systems (see the previous section), the structure of such an interface should correspond to the structure of the saturated monolayer with a unit cell $(14\times\sqrt{3})$). However, the saturated monolayer failed to reproduce the Moiré-like pattern at any angle of rotation. On the contrary, a simple model considering direct contact between the Ni(111) and hexagonal iodide lattices does reproduce the period and orientation of the observed superstructure.

The $NiI_2$ crystal consists of alternating trilayers, each consisting of two iodine planes and one nickel plane between them. It is likely that during the $NiI_2$ growth, conversion of the iodine monolayer in the lower plane of the I–Ni–I trilayer takes place.

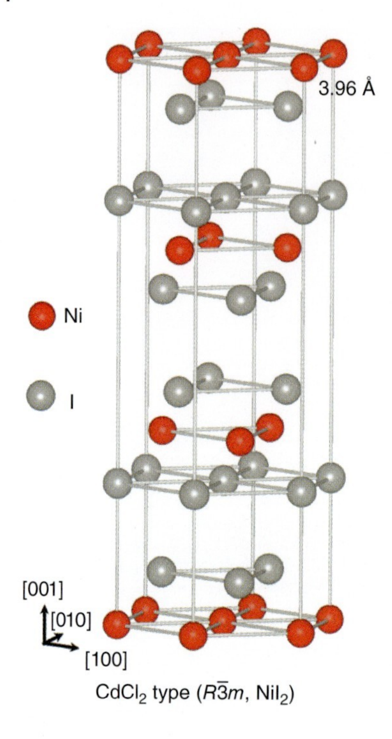


**Figure 35.36** Atomic structure of the bulk $NiI_2$ crystal. (Adapted from Ref. [181].)

## 35.3 Thick Halide Films

Little is known about the atomic structure of rather thick halide layers on metal surfaces. Here we present an interesting example of the Br/Cu(100) system. According to Nakakura and Altman [159], the surface of a thin CuBr film on Cu(100) corresponds well to the (111) plane of the $\gamma$-CuI (zinc blende phase).

However, after prolonged $Br_2$ exposures at room temperature, thicker CuBr multilayer films are formed. High-resolution STM images taken from an area more than three CuBr bilayers thick on Cu(100) are shown in Figure 35.37. Although the symmetry of the lattice remains hexagonal, the periodicity changes to twice that of bulk $\gamma$-CuBr(111).

The authors of Ref. [159] argued that the driving force for a (2×2) reconstruction is clear. Indeed, bulk-terminated $\gamma$-CuBr(111) is a polar surface containing a net surface charge of 1/4 per surface unit cell. The charge is positive for a Cu-terminated surface and negative for a Br-terminated surface. As shown in Figure 35.37b, a (2×2) ordered vacancy structure results in a surface unit cell with three Br and three Cu atoms missing the nearest neighbor, that is, a neutral surface. Such (2×2) structures had been observed for (111) zinc blende III–V semiconductor surfaces [182], and evidence for a (2×2) reconstruction in CuBr films on MgO(100) had been reported [183]. A (2×2) ordered adatom structure would also result in a neutral surface. In this case, however, the adatoms would be missing three nearest neighbors rather than one, and therefore the vacancy structure is expected to be favored.

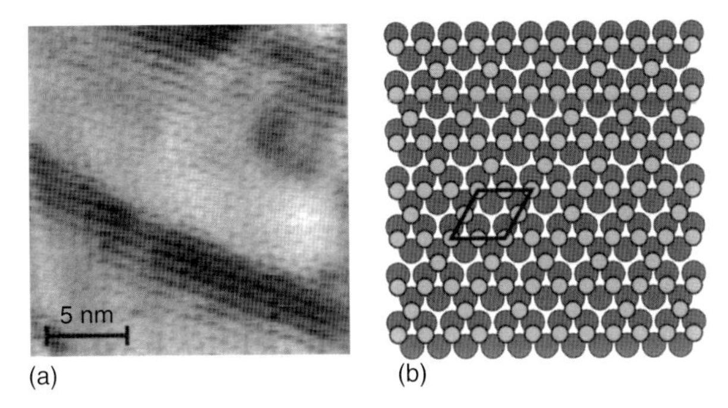


**Figure 35.37** (a) Atomically resolved image of multiple-layer CuBr film grown by room-temperature exposure to $Br_2$. The measured periodicity is 0.815 ± 0.016 nm, or double the bulk spacing for γ-CuBr(111). The step is a substrate step. (b) Model of (2×2) vacancy structure in γ-CuBr(111). The gray circles represent Br atoms and the black circles Cu atoms. By removing one-fourth of the surface atoms, the excess surface charge in bulk-terminated γ-CuBr(111) is removed. (From Ref. [159].)

## 35.4 Conclusions

In this chapter, we tried to give a personal view of the problem of halogen interaction with metal surfaces under UHV conditions. According to our analysis, for a number of systems the picture of halogen/metal interaction is far from complete. Indeed, for most systems recently reconsidered with STM in combination with DFT calculations, new findings have been reported. Therefore, further reconsideration of the systems studied in the 1970s–1980s is expected to provide a lot of new information.

In further research, special attention should be paid to LT-STM studies, which provide an opportunity to explore interactions of separate halogen atoms on the surface.

We have not considered here processes of dissociation of halogen molecules on metal surfaces and desorption of chlorine atoms. For detailed description of these processes, thorough DFT-based calculations are required. Many of the experimental results presented in this Chapter, however, are an ideal basis for a comparison with results obtained for similar halogen/metal interfaces in electrolyte solutions, subject of Volume 7 of this series of books.

## References

1. Jones, R.G. (1988) Halogen adsorption on solid surfaces. *Prog. Surf. Sci.*, **27**, 25–160.
2. Altman, E.I. (2001) The adsorption of halogens on metal and semiconductor surfaces, in *Adsorbed Layers on Surfaces. Part 1: Adsorption on Surfaces and Surface Diffusion of Adsorbates Landolt-Bornstein - Group III Condensed Matter*, vol. 42A1 (ed. H.P. Bonzel), Springer-Verlag, Berlin, pp. 420–442.
3. Andryushechkin, B.V. and Eltsov, K.N. (2003) Chemical and structural transformations on copper surface during chlorine action, in *Chemical State and*

*Atomic Structure of FCC Metal Surfaces in Chemical Reaction with Halogens, Proceedings of the Prokhorov General Physics Institute*, vol. 59 (eds V.I. Konov and K.N. Eltsov), Nauka, Moscow, pp. 106–133.

4. Eltsov, K.N., Zueva, G.Ya., Klimov, A.N., Martynov, V.V., and Prokhorov, A.M. (1991) Reversible coverage-dependent Cu + $Cl_{ads}$ → CuCl transition on Cu(111)/$Cl_2$ surface. *Surf. Sci.*, **251/252**, 753–758.
5. Andryushechkin, B.V., Eltsov, K.N., Shevlyuga, V.M., and Yurov, V.Yu. (1999) Direct STM observation of surface modification and growth of AgCl islands on Ag(111) upon chlorination at room temperature. *Surf. Sci.*, **431**, 96–108.
6. Andryushechkin, B.V., Eltsov, K.N., and Shevlyuga, V.M. (2003) Halide nucleation and growth on monocrystalline copper surface. *Phys. Low-Dim. Struct.*, **3/4**, 1–20.
7. Andryushechkin, B.V., Eltsov, K.N., and Shevlyuga, V.M. (2004) Atomic scale study of CuI film nucleation on copper under molecular iodine action. *e-J. Surf. Sci. Nanotechnol.*, **2**, 234–240.
8. Andryushechkin, B.V., Zhidomirov, G.M., Eltsov, K.N., Hladchanka, Y.V., and Korlyukov, A.A. (2009) Local structure of the Ag(100) surface reacting with molecular iodine: experimental and theoretical study. *Phys. Rev. B*, **80**, 125409.
9. Malinowski, E.R. and Howery, D.G. (1980) *Factor Analysis in Chemistry*, John Wiley & Sons, Inc., New York.
10. Gaarenstroom, S.W. (1982) Application of Auger line shapes and factor analysis to characterize a metal-ceramic interfacial reaction. *J. Vac. Sci. Technol.*, **20**, 458–461.
11. Koutecky, J. (1958) A contribution to the molecular-orbital theory of chemisorption. *Trans. Faraday Soc.*, **54**, 1038–1052.
12. Bradshaw, A. and Scheffler, M. (1979) Lateral interactions in adsorbed layers. *J. Vac. Sci. Technol.*, **16**, 447–454.
13. Zangwill, A. (1988) *Physics at Surfaces*, Cambridge University Press, Cambridge.
14. Einstein, T.L. (1996) Interactions between adsorbate particles, in *Handbook of Surface Science*, vol. 1 (ed. W.N. Unertl), Elsevier, Amsterdam, pp. 577–650.
15. Lau, K.H. and Kohn, W. (1978) Indirect long-range oscillatory interaction between adsorbed atoms. *Surf. Sci.*, **75**, 69–85.
16. Hildgaard, P. and Persson, M. (2000) Long-ranged adsorbate-adsorbate interactions mediated by a surface-state band. *J. Phys. Condens. Matter*, **12**, L13–L19.
17. Zheltov, V.V., Cherkez, V.V., Andryushechkin, B.V., Zhidomirov, G.M., Kierren, B., Fagot-Revurat, Y., Malterre, D., and Eltsov, K.N. (2014) Structural paradox in submonolayer chlorine coverage on Au(111). *Phys. Rev. B*, **89**, 195425.
18. Cerchez, V. (2011) Nano structures formed by molecular chlorine interaction with noble metal surfaces: scanning tunnelling microscopy/spectroscopy study. PhD thesis. Université Henri Poincaré, Nancy.
19. Andryushechkin, B.V., Zheltov, V.V., Cherkez, V.V., Zhidomirov, G.M., Klimov, A.N., Kierren, B., Fagot-Revurat, Y., Malterre, D., and Eltsov, K.N. (2015) Chlorine adsorption on Cu(111) revisited: LT-STM and DFT study. *Surf. Sci.*, **639**, 7–12.
20. Nanayakkara, S.U., Sykes, E.C.H., Fernández-Torres, L.C., Blake, M.M., and Weiss, P.S. (2007) Long-range electronic interactions at a high temperature: bromine adatom islands on Cu(111). *Phys. Rev. Lett.*, **98**, 206108.
21. den Nijs, M. (1988) The domain wall theory of two-dimensional commensurate-incommensurate phase transitions, in *Phase Transitions and Critical Phenomena*, vol. 12 (eds C. Domb and J.L. Lebowitz), Academic Press, New York, pp. 219–333.
22. Freimuth, H., Wiechert, H., Schildberg, H.P., and Lauter, H.J. (1990) Neutron-diffraction study of the commensurate-incommensurate phase transition of deuterium monolayers physisorbed on graphite. *Phys. Rev. B*, **42**, 587–603.

23. Cui, J. and Fain, S.C. (1989) Low-energy electron diffraction study of incommensurate $H_2$, HD, and $D_2$ monolayers physisorbed on graphite. *Phys. Rev. B*, **39**, 8628–8642.
24. Lyuksyutov, I.F., Naumovets, A.G., and Pokrovsky, V.L. (1992) *Two-Dimensional Crystals*, Academic Press, Boston, MA.
25. Kern, K. and Comsa, G. (1988) Physisorbed rare gas adlayers, in *Chemistry and Physics of Solid Surfaces VII*, Springer Series in Surface Sciences, vol. 10 (eds R. Vanselow and R. Howe), Springer-Verlag, Heidelberg, pp. 65–108.
26. Bak, P. (1982) Commensurate phases, incommensurate phases and the devil's staircase. *Rep. Prog. Phys.*, **45**, 587–629.
27. Persson, B.N.J. (1992) Ordered structures and phase transitions in adsorbed layers. *Surf. Sci. Rep.*, **15**, 1–135.
28. Patrykiejew, A., Sokolowski, S., and Binder, K. (2000) Phase transitions in adsorbed layers formed on crystals of square and rectangular surface lattice. *Surf. Sci. Rep.*, **37**, 209–344.
29. Goddard, P.J. and Lambert, R.M. (1977) Adsorption-desorption properties and surface structural chemistry of chlorine on Cu(111) and Ag(111). *Surf. Sci.*, **67**, 180–194.
30. Walter, W.K., Manolopoulos, D.E., and Jones, R.G. (1996) Chlorine adsorption and diffusion on Cu(111). *Surf. Sci.*, **348**, 115–132.
31. Andryushechkin, B.V., Eltsov, K.N., and Shevlyuga, V.M. (2000) Domain-wall mechanism of "$(n\sqrt{3} \times n\sqrt{3})R30°$" incommensurate structure formation in chemisorbed halogen layers on Cu(111). *Surf. Sci.*, **470**, L63–L68.
32. Jones, R.G. and Kadodwala, M. (1997) Bromine adsorption on Cu(111). *Surf. Sci.*, **370**, L219–L225.
33. Bardi, U. and Rovida, G. (1983) LEED, AES and thermal-desorption study of iodine chemisorption on the silver(100), silver(111) and silver(110) faces. *Surf. Sci.*, **128**, 145–168.
34. Cochran, S.A. and Farrell, H.H. (1980) The chemisorption of iodine on gold. *Surf. Sci.*, **95**, 359–366.
35. Shard, A.G. and Dhanak, V.R. (2000) Chlorine adsorption on silver (111) at low temperatures. *J. Phys. Chem. B*, **104**, 2743–2748.
36. Andryushechkin, B.V., Cherkez, V.V., Kierren, B., Fagot-Revurat, Y., Malterre, D., and Eltsov, K.N. (2011) Commensurate-incommensurate phase transition in chlorine monolayer chemisorbed on Ag(111): direct observation of crowdion condensation into a domain-wall fluid. *Phys. Rev. B*, **84**, 205422.
37. Andryushechkin, B.V., Eltsov, K.N., and Shevlyuga, V.M. (2001) Atomic scale observation of iodine layer compression on Cu(111). *Surf. Sci.*, **472**, 80–88.
38. Erley, W. and Wagner, H. (1977) Chlorine adsorption on Ni(111). *Surf. Sci.*, **66**, 371–375.
39. Holmes, D.J., Panagiotides, N., and King, D.A. (1989) Observation of a low-temperature incommensurate Ag(111)$(\sqrt{3} \times \sqrt{3})$R30° -Br phase. *Surf. Sci.*, **222**, 285–295.
40. Motai, K., Hashizume, T., Lu, H., Jeon, D., Sakurai, T., and Pickering, H.W. (1993) STM of the Cu(111)1×1 surface and its exposure to chlorine and sulfur. *Appl. Surf. Sci.*, **67**, 246–251.
41. Huang, L., Zeppenfeld, P., Horch, S., and Comsa, G. (1997) Determination of iodine adlayer structures on Au(111) by scanning tunneling microscopy. *J. Chem. Phys.*, **107**, 585–591.
42. Woodruff, D.P., Seymour, D.L., McConvile, C.F., Reley, C.E., Crapper, M.D., Prince, N.P., and Jones, R.G. (1988) A simple x-ray standing wave technique for surface-structure determination - theory and an application. *Surf. Sci.*, **195**, 237–254.
43. Kardar, M. and Berker, A.N. (1982) Commensurate-incommensurate phase diagrams for overlayers from a helical potts model. *Phys. Rev. Lett.*, **48**, 1552–1555.
44. Zeppenfeld, P., Kern, K., David, R., and Comsa, G. (1988) Diffraction from domain wall systems. *Phys. Rev. B*, **38**, 3918–3924.
45. Doll, K. and Harrison, M.N. (2000) Chlorine adsorption on the Cu(111)

surface. *Chem. Phys. Lett.*, **317**, 282–289.

46. Komarov, N.S., Pavlova, T.V., and Andryushechkin, B.V. (2014) Adsorption of molecular iodine on Ni(111), unpublished.
47. Andryushechkin, B.V., Cherkez, V.V., and Eltsov, K.N. (2014) Iodine adsorption on Ag(111), unpublished.
48. Tewordt, L. (1958) Distortion of the lattice around an interstitial, a crowdion, and a vacancy in copper. *Phys. Rev.*, **109**, 61–68.
49. Braun, O.M. and Kivshar, Yu.S. (2004) *The Frenkel-Kontorova Model: Concepts, Methods, and Applications*, Springer-Verlag, Berlin.
50. Xiao, W., Greaney, P.A., and Chrzan, D.C. (2003) Adatom transport on strained Cu (001): surface crowdions. *Phys. Rev. Lett.*, **90**, 156102.
51. Xiao, W., Greaney, P.A., and Chrzan, D.C. (2004) Pt adatom diffusion on strained Pt(001). *Phys. Rev. B*, **70**, 033402.
52. Andryushechkin, B.V., Cherkez, V.V., Gladchenko, E.V., Zhidomirov, G.M., Kierren, B., Fagot-Revurat, Y., Malterre, D., and Eltsov, K.N. (2010) Structure of chlorine on Ag(111): evidence of the (3×3) reconstruction. *Phys. Rev. B*, **81**, 205434.
53. Andryushechkin, B.V., Cherkez, V.V., Gladchenko, E.V., Zhidomirov, G.M., Kierren, B., Fagot-Revurat, Y., Malterre, D., and Eltsov, K.N. (2011) Atomic structure of Ag(111) saturated with chlorine: Formation of $Ag_3Cl_7$ clusters. *Phys. Rev. B*, **84**, 075452.
54. Lyuksyutov, I.F., Pfnür, H., and Everts, H.-U. (1996) Incommensurate-commensurate transition via domain wall evaporation in an overlayer. *Europhys. Lett.*, **33**, 673–678.
55. Villain, J. (1980) Commensurate-incommensurate transition of krypton monolayers on graphite: a low temperature theory. *Surf. Sci.*, **97**, 219–242.
56. Gao, W., Baker, T.A., Zhou, L., Pinnaduwage, D.S., Kaxiras, S., and Friend, C.M. (2008) Chlorine adsorption on Au(111): chlorine overlayer or surface chloride? *J. Am. Chem. Soc.*, **130**, 3560–3565.
57. Andryushechkin, B.V., Cherkez, V.V., Gladchenko, E.V., Pavlova, T.V., Zhidomirov, G.M., Kierren, B., Didiot, C., Fagot-Revurat, Y., Malterre, D., and Eltsov, K.N. (2013) Self-organization of gold chloride molecules on Au(111) surface. *J. Phys. Chem. C*, **117**, 24948–24954.
58. Erley, W. (1980) Chlorine adsorption on the (111) faces of Pd and Pt. *Surf. Sci.*, **94**, 281–292.
59. Bittner, A.M., Wintterlin, J., Beran, B., and Ertl, G. (1995) Bromine adsorption on Pt(111), (100), and (110) - an STM study in air and in electrolyte. *Surf. Sci.*, **335**, 291–299.
60. Schennach, R. and Bechtold, E. (1997) Chlorine adsorption on Pt(111) and Pt(110). *Surf. Sci.*, **380**, 9–16.
61. Song, M.B. and Ito, M. (2001) STM observation of Pt111(3×3)-Cl and c(4×2)-Cl structures. *Bull. Korean Chem. Soc.*, **22**, 267–270.
62. Xu, H., Yuro, R., and Harrison, I. (1998) The structure and corrosion chemistry of bromine on Pt(111). *Surf. Sci.*, **411**, 303–315.
63. Felter, T.E. and Hubbard, A.T. (1979) L.E.E.D. and electrochemistry of iodine on Pt(100) and Pt(111) single-crystal surfaces. *J. Electroanal. Chem.*, **100**, 473–491.
64. Farell, H.H. (1980) The coadsorption of I and Cl on Pt(111). *Surf. Sci.*, **100**, 613–625.
65. DiCenzo, S.B., Wertheim, G.K., and Buchanan, D.N.E. (1984) Site dependence of core-electron binding energies of adsorbates: I/Pt(111). *Phys. Rev. B*, **30**, 553–557.
66. Saidy, M., Mitchell, K.A.R., Furman, S.A., Labayen, M., and Harrington, D.A. (1999) Tensor LEED analyses for three chemisorbed structures formed by iodine on a Pt(111) surface. *Surf. Rev. Lett.*, **6**, 871–881.
67. Tysoe, W.T. and Lambert, R.M. (1982) Surface-chemistry of the metal-halogen interface - bromine chemisorption and dibromide formation on palladium (111). *Surf. Sci.*, **115**, 37–47.
68. Tysoe, W.T. and Lambert, R.M. (1988) Structure and reactivity at the halogen metal interface - chemisorption,

corrosion and reaction pathways in the Pd(111)-$Cl_2$ system. *Surf. Sci.*, **199**, 1–12.
69. Shard, A.G., Dhanak, V.R., and Santoni, A. (2000) Structures of chlorine on palladium (111). *Surf. Sci.*, **445**, 309–314.
70. Dhanak, V.R., Shard, A.G., D'Addato, S., and Santoni, A. (1999) The structure of (root 3 x root 3)R30 degrees iodine on Pd(111) surface studied by normal incidence X-ray standing wavefield absorption. *Chem. Phys. Lett.*, **306**, 341–344.
71. Göthelid, M., von Schenck, H., Weissenrieder, J., Åkermark, B., Tkatchenko, A., and Galván, M. (2006) Adsorption site, core level shifts and charge transfer on the Pd(111)-I(root 3 x root 3) surface. *Surf. Sci.*, **600**, 3093–3098.
72. Cox, M.P. and Lambert, R.M. (1981) Structural and kinetic aspects of the rhodium-chlorine system - chlorine chemisorption and surface chloride formation on Rh(111). *Surf. Sci.*, **107**, 547–561.
73. Shard, A.G., Dhanak, V.R., and Santoni, A. (1999) Structural studies of the (root 3 x root 3)R 30 degrees surfaces of chlorine and iodine on Rh (111). *Surf. Sci.*, **429**, 279–286.
74. Barnes, C.J., Wander, A., and King, D.A. (1993) A tensor LEED structural study of the coverage-dependent bonding of iodine adsorbed on Rh111. *Surf. Sci.*, **281**, 33–41.
75. Mesgar, M., Kaghazchi, P., Jacob, T., Pichardo-Pedrero, E., Giesen, M., Pichardo-Pedrero, E., Giesen, M., Ibach, H., Luque, N.B., and Schmickler, W. (2013) Chlorine-enhanced surface mobility of Au(100). *ChemPhysChem*, **14**, 233–236.
76. Wang, S.W. and Rikvold, P.A. (2002) Ab initio calculations for bromine adlayers on the Ag(100) and Au(100) surfaces: the c(2×2) structure. *Phys. Rev. B*, **65**, 155406.
77. Lamble, G.M., Brooks, R.S., Campuzano, J.C., King, D.A., and Norman, D. (1987) Structure of the c(2×2) coverage of Cl on Ag(100): a controversy resolved by surface extended x-ray-absorption fine-structure spectroscopy. *Phys. Rev. B*, **36**, 1796–1798.
78. Fu, H., Jia, L., Wang, W., and Fan, K. (2005) The first-principle study on chlorine-modified silver surfaces. *Surf. Sci.*, **584**, 187–198.
79. Citrin, P.H., Hamann, D.R., Mattheiss, L.F., and Rowe, J.E. (1982) Geometry and electronic structure of Cl on the Cu 001 surface. *Phys. Rev. Lett.*, **49**, 1712–1715.
80. Kenny, S.D., Pethica, J.B., and Edgell, R.G. (2003) A density functional study of Br on Cu(100) at low coverages. *Surf. Sci.*, **524**, 141–147.
81. Citrin, P.H., Eisenberger, P., and Hewitt, R.C. (1980) Adsorption sites and bond lengths of iodine on Cu111 and Cu100 from surface extended X-ray-absorption fine structure. *Phys. Rev. Lett.*, **45**, 1948–1951.
82. Lairson, B., Rhodin, T.N., and Ho, W. (1985) Adsorbate fluorescence EXAFS: determination of bromine bonding structure in c(2×2)Br-Ni(001). *Solid State Commun.*, **55**, 925–927.
83. Jones, R.G., Ainsworth, S., Crapper, M.D., Somerton, C., and Woodruff, D.P. (1987) A SEXAFS study of several surface phases of iodine adsorption on Ni100: I. Multi-shell simulation analysis. *Surf. Sci.*, **179**, 425–441.
84. Gross, A. (2013) Ab initio molecular dynamics study of $H_2$ adsorption on sulfur- and chlorine-covered Pd(100). *Surf. Sci.*, **608**, 249–254.
85. Gatwood, G.A. and Hubbard, A.T. (1981) Superlattices formed by interaction of hydrogen bromide and hydrogen chloride with Pt(111) and Pt(100) studied by LEED, Auger and thermal desorption mass spectroscopy. *Surf. Sci.*, **112**, 281–305.
86. Gatwood, G.A. and Hubbard, A.T. (1980) Superlattices formed by interaction of hydrogen iodide with Pt(111) and Pt(100) studied by LEED, Auger and thermal desorption mass spectroscopy. *Surf. Sci.*, **92**, 617–635.
87. Nakakura, C.Y. and Altman, E.I. (1998) Scanning tunneling microscopy study of the reaction of $Br_2$ with Cu(100). *Surf. Sci.*, **398**, 281–300.

88. Kiskinova, M. and Goodman, D.W. (1981) Modification of chemisorption properties by electronegative adatoms: $H_2$ and CO on chlorided, sulfided, and phosphided Ni(100). *Surf. Sci.*, **108**, 64–76.
89. Dowben, P.A., Mueller, D., Rhodin, T.N., and Sakisaka, Y. (1985) Molecular bromine adsorption and dissociation on iron and nickel surfaces. *Surf. Sci.*, **155**, 567–583.
90. Wang, Y.N., Marcos, J.A., Simmons, G.W., and Klier, K. (1990) Adsorption of dichloromethane and its interaction with oxygen on the Pd(100) surface - effect of chlorine layers on oxygen-chemisorption and oxidation of carbon residues. *J. Phys. Chem.*, **94**, 7597–7607.
91. Schimpf, J.A., Abreu, J.B., and Soriaga, M.P. (1994) Electrochemical regeneration of clean and ordered Pd(100) surfaces by iodine adsorption-desorption: evidence from low-energy electron diffraction. *J. Electroanal. Chem.*, **364**, 247–249.
92. Eltsov, K.N., Klimov, A.N., Yurov, V.Yu., Shevlyuga, V.M., Prokhorov, A.M., Bardi, U., and Galeotti, M. (1995) Surface atomic structure upon Cu(100) chlorination observed by scanning tunneling microscopy. *JETP Lett.*, **62**, 444–450.
93. Rowland, R.S. and Taylor, R. (1996) Intermolecular nonbonded contact distances in organic crystal structures: comparison with distances expected from van der Waals radii. *J. Phys. Chem.*, **100**, 7384–7391.
94. Andryushechkin, B.V., Eltsov, K.N., Shevlyuga, V.M., Bardi, U., and Cortigiani, B. (2002) Structural transitions of chemisorbed iodine on Cu(100). *Surf. Sci.*, **497**, 59–69.
95. Jones, R.G. and Woodruff, D.P. (1981) The adsorption of $I_2$ on Ni100 studied by AES, LEED and thermal desorption. *Vacuum*, **31**, 411–415.
96. Hermse, C.G.M., van Bavel, A.P., Koper, M.T.M., Lukkien, J.J., van Santen, R.A., and Jansen, A.P.J. (2006) Bridge-bonded adsorbates on fcc(100) and fcc(111) surfaces: a kinetic Monte Carlo study. *Phys. Rev. B*, **73**, 195422.
97. Suleiman, I.A., Radny, M.W., Gladys, M.J., Smith, P.V., Mackie, J.C., Kennedy, E.M., and Dlugogorski, B.Z. (2011) Chlorination of the Cu(110) surface and copper nanoparticles: a density functional theory study. *J. Phys. Chem. C*, **115**, 13412–13419.
98. Gladchenko, E.V. (2012) Halide structures on monocrystalline gold and copper surfaces. PhD thesis. A.M. Prokhorov General Physics Institute of Russian Academy of Sciences, Moscow (in Russian).
99. Dona, E., Cordin, M., Deisl, C., Bertel, E., Franchini, C., Zucca, R., and Redinger, J. (2009) Halogen-induced corrosion of platinum. *J. Am. Chem. Soc.*, **131**, 2827–2829.
100. Deisl, C., Dona, E., Penner, S., Gabl, M., Bertel, E., Zucca, R., and Redinger, J. (2009) The phase diagram of halogens on Pt(110): structure of the (4×1)-Br/Pt(110) phase. *J. Phys. Condens. Matter*, **21**, 134003.
101. Wang, Y., Sun, Q., Fan, K., and Deng, J. (2001) Interaction of halogen atom with Ag(110): ab initio pseudopotential density functional study. *Chem. Phys. Lett.*, **334**, 411–418.
102. Houssiau, L. and Rabalais, J.W. (1999) Scattering and recoiling imaging spectrometry (SARIS) study of chlorine chemisorption on Ni(110). *Nucl. Instrum. Methods Phys. Res., Sect. B*, **157**, 274–278.
103. Zhang, J., Diao, Z.Y., and Wang, Z.X. (2006) Adsorption and vibration of Cl atoms on Ni low-index surfaces. *Chem. Res. Chin. Univ.*, **22**, 488–492.
104. Amann, P., Cordin, M., Gotsch, T., Menzel, A., Bertel, E., Redinger, J., and Franchini, C. (2015) Halogen phases on Pd(110): compression structures, domain walls, and corrosion. *J. Phys. Chem. C*, **119**, 3613–3623.
105. Oberkalmsteiner, N., Cordin, M., Duerrbeck, S., Bertel, E., Redinger, J., and Franchini, C. (2014) Degenerate phases of iodine on Pt(110) at half-monolayer coverage. *J. Phys. Chem. C*, **118**, 29919–29927.
106. Bushell, J., Carley, A.F., Coughlin, M., Davies, P.R., Edwards, D., Morgan, D.J.,

and Parsons, M. (2005) The reactive chemisorption of alkyl iodides at Cu(110) and Ag(111) surfaces: a combined STM and XPS study. *J. Phys. Chem. B*, **109**, 9556–9566.
107. Göthelid, M., Tymczenko, M., Chow, W., Ahmadi, S., Yu, S., Bruhn, B., Stoltz, D., von Schenck, H., Weissenrieder, J., and Sun, C. (2012) Surface concentration dependent structures of iodine on Pd(110). *J. Chem. Phys.*, **137**, 204703.
108. Wang, Y., Wang, W.N., Fan, K.N., and Deng, J.F. (2001) The first-principle study of the iodine-modified silver surfaces. *Surf. Sci.*, **487**, 77–86.
109. Andryushechkin, B.V., Eltsov, K.N., and Shevlyuga, V.M. (2005) Atomic structure of chemisorbed iodine layer on Cu(110). *Surf. Sci.*, **584**, 278–286.
110. Denisenkov, V.S., Pavlova, T.V., and Andryushechkin, B.V. (2014) Iodine lattice compression on Ag(110): STM and DFT study, unpublished.
111. Erley, W. (1982) Chlorine adsorption on the (110) faces of Ni, Pd and Pt. *Surf. Sci.*, **114**, 47–64.
112. Fishlock, T.W., Pethica, J.B., Jones, F.H., Egdell, R.G., and Foord, J.S. (1997) Interaction of chlorine with nickel (110) studied by scanning tunnelling microscopy. *Surf. Sci.*, **377/379**, 629–633.
113. Fishlock, T.W., Pethica, J.B., Oral, A., Egdell, R.G., and Jones, F.H. (1999) Interaction of chlorine with nickel (110) studied by scanning tunnelling microscopy. *Surf. Sci.*, **426**, 212–224.
114. Zou, S., Gao, X., and Weaver, M.J. (2000) Electrochemical adsorbate-induced substrate restructuring: gold(110) in aqueous bromide electrolytes. *Surf. Sci.*, **452**, 44–57.
115. Gao, X. and Weaver, M.J. (1994) New-type of periodic long-range restructuring of ordered metal-surfaces induced by lateral adsorbate interactions - iodide on Au(110) electrodes. *Phys. Rev. Lett.*, **73**, 846–849.
116. Obliers, B., Anastasescu, M., Broekmann, P., and Wandelt, K. (2004) Atomic structure and tip-induced reconstruction of bromide covered Cu(110) electrodes. *Surf. Sci.*, **573**, 47–56.
117. Menzel, A., Swamy, K., Beer, R., Hanesch, P., Bertel, E., and Birkenheuer, U. (2000) Electronic structure of a catalyst poison: Br/Pt(110). *Surf. Sci.*, **454**, 88–93.
118. Swamy, K., Hanesch, P., Sandl, P., and Bertel, E. (2000) Halogen-metal interaction: bromine on Pt(110). *Surf. Sci.*, **466**, 11–29.
119. Deisl, C., Swamy, K., Penner, S., and Bertel, E. (2001) Molecular adsorption on the quasi-one-dimensional c(2×2)-Br/Pt(110) surface. *Phys. Chem. Chem. Phys.*, **3**, 1213–1217.
120. Swamy, K., Menzel, A., Beer, R., and Bertel, E. (2001) Charge-density waves in self-assembled halogen-bridged metal chains. *Phys. Rev. Lett.*, **86**, 1299–1302.
121. Blum, V., Hammer, L., Heinz, K., Franchini, C., Redinger, J., Swamy, K., Deisl, C., and Bertel, E. (2002) Structure of the c(2×2)-Br/Pt(110) surface. *Phys. Rev. B*, **65**, 165408.
122. Deisl, C., Swamy, K., Beer, R., Menzel, A., and Bertel, E. (2002) Structure tuning by Fermi-surface shifts in low-dimensional systems. *J. Phys. Condens. Matter*, **14**, 4199–4209.
123. Deisl, C., Swamy, K., Memmel, N., Bertel, E., Franchini, C., Schneider, G., Redinger, J., Walter, S., Hammer, L., and Heinz, K. (2004) (3×1)-Br/Pt(110) structure and the charge-density-wave-assisted c(2×2) to (3×1) phase transition. *Phys. Rev. B*, **69**, 195405.
124. Zhang, Z.R., Minca, M., Menzel, A., and Bertel, E. (2004) The surface electronic structure of Br/Pt(110) phases. *Surf. Sci.*, **566**, 476–481.
125. Menzel, A., Zhang, Z., Minca, M., Loerting, T., Deisl, C., and Bertel, E. (2005) Correlation in low-dimensional electronic states on metal surfaces. *New J. Phys.*, **7**, 102.
126. Bertel, E. and Dona, E. (2007) Fermi surface tuning in two-dimensional surface systems. *J. Phys. Condens. Matter*, **19**, 355006.
127. Dona, E., Loerting, T., Penner, S., Minca, M., Menzel, A., Bertel, E., Schoiswohl, J., Berkebile, S.,

Netzer, F.P., Zucca, R., and Redinger, J. (2007) Fluctuations and phase separation in Br/Pt(110). *Surf. Sci.*, **601**, 4386–4389.
128. Amann, P., Cordin, M., Braun, C., Lechner, B.A.J., Menzel, A., Bertel, E., Franchini, C., Zucca, R., Redinger, J., Baranov, M., and Diehl, S. (2010) Electronically driven phase transitions in a quasi-one-dimensional adsorbate system. *Eur. Phys. J. B*, **75**, 15–22.
129. Bertel, E. (2013) Quasi-critical fluctuations: a novel state of matter? *J. Nanopart. Res.*, **15**, 1407.
130. Cordin, M., Lechner, B.A.J., Duerrbeck, S., Menzel, A., Bertel, E., Redinger, J., and Franchini, C. (2014) Experimental observation of defect pair separation triggering phase transitions. *Sci. Rep.*, **4**, 4110.
131. Cordin, M., Lechner, B.A.J., Amann, P., Menzel, A., Bertel, E., Franchini, C., Zucca, R., Redinger, J., Baranov, M., and Diehl, S. (2010) Phase transitions driven by competing interactions in low-dimensional systems. *Europhys. Lett.*, **92**, 26004.
132. Foord, J.S. and Lambert, R.M. (1982) A structural and kinetic study of chlorine chemisorption and surface chloride formation on Cr(100). *Surf. Sci.*, **115**, 141–160.
133. Reed, A.P.C., Lambert, R.M., and Foord, J.S. (1983) Chemisorption and corrosion at the metal-halogen interface: overlayer growth and compound formation by bromine on Cr(100). *Surf. Sci.*, **134**, 689–702.
134. Hino, S. and Lambert, R.M. (1986) Chlorine chemisorption and surface chloride formation on iron: adsorption/desorption and photoelectron spectroscopy. *Langmuir*, **2**, 147–150.
135. Grunze, M. and Dowben, P.A. (1982) A review of halocarbon and halogen adsorption with particular reference to iron surfaces. *Appl. Surf. Sci.*, **10**, 209–239.
136. Stott, Z.T. and Hughes, H.P. (1983) Chlorine and iodine adsorption on the Ta(100) and Ta(110) surfaces. *Surf. Sci.*, **126**, 455–462.
137. Reed, A.P.C., Lambert, R.M., and Foord, J.S. (1983) Chemisorption and epitaxial growth: structural and kinetic studies of halogen interactions with vanadium and chromium surfaces. *Vacuum*, **33**, 707–714.
138. Davies, P.W. and Lambert, R.M. (1980) Surface chemistry of the metal-halogen interface: bromine chemisorption and related studies on vanadium (100). *Surf. Sci.*, **95**, 571–586.
139. Kramer, H.M. and Bauer, E. (1981) The adsorption of chlorine on W(100). *Surf. Sci.*, **107**, 1–19.
140. Rawlings, K.J., Price, G.G., and Hopkins, B.J. (1980) A LEED and AES study of the adsorption of bromine on W(100) at room temperature. *Surf. Sci.*, **100**, 289–301.
141. Rawlings, K.J., Price, G.G., and Hopkins, B.J. (1980) A LEED and AES study of the adsorption of iodine on W(100) at room temperature. *Surf. Sci.*, **95**, 245–256.
142. Foord, J.S. and Lambert, R.M. (1987) Halogen adsorption and halogen-induced surface phase transitions on Cr(110). *Surf. Sci.*, **185**, L483–L488.
143. Linsebigler, A.L., Smentkowski, V.S., Ellison, M.D., and Yates, J.T. Jr. (1992) Interaction of chlorine with iron(110) in the temperature range 90-1050 K. *J. Am. Chem. Soc.*, **114**, 465–473.
144. Mueller, D.R., Rhodin, T.N., Sakisaka, Y., and Dowben, P.A. (1991) Chemisorbed halogen adlayers on Fe(110): electronic band structure and surface geometry. *Surf. Sci.*, **250**, 185–197.
145. Preyss, W., Ebinger, H.D., Fick, D., Polenz, C., Polivka, B., Saier, V., Veith, R., Weindel, Ch., and Jänsch, H.J. (1997) Adsorption of chlorine on Ru(001) and codesorption with lithium. *Surf. Sci.*, **373**, 33–42.
146. Hofmann, J.P., Rohrlack, S.F., Heß, F., Goritzka, J.C., Krause, P.P.T., Seitsonen, A.P., Moritz, W., and Over, H. (2012) Adsorption of chlorine on Ru(0001)–A combined density functional theory and quantitative low energy electron diffraction study. *Surf. Sci.*, **606**, 297–304.
147. Fishlock, T.W., Pethica, J.B., and Egdell, R.G. (2000) Observation of a nanoscale

chessboard superstructure in the Br-Cu(100) adsorbate system. *Surf. Sci.*, **445**, L47–L52.

148. Bowker, M. and Waugh, K.C. (1983) The adsorption of chlorine and chloridation of Ag(111). *Surf. Sci.*, **134**, 639–664.
149. Rovida, G. and Pratesi, F. (1975) Chlorine monolayers on the low-index faces of silver. *Surf. Sci.*, **51**, 270–282.
150. Tu, Y.Y. and Blakely, J.M. (1978) Chlorine adsorption on silver surfaces. *J. Vac. Sci. Technol.*, **15**, 563–567.
151. Wu, K., Wang, D., Deng, J., Wei, X., Cao, Y., Zei, M., Zhai, R., and Guo, X. (1992) Chlorine on Ag(111): the intermediate coverage case. *Surf. Sci.*, **264**, 249–259.
152. Andryushechkin, B.V., Eltsov, K.N., Shevlyuga, V.M., and Yurov, V.Y. (1998) Atomic structure of saturated chlorine monolayer on Ag(111) surface. *Surf. Sci.*, **407**, L633–L639.
153. Endo, O., Kondoh, H., and Ohta, T. (1999) Scanning tunneling microscope study of bromine adsorbed on the Ag(111) surface. *Surf. Sci.*, **441**, L924–L930.
154. Schardt, B.C., Yau, S.L., and Rinaldi, F. (1989) Atomic resolution imaging of adsorbates on metal-surfaces in air - iodine adsorption on Pt(111). *Science*, **243**, 1050–1053.
155. Tkatchenko, A., Batina, N., and Galván, M. (2006) Potential energy landscape of monolayer-surface systems governed by repulsive lateral interactions: the case of (3×3)-I-Pt(111). *Phys. Rev. Lett.*, **97**, 036102.
156. Andryushechkin, B.V., Cherkez, V.V., Pavlova, T.V., Zhidomirov, G.M., and Eltsov, K.N. (2013) Structural transformations of Cu(110) surface induced by adsorption of molecular chlorine. *Surf. Sci.*, **608**, 135–145.
157. Galeotti, M., Cortigiani, B., Torrini, M., Bardi, U., Andryushechkin, B., Klimov, A., and Eltsov, K. (1996) Epitaxy and structure of the chloride phase formed by reaction of chlorine with Cu(100). A study by X-ray photoelectron diffraction. *Surf. Sci.*, **349**, L164–L168.
158. DiCenzo, S.B., Wertheim, G.K., and Buchanan, D.N.E. (1982) Epitaxy of CuI on Cu(111). *Appl. Phys. Lett.*, **40**, 888–890.
159. Nakakura, C.Y. and Altman, E.I. (1999) Scanning tunneling microscopy study of halide nucleation, growth, and relaxation on singular and vicinal Cu surfaces. *Surf. Sci.*, **424**, 244–261.
160. Besenbacher, F., Nielsen, L.P., and Sprunger, P.T. (1997) Surface alloying in heteroepitaxial metal-on-metal growth, in *Growth and Properties of Ultrathin, Epitaxial Layers*, vol. 8 (eds D.A. King and D.P. Woodruff), Elsevier, Amsterdam, pp. 207–257.
161. Wiederholt, T., Brune, H., Wintterlin, J., Behm, R.J., and Ertl, G. (1994) Formation of two-dimensional sulfide phases on Al(111): an STM study. *Surf. Sci.*, **324**, 91–105.
162. Weiss, W. and Ranke, W. (2002) Surface chemistry and catalysis on well-defined epitaxial iron-oxide layers. *Prog. Surf. Sci.*, **70**, 1–151.
163. Barth, J., Behm, R.J., and Ertl, G. (1995) Adsorption, surface restructuring and alloy formation in the NaAu(111) system. *Surf. Sci.*, **341**, 62–91.
164. Joyce, B.A., Sudijono, J.L., Belk, J.G., Yamaguchi, H., Zhang, X.M., Dobbs, H.T., Zangwill, A., Vvedensky, D.D., and Jones, T.S. (1997) A scanning tunneling microscopy-reflection high energy electron diffraction-rate equation study of the molecular beam epitaxial growth of InAs on GaAs(001), (110) and (111)A-Quantum dots and two-dimensional modes. *Jpn. J. Appl. Phys.*, **36**, 4111–4117.
165. Belk, J.G., Pashley, D.W., Joyce, B.A., and Jones, T.S. (1998) Dislocation displacement field at the surface of InAs thin films grown on GaAs(110). *Phys. Rev. B*, **58**, 16194–16201.
166. Guha, S., Madhukar, A., and Rajkumar, K.C. (1990) Onset of incoherency and defect introduction in the initial stages of molecular beam epitaxical growth of highly strained $In_xGa_{1-x}As$ on GaAs(100). *Appl. Phys. Lett.*, **57**, 2110–2112.
167. Mo, Y.-W., Savage, D.E., Swartzentruber, B.S., and Lagally, M.G.

(1990) Kinetic pathway in Stranski-Krastanov growth of Ge on Si(001). *Phys. Rev. Lett.*, **65**, 1020–1023.
168. Kitson, M. and Lambert, R. (1980) A structural and kinetic study of chlorine and silver chloride on Ag(100): the transition from overlayer to bulk halide. *Surf. Sci.*, **100**, 368–380.
169. Andryushechkin, B.V., Eltsov, K.N., Shevlyuga, V.M., Tarducci, C., Cortigiani, B., and Bardi, U. (1999) Epitaxial growth of AgCl layers on the Ag(100) surface. *Surf. Sci.*, **421**, 27–32.
170. Andryushechkin, B.V. and Eltsov, K.N. (1992) Local structure of a copper surface chlorinated at a low temperature. *Surf. Sci.*, **265**, L245–L247.
171. Andryushechkin, B.V., Eltsov, K.N., and Shevlyuga, V.M. (1995) Local structure determination for surface chlorination with EELFS. *Physica B*, **208/209**, 471–473.
172. Andryushechkin, B.V., Cherkez, V.V., Gladchenko, E.V., Zhidomirov, G.M., Kierren, B., Fagot-Revurat, Y., Malterre, D., and Eltsov, K.N. (2013) New insight into the structure of saturated chlorine layer on Ag(111): LT-STM and DFT study. *Appl. Surf. Sci.*, **267**, 21–25.
173. Wyckoff, R.W.G. (1963) *Crystal Structures*, vol. 1, John Wiley & Sons, Inc., New-York and London.
174. Andryushechkin, B.V., Eltsov, K.N., and Shevlyuga, V.M. (2004) CuI growth on copper surfaces under molecuar iodine action: influence of the surface anisotropy in the iodine monolayer. *Surf. Sci.*, **566/568**, 203–209.
175. Hai, N.T.M., Huemann, S., Hunger, R., Jaegermann, W., Wandelt, K., and Broekmann, P. (2007) Combined scanning tunneling microscopy and synchrotron X-ray photoemission spectroscopy results on the oxidative CuI film formation on Cu(111). *J. Phys. Chem. C*, **111**, 14768–14781.
176. Broekmann, P., Hai, N.T.M., and Wandelt, K. (2006) Copper dissolution in the presence of a binary compound: CuI on Cu(100). *J. Appl. Electrochem.*, **36**, 1241–1252.
177. Andryushechkin, B.V., Eltsov, K.N., and Cherkez, V.V. (2006) Epitaxial growth of semiconductor thin films on metals in the halogenation process. Atomic structure of copper iodide on the Cu(110) surface. *JETP Lett.*, **83**, 162–166.
178. Lundgren, E., Mikkelsen, A., Andersen, J.N., Kresse, G., Schmid, M., and Varga, P. (2006) Surface oxides on close-packed surfaces of late transition metals. *J. Phys. Condens. Matter*, **18**, R481–R499.
179. Kleinherbers, K.K., Janssen, E., Goldmann, A., and Saalfeld, H. (1989) Submonolayer adsorption of halogens on Ag(001) and Ag(011) studied by photoemission. *Surf. Sci.*, **215**, 394–420.
180. Andryushechkin, B.V., Eltsov, K.N., and Shevlyuga, V.M. (1999) Atomic structure of silver chloride formed on Ag(111) surface upon low temperature chlorination. *Surf. Sci.*, **433/435**, 109–113.
181. Kurumaji, T., Seki, S., Ishiwata, S., Murakawa, H., Kaneko, Y., and Tokura, Y. (2013) Magnetoelectric responses induced by domain rearrangement and spin structural change in triangular-lattice helimagnets $NiI_2$ and $CoI_2$. *Phys. Rev. B*, **87**, 014429.
182. Haberern, K.W. and Pashley, M.D. (1990) GaAs(111)A-(2×2) reconstruction studied by scanning tunneling microscopy. *Phys. Rev. B*, **41**, 3226–3229.
183. Yanase, A. and Segawa, Y. (1995) Two different in-plane orientations in the growths of cuprous halides on MgO(001). *Surf. Sci.*, **329**, 219–226.

# 36
# Adsorption of Hydrogen

*Klaus Christmann*

## 36.1
## Introduction

"Am Anfang war der Wasserstoff" (In the beginning was – "HYDROGEN"). This is the title of a book written by the German author and journalist Hoimar von Ditfurth that appeared in 1979 and points to the most important role of the element hydrogen as an apparently extremely simple building block in the development of the inorganic and organic matter in our universe [1]. Despite the fact that this book appeared quite a long time ago, it manifests the modern scientific view of a continuous, evolutionary, biogenesis of humankind, and puts the role of "homo sapiens" into perspective. Latest, since the ending of the twentieth and the beginning twenty-first century, and in view of the fact that the population on our planet is still growing exponentially, it is one of the most prominent economic and political tasks to appease human's hunger for energy. While the resources based on fossil fuels are assumed to get exhausted soon, hydrogen has certainly become one of the most valuable energy carriers [2], and a lot of effort is invested to efficiently produce and store hydrogen gas [3, 4]. At first glance, this appears self-contradictory, since the chemical element "hydrogen" is the most abundant species in our universe. However, the occurrence of elemental hydrogen on earth is vanishingly small, and the overwhelming amount is oxidized and available only in the form of water. To cleave the first hydrogen–oxygen bond of $H_2O$ according to

$$\text{H–O–H} \rightarrow \text{H} + \text{O–H}$$

requires the considerable energy of 498 kJ/mol (5.16 eV), and the cleavage of the remaining hydroxyl bond

$$\text{O–H} \rightarrow \text{H} + \text{O}$$

still needs 424 kJ/mol (4.39 eV).

Since this chapter is devoted to hydrogen in its interaction with *surfaces*, we desist from expanding too much on the chemical and physical properties of this element, a description of which can be found in many textbooks of chemistry and physics [5–8] as well as on the Internet [9]. However, some remarks are nevertheless useful,

*Surface and Interface Science: Solid-Gas Interfaces I*, First Edition. Edited by Klaus Wandelt.

because they help us to understand the specific reactive behavior of hydrogen with surfaces.

Hydrogen was officially discovered by Henry Cavendish in 1766, although chemists have certainly encountered this element before, among others Robert Boyle in his reaction experiments with iron and acids. Its name comes from the combined Greek words hydro and genes, meaning, "water forming." It is the first element in the periodic table (PT) and, in the form of the isotope, "hydrogen" (sometimes also called "protium") ${}^{1}_{1}H$, consists of a proton and a single electron. The two other known isotopes, deuterium ${}^{2}_{1}H$ and tritium ${}^{3}_{1}H$, have one and two "extra" neutrons, respectively, and, hence, possess the two and three times the atomic mass. This has a strong impact on hydrogen's kinetic behavior, especially concerning surface diffusion and vibrational phenomena, which show relatively large "isotope effects," that is, differences between H, D, and T, compared with the heavier chemical elements. Together with the noble gas helium, hydrogen forms the unique first period of the PT of the chemical elements; the filled 1s shell contains only two electrons, while all following periods contain elements with at least eight outer-shell electrons. In other words, the H atom unites the properties of the first and the last-but-one family of the PT and can, therefore, resemble both an alkali metal atom – when it loses its valence electron to form a proton ($H^+$ ion) – and a halogen atom – when it accepts an electron and builds a hydride [$H^-$] ion.

The single outer unpaired valence electron of an H atom is responsible for the chemical instability of isolated H atoms. If two H atoms come together, they immediately recombine to form a hydrogen molecule, $H_2$, provided the excess energy released during the recombination reaction can be transferred to the wall of a reaction vessel or, more generally, dissipated to any surface. Accordingly, on earth and at temperatures below ~2000 K, the thermodynamically stable form of hydrogen is di-hydrogen, $H_2$. From the viewpoint of theoretical chemistry, this simplest homonuclear diatomic molecule exhibits a strong chemical bond (bond dissociation energy 435 kJ/mol), which can be regarded as the prototype of an atomic or covalent bond. The chemical and physical properties of both the hydrogen molecule and the hydrogen atom are well known and listed in numerous text and reference books [10]. In Table 36.1, we provide the reader with some physical and chemical data of the natural element "hydrogen" (in the form of the $H_2$ molecule).

Besides its simple electronic structure, a few other properties of the hydrogen molecule make it really unique and distinguish it from any other homo or heteronuclear molecular species. As will be shown later, these properties have impact also on hydrogen's interaction with surfaces (see Section 36.2) and are therefore of importance in our context.

There is, first, the very small geometrical size of $H_2$ due to the extremely short internuclear distance of only $7.4 \times 10^{-11}$ m. However, the hydrogen molecule must not be envisaged as a dumpbell, but it rather has an almost perfect spherical shape, as is evident from Figure 36.1 taken from an article by Silvera [11]. In its dynamical interaction with surfaces, however, the molecular orientation with respect to the surface plane can have some significance, for example, for hydrogen trapping, sticking, and dissociation phenomena; numerous theoretical considerations have been

**Table 36.1** Some physical and chemical properties of molecular hydrogen.

| Property | Unit | $H_2$ ($^1_1H$) | $D_2$ ($^2_1H$) | $T_2$ ($^3_1H$) |
|---|---|---|---|---|
| Relative abundance | % | 99.985 | 0.015 | $\sim 10^{-15}$ |
| Absolute mass | kg | $3.34706538 \times 10^{-27}$ | $6.68898878 \times 10^{-27}$ | $10.01653442 \times 10^{-27}$ |
| Molar mass | kg/mol | $2.01565006 \times 10^{-3}$ | $4.028203552 \times 10^{-3}$ | $6.032098551 \times 10^{-3}$ |
| Binding energy | eV/molecule | 4.519 | 4.602 | 4.636 |
| (heat of dissociation) | kJ/mol | 436.002(4) | 443.546(4) | 446.9 |
| Internuclear distance | m | $0.7416 \times 10^{-10}$ | $0.7416 \times 10^{-10}$ | $0.7416 \times 10^{-10}$ |
| Ionization potential | eV | 15.427 | 15.46 | |
| Moment of inertia | kg m$^2$ | $4.60 \times 10^{-48}$ | $9.19 \times 10^{-48}$ | $6.11 \times 10^{-48}$ |
| Rotational constant B | cm$^{-1}$ | 60.809 | 30.4 | |
| Ground-state configuration | | $^1\Sigma_{g+}$ | | |
| Ground state vibration $\tilde{\nu}$ | cm$^{-1}$ | 4395.2 | 3118.4 | 3817 |
| Density at 20 °C $\rho$ | $10^3$kg/m$^3$ | 0.00008988 | 0.000168 | 0.000269 |
| Triple point | K | 13.95 | | |
| Melting point | K | 14.025 | 18.73 | 20.62 |
| Boiling point | K | 20.39 | 23.67 | 25.04 |
| Heat of fusion at 13.9 K | J/mol | 117.2 | 197 | 250 |
| Heat of vaporization | J/mol | 994 | 1.126 | 1.393 |
| Molar volume at 13.9 K | m$^3$ | $26.15 \times 10^{-6}$ (liquid)<br>$23.31 \times 10^{-6}$ (solid) | $23.14 \times 10^{-6}$ (liquid)<br>$20.48 \times 10^{-6}$ (solid) | |
| Crystal structure of solid | | h.c.p./f.c.c. | | |
| Lattice constants (hcp) | $10^{-10}$ m | 3.75 (*a*)<br>6.12 (*c*)<br>1.632 (*c*/*a*) | | |

devoted to this topic [12] (see Section 36.2). From the co-volume of hydrogen (van der Waals constant $b = 0.02661$ dm$^3$ mol$^{-1}$ [13]), one can estimate a van der Waals radius of $1.38 \times 10^{-10}$ m. The distance dependence of the electron density around the H–H axis is also illustrated in Figure 36.1 and provides information on the diameter and shape of the $H_2$ molecule. Compared to almost any other molecular species, hydrogen has a lack of electrons, which makes it a very weak electron scatterer. This has implications on the visibility and detectability of adsorbed hydrogen on surfaces, for example, by low-energy electron diffraction or other scattering methods, or techniques that are based on multielectron excitations such as Auger electron spectroscopy (AES) or X-ray photoelectron spectroscopy (XPS).

Another peculiar property of the $H_2$ entity is its tiny atomic mass of only 2.016 g/mol = $3.348 \times 10^{-27}$ kg/molecule, a value that is smaller than for any other chemical element and responsible for the apparent quantum behavior of hydrogen and its isotopes. For a given velocity *v*, the corresponding de Broglie wavelength exceeds that of any other element: consequently, hydrogen can easily reveal its wave nature. The respective particle–wave dualism manifests itself in hydrogen

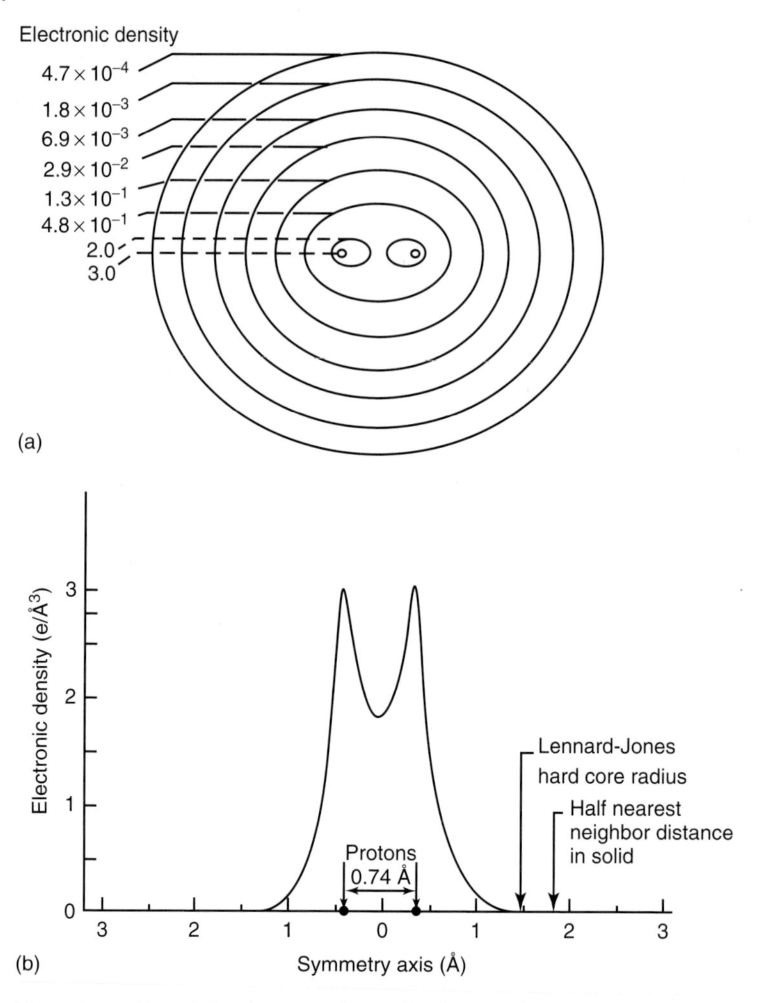


**Figure 36.1** Map of the electronic charge distribution of the $H_2$ molecule. (a) Contour lines of equal charge density. (b) Density distribution along the internuclear molecular axis. (After Silvera [11].)

scattering, diffraction, and tunneling phenomena, and provides the experimentalist with a variety of powerful tools that help to elucidate the details of hydrogen's interaction with surfaces, as will be shown in more detail later. We recall in this context a phenomenon known as *quantum delocalization*, which is a consequence of hydrogen's light mass and responsible for the fact that a band model is appropriate to describe the motion of H atoms parallel and perpendicular to a surface [14–16].

Especially interesting is the hydrogen molecule with respect to the nuclear spins of the two protons. According to quantum chemistry, the spin quantum number of each of the protons can be $s = +\frac{1}{2}$ or $-\frac{1}{2}$; if both protons rotate in the same direction (parallel nuclear spins), they form the compound "*ortho*-hydrogen," while the species with antiparallel nuclear spins is known as "*para*-hydrogen" [17].

Actually, the two modifications differ somewhat with respect to their energy content, with the *para*-hydrogen being the slightly more stable species. For a given number of molecules, both forms exist in a temperature-dependent equilibrium, with the equilibrium being shifted to pure *para*-hydrogen at −273 °C:

$$o\text{-H}_2 \leftrightarrow p\text{-H}_2; \quad \Delta H_R = -80\,\text{J/mol}.$$

Increasing the temperature supports the conversion to ortho-hydrogen, and at room temperature the equilibrated mixture contains 75% *o*-hydrogen and 25% *p*-hydrogen. This is just the highest possible content of *o*-hydrogen, because the multiplicity for a singlet state (total spin quantum number $S = 0$) is just 1, while for a triplet state (total spin quantum number $S = 1$) one has a multiplicity of 3. Because of the weakness of the spin–spin interaction, the equilibration is a very slow process and can last several months. However, by means of an appropriate catalyst (charcoal surfaces, for example), the equilibration can be very much accelerated. This bears some implications for the industrial liquefaction of hydrogen, where one adds a catalyst in order to avoid the spontaneous transformation of $o$-$H_2$ to $p$-$H_2$, which releases some energy that would cause unwanted vaporization of liquid hydrogen. The catalytic transformation can be reckoned as a surface-mediated process, in which the spin conversion is accomplished by the magnetic interaction of the $H_2$ nuclear spins with paramagnetic centers at the surface, for example, unpaired electron spins. In a recent publication, Buntkowsky *et al.* showed that also the interaction with diamagnetic substances could induce the respective spin transformation [18], which may have some implications for the future use of spin-marked hydrogen in magnetic resonance tomography (MRT). Also very recently, Fukutani and Sugimoto published a comprehensive work on physisorption and *o*–*p* conversion of hydrogen on surfaces [19].

Very briefly, we address another topic that makes molecular hydrogen a desired "chemical" in industry and technology. There has long been a worldwide demand for hydrogen as a valuable raw material to produce and synthesize all kind of chemical products – if we confine ourselves just to heterogeneously catalyzed reactions, we could already mention hundreds of chemical processes including all kinds of hydrogenations, especially of carbon monoxide via the Fischer–Tropsch synthesis [20–22], hydrotreating and hydrocracking [23, 24], or hydrodesulfurization [25]. In all these reactions, the interaction between hydrogen and the respective catalyst surfaces is a key issue. Goodman has emphasized the role of surface and catalytic model studies to understand the respective mechanisms [26]; for an overview we recommend the monograph by Ertl [27]. Today, the use of hydrogen has reached an entirely new dimension, owing to its use in fuel cells and batteries [28, 29]. Schlögl has recently given a comprehensive survey about the general role of chemistry in the energy challenge (with some emphasis on hydrogen) [30].

So far, we have dealt preferentially with the hydrogen *molecule*, and a few notes on the H *atom* are certainly useful. The H atom with its proton and its 1s electron can be quantitatively described by the stationary Schrödinger equation [31], and the probability to encounter the electron is highest at a distance of $0.529 \times 10^{-10}$ m, which represents the classical Bohr radius. It requires an ionization energy of 13.59 eV to completely remove the 1s electron and to form a "naked" proton. On the other hand, the H atom can also accept an additional electron to form a (negatively charged) closed-shell unit, the hydride ion, $H^-$, the electron affinity for this process being 0.754 eV. Some other relevant properties of the H atom may be seen in Table 36.2.

Because of its unpaired valence electron, H can act as a strong nucleophilic agent and form stable compounds with the nonmetallic main-group elements of the PT. Depending on the electronegativity of the binding partner, the character of the chemical bond to the H atom can vary from almost complete covalency (e.g., in H–carbon or H–silicon bonds) to strong ionicity (in bonding to halogens or alkali metals). The Pauling electronegativity of H is 2.1 [32], and many elements on the left-hand side of the PT have electronegativities smaller than (or at most equal to) 2. Accordingly, the respective elements can donate electrons to the hydrogen atom, resulting in bonds with pronounced ionic character. This holds, in a reverse sense, also for the (nonmetallic) elements on the right-hand side of the PT with their comparatively large electronegativities, and the hydrogen atom now acts as an electron donor. The electronegativities of most of the transition metals (TMs), however, vary between 1.5 and 2.5 and resemble that of hydrogen, and no clear-cut direction of electron transfer can be predicted in each case.

**Table 36.2** Some physical and chemical properties of the H atom.

| Property | Unit | $^1_1$H (=H) | $^2_1$H (=D) | $^3_1$H (=T) |
|---|---|---|---|---|
| Absolute mass | kg | $1.67353269 \times 10^{-27}$ | $3.344494392 \times 10^{-27}$ | $5.008267212 \times 10^{-27}$ |
| Molar mass | kg/mol | $1.00782503207 \times 10^{-3}$ | $2.0141017778 \times 10^{-3}$ | $3.0160492777 \times 10^{-3}$ |
| Natural isotope mixture | % | 99.9885(70) | 0.0115(70) | Traces |
| Nuclear spin $I$ | $h/2\pi$ | ½+ | 1+ | ½+ |
| Nuclear magnetic moment $\mu$ (in nuclear magnetrons $\mu_N$) | | 2.792847351(28) | 0.857438230(24) | 2.97896248(7) |
| Atomic radius (from internuclear separation) | m | $0.3707 \times 10^{-10}$ | | |
| Bohr radius $a_0$ | m | $0.52917721092(17) \times 10^{-10}$ | | |
| Velocity of electron in the innermost orbit | m/s | $2.187 \times 10^6$ | | |
| Ground-state configuration | — | $^2S_{1/2}$ | | |
| First ionization potential I | eV | 13.595 | | |
| Electron affinity A ($1s^2$) | eV | 0.756 | | |
| Pauling's electronegativity | | 2.20 | | |

It is known that various TMs can, at least physically, take up hydrogen atoms, simply by dissolution; they *absorb* hydrogen. This sorption process depends, among others, on temperature and hydrogen gas pressure, and one obtains absorption isotherms and can relate the dissolved amount of hydrogen with the number of metal atoms. The H-to-metal atom ratio can (and in many cases will) be nonstoichiometric and vary with the applied hydrogen gas pressure. In some other cases, at least in a certain pressure–temperature regime, the H-to-metal ratio remains constant, a hint to a stoichiometric hydride, but it is not always easy to distinguish hydridic phases from simply dissolved (absorbed) hydrogen. In Figure 36.2, we give a schematic overview of the types of hydrides formed by the elements of the PT, along with the activity of the metallic elements to dissociate molecular hydrogen. However, since it is not our intention to expand on hydride chemistry and physics, we refer the reader to the pertinent literature and mention explicitly the early exhaustive review article on metal hydrides by Gibb [33]. More information about the topic "absorption of hydrogen" can be obtained from the numerous respective monographs and original publications [34–37].

A final issue here may be the actual experimental production of *atomic* hydrogen, which is of some relevance in the context of surface experiments that probe the interaction between H *atoms* and solid surfaces. The homolytic scission of the covalent H–H bond requires the supply of the binding energy of 435 kJ/mol (which in this case is better called *dissociation energy*). Accordingly, this energy has to be spent on the system – either chemically, by electromagnetic radiation, electrically,

| | | | | | | | | | | | | | | | | | |
|---|---|---|---|---|---|---|---|---|---|---|---|---|---|---|---|---|---|
| H | | | | | | | | | | | | | | | | | He |
| Li | Be | | | | | | | | | | | B | C | N | O | F | Ne |
| Na | Mg – | | | | | | | | | | | Al – | Si (+) | P | S | Cl | Ar |
| K – | Ca + | Sc + | Ti + | V + | Cr + | Mn + | Fe + | Co + | Ni + | Cu (+) | Zn – | Ga – | Ge (+) | As | Se | Br | Kr |
| Rb | Sr | Y + | Zr + | Nb + | Mo + | Tc + | Ru + | Rh + | Pd + | Ag – | Cd – | In – | Sn – | Sb | Te | I | Xe |
| Cs | Ba + | La + | Hf + | Ta + | W + | Re + | Os + | Ir + | Pt + | Au – | Hg | Tl | Pb – | Bi | Po | At | Rn |
| Fr | Ra | Ac | Th | Pa | U | | | | | | | | | | | | |
| Ionic (salt-like) hydrides | | Transition metal hydrides ("metallic" hydrides) | | | | | | | | Border line cases | | Covalent hydrides | | | | | |

**Figure 36.2** Schematic overview of the types of hydrides formed by the elements of the PT. In addition, those metals that chemisorb hydrogen dissociatively with moderate or high sticking probability are indicated by a plus (+) sign. A minus (–) sign denotes elements where an activation barrier for dissociation has been verified.

or thermally. In 1981, Slevin and Stirling reported on a beam source that provided atomic hydrogen by radio frequency (RF) excitation at 35 MHz [38]. Ten years later, Hodgson and Haasz developed a glow-discharge RF source to efficiently produce H atoms [39]. In most cases, however, atomic hydrogen is produced thermally. From simple equilibrium thermodynamics it is known that one can produce atomic hydrogen simply by heating $H_2$ gas to very high temperatures in order to shift the equilibrium to the side of dissociation; an early overview has been given by Moore and Unterwald [40]. One of the pioneers here is Langmuir, who investigated the thermal hydrogen dissociation as early as 1912 [41]. He showed that the $H_2$ dissociation is practically complete at $T = 8000$ K, while already at $T = 2000$ K only 0.1% of the $H_2$ molecules are dissociated. At $T = 2900$ K, about 7% H atoms exist, and at $T = 3800$ K, a degree of dissociation of 40% is reached. Since tungsten has the highest melting point of all elements, $T_m = 3683$ K, this sets a benchmark regarding practical surface-chemical experiments in that H atoms can be produced in reasonable yields by allowing a flux of $H_2$ molecules to pass over an incandescent W filament in vacuum; thereafter, the beam of H atoms is directed toward the specimen target. The vacuum therefore is necessary to avoid oxidative corrosion of the W filament and, more importantly, to prevent the H atoms from recombination. Besides their thermal velocity, the H atoms carry a lot of chemical energy and can strongly attack even surfaces that are normally inert to molecular hydrogen. In this way, solid surfaces hit by H atoms may undergo structural changes (reconstructions), dissolve H atoms, or form chemical compounds (hydrides) with hydrogen. In case these hydrides are volatile, the solid material can be partially removed and transferred to the gas phase, particularly at elevated temperatures. An example is the interaction of H atoms with Si surfaces, in which silanes can be formed after prolonged hydrogen exposure [42]. Some other semiconductors containing gallium, arsenic, or antimony also fall in this category.

## 36.2 Hydrogen Interacting with Surfaces

### 36.2.1 General Remarks and Clarification of Terms

As we have indicated on various occasions in the preceding section, the interaction of the $H_2$ molecule with *solid surfaces* (preferentially *metallic* surfaces) has attracted and still attracts much attention in science and technology, for a variety of reasons (hydrogen's role in heterogeneous catalysis, battery and fuel cell technology, materials science, plasma physics). Groß recently underlined this in a "surface science perspective" article [43].

Whenever hydrogen gas interacts with solid or liquid matter, it must first *come into contact* with the *surface* of the condensed material. The penetration into the interior of the bulk and possible chemical reactions are then only *subsequent* steps of the respective *surface* interaction. Accordingly, the first decisive reaction events have to do with the encounter of the $H_2$ molecule with the surface, and we summarize this in a wide sense as "adsorption." This adsorption comprises all interactions of the hydrogen molecule (or of the H atom, in case predissociated hydrogen is used) with the force fields of the respective surface. Only once the $H_2$ molecule has been trapped and adsorbed, consecutive reactions (dissociation, chemisorption, transient or permanent subsurface state population, bulk absorption, hydride formation, etc.) can proceed. There exists a vast literature concerning the interaction of hydrogen with solid surfaces, in particular with metallic surfaces. Only some references are listed here comprising both experiment and theory where the reader can plunge into the extensive literature [44–51].

In order to clearly specify the interaction scenario of the adsorption process, it is useful to distinguish two main situations: in the first case, we select a *single $H_2$ molecule* (or an isolated H atom) and follow it on its way toward the solid surface while watching and describing all chemical/physical changes that occur during this approach on the adsorbing particle *and* on the surface. Let us call this scenario the *single-particle interaction*; it may be treated theoretically by representing the solid as a (small) cluster of metal atoms that interacts with the hydrogen molecule (or atom), or by really taking into account a semi-infinitely extended crystalline surface instead of the cluster. A careful (experimental and theoretical) analysis of all individual reaction steps can then disclose the details of the interaction, including the energetics, adsorption geometry (bond lengths and bond angles), or vibrational or electronic excitations of the H-containing surface complex.

Once we have understood the respective sequence of individual reaction steps, which one usually subsumes as the "interaction mechanism," we can move on to the second (and often equally difficult) scenario that one encounters when not a *single* molecule is approaching the surface, but a whole *ensemble* of molecules are. In other words, we now consider a vast number of undistinguishable species that hit the surface in a given time interval. This corresponds exactly to the situation when the surface is, in a practical experiment, exposed to a certain finite hydrogen gas pressure. We will call this scenario *multiparticle interaction*, which is, from the viewpoint of *practical* chemistry and catalysis, certainly especially important. Unfortunately, the accurate details of any individual chemical reaction (that one can, at least in principle, still follow in the single-particle approach) are no longer separately accessible under these conditions, owing to the vast number of particle–particle interactions that come now into play. Rather, one has to treat the *combined* system, that is, the surface plus, say, a mole of hydrogen molecules or atoms, and all the possible mutual interactions by means of appropriate (and usually complicated) quantum statistics and statistical thermodynamics [52].

In both the single- and the multiparticle concept, it is – as generally in physical chemistry – useful to distinguish stationary states, that is, time-*independent* phenomena, representing chemical equilibrium, and time-*dependent* processes, known as *interaction kinetics* or, more generally, as *interaction dynamics*. In the first case, the chemical interaction, that is, the breaking and making of chemical bonds involving hydrogen, can be described by the *stationary* Schrödinger equation and calculated using the Born–Oppenheimer approximation [51, 52]. The energetics of the dissociation of the hydrogen molecule falls in this category, but also the formation of new chemical phases with long-range order or rearrangement processes within a given phase, which can be driven by energy and entropy. Chemical thermodynamics predicts the direction and extent as well as the free energy of the reaction(s) that occur during the interaction with the surface [53]. In this respect, the energetics of an interaction process and the final structural "state" of the reaction product(s) are *stationary* properties and should be *independent* of time. However, as known from ordinary chemical reaction kinetics, thermodynamic predictions often fail because they neglect the existence of kinetic activation barriers, which can slow down the reaction rate to almost zero. In other words, time-dependent processes very often determine the reaction scenario, and the knowledge and appropriate consideration of the involved activation barriers are decisive prerequisites to correctly treat the overall reaction cycle and obtain reliable information about the final reaction *rates*. In all these cases and especially when one is interested in the behavior of adsorption systems under the influence of electromagnetic radiation (photochemical processes), it is essential to consider the interaction dynamics based on the *time-dependent* Schrödinger equation. Modern experimental means to monitor the chemical and physical changes of the quantum-state populations on very short time scales (femto- or even attoseconds) include molecular beam methods combined with laser pulse techniques and various (sometimes sophisticated pump-and-probe type) spectroscopies. Especially for the interaction of hydrogen with metal surfaces with emphasis on the dissociation reaction, this procedure has led to fundamental new insights [54–59].

Quite an important experimental parameter to distinguish single and multiple particle interaction is the concentration of adsorbed species at the surface, $\sigma$ (particles/m$^{-2}$). While $\sigma$ tends to zero in case of single-particle interaction, it can usually reach values close to the number of *adsorption sites* per unit area provided by the solid surface. A common single-crystal surface of a close-packed metal contains approximately $10^{19}$ atoms/m$^2$. One can conveniently relate the number of actually occupied adsorption sites, $N_{ad}$, to the total number of available adsorption sites, $N_{max}$, and obtains a dimensionless number that varies between 0 and 1, which is called "coverage" $\Theta$:

$$\text{coverage}\,\Theta = \frac{N_{ad}}{N_{max}}; \quad 0 \le \Theta \le 1, \tag{36.1}$$

based on the idea that the adsorption capacity of a given surface atom is exhausted once it carries an adsorbed particle. This is certainly so when a strong chemical bond is formed between the adsorbed particle and the surface atom, the underlying process being called *chemisorption*, see below. Especially in case of hydrogen chemisorption, due to the smallness of a H atom or a $H_2$ molecule, the number of these adsorption sites can easily reach the number of exposed surface atoms [60, 61]. In some cases, it may even exceed $\Theta = 1$ [62]. This is quite in contrast to, for example, the adsorption of voluminous, heavy noble gas or alkali metal atoms or large organic molecules, where maximum coverage at or below 0.5 is the rule [63].

Since the adsorbed particles can and usually will interact with each other, depending on their mutual proximity, the physical properties describing the adsorbed state are usually coverage-dependent. The underlying interaction forces can have long-range character (in case of dipole–dipole interactions) or act on the short-range scale (in case of direct orbital–orbital overlapping, i.e., Pauli repulsion). The mutual adsorbate–adsorbate interactions can have great influence, for example, in modifying the energy released during the adsorption process (called the *adsorption energy*) or the efficiency of impinging particles to dissipate their kinetic energy to the surface and to reside in a bound state at the surface. The latter efficiency can be expressed $s$ using the term *sticking probability.* This sticking probability or sticking coefficient is defined as the ratio of the number of atoms or molecules that "adsorb" to the total number of particles that impinge upon the surface within the same period. In this definition, "adsorption" requires that the incoming particle really gets accommodated on the surface, that is, loses its "memory" of momentum and energy after the collision. Generally, $s$ can vary between 0 and 1: $s = 0$ means that the particle is simply reflected from the surface, while it is completely accommodated if $s = 1$. The sticking coefficient is usually strongly coverage-dependent, with the simple idea behind being that an atom or molecule arriving from the gas phase needs an empty site in order to form an adsorptive bond to the surface. It can also depend on the temperature and (to some extent) on the morphology (roughness, number of crystallographic defects, etc.) of the surface.

Before we continue and expand on the interaction of a single hydrogen molecule with a solid surface (see, Section 36.2.2), we have to briefly mention some other important terms that we will encounter in the following. Figure 36.3 may help to illustrate the principal situation. The respective two-dimensional Lennard–Jones representation [64] shows the potential energy balance of an incoming hydrogen molecule as a function of its distance to the surface, where the shaded area indicates the location of the surface. The $H_2$ molecule is approaching the surface from right to left, and the first step of the aforementioned sticking event is the (often transient) trapping of the intact hydrogen molecule in a weak van der Waals-like potential with a depth of a few kilojoules only, from which the molecule can easily escape again into the gas phase. We call this process *physisorption*, because only physical interaction forces of the van der Waals type prevail. The time a molecule spends in this bound state is called the *surface residence time* $\tau_{surf}$, which is determined by Boltzmann statistics via

$$\tau_{surf} = \tau_0 \exp\left(\frac{E_{phys}}{RT}\right) \tag{36.2}$$

$\tau_0$ being the pre-exponential factor, $E_{phys}$ denoting the depth of the potential, and $RT$ the thermal energy of the system. Because of the weakness of the van der Waals forces, the position of the minimum of the physisorption potential is located fairly far away from the surface. Closer to the surface, there exists a usually much deeper potential well that is provided only if the $H_2$ molecule falls apart into its atoms. The larger depth of this potential is due to the fact that individual H atoms can interact much more strongly with the surface leading to the buildup of real chemical bonds. We recall that *dissociation* of the $H_2$ molecule into the atoms is urgently required in the course of this process (see Section 36.2.3.1)! The overall process

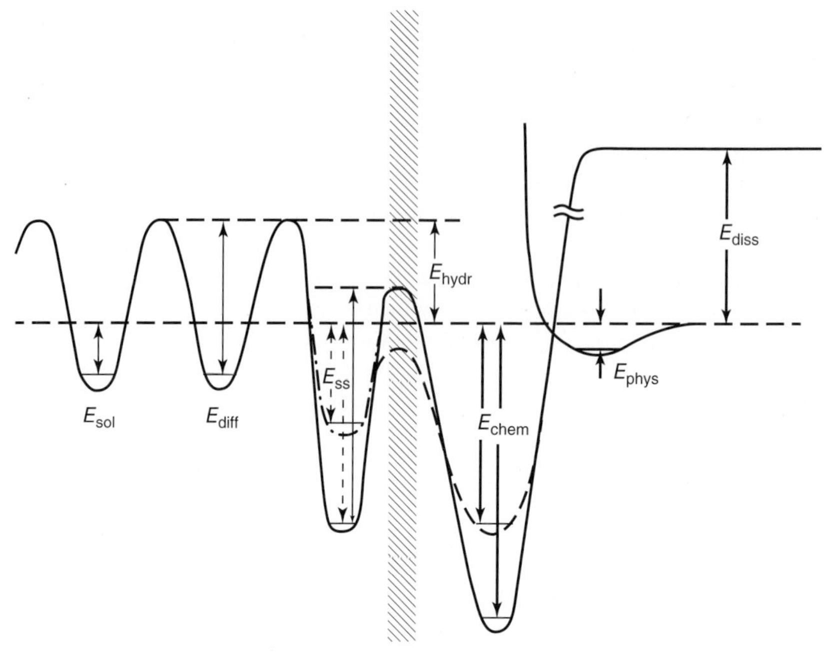


**Figure 36.3** One-dimensional Lennard–Jones diagram showing the change in potential energy of a hydrogen molecule approaching a metal surface. Hatching indicates the position of the surface. The following possible partial reactions can happen: (i) physisorption of the $H_2$ molecule in a weak potential of depth $E_{phys}$; (ii) dissociation of the $H_2$ molecule resulting in a stable chemisorptive bond with adsorption energy $E_{chem}$. The depth of this potential depends on coverage as indicated by the broken line; (iii) transport of H atoms into subsurface sites, with a (likewise coverage-dependent) sorption energy $E_{ss}$; (iv) absorption of H atoms in interstitial sites with the heat of solution $E_{sol}$. $E_{hydr}$ and $E_{diff}$ denote the activation energies for hydrogenation and lattice diffusion, respectively.

including dissociation is called *chemisorption*, which usually involves adsorption energies $E_{chem}$ of more than ~20 kJ/mol and in some cases (hydrogen interaction with tungsten or molybdenum surfaces) even more than 150 kJ/mol. The two potentials are separated by an activation barrier (called the *activation energy* for adsorption, $\Delta E^*_{ad}$, if the barrier has a positive sign (see Section 36.2.3, Figure 36.9). Depending on $\Delta E^*_{ad}$ in relation to the thermal energy, the incoming hydrogen molecule may dissociate spontaneously (*non-activated* dissociative *adsorption*), or it must have an appropriate surplus of external or internal energy to overcome this barrier before it can dissociate into the atoms (*activated adsorption*). On TM surfaces with their unfilled d electron states, the spontaneous dissociative chemisorption is the rule, but this does not hold for the free-electron metals of the left side of the PT for reasons that will be explained further below. The dissociation event heavily influences both the adsorption kinetics and energetics. It will therefore be a major issue of our considerations to dwell on the (homolytic) dissociation process in view of both interaction dynamics and energetics (see Section 36.2.3).

Figure 36.3 discloses some further details: left to the (hatched) surface region of the crystal there exists a potential of depth $E_{SS}$, which refers to the population of the so-called *subsurface* sites, which are – in a strict sense – located just between the first and second atomic layer of the crystal. The small physical size of a hydrogen atom can substantially facilitate the population of these subsurface sites. Moving further to the left toward the interior of the solid often reveals another, somewhat weaker, periodic potential (depth $E_{sol}$), which indicates the ability of some metals (Pd!) to *absorb* and dissolve H atoms in the bulk including the possibility to form hydridic phases or compounds. This dissolution process can be exothermic (in case the periodic potential is below the energy zero line) or endothermic (if positive energy barriers are involved). Again, a (concentration-dependent) activation barrier may separate the subsurface from the sorption sites.

## 36.2.2 The Associative Adsorption of $H_2$ *Molecules* on Surfaces

### 36.2.2.1 Trapping and Sticking; Energy Accommodation

For the following considerations, we "accompany" a hydrogen molecule on its way toward a given single-crystalline solid surface – under conditions of thermal equilibrium, it has, at 300 K temperature, a mean velocity of 1775 km/s – almost four times as fast as a nitrogen molecule at the same temperature. This means that the molecule is approaching the surface at a rate of 18 Å/ps. As a *closed-shell entity*, the $H_2$ molecule does not experience any "chemical" interaction with the surface, but perceives at first only the aforementioned "physical" interaction forces. There exist now three main "reaction channels."

1) The $H_2$ molecule hits the surface, but it is unable to dissipate sufficient kinetic energy during the very collision event. It is either immediately reflected to the gas phase (and retains its memory of impact momentum and energy, a process called *elastic scattering*), or it leaves the physisorption potential already after a few vibrational periods. There may, however, occur some changes in its internal distribution of quantum states even if it does not really exchange much energy with the surface. We describe the respective process as *direct inelastic scattering*; it results in "transient trapping."
2) The molecule may become trapped in the physisorption potential for a much longer time ($>10^{-6}$ s), provided its thermal energy is low enough. We assume complete energy accommodation; only occasionally the $H_2$ molecule may leave the surface by thermal desorption. Its residence time on the surface is determined by the interplay between the thermal and potential energy according to Boltzmann statistics, (Equation 36.2). The overall situation corresponds to true physisorption.
3) The molecule utilizes its initial thermal energy to overcome the activation energy barrier that separates the physisorption and chemisorption potential

energy well (Figure 36.3). This will automatically lead to dissociation, and the two H atoms may then form more or less strong bonds to the surface, resulting in true chemisorption. This (practically very important) case will be the subject of Section 36.2.3.

**Case 1) Transient Trapping** This (from a *chemical* viewpoint certainly not much interesting) phenomenon has been thoroughly investigated by many researchers using molecular beam experiments. We refer to original work by Robota *et al.*, who studied $H_2$ and $D_2$ scattering on Ni(111) and Ni(110) surfaces [54]. They measured, among others, the angular distribution of the off-scattered molecules and found, particularly for the (110) surface orientation along the [001] direction, pronounced diffraction effects due to scattering from the molecular potential well. In addition, the authors observed a strong background scattering intensity peaking in specular direction, which originates from energy exchange between the incident molecules and the surface. This is illustrated by Figure 36.4 for deuterium molecules from a Ni(110) surface. Cowin *et al.* examined the scattering of HD molecules at a Pt(111) surface and found evidence of rotationally inelastic diffractive scattering effects that could be explained by a simple model in which an eccentrically weighted sphere collided with a hard wall [65].

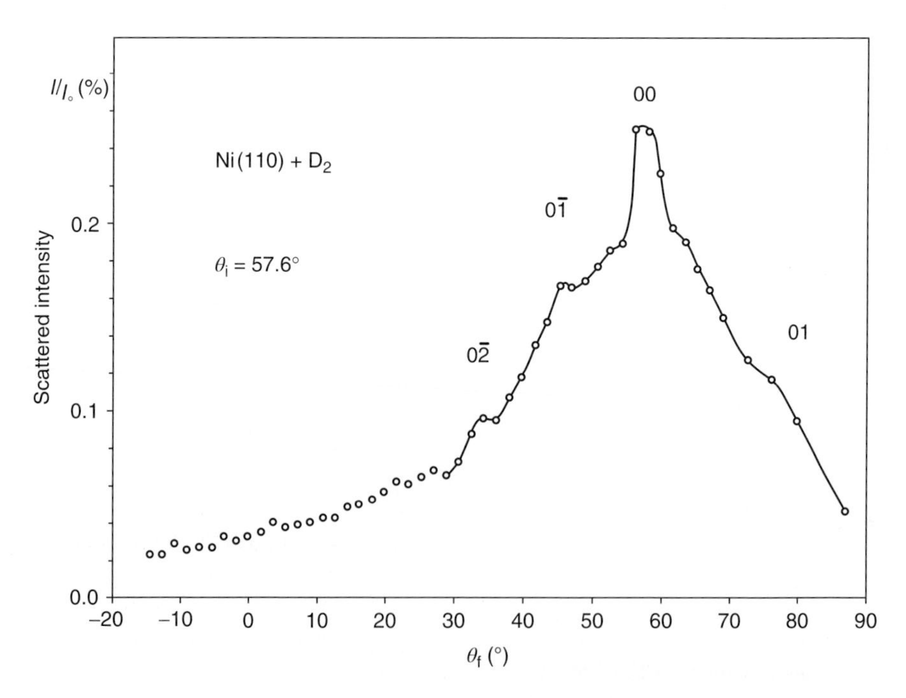


**Figure 36.4** Angular distribution of deuterium molecules scattered from the Ni(110) surface at a temperature of 400 K. The angle of incidence is 57.6° along the [001] direction. The diffraction maxima of $n = 1$ and 2 are superimposed on the diffuse background of inelastic scattering. (After Robota *et al.* [54].)

**Case 2) Physisorption** If the temperature of the surface is low enough, the hydrogen molecule can and will stay in the physisorption potential, and we may consider this case as true sticking. Since, especially, on metal surfaces the aforementioned reaction channel 3 is significant, it is difficult to determine the true sticking coefficients for $H_2$ *molecules*, since in many cases there immediately occurs hydrogen dissociation with instantaneous population of chemisorption sites, thereby simulating an effective sticking of H *atoms*. Accordingly, there are not very many reports on *molecular* sticking coefficients, one of the rare examples being the study by Frieß *et al.*, who investigated, at temperatures as low as 4–6 K, the $H_2$ and $D_2$ adsorption on a Ru(0001) surface that was passivated before by a layer of chemisorbed H atoms in order to suppress the dissociation reaction [66]. In Figure 36.5, we present their coverage dependence of the molecular sticking coefficient, which quite interestingly exhibits an initial increase with $\Theta$ from almost 0 to ~0.7 close to the monolayer coverage. This effect is explained by the improved energy transfer on adsorbate islands compared to the bare surface [67] and demonstrates the importance of nonclassical effects in the accommodation and sticking of light particles such as hydrogen.

In this context, there is an almost hydrogen-specific problem concerning the energy transfer of an incoming hydrogen molecule, namely, the effective dissipation of its kinetic energy to the solid surface. This requires that the incoming particle couples to the thermal vibrations of the crystal or, in other words, to the phonon bath of the solid. If $m$ denotes the mass of the impinging adsorbate particle and $M$ that of the surface atom hit by the particle, the efficiency of this kind of energy transfer scales with the mass ratio $\mu = m/M$. If we describe the binary elastic collision between the $H_2$ molecule and the surface atom by the hard-sphere approximation

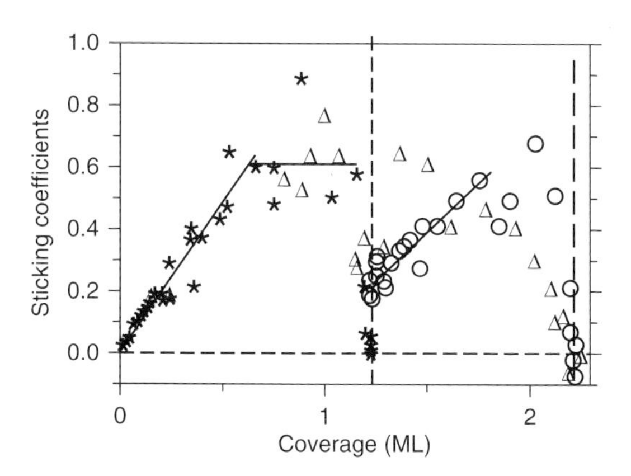


**Figure 36.5** Coverage dependence of the molecular sticking coefficient of hydrogen on the Ru(0001) surface passivated with a $(1 \times 1)$ layer of chemisorbed H atoms. Triangles and stars denote the results of direct dosing at 4.8 and 5.5 K, respectively; circles indicate adsorption onto an annealed, compressed first layer at $T = 4.8$ K. (After Frieß *et al.* [66].)

and apply the rules of classical conservation of energy and momentum, one can express the amount of energy transfer $\Delta E = E_i - E_f$ by the Baule formula [68, 69]

$$\Delta E = \frac{4\mu}{(1+\mu)^2} E_i. \tag{36.3}$$

The energy $\Delta E$ is then submitted to the phonon bath of the solid and finally dissipated. For hydrogen with $\mu \ll 1$, the amount of transferred energy is usually small, and therefore this type of energy accommodation is believed to be not very effective. Accordingly, other mechanisms have been invoked to account for the experimental observation that hydrogen molecules quite often do stick and dissociate on surfaces with an appreciable initial sticking probability. In another view, the time during the collision is simply too short for transferring sufficient energy to the surface. A frequently discussed possibility especially for metal surfaces is the rapid excitation of electron–hole pairs right at the Fermi level, a mechanism also called *electronic friction* [70]. Various theoreticians have studied the details of the hydrogen accommodation problem [71–73]: among others, the nonadiabatic energy dissipation during adsorption at TM surfaces [74]. Andersson *et al.* [75] have analyzed the sticking of light inert particles on surfaces, especially of $H_2$ molecules on the Cu(100) surface at $T = 10$ K, and concluded (somewhat in contrast to the classical picture discussed above) that the sticking processes are inherently quantum in nature. A theory based on perturbative coupling to phonons can satisfactorily account for most of their measured data. This is taken as evidence that phonon excitation should still be the dominant energy-transfer mechanism in such systems.

At finite temperatures, the trapped and weakly held molecule is usually very mobile and can easily diffuse across the surface while looking for appropriate conditions to still overcome the dissociation barrier. In this configuration, the trapped molecule is considered a *precursor*, with interesting consequences for the overall adsorption kinetics [76]. Unless experiments are carried out at extremely low temperatures, the existence of such a precursor manifests itself mainly in the coverage dependence of the sticking probability, $s(\Theta)$, where $s$ now refers to *atomic* H adsorption. We will return to this topic in Section 36.2.3.

In some of our own hydrogen adsorption experiments with Pd, Ni, and Rh(210) surfaces, we came across an interesting interplay between molecular and dissociative trapping, sticking, and adsorption phenomena [77]. The starting point was an observation by Mårtensson *et al.* who discovered, by means of vibrational spectroscopy, the formation of a molecularly adsorbed hydrogen species near the atomic steps of a Ni(510) single-crystal surface even at temperatures as high as 100 K [78]. The "survival" of molecular hydrogen was explained by a peculiar configuration of the electronic structure near the steps that stabilized Ni–$H_2$ complexes. On Pd(210), it could be shown experimentally [79] and by theory [80] that there is a ($T$-dependent) competition between molecular and dissociative hydrogen adsorption, with a preference for molecular adsorption at temperatures <120 K. Sites occupied with H *atoms* seem to stabilize the adjacent $H_2$/Pd complexes. Quite interestingly, the uptake of molecular hydrogen (as monitored by a negative work function (WF) change) occurs with a high probability, pointing to a large sticking coefficient for $H_2$ on the (210) surfaces of Pd, Ni, and Rh [77].

**Case 3) Dissociative Chemisorption** It can, as indicated above, also happen that the impinging $H_2$ molecule utilizes its initial thermal energy to overcome or penetrate the activation energy barrier to the chemisorption potential. This inevitably requires its dissociation, and the two H atoms may then form more or less strong bonds to the surface resulting in true chemisorption. The dissociation reaction of hydrogen is a key issue in the course of the interaction and will be the subject of Section 36.2.3.

#### 36.2.2.2 Energetics and Kinetics of Hydrogen Physisorption

We have seen that the nondissociative adsorption of hydrogen molecules requires experiments at low or even very low temperatures. In view of the boiling point of liquid hydrogen of $T_b = 20.4$ K, it is evident that hydrogen molecules can only be condensed on surfaces at or below this temperature; practically, cooling with liquid helium is mandatory, and some care has to be taken to accurately measure temperatures and pressures and keep the samples free of contaminating heavier condensates [81]. A more recent review article on the physisorption of hydrogen on various surfaces has been published by Ptushinskii [82].

Two hydrogen molecules attract each other weakly at larger distances as a result of mutual electron–electron polarization, the origin of the van der Waals forces; if they are brought very close together, strong direct orbital–orbital repulsions become effective, which rise very steeply with decreasing separation. The already mentioned Lennard–Jones potential is usually written as

$$V(r) = C_n/r^n - C_6/r^6, \tag{36.4}$$

where $n$ is a large integer, which is often chosen as $n = 12$, and we have the well-known (12,6)-potential. In many cases, $V(r)$ is given in the parameterized form

$$V(r) = 4\varepsilon \left\{ \left(\frac{\sigma}{r}\right)^{12} - \left(\frac{\sigma}{r}\right)^{6} \right\}, \tag{36.5}$$

in which $\varepsilon$ is the depth of the minimum of the curve, which occurs at $r_{ph} = \sqrt[6]{2}\sigma$. The values for gaseous $H_2$ are [83] $\varepsilon = 4.597 \times 10^{-22}$ J/mol (=276.8 J/mol) and $\varepsilon/k_B = 33.3$ K; if $\varepsilon$ is divided by Boltzmann's constant $k_B$ for convenience, $\sigma = 2.97 \times 10^{-10}\ m$ or $r_{ph} = 3.33 \times 10^{-10}\ m$. Note that these values hold only for the mutual interaction of *two* $H_2$ molecules in the gas phase. On surfaces, there exists a new situation in that a $H_2$ molecule directly in front of a given surface site will interact with all accessible surface atoms within a certain range of action. At finite coverages, additional mutual $H_2$–$H_2$ interactions come additionally into play. The resulting interaction energies may then be determined by summation over all possible geometrical configurations and will certainly substantially exceed the bimolecular van der Waals energy mentioned above. An example of the principal procedure can be found in the work by Küppers and Seip [84], who determined physisorption energies for xenon interacting with Pd surfaces using the Lennard–Jones (12,6) potential and could predict the experimental values quite well.

Since there are not many surface studies dealing with the energetics of physisorbed hydrogen, we again invoke the already cited work by Frieß *et al.* [66], in which the energies to remove physisorbed hydrogen and deuterium molecules from a passivated Ru(0001) surface have been determined as a function of surface coverage with remarkable accuracy. The technique used here was thermal desorption spectroscopy (TDS), where the covered surface is heated with a linear ramp while the molecules leaving the surface are collected with a mass spectrometer in the pumped vacuum system [85–87]. From the shape and temperature position of the resulting hydrogen partial pressure signal, the desorption energy $E_{des}$ could be evaluated (which equals the *adsorption* energy for nonactivated adsorption); it corresponds exactly to the energy required to break the physisorptive bond to the surface. We present in Figure 36.6 the $D_2$ thermal desorption spectra from Frieß *et al.* taken at linear heating rates between 0.1 and 0.2 K/s from a Ru(0001)(1 × 1)D surface that was exposed to increasing doses of deuterium at ~4 K. The TD peak at ~15 K reflects deuterium molecules that are adsorbed within the first layer, that is, have direct contact to the chemisorbed layer of D adatoms on the Ru surface. Remarkable is the invariance of the desorption temperature with coverage, which clearly points to a first-order desorption kinetics as expected for a situation in which the rate-determining step is the cleavage of an individual $D_2$–surface bond.

$$-\frac{d\Theta}{dt} = \nu_1 \Theta^1 \exp\left(-\frac{E_{des}}{RT}\right), \tag{36.6}$$

where $\Theta$ = coverage, $\nu_1$ = first-order frequency factor (1/*s*), and $E_{des}$ = desorption energy. Note that for dissociative adsorption and associative desorption (two

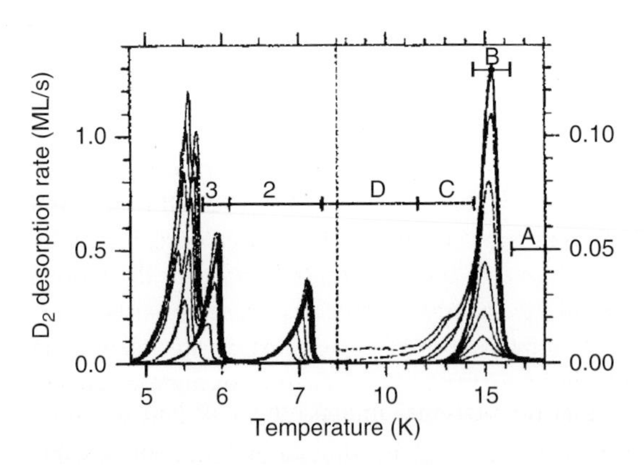


**Figure 36.6** Thermal desorption spectra monitoring deuterium molecules leaving the Ru(0001) (1 × 1)D surface after increasing exposures in the temperature range 4–18 K. The heating rate was 0.2 K/s in the monolayer range and 0.1 K/s in the multilayer range. According to the authors, at least five layers (indicated by labels A–D) can be distinguished for the monolayer range and more than two layers (labeled 2 and 3) for the multilayer range, whereas still higher multilayers are not indicated anymore. Note the different intensity scale for the monolayer (right axis) and multilayer desorption (left axis). See text for further details. (After Frieß *et al.* [66].)

individual adsorbed D atoms have to recombine prior to desorption as a $D_2$ entity), a second-order desorption kinetics applies:

$$-\frac{d\Theta}{dt} = \nu_2 N_{max} \Theta^2 \exp\left(-\frac{E_{des}}{RT}\right), \tag{36.7}$$

where $\nu_2$ = second-order frequency factor ($m^2/s$) and $N_{max}$ = total number of available adsorption sites ($m^{-2}$). In this case, the peak positions of the TD maxima shift with coverage to lower temperatures, and we will return to this when discussing the case of dissociative adsorption (Section 36.2.3).

The TD spectra of Figure 36.6 clearly reveal the layer-by-layer growth of physisorbing deuterium, in line with a perfect surface wetting of molecular deuterium on the Ru substrate. For higher coverages, the formation of at least four more layers can be distinguished. The desorption contribution stemming from the second layer appears around 7 K, while the third and following layers cause desorption peaks between 5 and 6 K. The marked difference in energies with respect to the first monolayer reflects the rapidly decreasing contribution of the direct $D_2$–Ru interaction. Note that all respective TD maxima exhibit a common leading edge and, hence, obey zero-order kinetics, where the rate is independent of the surface concentration – as expected for a true evaporation or sublimation process:

$$-\frac{d\Theta}{dt} = \frac{\nu_0}{N_{max}} \exp\left(-\frac{E_{des}}{RT}\right), \tag{36.8}$$

in which $\nu_0$ stands for the zero-order pre-exponential factor [$1/(s\,m^2)$]. The authors evaluated desorption energies for the various states: for the first-layer peak at 15 K, they report a constant value of 340 K (2.83 kJ/mol), which decreases by a factor of 2 for the second-layer state ($E_{phys}$ = 170 K (=1.43 kJ/mol)). Note that, due to compression effects within the second adsorbed layer, the coverages within the individual layers are not identical, the reason being zero-point vibrational effects. It is unnecessary to say that the just discussed phenomena hold in the same way (with light modifications) also for molecular *hydrogen*.

Zero-point energy effects also dominate physisorbed $H_2$ layers formed on other solid surfaces. Extensive studies were performed with graphite surfaces, and depending on temperature and surface coverage, interesting phases of molecular hydrogen with a rich structural variety could be observed. Just as with physisorbed diatomic molecules, a new phenomenon, namely *orientational ordering*, can occur in that the molecular axis, which is oriented parallel to the surface plane, can take on a variety of relative azimuthal orientations, among others, of herringbone or pinwheel character [88–90]. The reason is that, in view of the extremely weak molecule–surface binding, the adsorbate–adsorbate interactions dominate the adsorbate–substrate forces. For hydrogen physisorbed on graphite, Cui *et al.* [91] examined the orientational ordering of physisorbed deuterium as a function of temperature in the range between 2 and 5 K by means of neutron and electron diffraction and found three incommensurate solid phases and, in addition, the same commensurate ($\sqrt{3} \times \sqrt{3}$)R30° structure that had been reported earlier in low-energy electron diffraction (LEED) experiments by Seguin and Suzanne [92].

This latter study is of particular interest for surface scientists because it was the first time that a LEED pattern from physisorbed hydrogen could be measured, thus allowing the judgment directly of the diffractive power of hydrogen molecules for low-energy electrons. In Figure 36.7, we present the respective LEED pattern from Ref. [92]. One has to look quite carefully at the image in order to realize the very weak, but sharp, $H_2$-induced "extra" spots in the $\sqrt{3}$-positions, and it is evident that the electron scattering power of hydrogen is indeed quite low, at least in the absence of strong interactions with the underlying surface. As we will see in Section 36.2.6, LEED beams produced by H atoms chemisorbed at metal surfaces can have considerably more intensity because of partial charge-transfer effects and possible H-induced displacements of substrate atoms (H-induced surface reconstruction); respective considerations can, for example, be taken from Refs [14, 93, 94].

A final remark may be added concerning the question how one can actually distinguish physisorbed $H_2$ molecules from chemisorbed H atoms – besides perhaps the already discussed thermal desorption contributions at very low temperatures. An unambiguous criterion for the presence of molecular hydrogen is certainly the observation of the inner-molecular H–H stretching vibration $\nu$, which appears for the unperturbed gas phase $H_2$ molecule at 4153 cm$^{-1}$ (=516 meV) [78]. (More details of the vibrational spectroscopy can be found in Section 36.2.7.)

**Figure 36.7** LEED pattern of the $(\sqrt{3} \times \sqrt{3})$ R30° structure formed by molecular hydrogen adsorbed on a graphite(0001) surface at ~10 K and under an equilibrium pressure of $1.2 \times 10^{-8}$ Torr. The electron energy was 142 eV. The intense spots are due to the hexagonal graphite and the faint spots due to the hydrogen layer. (After Seguin and Suzanne [92].)

For a physisorbed molecule that is largely unperturbed by the adsorption, surface vibrational modes close to this value should be observed. Mårtensson *et al.* reported for their molecular species on Ni(510) $\nu = 4339\,\text{cm}^{-1}$ (=538 meV) [78], and in our own HREELS study (high-resolution electron energy loss spectroscopy) of $H_2$ on Pd(210) we observed the H–H vibrational band around $3347\,\text{cm}^{-1}$ (415 meV) [77], already indicating a nonnegligible interaction with the underlying surface. For Ag(111) between 10 and 20 K, Gruyters and Jacobi performed HREELS experiments and could excite and identify rotational transitions of *ortho*- and *para*-hydrogen species [95]. The H–H stretching mode $\upsilon = 0 \rightarrow 1$ was reported for Ag(111) at $4150\,\text{cm}^{-1}$ (=514.5 meV), quite similar to the gas-phase value. Corresponding results had been obtained earlier by Avouris *et al.* for $H_2$/Ag(111) at 10 K, and again rotational transitions could be distinguished for *ortho*- and *para*-hydrogen [96]. Cu(100) surfaces interacting with hydrogen molecules at 10 K were examined by the group of Andersson [97], and we refer to work in which a vibrational analysis of the physisorbed $H_2$ molecules has been performed, whereby the rotational transitions $J = 0 \rightarrow 2$ and $1 \rightarrow 3$ could by excited by the low-energy electrons of the HREELS experiment. The inner-molecular stretching vibration $\upsilon = 0 \rightarrow 1$ was found at $4178\,\text{cm}^{-1}$ (=518 meV), pointing again to only negligible perturbations of the $H_2$ molecule by the Cu surface.

Another elegant possibility to differentiate between adsorbed $H_2$ and H consists in performing $H_2/D_2$ exchange experiments and check on isotopic scrambling, that is, the formation of hydrogen deuteride (HD) in the course of the adsorption reaction. The HD molecule can, of course, be formed only if either $H_2$ or $D_2$ or both species dissociate into atoms. The absence of a mass 3 contribution in certain low-temperature states in our hydrogen TD spectra from the Pd(210) surface [79, 80] proved the molecular nature of the respective states. It is on the same line that experiments probing the vibrational modes of adsorbed molecules are often performed also with deuterium instead of hydrogen, and the influence of the dual mass of $D_2$ makes the respective vibration appear at a frequency that is lower by the factor $\frac{1}{2}\sqrt{2}$. This can greatly help in identifying the respective vibrational modes.

### 36.2.3 The Dissociative Chemisorption of Hydrogen (Adsorption of H Atoms)

Physisorption with its shallow potential well is not the only way to bind hydrogen to surfaces. As we saw from Figure 36.3 (Section 36.2.1), there often exists a second (usually much deeper) binding potential closer to the surface, which is populated with H *atoms* – the respective process being called *chemisorption*, which is now in our focus. In the preceding section we listed this already as case 3 "dissociative chemisorption", and we use Figure 36.8 to illustrate how the dissociation process could actually take place: the $H_2$ molecule (for simplicity drawn as a dumpbell-like entity parallel to the surface) falls more and more apart into its two H atoms as it approaches the surface. Whenever hydrogen molecules come into contact with active surfaces at noncryogenic temperatures, this dissociative adsorption represents the usual and practically important interaction channel; it is certainly the

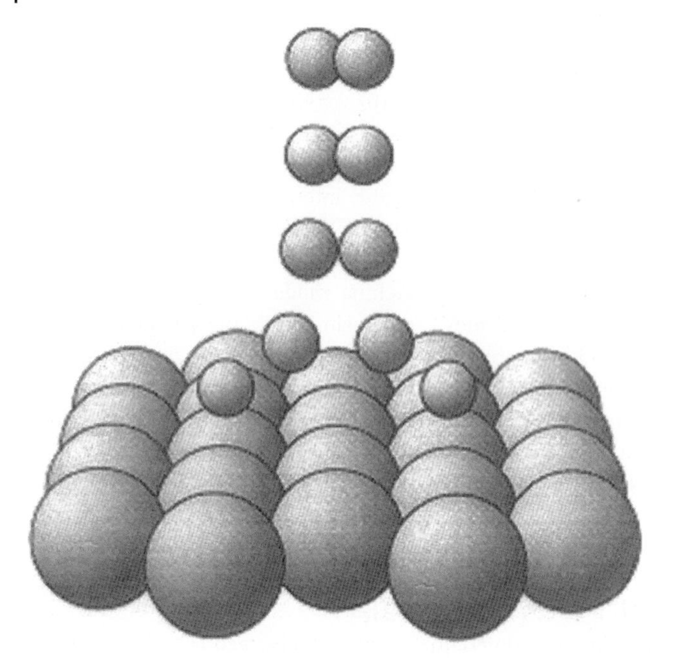

**Figure 36.8** Pictorial illustration showing how the dissociation of a hydrogen molecule could occur over a (111) metal surface.

decisive process in all heterogeneously catalyzed chemical conversions that imply transfer of hydrogen. At least on most of the TM surfaces the $H_2$ entity will break apart into the atoms *spontaneously*, while on the "free-electron" metals of groups 1–3 of the PT including the coinage metals Cu, Ag, and Au an additional activation energy is required to induce the dissociation reaction (activated adsorption, see Section 36.2.1). In Figure 36.9, we redraw and summarize these two cases, viz., activated and nonactivated hydrogen adsorption, using the one-dimensional Lennard–Jones potential energy diagram for convenience. Note that in case (a), that is, spontaneous dissociation, the barrier is below the energy-zero level, whilst in case (b), that is, activated dissociation, the barrier has the positive height of $E^*_{ad}$ above the energy-zero level.

#### 36.2.3.1 Some Fundamentals about the Dissociation Reaction

The orientation of the $H_2$ molecular axis parallel to the surface plane of Figure 36.8 is certainly a constraint, since the incoming molecule can be oriented with respect to the surface plane in any way; it could, for example, hit the surface "head on" while rotating in a kind of a cartwheel configuration, or it may indeed "land" there rotating parallel to the surface in a "helicopter"-like fashion. Actually, the hydrogen molecule has six degrees of freedom of translational, rotational, and vibrational motion while it is approaching the surface. The task now consists in appropriately considering the quantum mechanical interaction for all these possible molecular orientations relative to the surface (including its internal rotational and vibrational excitations) in

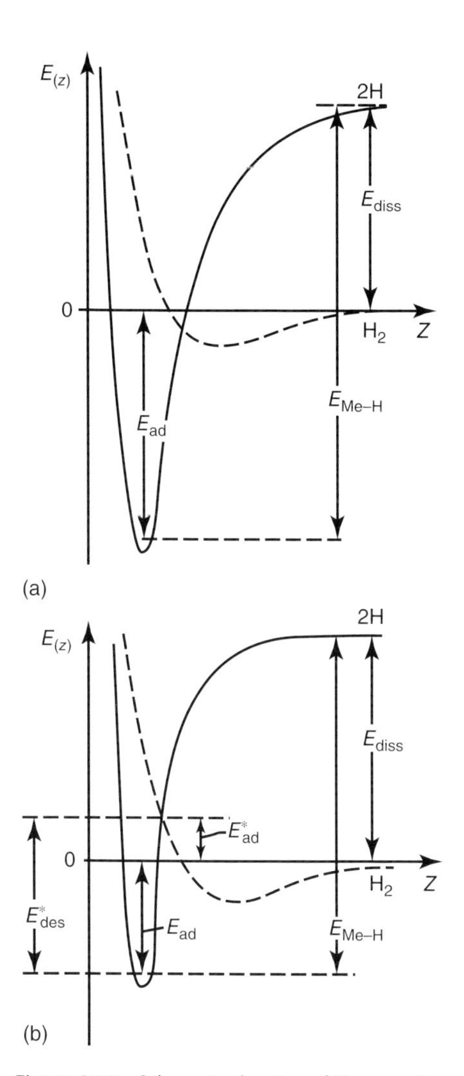


**Figure 36.9** Schematic drawing of the one-dimensional Lennard–Jones potential energy diagram for the cases of (a) spontaneous dissociation and (b) activated dissociation with barrier height $E^*_{ad}$.

order to find out the most favorable path for the dissociation to take place. This complicated problem is the essence of quite a number of theoretical considerations, in which the potential energy of hypersurfaces of the $H_2$–surface system are calculated, usually under the assumption of the Born–Oppenheimer approximation, where the electronic configurations instantaneously follow the nuclear coordinates. For more details about the various theoretical approaches and computational procedures, we recommend the book by Groß [52] as well as several review articles [47, 58, 98–100] and original papers, from which only a few are listed here [101–106].

In order to somewhat simplify the problem, we start off with the $H_2$ molecule axis oriented parallel to the metal surface (similar to the trapezoidal configuration of Figure 36.8) and use transition-state theory (TST) [107] to describe the dissociation reaction; that is, we follow the potential energy of the reaction

$$H_2 + 2\,Me \rightarrow \begin{pmatrix} H_2 \\ -Me - Me- \end{pmatrix}^{\ddagger} \leftrightarrow 2\,Me\text{–}H \qquad (36.9)$$

and confine our consideration to only two coordinates, namely the internuclear H–H bond extension denoted **y** and the vertical distance of the H–H bond to the metal surface, **x**. Far away from the surface, that is, for large $x$-values, $y$ equals, of course, the hydrogen gas phase value (0.74 Å). As $x$ gets smaller, the interaction forces become more and more noticeable; they kind of lug at the H–H bond and try to increase $y$ with the consequence that the H–H bond separation becomes larger for smaller $x$ until $x$ finally equals the (short) Me–H bond length, resulting in two independent H–Me surface complexes. The potential energy of the system as a function of both coordinates $x$ and $y$, $V(x,y)$, could best be illustrated by a three-dimensional plot with $V(x,y)$ as the $z$ coordinate and the distances $x$ and $y$ usually chosen as ordinate and abscissa, respectively. Our point of departure may be the physisorbed molecule trapped in the shallow van der Waals minimum far outside the surface. We approach the surface from the right to left, that is, beginning with large $x$ values. In moving to the left, the $H_2$ molecule first feels the attractive van der Waals forces, passes the respective minimum, but then has to go uphill along the broken line and reach a more or less pronounced barrier while always staying in the energetic "valley" like a hiker climbing a mountain pass. The height of this pass P actually represents the aforementioned transition state ‡. There the molecule has to "decide" to either pass the barrier and continue to move downhill until it reaches the deep minimum of the chemisorption potential (which means successful dissociation), or it may return to the physisorption minimum and remain intact as a molecule – the latter case corresponding to a reflection at the dissociation potential barrier. The path the reaction system takes can be more conveniently sketched in the paper plane by taking the projection of $V(x,y)$ in an $x,y$-diagram yielding the well-known "elbow" plots, in which the various energy values $V$ are indicated as contours of constant height. In Figure 36.10 we provide a schematic summary.

Incidentally, this figure discloses another complication, since the transition state P may be located either as a "late" barrier in the so-called "exit" channel quite close to the surface, that is, at small $x$ (Figure 36.10a), or as an "early" barrier in the "entrance" channel on the far right side (Figure 36.10b). Actually, the position of the activation barrier has major implications on how the molecule can best overcome the barrier. A high kinetic (translational) energy (along the $x$-coordinate) is certainly favorable if the barrier appears in the entrance channel, while surmounting a late barrier (requiring larger $y$-distances) will profit from an excitation of the intramolecular H–H stretching vibration. A high surface temperature would certainly support this excitation and improve the dissociative sticking, but not help to overcome the "early" barrier. Accordingly, the temperature dependence of the hydrogen sticking probability could provide some hints to the location of the activation barrier of dissociation.

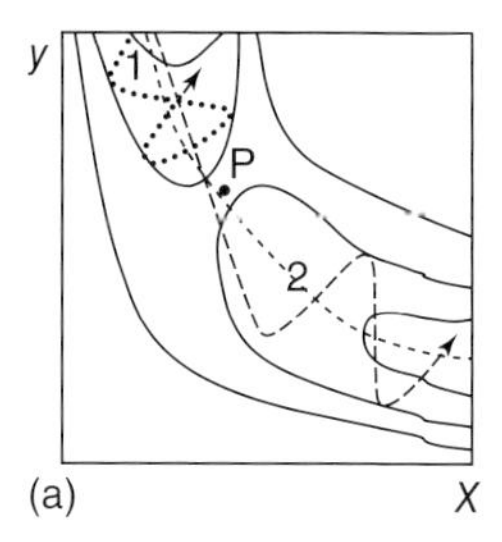

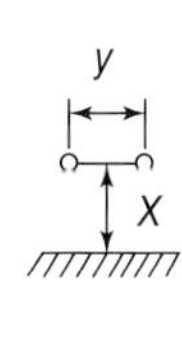

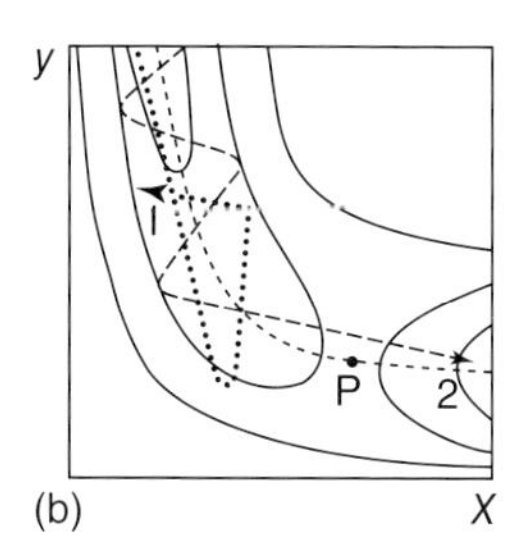


**Figure 36.10** Summary of the $H_2$ dissociation process in terms of transition-state theory using a two-dimensional representation of the interaction potential. Plotted is the intranuclear separation of the two H atoms ($y$) as a function of the distance ($x$) of the molecule from the surface plane. Some typical trajectories are indicated: (1) denotes an aborted dissociation attempt, where the $H_2$ molecule is reflected at the barrier and remains intact and (2) a successful dissociation event. The saddle point $P$, that is, the mountain pass of the hypersurface, can lie in the exit channel only close to the surface (small $x$, already stretched $y$), or in the entrance channel (at large $x$ and still small $y$ values) as indicated in the subfigures (a) and (b), respectively. The potential energy $V(x,y)$ is plotted as a projection in the plane of the paper.

A large number of experimental investigations have been carried out during the last decades in which supersonic molecular beams of hydrogen molecules were directed to metal single-crystal surfaces and the sticking probability for dissociative adsorption was followed as function of the translational, rotational, or vibrational energy of the impinging molecules [54–57, 108–111]. The first studies of this kind were performed already in the 1970s by Balooch, Stickney, and Cardillo [112, 113], who did hydrogen beam scattering on copper single-crystal surfaces and provided the first convincing evidence of an activation barrier for $H_2$ dissociation on this metal.

On the theoretical side, many researchers have studied the dynamics and energetics of the dissociation reaction over metal surfaces, in particular over Mg, Ni, Cu, and Pd surfaces [102–106, 114–116]. In most of this work, the potential energy surface of the system (consisting of a $H_2$ molecule approaching a surface section of defined geometry) is calculated as a function of the intranuclear H–H distance (mostly called "$r$" here and given in atomic units) and the molecule's distance to the surface, often named $z$. Again, the potential energy values are associated with the contour lines. Here, the convention is to plot $r$ on the abscissa and make $z$ the ordinate, with the consequence that one obtains indeed again an "elbow" plot, but it is turned around compared with Figure 36.10: now the trajectory of the molecule begins far on the left-hand side at short $r$ and large $z$ values and continues to the right, until the dissociation is complete at small $z$ and large $r$ values. Correspondingly, the "entrance channel" is now located in the upper left, where also an "early" barrier would be located, and the "exit channel" appears more on the right (as would a "late" barrier). To give an example, we reproduce two theoretical elbow plots from work by Engdahl *et al.* [102] in Figure 36.11, one for the Ni(110) and the other for the Cu(110) surface. Clearly, there is a small activation barrier in the entrance channel for $H_2$/Ni(110) that is easy to overcome, but a larger barrier in the exit channel for the Cu(110) surface – totally in line with the experimental observation of a nonactivated hydrogen

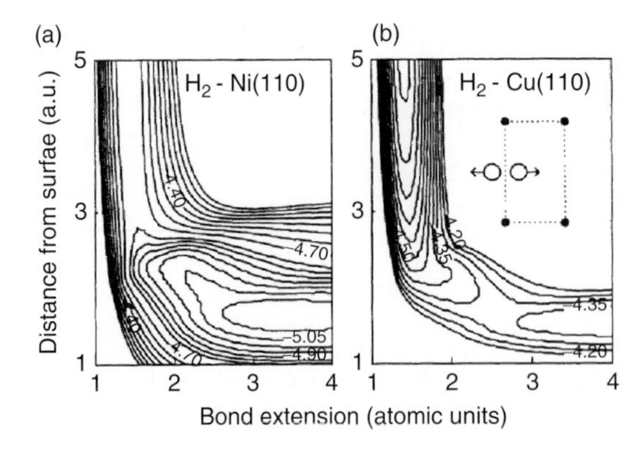


**Figure 36.11** Theoretical elbow plot from work by Engdahl *et al.* [102] for $H_2$ interacting with a Ni(110) (a) and a Cu(110) surface (b). The potential energy $V(r,z)$ is represented by contour lines. The hydrogen molecule approaches the surface with the axis parallel to the surface over a twofold bridge site and dissociates into the adjacent center sites.

adsorption with a high and $T$-independent sticking coefficient in the Ni(110) case, and thermally strongly activated H adsorption on Cu(110) with very low sticking probabilities for thermal hydrogen molecules.

#### 36.2.3.2 Spontaneous Dissociation and Sticking

We now turn to the practical consequences of the dissociation dynamics discussed above. The experimental quantity that reflects most of the dynamic properties is the sticking coefficient $s$ for *dissociative* adsorption. A high value of this quantity will lead to a rapid occupation of surface sites with H atoms, while a low sticking coefficient and a strong $s(T)$ dependence always indicate a large positive activation energy for adsorption, $\Delta E^*$. From the viewpoint of reaction kinetics, the dissociation is a second-order process, since a $H_2$ molecule breaking apart produces two individual H atoms, each of them requiring a separate adsorption site. Accordingly, the increase of coverage with time should, for nonactivated adsorption, obey the relation

$$+\frac{d\Theta}{dt} = \frac{\Phi s_0}{N_{\max}}(1-\Theta)^2, \tag{36.10}$$

in which $\Phi$ stands for the collision rate $= P(2\pi mkT)^{-\frac{1}{2}}$ ($P$ = gas pressure, $m$ = molecular mass, $k$ = Boltzmann's constant). In case of activated adsorption, the right side of Equation 36.10 has to be multiplied by a Boltzmann factor $f = \exp\{-\Delta E^*/(\mathrm{RT})\}$, which, depending on the magnitude of $\Delta E^*$, can substantially reduce the rate of adsorption. The main assumption in Equation 36.10 is that the two H atoms formed by dissociation statistically occupy an adsorption site until the whole surface is covered, which is often far from reality for several reasons. First of all, two H atoms adsorbed close to each other interact and may modify the potential energy hyperface, leading to changes in the probability of dissociation. Second, the appreciable lifetime of a molecular precursor state can lead to an almost constantly high $s$ value for small and medium coverages according to the Kisliuk model [76].

As pointed out in Section 36.2.2.1, the $H_2$ molecule can initially almost freely migrate across the surface and successfully "search" for a dissociation site. Only as the number of these sites decreases beyond a critical value, the sticking probability bends over and rapidly turns to zero, resulting in a convex shape of the $s(\Theta)$ curve. As an example, we present in Figure 36.12 the respective $s(\Theta)$ function for hydrogen adsorbing on a Pd(100) surface as determined by Behm *et al.* [117]. In the context of hydrogen interaction with palladium, it is perhaps worth mentioning that modern developments in low-temperature scanning tunneling microscopy (STM) have enabled scientists to directly watch $H_2$ molecules dissociating on a Pd(111) surface, with the somewhat surprising insight that the dissociation reaction requires at least three adjacent sites in order to be successful [118].

Furthermore, it is a well-known experimental fact that just the hydrogen adsorption rate can depend extremely sensitively upon the crystallographic and chemical cleanliness of a surface, resulting, for example, in self-poisoning effects due to repulsive H–H long-range interactions even for a sparsely covered surface. In these cases, the $s(\Theta)$ curve declines much more strongly with coverage and can take on a more concave shape. The strong influence of structural defects on hydrogen sticking has been convincingly demonstrated for the system H/Pt(111) by Poelsema *et al.* [119]. They took every effort to prepare a largely defect-free surface and discovered a steady decrease of the initial sticking coefficient with increasing crystallographic quality. In addition, they found an increase of $s$ with temperature, pointing to an activation barrier for adsorption at least for the defect-free Pt(111) surface. One may therefore conclude that the dissociation probability of hydrogen molecules is in many cases also tied to the presence of defects such as dislocations, steps, kinks, and so on. Concerning "chemical" surface impurities such as foreign atoms of metals or carbon, sulfur, oxygen, there have also been numerous studies in this area showing that

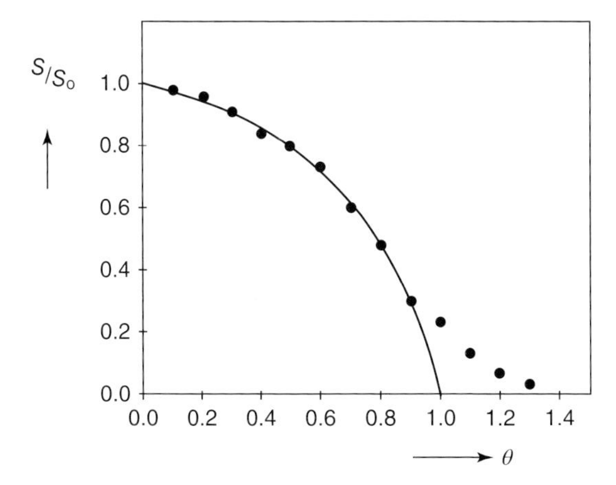


**Figure 36.12** Coverage dependence of the hydrogen sticking probability on a Pd(100) surface that was exposed to hydrogen molecules at $T = 170$ K. Indicated is a first-order precursor model (full line) that describes the reality up to about 1 ML coverage. (After Behm *et al.* [117].)

the adsorption rate (and binding energy) of hydrogen adsorbed in neighboring sites can be strongly modified – a phenomenon that is of great importance in catalytic chemistry.

Another issue in this context is worth mentioning, the so-called spill-over effect. On bimetallic surfaces consisting, for example, of small islands of an active metal embedded in larger areas of an inactive metal, the H atoms are more strongly bound to the active metal but merely weakly held on the inactive metal. Prior to adsorption, the $H_2$ molecules have undergone dissociation preferentially on the active islands, but once the chemically reactive H atoms are formed they can diffuse, that is, spill over, to the areas of the inactive metal where they are at disposition for surface reactions. Because of their lower adsorption energy on the respective sites, they are more mobile and reactive than the H atoms bound to the active metal. Examples of spill-over systems could be alloy surfaces of a TM mixed with Cu or Ag, such as $PtCu_3$ [120], or bimetallic surfaces of immiscible metals such as Cu on Ru surfaces [121] where electronic effects such as the so-called ensemble effect could be studied. Here, only an *ensemble* of coherent surface atoms, often with a certain geometrical structure, provides an appropriate basis for a chemical surface reaction to take place – in the simplest case, this "reaction" could be just the dissociation of a molecule. Particularly revealing studies have been performed in Wandelt's laboratory utilizing the $Cu_3Pt(111)$ surface [120]. In this ordered alloy, individual Pt atoms are surrounded only by Cu atoms. It could be shown that molecular hydrogen dissociates exclusively at the isolated Pt atoms. After complete occupation of the Pt sites, also Cu atoms became covered with hydrogen. A later theoretical investigation of the same system basically confirmed the experimental observations [122]. For more details on spill-over phenomena and their catalytic relevance, we refer to the pertinent literature [123–125].

In concluding this section on hydrogen sticking, we would like to point to a very helpful survey article on gas adsorption kinetics, sticking coefficients, precursor states, and so on, by Morris, Bowker, and King, in which the reader can also find many data about experimental sticking coefficients along with illustrating figures that provide all kinds of sticking (coverage) functions $s(\Theta)$ [126]. In Table 36.3 we list some experimentally determined sticking coefficients for dissociative hydrogen chemisorption on various metal single-crystal surfaces.

#### 36.2.3.3 **Activated Dissociative Adsorption**

On surveying the extensive literature concerning activated dissociative hydrogen adsorption, one immediately realizes that most of the work has been performed with copper surfaces, and at least for the low-Miller-index planes, activated adsorption seems to be safely confirmed. An excellent overview of the relevant publications until 1991 is provided by Michelsen and Auerbach [149].

The molecular beam experiments by Balooch *et al.* [112, 113] had clearly confirmed that hydrogen adsorption on copper surfaces is thermally activated, and we recall again Figure 36.9b, which displays the respective positive activation barrier. From their measurements with variable beam energy, the authors concluded activation barrier heights of about 10–20 kJ/mol. In the preceding section, we stated

**Table 36.3** Some experimental values for the initial sticking coefficient of hydrogen on metal surfaces.

| Surface | Surface temperature (K) | Initial sticking coefficient $s_0$ | References |
|---|---|---|---|
| Fe(100) | 300 | $1.5 \times 10^{-3}$ | [127] |
| Co(0001) | 110 | ~0.1 | [128] |
| Co(10-10) | 100 | ≥0.8 | [129] |
| Ni(100) | 120 | 0.06–0.1 | [130] |
| Ni(110) | 100 | 0.9–1.0 | [131] |
| Ni(110) | 140 | 0.96 | [132] |
| Ni(111) | 155 | 0.1 | [54] |
| Cu(100) | 300 | 0.00001 | [55] |
| Cu(110) | 90 | 0.05 | [133] |
| Cu(111) | 300 | $<10^{-7}$ | [56] |
| Ru(0001) | 95 | 0.58 | [134] |
| Ru($10\bar{1}0$) | 100 | ~1.0 | [135] |
| Rh(100) | 100 | 0.53 | [136] |
| Rh(110) | 80 | 0.97 | [137] |
| Rh(111) | 100 | 0.65 | [138] |
| | 180 | 0.01 | [139] |
| Pd(100) | 170 | 0.5 | [117] |
| Pd(110) | 130 | 0.7 | [140] |
| Pd(111) | 100 | 0.5–0.6 | [141] |
| Ag(111) | 110 | ~0 | [142] |
| Re($10\bar{1}0$) | 100 | 0.7 | [143] |
| W(100) | 100 | 0.9 | [127] |
| W(110) | 90 | 0.22 | [144] |
| W(111) | 80 | 0.85 | [127] |
| Ir(111) | 100 | $7 \cdot 10^{-3}$ | [145] |
| Pt(100)(1 × 1) | 300 | 0.17 | [146] |
| Pt(100)-hex | 150 | 0.06 | [146] |
| Pt(110)(1 × 2) | 170 | 0.8 | [147] |
| Pt(111) | 240 | $6 \cdot 10^{-2}$ | [148] |

that an activation barrier can be located in the "entrance" or "exit" channel, and the $H_2$ molecule needs translational or vibrational excitation to overcome the barrier. An early barrier is likely if an increasing thermal velocity of the incident beam of $H_2$ molecules leads to an improvement of the sticking. In the following, we refer to the work by Anger *et al.* [55], who produced supersonic hydrogen beams of different temperatures that were scattered at the three low-index Cu surfaces (100), (110), and (111). The sticking coefficient was measured as well as the angular distribution of the scattered molecules. With the nozzle kept at 300 K (corresponding to a beam energy of ~70 meV), a hardly measurable sticking coefficient of the order of $10^{-5}$ was found for all three Cu surfaces. Only when the beam energy reached values >200 meV, noticeable hydrogen sticking occurred and $s$ increased rapidly with increasing beam energy, as shown in Figure 36.13, with slight differences

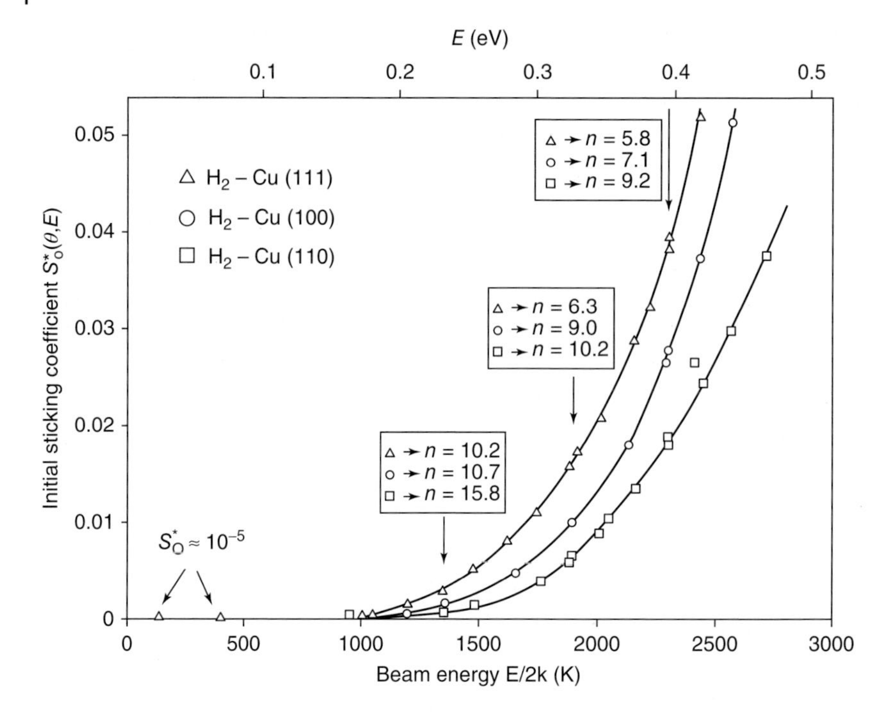


**Figure 36.13** Initial sticking probability for a monoenergetic hydrogen beam on low-index Cu surfaces as a function of the beam energy. The temperature of the surface was 190 K. The values *n* indicate the exponent (and hence, the sharpness) of the cosine distribution of the off-scattered $H_2$ molecules for the three Cu surfaces with (111) orientation (triangle symbols), (100) orientation (open circles), and (110) orientation (open squares) at the beam energies indicated by the arrows. At beam energies <0.1 eV, the sticking coefficient is merely ~$10^{-5}$. (After Anger *et al.* [55].)

between the three crystal face orientations. Figures of this kind can also be found in Michelsen's and Auerbach's review article [149].

Somewhat later, measurements of $H_2$ interaction with Cu(100) were performed by Rasmussen *et al.* [150]. They deserve particular attention, because here hydrogen pressures up to 5 bar were applied while the Cu surface was kept at temperatures between 218 and 258 K. The barrier heights for $H_2$ and $D_2$ adsorption were determined as $48 \pm 6$ and $56 \pm 8$ kJ/mol, and values of $2.5 \times 10^{-13}$ and $7.8 \times 10^{-14}$ were found for $H_2$ and $D_2$ sticking at 242 K, respectively. Furthermore, from the pressure dependence of the sticking coefficient the authors argued qualitatively that vibrational energy played an important role in overcoming the barrier.

From H-induced WF change measurements with *stepped* Cu surfaces, in particular the Cu(311) surface, Pritchard *et al.* [151] deduced at least a partial spontaneous dissociative adsorption with a nonnegligible sticking probability. For the low-index surfaces, on the other hand, the authors confirmed the activated nature of adsorption. Using isotherm measurements, isosteric heats of H chemisorption with values ~40 kJ/mol were determined.

Because of their chemical similarity with copper, the two other coinage metals (Ag and Au) have also frequently been investigated concerning their affinity to adsorb hydrogen. Indeed, also on Ag and Au surfaces the hydrogen dissociation is strongly activated, and a noticeable sticking occurs only for higher energies of a molecular hydrogen beam. However, there is one important difference with respect to Cu: while the depth of the chemisorption potential of the H/Cu system is still large compared to the physisorption potential and located well below the energy zero line (meaning that hydrogen adsorption on Cu surfaces is exothermic), the binding potential of the H + Ag and H + Au interaction appears to be much smaller (although still closer to the surface) than the van der Waals minimum. This is schematically sketched in Figure 36.14. Thermodynamically, the formation of a Ag–H and a Au–H adsorptive bond is endothermic. Therefore, most of the experimentalists studying these systems have utilized atomic hydrogen sources to bring the hydrogen on the surface and keep it there at sufficiently low temperatures. For more details, we refer to the work by Sprunger and Plummer, who carefully studied the H-on-Ag(110) system [152] and Iwai *et al.* who examined the Au(100) surface [153].

In summary, it appears as if activated adsorption is characteristic of smooth, defect-free, low-index Cu surfaces, though on stepped or defected surfaces also spontaneous dissociation may occur to some extent. This is also in line with the already mentioned Helium atom scattering experiments reported by Poelsema for the H/Pt(111) system [119], where very low sticking coefficients were found for the defect-free Pt surface, which could, however, increase by orders of magnitude for surfaces having defects.

Although Ni is certainly considered the TM with a high tendency for spontaneous dissociative hydrogen chemisorption (proven by its role in catalytic hydrogenation reactions), surprisingly small (and energy-dependent) sticking probabilities have

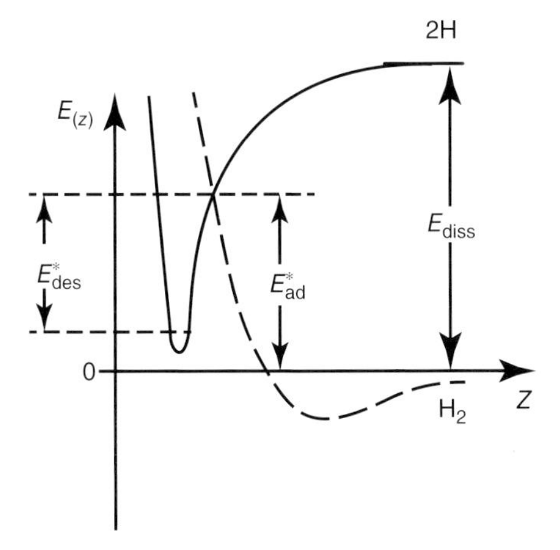


**Figure 36.14** Schematic drawing of the one-dimensional Lennard–Jones potential energy diagram for the cases of activated $H_2$ dissociation (barrier height $E^*_{ad}$) over silver and gold surfaces. Note that the depth of the chemisorption potential is above $E = 0$, which makes the adsorption endothermic.

been reported in some cases for the crystallographically smooth (100) and (111) surface orientations [130, 154]. Molecular beam (MB) experiments revealed that increasing the beam energies of the hydrogen molecules led to a pronounced rise of the sticking coefficient. In their MB study, Robota *et al.* [54] arrived at the same conclusions for the Ni(111) surface, whereas the Ni(110) surface orientation with its more corrugated surface revealed high and energy-independent sticking coefficients indicative of spontaneous dissociative adsorption. In summary, it appears as if, for the close-packed Ni and Pt surfaces, indeed a (small) activation barrier exists for dissociative hydrogen adsorption. From a chemical-catalytic viewpoint, these activation barriers are, however, not of much significance, since on any practical Ni catalyst surface there are certainly countless patches with (110) or any high-Miller-index orientation, if not crystallographic defects, that promote spontaneous dissociation with sufficient velocity.

### 36.2.4
### The Energetics of Hydrogen Adsorption and Desorption

We now depart from the following scenario: the dissociation of the $H_2$ molecule is accomplished, and the H atoms reside in their most favorable adsorption sites at the bottom of the chemisorption well, and the associated adsorption energy has been released. To consider the overall energetics, we refer again to Figure 36.9a. Let us imagine that we supply the $H_2$ molecule far away from the surface (large $z$) with the dissociation energy of $E_{diss} = 435\,kJ/mol$ and bring the two individual H atoms then close to the surface. Because of their unpaired electron, they interact strongly with the surface atom(s) underneath and form, in each case, a Me–H bond, whose energy is denoted as $E_{Me-H}$. As can be seen, the overall depth of the potential associated with the H–Me bond energy just equals the sum of the dissociation energy and the depth of the chemisorption potential $E_{chem}$ (the latter is referred to the energy zero line). Since the Me–H bond energy is gained twice, the energy balance yields immediately

$$2E_{Me-H} = E_{chem} + E_{diss}, \qquad (36.11)$$

from which the energy of a single metal–H bond can easily be determined, provided the chemisorption (= adsorption) energy $E_{chem}$ is known. We shall see further below how this quantity is experimentally accessible. It may be of interest to compare the dissociation energy of an Me–H surface complex with the bond energy of the respective binary hydride molecule. Both kinds of data are available in the literature. While Greeley and Mavrikakis [48] and Ferrin *et al.* [51] have compiled the latest theoretical and experimental values of H–surface binding energies, the respective experimental and theoretical energies for binary hydride molecules (either in neutral or singly ionized form) can be found in Refs [155, 156]. In Table 36.4 we compare the data for some selected metal–hydrogen systems.

At a first glance, the differences between the various bond energies of the binary hydrides scatter from 86 up to 352 kJ/mol, while the metal–H bond energies for the chemisorption systems range only from 233 to 266 kJ/mol. This means that the H–surface bond energies and bond types are chemically much more

**Table 36.4** Comparison of H chemisorption energies and bond dissociation energies for binary hydride molecules.

| System | Bond dissociation energy of the binary hydride molecule at 298 K (kJ/mol); Refs [155, 156] | Energy of the metal–hydrogen bond in the chemisorption complex at 298 K (kJ/mol) | References |
|---|---|---|---|
| Ag-H | 226 | 239 | [152] |
| Al-H | 285 | 253 | [157] |
| Au-H | 314 | 240 | [158] |
| Be-H | 226 | <264 | [159] |
| Ca-H | 168 | — | [155, 156] |
| Cr-H | 280 | — | [155, 156] |
| Cu-H | 280 | 243 | [150] |
| Ga-H | <274 | — | [155, 156] |
| K-H | 183 | — | [155, 156] |
| Mg-H | 197 | — | [155, 156] |
| Na-H | 201 | — | [155, 156] |
| Ni-H | 289 | 266 | [160] |
| Pt-H | 352 | 251 | [161] |
| Ti-H | ~159 | 233 | [162] |
| Zn-H | 86 | 241 | [163] |

similar than the bond situation predominating in the isolated hydride molecules. Furthermore, a direct correspondence between the Me–H surface bond and the hydride dissociation energy exists only in a few cases. At least Ni, Cu, Ag, and Al show some similarity, while larger deviations are evident for Pt, Au, or Zn. These discrepancies are likely caused by the fact that, in the case of a metal–hydrogen adsorption complex that is embedded in a surface, usually a number of metal atoms take part in the formation of a H–Me bond. Accordingly, the electronic configuration of the chemisorption complex will probably differ from that of an isolated hydride molecule. Nevertheless, at least the principal trends and the gross order of magnitude are more or less reflected in the energy values of Table 36.4.

We recall that it is necessary to distinguish between the single-particle and the multiparticle interaction also when considering the energetics of adsorption. This means that the first $H_2$ molecules arriving and dissociating at a metal surface will usually face a different local environment than a molecule that arrives on an already partially or largely covered surface – the main reason being lateral interactions between neighboring adsorbed H atoms that come increasingly into play at higher H surface concentrations. In the following we will, therefore, first consider the single-particle interaction and then turn to the coverage dependence of the adsorption energy.

#### 36.2.4.1 **The Adsorption Energy in the Low-Coverage Limit**

The determination of adsorption energies of hydrogen on surfaces can be performed in many different ways. Because of spatial limitations, we cannot devote too much attention to this experimental matter. We simply state that either thermodynamic (equilibrium) measurements can be employed or the adsorption energy is deduced from the adsorption/desorption kinetics (more details can be seen in Ref. [164] and other textbooks on surface chemistry).

In the first case, usually adsorption *isotherms* are measured, that is, the hydrogen coverage is monitored as a function of pressure for constant temperatures: $\Theta_H = f(p)_T$ (equally well isobars can be taken: $\Theta_H = f(T)_P$). Under dynamic equilibrium, the chemical potentials $\mu_I$ of the gas phase and the adsorbate phase are equal and remain equal, and the thermodynamic analysis based on the Clausius–Clapeyron formalism [164, 165] yields the so-called isosteric heat of adsorption $Q_{st}$ as a differential quantity, which for nonactivated adsorption is equal to the depth of the chemisorption potential $E_{chem}$. The extrapolation of $Q_{st}$ to $\Theta_H = $ zero yields the initial adsorption energy $E_0$ (zero-coverage limit). The experimental difficulty lies in reliably measuring the associated hydrogen coverages if small-area single-crystal surfaces ($A \approx 1\,cm^2$) are used: because of their well-defined geometry, these samples provide the most reliable information, but they have the disadvantage that the number of H atoms at the monolayer concentration amounts only to about $10^{15}$ (equivalent to merely $\sim 10^{-9}$ mol), and extremely sensitive experimental techniques are required to directly determine the amount of adsorbate particles. Therefore, mostly *indirect* methods are used in which a physical property of the adsorbed particle is measured that correlates uniquely with its concentration. A typical quantity of this kind is the dipole moment of the adsorbate complex. The sum of the surface dipoles then correlates with the overall adsorbate-induced WF change ($\Delta\Phi$) often in a unique way, and WF changes can often be sensitively measured. In Section 36.2.8 we will return to the issue of H-induced WF changes. Actually, in many cases the H-induced $\Delta\Phi$ has been used for the determination of $Q_{st}$. We refer to the early work by Conrad *et al.* for Pd [166] and by Christmann *et al.* [160] for Ni single-crystal surfaces. Figure 36.15 shows an example for experimental adsorption isotherms of hydrogen on Ni(100) measured by the H-induced WF change. Horizontal cuts represent constant coverages that are achieved at different pairs of hydrogen pressure and sample temperature. The initial adsorption energy (equal to the isosteric heat of adsorption) for hydrogen on Ni(100) came out as 96.2 kJ/mol [160]. Of course, it is also possible to *directly* measure the adsorption energy, for example, by calorimetry (in this case the samples are exposed to larger "portions" of hydrogen gas, and the *integral* heats that are released during the adsorption are collected). However, according to what has been said above, one needs here an extremely sensitive setup. This can be realized with evaporated polycrystalline films, which have a larger surface area and a higher internal roughness than single-crystal samples; the respective calorimetric technique had been described already in the 1970s by Wedler [167]. In their experiments, they used a glass calorimeter and low temperatures to measure the integral heat of hydrogen adsorption on polycrystalline Ni films [168]. Only in recent years has it become possible to measure calorimetric heats also from small-area single-crystal surfaces using sophisticated sensitive techniques, where, among others, a pyroelectric foil is pushed against the single-crystal sample [169, 170] and the change in the charge distribution is measured.

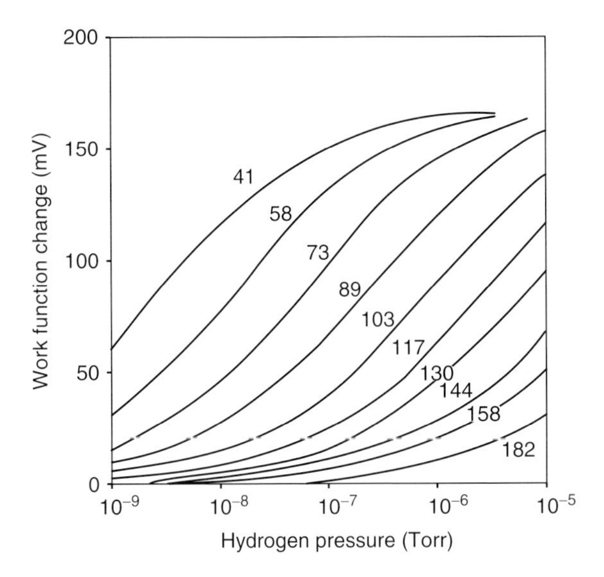


**Figure 36.15** Example of experimental adsorption isotherms for the adsorption of hydrogen on the Ni(100) surface. Plotted is the H-induced work function change $\Delta\Phi$ against the $H_2$ pressure (Torr) for various constant temperatures ranging from 41 to 182 °C. In the coverage range studied, $\Delta\Phi$ is proportional to the coverage $\Theta$. (After Ref. [160].)

In the second case of deducing adsorption energies from *kinetic* measurements, mostly TDS has been used (see Section 36.2.2.2 and Figure 36.6). The position of the desorption maxima on the temperature scale correlates with the desorption energy, and the number of TD peaks indicate different hydrogen binding states. For hydrogen-on-metal systems, various thermal desorption spectra have been collected and reproduced in the article by Morris *et al.* [126]. In order to determine the desorption energy accurately, a careful experimental setup as well as a reliable procedure to evaluate the data is required; for more details, we refer to the fundamental articles by Redhead [85] treating the desorption energy as coverage-independent, and by King [171] who explicitly considered coverage dependences by using a complete line shape analysis of the TD curves. Furthermore, we recommend the methodological comparison by de Jong and Niemantsverdriet [87]. A typical experimental example for a series of hydrogen thermal desorption spectra is presented in Figure 36.16, again for the system H/Ni(100); the desorption energy determined from a second-order analysis agreed well with that from the equilibrium measurements [172]. The second-order kinetics is indeed expected for associative hydrogen desorption, where the rate-limiting step is the recombination of individual H atoms and their fusion to a molecular transition state. However, quite a number of hydrogen desorption spectra reveal deviations from a clear second-order kinetics, for which the reasons are not always known. The existence of crystallographic surface defects may be playing a similarly crucial role as during the sticking, that is, the adsorption process, and according to microscopic reversibility,

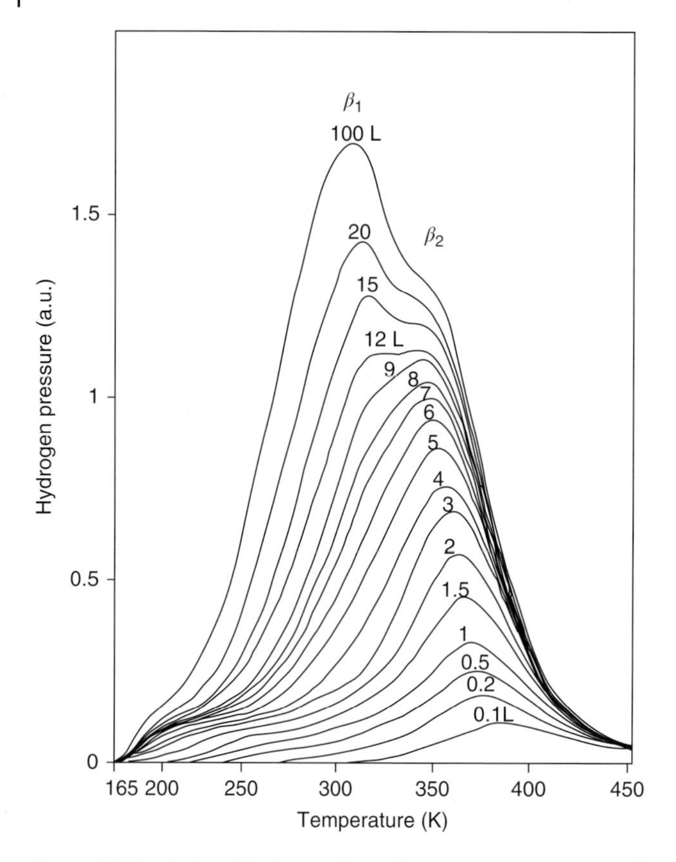


**Figure 36.16** Series of hydrogen thermal desorption spectra from a Ni(100) surface that was exposed to $H_2$ at 120 K and then linearly heated at a rate of 10 K/s. Parameter of the curves is the hydrogen exposure in (Langmuir) (1 L = $10^{-6}$ Torr × 1 s). The second-order $\beta_2$ state between 350 and 400 K saturates after ~12 L and is followed by a $\beta_1$ state at ~300 K. (After Ref. [172].)

also the desorption kinetics could be affected [173]. Practical surfaces, regardless of whether they have single-crystalline or polycrystalline structure, usually exhibit a number of such defects (steps, kinks, dislocations, point defects, etc.), which easily act as traps or active centers for hydrogen dissociation and chemisorption. In most cases, these sites provide somewhat higher adsorption energy than the "nominal" sites of the plane surface. In the presence of defects, one would actually measure higher, sometimes substantially higher, values for the initial adsorption energy. We refer to experiments in which the adsorption energies of hydrogen on flat (111) and periodically stepped (111) surfaces of Pd [166] and Pt [174, 175] have been determined. For both metals, a marked initial increase of $E_{\text{chem}}$ has been found by about ~10 kJ/mol compared to the "flat" (111) surfaces, the reason being the higher coordination number of hydrogen adsorbed at steps or kinks. The actual distribution of H atoms adsorbed on the plane surface and at the step sites is then governed by the energy difference between the sites and the thermal energy $kT$

according to Boltzmann statistics. In order to accurately and reliably determine the initial heats of hydrogen adsorption on a clean surface, it is essential to carefully control its structural (crystallographic) quality. From numerous studies of hydrogen adsorption, it also turns out that this particular interaction is particularly sensitive also to chemical surface impurities, for example, foreign atoms such as carbon, sulfur, or oxygen [176].

We now turn to a brief compilation and discussion of the initial values of hydrogen adsorption energies. In Table 36.5, we first present a listing of some experimental values of this quantity, organized in the form of the PT; more extensive collections of respective data can be found in Refs [48, 50].

Table 36.5 reveals that there is a fairly large range over which the $E_{chem}$ values fluctuate. Quite low values (<50 kJ/mol) are found for Ag or Zn, medium values (around 80–100 kJ/mol) for most of the group 8–10 metals, wherein the group 8 elements usually yield somewhat higher values), and quite high energies for the group 5–7 elements with values ranging from 120 to 176 kJ/mol for Mo and W, respectively. However, a gross interpretation of the overall numbers is not particularly illustrating for the following reasons. Because of the different mechanisms of the H–surface interactions (activated or nonactivated chemisorption) and the appearance of different binding states, one should preferably compare chemically similar materials and, if possible, analogous crystal face orientations. In this way, TMs of the platinum group could be compared, the coinage metals Cu, Ag, and Au; or the "free-electron" metals of the groups 1, 2, 3, 12, and 13 of the PT (which form adsorptive bonds only if hydrogen is offered in the reactive atomic form, the reason being the (sometimes substantial) activation barrier for dissociation (see Section 36.2.3.3). Another class of interest is elemental semiconductors, Si and Ge in the first place, but also C (in the form of graphite, graphene, or diamond), which exhibit an often specific hydrogen–surface interaction scenario compared to metals, owing to their different electronic structure and the mostly *covalent* nature of the bonds formed. With these materials, quite high adsorption energies have been reported: 260 kJ/mol for Si and 145 kJ/mol for Ge [50]. Furthermore, the high reactivity of H atoms

**Table 36.5** Some selected (average) values for experimental initial adsorption energies of hydrogen (kJ/mol).

| Group 3 | Group 4 | Group 5 | Group 6 | Group 7 | Group 8 | Group 9 | Group 10 | Group 11 | Group 12 |
|---|---|---|---|---|---|---|---|---|---|
| 21 | 22 | 23 | 24 | 25 | 26 | 27 | 28 | 29 | 30 |
| Sc | Ti | V | Cr | Mn | Fe | Co | Ni | Cu | Zn |
| | | | | | 88–110 | 60–80 | 81–96 | 45–96 | 47 |
| 39 | 40 | 41 | 42 | 43 | 44 | 45 | 46 | 47 | 48 |
| Y | Zr | Nb | Mo | Tc | Ru | Rh | Pd | Ag | Cd |
| | | 62–110 | 83–145 | | 80–125 | 70–109 | 71–102 | 25–45 | |
| 57 | 72 | 73 | 74 | 75 | 76 | 77 | 78 | 79 | 80 |
| La | Hf | Ta | W | Re | Os | Ir | Pt | Au | Hg |
| | | | 130–167 | 90–130 | | 75–100 | 50–70 | <50 | |

Values taken from Ref. [50].

with these materials often causes accompanying surface rearrangement processes (reconstructions), where the surfaces take on new lateral periodicities (see Section 36.2.4.3).

And finally, as we have seen, for one and the same material also the surface *geometry* can significantly influence the strength of an adsorptive bond, a phenomenon known as *crystal face specificity*. Usually, crystallographically more "open" surfaces bind H atoms somewhat more strongly than the densely packed surfaces. Note that, on some of the latter surfaces, even small activation barriers for dissociation can exist, which can sometimes seriously reduce the rate of adsorption. On the other hand, an inspection of the adsorption energy values for H on Ni, Pd, or Pt reveals that the differences between the densely packed faces (100) and (111) are not pronounced, and it appears as if in many cases there is no particularly strong crystal face specificity for hydrogen chemisorption.

In order to better understand the distinct catalytic properties of surface reactions involving hydrogen, it is common to plot the adsorption energy $E_{chem}$ or the metal–hydrogen bond energy $E_{Me-H}$ versus the atomic number of the respective element and to correlate these quantities, for example, with the cohesive energy of the solids and their heats of fusion or sublimation, respectively. Often, these systematic comparisons can shed light on peculiar catalytic properties of the individual chemical elements and might explain why Ni and Pd exhibit different methanation activity or why it is advantageous to use Co instead of Ni in order to preferentially obtain oxygen-containing hydrocarbons in a Fischer–Tropsch synthesis. In order to facilitate a surface transfer of H atoms to co-adsorbing target molecules (e.g., unsaturated hydrocarbons), the adsorption energy of the H atom should neither be too high nor too low – in the first case, its lateral mobility would be restricted, thus inhibiting the transfer; and in the second case, the reduced thermal stability of the adsorptive bond causes competing thermal desorption and, hence, a loss of reactive species. Quite often, so-called volcano curves are obtained, if the turnover numbers (TONs) for a certain catalytic reaction (hydrogenation, hydrodesulfurization, etc.) are plotted against the atomic number $Z$ of the TM under consideration (TON is a measure for the number of moles of reactant that a mole of catalyst can convert into the desired product before becoming inactive). Apparently, metals with lower $Z$ bind too weakly and those with higher $Z$ too strongly, while the optimum adsorptive bond strength is somewhere in between. Accordingly, there is often a maximum of the TON in the regime of the metals of group 8–10 of the PT. Ni, Ru, Rh, Pd, and Pt are known to be catalytically more active for hydrogenation reactions than metals such as Cu, Cr, W, and Re. For more details about this interesting subject, we recommend some more basic publications ranging from the early 1970s till today [48, 177–180].

#### 36.2.4.2 **The Coverage Dependence of the Adsorption Energy**

We recall the distinction between single- and multiparticle interactions made in Section 36.2.1. We now explicitly consider the adsorption of an ensemble of H atoms on a single-crystal surface. In the simple Langmuir model, there should be no difference in the overall energetics as long as each H atom can occupy an identical adsorption site, since the very first and the very last adsorbing particle should experience exactly the same interaction. In a simple view, we now consider the variation of the chemisorption potential parallel to the surface, and would at first expect a homogeneous variation in the $x$ and $y$ directions with the periodicity of the adsorption sites, according to the crystallography of the surface. This is schematically illustrated in Figure 36.17a. In order to move an adsorbed particle from one site to an adjacent one, the diffusion energy $E_{\mathrm{diff}}$ must be spent, unless tunneling provides an easier diffusion path (see Section 36.2.5).

However, in reality one finds that even for sparsely covered surfaces the available adsorption sites are no longer equivalent, the reason being particle–particle interactions, leading to a situation in which adjacent adsorbed particles repel or attract each other. This is depicted in Figure 36.17b, where the potential energy situation of Figure 36.17a is superimposed and modulated by a *repulsive* pairwise interaction potential of two adsorbate particles. Figure 36.17c depicts the same situation for *attractive* interactions. With voluminous adsorbates such as the noble gases Kr or Xe, *direct* orbital–orbital repulsions usually lead to a reduction of the potential energy, because it costs energy to surmount the Pauli repulsion between the two closed-shell entities. With the comparatively small H atom, however, another type of interaction is common, the so-called *indirect* interactions, which likewise have a quantum chemical origin. Naively speaking, an already adsorbed H atom changes the electronic charge distribution of the metallic substrate around its adsorption site due to utilization of the metal electrons for the formation of the Me–H chemisorption bond. In many cases, the respective disturbances are not restricted to the direct neighborhood but may affect the adsorption sites relatively far away from the occupied site in question. The responsible indirect interactions have been dealt with in many chemisorption theories. Depending on the distance from the site in question, they can be attractive or repulsive and therefore lead to a reinforcement or a weakening of the adsorptive bond. The existence of the indirect interactions was first concluded by Koutecký in the 1958 [181] and later confirmed many times, among

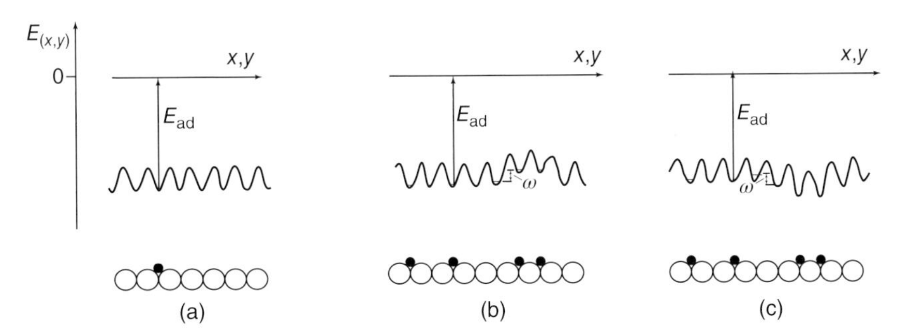


**Figure 36.17** Modulation of the potential energy of adsorption $E(x,y)$ parallel to the surface by the pairwise interaction potential between neighboring adsorbed particles. Panel (a) shows the situation for a single particle (no lateral interactions), while Panels (b, c) refer to repulsive and attractive particle–particle interactions, respectively.

others by Grimley [182, 183], Einstein and Schrieffer [184], Lundqvist *et al.* [185], and Muscat [186, 187].

Lateral adsorbate–adsorbate interactions manifest themselves in the coverage dependence of the heat of adsorption: Other than predicted by the Langmuir model, the adsorption energy is not independent of coverage but usually decreases with $\Theta$ – slowly at small or medium coverages and strongly for elevated adsorbate concentrations, reaching only small values near saturation. As an example, we present in Figure 36.18 the $E_{\mathrm{chem}}$ ($\Theta$) dependence for the H-on-Ni(111) system measured volumetrically by Rinne [188]. Clearly, the decrease of $E_{\mathrm{chem}}$ with increasing coverage is evident, wherein, within the first ~25% of $\Theta_{\mathrm{max}}$, the adsorption energy remains constant (the H atoms are sufficiently far away from each other to be able to interact), but steadily decreases at higher surface concentrations. In some (rare) cases, attractive interactions can predominate, and one finds an initial increase of the adsorption energy with coverage, resulting in phenomena such as condensation and island formation. The interaction of hydrogen with the Ni(110) surface is an example here [160, 189]. Within a certain pressure interval, at a constant temperature, the uptake of H atoms increases steeply due to 2D condensation. In this situation, it is energetically more favorable for an arriving H atom to "dock" to an already existing island or chain rather than to build a new individual adsorption complex.

We just mention here that it is also the lateral H–H interactions that are responsible for the formation of hydrogen phases with long-range order, and from thermodynamic measurements and analyses of the stability of such phases it is possible to get distinct estimates on their magnitude. It turns out that they range around values that amount to about 1/10 of the adsorption energy (see Section 36.2.6).

Another aspect is to be taken into account when dealing with coverage dependences, namely the occurrence of distinct binding states (which are often apparent

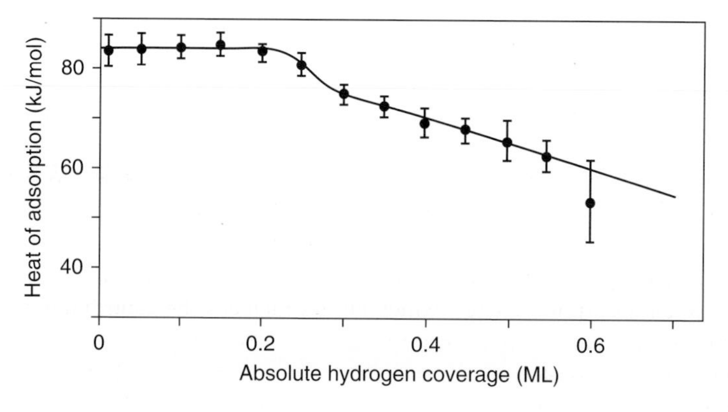


**Figure 36.18** Coverage dependence of the isosteric heat of adsorption of hydrogen on the Ni(111) surface. The respective measurements have been performed volumetrically in an all-glass ultrahigh-vacuum apparatus [188], and coverages are given on the absolute scale. Only at lower coverages the heat remains constant; at higher coverages it decreases as a result of repulsive H–H interactions.

especially in thermal desorption spectra as peaks or shoulder states). While the Langmuir model suggests an energetically homogeneous surface, lateral interactions or inherently different crystallographic adsorption sites make the adsorption energy of the surface *heterogeneous*. We distinguish *a priori heterogeneity* (in case sites with initially different adsorption energy exist) and *a posteriori heterogeneity* that is *induced* only by the aforementioned lateral adsorbate–adsorbate interactions on an initially homogeneous surface. Surfaces with periodic steps often provide geometrically different sites from the very beginning, leading to distinct adsorption states and hence separate thermal desorption peaks. Flat, densely packed, defect-free surfaces are initially quite homogeneous; only as the adsorption sites become occupied do lateral interactions between the adsorbate particles come gradually into effect, leading to favorable and less favorable adsorption sites. Consequently, additional TD states develop (which often show up only as shoulders; see Figure 36.16). In a number of cases, particularly with corrugated surfaces and close to saturation coverage, a hydrogen species with surprisingly low adsorption energy emerges, and the corresponding TD states are well separated from the respective peak(s) of the low-coverage regime. The H-on-Rh(110) system is an example where a H coverage of $\Theta = 2$ (corresponding to twice the number of Rh surface atoms) produces a $(1 \times 1)$-2H phase with a clear split-off TD state appearing at 150 K and reflecting H atoms with a binding energy of merely 33 kJ/mol [137]. Another example is the H-on Ru($10\bar{1}0$) system, where likewise a total coverage of $\Theta = 2$ is attained; here the mutual H–H interactions are somewhat less pronounced due to the somewhat larger Ru–Ru spacing, and the split-off state appears at 220 K, reflecting a desorption energy of 56 kJ/mol, compared to the 80 kJ/mol of the low-coverage state [135].

#### 36.2.4.3 Surface Relaxation and Surface Reconstruction Effects

Any thermodynamic system containing a surface is provided with a surplus of free energy and strives for a minimization of this "extra" energy. For clean metal surfaces, this tendency causes a phenomenon known as *multilayer relaxation*: the system can gain energy by somewhat contracting the outermost surface layer with respect to the second and third layer, whereby the lateral translational symmetry within both layers remains unaffected. The changes of the perpendicular layer distances usually amount to a couple of percent of the distances of the unperturbed system [190]. Any adsorption of chemically active gases such as hydrogen then introduces a perturbation as the electron cloud of the 1s orbital interacts with the localized or delocalized wave functions of the solid surface. In mild cases, only the inherent layer relaxation will be modified (usually it is reduced), and one does not expect significant alterations of the overall adsorption energy, which always includes the contributions associated with the geometrical changes of the surface atomic layers. Experimentally, it is impossible to separately measure the contribution due to the formation of the individual adsorptive bond and that due to the lifting of an existing layer relaxation. One can, however, follow the changes of the multilayer relaxation as the surface becomes covered with H atoms. A good example here is the H-on-Rh(110) system, where various ordered hydrogen LEED phases are formed at low, medium, and high coverages, all of which allow accurate structure analyses from which, in a series of meticulous studies, not only the position of the adsorbed H atoms but also coverage-dependent information on the extent of the multilayer relaxation could be inferred [191–195].

If the interaction is more vigorous, the so-called *surface reconstruction* sets in, where the topmost atoms of the solid take on a new periodicity, usually in addition to the relaxation changes mentioned above. Numerous experimental [196] and theoretical [197] investigations have been devoted to the phenomenon of adsorption-induced surface reconstruction, from both energetic and mechanistic viewpoints. A reconstructed surface exhibits distinct "extra" spots in the LEED pattern due to changes in the translational symmetry. Some of the topmost surface atoms experience translational movements from their former equilibrium positions, where displacive reconstructions comprise smaller lateral shifts of atoms, without the need of lateral diffusion, while heavier interactions trigger explicit site changes in that atoms leave their former and take on totally new positions, which usually requires extensive diffusion. Examples are missing-row (MR) or added-row (AR) reconstructions, which frequently occur with the anisotropic face-centered cubic (fcc) (110) or hexagonal closed-packed (hcp) $(10\bar{1}0)$ surface orientations [196, 131]. Likewise, the so-called registry shifts have been reported, especially with base-centered cubic (bcc) (211) surfaces [198]. Titmuss *et al.* have reviewed the topic "Reconstruction of clean and adsorbate-covered metal surfaces" until 1996, and the reader is referred to this article for further details [199].

Concerning the overall energetics of adsorption, it is self-evident that, especially in the case of the latter type of reconstructions, the energy required for the structural rearrangement can be considerable and must be spent from the pool of the adsorption energy. Figure 36.19 shows the principal situation. Assume that a given surface can exist in two different crystallographic forms A and B, with minima $E_A$ and $E_B$ in the free energy function. For a clean surface, A should be more favorable than B. An activation barrier prevents the conversion of one configuration into the other. However, as the surface A becomes gradually covered with the adsorbate (in our case atomic hydrogen), the free energy changes continuously in favor of configuration B, and at a certain critical coverage $\Theta_{crit}$ the total energy minimum of configuration B will extend the minimum A in depth. Provided that the thermal energy of the surface is sufficiently high, the system will slowly or rapidly turn from configuration A into configuration B. There are quite a number of reports in which just this principal scenario has been studied, especially for the (100) surfaces of Pt and Ir [200, 201]; see also Chapter 4 in Volume 2. Tománek and Bennemann developed an electronic model for the relaxation and reconstruction trends at metal surfaces that provides further useful information on this issue [202]. However, the simple thermodynamic view does certainly not really allow conclusions on the *mechanism* of a given reconstruction. The structural rearrangement may initially happen only locally, that is, in the direct vicinity of the adsorbed H atom, at a coverage that is still below the critical value $\Theta_{crit}$. As more atoms get adsorbed, the surface becomes more and more unstable, until the overall restructuring occurs. It may be, on the other hand, that already quite a few adsorbed atoms make the whole surface instable and flip over to the new crystallographic configuration.

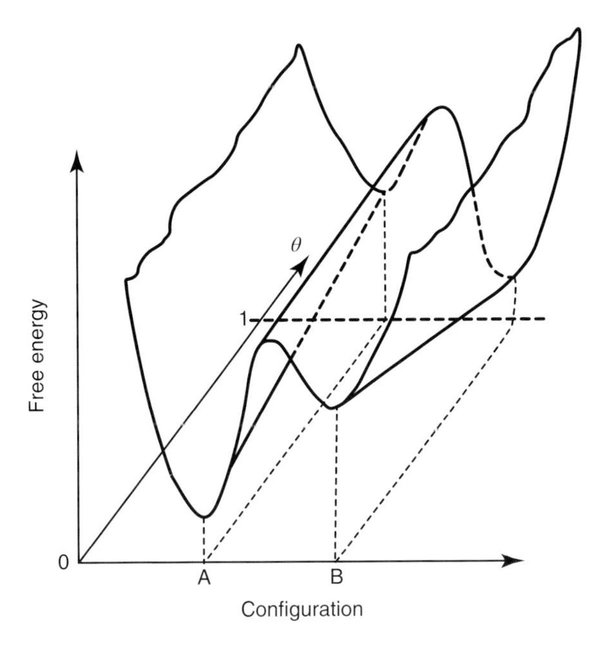


**Figure 36.19** Schematic view of the dependence of the surface free energy on the surface configuration and the adsorbate coverage. In the clean state, the surface exists in the energetically more favorable configuration "A," which is separated from the less stable configuration "B." However, as the adsorbate coverage Θ increases, "A" loses and "B" gains stability and the depths of the minima revert, until configuration "B" is favored over configuration "A."

An almost "classical" case of a reconstruction induced by hydrogen is the Ni(110) surface. Already in 1962 Germer and McRae studied the hydrogen adsorption on Ni(110) and reported on a H-induced rearrangement of the Ni surface atoms [203]. In 1974, Taylor and Estrup revisited this system and observed temperature-dependent rearrangement effects [204]. In the 1980s, Christmann *et al.* performed detailed LEED, WF, and thermal desorption experiments and showed that actually two types of reconstructions occurred – one of the pairing-row (PR) displacive type below 220 K at hydrogen coverages $1 < \Theta < 1.5$ monolayers, and another one of the MR type taking place only above 220 K [205, 206]. The PR reconstruction below 220 K creates additional H adsorption sites beyond $\Theta = 1.0$, but at 220 K the surface becomes unstable and returns instantaneously to the unreconstructed $(1 \times 1)$ configuration. Accordingly, the respective surplus of formerly adsorbed hydrogen atoms must leave the surface within a very short time, thereby producing a sharp TD peak that has been addressed as "surface explosion" [207]. In Figure 36.20 we present a respective series of hydrogen TD spectra showing the sharp $\alpha$-state that exhibits a zero-order desorption kinetics. After the invention of the tunneling microscope, various groups have studied the H-on-Ni(110) system by scanning tunneling microscopy (STM) and could basically confirm and deepen the structural conclusions deduced from the previous work, especially concerning the MR

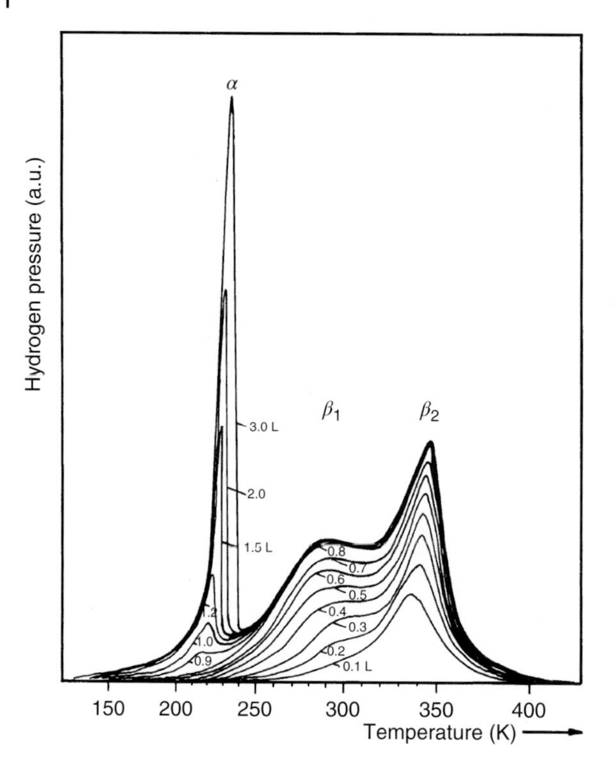


**Figure 36.20** Series of thermal desorption spectra obtained after adsorption of hydrogen on the Ni(110) surface at 120 K. The heating rate was 9 K/s, and the various desorption states are indicated (in their sequence of population) as $\beta_2$, $\beta_1$, and $\alpha$. Note the extreme sharpness and peculiar desorption kinetics of the $\alpha$ state. (After Ref. [207].)

reconstruction of the high-temperature (1 × 2) phase [189, 208, 209]. It may be added that also the Pd(110) surface exhibits a strong H-induced reconstruction [210–214]. Upon hydrogen exposure at 120 K, a PR-type of reconstruction was confirmed by ion scattering experiments by Niehus *et al.* [211], whereas at room temperature rather an MR reconstruction seems to prevail as revealed by STM studies [212, 213]. It appears therefore that even very small H exposures are sufficient to induce the reconstruction. The Cu(110) surface likewise reconstructs in an MR configuration if it is exposed to atomic hydrogen [215].

#### 36.2.4.4 **Subsurface Hydrogen**

H-induced surface relaxations and reconstructions discussed above may be considered a preceding step for the occupation of subsurface sites and/or interstitial sites in the bulk of the solid. Relaxation and especially reconstructions open the way for the H atoms to permeate through the top layer and to reside in subsurface sites. Unfortunately, the usage of the term "subsurface" hydrogen in the literature is not always consistent. In a general sense, "subsurface" hydrogen could refer to all H atoms that are located below the surface and reside not only between the first and second, but also between the second and third, third and fourth, and so on, layer, however still restricted to the surface region of the crystal. In a more rigorous definition, the term "subsurface" refers only to H atoms that adsorb in sites between the first and second atomic layer of the surface. In a recent extensive density functional theory (DFT) calculation, Ferrin *et al.* [51] compared for a large number of H-on-metal adsorption systems the binding energies of H atoms located in surface sites with those of first- and second-layer subsurface sites. The authors concluded that the formation of *surface* hydrogen is exothermic with respect to gas-phase $H_2$ on all metals studied with the exception of Ag and Au. For each metal investigated, hydrogen in its preferred subsurface state is always less stable than in its preferred surface site. This is in line with most experimental observations of subsurface hydrogen systems, but puts some question marks behind earlier reports that claimed subsurface hydrogen also for the Ni(111) [216] or the Cu(110) surface [217]. Subsurface hydrogen is well documented for Pd surfaces and identified mostly by TD experiments [79, 117, 210, 218] in which the hydrogen desorption peaks due to *subsurface* hydrogen appear only on the low-temperature tail of the TD states associated with *surface* hydrogen. For a better understanding of the energetics of the formation of subsurface hydrogen, we refer again to Figure 36.3 to illustrate the origin of these states, which usually depend on the coverage of the subsurface sites because H atoms trapped in these states interact with each other in a way similar to that by the atoms chemisorbed at the surface.

In a first study, dating back to 1983 [218], the existence of subsurface hydrogen on the Pd(110) surface was deduced from combined TDS, LEED, and WF change ($\Delta\Phi$) experiments. The scenario is, however, relatively complicated and requires a closer inspection of the TD spectra displayed in Figure 36.21. The sharp $\alpha_2$ state is obviously associated with a temperature-driven $(1 \times 2)$-3H $\rightarrow$ $(2 \times 1)$-2H phase transition of the Pd surface, since parallel LEED experiments reveal that the H-induced $(1 \times 2)$ reconstruction is lifted just as the $\alpha_2$ species begins to desorb. The respective phase transformation induces an annihilation of ~0.5 ML of the H adsorption sites, similar to the H-on-Ni(110) system (see Figure 36.20). It is now most remarkable that the $\alpha_2$ peak appears only for coverages $\Theta > 1.5$, whereas the $(2 \times 1) \rightarrow (1 \times 2)$ reconstruction sets in already at $\Theta > 1.0$. This was taken as a serious hint to subsurface hydrogen. Simultaneous WF change measurements taken for coverages <1.5 ML while the H-covered sample was heated indicated a loss of hydrogen from the surface at *lower* temperatures than the onset of the $\alpha_2$ desorption. Obviously, part of the hydrogen can move to below the surface into the subsurface region from where it can escape only at somewhat higher temperatures. This was documented in Ref. [218] by the TD spectra spanning the coverage range $1 \leq \Theta \leq 1.5$, which did *not* show the $\alpha_2$ peak during the temperature-driven phase transition. The respective H atoms rather desorbed in a broad and not very significant $\beta_1$ shoulder on the low-temperature tail of the $\beta_2$ state, the latter reflecting as usual the chemisorbed *surface* hydrogen. For initial coverages >1.5 ML, on the other hand, the subsurface region becomes well populated during adsorption at 130 K, and most of the respective subsurface hydrogen species then contributes to the $\alpha_2$ peak. How complex the actual H-on-Pd(110) system really is may be inferred from a very recent and careful combined TDS and

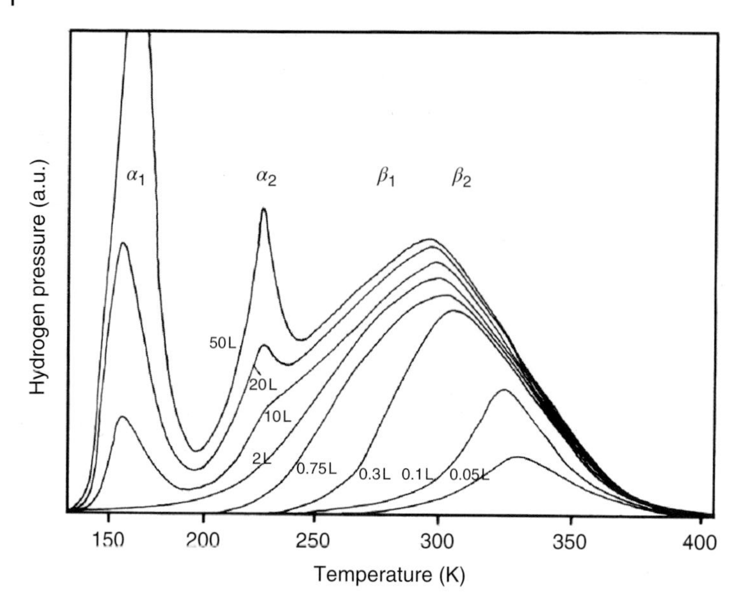


**Figure 36.21** Family of thermal desorption traces obtained from a Pd(110) surface at 130 K after $H_2$ exposures between 10 and 500 L. The intense $\alpha_1$ peak is interpreted as stemming from subsurface hydrogen that has accumulated in the near-surface region; the $\alpha_2$ state is associated with a reconstructive phase transformation and the $\beta$ states reflect chemisorbed hydrogen in different sites. (After Ref. [218].)

depth-profiling study using the nuclear-reaction analysis (NRA) with $^{15}$N nuclei by Ohno *et al.* [219]. Although the main spectroscopic features of our previous work could be confirmed, the depth profiling allowed a somewhat better assignment of the observed TD peaks and their correlation with the subsurface hydrogen and surface-near hydride species.

Especially with palladium, the subsurface hydrogen itself is only a precursor to the population of sites that are located in deeper layers, with a gradual transition to the filling of bulk sites (H sorption or dissolution). In a hydrogen uptake experiment, this process shows up by TD states (or other phenomena) that do no longer reach saturation; rather the uptake continues steadily. In the TD spectra shown in Figure 36.21, the *absorbed* hydrogen produces a huge, nonsaturable desorption state at even lower temperature, which is denoted as $\alpha_1$. Another example is provided by the Pd(210)/H system, where likewise a nonsaturable TD state is associated with the disappearance of H atoms from the near-bulk surface [79].

### 36.2.5
### Surface Diffusion of Hydrogen

Surface diffusion is essential for a given adsorbed particle to reach the most favorable adsorption site on a surface or to react with co-adsorbed species in a Langmuir–Hinshelwood (LH) type of conversion. (In an LH reaction, both species react with each other from the adsorbed state [180]). At a given temperature, "classical" surface diffusion or migration implies both rapid motion in a weakly bound molecular precursor state or the (usually slower) diffusion within the chemisorbed atomic state, where the adsorbed particles rather "jump" or "hop" from one site to a neighboring site. We refer to Figure 36.17a, which displays the periodic variation of the potential energy *parallel* to the surface, that is, $E_{chem}$ $(x,y)$, at the bottom of the chemisorption potential well $E_{chem}$ $(z)$. The periodicity in this case is determined by the regular position of the surface atoms and adsorption sites, respectively. Similar to Equation 36.2, we may define a residence time (or lifetime) of an adsorbed particle in a given site, $\tau_{site}$:

$$\tau_{site} = \tau_0' \exp\left(\frac{E_{diff}}{RT}\right), \tag{36.12}$$

in which $E_{diff}$ stands for the classical activation energy for diffusion, and $\tau_0'$ stands for the pre-exponential. Usually, $E_{diff}$ is merely 10% or 20% of the adsorption energy $E_{chem}$; therefore the adsorbed atoms start to move or hop from one site to another long before they can leave the chemisorption potential by thermal desorption. The lifetime of a particle in a given site depends sensitively on the thermal energy (temperature) of the surface: $E_{diff} > \sim 10\, k_B T$ corresponds to basically immobile particles, while $E_{diff} < \sim k_B T$ means practically unhindered motion across the surface. Concerning diffusion of adsorbates on surfaces *in general*, there is a wealth of literature from which we just cite some useful review articles [219–222].

One can try to describe the macroscopic surface diffusion by a continuum model, with the thermodynamic driving force being the chemical potential $\mu$ or the concentration (coverage) gradient in one direction, $d\Theta/dx$. Depending on the surface properties and the boundary conditions of the diffusion problem, one first applies and modifies Fick's first and second law, respectively. As pointed out by Ertl [27], the Fickian diffusion coefficient $D_x$ can be related to the diffusion coefficient $D_x^*$ of the Einstein–Smoluchowski relation

$$D_x^* = \frac{\langle x^2 \rangle}{2t}, \tag{36.13}$$

in which $\langle x^2 \rangle$ denotes the mean square displacement due to the random walk of an individual diffusing particle within time interval $t$. The Fickian diffusion coefficient $D_x$ is usually measured by watching the decay of the concentration gradient of the adsorbed species with time. In the initial concentration profile, the particles have, however, close mutual distances and hence interact with each other, leading to a relation between $D_x$ and $D_x^*$ that contains an unknown differential factor, which becomes unity only for the limit $\Theta \rightarrow 0$; only then is $D_x = D_x^*$:

$$D_x = D_x^* \left( \frac{\partial \frac{\mu}{kT}}{\partial \ln \Theta} \right)_T. \tag{36.14}$$

Therefore, the determination of Fickian diffusion coefficients based on the time-dependent dispersion of initially dense concentration profiles is somewhat hampered and often provides not more than an estimation of this quantity. By contrast, the measurement of the individual particle hopping either by field-emission microscopy or STM experiments is much more reliable.

The temperature dependence of the classical diffusion rate obeys the Boltzmann statistics and follows an Arrhenius-type of relation

$$D(T) = D_0 \exp\left(-\frac{E_{\text{diff}}}{\text{RT}}\right), \tag{36.15}$$

with $D_0$ = pre-exponential factor ($\text{cm}^2/\text{s}$), and $E_{\text{diff}}$ = activation energy for diffusion (kJ/mol). Accordingly, the activation energy for diffusion is accessible from measured diffusion coefficients by constructing semilogarithmic plots of $D$ against reciprocal temperature.

Focusing now on diffusion of *hydrogen* on surfaces, we have to mention the work by Gomer in the first place. Gomer has been one of the pioneers who drew the attention of surface scientists to surface diffusion phenomena. Beginning in the 1950s and 1960s, he utilized mainly field-emission techniques and first followed the propagation of hydrogen diffusion fronts by direct observation in a field-electron microscope [223, 224]. Later, he developed and applied the field-emission fluctuation technique, among others, to study the diffusion hydrogen of on W(110) surfaces [225, 226].

Beginning perhaps in the 1970s, also other experimental techniques were gradually introduced. Ertl and Neumann were probably the first who used laser-induced thermal desorption [227], a method that was further improved in the following years [228–230] and that is based on a "hole burning" and "refilling" technique. A well-focused laser beam of cross section A is directed onto the hydrogen-covered surface and locally heats the hit area just to a temperature at which all H atoms desorb. The refilling of the empty hole of area A by the surrounding adsorbed H atoms is then followed and monitored with time by firing subsequent laser pulses onto the same area. The H signal is then plotted as a function of time and evaluated using Fick's second law. One of the main difficulties consists in correctly adjusting (and calibrating) the laser power in order not to melt or destroy the hit surface area but to still safely desorb the adsorbed H atoms.

STM then opened the field for a direct observation of surface diffusion phenomena even in real time since the mid-1980s. In this way, direct particle counting from the STM frames as a function of real time could monitor the hopping and displacements of oxygen and nitrogen atoms, and diffusion parameters were subsequently determined from a statistical analysis [231–233]. It is, however, not straightforward to apply this technique to diffusing *hydrogen* atoms on surfaces, since their diffusion rate is substantially higher – unless one reduces the temperature of the sample to very low values. The introduction of the low-temperature (4 K) STM in conjunction with inelastic electron tunneling [234, 235] paved the way to the study of surface diffusion in conjunction with vibrational excitations even of single molecules on surfaces [236]. Concerning other STM studies, a more recent publication by Mitsui *et al.* is worth mentioning. The authors studied the dissociation, adsorption, and diffusion of hydrogen on a Pd(111) surface by means of STM in the temperature range between 37 and 90 K [118, 237]. It had been known from previous LEED work that hydrogen forms two superstructures on Pd(111): a $(\sqrt{3}\times\sqrt{3})$-H phase at $\Theta = 1/3$, and a $(\sqrt{3}\times\sqrt{3})$-2H phase at $\Theta = 2/3$ [238, 239]. Depending on hydrogen coverage, Mitsui *et al.* could follow single H atom diffusion at low coverage, vacancy, and interstitial diffusion in the two $\sqrt{3}$-phases, and H vacancy diffusion in the final $(1\times 1)$ phase that forms at saturation. For the latter kind of diffusion, a hopping rate of $4\times 10^{-4}\,s^{-1}$ at 65 K was reported, and under the assumption of $D_0 = 10^{13}\,s^{-1}$, the activation barrier for vacancy diffusion was estimated as 0.21 eV = 20.3 kJ/mol [237].

Compared to that of heavier adsorbates such as N or O atoms or CO molecules, surface diffusion of adsorbed hydrogen exhibits again some peculiarities that have to do with hydrogen's small mass and/or size and results in quantum effects. The two manifestations of these effects are (i) tunneling and (ii) quantum motion. In addition to the discrete surface diffusion discussed above, these quantum effects can, under certain circumstances, come into play and even dominate the diffusion. (iii) Finally, the small size of the H atoms can additionally facilitate a permeation through the topmost surface layers to the second, third, and so on, layers and diffusion via interstitial sites, leading to the well-known solubility of hydrogen in the bulk of a number of metals.

1) *Tunneling effects.* Again, it was Gomer who showed as one of the first researchers that that the diffusion of H, D, and T atoms on surfaces can proceed in two different ways, namely, via the "classical" diffusion discussed above consisting of thermally activated discrete particle hopping events, and by tunneling, whereby the respective atoms behave wave-like and partially propagate through the activation barriers of Figure 36.17a. This tunneling becomes apparent as one follows the temperature dependence of the diffusion coefficient $D(T)$. Whereas the classical diffusion usually shows an Arrhenius-type of T dependence (see Equation 36.15) and usually increases strongly with temperature, tunneling is practically independent of temperature, because no activation barrier has to be surmounted. This could be verified by DiFoggio and Gomer [225, 226] and Auerbach *et al.* [240] for hydrogen diffusion on W field-emitter surfaces. A summary of their diffusion data is presented in Figure 36.22.
2) *Quantum delocalization.* We mentioned this phenomenon (which is often called also *quantum motion* [14–16]) already in the introduction. It is characterized by adsorbed H atoms that are vibrationally excited from the ground state to higher lying vibrational band states, where they are largely delocalized and actually appear anywhere on the surface. Convincing experimental evidence was obtained for the H-on-Pt(111) system, where vibrational loss (HREELS) measurements by Badescu *et al.* [241, 242] revealed systematic vibrational features that could be associated with the various possible excitations of H atoms into the ladder of band states. The prerequisite is apparently an energetically smooth and homogeneous flat surface, which is apparently provided by the Pt(111) surface.
3) *Small size.* A unique property of the H atom is its small atomic radius, which often provides easy permeation and diffusion through interstitials toward the

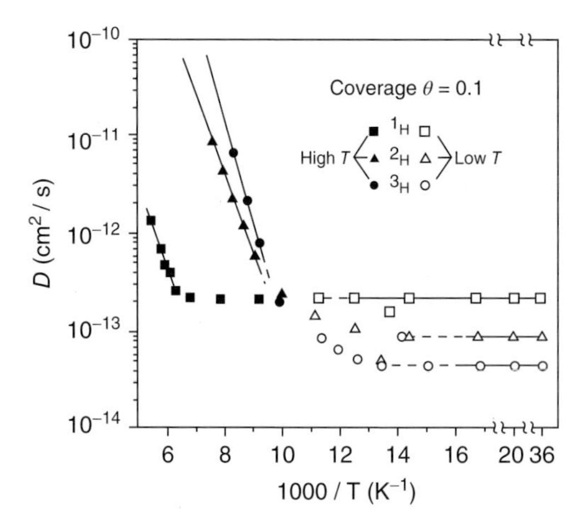


**Figure 36.22** Logarithm of the diffusion coefficient $D$ ($cm^2/s$) plotted versus the inverse absolute temperature (1/K) for the diffusion of the three hydrogen isotopes ${}_1^1H$, ${}_1^2H$, and ${}_1^3H$ on a sparsely covered W(110) surface. Note the different behavior in the low (open symbols) and the high (full symbols) temperature regimes. The data points were fitted by straight lines developed from theoretical diffusion models. (After Ref. [240].)

interior of a crystal. Because of space limitations, we refrain from expanding on the respective bulk diffusion phenomena, although these are certainly prominent properties of certain metals such as V, Ti, Zr, Nb, Pd, or Ta, which could act (within a modest frame) as storage materials for hydrogen. Especially palladium surfaces show a variety of diffusion-related phenomena including overlayer–subsurface layer transitions as well as bulk absorption and formation of hydridic phases. More about the sorption of hydrogen in bulk materials can be found in the monograph by Alefeld and Völkl [243] and other related publications [244, 245].

One particular aspect should still be mentioned here, namely the effect that hydrogen atoms permeating through surfaces can have dramatic consequences with respect to the chemical and mechanical stability of metals, an effect known as *hydrogen embrittlement*. H atoms can, for example, diffuse into the interior of metals via crystallographic defects such as screw dislocations, but also via adsorption into and subsequent diffusion on the internal surface of crack tips. Therefore, metallurgists are highly interested in exploring the origin of the H-induced embrittlement including the hydrogen diffusion phenomena [246, 247]. For further information, we recommend the monograph *Atomistics of Fracture* by Latanision and Pickens [248].

In concluding this section, we would like to point to a more recent theoretical publication in which the present state of the art concerning hydrogen diffusion on TM surfaces is concisely presented. Kristinsdóttir and Skúlason report systematic DFT calculations of hydrogen diffusion on 23 close-packed TM surfaces [249]. Specifically, potential energy surfaces of H atoms on the d metals of the fourth,

**Table 36.6** Some experimentally determined diffusion parameters for hydrogen (deuterium) atom diffusion.

| Surface | Temperature range (K) | Diffusion coefficient $D_0$ ($cm^2/s$) | Diffusion energy (kJ/mol) | Experimental method | References |
|---|---|---|---|---|---|
| Ni(100) | 211 | $2.1 \times 10^{-7}$ | 17.6 | Laser-induced thermal desorption of D atoms | [250] |
| Ni(100) | 140–250 | $8 \times 10^{-6}$ | 13.4 | Field emission fluctuation (FEF) technique | [251] |
| Ni(111) | 110–240 | $2.8 \times 10^{-3}$ | 18.9 | Laser-induced thermal desorption | [252] |
| Ru(0001) | 260–330 | $6.3 \times 10^{-4}$ | 16.7 at low coverage | Laser-induced thermal desorption | [229] |
| Rh(111) | 150–300 | $8 \times 10^{-2}$ | 15.5 at low coverage | Laser-induced thermal desorption | [253] |
| W(100) | >220 (thermally activated $T$ regime) | $10^{-5}$–$10^{-7}$ | 16.7–29.3 | FEF | [254] |
| | 140–220 (tunneling r.) | $10^{-9}$–$10^{-10}$ | 4.2–8.4 | | |
| W(110) | 143–300 | $5 \times 10^{-5}$ | $20.0 \pm 1.5$ | FEF | [225] |
| W(211) | 80–250 | $9 \times 10^{-5}$ (H) | 31 (H) | Along surface channels | [255] |
| | | $5 \times 10^{-6}$ (D) | 27.6 (D) | Along surface channels | |
| | | $8 \times 10^{-6}$ (H) | 27.2 (H) | Across surface channels | |
| | | $2 \times 10^{-6}$ (D) | 25.5 (D) | Across surface channels (FEF) | |
| Pt(111) | 250–250 | $5 \times 10^{-1}$ (D) | 29.3 | LID; H (D) coverage ~0.3 | [228] |

fifth, and sixth periods have been constructed. Hydrogen was moved on the surfaces in the $x$ and $y$ directions with a certain interval, which resulted in a grid. The potential energy was then calculated in each position, whereby the H atom was located slightly above the surface at a distance of ~1.5 Å and the minimum energy calculated in $z$-direction for fixed $x$ and $y$ directions. All the activation energies for hydrogen diffusion are comparatively low, and range from 0.04 eV (3.86 kJ/mol) for platinum to 0.28 eV (27 kJ/mol) for yttrium and zirconium. In addition, the temperatures are estimated at which tunneling phenomena start to become apparent.

Table 36.6 presents a selection of some diffusion energies of hydrogen and deuterium on metal surfaces.

### 36.2.6 The Structure of Chemisorbed Hydrogen Phases

In the preceding sections we addressed mainly phenomena related to the kinetics and energetics of hydrogen adsorption, the final result being that an H atom (or a $H_2$ molecule) has found a fixed adsorption site on the surface where it rests under equilibrium conditions, provided the temperature is sufficiently low. We shall now direct our attention to the local geometry of this site, which is characterized by a coordination number $Z$, counting the surface atoms the H adatom is attached to. Concerning the local geometry of the respective site (for simplicity we assume a common metal single-crystal surface), one distinguishes terminal, long or short bridge, fourfold and threefold hollow (octahedral or tetrahedral) sites. Decisive parameters are then the bond lengths of the H atom to all of its neighbors, including long-range effects, that is, any changes induced on the distances to the next-nearest neighbors, and so on, that may lead to an overall layer relaxation or to reconstruction as discussed above. It is often helpful to think of a cluster-like adsorption complex (with the H atom located in the aforementioned site on top) that is embedded in the semi-infinite surface. In the following, we will first be interested in the structure of these *individual* adsorption complexes (Section 36.2.6.1) and return thereby, in a sense, to the single-particle consideration. In Section 36.2.6.2 we will then, following the multiparticle view, couple all adsorbed individual complexes and investigate their lateral interactions that often lead to long-range ordering effects and thus to the formation of well-ordered H phases. Among others, we will show that the coverage and temperature dependence of the long-range order provides valuable thermodynamic information about phase stabilities and lateral interaction forces operating between adsorbed H atoms.

In the beginning, it is useful to make some introductory remarks about experimental techniques that are particularly suited for probing the structure of hydrogen adsorbates. Low-energy electron diffraction must certainly be reckoned among the most important tools also for hydrogen surface structure determination, although, as mentioned before, the diffractive power of H atoms for low-energy electrons is small [256, 257]. Heinz and Hammer have prepared a useful compilation of problems that one encounters when LEED is employed for structure determination of adsorbed hydrogen [258]. Since the early experiments by Davisson and Germer, who studied adsorbate-induced changes of the surface structure of Ni samples by LEED already in 1927 [259], it took about 30–40 years until the LEED method could be routinely performed in surface science laboratories. In their monograph *Low-energy Electrons and Surface Chemistry*, Ertl and Küppers give a brief historical summary about how the LEED method developed [260]. Soon it became clear that the (strong) physical interaction of low-energy electrons with a solid surface is a complicated scattering problem that requires a profound theoretical treatment in order to be of any use for quantitative surface structure determination. Over the years, powerful LEED theories have been developed [256, 257, 260–265] including special computational techniques such as "renormalized forward scattering" and "layer doubling", which are helpful to treat, for example, the comparatively strong forward scattering properties of the H atom and the convergence problem caused by the unusually short distances between the H layer and the topmost surface plane. In one of the first studies of this kind, the geometry of H on a Ni(111) surface was addressed [14] and later refined with respect to concomitant reconstruction effects, which led to a greatly improved quality of the geometrical parameters [94].

Besides LEED, also other scattering techniques with light particles can provide structural information. Helium atom scattering (HAS) is a frequently used method that was independently developed in various laboratories [266] and successfully employed also for the structure determination of H adsorbed layers [267, 268]. Although this method is very sensitive, it does not image the H atom directly, but rather probes the integral surface corrugation, and the correct choice of the corrugation potential that governs the scattering is essential for a meaningful structure determination [266–269]. Valuable structural information on the location of the adsorbed H atoms could also be obtained from X-ray photoelectron diffraction [270] as well as from ion scattering experiments combined with transmission channeling [271].

At a first glance, STM should be able to directly image adsorbed hydrogen species and is therefore expected to significantly contribute to a structural analysis of H adsorption systems. However, STM has some problems that have to do with the small charge density around an H atom along with a reduced tunneling probability [118] and the usually high mobility of adsorbed H at room temperature that requires very low temperatures in order to fix the H atoms on the surface. In addition, it must be checked in each case whether the tunneling current induced or modified by an adsorbed H atom indicates protrusions or depressions in the respective STM image. Most of the previous STM investigations with adsorbed hydrogen have been performed at or around room temperature, and the structural changes that appeared in the images pertained only to H-induced restructuring effects of the metal surface. Furthermore, the accuracy achieved in STM experiments concerning horizontal and vertical distances cannot always compete with the precision of a quantitative LEED analysis. This is why most of the reliable surface structure information is still gained from diffraction experiments. However, one should not conceal the fact that a LEED analysis is possible only if the adsorbed H species forms a periodic grating with long-range order that produces LEED "extra" spots or at least sufficiently modifies the intensity–voltage [$I(V)$] dependence of the integer-order beams of the substrate. These LEED $I(V)$ curves are the heart of any structure determination because – as in X-ray diffraction – the structural information is hidden in the *intensity* of the diffracted beams.

Often, quite helpful supporting structural information comes from vibrational spectroscopy (HREELS or infrared spectroscopy), a subject that we briefly touched already in Section 36.2.2.2 in the context of the distinction between molecular and atomic adsorption. Since both the number and frequencies of hydrogen vibrational modes, in conjunction with the electron scattering conditions of an HREELS experiment, reflect the local geometry or at least *symmetry* of an adsorption complex [270, 271], there are several publications in which hints for the correct hydrogen adsorption site(s) came from a vibrational analysis. This will be the topic of Section 36.2.7.

#### 36.2.6.1 The Local H–Metal Adsorption Complex

A fairly complete and consistent picture has emerged for the structure of adsorbed hydrogen layers on *metallic* surfaces thanks to numerous structure analyses that were performed during the 1980s and 1990s of the past century. It is virtually impossible to list all the relevant publications pertinent to this field; instead, we rather select a few especially revealing studies here and ask the reader to consult the respective book chapters or review articles [14, 45–51, 262–265, 269, 270]. From all these studies, certain H-specific trends can be deduced.

1) *Defects and impurities.* Hydrogen adsorption by responds extremely sensitively to structural defects and impurities. Structural defects often act as preferred adsorption centers, as we have seen in the context with stepped surfaces in Section 36.2.4.1. Certain chemical impurities such as carbon or sulfur often suppress the hydrogen uptake dramatically because they utilize similar bonding orbitals for bonding and/or modify the surface charge distribution in their vicinity in a way that the bonding conditions for H atoms are impaired.
2) *Local adsorption sites.* In most structure analyses, it turns out that H atoms prefer adsorption sites with high local coordination. As is evident from Table 36.7, highly coordinated sites are usually chosen, with only very few exceptions. The fourfold hollow site is occupied on the (100) surface and the threefold hollow site on the (111) surface. For the fcc (110) and hcp $(10\bar{1}0)$ surfaces, there is at least a preference for quasi-threefold coordinated sites of the fcc or hcp type, as illustrated in Figure 36.23. The exceptions actually concern a few fcc (110) surfaces such as Pt(110) [295, 297] or hcp $(10\bar{1}0)$ surfaces such as Ru $(10\bar{1}0)$ [297], where under certain coverage conditions H atoms became adsorbed also in the less coordinated bridge sites.
3) *Local geometrical effects.* There is hardly any case where chemisorbed H atoms do not induce at least slight alterations of the atomic distances in their direct environment. It appears that the small H atom somewhat attracts their direct neighbors upon chemisorption, thereby causing slight local geometrical changes that usually consist of a reduction or even lifting of the inherent multilayer relaxation of the clean surface. In summary, then, a certain surface buckling results [258].
4) *Hydrogen – metal bond lengths.* Concerning the bong lengths $d_{\mathrm{Me-H}}$ within the H-to-metal complex, it is safe to say that the measured values are usually close to the sum of the hard-sphere radii of the metal surface atom and the H atom, the latter being around 0.5–0.6 Å compatible with the Bohr radius of $a_0 = 0.53$ Å. Accordingly, Hammer *et al.* report for the (2 × 2)-2H honeycomb phase formed by H on the Ni(111) surface a Ni–H bond length of $d_{\mathrm{Ni-H}} = 1.73$ Å,

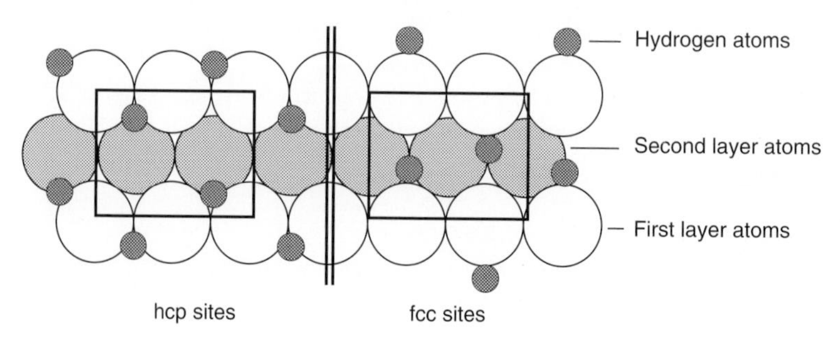


**Figure 36.23** Cutout of a ball model of an hcp $(10\bar{1}0)$ surface showing the hcp and fcc type of adsorption sites. The H atoms are indicated as gray circles.

**Table 36.7** Ordered H phases, critical temperatures, adsorption sites, and bond distances for some H-on-metal systems.

| Surface | Ordered H phase | Critical T (K) | H coverage | H-substrate distance | Radius of H atom | Coordination | Experimental methods | Remarks | References |
|---|---|---|---|---|---|---|---|---|---|
| Be(0001) | $(\sqrt{3}\times\sqrt{3})R30°$ | <270 | 0.6–1.0 | 1.53 ± 0.2 | | Threefold | LEED I,V | Exposure to H atoms | [272] |
| Al(100) | p(1 × 1) | | 1.0 | | | Onefold | LEED, HREELS | Exposure to H atoms | [273] |
| Ti(0001) | (1 × 1) | | | 1.8–1.9 | | | LEED, ARUPS | | [274] |
| Fe(100) | | | | | | Fourfold | HREELS | | [275] |
| Fe(110) | p(2 × 1) = c(2 × 2) | 245 | 0.5 | 1.75 ± 0.05 | 0.47 ± 0.05 | Twofold long-bridge | LEED | H-induced reconstruction included | [276] |
| | (2 × 2)-2H | | 0.5 | 1.84 | 0.58 | Threefold | LEED | | [94] |
| | (3 × 3)-6H | 265 | 0.67 | | | | | | [277] |
| Co(10-10) | c(2 × 4)-4H | 270 | 0.5 | | | Threefold | LEED, HREELS | H-induced reconstruction | [129] |
| | (2 × 1)-2H p2mg | 290 | 1.0 | | | | | | |
| | (1 × 2) | | | | | | | | |
| Ni(100) | (1 × 1) | | 1.0 | 1.95–2.0 | | Fourfold | He diffraction | | [278] |
| Ni(110) | c(2 × 6) | <175 | 0.33 | 1.72 ± 0.05 | 0.47 ± 0.05 | Threefold | LEED, He diffraction | H-induced PR reconstruction | [206] |
| | c(2 × 4) | | 0.5 | | | | | | |
| | c(2 × 6) | <175 | 0.67 | | | | LEED | | [279] |
| | c(2 × 6) | | 0.83 | | | | | | |
| | (2 × 1)-2H p2mg | <220 | 1.00 | | | | LEED | | [280] |
| | (1 × 2)-3H | >220 | 1.5 | | | | | | |
| Ni(111) | (2 × 2)-2H | 270 | 0.5 | 1.73 ± 0.08 | 0.49 ± 0.08 | Threefold honeycomb | LEED | Slight H-induced reconstruction included | [94] |
| Cu(110) | (1 × 2)-rec | | 0.5 | 1.54 ± 0.09 | | Threefold | Low-energy ion scattering | Exposure to H atoms, MR-reconstruction | [281] |

(continued overleaf)

**Table 36.7** (*Continued*)

| Surface | Ordered H phase | Critical T (K) | H coverage | H-substrate distance | Radius of H atom | Coordination | Experimental methods | Remarks | References |
|---|---|---|---|---|---|---|---|---|---|
| Mo(110) | $(2\times2)$-2H | 200 | 0.5 | 1.93 | 0.57 | Threefold hollow | LEED | | [282] |
| | $(1\times1)$ | | 1.0 | | 0.65 | Threefold hollow | | | |
| Ru(0001) | $(\sqrt{3}\times\sqrt{3})$R30° | 74.5 | 0.33 | $2.0\pm0.2$ | 0.56 | Threefold f.c.c. | LEED | | [283] |
| | p$(2\times1)$ | 68 | 0.5 | | | | LEED | | [284, 285] |
| | $(2\times2)$-3H | 71 | 0.75 | $1.91\pm0.06$ | | Threefold | VLEED | | |
| | $(1\times1)$ or disordered | | 1.0 | | | | | | |
| Ru(10-10) | c$(2\times2)$-3H | | 1.5 | $2.01\pm0.4$ | 0.65 | Quasi-threefold | LEED | Slight buckling included | [286] |
| | $(1\times1)$-2H | | 2.0 | $1.95\pm0.6$ | 0.70 | Quasi-threefold | LEED | | |
| Rh(110) | $(1\times3)$-H | 130 | 0.33 | $1.86\pm0.1$ | $0.52\pm0.1$ | Quasi-threefold | LEED | H-induced shift-buckling | [192] |
| | $(1\times2)$-H | 140 | 0.5 | $1.87\pm0.1$ | $0.53\pm0.1$ | Quasi-threefold | LEED | | [194] |
| | $(1\times3)$-2H | >180 | 0.67 | | | Quasi-threefold | | | |
| | $(1\times2)$-3H | >160 | 1.5 | 1.87–1.93 | $0.5\pm0.1$ | | LEED | | [193] |
| | $(1\times1)$-2H | | 2.0 | $1.84\pm0.2$ | $0.5\pm0.2$ | | LEED | | [191] |
| Pd(100) | c$(2\times2)$ | 260 | 0.5 | 1.97 | | Fourfold hollow | LEED; transmission channeling | | [117, 287] |
| Pd(110) | $(2\times1)$-2H | | 1.0 | $2.00\pm0.1$ | $0.6\pm0.1$ | Quasi-threefold | LEED | MR-reconstruction | [288] |
| | $(1\times2)$-3H | | 1.5 | | | | STM | | [212] |

| | | | | | | | | | |
|---|---|---|---|---|---|---|---|---|---|
| Pd(111) | $(\sqrt{3}\times\sqrt{3})R30^\circ$-H | 85 | 0.33 | 1.78–1.80 | | Threefold fcc | LEED | Partial occupation of subsurface sites | [238, 289] |
| | $(\sqrt{3}\times\sqrt{3})R30^\circ$-2H | 105 | 0.67 | | | | LEED | | |
| Ta(100) | $(1\times1)$-H | | 1.0 | 1.92–1.94 | 0.5–0.6 | Threefold hollow | LEED, HREELS | | [290] |
| W(100) | $c(2\times2)=(\sqrt{2}\times\sqrt{2})R45^\circ$ | | 0 | $1.97\pm0.04$ | | Zig-zag model | LEED | Clean surface is reconstructed | [291] |
| | $c(2\times2)$-H, followed by complex series of LEED patterns indicative of IC phases | | 0.25 | | | Twofold bridge | LEED, HREELS | Dimer model | [292] |
| | $(1\times1)$-2H | | 2.0 | | | Twofold bridge | LEED | | |
| W(110) | $p(2\times1)$ | | 0.4–0.5 | | | Quasi-threefold | LEED | | [293] |
| | $p(2\times2)$ | | 0.4–0.8 | | | | | | |
| | $p(1\times1)$ | | 1.0 | | | | | | |
| Re(10-10) | $c(2\times2)-3H$ | | 1.5 | $1.85\pm0.4$ | $0.47\pm0.4$ | Two H atoms in quasi-threefold, one H atom in twofold site | LEED | | [286] |
| Ir(111) | $(1\times1)$ | | 1.0 | | | Onefold (terminal) site; quantum delocalization | HREELS | Exposures of >100 l | [294] |
| Pt(110)-$(1\times2)$ | | | | | | Twofold (bridge) site | He diffraction | | [295] |
| Pt(111) | $(1\times1)$ | | | $1.9\pm0.1$ | | Threefold hollow (fcc) | Ion channeling | | [296] |

with an atomic radius of the adsorbed H atom of $r_H = 0.49$ Å [94]. For another (nonprimitive) (2 × 2)-2H phase formed by hydrogen on Fe(110) [277], the same authors arrive at $d_{Fe-H} = 1.84$ Å, with the H part amounting to $r_H = 0.58$ Å, in good agreement with a previous analysis by Moritz *et al.* [276]. Further reliable LEED bond length determinations are available for the H-on-Ni(110) (2 × 1) phase by Reimer *et al.* [280] and for Pd(110) by Skottke *et al.* [288]. Here, the bond lengths came out as $d_{Ni-H} = 1.72$ Å and $r_H = 0.48$ Å for the nonprimitive, unreconstructed (2 × 1) phase on Ni(110) and $d_{Pd-H} = 2.0$ Å of the H atoms adsorbed in the analog (2 × 1) phase on Pd(110), leading to an effective H radius of $r_H = 0.6$ Å. For hydrogen atoms adsorbed on the Ru(0001) surface, a Ru–H bond distance of $d_{Ru-H} = 2.0$ (±0.2) was reported, with $r_H = 0.75$ (±0.2) A [284]. A collection of values for several H-on-TM systems is listed in Table 36.7.

The usually observed high coordination of adsorbed hydrogen atoms suggests that more than just one atom of the solid participates in and provides for an appropriate chemical bonding to the surface. A particularly illustrative example is the (2 × 2)-2H structure formed by half a monolayer of H atoms on the Ni(111) surface. This structure will be dealt with in some detail also in Section 36.2.6.2, and we refer to the respective ball model of Figure 36.25. The H atoms shown as small full circles form a graphite-like hexagonal grating, which is often called the *honeycomb* phase. The (nonprimitive) oblique unit mesh is indicated in Figure 36.25; it contains two H atoms that are both located in threefold-coordinated sites with trigonal symmetry, A and B. However, these sites are not really equivalent: if the second layer of Ni atoms underneath the top layer is considered, then site A has a Ni atom directly underneath (an hcp site), while site B has not (an fcc site). Because of their close proximity to the surface plane, the H atoms actually "feel" the difference, which causes, as detailed LEED investigations proved [94], a slight undulation of the H honeycomb layer.

The participation of more than a single surface atom in the adsorptive binding of an H atom suggested by the experimental studies is corroborated by the results of various theoretical calculations. As an example, we quote the work by Kresse and Hafner [309], who performed DFT calculations for the H adsorption on the three low-index Ni planes and devoted particular attention to the bonding situation of H in the honeycomb structure. They found $d_{Ni-H} = 1.67$ Å for both hcp and fcc sites, and correctly predicted the buckling. Greeley and Mavrikakis focused in their DFT calculations on the H/Ni(111) system and stated that the hydrogen–metal bond distances on threefold coordinated sites ranged between 1.73 and 2.03 Å, with the perpendicular hydrogen–surface distance varying between 0.67 and 1.05 Å [48]. We just add that in these and many other calculations also, the adsorption energies (see Section 36.2.4) and the H-induced WF changes (see Section 36.2.8) are determined. In most cases, the energies come out correctly to within 10%, and the agreement between the computed WF changes with the experimental data is also quite satisfactory. This provides some confidence regarding the quality and reliability of the presently utilized computing techniques, especially those based on DFT (see Section 36.3).

#### 36.2.6.2 Adsorbed Hydrogen Phases with Long-Range Order

On a periodic surface lattice, a mobile adsorbate *without* lateral interactions $\omega$ would occupy the available sites, that is, the potential minima provided by the surface lattice, entirely randomly, forming a so-called lattice gas. In a diffraction experiment, only diffuse background and no "extra" spots would appear [260]. Turning on the lateral interaction forces $\omega$ introduces immediately an energetic heterogeneity and creates favorable and less favorable sites that are preferentially occupied and avoided, respectively. We distinguish two simple cases: (i) attractive interactions, and (ii) repulsive interactions.

(i) *Attractive interactions*. Even with sparsely covered surfaces, the adsorbing particles attract each other and form, after a series of diffusion steps, more or less dense islands resulting in the formation of locally ordered patches that produce sharp "extra" spots in a LEED pattern even at a fairly early stage of coverage $\Theta$. As $\Theta$ increases, the empty space between the islands becomes filled until the entire surface is covered with an ordered phase. The intensity of the "extra" spots varies roughly linearly with coverage until the optimum order is attained [310].

(ii) *Repulsive interactions*. In this case, the adsorbed particles try to avoid nearness to the neighboring species. At low coverages, the mutual distances are as large as possible, and the "extra" spot intensity increases at a rate greater than that of first power of coverage. At certain coverages, the particles reside in equivalent sites of a grid with lateral distances that are optimized with respect to minimum interaction energy.

For the formation of adsorbate phases with long-range order, these adsorbate–adsorbate interactions $E_\omega$ are decisive, but it is even more important to relate $E_\omega$ to the depth of the chemisorption potential $E(z)$ (see Figure 36.17). For weakly adsorbed (physisorbed) species – hydrogen molecules, for example – $E_\omega$ is not too much different from $E(z)$, with the consequence that a rich variety of phases can form as we have briefly noted in the context of hydrogen physisorption on graphite surfaces in Section 36.2.2. The other case is *chemisorption* of H atoms, where $E(z)$ is usually much larger than $E_\omega$. Then phases with long-range order can and will appear only at sufficiently low temperatures at which the mobility of the H atoms is limited and the adsorbed atoms stay long enough in the shallow periodic potential wells created by the induced heterogeneity.

Assuming temperature and coverage conditions that lead to the optimum development of a hydrogen superlattice, the *intensity* of the resulting H-induced LEED beams deserves some attention. As various LEED experiments reveal, one can distinguish patterns with very weak "extra" spot intensity from those where the fractional-order beams are as intense as the integer-order beams of the substrate, the intensity ratio ranging around 1/100. This suggests subdividing the H adsorption systems into two categories: the one showing the weak "extra" intensities comprises all systems in which the ordered H phase resides on a nondistorted rigid substrate lattice (i.e., systems without reconstructions), and the idea behind that the "extra" LEED intensity stems exclusively from ordered H atoms with their weak diffractive power. The other class then contains all systems in which the adsorbed H atoms induce substantial geometrical distortions of the underlying lattice, owing to surface restructuring effects (i.e., systems with reconstruction). Here, ordered H atoms do not really contribute to the integral LEED intensity, whose high value is rather caused by the H-induced displacement of substrate atoms with their comparatively strong scattering power. Most of the LEED structure analyses of H chemisorption systems have revealed that the substrate lattice is indeed not really rigid, with the consequence that even in cases with weak "extra" spots a certain intensity contribution is due to (slightly) displaced substrate atoms. Nevertheless, for a coarse distinction, we find it useful to keep the simple two-group category here and present first examples.

**36.2.6.2.1 Ordered H Phases without Surface Reconstruction** In the past ~50 years, a whole variety of ordered hydrogen phases have been observed, mostly on metallic surfaces, but some also on semiconductors, silicon in particular. However, on these latter materials, the binding situation is quite different from that on metal surfaces, because semiconductors with their predominantly covalent binding forces exhibit *directed* valencies leading to the usual picture of "dangling bonds" that "stick out" of the surfaces, resulting in an unfavorable energetic situation. Often, the surface free energy can be minimized by pronounced reconstructions, a good example being the famous $(7 \times 7)$ phase of the Si(111) surface [311]. If the reconstructed surfaces of this or similar kind are exposed to atomic or reactive hydrogen (either in the gas phase or under electrochemical conditions in solution), a lifting of the reconstruction can be achieved. Hence, the "extra" spots in the diffraction pattern disappear and give way for a perfect $(1 \times 1)$ pattern in which each dangling bond carries a single hydrogen atom and provides a peculiar chemical reactivity of these $(1 \times 1)$-H surfaces [42, 312, 313]. It is unnecessary to underline that the respective ordered hydrogen phases are not due to lateral H–H interactions but rather reflect the specific binding situation on semiconductor surfaces. In the following considerations, we are, however, more interested in the lateral interactions between H atoms adsorbed on metallic solids and will therefore not further expand on H-induced ordering on semiconductor, oxide, or insulator surfaces.

A good starting point is to compare the tendency of adsorbed H atoms to form ordered phases on geometrically similar metal surfaces, for example, on close-packed (111) surfaces of the fcc lattice or (0001) surfaces of the hcp crystal system. These surfaces possess low surface free energies and do not reconstruct in their clean state. They are geometrically flat, that is, do not exhibit a pronounced corrugation or site heterogeneity. This makes the adsorbed H atoms quite mobile even at room temperature. The respective high mobility or diffusivity is manifested by the fact that ordered H phases could never be observed for these surfaces at or above 300 K; in other words, the critical temperatures of the ordered H phases are well below room temperature, and in order to achieve H atom ordering, the surface temperatures must at least be lowered to below ~270 K [314]. To illustrate this, we present in Table 36.8 a summary of H-induced superstructures reported for the hexagonal surfaces of the group 8–11 elements of the PT, which is adapted from our recent work on the H-on-Co(0001) system [128].

**Table 36.8** Ordered hydrogen phases observed on TM surfaces with hexagonal symmetry.

| **Fe(111)** | **Co(0001)** | **Ni(111)** | **Cu(111)** |
|---|---|---|---|
| No superstructure | c(2 × 2) at $\Theta = 0.5$ | (2 × 2)-2H at $\Theta = 0.5$ | (2 × 2) at $0.3 < \Theta < 0.56$ |
| | $T_c \approx 240$ K | $T_c = 270$ K | $T_c = 100$ K |
| [298] | [299, 300] | [14] | [301][a)] |
| **Ru(0001)** | **Rh(111)** | **Pd(111)** | **Ag(111)** |
| (1) ($\sqrt{3}$x$\sqrt{3}$)R30° at $\Theta = 0.33$ | No ordered phase for $T > 80$ K | (1) ($\sqrt{3}$x$\sqrt{3}$)R30° – H at $\Theta = 0.3$ | (1) (2 × 2) at $\Theta = 0.5$ |
| $T_c = 74.5$ K | | $T_c = 85$ K | $T_c = 140$ K |
| (2) p(2 × 1) at $\Theta = 0.50$ | | (2) ($\sqrt{3}$x$\sqrt{3}$)R30° – 2H at $\Theta = 0.67$ | (2) mixed (2 × 2) + (3 × 3) at $\Theta > 0.5$ |
| $T_c = 68.0$ K | | $T_c = 105$ K | |
| (3) (2 × 2)-2H at $\Theta = 0.75$ | | | |
| $T_c = 71.0$ K | | | |
| [283] | [302] | [239] | [303] |
| | [304] | [289] | [142] |
| **Os(0001)** | **Ir(111)** | **Pt(111)** | **Au(111)** (thin film) |
| Not investigated | No ordered phase for $T > 130$ K | No ordered phase for $T > 80$ K | No ordered phase |
| | [145] | [305] | [306] |
| | [307] | [241] | |

a) Note: The (3 × 3) phase at $0.5 < \Theta < 0.67$ reported by [303, 308] has not been analyzed yet and may reflect a surface reconstruction.

A visual inspection reveals that for the 3d metals Ni and Co, a single phase with (2 × 2) symmetry is observed, the optimum coverage likely being $\Theta_H = 0.5$, although only for Ni(111) this coverage (and the adsorption site) has been confirmed by a dynamical LEED calculation. The Ni surface also exhibits the highest critical temperature of $T_c = 270$ K. On Cu(111), a (2 × 2) phase with weak "extra" spots has been reported around $\Theta_H = 0.5$, too. At still higher coverage, an additional (3 × 3) LEED pattern appears with more intense spots and existing up to ~240 K, whereby the disappearance of the (3 × 3) structure for $T > 250$ K is not caused by disordering effects but rather by thermal desorption [301, 308]. These findings may indicate that a H-induced surface reconstruction is at least involved in the respective phase formation. As far as the 4d metals with their somewhat larger lattice spacing are concerned, more than a single ordered H phase is common. On Ru(0001), a $\sqrt{3} \times \sqrt{3}$ phase appears already at a coverage of 1/3, in addition to a (2 × 2)-2H phase at $\Theta_H = 0.5$, and even a third ordered structure, namely a (2 × 2)-3H, exists at $\Theta_H = 0.75$. The critical temperatures of all phases are below 80 K [283]. A similar situation holds for the Pd(111) surface, where two $\sqrt{3} \times \sqrt{3}$ phases have been reported, at coverages of $\Theta_H = 0.3$ and 0.67 and with $T_c$ at 85 and 105 K, respectively [238, 239, 289]. Interestingly, no (2 × 2) phase at $\Theta_H = 0.5$ exists – at least not for temperatures >80 K, on the contrary, calculations using the formalism of the "embedded atom method" (EAM) predict the absence of an ordered hydrogen phase at half a monolayer coverage [239], which may be due to the partial occupation of subsurface sites [289]. However, no matter which ordered H phase is considered on these hexagonal surfaces, on all the respective structures with trigonal or hexagonal symmetry the H atoms were found to reside in threefold hollow sites [289].

Turning to similar considerations for (100)-oriented surfaces of TMs with fourfold symmetry, ordered hydrogen phases without a distinct reconstruction are much less frequently observed; at and above 140 K a c(2 × 2) superstructure because adsorbed hydrogen has been found only for the Pd(100) surface [117]. Neither for Fe(100) nor for Ni(100), Cu(100), and Rh(100) have such phases been reported, although hydrogen coverages up to 1 ML can be achieved on some of these surfaces giving rise to well-ordered (1 × 1) phases. The variety of H-induced superstructures on Mo(100) and W(100) involve marked substrate reconstructions (see Section 36.2.6.2.2), and for the (100) surfaces of Pt and Ir, which are reconstructed in the clean state, adsorbed hydrogen is not really capable of lifting this inherent reconstruction.

The situation is again different if the fcc and hcp surfaces of twofold symmetry [(110) and $(10\bar{1}0)$] are considered. Here, hydrogen adsorption is responsible for a whole variety of ordered phases without noticeable reconstruction in the submonolayer coverage regime; reconstructions definitely occur, too, but require higher coverages and/or temperatures [131] (see below). An almost classical example is the Ni(110) surface, where six different H phases appear in the coverage range from $0.33 \leq \Theta_H \leq 1.0$, that is, a (2 × 3)-1d at $\Theta_H = 0.3$, a c(2 × 4) at $\Theta_H = 0.5$, a c(2 × 6) at $\Theta_H = 0.67$, a second c(2 × 6) at $\Theta_H = 0.83$, and a (2 × 1)-2H *p*2*mg* phase at $\Theta_H = 1.0$. Especially for the c(2 × 6) phases, the surface unit cells are comparatively large [206, 268, 315], indicating long-range periodicities, which are, however, not due to direct wide-range interaction but are rather a consequence of indirect "through-bond" interactions with oscillatory character [316], or indicate elastic surface strain that is compensated by slight rumpling in the direction perpendicular to the rows and troughs. On the bcc (211) surface of iron also, a rich variety of ordered H phases has been observed. In the submonolayer coverage range, the surface remains unreconstructed; however, as $\Theta_H$ exceeds 1 ML, a clear surface reconstruction sets in with partly huge unit cells containing up to 20 H atoms [317]. On the Rh(110) surface, too, several ordered H phases exist in the coverage range $0.33 \leq \Theta_H \leq 2.0$, whereby restructuring effects remain moderate [195]. Interestingly, unlike Ni(110)/H, no ordered (2 × 1) phase appears around $\Theta_H = 1.0$. This phase definitely exists on the Pd(110) surface, where in addition subsurface sites are populated under appropriate temperature and coverage conditions [210, 218].

Worthy of mention is the formation of ordered H phases on the densely packed bcc (110) surfaces, a prominent example being the Fe(110) surface, where two ordered H phases could be ensured, namely a c(2 × 2) and a (2 × 2)-2H structure [94, 277, 276], which could reversibly be transformed into each other. At sufficiently high temperatures, the H atoms are fluctuating between neighboring threefold-coordinated long-bridge sites [318]. As dynamical LEED analyses have shown, the H phase formation is accompanied by a slight restructuring of the Fe surface in that the H atoms push Fe atoms toward the bulk [94]. On the geometrically similar W(110) surface, two ordered H structures were reported for the submonolayer coverage range, a p(2 × 1) phase reflecting $\Theta_H = 0.5$, and a (2 × 2) phase associated with $\Theta_H = 0.75$ [293]. The critical temperatures for both phases are below 270 K, thus supporting the assumption that a H-induced reconstruction is not involved – as expected for a densely packed bcc surface.

We conclude this section by simply mentioning that numerous theoretical investigations have been devoted to the formation of ordered hydrogen phases on metal surfaces with the aim to relate the phase formation with the operation of lateral indirect interaction forces. On the many computations, we mention the extensive work by Newns and Muscat who calculated, for a variety of surface geometries of TM surfaces, possible hydrogen superstructures using an embedded cluster approach and considering the interaction energies of groups of neighboring atoms [319–321, 187], often in good agreement with the experiment. Later, Kresse and Hafner modeled the adsorption of H atoms on the low-index surfaces of Ni by means of first-principles calculations and considered especially the ordered (2 × 2)-2H phase [309]. More recent DFT-GGA work by Greeley and Mavrikakis was likewise concerned with the energetics and the (2 × 2)-2H phase formation on the Ni(111) surface [322], and the experimental behavior could be well predicted by theory.

**36.2.6.2.2 Ordered H Phases with Reconstruction** As we have repeatedly emphasized, the tiny adsorbing H atoms can really "dip" into the surfaces and at least locally perturb the substrate's electronic structure markedly. This often results in heavy restructuring of the surface atomic layers, which usually produces intense "extra" diffraction spots in a LEED experiment. These are a good indicator of the degree of the restructuring phenomena, and a dynamical LEED analysis then helps to assess the displacements of the substrate's surface atoms. However, it is very difficult to reliably determine the location of the merely weakly scattering H atoms.

From the wealth of systems that reconstruct under hydrogen, we select the famous H-on-W(100) system that has kept surface scientists busy for quite a number of years [291, 323–326]. The clean W(100) surface is somewhat unstable; in order to reduce the surface free energy, the W atoms tend to get closer together and form zig-zag chains in the [11] direction, thereby causing a c($\sqrt{2} \times \sqrt{2}$) R45° LEED pattern with *p2mg* symmetry (in another notation, the superstructure is assigned as c(2 × 2)). Upon hydrogen adsorption, this reconstruction changes in that the zig-zag chains dissolve and give way for a dimer structure in [01] direction, which then produces a ($\sqrt{2} \times \sqrt{2}$) R45° pattern with *c2mm* symmetry. Figure 36.24, taken from Ref. [326], displays the situation.

The other frequently studied adsorption system with strong H-induced reconstruction is the H-on-Ni(110)-system, where actually two different types of reconstructions occur as repeatedly mentioned earlier. The first PR reconstruction takes place on the *cold* Ni surface at $T < 220$ K and shows up as sharp and fairly bright "extra" spots of a (1 × 2) structure. The critical coverage here is $\Theta_H = 1.0$, because as soon as the coverage reaches one monolayer, the former unreconstructed (2 × 1)-2H phase transforms (via appropriate island growth) to the (1 × 2) structure with paired rows in [$1\bar{1}0$] direction. At $T > 220$ K, prolonged hydrogen exposure causes a missing/AR reconstruction with less pronounced ordering, as

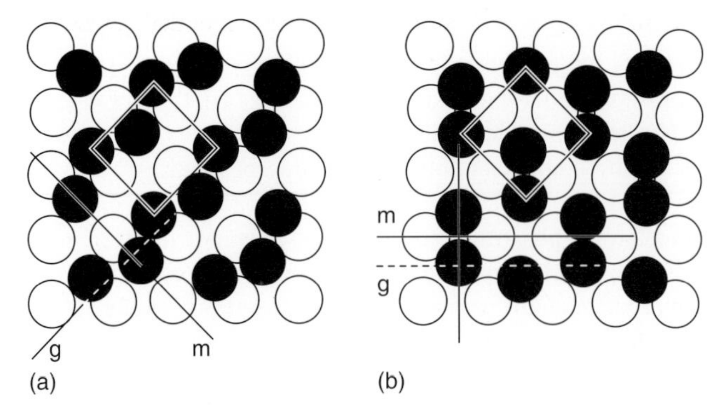


**Figure 36.24** Structure models of the reconstructing W(100) surfaces, with the W *surface* atoms shown as dark circles. (a) Clean surface with pair formation along the (1,1) glide plane (g), leading to a zig-zag model with *p2mg* symmetry. (b) W(100) surface with adsorbed hydrogen (H atoms not shown). The W dimers now form along (1,0) direction, resulting in the dimer model with *c2mm* symmetry. Unit cells and glide and mirror planes are indicated. (After Barker and Estrup [291, 323].)

indicated by intense but always streaky LEED $(1 \times 2)$ "extra" spots. The STM work performed on this system has provided valuable supporting structural information as mentioned already in Section 36.2.4.2 [189, 208, 209].

Besides the just discussed cases, there are a number of other metal surfaces that do reconstruct under hydrogen: molybdenum with its (100) surface in the first place [327], but also Cu(110), Pd(110), Ag(110), and so on. For more details, we refer the reader to the listing of systems presented in Ref. [50]. Finally, we emphasize again that the driving force for all these restructuring processes is the overall energetics, that is, the tendency of the systems to reduce their surface free energy under the influence of adsorbed H atoms. However, the presence of kinetic activation barriers may in some cases require certain temperature and coverage conditions for the reconstructions to actually occur.

#### 36.2.6.3 Phase Transitions within Adsorbed H Layers

The thermodynamic behavior of the ordered H phases provides a key to getting access to the kind and strength of the lateral H–H interactions. This has been demonstrated in quite a number of cases, and a voluminous literature has been accumulated regarding both experimental and theoretical contributions. The degree of long-range order within the hydrogen phases reflects primarily the configurational interaction (CI), which depends on the sign, magnitude, and range of the H–H interaction forces as well as on the lattice geometry and number of neighbors. For more details, the reader is referred to the pertinent literature on two-dimensional phase transitions within adsorbed phases [328–337].

Usually, the intensity of electron beams diffracted at the grating of the adsorbate is a good measure of the long-range order, and, hence, measurements of the LEED intensity of a fractional-order beam as a function of coverage and temperature provide valuable insight into the CI [331]. For fixed experimental settings of the LEED parameters (primary beam energy, polar, and azimuthal angles of incidence), constant atomic scattering factors $f$ of the adsorbed particles, and a given coverage $\Theta$, the fractional-order beam intensity $I$ is solely a function of the degree of long-range order within the adsorbate layer:

$$I \propto G(\text{order}) \exp(-2\,W), \tag{36.16}$$

where $\exp(-2W)$ stands for the Debye–Waller factor and accounts for the temperature dependence. This configurational part of the LEED intensity, $I_{\text{conf}}$, is obtained by the summation over the individual scattering contributions of all (hydrogen) adatoms within the coherence zone of the LEED experiment containing $M^2$ adsorption sites:

$$I_{\text{conf}} = \frac{C}{M^4} \sum_{r} \sum_{r'} \langle f_r f_{r'} \rangle \exp\left(i\Delta k(r - r')\right), \tag{36.17}$$

where $C$ is a constant, $\Delta\mathbf{k} = \mathbf{k} - \mathbf{k}'$ is the difference between the wave vectors of the scattered and the primary electrons, and $\mathbf{r}$ is the location of the adsorption site with scattering factor $f_{\text{r}}$. $f_{\text{r}}$ is set equal to unity if the site $\mathbf{r}$ is actually occupied by an adsorbed atom and equal to zero if the site is empty. The brackets $\langle\ \rangle$ denote the ensemble average.

In the simplest approximation, a lattice gas model with only pairwise interactions is assumed, and at low temperatures the lateral mobility (i.e., thermal disorder) is sufficiently suppressed so as to keep the majority of the H atoms in their "correct" sites of the superlattice, reasonably contributing to the fractional-order beam intensity. Higher temperatures, however, improve the mobility of the adsorbed atoms, which then spend less time in their "correct" sites. Right at the border between the ordered and disordered (i.e., lattice gas) phase, a further temperature increase produces overall disorder, which is indicated by a sudden breakdown of the fractional-order beam intensity. In turn, a decrease of $T$ to below the order–disorder borderline makes the respective intensity to reappear. Concerning the H coverage, it is clear that the "extra" intensity becomes higher when $\Theta_{\text{H}}$ gets nearer to the optimum coverage of the ordered phase. In summary, a measurement of the degree of order (which is peoportional to fractional-order beam intensity $I$) as a function of coverage and temperature enables one to construct a two-dimensional phase diagram, in which the range of existence of the ordered phase(s) is plotted in a $T$ versus $\Theta$ or a $T$ versus $\mu$-plane diagram, with $\mu$ denoting the chemical potential of the adsorbate.

In the following, we select again the H-on-Ni(111) system to illustrate what has been said above. At temperatures below 270 K, adsorbed H atoms form the frequently mentioned ordered $(2 \times 2)$-2H phase with its optimum coverage of $\Theta_{\text{H}} = 0.5$. The structure model of Figure 36.25 shows the well-known graphitic or honeycomb-like array of the H atoms, which alternatively occupy the fcc and hcp sites. We have followed the fractional-order LEED beam intensity $I_{1/2,1/2}$ at a given coverage $\Theta_{\text{H}}$ as a function of temperature and obtained the $I(T)$ curves displayed in Figure 36.26. They exhibit an S-like shape, and as pointed out in Ref. [14], the inflection points can be taken as an indicator of the temperature at which the order–disorder phase

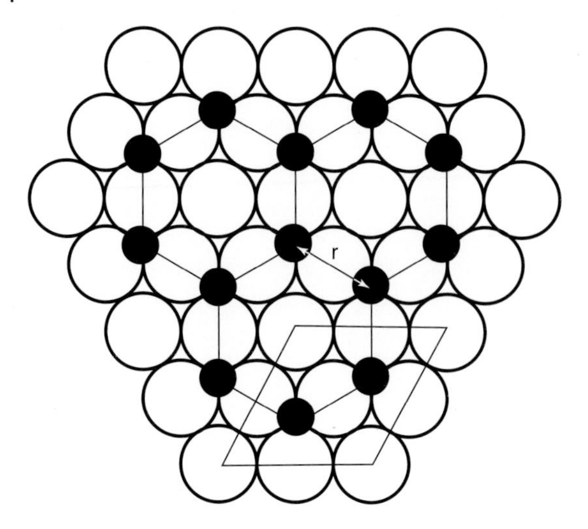


**Figure 36.25** Structure model of the graphitic (2 × 2) phase formed by half a monolayer of adsorbed H atoms on the Ni(111) surface. The "honeycomb" geometry is indicated as well as the (nonprimitive) unit mesh with the H atoms (dark circles) being adsorbed in nonequivalent sites of trigonal symmetry. The nearest distance between two H atoms in the graphitic phase is $r = \frac{a_0}{3}\sqrt{6} = 2.87$ Å with $a_0$(Ni) = 3.52 Å.

transition occurs. The parameter of the curves is the hydrogen coverage $\Theta_H$. Putting the information together yields the phase diagram shown in Figure 36.27, which is not symmetric with respect to $\Theta_H = 0.5$. Monte Carlo studies with lattice gas systems of square [333] and triangular geometry [334] having only pairwise mutual interactions yield, instead, phase diagrams that are *symmetric* with respect to $\Theta = 0.5$. The asymmetry of the H-on-Ni(111) phase diagram is, however, evident and could be naively understood as follows: at $\Theta_H = 0.25$, long-range order persists up to about 150 K and disorder sets in for $T > 150$ K. As more H atoms get adsorbed, they can gradually populate the "correct" sites of the honeycomb lattice until at $\Theta_H = 0.5$ the optimum order is achieved and the fractional-order beam intensities have reached their maximum value. In this configuration, a homogeneous mutual distance between adjacent H atoms of 2.87 Å is reached. At the same time, the temperature at which the order just persists, that is, the *critical* temperature $T_c$, reaches a maximum value of 270 K in this particular case. Beyond $\Theta_H = 0.5$, however, there are no longer equivalent adsorption sites available where H atoms could avoid having a smaller distance than 2.87 Å – at least not without entirely destroying the graphitic structure, since for distances $d < 2.87$ Å repulsive interactions come increasingly into play. This easily explains why for $\Theta_H > 0.5$ the overall order breaks down rapidly with coverage and the fractional-order beam intensity tends to zero already at $\Theta_H = 0.6$.

In statistical mechanics, there are various interaction models that can predict the degree of long-range order, that is, the configuration of the adatoms, as a function of temperature and coverage for a set of fixed lateral interaction energies [332]. For strictly pairwise interaction energies $E_\omega$ and for square lattice geometry, the famous

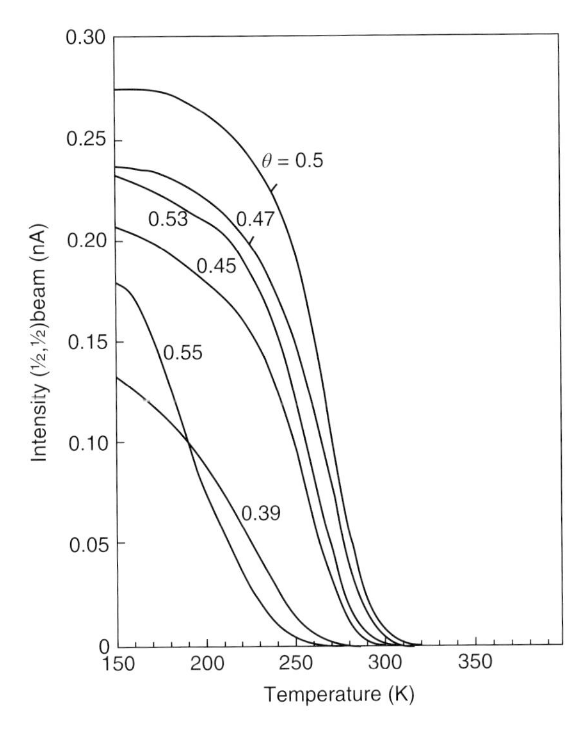


**Figure 36.26** Temperature dependence of the fractional-order beam intensity $I_{1/2,1/2}$ of the $(2\times2)$ structure formed by hydrogen at $\Theta = 0.5$ on Ni(111). The curves have been measured at various initial H coverages, as indicated in the figure. (After Ref. [14].)

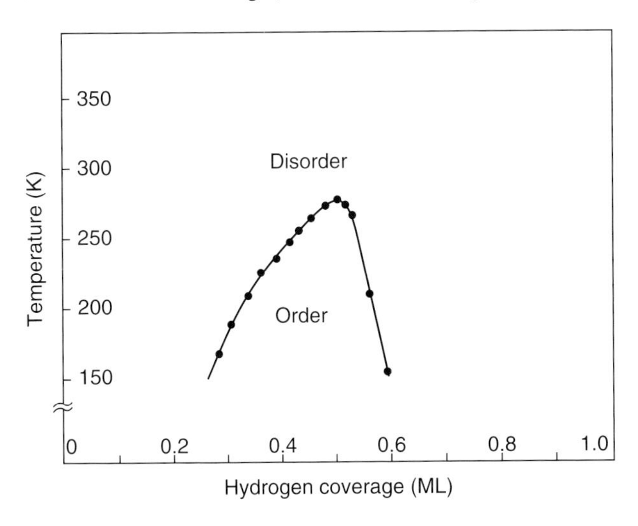


**Figure 36.27** Phase diagram for the $(2\times2)$-2H phase on Ni(111). The temperatures at which the order breaks down were derived from the points of inflection of the intensity (temperature) curves of Figure 36.26. The critical temperature could be determined as $T_c = 270\,\text{K}$. Note that the diagram is asymmetric with respect to $\Theta = 0.5$.

two-dimensional Ising model (initially developed for a magnetic spin lattice [335]) provides a relation between $E_\omega$ and the critical temperature $T_c$:

$$E_\omega = 2 \cdot \ln(1 + \sqrt{2}) k_B T_c = 1.76 k_B T. \quad (36.18)$$

For lattices with a different (e.g., trigonal) geometry, similar relations will hold, although the treatment of the honeycomb lattice symmetry requires a little more effort [330]. Nevertheless, Equation 36.18 allows at least a rough estimate of the (repulsive) pairwise lateral interaction energies. Inserting $T_c = 270$ K yields $E_\omega \approx 4$ kJ/mol, which appears quite reasonable in terms of the experience that $E_\omega$ usually ranges around 1/10 to 1/20 of the heat of adsorption.

More quantitative information concerning the sign and magnitude of the H–H interactions can be obtained from several theoretical publications that were explicitly devoted to a description and interpretation of the experimental phase diagram of H on Ni(111) [336, 337]. Domany, Schick, and Walker proposed the honeycomb structure also from a theoretical viewpoint and explained the asymmetry of the phase diagram by the fact that there existed two threefold sites per Ni surface atom. In this case, the lattice gas density $n$ (which is the decisive quantity in the statistical consideration) is related to the H coverage as $n = 0.5\ \Theta$. The authors further concluded that, in a hexagonal array, a given H atom interacts with its nearest and second-nearest neighbors via repulsive forces, whereby the repulsion $\omega_1$ between nearest neighbors at a distance of 1.44 Å is much larger than that between second neighbors, $\omega_2$, at a distance of 2.49 Å. The exclusive operation of these repulsive interactions, however, yields a markedly narrower phase diagram than that observed in the experiment, and the authors argue that additional third-neighbor attractions must also be involved. For the second-neighbor repulsions, they estimate $\omega_2 = 0.066$ eV = 0.64 kJ/mol at a mutual distance of 2.49 Å, a similar value that was reported for the H-on-W(100) system [291, 323].

Another example for a phase diagram of adsorbed hydrogen displaying only a single ordered phase is the H-on-Pd(100) system, where a c(2 × 2) structure appears around $\Theta_H = 0.5$ and below $T = 260$ K [117]. From the temperature dependence of the fractional-order LEED beam intensity, again the phase diagram was constructed, which is shown in Figure 36.28. The ordered c(2 × 2) phase is again not entirely symmetric with respect to $\Theta = 0.5$, but not as asymmetric as in the Ni(111) case. Conspicuous is its large width as compared to theory, which points to additional attractive interactions [333]. From arguments presented in Ref. [117], one can summarize that the first-neighbor repulsions $\omega_1$ are about 2.5 kJ/mol, while the attractions between second neighbors is assumed to range between 0.4 and 1.7 kJ/mol. Furthermore, the slight asymmetry of the phase diagram could point to the fact that the H–H interactions are not strictly pairwise but contain contributions of three-particle ("trio") interactions [338].

We conclude by recalling the remarks made in Section 36.2.6.2.2 about H phases without surface reconstructions. In most of the experiments, the lowest attainable surface temperature was about 60–70 K, and in the majority of cases more than just a single ordered H phase was found: three phases appeared on Ru(0001) [283], two on Pd(111) [238, 239], and two on Fe(110) [277]. The anisotropic surfaces with twofold

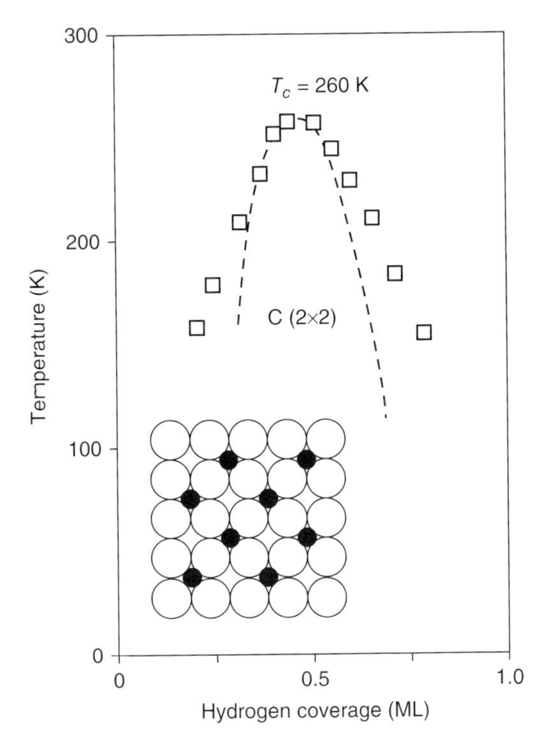


**Figure 36.28** Phase diagram for the c(2 × 2) phase formed by hydrogen on the Pd(100) surface around a coverage of $\Theta = 0.5$. Open squares denote the experimental data, and the dashed line describes a theoretical prediction based on repulsive nearest-neighbor interactions by Binder and Landau [333]. Apparently, the experimental PD is slightly asymmetric with respect to $\Theta = 0.5$ and has an extended width, which points to additional attractive and not strictly pairwise interactions. The critical temperature is $T_c = 260$ K. In the c(2 × 2) phase, the H atoms are located in the fourfold hollow sites, as indicated in the ball model of the inset.

symmetry regularly show a larger variety of phases, good examples being the Fe(211) [317], the Ni(110) [206], the Ru(10$\bar{1}$0) [135], and the Rh(110) [195] hydrogen systems. The lateral interaction energies (in case where they have been determined) are of similar order of magnitude as discussed above; typically, the first neighbor repulsion is dominant.

### 36.2.7
### Surface Vibrations of Adsorbed Hydrogen

Vibrations of adsorbed particles contain a lot of extremely valuable information. The vibrational frequency maps the strengths of the adsorptive chemical bond and is determined by the vibrating masses and the force constant(s) of the involved bonds. For minor elongations of the vibrational amplitude, the nuclei will stay near the bottom of the chemisorption potential $U(z)$ and "feel" a harmonic potential. Consequently, the harmonic oscillator formalism can be successfully applied, and one observes a single characteristic frequency $\nu_0$ only. For larger elongations, anharmonicity effects have to be taken into account, with the consequence that overtones $\nu_i \neq \nu_0$ can appear. During irradiation with an electromagnetic field, there is interaction between the electrical field vector and the dynamic dipole moment(s) of the vibrating molecule, leading to excitation/de-excitation of vibrations and absorption/emission of IR radiation.

Typically, molecular force constants are $\sim$1 eV/Å, and vibrational frequencies between 500 and 3500 $cm^{-1}$ are expected. The respective wavelengths range between $2 \times 10^{-3}$ and $3 \times 10^{-4}$ cm, and the associated energies are typically in the infrared region of the electromagnetic spectrum. Accordingly, both infrared and Raman spectroscopy are appropriate tools to determine surface vibrational modes. With small single crystal samples, however, one often faces an intensity problem, and various improvements have been developed to increase the detection sensitivity. Greenler was the first to point to the fact that the absorption per reflection for the parallel component of the incident radiation should be very sensitive to the angle of incidence, with a sharp peak around 88° to the surface normal [339]. This was the starting point for the IRAS (infrared reflection–absorption spectroscopy) technique, where the absorption of infrared radiation due to the excitation of surface vibrations of the adsorbed species is measured after reflection from a flat (usually metal) substrate surface. Other optical vibrational spectroscopies are infrared emission spectroscopy (IES) [340] and inelastic electron tunneling spectroscopy (IETS) (see Section 36.2.5) [234, 235]. For more details about vibrational spectroscopy methods, we refer to the literature [341–343] and recommend a more recent overview of the present state of the art by Hirschmugl [344].

While optical infrared spectroscopy has the advantage that it affords a very high spectral resolution of a few wavenumbers only, there is still a lack of sensitivity and detection problem for low and very low vibrational frequencies. Here, HREELS has become one of the most successful experimental tools to detect surface vibrational modes, especially of adsorbates on metal surfaces. For more details about the physics of low-energy electron scattering, selection rules, and instrumentation, we refer the reader to the monograph *Electron-energy Loss Spectroscopy and Surface Vibrations* by Ibach and Mills [345] and a comprehensive article by Froitzheim [346]. Concerning the physics of HREELS, it suffices to say here that two main scattering mechanisms operate, namely dipole scattering (very similar to what has been said above for optical IR spectroscopy), and impact scattering where the impinging electron is captured for a short period in an orbital of the adsorbate, leading to a short-lived negative ion resonance that also contributes to the scattering intensity.

The experimental setup of an HREELS experiment is as follows. A highly monochromatized beam of low-energy electrons ($1 < E_p < 10$ eV) is directed to the (metallic) surface at an angle $\alpha$ (often chosen as 45°). As with LEED, there is a high cross section for interaction of these electrons with matter, making HREELS highly surface sensitive. In the dipole scattering mechanism indicated above, the Coulomb field associated with the impinging electrons interacts with the dipole moments of the vibrating surface particles, whereby the electric field vector couples to the perpendicular component of the surface dipole moment and excites appropriate vibrations $\nu_i$. The required excitation energy $E_i$ is quite small, that is, between 0.01 and 0.5 eV, and withdrawn from the overall energy distribution of the scattered electrons during the absorption process. The electrons backscattered in specular direction are then collected in an electrostatic analyzer and sorted with high resolution with respect to their kinetic energy. The inelastic electrons that have lost $E_i$ now appear in the energy distribution at an energy $E_p - E_i$ as small loss peaks usually quite close to the intense elastic peak of energy $E_p$, which requires the aforementioned high resolution. In contrast to the IRAS technique, HREELS is capable of detecting low-energy vibrations with low frequencies; in addition, it is quite sensitive and can detect still a couple of percent of a monolayer of adsorbed particles provided there is sufficient cross section for vibrational excitation. A textbook example is carbon monoxide adsorbed on metal surfaces, where both the (high-frequency) C–O stretching vibration and the (low-frequency) CO–metal vibration can be detected even for surface concentrations of <1% of a monolayer.

HREELS can be well applied also to adsorbed hydrogen, and our present knowledge about H-surface vibrational modes on metal and semiconductor surfaces has greatly benefited from the HREELS technique, beginning perhaps with the early work on H adsorbed on the W(100) surface by Probst and Piper [347] and subsequent studies on the same system by Froitzheim *et al.* [348] and Ho *et al.* [349]. More detailed information is available from several review articles and monographs [341, 346, 350–352]. Hamann has performed LAPW (linear augmented plane wave) calculations for hydrogen vibrations at TM surfaces [351]. For adsorbed hydrogen, a relatively wide range of frequencies must be considered – from 400 $cm^{-1}$ (for deuterium atoms adsorbed on Pt(111) [349]) to more than 4000 $cm^{-1}$ for the H–H stretching vibration of molecularly adsorbed $H_2$, (see Section 36.2.2.2). Compared to CO on TM surfaces, which is detected by HREELS with high sensitivity, the cross section for exciting H atoms is comparatively small because of the smaller dipole moment associated with the hydrogen–substrate vibrations caused by the close proximity of the adsorbed H atom to the surface and the usually smaller charge separation. The fact that only the perpendicular component of the dynamic dipole moment can couple with the impinging electrons provides an important symmetry condition, which is known as the *surface selection rule*. It states that only those vibrations can be dipole-excited and seen in the specular electron beam that possess a component of the dynamic dipole moment perpendicular to the surface. Recalling what we have said before, one can delineate dipole scattering from impact scattering contributions by measuring the angular dependence of the scattered electron intensity, for only the "dipole peaks" have a sharp maximum in the specular direction. Concerning the optimum sensitivity for *impact scattering*, it is advisable to adjust the primary electron beam energy appropriately, since in contrast to dipole scattering the cross section for electron absorption can strongly depend on energy. This has been first observed by Baró *et al.* [353] for H on Pt(111) and later confirmed by Conrad *et al.* for H on Ru(0001) [354] and on Pd(111) [355] as well as by Mate and Somorjai [302] for H on Rh(111). Apparently, the electron resonances during the short-range impact process are strongly energy-dependent. Impact scattering provides valuable symmetry selection rules that can be summarized as follows [350]:

1) The inelastic scattered intensity is zero for a vibration that is odd with respect to a mirror symmetry plane if the trajectories of the incident and the scattered electrons are both in this mirror plane.
2) The inelastic scattered intensity also vanishes in the specular direction for a vibration that is odd with respect to a mirror symmetry plane. The plane of the incident and the scattered beam must be perpendicular to the mirror plane and the surface.
3) The inelastic scattering intensity goes to zero in the specular direction if the considered vibration is odd with respect to twofold rotation symmetry.

Without going into details, the application of these selection rules (by varying the geometrical incidence and detection planes as well as the detection angle) supports the assignment of the H vibrations particularly for surfaces with twofold symmetry such as Ni(110) or Ru($10\bar{1}0$) [135, 356] and allows deriving conclusions on the H adsorption site that is taken at different coverages.

Another big advantage is that deuterium can be adsorbed instead of hydrogen, and the 100% mass difference causes easily detectable red shifts in frequency by a $\sqrt{2}$ factor, which often allows an unambiguous assignment of a certain vibrational mode.

Within a simple spring model and the harmonic approximation, the number and frequencies of the hydrogen vibrational modes can be correlated with the local geometry of the adsorption site, and hence HREELS provides invaluable information about the nature of this site. Following Hamann [351], we show in Figure 36.29a the high-symmetry sites that can be occupied by hydrogen atoms on a low-index

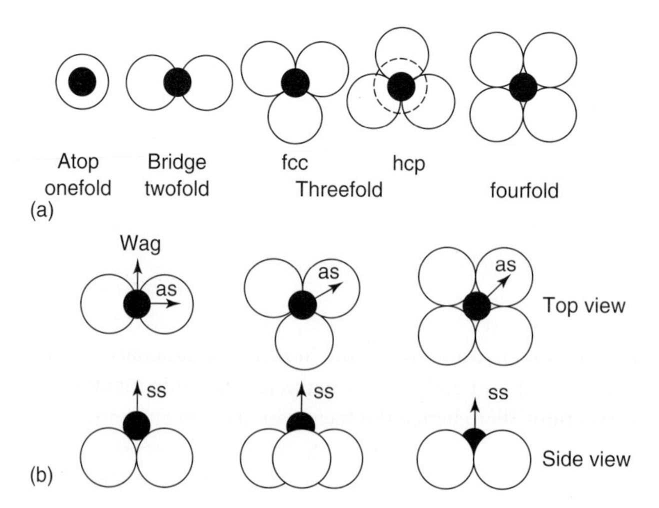


**Figure 36.29** Correlation between site symmetry and vibrational modes after Hamann [351]: (a) High-symmetry sites for hydrogen adsorption on low-index surfaces. (b) Vibrational modes of the high-symmetry sites of (a). *ss* denotes the symmetric stretch, *as* the asymmetric stretch, and *wag* the wagging mode.

metal surface and in Figure 36.29b the possible vibrational modes of H atoms adsorbed in these sites. Symmetry rules predict that three and fourfold sites possess only two inequivalent vibrational modes: the vertical mode is called the "symmetric stretch" (ss), and the two degenerate parallel modes the "asymmetric stretch" (as). Note that, within the harmonic spring model, any direction of horizontal motion has the same force constant for these sites. For the bridge-bonded H atom, the asymmetric stretch (as) mode, however, is inequivalent to the wagging mode, and three fundamental frequencies can be observed, that is, the ss-, as-, and wag mode. This can be demonstrated using the classical example of "H on W(100)," and we refer explicitly to the early work by Ho *et al.* [349]. In Figure 36.30, typical loss spectra obtained for the saturation coverage of hydrogen on W(100) are displayed. This surface with its fairly large squared unit mesh can accommodate two H atoms per W surface atom, leading to a saturation coverage of $\sim 2 \times 10^{19}$ H atoms/$m^2$ [61, 357], with the H atoms being located in bridge sites. Figure 36.30a presents the loss spectrum in the specular direction (primary energy $E_p = 9.65$ eV, angle of incidence $\alpha = 23°$). Only a single loss is observed at $h\nu_1 = 130$ meV (1048 $cm^{-1}$) and identified as the symmetric stretching vibration normal to the surface ($A_1$ mode). However, if the electrons are detected in off-specular direction ($\Delta\alpha = 17°$), three additional modes appear at 80, 160, and 260 meV (= 645, 1290, and 2096 $cm^{-1}$), while the main 130 meV loss remains dominant. This is shown in Figure 36.30b, together with the vibrational assignments made by the authors. All modes shift by $\sim\sqrt{2}$ when deuterium is used instead of H and are thus unequivocally identified as H vibrations. The 260 meV mode is taken as the first overtone ($\upsilon = 0 \rightarrow 2$) of the stretching mode $\nu_1$, while the 80 meV loss $\nu_2$ is due to the out-of-plane bending vibration. The $\nu_3$ shoulder mode at 160 meV is denoted as the asymmetric stretch. Apparently, all fundamental modes of the hydrogen–surface complex can be excited, and since the 130 meV mode has its intensity maximum in the specular direction, it is clearly associated with the dipole-active symmetric stretching vibration $\nu_{ss}$ as stated above.

In this way, the vibrational spectra of many H-on-TM adsorption systems have been measured and analyzed in the past, and we present in Table 36.9 a collection of the respective vibrational data. A much more extensive compilation can be found in Ref. [50]. From the overall body of data, one can roughly correlate the frequency ranges with the geometry of H adsorption sites: H atoms bound to a single atom have vibrations in the 2200–1600 $cm^{-1}$ range; edge-bridging H atoms lead to absorption between 1400 and 800 $cm^{-1}$; while triply coordinated adsorbed H yields absorption features even below 800 $cm^{-1}$ [377]. In addition to the "normal" vibrational behavior, however, various peculiar effects can additionally happen – we only mention H-induced surface reconstructions that change the local geometry of the adsorption site(s) and greatly affect the vibrational spectra. Furthermore, the issue of subsurface hydrogen has gained considerable interest also in the analysis of vibrational modes – depending on the coverage and temperature conditions, H atoms can move to subsurface sites and get back to surface positions where they are detectable – with the consequence that the respective vibrational modes disappear or reappear. An example may be the H + Cu(111) system, which has been revisited recently [378].

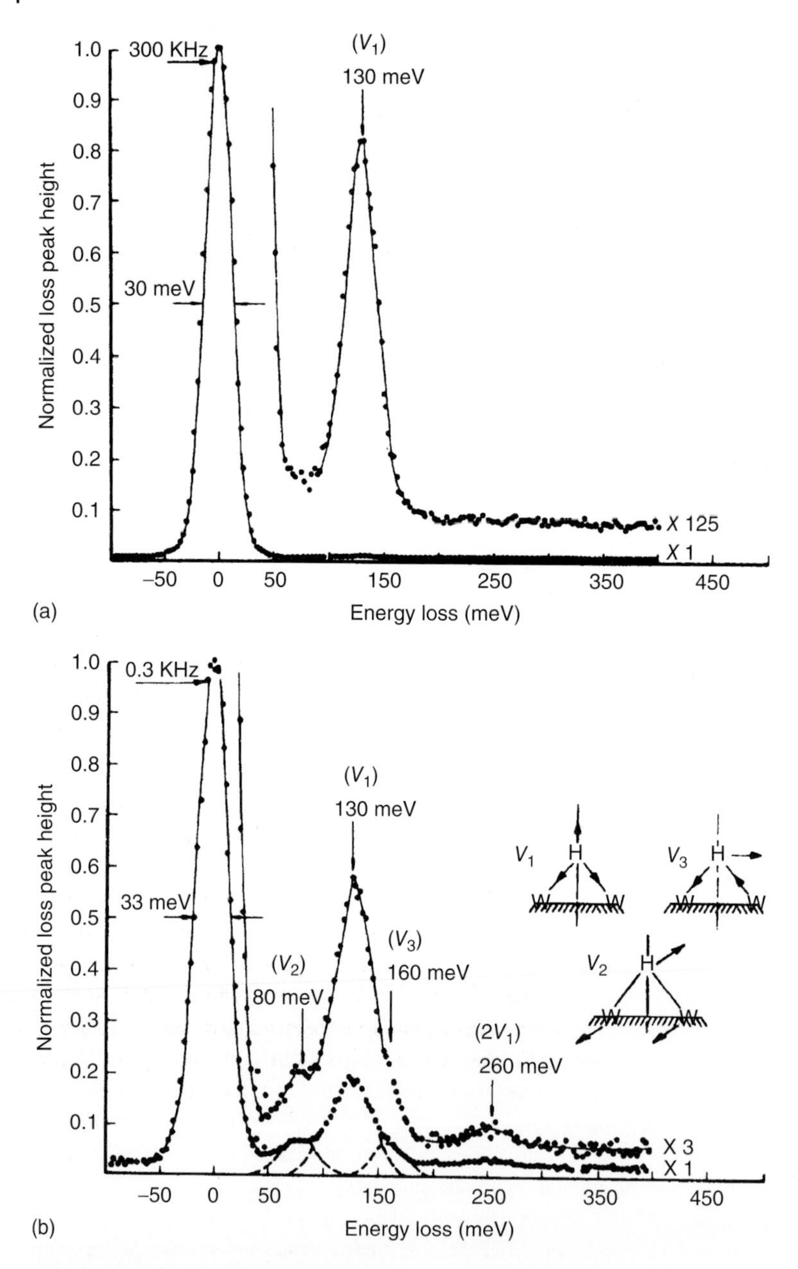


**Figure 36.30** Normalized high-resolution electron-energy loss spectra for the saturation coverage of H chemisorbed on the W(100) surface, with $\alpha_I = 23°$ electron incidence angle and primary beam energy $E_p = 9.65$ eV. The incident beam is along the [100] crystal direction. (After Ho *et al.* [349].) (a) Specular direction and (b) +17° off the specular direction toward the surface. The fundamental vibrational modes are indicated in the inset and are compatible with bridge-site bonding ($C_{2v}$ symmetry).

**Table 36.9** Selection of vibrational modes of hydrogen adsorbed on various solid surfaces.

| Surface | H coverage (ML) | *T* (K) | Experimental frequencies or loss bands | | Mode assignment (ss = symmetry stretch; as = asymmetry stretch) | Remarks | References |
|---|---|---|---|---|---|---|---|
| | | | ($cm^{-1}$) | (meV) | | | |
| Be(0001) | <0.38 | 80 | 1492 | 185 | Be-H stretch (bridge site) | Exposure to H atoms | [159] |
| C(0001) | 0.5 (sat'n) | 300 | 1210 | 150 | C-H bending | HOPG was exposed to H atoms | [358] |
| | | | 2650 | 329 | C-H stretching | | |
| Al(111) | 1.0 | 85 | 800 | 99 | Bending (onefold) | Exposure to H atoms; preformed $AlH_3$ molecules at surface | [359] |
| | | | 1700 | 211 | Stretching (onefold) | | |
| | | | 1200 | 149 | Stretching (twofold) | | |
| Si(100)-(1 × 1)H | 1.0 | 500 | 630 | 78 | Monohydride scissor mode | Exposure to H atoms | [360] |
| | | | 2080 | 258 | Monohydride stretch | | |
| Si(111)-(1 × 1)H | 1.0 | 300 | 627 | 78 | as (bending mode) | Exposure to H atoms | [361] |
| | | | 2084 | 258 | ss (Si-H sym. stretch) | | |
| Fe(100) | Low covererages | | 700 | 87 | ss (fourfold-hollow) | | [275] |
| | High coverages | | 1000 | 124 | as (fourfold hollow) | | |
| Fe(110) | | 300 | 1060 | 132 | ss (short-bridge) | Specular | [362] |
| | | | 880 | 109 | | nonspecular | |
| Ni(100) | 1.0 | | 645 | 80 | ss (fourfold hollow) | Collective as mode with large dispersion | [363] |
| | | | 700–850 | 87–105 | as (fourfold hollow) | | |
| Ni(110) | 1.0 | 100 | 637 | 79 | Low-symmetry short bridge sites | (2 × 1)-2H phase | [364] |
| | 1.5 | | 1049 | 130 | Low-symmetry + high-symmetry short bridge sites | (1 × 2)-3H phase | |
| | | | 613 | 76 | | | |
| | | | 944 | 117 | | | |
| | | | 1210 | 150 | | | |

(*continued overleaf*)

**Table 36.9** *(Continued).*

| Surface | H coverage (ML) | *T* (K) | Experimental frequencies or loss bands | | Mode assignment (ss = symmetry stretch; as = asymmetry stretch) | Remarks | References |
|---|---|---|---|---|---|---|---|
| | | | ($cm^{-1}$) | (meV) | | | |
| Ni(111) | 0.05–0.5 | 100 | 726 | 90 | as (threefold) | | [365] |
| | >0.5–1.0 | | 1048 | 130 | ss (threefold) of honeycomb phase | | |
| | | | 726 | 90 | Additional modes due to (1 × 1)-H phase at Θ = 1.0 | | |
| | | | 927 | 115 | | | |
| | | | 1129 | 140 | | | |
| Cu(100) | Low Θ | 83 | 565 | 70 | ss (fourfold hollow) | Exposure to H atoms; authors report on H-driven reconstruction with *p4g* or *pgg* symmetry | [366] |
| | High Θ, up to 1.03 ML | | 565 | 70 | Modified fourfold hollow | | |
| | | | 468 | 58 | ss and as twofold (bridge-bonded) | | |
| | | | 948 | 118 | | | |
| Cu(111) | <0.5 | 100 | 806 | 100 | as (threefold-hollow) | Exposure to H atoms | [367] |
| | >0.5 | | 1040 | 129 | ss (threefold-hollow) | | |
| | | | 927 | 115 | New site due to (3 × 3) phase | | |
| Ge(100)-(2 × 1)H | | | 532 | 66 | as (Ge-H bending) | Exposure to H atoms | [368] |
| | | | 1976 | 245 | ss (Ge-H stretch) | | |
| Ru(0001) | 0–1.0 | 170 | 820 | 102 | as (threefold hollow) | | [369] |
| | | | 1137 | 141 | ss (threefold hollow) | | |
| Rh(100) | 1.0 | 90 | 528 | 65 | as (fourfold-hollow) | Various overtones reported | [370] |
| | | | 661 | 82 | ss (fourfold-hollow) | | |
| Pd(100) | 0.08 | 80 | 486 | 60 | ss (fourfold-hollow) | | [371] |

| | | | | | | | |
|---|---|---|---|---|---|---|---|
| | 0.59 | | 502 | 62 | ss (fourfold-hollow) | c(2 × 2) phase | |
| | 0.96 | | 512 | 63.5 | ss (fourfold-hollow) | (1 × 1)-H phase | |
| Pd(110) | 1.0 | 100 | 790 | 98 | as (quasi-threefold sites) | (2 × 1)-2H phase | [372] |
| | | | 968 | 120 | ss (quasi-threefold sites) | | |
| | >1.0 | | 790 | 98 | ,, | (1 × 2) reconstruction causes only line broadening | |
| | | | 968 | 120 | ,, | | |
| Pd(111) | ~0.4–0.6 | 120 | 774 | 96 | as (threefold-hollow) | | [355, 373] |
| | | | 1000 | 124 | ss (threefold-hollow) | | |
| Ag(110) | 0.1–1.0 | 100 | 484 | 60 | as (tilted trigonal sites) | Exposure to H atoms | [152] |
| | | | 847 | 105 | ss (tilted trigonal sites) | | |
| W(100) (1 × 1)-2H | 1.4 | 100 | 1100 | 136 | ss (twofold bridge) | RAIRS expts | [374] |
| | 1.65–2.0 | | 1070 | 133 | ss (twofold bridge) | | |
| | | | 1260 | 156 | 2 $\nu_2$ overtone wagging | | |
| Ir(111) | 0.4–0.6 | 170 | 560 | 69 | ss (threefold hollow site) | High-frequency mode falsely attributed to CO contamination | [375] |
| | | | 2025 | 251 | | | |
| Ir(111) | <0.4 | 90 | 525 | 65 | ss (threefold-hollow) | Excitation into protonic bands | [294] |
| | >0.44 | 90 | 2030 | 252 | ss (onefold terminal site) | | |
| Pt(111) (1 × 1)-H | ~1.0 | 85 | 540 | 67 | as (threefold hollow) | | [376] |
| | | | 903 | 112 | ss (threefold hollow) | | |
| | | | 1234 | 153 | Overtone + combination loss | | |

A final note should be devoted to vibrational excitation in the delocalized "hydrogen bands" that we mentioned already in Section 36.2.5 in the context of hydrogen diffusion and addressed as quantum delocalization [16]. The vertical bandgaps of the delocalized vibrational bands are separated by a couple of milli-electron volts only, and sophisticated HREELS spectrometers with an energy resolution $>1$ meV are required to obtain the corresponding information. As an example, we merely point to recent HREELS measurements performed with the H-on-Pt(111) system by Badescu *et al.* [241, 242], which showed for the hydrogen $(1 \times 1)$ phase at one monolayer H coverage various vibrational peaks that were – based on and supported by parallel first-principles calculations – associated with discrete adsorption sites near the ground state and delocalized band-like features at higher excitations.

### 36.2.8
### Electronic Interactions: UV-Photoemission and Work Function Effects

The nature of the adsorption site, the bond lengths, and the bond angles are certainly a consequence of the electronic interaction between the adsorbing H atoms or $H_2$ molecule and the underlying substrate surface. We have already touched upon the phenomenon of hydrogen physisorption in Section 36.2.2, and stated that physisorption is dominated by van der Waals forces that do not change the charge distribution at the surface significantly and do little more than concentrating the adsorbate "hydrogen" from the gas phase at the surface. In the following, we will consider some details of the "post-dissociation" situation, after the H atoms have been produced by dissociation and become bound to the (metal) surface, whereby some charge is exchanged between the adsorbate and substrate. We recall that the dissociation can be initiated close to the surface by the possibility of electron tunneling between the occupied $1\sigma$ molecular orbital (MO) of the approaching $H_2$ molecule and those wave functions of the substrate that have the appropriate symmetry for interaction and electron exchange. Upon dissociation, the adatom electron states become shifted and broadened, and both the ionization and the electron affinity level degenerate to narrow bands and usually change their energetic positions. As photoemission studies have revealed, this holds true in particular for a H *atom* that can almost immerse in the electronic sphere of the substrate metal – with the consequence that its 1s atomic orbital strongly interacts with the metal electron wave functions. In a band-like view, the tails of both the filled H 1s and the empty electron affinity bands reach into the Fermi level of the surface, and there occurs a net charge flow, resulting in a partial filling of these bands, which then contribute to the chemisorptive bonding. Already in the 1980s, Nørskov and Stoltze performed exemplary calculations for $H_2$ interaction with a Mg(0001) surface and verified the aforementioned shift and broadening of the involved electron states [114, 379], and in many subsequent calculations similar effects were predicted (see Section 36.3). In summary, the antibonding $2\sigma_u^*$ molecular orbital (MO) of the $H_2$ molecule becomes increasingly filled by metal electrons as the $H_2$ molecules approach the surface, thus weakening the H–H intermolecular bond. At the same time, the MO is broadened and pulled down to below the Fermi level of the substrate metal und experiences an attraction, which bends the (originally repulsive) potential energy curve over and produces a minimum for providing a first chemisorbed molecular state. At this stage, the covalent H–H bond is not yet completely cleaved, and some additional activation energy is required for completing the dissociation, with the corresponding barrier being located more closely to the surface. The dissociated H state with its strong 1s character then produces a single electron resonance level between 6 and 10 eV below the Fermi level, which can be detected by UV photoelectron spectroscopy (UPS) as "extra" emission. Although the overall scenario was originally applied to an electron-rich sp metal, the general conclusions may also hold (certainly with some modifications) for TMs with their unfilled d electron states, whereby just these empty d states are essential for reducing or annulling the aforementioned activation barrier for dissociation but are less important for the actual bonding to the surface. This has been made plausible especially by Harris [380–382], who clearly associated the dissociation barrier with the existence of a Pauli repulsion between the closed-shell entity of a $H_2$ molecule and the occupied sp orbitals of electron-rich metals such as Cu, Ag, or Au. While the sp-like orbitals of the TMs (Ni, Pd, Pt, etc.) extend somewhat into space and tend to repel the incoming hydrogen molecule anyway, there are, however, also empty d states available at the Fermi level, and it does not cost energy for such a system to rehybridize and allow the sp electrons to avoid the Pauli repulsion because they are promoted to the empty d electron states at the Fermi level. As Harris states it, "the mere existence of 'holes' in the d-band opens up an escape route for the s electrons that is not available to those of a simple or noble metal" [382]. At any rate, the Pauli repulsion between a hydrogen molecule and a TM with empty d states will be reduced compared to a simple or noble metal with a similar s electron density but no d holes. In this way, one can easily understand why typical TM surfaces such as Ni or Pt spontaneously dissociate hydrogen molecules, but activation barriers exist for the same process on electron-rich metals such as Cu, Ag, or Au. Of course, in a more expanded view, the spatial orientation of the $H_2$ molecule relative to the surface atoms and its internal energy must be carefully considered. As pointed out in Section 36.2.3.1, this requires a full multidimensional quantum dynamical treatment in order to obtain the appropriate potential energy hypersurface. During the last decades, an increasing number of theoretical publications have appeared for the H-on-Pd(100) system, among others by Wilke, Groß, and Scheffler [115, 383], who were probably the first to point out the dynamical steering effects that "conduct" the molecules to surface sites with a preferred dissociation probability.

The aim of an electronic theory describing the interaction of H with metal surfaces is to correctly predict the adsorption energy, the binding sites (including subsurface sites), the electron binding states, vibrational frequencies, and the charge transfer associated with the H–metal bond formation. While there have been multiple methodical approaches for chemisorption theories in the past, ranging from *ab initio* to semiempirical tight-binding or cluster approximations, theories based on DFT nowadays dominate the theoretical treatments also of the hydrogen–surface interaction (see Section 36.3), and we recommend the comprehensive overviews given by Greeley and Mavrikakis [48] and Ferrin *et al.* [51].

*Experimental* access to some aspects and details of the electronic interaction between H and (metal) surfaces can be obtained from electron spectroscopies, UPS in the first place [384, 385]. UPS probes the electron density of states and their adsorption-induced changes within the valence band region up to ~12 eV below the Fermi level $E_F$. In contrast, Bremsstrahlung isochromat spectroscopy (BIS) (="inverse" photoemission IPE) provides information about the density of states for $E > E_F$, that is, above the Fermi level [386, 387]. The knowledge of the distribution of electronic states above and below $E_F$ is a cornerstone for understanding the quantum chemistry of H-to-metal bonding [114, 379, 385–391]. Most important are also measurements of H-induced WF changes, although the physical information hidden in this quantity is in some respect more difficult to extract [392–394].

Somewhat contrary to the well-known, relatively sharp by: MO levels that appear in the valence band UP spectra of CO on TM surfaces, the respective photoemission data of adsorbed hydrogen systems often reveal merely broad single peaks in the energy range 6–10 eV below $E_F$ that mirror the H-induced changes in the electron density of states in the valence band region according to the theory of Nørskov *et al.* [114, 381, 388] mentioned above. A listing of available hydrogen UPS studies can be found again in Ref. [50]. UV photoemission measurements performed in an energy-dependent and angle-resolved manner ("angle-resolved UV photoelectron spectroscopy (ARUPS)") can disclose the density of states within the surface Brillouin zone (SBZ), as was shown several times for various TM surfaces [395–397]. Conveniently, these studies are performed at a synchrotron facility with special display analyzers [398, 399]. Some time ago, we studied the H-on-Ni(110) system by means of angle-resolved UV photoemission and found, among others, a broad H-induced "extra" emission due to hydrogen around 8 eV below the Fermi level [45]. Komeda *et al.* [400] performed ARUPS measurements on the hydrogen-on-Ni(110) system, too, and could map the (1 × 2) SBZ of the PR-reconstructed phase. On the same line are angle-resolved UV photoemission studies of H adsorbed on the Fe(110) surface by Maruyama *et al.* [401]. They probed both the (2 × 1)-H and (3 × 1)-H phases in a synchrotron study and found the H 1s-derived split-off state around 7.9 and 8.2 eV for the aforementioned phases, respectively. An example from their work is given as Figure 36.31, showing the angle-resolved UP spectra of the (3 × 1) phase at 80 K along the $\overline{\Gamma N'}$ (=[1$\bar{1}$0] azimuth) at a photon energy of 40 eV. A more general compilation of how photoemission experiments at modern synchrotrons can contribute to the understanding of surface physical (magnetism, high-$T_c$ superconductors) and chemical solid-state properties (imaging of chemical waves, structure of adsorbate layers near atmospheric pressure) has been published by Ferrer and Petroff [402].

We now turn to and expand somewhat on the other quite frequently employed technique to follow changes in the surface charge density distribution, namely, measurements of the adsorbate-induced WF change ($\Delta\Phi$). $\Delta\Phi$ effects can be measured with high sensitivity and precision, and this is the reason why WF effects have been studied since the early days of surface science. Particularly on metal surfaces, WFs and their changes can be measured by UV photoemission (using the cut-off edge of the secondary electrons in the energy distribution curve (EDC)), by field emission

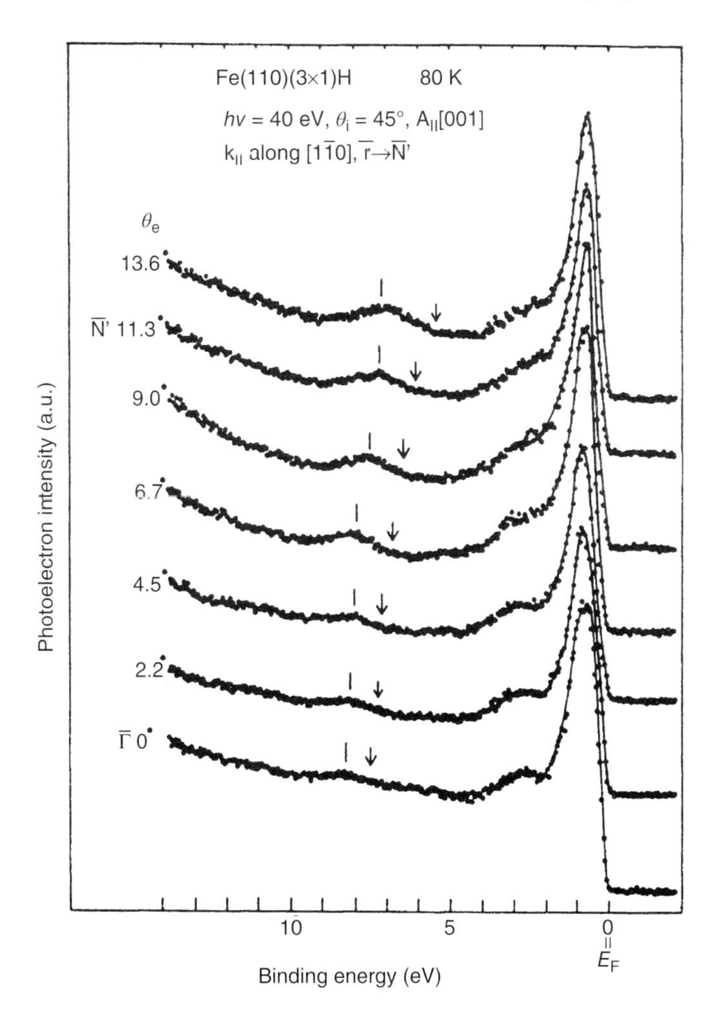


**Figure 36.31** Angle-resolved UV-photoelectron spectra of Fe(110)-(3 × 1)H recorded at $T = 80$ K along the [1$\bar{1}$0] azimuth and with a photon energy of $h\nu = 40$ eV, incidence angle $\Theta_i = 45°$, and vector $A_{||}$ along [001] direction. The emission angle $\Theta_e$ changes along the [1$\bar{1}$0] azimuth, as indicated in each curve. The H-induced "extra" emission is also indicated by tick marks and the expected position of the Fe 4s band emission by arrows. (After Ref. [401].)

techniques [224], by the diode [403, 404] method, or by the Kelvin method using a vibrating capacitor [405, 406], conveniently in a self-compensating mode using a lock-in amplifier technique [407]. Modern developments include the use of scanning probe microscopy methods with the advantage that even local WFs can be determined by taking current–voltage curves or by measuring the distance dependence of the tunnel current [408, 409]. An overview of the various integral methods and some theoretical background can be found in the book by Woodruff and Delchar [410].

WF changes due to adsorption can span a considerable range from a couple of millivolts (in the case of hydrogen on Co(0001) [128] to 2–3 V (in case of alkali metal adsorption) and can be positive (increase of $\Phi$) or negative (decrease of $\Phi$). Although it can be measured relatively simply, an inherent problem of WF measurements is that the $\Delta\Phi$ contributions can stem from the surface *and* from the bulk of the solid, which are not always easy to separate.

The electrochemical potential $\eta$ of the metal valence band electrons is composed of the chemical potential $\overline{\mu}$ of electrons inside the metal (responsible for the Fermi level $E_F$) and the electrostatic potential $e_0\varphi$:

$$\eta = \overline{\mu} - e_0\varphi. \tag{36.19}$$

$\overline{\mu} = (\partial G/\partial n_e)$, with $G$ = Gibbs energy and $n_e$ = mole number of conduction electrons; and $\varphi$ is the inner potential, which is equivalent to the work necessary to transfer an electron from infinity to the Fermi level of the metal. The inner potential $\varphi$ contains two independent contributions: the outer electric potential $\psi$ of the solid surface (= 0 in the absence of external electric fields), and the so-called surface potential (SP) often denoted as $\chi$, that is

$$\varphi = \psi + \chi. \tag{36.20}$$

The SP reflects inherent charge asymmetries at the surface (which are common for metals with their free electrons) since these electrons tend to redistribute in order to moderate the sharp termination of the bulk properties by the surface. This is known as the *Smoluchowski effect* [411], which makes the electrons spill out somewhat into vacuum, thereby causing asymmetries in the charge density distribution between electrons and positive ion cores at the surface. It is then evident that all adsorbates that influence this surface charge distribution produce changes of the SP. For a neutral surface with $\psi = 0$, Equation 36.20 states that the inner potential $\varphi$ is equal to $\chi$, and an external amount of work $W = e_0\Phi$ is required to transport an electron from the Fermi level to about $10^{-6}$ m outside the surface. At this distance, the image charge forces acting on the transferred electron are no longer perceptible and do not affect $W$ anymore. $W$ represents the classical WF, and is apparently composed of the chemical potential of the metal electrons plus the energy contribution required to move the test electron through the SP $\chi$:

$$e_0\Phi = \overline{\mu} - e_0\chi. \tag{36.21}$$

From Equation 36.21, it is immediately apparent that changes of the SP contribute directly to the WF, whereby the WF change is equal to $-\chi$. Usually, adsorbing particles, which more or less alter the charge distribution at the surface, are responsible for these changes. Considering a layer of adsorbed dipoles with initial dipole moment $\mu_0$ and surface concentration $\sigma$ (=number of dipoles/unit area), the WF change produced by the dipoles can be expressed for the low-coverage limit with negligible dipole–dipole interactions by the Helmholtz equation

$$\Delta\Phi = \mu_0\sigma/\varepsilon_0, \tag{36.22}$$

where $\varepsilon_0$ is the vacuum dielectric constant = $8.85 \times 10^{-12}$ As/(Vm). Equation 36.22 can provide information about charge separations of adsorbed dipoles and surface concentrations in the low-coverage limit, respectively. From Equation 36.22, it follows that the initial dipole moment $\mu_0$ is accessible from the initial slope of the $\Delta\Phi$ ($\sigma$) (i.e., the $\Delta\Phi$ ($\Theta$) dependence) via

$$\mu_0 = C\left|\frac{\partial \Delta\Phi}{\partial \sigma}\right|_{\sigma \to 0} \tag{36.23}$$

($C$ is the conversion factor to change electrostatic units to SI units).

For a sparsely covered surface, the adsorbate-induced WF change can be taken as a convenient monitor of the adsorbate coverage, thus enabling isotherm and equilibrium measurements, as already mentioned in Section 36.2.4. At higher concentrations, however, dipole–dipole interactions become effective, causing depolarization effects, which suspend the linear dependence between the number of adsorbed dipoles and the WF change predicted by Equation 36.22 and lead to a relation between $\Delta\Phi$ and coverage $\Theta$ of the form

$$\Delta\Phi \propto \Theta^n, \tag{36.24}$$

with $n < 1$. At higher coverages, one can try to adopt the Topping model to account for the electrostatic interaction and depolarization effects [412].

Turning now to the actual WF phenomena observed during dissociative adsorption of hydrogen, several specific properties can be mentioned:

(i) *Adsorbed H atoms mostly increase the work function.* In most cases, adsorption of H atoms causes an initial WF increase, which means that the first adsorbing H atoms build up a negative dipole barrier at the surface, that is, the negative ends of the metal–H dipoles point away from the surface and an extra amount of work has to be done for moving an electron from the Fermi level of the metal through the dipole layer into the vacuum. For systems in which the number of adsorbed dipoles can be determined independently, the amount of charge $q$ localized on the dipole can be estimated. For hydrogen adsorbate systems, especially those containing smooth low-index surfaces, the $\Delta\Phi$ values seldom exceed 0.5 V and $q$ is generally less than a tenth, often only a hundredth, of the elementary charge $e_0$ with the consequence that the ionicity of a metal–hydrogen adsorptive bond is always quite small. Hydrogen atoms adsorbing on more open higher index surfaces can produce considerably largerWF changes, which sometimes reach almost 1 V [137].

Only a few H adsorption systems exhibit an opposite behavior, in that they show an initial *decrease* of $\Delta\Phi$. Among these are the bcc (110) surfaces of Fe [413] and W [144], as well as the fcc (111) surface of Pt [175]. Recently, also for the hcp (0001) surface of Co a small initial decrease of merely 20 mV was reported [128]. One could only speculate that on these surfaces the H atoms have a particular short distance to the topmost surface layer and are more immersed in the "sea" of valence electrons, so that image charge effects can possibly invert the dipole polarity. For the interpretation of the sign of the H dipole moment, the overall charge distribution around the H–surface complex is decisive. Especially, local restructuring processes occurring at the metal surface can have a substantial influence on the sign and magnitude of the experimental dipole moment (see below).

(ii) *The work function–coverage relation can be complicated.* Various H-on-TM adsorption systems show a complicated $\Delta\Phi(\Theta)$ relation where the WF runs through a maximum at medium or high coverages. A typical example is again provided by the H-on-Ni(111) system, where $\Delta\Phi$ reaches a maximum of ~165 mV at half a monolayer coverage, that is, when the H atoms arrange themselves in the honeycomb phase of Figure 36.25. As $\Theta$ increases to beyond 0.5, WF decreases again and reaches values below 90 mV around $\Theta = 0.9$. Interestingly, an inverse behavior has recently been reported for the Co(0001) surface, where conjecturally the same graphitic H phase is formed, which coincides with a $\Delta\Phi$ minimum of approximately −20 mV, followed by a slight re-increase as $\Theta > 0.5$ ML. The WF behavior of H adsorbing on the Ru(0001) surface is even more complicated. In their early publication, Feulner and Menzel described WF effects during hydrogen exposure of the clean Ru(0001) surface [134], which also exhibited a small $\Delta\Phi$ maximum around +30 mV followed by a continuous (and slightly $T$-dependent) decrease to values of approximately −30 mV near saturation. The authors initially assumed parallel adsorption of negatively and positively polarized H species, but the origin of this behavior remained unclear. Some authors even assumed a parallel occupation of subsurface sites [414]. Elucidation came from a more recent careful HREELS study by Kostov *et al.* [369], which not only ruled out the subsurface H explanation but also gave hints to the reasoning for the WF effects. It could be shown that the H atoms adsorb practically at all coverages in threefold-coordinated positions, leading to a very homogeneous $(1 \times 1)$ H phase at saturation. In the intermediate coverage range, however, the properties of the well-defined H state change slightly with coverage, around $\Theta = 0.5$ even in a stepwise fashion, with coexistence of Ru–H complexes with different bond lengths, ranging from 0.7 to 0.8 Å at $\Theta = 0.2$ ML to 1.1 Å at saturation [415]. However, even with these length differences, the turnaround of the WF is difficult to explain – it seems as if marked effects can arise also from an accompanying H-induced redistribution of the surface charge density due to layer relaxation/restructuring, as proposed by Menzel [416]. It is this sensitivity to the local surface structure that makes predictions of H-induced WF effects and their coverage dependence difficult to predict even in modern DFT computations. We refer to a long list of WF changes theoretically forecasted by Greeley and Mavrikakis [48], which contain mostly small, positive but also some negative $\Delta\Phi$ values, though most of them remain to be verified experimentally.

The sensitivity with respect to the local charge density becomes also apparent from the fact that H atoms adsorbed in different adsorption sites can give rise to entirely different WF changes. A convincing example is perhaps H adsorption on a stepped Pt(111) surface, where a preferential adsorption of H atoms at certain steps could be deduced from WF effects [174]. While hydrogen adsorption on the flat Pt(111) terraces produced the expected ΔΦ decrease, the preferential occupation of step sites in the very beginning of the adsorption led to an initial ΔΦ increase, which could even be associated with the population of two different step sites. This is illustrated in Figure 36.32.

(iii) Electron-rich metal surfaces exposed to pre-dissociated H atoms often show a decrease of the WF change. In many of these cases, the formation of a surface hydride could be verified, for example, for the H + Be(0001) [159] or the H + Mg(0001) [417] adsorption systems. In this respect, hydrogen atom interaction with Cu(100) [366] or Au(100) [113] also leads to a decrease of ΔΦ, although there are no hints to the formation of surface hydrides in these latter cases.

A final remark concerns the WF change associated with the physisorption or weak chemisorption of hydrogen molecules on surfaces. One of the few examples here

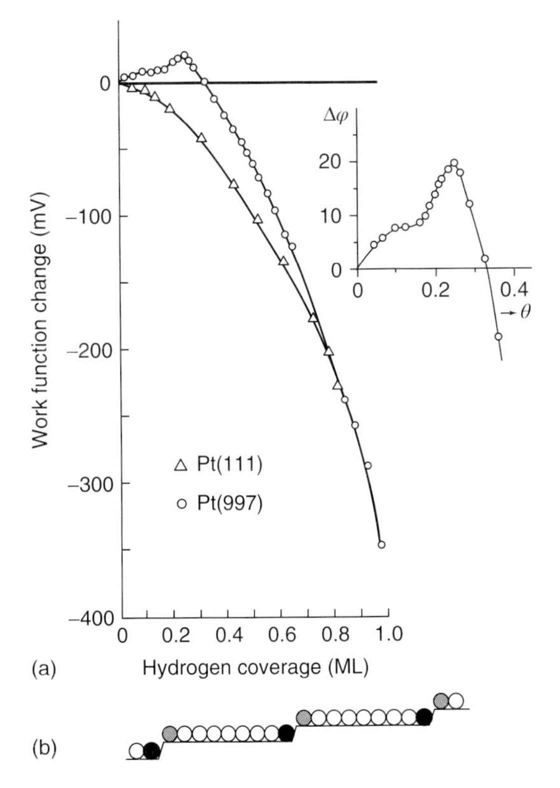


**Figure 36.32** Hydrogen adsorption on a clean Pt(111) and a Pt(997) surface with periodic steps. (a) H-induced change of the work function (mV) as a function of H coverage (ML) at 120 K for the Pt(997) surface (upper curve) and the (111) surface without steps. Only the stepped surface shows an initial ΔΦ increase, while ΔΦ decreases from the beginning on the flat surface. As the inset reveals, the structured initial increase suggests the preferential occupation of two kinds of step sites that are illustrated in (b). (b) Schematic side view of the Pt(997) surface, in which the different step sites for H adsorption are indicated as full and hatched circles, respectively. (After Ref. [174].)

is the chemisorption of $H_2$ molecules on the Pd(210) surface around 60 K, which showed an instantaneous fall of $\Delta\Phi$ by almost 400 mV during $H_2$ exposure, thus suggesting a positively polarized molecular hydrogen species [79, 80]. In concluding this section, we present in Table 36.10 some experimental data of H-induced WF changes.

## 36.3 Some Remarks about Theories Describing the Chemisorption of Hydrogen

In the beginning of this (short) compilation, we recall that the main task of theoretically treating an adsorption system that consists of an adsorbate such as a hydrogen molecule or a H atom and an extended metal surface is actually relatively straightforward. It always aims at the determination of the system's total energy $E_{tot}$ and the associated total wave function $\Psi_{tot}$. From these quantities, many other properties can then be deduced. In the time-independent Schrödinger equation, the Hamiltonian acting on $\Psi_{tot}$ yields the product between $E_{tot}$ and $\Psi_{tot}$. In practice, the adsorption systems consist of $M$ nuclei and $N$ electrons, and even within the Born–Oppenheimer approximation for resting nuclei the Schrödinger equation can no longer be solved as soon as $N$ exceeds 1 because of the electron–electron repulsion and correlation. Numerous quantum chemical procedures have been developed to solve the Schrödinger equation at least approximately. Conceptually, relatively simple is the Hartree–Fock (HF) approach, where one departs from one-electron wave functions in the mean field of all other electrons and constructs the total wave function as a product of all these one-electron wave functions self-consistently (SCF method). For practical purposes, this is, however, a long-lasting and costly procedure. Accordingly, numerous other treatments of the many-body problem have been pursued in the past. In a review article on hydrogen interaction with metal surfaces written in 1988 [45], various types of theories had been introduced that had so far well contributed to a better understanding of the chemisorption of hydrogen on metal surfaces. In a more recent compilation, one can find a listing of some system-specific theoretical treatments of H-on-metal systems [50]. As pointed out in Ref. [45], one can subdivide the current chemisorption theories into three main groups (which have some overlap though), reflecting the solid-state physics view: an infinite solid with largely delocalized wave functions is considered in the first group; the more localized treatment of the chemisorption complex in the second category, where a limited number of interacting atoms is taken into account; and a third group which combines the advantages of (i) and (ii) in that they focus on calculations of the local charge density and its perturbation by the adsorbing particles. In detail, we have

(i) band structure models including pseudopotential, $X_\alpha$ band theory, or *ab initio* HF methods,
(ii) cluster models, among others all-electron *ab initio* calculations with multi-CI, the SCF $X_\alpha$ approach, or various pseudo-potential cluster methods,

**Table 36.10** Hydrogen-induced changes of the work function ($\Delta\Phi$) for some H–metal adsorption systems.

| Surface | Adsorption temperature (K) | Work function change at sat'n coverage (mV) | Coverage dependence | Initial dipole moment (D); note: 1 $D = 3.33 \cdot 10^{-30}$ (As·m) | Experimental method | Remarks | References |
|---|---|---|---|---|---|---|---|
| Be(0001) | 90 | −440 | Minimum at −560 mV at $\Theta = 0.4$ | | HREELS analyzer in retarding field mode | Exposure to H atoms | [159] |
| Mg(0001) | 110 | $-950 \pm 70$ | | | HREELS analyzer in retarding field mode | Exposure to H atoms; formation of surface hydride | [417] |
| Fe(100) | 140 | +75 | | | Kelvin probe | | [413] |
| Fe(110) | 140 | $-85 \pm 5$ | Small initial minimum of −55 mV at completion of $\beta_2$-TD state | −0.032 | Kelvin probe | | [413] |
| Fe(111) | 140 | +310 | | | Kelvin probe | Annealing required | [413] |
| Co(0001) | 90 | −12 | Decrease to minimum of −18 mV at $\Theta \approx 0.55$, then re-increase | −0.0048 | Kelvin probe | | [128] |
| Ni(100) | 150 | +45 | Broad maximum of 96 mV around $\Theta = 0.5$ | +0.049 | Kelvin probe | | [172] |
| Ni(110) | 120 | +510 | $\Delta\Phi$ linear with positive break at $\Theta = 1$ due to reconstruction | +0.06 | Kelvin probe | (1 × 2)PR reconstruction | [131] |
| Ni(111) | 273 | +195 | $\Delta\Phi$ linear with $\Theta$ | | Kelvin probe | Saturation coverage at 273 K $\Theta_{sat} \approx 0.5$ | [160] |
| Ni(111) | 110 | +80 | $\Delta\Phi$ linear with $\Theta$ up to maximum of 165 mV at $\Theta = 0.5$, then decrease by 85 mV at sat'n | +0.029 | Kelvin probe | Formation of (2 × 2)-2H phase at $\Theta = 0.5$ | [14] |

(*continued overleaf*)

Table 36.10 (Continued)

| Surface | Adsorption temperature (K) | Work function change at sat'n coverage (mV) | Coverage dependence | Initial dipole moment (D); note: 1 $D = 3.33 \cdot 10^{-30}$ (As·m) | Experimental method | Remarks | References |
|---|---|---|---|---|---|---|---|
| Cu(100) | 10 | −120 | Linear decrease of $\Delta\Phi$ with $\Theta$ | | $e^-$ beam retardation method | Physisorption of $H_2$ molecules | [418] |
| Cu(100) | 170 | −250 | | | Kelvin probe | Exposure to H atoms; surface reconstructs to *p4g* phase | [366] |
| Mo(100) | 160–240 | +1400 | Linear increase to 600 mV (c(2 × 2) phase), then plateau, followed by re-increase | | | H-induced reconstruction involved | [419] |
| Ru(0001) | 100–150 | −10 to 30 | $\Theta_{sat'n} = 1.0$ | +0.06 to 0.01 | Kelvin probe | 1. +25 mV first increase at small $\Theta$; 2. −10 to −30 mV near $\Theta = 1.0$ | [134] |
| Rh(111) | 130 | +50 | | | UV photoemission | | [420] |
| Rh(110) | 100 | +930 | $\Theta_{sat'n} = 2.0$ | 0.027 | Kelvin probe | Almost linear increase with $\Theta$ up to ~80% of $\Theta_{sat'n}$ | [137] |
| Pd(100) | 170 | +200 | $\Theta_{sat'n} = 1.4$ | 0.021 | Kelvin probe | Almost linear increase with positive break at $\Theta = 1.0$ | [117] |
| Pd(110) | 300 | +360 | | | Kelvin probe | (1 × 2) reconstruction included | [166] |

| | | | | | | | |
|---|---|---|---|---|---|---|---|
| Pd(110) | 130 | +325 | $\Theta_{sat'n} = 1.5$ | 0.071 | Kelvin probe | (1 × 2) reconstruction included | [140] |
| Pd(111) | 300 | +180 | | | Kelvin probe | | [140] |
| Pd(210) | 40 | −400 | | | Kelvin probe | Weak chemisorption of $H_2$ molecules | [80] |
| Pd(210) | 120 | +160 | $\Theta_{sat'n} \approx 3.0$ | | Kelvin probe | H atom chemisorption | [79] |
| Ag(110) | 100 | +220 | $\Theta_{sat'n} = 1.0$ | | HREELS analyzer in RF mode | Exposure to H atoms | [152] |
| Ag(111) | 100 | +320 | $\Theta_{sat'n} = 0.5–0.6$ | 0.18 | HREELS analyzer in RF mode | Exposure to H atoms | [421] |
| W(100) | 330 | +900 | $\Theta_{sat'n} \approx 2.0$ | 0.21 | Contact potential difference, via RF mode | | [61] |
| W(110) | 90 | −500 | $\Theta_{sat'n} = 1.0$ | | Kelvin probe | | [144] |
| Pt(100)-(5 × 20) | 100 | +70 (state b) | $0 < \Theta < 1.0$ | | Kelvin probe | $\Delta\Phi(\Theta)$ passes a maximum. Complex temperature dependence due to thermally activated reconstruction/ deconstructuction processes | [422] |
| | | −370 (states $a_1$ and $a_2$) | | | | | |
| Pt(111) | 150 | −230 | $0 < \Theta < 1.0$ | −0.036 | Kelvin probe | $\Delta\Phi \sim \Theta^{1.33}$ | [305] |
| Au(100) | 100 | −200 | $\Theta_{sat'n} \approx 0.3$ | | Contact potential difference, via RF mode | Exposure to H atoms | [153] |

(iii) models that aim at calculations of electron charge densities, among which are the effective medium theory (EMT), the (somewhat related) EAM, and the whole group of DFT.

The underlying concept of these latter theories (iii) is based on the theorem of Hohenberg and Kohn, in which the ground state of a system of $N$ electrons is uniquely correlated with the local electron density $n(\vec{r})$ [423]. In EMT, the adsorption of a hydrogen or adsorbate atom is then treated as perturbation on a host substrate, in the simplest case a homogeneous electron gas. The binding energy of an H atom on a TM surface with a given electron density profile $n(\vec{r})$ is then estimated from the binding energy of the H atom in the homogeneous electron gas. There exist a multitude of publications on the EMT method, which was originally proposed by Stott and Zaremba [424] and developed further by Nørskov and Lang [425]. Surprisingly, there are various examples where EMT could correctly predict the binding energies of H on metals despite its relative simplicity [15, 114, 426, 427]. We refer especially to Ref. [427], which summarizes some data for hydrogen adsorption. The adsorption of H atoms on a metal surface resembles in some respect the bonding of a metal to a metal, which manifests itself, for example, in the tendency of H atoms to prefer highly coordinated surface sites. This explains perhaps the great success of the EMT just for hydrogen adsorption systems. All in all, EMT has developed into a powerful instrument to calculate the geometry of the H adsorption site, H-metal layer distances, binding energies, vibrational ground state frequencies, and the energetic position of the hydrogenic 1s-like orbitals in the density of electronic states in valence band spectra [388]. For further details on this method, we recommend the monograph by Choy [428]. EAM was introduced by Daw and Baskes [429]. It pursues again the concept of embedding impurity atoms (e.g., an H atom) in a substrate of host atoms and considers among others the "embedding" energy required to embed the atom in question in the electron density of the host. It has been proven to predict various bulk properties quite well: for example, cohesive energies, heats of solution, or embrittlement phenomena. EAM and EMT may be considered highly simplified (but anyway quite useful) versions of the general DFT, which is almost exclusively used these days when hydrogen chemisorption is to be theoretically modeled.

To underline this, we would like to point again to the recent "surface science perspective" article by Groß [43] entitled "Hydrogen on metal surfaces: Forever young." In this publication, the author briefly recapitulates the present status of the theoretical efforts to treat the metal–hydrogen interaction and emphasizes that there is still a need for comprehensive studies that address the H–metal interaction for a broad variety of systems by means of the same comparable theoretical approach in order to be able to establish general trends among the metals. He repeatedly commends therein recent DFT work by Ferrin *et al.*, which deals with the stability of atomic hydrogen on the surface and in the subsurface region of a variety of TMs, with emphasis on binding energies, magnetic properties, and adsorption and absorption geometries for each surface facet studied. The authors themselves designate their work as "the first systematic DFT study of the energetics of hydrogen and its diffusion in the near surface region on multiple facets of a large number of transition metals" [51]. This statement accentuates the rapid and very promising development of DFT-type calculations during the last decades. For further reading on DFT, we recommend the book by Chong *et al.* [430], where further mathematical and technical details can be found. For a modeling of chemisorption systems, calculations using slab geometries (supercells) have proven very useful, because they reduce the computational effort substantially [431].

The main difficulty in the DFT-based calculations must not be concealed – it is adequate consideration of the electron exchange and correlation energy term. One must look for an appropriate respective functional, for which several strategies can be followed. On one side, one can try to determine the exchange energy by departing from the local density of the homogeneous electron gas – this kind of approximation is known as the *local density approximation* (LDA) [432]. More refined LDAs take into account inhomogeneities in the local density by considering local gradients – these nonlocal procedures are known as the *generalized gradient approximations* (GGAs). Various approaches have been developed that use different analytic forms for the correlation energy. These have then generated different LDAs for the correlation functional, leading, for example, to the Perdew-Wang (PW91) [433] or the DACAPO code [434, 435].

Finally, we return again to the article by Groß [43] and underline the excellent systematic DFT work on hydrogen adsorption systems performed by the group of Mavrikakis. The authors use periodic, self-consistent DFT-GGA (PW91) calculations and are, as mentioned already above, able to handle not only the problem of H chemisorption but also the occupation of subsurface sites in the second and third atomic layer of selected TMs. An especially convincing example is the verification of a quite unusual terminal adsorption site for hydrogen on the Ir(111) surface. Recently, Hagedorn *et al.* have studied hydrogen adsorption on the Ir(111) surface by means of HREELS and found that, as the coverage went toward the $(1 \times 1)$ saturation phase, there appeared a vibrational band at $2030\,\text{cm}^{-1}$, which urged the authors to assume terminal adsorption sites for the H atoms [294]. In a subsequent DFT calculation, Krekelberg *et al.* [436] could indeed correctly predict just this site. To show the power of their DFT calculations, we may add another example, which concerns the prediction of H-induced WF changes. In a recent study of hydrogen adsorption on the Co(0001) surface [128], we measured the exposure and coverage dependence of the H-induced WF change $\Delta\Phi$ and found a small initial decrease by about $-20$ mV, followed by a slight re-increase to $-10$ mV. More than 10 years ago, Klinke and Broadbelt [437] had performed combined full-potential-linear-augmented-plane-wave (FP-LAPW) calculations on the same system. The FP-LAPW method combines both the plane wave and the DFT procedure with the benefit that both delocalized and localized bonding effects can be accurately described. The authors predicted, for a coverage of $\Theta = 0.5$, occupation of either the fcc or hcp threefold hollow sites with binding energies of 2.88 and 2.87 eV, respectively. The associated WF changes were given as $-45$ and $-29$ mV, respectively, in remarkable agreement with our experiment. In subsequent DFT calculations using the DACAPO code, Greeley and Mavrikakis [48] predicted, likewise for the H + Co(0001) system, occupation of the same sites with binding energies of 2.89 and 2.86 eV, both associated with a $\Delta\Phi$ of −40 mV, which is not only in very good agreement with the experiment but also shows in addition that different theoretical treatments of the same H adsorption system can lead to practically identical results.

In conclusion of this brief overview, these two latter examples may illustrate how sophisticated and how reliable many hydrogen-on-metal chemisorption data can be forecasted or verified by appropriate DFT calculations – certainly with some exceptions here and there, but the overall trend is definitely that especially adsorption energies can be predicted to within a couple of percent. Future tasks here may focus on modeling the electron exchange/correlation problem by even better functionals and a better inclusion of the long-range dispersion (van der Waals) forces that pose even now many difficulties.

## References

1. von Ditfurth, H. (2005) *Am Anfang war der Wasserstoff*, Deutscher Taschenbuch-Verlag, München.
2. Bockris, J.O.'M. and Justi, E.W. (1990) *Wasserstoff, Energie für alle Zeiten*, Augustus-Verlag, Augsburg.
3. Schlapbach, L. and Züttel, A. (2001) *Nature*, **414**, 353.
4. *http://de.wikipedia.org/wiki/Wasserstoffspeicherung* (accessed 23 March 2015).
5. Mackey, K.M. (1973) *The Element Hydrogen in Comprehensive Inorganic Chemistry*, Pergamon Press, Oxford.
6. Remy, H. (1973) *Lehrbuch der Anorganischen Chemie*, 13. Aufl., Geest & Portig, Leipzig.
7. Huheey, J.E. (1983) *Inorganic Chemistry – Principles of Structure and Reactivity*, Longman, London.
8. Hollemann, A.F. and Wiberg, N. (2007) *Lehrbuch der Anorganischen Chemie*, 102nd edn, de Gruyter, Berlin.
9. *http://de.wikipedia.org/wiki/Wasserstoff* (accessed 23 March 2015).
10. Royal Society of Chemistry *http://www.rsc.org/periodic-table/element/1/hydrogen* (accessed 23 March 2015).
11. Silvera, I.F. (1980) *Rev. Mod. Phys.*, **52**, 393.
12. See, for example: Groß, A. (2009) *Theoretical Surface Science – A Microscopic Perspective*, 2nd edn, Springer, Berlin.
13. Wedler, G. and Freund, H.-J. (2012) *Lehrbuch der Physikalischen Chemie*, 6. Aufl., Wiley-VCH Verlag GmbH, Weinheim, p. 269.
14. Christmann, K., Behm, R.J., Ertl, G., Van Hove, M.A., and Weinberg, W.H. (1979) *J. Chem. Phys.*, **70**, 4168.
15. Puska, M.J., Nieminen, R.M., Manninen, M., Chakraborty, B., Holloway, S., and Nørskov, J.K. (1983) *Phys. Rev. Lett.*, **51**, 1081.
16. Nishijima, M., Okuyama, H., Takagi, N., Aruga, T., and Brenig, W. (2005) *Surf. Sci. Rep.*, **57**, 113.
17. (a) Bonhoeffer, K.F. and Harteck, P. (1929) *Z. Elektrochem.*, **35**, 621; (b) Farkas, A. (1935) *Orthohydrogen, Parahydrogen and Heavy Hydrogen*, 1st edn, Cambridge University Press.
18. Buntkowsky, G., Walaszek, B., Adamczyk, A., Xu, Y., Limbach, H.-H., and Chaudret, B. (2006) *Phys. Chem. Phys.*, **8**, 1929.
19. Fukutani, K. and Sugimoto, T. (2013) *Prog. Surf. Sci.*, **88**, 279.
20. Fischer, F. and Tropsch, H. (1926) *Ber. Dtsch. Chem. Ges.*, **59**, 830.
21. P.N. Rylander, Catalytic processes in organic conversions, in: *Catalysis, Science and Technology* eds. J.R. Anderson and M. Boudart, vol. 4, Springer Berlin.
22. Khodakov, A.Y., Chu, W., and Fongarland, P. (2007) *Chem. Rev.*, **107**, 1692.

23. Prins, R., de Beer, V.H.J., and Somorjai, G.A. (1989) *Catal. Rev. Sci. Eng.*, **31**, 1.
24. Topsoe, H., Clausen, B.S., and Massoth, F.E. (1996) *Hydrotreating Catalysis, Science and Technology*, Springer, Berlin.
25. Chianelli, R.R., Berhault, G., Raybaud, P., Kasztelan, S., Hafner, J., and Toulhoat, H. (2002) *Appl. Catal., A*, **227**, 83.
26. Goodman, D.W. (1995) *Chem. Rev.*, **95**, 523.
27. Ertl, G. (2009) *Reactions at Solid Surfaces*, John Wiley & Sons, Inc., Hoboken, NJ.
28. *http://en.wikipedia.org/wiki/Fuel_cell* (accessed 23 March 2015).
29. Thomas, C.E. (2009) *Int. J. Hydrogen Energy*, **34**, 6005.
30. Schlögl, R. (2010) *ChemSusChem*, **3**, 209.
31. Wedler, G. and Freund, H.-J. (2012) *Lehrbuch der Physikalischen Chemie*, 6. Aufl., Wiley-VCH Verlag GmbH, Weinheim, p. 1090.
32. Pauling, L. (1969) *Grundlagen der Chemie*, 8. Aufl., Verlag Chemie, Weinheim, p. 159.
33. Gibb, T.R.P. Jr., (1962) *Primary Solid Hydrides*, Progress in Inorganic Chemistry, vol. 3, Interscience, New York.
34. Wipf, H. (ed.) (1997) *Hydrogen in Metals III*, Springer Series 'Topics in Applied Physics', vol. 73, Springer-Verlag, Berlin, Heidelberg.
35. Noreus, D., Rundqvist, S., and Wicke, E. (eds) (1992) *Metal-Hydrogen Systems: Fundamentals and Applications*, vol. 2, Oldenbourg-Verlag, München.
36. Fukai, Y. (2005) *The Metal – Hydrogen System, Basic Bulk Properties*, Springer Series in Materials Science, 2nd edn, vol. 21, Springer-Verlag, Berlin.
37. Pundt, A. and Kirchheim, R. (2006) *Annu. Rev. Mater. Res.*, **36**, 555, and references therein.
38. Slevin, J. and Stirling, W. (1981) *Rev. Sci. Instrum.*, **52**, 1780.
39. Hodgson, J.A.B. and Haasz, A.A. (1991) *Rev. Sci. Instrum.*, **62**, 96.
40. Moore, G.E. and Unterwald, F.C. (1964) *J. Chem. Phys.*, **40**, 2639.
41. Langmuir, I. (1912) *J. Am. Chem. Soc.*, **34**, 860.
42. Wei, Y., Li, L., and Tsong, I.S.T. (1995) *Appl. Phys. Lett.*, **66**, 1818.
43. Groß, A. (2012) *Surf. Sci.*, **606**, 690.
44. Knor, Z. (1982) in *Catalysis*, vol. 3 (eds J.R. Anderson and M. Boudart), Springer, Berlin, p. 231.
45. Christmann, K. (1988) *Surf. Sci. Rep.*, **9**, 1.
46. J.W. Davenport and P.J. Estrup, in: *The Chemical Physics of Solid Surfaces and Heterogeneous Catalysis*, (D.A. King and D.P. Woodruff eds), Vol. 3A, Elsevier, Amsterdam 1990, p. 1
47. Groß, A. (1998) *Surf. Sci. Rep.*, **32**, 291.
48. Greeley, J. and Mavrikakis, M. (2005) *J. Phys. Chem. B*, **109**, 3460, and references given therein.
49. Hynes, J.T., Klinman, J.P., Limbach, H.-H., and Schowen, R.L. (eds) (2006) *Hydrogen Transfer Reactions*, Wiley-VCH Verlag GmbH, Weinheim.
50. Christmann, K. (2006) Adsorbate properties of hydrogen on solid surfaces, in *Landolt-Börnstein, Group III*, Physics of Covered Solid Surfaces, Subvolume A, Part 5, vol. 42 (ed. H.-P. Bonzel), Springer, Berlin.
51. Ferrin, P., Kandoi, S., Nilekar, A.U., and Mavrikakis, M. (2012) *Surf. Sci.*, **606**, 679.
52. Groß, A. (2009) *Theoretical Surface Science*, 2nd edn, Springer, Berlin.
53. Spaarnay, M.J. (1985) *Surf. Sci. Rep.*, **4**, 101.
54. Robota, H.J., Vielhaber, W., Lin, M.C., Segner, J., and Ertl, G. (1985) *Surf. Sci.*, **155**, 101.
55. Anger, G., Winkler, A., and Rendulic, K.D. (1989) *Surf. Sci.*, **220**, 1.
56. Rettner, C.T., Auerbach, D.J., and Michelsen, H.A. (1992) *Phys. Rev. Lett.*, **68**, 1164.
57. Rettner, C.T., Michelson, H.A., and Auerbach, D.J. (1995) *J. Chem. Phys.*, **102**, 4625.
58. Darling, G.R. and Holloway, S. (1995) *Rep. Prog. Phys.*, **58**, 1595.
59. Wetzig, D., Rutkowski, M., David, R., and Zacharias, H. (1996) *Europhys. Lett.*, **36**, 31.
60. Estrup, P.J. and Anderson, J. (1966) *J. Chem. Phys.*, **45**, 2254.
61. Madey, T.E. (1973) *Surf. Sci.*, **36**, 281.

62. Christmann, K., Ehsasi, M., Block, J.H., and Hirschwald, W. (1986) *Chem. Phys. Lett.*, **131**, 192.
63. Diehl, R.D. and McGrath, R. (1996) *Surf. Sci. Rep.*, **23**, 43.
64. Lennard-Jones, J.E. (1932) *Trans. Faraday Soc.*, **28**, 333.
65. Cowin, J.P., Yu, C., Sibener, S.J., and Wharton, L. (1983) *J. Chem. Phys.*, **79**, 3537.
66. Frieß, W., Schlichting, H., and Menzel, D. (1995) *Phys. Rev. Lett.*, **74**, 1147.
67. Schlichting, H., Menzel, D., Brunner, T., and Brenig, W. (1988) *Phys. Rev. Lett.*, **60**, 2515.
68. Baule, B. (1914) *Ann. Phys.*, **44**, 145.
69. Goodman, F.O. (1974) *Prog. Surf. Sci.*, **5**, 261.
70. Knowles, T.R. and Suhl, H. (1977) *Phys. Rev. Lett.*, **39**, 1417.
71. Schönhammer, K. and Gunnarsson, O. (1980) *Phys. Rev.*, **B22**, 1629; *Surf. Sci.* **117** (1982) 53.
72. Brenig, W. (1982) *Z. Phys.*, **B48**, 127.
73. Mengel, S.K. and Billing, G.D. (1997) *J. Phys. Chem. B*, **101**, 10781.
74. Nienhaus, H., Bergh, H.S., Gergen, B., Majumdar, A., Weinberg, W.H., and McFarland, E.E. (1999) *Phys. Rev. Lett.*, **82**, 446.
75. Andersson, S., Wilzén, L., Persson, M., and Harris, J. (1989) *Phys. Rev.*, **B40**, 8146.
76. (a) Kisliuk, P. (1957) *J. Phys. Chem. Solids*, **3**, 95; (b) Kisliuk, P. (1958) *J. Phys. Chem. Solids*, **5**, 78.
77. Christmann, K. (2009) *Surf. Sci.*, **603**, 1405.
78. Mårtensson, A.S., Nyberg, C., and Andersson, S. (1986) *Phys. Rev. Lett.*, **57**, 2045.
79. Muschiol, U., Schmidt, P.K., and Christmann, K. (1998) *Surf. Sci.*, **395**, 182.
80. Schmidt, P.K., Christmann, K., Kresse, G., Hafner, J., Lischka, M., and Groß, A. (2001) *Phys. Rev. Lett.*, **87**, 096103.
81. Schlichting, H. and Menzel, D. (1993) *Surf. Sci.*, **285**, 209.
82. Ptushinskii, Y.G. (2004) *Low Temp. Phys.*, **30**, 1.
83. Hirschfelder, J.O., Curtiss, C.F., and Bird, R.B. (1954) *Molecular Theory of Gases and Liquids*, 1st edn, John Wiley & Sons, Inc., New York.
84. Küppers, J. and Seip, U. (1982) *Surf. Sci.*, **119**, 291.
85. Redhead, P.A. (1962) *Vacuum*, **12**, 203.
86. Pétermann, L.A. (1972) *Prog. Surf. Sci.*, **3**, 1.
87. de Jong, A.M. and Niemantsverdriet, J.W. (1990) *Surf. Sci.*, **233**, 355.
88. McTague, J.P. and Novaco, A.D. (1979) *Phys. Rev.*, **B19**, 5299.
89. Harris, A.B. and Berlinsky, J. (1979) *Can. J. Phys.*, **57**, 1852.
90. Freimuth, H. and Wiechert, H. (1986) *Surf. Sci.*, **178**, 716.
91. Cui, J., Fain, S.C., Freimuth, H., Wiechert, H., Schildberg, H.P., and Lauter, H.J. (1988) *Phys. Rev. Lett.*, **60**, 1848.
92. Seguin, J.L. and Suzanne, J. (1982) *Surf. Sci.*, **118**, L241.
93. van Hove, M.A. and Tong, S.Y. (1979) *Surface Crystallography by LEED*, Springer, New York.
94. Hammer, L., Landskron, H., Nichtl-Pecher, W., Fricke, A., Heinz, K., and Müller, K. (1993) *Phys. Rev.*, **B47**, 15969.
95. Gruyters, M. and Jacobi, K. (1994) *Chem. Phys. Lett.*, **225**, 309.
96. Avouris, P., Schmeisser, D., and Demuth, J.E. (1982) *Phys. Rev. Lett.*, **48**, 199.
97. Andersson, S. and Harris, J. (1982) *Phys. Rev. Lett.*, **48**, 545.
98. Groß, A. (1996) *Surf. Sci.*, **363**, 1.
99. Kroes, G.-J. (1999) *Prog. Surf. Sci.*, **60**, 1–85.
100. Groß, A. (2010) *ChemPhysChem*, **11**, 1374.
101. Nørskov, J.K., Stoltze, P., and Nielsen, U. (1991) *Catal. Lett.*, **9**, 173.
102. Engdahl, C., Lundqvist, B.I., Nielsen, U., and Nørskov, J.K. (1992) *Phys. Rev.*, **B45**, 11362.
103. Mowray, R.C., Kroes, G.D., Wiesenekker, G., and Baerends, E.J. (1997) *J. Chem. Phys.*, **106**, 4248 (4D calculations).
104. Eichler, A., Kresse, G., and Hafner, J. (1998) *Surf. Sci.*, **397**, 116 (Rh, Pd, Ag).
105. Eichler, A., Hafner, J., Groß, A., and Scheffler, M. (1999) *Phys. Rev.*, **B59**, 13297.

106. Nobuhara, K., Kasai, H., Diño, W.A., and Nakanishi, H. (2004) *Surf. Sci.*, **566 568**, 703 (elbow plots of Mg, Ti, La, Ni, Pd).
107. Eyring, H. and Polanyi, M. (1931) *Z. Phys. Chem. Abt. B*, **12**, 279.
108. Kubiak, G.D., Sitz, G.O., and Zare, R.N. (1985) *J. Chem. Phys.*, **83**, 2538.
109. Rendulic, K.D. (1988) *Appl. Phys.*, **A47**, 55.
110. Rendulic, K.D. and Winkler, A. (1989) *Int. J. Mod. Phys.*, **B3**, 941.
111. Zacharias, H. (1990) *Int. J. Mod. Phys.*, **B4**, 45.
112. Balooch, M., Cardillo, M.J., Miller, D.R., and Stickney, R.E. (1974) *Surf. Sci.*, **46**, 358.
113. Cardillo, M.J., Balooch, M., and Stickney, R. (1975) *Surf. Sci.*, **50**, 263.
114. Nørskov, J.K., Houmøller, A., Johansson, P.K., and Lundqvist, B.I. (1981) *Phys. Rev. Lett.*, **46**, 257.
115. Groß, A., Wilke, S., and Scheffler, M. (1995) *Phys. Rev. Lett.*, **75**, 2718.
116. Forni, A. and Tantardini, G.F. (1996) *Surf. Sci.*, **352–354**, 142.
117. Behm, J.R., Christmann, K., and Ertl, G. (1980) *Surf. Sci.*, **99**, 320.
118. Mitsui, T., Rose, M.K., Fomin, E., Ogletree, D.F., and Salmeron, M. (2003) *Nature*, **422**, 705.
119. Poelsema, B., Verheij, L.K., and Comsa, G. (1985) *Surf. Sci.*, **152/153**, 496.
120. Linke, R., Schneider, U., Busse, H., Becker, C., Schröder, U., Castro, G.R., and Wandelt, K. (1994) *Surf. Sci.*, **307–309**, 407.
121. Shimizu, H., Christmann, K., and Ertl, G. (1980) *J. Catal.*, **61**, 412.
122. Diño, W.A., Kasai, H., and Okiji, A. (2001) *Surf. Sci.*, **482–485**, 318.
123. Conner, W.C. Jr., (1988) in *Hydrogen Effects in Catalysis*, Chapter 12 (eds Z. Páal and P.G. Menon), Marcel Dekker, New York, p. 311 ff.
124. Conner, W.C. and Falconer, J.L. (1995) *Spillover in Heterogeneous Catalysis. Chem. Rev.*, **95**, 759.
125. Jung, K.D. and Bell, A.T. (2000) *J. Catal.*, **193**, 207.
126. Morris, M.A., Bowker, M., and King, D.A. (1984) in *Simple Processes at the Gas – Solid Interface*, Chemical Kinetics, vol. 19, Chapter 1 (eds C.H. Bamford, C.F.H. Tipper, and R.G. Compton), Elsevier, Amsterdam.
127. Berger, H.F. *et al.* (1992) *Surf. Sci. Lett.*, **275**, L627.
128. Hüesges, Z. and Christmann, K. (2013) *Z. Phys. Chem.*, **227**, 881.
129. Ernst, K.H. *et al.* (1994) *J. Chem. Phys.*, **101**, 5388.
130. Hamza, A.V. and Madix, R.J. (1985) *J. Phys. Chem.*, **89**, 5381.
131. Christmann, K. (1989) *Mol. Phys.*, **66**, 1.
132. Winkler, A. and Rendulic, K.D. (1982) *Surf. Sci.*, **118**, 19.
133. Rohwerder, M. and Benndorf, C. (1994) *Surf. Sci.*, **307**, 789.
134. Feulner, P. and Menzel, D. (1985) *Surf. Sci.*, **154**, 465.
135. Lauth, G., Schwarz, E., and Christmann, K. (1989) *J. Chem. Phys.*, **91**, 3729.
136. Kim, Y. *et al.* (1982) *Surf. Sci.*, **114**, 363.
137. Ehsasi, M. and Christmann, K. (1988) *Surf. Sci.*, **194**, 172.
138. Yates, J.T. *et al.* (1979) *Surf. Sci.*, **84**, 427.
139. Colonell, J.L. *et al.* (1996) *Surf. Sci.*, **366**, 19.
140. Cattania, M.-G. *et al.* (1983) *Surf. Sci.*, **126**, 382.
141. Beutl, M. *et al.* (1995) *Chem. Phys. Lett.*, **247**, 249.
142. Healey, F. *et al.* (1995) *Surf. Sci.*, **328**, 67.
143. Muschiol, U. *et al.* (1995) *Surf. Sci.*, **331–333**, 127.
144. Nahm, T.-U. and Gomer, R. (1997) *Surf. Sci.*, **380**, 434.
145. Engstrom, J.R. *et al.* (1987) *J. Chem. Phys.*, **87**, 3104.
146. Dixon-Warren, S.J. *et al.* (1995) *J. Chem. Phys.*, **103**, 2261.
147. Shern, C.S. (1992) *Surf. Sci.*, **264**, 171.
148. Poelsema, B. *et al.* (1985) *Surf. Sci.*, **151/152**, 496.
149. Michelsen, H.A. and Auerbach, D.J. (1991) *J. Chem. Phys.*, **94**, 7502.
150. Rasmussen, P.B., Holmblad, P.M., Christoffersen, H., Taylor, P.A., and Chorkendorff, I. (1993) *Surf. Sci.*, **287/288**, 79.
151. Pritchard, J., Catterick, T., and Gupta, R.K. (1975) *Surf. Sci.*, **53**, 1.

152. Sprunger, P.T. and Plummer, E.W. (1993) *Phys. Rev.*, **B48**, 14436.
153. Iwai, H., Fukutani, K., and Murata, Y. (1996) *Surf. Sci.*, **357/358**, 663.
154. Rendulic, K.D., Anger, G., and Winkler, A. (1989) *Surf. Sci.*, **208**, 404.
155. Cottrell, T.L. (1958) *The Strength of Chemical Bonds*, 2nd edn, Butterworth, London.
156. Schilling, J.B., Goddard, W.A. III, and Beauchamp, J.L. (1986) *J. Am. Chem. Soc.*, **108**, 582.
157. Winkler, A. *et al.* (1991) *Surf. Sci.*, **251–252**, 886.
158. Sault, A.G. *et al.* (1986) *Surf. Sci.*, **169**, 347.
159. Ray, K.B., Hannon, J.B., and Plummer, E.W. (1990) *Chem. Phys. Lett.*, **171**, 469.
160. Christmann, K., Schober, O., Ertl, G., and Neumann, M. (1974) *J. Chem. Phys.*, **60**, 4528.
161. Norton, P.R. *et al.* (1982) *Surf. Sci.*, **121**, 103.
162. Nowicka, E. and Dus, R. (1996) *Langmuir*, **12**, 1520.
163. Chan, L. and Griffin, G.L. (1984) *Surf. Sci.*, **145**, 165.
164. Roberts, M.W. and McKee, C.S. (1978) *Chemistry of the Metal – Gas Interface*, Clarendon Press, Oxford.
165. Christmann, K. (1991) *Introduction to Surface Physical Chemistry*, Steinkopff-Verlag, Darmstadt.
166. Conrad, H., Ertl, G., and Latta, E.E. (1974) *Surf. Sci.*, **41**, 435.
167. Wedler, G. (1976) *Chemisorption – An Experimental Approach*, Butterworths, London, and references therein.
168. Wedler, G. and Fisch, H. (1972) *Ber. Bunsen. Phys. Chem.*, **76**, 1160.
169. Stuckless, J.L., Frei, N.A., and Campbell, C.T. (2000) *Sens. Actuators, B*, **62**, 13.
170. Borroni-Bird, C.E. and King, D.A. (1991) *Rev. Sci. Instrum.*, **62**, 2177.
171. King, D.A. (1975) *Surf. Sci.*, **47**, 384.
172. Christmann, K. (1979) *Z. Naturforsch.*, **34a**, 22.
173. Madix, R.J., Ertl, G., and Christmann, K. (1979) *Chem. Phys. Lett.*, **62**, 38.
174. Christmann, K. and Ertl, G. (1976) *Surf. Sci.*, **60**, 365.
175. Christmann, K., Ertl, G., and Pignet, T. (1976) *Surf. Sci.*, **54**, 365.
176. Christmann, K. (1990) *J. Vac. Soc. Jpn.*, **33**, 549.
177. Balandin, A.A. (1969) *Adv. Catal. Rel. Subj.*, **19**, 1.
178. Bond, G.C. (1969) *Surf. Sci.*, **18**, 11.
179. Shustorovich, E. (ed.) (1991) *Metal – Surface Reaction Energetics: Theory and Applications to Heterogeneous Catalysis, Chemisorption, and Surface Diffusion*, Verlag Chemie, Weinheim.
180. Chorkendorff, I. and Niemantsverdriet, J.W. (2003) *Concepts of Modern Catalysis and Kinetics*, Wiley-VCH Verlag GmbH, Weinheim.
181. Koutecký, J. (1958) *Trans. Faraday Soc.*, **54**, 1038.
182. Grimley, T.B. (1967) *Proc. Phys. Soc. London*, **90**, 751.
183. Grimley, T.B. and Torrini, M. (1973) *J. Phys. C*, **6**, 868.
184. Einstein, T.L. and Schrieffer, J.R. (1973) *Phys. Rev.*, **B7**, 3629.
185. Lundqvist, B.I., Gunnarsson, O., Hjelmberg, H., and Nørskov, J.K. (1979) *Surf. Sci.*, **89**, 196.
186. Muscat, J.P. (1981) *Surf. Sci.*, **110**, 85.
187. Muscat, J.P. (1985) *Surf. Sci.*, **152/153**, 684.
188. Rinne, H. (1974) Absolutmessungen der Adsorption von Wasserstoff an der Nickel(111)-Fläche und an ultradünnen Nickelfilmen. PhD thesis. Technische Universität Hannover.
189. Alemozafar, A.R. and Madix, R.J. (2004) *Surf. Sci.*, **557**, 231.
190. Müller, K. (1986) *Ber. Bunsen Ges. Phys. Chem.*, **90**, 184.
191. Oed, W., Puchta, W., Bickel, N., Heinz, K., Nichtl, W., and Müller, K. (1988) *J. Phys. C*, **21**, 237.
192. Lehnberger, K., Nichtl-Pecher, W., Oed, W., Heinz, K., and Müller, K. (1989) *Surf. Sci.*, **217**, 511.
193. Michl, M., Nichtl-Pecher, W., Oed, W., Landskron, H., Heinz, K., and Müller, K. (1989) *Surf. Sci.*, **220**, 59.
194. Puchta, W., Nichtl, W., Oed, W., Bickel, N., Heinz, K., and Müller, K. (1989) *Phys. Rev.*, **B39**, 1020.
195. Nichtl-Pecher, W., Oed, W., Landskron, H., Heinz, K., and Müller, K. (1990) *Vacuum*, **41**, 297.

196. Christmann, K. (1987) *Z. Phys. Chem. (NF)*, **154**, 145.
197. Nørskov, J.K. (1994) *Surf. Sci.*, **299/300**, 690.
198. Hassold, E., Löffler, U., Schmiedl, R., Grund, M., Hammer, L., Heinz, K., and Müller, K. (1995) *Surf. Sci.*, **326**, 93.
199. Titmuss, S., Wander, A., and King, D.A. (1996) *Chem. Rev.*, **96**, 1291.
200. Müller, K., Lang, E., Endriss, H., and Heinz, K. (1982) *Appl. Surf. Sci.*, **11/12**, 625.
201. Müller, K. (1986) *Phys. Bl.*, **42**, 69.
202. Tománek, D. and Bennemann, K.-H. (1985) *Surf. Sci.*, **163**, 503.
203. Germer, L.H. and MacRae, A.U. (1962) *J. Chem. Phys.*, **37**, 1382.
204. Taylor, T.N. and Estrup, P.J. (1974) *J. Vac. Sci. Technol.*, **11**, 244.
205. Christmann, K., Penka, V., Behm, F.J., Chehab, F., and Ertl, G. (1984) *Solid State Commun.*, **51**, 487.
206. Penka, V., Christmann, K., and Ertl, G. (1984) *Surf. Sci.*, **136**, 307.
207. Christmann, K., Chehab, F., Penka, V., and Ertl, G. (1985) *Surf. Sci.*, **152/153**, 356.
208. Nielsen, L.P., Besenbacher, F., Laegsgaard, E., and Stensgaard, I. (1991) *Phys. Rev.*, **B44**, 13156.
209. Besenbacher, F., Stensgaard, I., Ruan, L., Nørskov, J.K., and Jacobsen, K.W. (1992) *Surf. Sci.*, **272**, 334.
210. Cattania, M.-G., Penka, V., Behm, R.J., Christmann, K., and Ertl, G. (1983) *Surf. Sci.*, **126**, 126.
211. Niehus, H., Hiller, C., and Comsa, G. (1986) *Surf. Sci.*, **173**, L599.
212. Kampshoff, E., Waelchli, N., Menck, A., and Kern, K. (1996) *Surf. Sci.*, **360**, 55.
213. Kralj, M., Becker, C., and Wandelt, K. (2006) *Surf. Sci.*, **600**, 4113.
214. Shuttleworth, I.G. (2013) *Surf. Sci.*, **615**, 119.
215. Goerge, J., Zeppenfeld, P., David, R., Büchel, M., and Comsa, G. (1993) *Surf. Sci.*, **289**, 201.
216. Eberhardt, W., Greuter, F., and Plummer, E.W. (1981) *Phys. Rev. Lett.*, **46**, 1085.
217. Rieder, K.H. and Stocker, W. (1986) *Phys. Rev. Lett.*, **57**, 2548.
218. Behm, R.J., Penka, V., Cattania, M.-G., Christmann, K., and Ertl, G. (1983) *J. Chem. Phys.*, **78**, 7486.
219. Ohno, S., Wilde, M., and Fukutani, K. (2014) *J. Chem. Phys.*, **140**, 134705.
220. King, D.A. (1980) *J. Vac. Sci. Technol.*, **17**, 241.
221. Naumovets, A.G. and Vedula, Y.S. (1985) *Surf. Sci. Rep.*, **4**, 365.
222. Gomer, R. (1990) *Rep. Prog. Phys.*, **53**, 917.
223. Wortman, R., Gomer, R., and Lundy, R. (1957) *J. Chem. Phys.*, **27**, 1099.
224. Gomer, R. (1961) *Field Emission and Field Ionization*, Harvard University Press, Cambridge.
225. DiFoggio, R. and Gomer, R. (1980) *Phys. Rev. Lett.*, **44**, 1258.
226. DiFoggio, R. and Gomer, R. (1982) *Phys. Rev.*, **B25**, 3490.
227. Ertl, G. and Neumann, M. (1972) *Z. Naturforsch., A*, **27**, 1607.
228. Seebauer, E.G. and Schmidt, L.D. (1986) *Chem. Phys. Lett.*, **123**, 129.
229. Mak, C.H., Brand, J.L., Deckert, A.A., and George, S.M. (1986) *J. Chem. Phys.*, **85**, 1676.
230. Mak, C.H., Koehler, B.G., Brand, J.L., and George, S.M. (1987) *J. Chem. Phys.*, **87**, 2340.
231. Zambelli, T., Trost, J., Wintterlin, J., and Ertl, G. (1996) *Phys. Rev. Lett.*, **76**, 795.
232. Trost, J., Zambelli, T., Wintterlin, J., and Ertl, G. (1996) *Phys. Rev.*, **B54**, 17850.
233. Wintterlin, J., Trost, J., Renisch, S., Schuster, R., Zambelli, T., and Ertl, G. (1997) *Surf. Sci.*, **394**, 159.
234. Stipe, B.C., Rezaei, M.A., and Ho, W. (1997) *J. Chem. Phys.*, **107**, 6443.
235. Stipe, B.C., Rezaei, M.A., and Ho, W. (1998) *Science*, **280**, 1732.
236. *http://www.physics.uci.edu/~wilsonho/stm-iets.html* (accessed 23 March 2015).
237. Mitsui, T., Rose, M.K., Fomin, E., Ogletree, D.F., and Salmeron, M. (2003) *Surf. Sci.*, **540**, 5.
238. Felter, T.E. and Stulen, R.H. (1985) *J. Vac. Sci. Technol., A*, **3**, 1566.
239. Felter, T.E., Foiles, S.M., Daw, M.S., and Stulen, R.H. (1986) *Surf. Sci. Lett.*, **171**, L379.

240. Auerbach, A., Freed, K.F., and Gomer, R. (1987) *J. Chem. Phys.*, **86**, 2356.
241. Badescu, S.C., Salo, P., Ala-Nissila, T., Ying, S.C., Jacobi, K., Wang, Y., Bedürftig, K., and Ertl, G. (2002) *Phys. Rev. Lett.*, **88**, 136101.
242. Badescu, S.C., Jacobi, K., Wang, Y., Bedürftig, K., Ertl, G., Salo, P., Ala-Nissila, T., and Ying, S.C. (2003) *Phys. Rev.*, **B68**, 205401.
243. Alefeld, G. and Völkl, J. (eds) (1978) *Hydrogen in Metals*, vol. 1 and 2, Springer-Verlag, Berlin.
244. Lewis, F.A. (1967) *The Palladium – Hydrogen System*, Academic Press, London.
245. Burch, R. (1980) in *Chemical Physics of Solids and Their Surfaces* (eds M.W. Roberts and J.M. Thomas), Royal Society of Chemisty, London, p. 1.
246. Bernstein, I.M. and Thompson, A.W. (eds) (1974) *Hydrogen in Metals*, American Society of Metal, Metals Park, OH.
247. Louthan, M.R. Jr., (2008) *J. Fail. Anal. Prev.*, **8**, 289.
248. Latanision, R.M. and Pickens, J.R. (eds) (1983) *Atomistics of Fracture*, Plenum Press, New York.
249. Kristinsdóttir, L. and Skúlason, E. (2012) *Surf. Sci.*, **606**, 1400.
250. Mullins, D.A. *et al.* (1986) *Chem. Phys. Lett.*, **129**, 511.
251. Lin, T.-S. and Gomer, R. (1991) *Surf. Sci.*, **255**, 41.
252. Cao, G.X. *et al.* (1997) *Phys. Rev. Lett.*, **79**, 3696.
253. Seebauer, E.G. *et al.* (1988) *J. Chem. Phys.*, **88**, 6597.
254. Daniels, E.A. and Gomer, R. (1995) *Surf. Sci.*, **336**, 245.
255. Daniels, E.A. *et al.* (1988) *Surf. Sci.*, **204**, 129.
256. Pendry, J.B. (1974) *Low-Energy Electron Diffraction*, Academic Press, New York.
257. Tong, S.Y. and van Hove, M.A. (1977) *Phys. Rev. B*, **16**, 1459.
258. Heinz, K. and Hammer, L. (1996) *Z. Phys. Chem.*, **197**, 173.
259. Davisson, C.J. and Germer, L.H. (1927) *Phys. Rev.*, **30**, 705.
260. Ertl, G. and Küppers, J. (1985) *Low-Energy Electrons and Surface Chemistry*, 2nd edn, Chapter 9, VCH-Verlagsgesellschaft, Weinheim.
261. Marcus, P.M. and Jona, F. (eds) (1984) *Determination of Surface Structure by LEED*, Plenum Press, New York.
262. van Hove, M.A., Weinberg, W.H., and Chan, C.-M. (1986) *Low-Energy Electron Diffraction: Experiment, Theory and Surface Structure*, Springer, Berlin.
263. Müller, K. (1993) *Prog. Surf. Sci.*, **42**, 245.
264. Heinz, K. (1995) *Rep. Prog. Phys.*, **58**, 637.
265. Woodruff, D.P. (2002) *Surf. Sci.*, **500**, 147.
266. Hulpke, E. (ed.) (1992) *Helium Atom Scattering from Surfaces*, Springer Series in Surface Science, vol. 27, Springer, Berlin, and references given therein.
267. Rieder, K.H. and Engel, T. (1980) *Phys. Rev. Lett.*, **45**, 824.
268. Engel, T. and Rieder, K.H. (1982) *Structural Studies of Surfaces*, Springer Tracts in Modern Physics, vol. 91, Springer, Berlin, p. 55.
269. Rieder, K.H. (1994) *Surf. Rev. Lett.*, **1**, 51.
270. Knauff, O., Grosche, U., Wesner, D.A., and Bonzel, H.P. (1992) *Surf. Sci.*, **277**, 132 [structure determination of H on Ni(110)].
271. Foss, M., Besenbacher, F., Klink, C., and Stensgaard, I. (1993) *Chem. Phys. Lett.*, **215**, 535 [structure determination of H on the Cu(100) surface].
272. Pohl, K. and Plummer, E.W. (1999) *Phys. Rev.*, **B59**, R5324.
273. Paul, J. (1988) *Phys. Rev.*, **B37**, 6164.
274. Feibelman, P.J. *et al.* (1980) *Phys. Rev.*, **B22**, 1734.
275. Merrill, P.B. and Madix, R.J. (1996) *Surf. Sci.*, **347**, 249.
276. Moritz, W., Imbihl, R., Behm, R.J., Ertl, G., and Matsushima, T. (1985) *J. Chem. Phys.*, **83**, 1959.
277. Imbihl, R., Behm, R.J., Christmann, K., Ertl, G., and Matsushima, T. (1982) *Surf. Sci.*, **117**, 257.
278. Rieder, K.H. and Wilsch, H. (1983) *Surf. Sci.*, **131**, 245.
279. Rieder, K.H. (1983) *Phys. Rev.*, **B27**, 7799.

280. Reimer, W., Penka, V., Skottke, M., Behm, J.J., Ertl, G., and Moritz, W. (1987) *Surf. Sci.*, **186**, 45.
281. Mijiritskii, A.V. *et al.* (1998) *Phys. Rev.*, **B57**, 9255.
282. Arnold, M. *et al.* (1997) *J. Phys.*, **C9**, 6481.
283. Sokolowski, M., Koch, T., and Pfnür, H. (1991) *Surf. Sci.*, **243**, 261.
284. Held, G., Pfnür, H., and Menzel, D. (1992) *Surf. Sci.*, **271**, 21.
285. Feulner, P. *et al.* (1986) *Surf. Sci.*, **173**, L576.
286. Döll, R., Hammer, L., Heinz, K., Bedürftig, K., Muschiol, U., Christmann, K., Seitsonen, A.P., Bludau, H., and Over, H. (1998) *J. Chem. Phys.*, **108**, 8671.
287. Besenbacher, F. *et al.* (1987) *Surf. Sci.*, **191**, 288.
288. Skottke, M., Behm, R.J., Ertl, G., Penka, V., and Moritz, W. (1987) *J. Chem. Phys.*, **87**, 6191.
289. Felter, T.E., Sowa, E.C., and van Hove, M.A. (1989) *Phys. Rev.*, **B40**, 891.
290. Yamazaki, H. *et al.* (2004) *Surf. Sci.*, **563**, 41.
291. Barker, P.A. and Estrup, P.J. (1981) *J. Chem. Phys.*, **74**, 1442.
292. Passler, M. *et al.* (1985) *Surf. Sci.*, **150**, 263.
293. Nahm, T.-U. and Gomer, R. (1997) *Surf. Sci.*, **375**, 281.
294. Hagedorn, C.J., Weiss, M.J., and Weinberg, W.H. (1999) *Phys. Rev.*, **B60**, R14016.
295. Kirsten, E., Parschau, G., Stocker, W., and Rieder, K.H. (1990) *Surf. Sci. Lett.*, **231**, L183.
296. Lui, K.M. *et al.* (1999) *J. Appl. Phys.*, **86**, 5256.
297. Zhang, Z., Minca, M., Deisl, C., Loerting, T., Menzel, A., and Bertel, E. (2004) *Phys. Rev.*, **B70**, 121401.
298. Boszo, F. *et al.* (1977) *Appl. Surf. Sci.*, **1**, 101.
299. Huesges, Z. and Chr, K. (2013) *Z. Phys. Chem.*, **227**, 808.
300. (a) Habermehl-Cwirzen, K.M.E. *et al.* (2004) *Phys. Scr.*, **T108**, 28; (b) Sicot, M. *et al.* (2008) *Surf. Sci.*, **603**, 3667.
301. Lee, G., Poker, D.B., Zehner, D.M., and Plummer, E.W. (1996) *Surf. Sci.*, **357/358**, 717.
302. Mate, C.M. and Somorjai, G.A. (1986) *Phys. Rev.*, **B34**, 7417.
303. Lee, G. *et al.* (1994) *J. Vac. Sci. Technol.*, **A12**, 2119.
304. Yanagita, H. *et al.* (1999) *Surf. Sci.*, **441**, 507.
305. Chr, K. *et al.* (1975) *Surf. Sci.*, **54**, 365.
306. Okada, M. *et al.* (2003) *Surf. Sci.*, **523**, 218.
307. Hagedorn, C.J. *et al.* (1999) *Phys. Rev.*, **B60**, R13941.
308. McCash, E.M., Parker, S.F., Pritchard, J., and Chesters, M.A. (1989) *Surf. Sci.*, **215**, 363.
309. Kresse, G. and Hafner, J. (2000) *Surf. Sci.*, **459**, 287.
310. Estrup, P.J. and McRae, E.G. (1971) *Surf. Sci.*, **25**, 1.
311. Chabal, Y.J. (1986) *Surf. Sci.*, **168**, 594.
312. Higashi, S., Chabal, Y.J., Trucks, G.W., and Raghavachari, K. (1990) *Phys. Rev. Lett.*, **65**, 1124.
313. Becker, R.S., Higashi, G.S., Chabal, Y.J., and Becker, A.J. (1990) *Phys. Rev. Lett.*, **65**, 1917.
314. Christmann, K. (1986) *Ber. Bunsen Ges. Phys. Chem.*, **90**, 307.
315. Rieder, K.H. and Stocker, W. (1985) *Surf. Sci.*, **164**, 55.
316. Grimley, T.B. (1967) *Proc. Phys. Soc. London*, **92**, 776.
317. Schmiedl, R., Nichtl-Pecher, W., Hammer, L., Heinz, K., and Müller, K. (1995) *Surf. Sci.*, **324**, 289.
318. Nichtl-Pecher, W., Gossmann, J., Hammer, L., Heinz, K., and Müller, K. (1992) *J. Vac. Sci. Technol., A*, **10**, 501.
319. Muscat, J.-P. and Newns, D.M. (1979) *Surf. Sci.*, **89**, 282.
320. Muscat, J.-P. (1985) *Prog. Surf. Sci.*, **18**, 59.
321. (a) Muscat, J.-P. (1984) *Surf. Sci.*, **139**, 491; (b) Muscat, J.-P. (1984) *Surf. Sci.*, **148**, 237.
322. Greeley, J. and Mavrikakis, M. (2003) *Surf. Sci.*, **540**, 215.
323. Barker, R.A. and Estrup, P.J. (1978) *Phys. Rev. Lett.*, **41**, 1307.
324. Debe, M.K. and King, D.A. (1979) *Surf. Sci.*, **81**, 193.
325. King, D.A. (1983) *Phys. Scr.*, **T4**, 34.

326. Schmidt, G., Zagel, H., Landskron, H., Heinz, K., Müller, K., and Pendry, J.B. (1992) *Surf. Sci.*, **271**, 416.
327. Barker, R.A., Semancik, S., and Estrup, P.J. (1980) *Surf. Sci.*, **94**, L162.
328. Ertl, G. and Küppers, J. (1970) *Surf. Sci.*, **21**, 61.
329. Bauer, E. (1980) in *Phase Transitions in Surface Films* (eds J.G. Dash and J. Ruvalds), Plenum, New York.
330. Schick, M. (1981) *Prog. Surf. Sci.*, **11**, 245.
331. Doyen, G., Ertl, G., and Plancher, M. (1975) *J. Chem. Phys.*, **62**, 2957.
332. Hill, T.L. (1960) *Introduction to Statistical Mechanics*, Addison-Wesley, Reading, MA, 2nd printing.
333. Binder, K. and Landau, D.P. (1976) *Surf. Sci.*, **61**, 577.
334. Mihura, B. and Landau, D.P. (1977) *Phys. Rev. Lett.*, **38**, 977.
335. Ising, E. (1925) *Z. Phys.*, **31**, 253.
336. Domany, E., Schick, M., and Walker, J.S. (1979) *Solid State Commun.*, **30**, 331.
337. Nagai, K., Ohno, Y., and Nakamura, T. (1984) *Phys. Rev.*, **B30**, 1461.
338. Einstein, T.L. (1979) *Surf. Sci.*, **84**, L497.
339. Greenler, R. (1975) *J. Vac. Sci. Technol.*, **12**, 1410, and references therein.
340. Chiang, S., Tobin, R.G., and Richards, P.L. (1984) *J. Vac. Sci. Technol.*, **A2**, 1069.
341. Avouris, P. and Demuth, J.E. (1984) *Annu. Rev. Phys. Chem.*, **35**, 49.
342. Hollins, P. and Pritchard, J. (1985) *Prog. Surf. Sci.*, **19**, 275.
343. Ueba, H. (1986) *Prog. Surf. Sci.*, **22**, 181.
344. Hirschmugl, C.J. (2002) *Surf. Sci.*, **500**, 577.
345. Ibach, H. and Mills, D.L. (1982) *Electron Energy Loss Spectroscopy and Surface Vibrations*, Academic Press, New York.
346. Froitzheim, H. (1977) in *Electron Spectroscopy for Surface Analysis* (ed. H. Ibach), Springer, Berlin, p. 205.
347. Probst, F.M. and Piper, T.C. (1967) *J. Vac. Sci. Technol.*, **4**, 53.
348. Froitzheim, H., Ibach, H., and Lehwald, S. (1976) *Phys. Rev. Lett.*, **36**, 1549.
349. Ho, W., Willis, R.F., and Plummer, E.W. (1978) *Phys. Rev. Lett.*, **40**, 1463.
350. *Vibrational Spectroscopy of Adsorbates*, R.F. Willis ed., vol. 15 Springer Series in Chemical Physics (Springer, Berlin, 1980)
351. Hamann, D.R. (1987) *J. Electr. Spectrom. Relat. Phenom.*, **44**, 1.
352. Mate, C.M., Bent, B.E., and Somorjai, G.A. (1988) Vibrational spectroscopy of hydrogen adsorbed on metal surfaces, in *Hydrogen Effects in Catalysis* (eds Z. Paàl and P.G. Menon), Dekker, New York, p. 57 ff.
353. Baró, A.M., Ibach, H., and Bruchmann, D. (1979) *Surf. Sci.*, **88**, 384.
354. Conrad, H., Scala, R., Stenzel, W., and Unwin, R.J. (1984) *J. Chem. Phys.*, **81**, 6371.
355. Conrad, H., Kordesch, M.E., Scala, R., and Stenzel, W. (1986) *J. Electr. Spectrom. Relat. Phenom.*, **38**, 289.
356. Voigtländer, B., Lehwald, S., and Ibach, H. (1989) *Surf. Sci.*, **208**, 113.
357. Schmidt, L.D. (1975) in *Interaction on Metal Surfaces*, Topics in Applied Physics, vol. 4 (ed. R. Gomer), Springer, Berlin, p. 82.
358. Zecho, T. *et al.* (2002) *J. Chem. Phys.*, **117**, 8486.
359. Kondoh, H. *et al.* (1991) *Chem. Phys. Lett.*, **187**, 466.
360. Butz, R. *et al.* (1984) *Surf. Sci.*, **147**, 343.
361. Caudano, Y. *et al.* (2002) *Surf. Sci.*, **502–503**, 91.
362. Baró, A.M. and Erley, W. (1981) *Surf. Sci.*, **112**, L759.
363. Okuyama, H. *et al.* (2002) *Phys. Rev.*, **B66**, 235411.
364. Jo, M. *et al.* (1985) *Surf. Sci.*, **154**, 417.
365. H. Yanagita *et al.*, *Phys. Rev.* .**B56** (1997) 14952
366. Chorkendorff, I. and Rasmussen, P.B. (1991) *Surf. Sci.*, **248**, 35.
367. Lee, G. and Plummer, E.W. (2002) *Surf. Sci.*, **498**, 229.
368. Papagno, L. *et al.* (1986) *Phys. Rev.*, **B34**, 7188.
369. Kostov, K.L., Widdra, W., and Menzel, D. (2004) *Surf. Sci.*, **560**, 130.
370. Richter, L.J. *et al.* (1988) *Phys. Rev.*, **B38**, 10403; and Richter, L.J. *et al.* (1988) *Surf. Sci.* **195** L182.
371. Nyberg, C. and Tengstål, C.G. (1983) *Phys. Rev. Lett.*, **50**, 1680.

372. Ellis, T.H. and Morin, M. (1989) *Surf. Sci.*, **216**, L351.
373. Conrad, H. *et al.* (1986) *Surf. Sci.*, **178**, 578.
374. Riffe, D.M. and Sievers, A.J. (1989) *Surf. Sci.*, **210**, L215.
375. Chakarov, D.V. and Marinova, T.S. (1988) *Surf. Sci.*, **204**, 147.
376. Richter, L.J. and Ho, W. (1987) *Phys. Rev.*, **B36**, 9797.
377. Kaesz, H.D. and Saillant, R.B. (1972) *Chem. Rev.*, **72**, 231.
378. Mudiyanselage, K., Yang, Y., Hoffmann, F.M., Furlong, O.J., Hrbek, J., White, M.G., Liu, P., and Stacchiola, D.J. (2013) *J. Chem. Phys.*, **139**, 044712.
379. Nørskov, J.K. and Stoltze, P. (1987) *Surf. Sci.*, **189/190**, 91.
380. Harris, J. and Andersson, S. (1985) *Phys. Rev. Lett.*, **55**, 1583.
381. Harris, J., Andersson, S., Holmberg, C., and Nordlander, P. (1986) *Phys. Scr.*, **T 13**, 155.
382. Harris, J. (1988) *Appl. Phys.*, **A47**, 63.
383. Wilke, H. and Scheffler, M. (1996) *Phys. Rev.*, **B53**, 4926.
384. Feuerbacher, B., Fitton, B., and Willis, R.F. (eds) (1978) *Photoemission and the Electronic Properties of Surfaces*, John Wiley & Sons, Inc., Chichester, New York.
385. Feuerbacher, B. and Fitton, B. (1977) in *Electron Spectroscopy for Surface Analysis*, Topics in Current Physics, vol. 4 (ed. H. Ibach), Springer, Berlin, p. 151.
386. Dose, V. (1983) *Prog. Surf. Sci.*, **13**, 225.
387. Rangelov, G., Memmel, N., Bertel, E., and Dose, V. (1990) *Surf. Sci.*, **236**, 250.
388. Nørskov, J.K. (1982) *Phys. Rev. Lett.*, **48**, 1620.
389. Siegbahn, P.E.M., Blomberg, M.R.A., and Bauschlicher, C.W. (1984) *Chem. Phys.*, **81**, 2103.
390. Panas, I., Siegbahn, P., and Wahlgren, U. (1987) *Chem. Phys.*, **112**, 325.
391. Weinert, M. and Davenport, S.W. (1985) *Phys. Rev. Lett.*, **54**, 1547.
392. Herring, C. and Nichols, P. (1949) *Rev. Mod. Phys.*, **21**, 85.
393. Culver, R.V. and Tompkins, F.C. (1959) *Adv. Catal. Relat. Subj.*, **11**, 67.
394. Hölzl, J. and Schulte, F.K. (1979) in *Springer Tracts in Modern Physics*, vol. 85 (ed. G. Höhler), Springer, Berlin, pp. 1–150.
395. Himpsel, F.J., Knapp, J.A., and Eastman, D.E. (1979) *Phys. Rev.*, **B19**, 2872.
396. Greuter, F., Strathy, I., Plummer, E.W., and Eberhardt, W. (1986) *Phys. Rev.*, **B33**, 736.
397. Hofmann, P. and Menzel, D. (1985) *Surf. Sci.*, **152/153**, 382.
398. Kevan, S.D. (ed.) (1992) *Angle-Resolved Photoemission: Theory and Current Applications*, Series: Studies in Surface Science and Catalysis, vol. 74, Elsevier, Amsterdam.
399. Himpsel, F.J., Heimann, P., and Eastman, D.E. (1981) *Phys. Rev.*, **B24**, 2003.
400. Komeda, T., Sakisaka, Y., Onchi, M., Kato, H., Masuda, S., and Yagi, K. (1987) *Phys. Rev.*, **B36**, 922.
401. Maruyama, T., Sakisada, Y., Kato, H., Aiura, Y., and Yanashima, H. (1991) *Surf. Sci.*, **253**, 147.
402. Ferrer, S. and Petroff, Y. (2002) *Surf. Sci.*, **500**, 605.
403. Knapp, A.G. (1973) *Surf. Sci.*, **34**, 289.
404. Christmann, K. and Herz, H. (1979) *Rev. Sci. Instrum.*, **50**, 988.
405. Mignolet, J.C.P. (1950) *Spec. Discuss. Faraday Soc.*, **8**, 326.
406. Palmberg, P.W. and Tracy, J.C. (1969) *J. Chem. Phys.*, **51**, 4852.
407. Ertl, G. and Küppers, J. (1971) *Ber. Bunsen Ges. Phys. Chem.*, **75**, 1017.
408. Sharma, R.B., Vinod, C.P., and Kulkarni, G.U. (2002) *Bull. Mater. Sci.*, **25**, 247.
409. *http://en.wikipedia.org/wiki/Scanning_tunneling_microscope* (accessed 24 March 2015).
410. Woodruff, D.P. and Delchar, T.A. (1985) *Modern Techniques of Surface Science*, Cambridge University Press, Cambridge.
411. Smoluchowski, R. (1941) *Phys. Rev.*, **60**, 661.
412. Topping, J. (1927) *Proc. R. Soc. London, Ser. A*, **114**, 67.
413. Bozso, F., Ertl, G., Grunze, M., and Weiss, M. (1977) *Appl. Surf. Sci.*, **1**, 103.
414. Peden, C.H.F., Goodman, D.W., and Houston, J.E. (1988) *Surf. Sci.*, **194**, 92.
415. Feibelman, P.J. and Hamann, D.R. (1987) *Surf. Sci.*, **179**, 153.

416. Menzel, D. (1999) *Surf. Rev. Lett.*, **6**, 835.
417. Sprunger, P.T. and Plummer, E.E. (1991) *Chem. Phys. Lett.*, **187**, 559.
418. Wilzén, L. *et al.* (1988) *Surf. Sci.*, **205**, 387.
419. Estrup, P.J. (1979) *J. Vac. Sci. Technol.*, **16**, 635.
420. Witte, G. *et al.* (1995) *Surf. Sci.*, **323**, 228.
421. Lee, G. and Plummer, E.W. (1995) *Phys. Rev.*, **B51**, 7250.
422. (a) Pennemann, B. *et al.* (1991) *Surf. Sci.*, **249**, 35; (b) Pennemann, B. *et al.* (1991) *Surf. Sci.*, **251/252**, 877.
423. Hohenberg, P. and Kohn, W. (1964) *Phys. Rev.*, **136**, B864.
424. Stott, M.J. and Zaremba, E. (1980) *Phys. Rev.*, **B22**, 1564.
425. Nørskov, J.K. and Lang, N.D. (1980) *Phys. Rev.*, **B21**, 2136.
426. Nørskov, J.K. (1981) *J. Vac. Sci. Technol.*, **18**, 420.
427. Nordlander, P., Holloway, S., and Nørskov, J.K. (1984) *Surf. Sci.*, **136**, 59, and references given therein.
428. Choy, T.C. (1999) *Effective Medium Theory: Principles and Applications*, International Series of Monographs on Physics, vol. 102, Oxford University Press, Oxford.
429. Daw, M.S. and Baskes, I.M. (1984) *Phys. Rev.*, **B29**, 6443.
430. Chong, D.P. (ed.) (1995) *Recent Advances in Density Functional Methods*, Series: Recent Advances in Computational Chemistry, vol. 1, World Scientific Publication Co., Singapore.
431. Neugebauer, J. and Scheffler, M. (1992) *Phys. Rev.*, **B46**, 16067.
432. *http://en.wikipedia.org/wiki/Local-density_approximation GGA* (accessed 24 March 2015).
433. Perdew, J.P., Chevary, J.A., Vosko, S.H., Jackson, K.A., Pederson, M.R., Singh, D.J., and Fiolhais, C. (1992) *Phys. Rev. B*, **46**, 6671.
434. Hammer, B., Hansen, L.B., and Nørskov, J.K. (1999) *Phys. Rev. B*, **59**, 7413.
435. Greeley, J., Nørskov, J.K., and Mavrikakis, M. (2002) *Annu. Rev. Phys. Chem.*, **53**, 319.
436. Krekelberg, W.P., Greeley, J., and Mavrikakis, M. (2004) *J. Phys. Chem. B*, **108**, 987.
437. Klinke, D.J. II, and Broadbelt, L.J. (1999) *Surf. Sci.*, **429**, 169.

# 37
# Adsorption of Water

*Sabine Maier and Miquel Salmeron*

## 37.1
## Introduction

Water is the most abundant molecule on Earth's surface. It is a small molecule with interesting physical properties, which arise as a consequence of its net dipole moment and the ability to form hydrogen-bond networks. Hydrogen bonding in ice and liquid water is responsible for their unusually high melting and boiling points, respectively. The phase diagram of water is complex, having a number of triple points resulting in many solid phases, that is, different ice phases. All the crystalline phases of ice have in common that the water molecules are hydrogen-bonded to four neighboring molecules, with the oxygen atoms in fixed positions relative to each other. The hydrogen atoms may or may not be ordered. The disorder in the orientation of the $H_2O$ molecules is referred to as *proton disorder*.

Hexagonal ice (ice Ih) is the form of natural snow and ice on earth, giving rise to the sixfold symmetry in ice crystals formed from water vapor. Hexagonal ice is therefore also the phase we expect to form if water adsorbs on cold surfaces at atmospheric pressure or in vacuum. In an ice Ih crystal, the oxygen atoms are arranged in a hexagonal lattice with the wurtzite structure (see also Chapter 17 in Volume 3). However, at the interface with a surface, a multitude of structures have been observed as the water molecules adapt to the structure of the underling surface while simultaneously optimizing its hydrogen-bonded network.

The adsorption of water on surfaces plays an important role in many phenomena in nature, for example, in the formation of raindrops, and in interstellar ice. In the same way, solid/liquid interfaces are ubiquitous in technological applications, including electrochemistry, corrosion, and environmental chemistry, and in heterogeneous catalysis where water is present as a reactant, product, or an intermediate species. Hence, numerous experimental and theoretical studies have been performed with the aim to understand the adsorption structures of water at the molecular level on a variety of surfaces. Also, the dissociation of water in the presence of a catalytically active surface attracts considerable attention as part of an economically attractive way to split water in oxygen and hydrogen, which is a key component in a sustainable hydrogen economy. New techniques and

*Surface and Interface Science: Solid-Gas Interfaces I*, First Edition. Edited by Klaus Wandelt.

methods are currently being explored to use the power of sunlight to efficiently split water into its components, paving the way for a broad use of hydrogen as a clean, green fuel.

In the 1980s and 1990s, the bilayer model was essentially the "standard model" of water adsorption on close-packed metals. The bilayer model proposed by Doering and Madey [1] consists of an epitaxial two-dimensional arrangement of water molecules forming a hexagonal network similar to the (001) basal plane of ice Ih. In this bilayer, half the water molecules are bound to the metal via the oxygen lone-pair orbital, while the other half are lifted from the surface and H-bonded to the first half in the tetrahedral bonding configuration of ice Ih. However, our understanding of the molecular structure of the water overlayer has progressed substantially through the application of local probes such as scanning tunneling microscopy (STM) in combination with extensive first-principles density functional theory (DFT) calculations. Through these advances, it is now well established that, in order to maximize simultaneously hydrogen-bonding between water molecules and bonding to the metal, the water molecules adopt adsorption configurations different from the ice-like patterns. This has been confirmed by the observation of water structures on hexagonal close-packed surfaces that are not exclusively built from hexagons but includes also pentagons or combinations of heptagons and pentagons. Similar to the richness of bulk ice structures, a multitude of metastable water structures on surfaces have been observed.

This chapter will first focus on water adsorption on metal surfaces, where the largest amount of molecular-level results exists. A brief outline of water adsorption on other surfaces, such as metal oxides, salts, and graphene, will be given toward the end of the chapter. We will start with the adsorption and diffusion properties of water monomers, followed by a discussion of the peculiar diffusion of the water dimers. We next discuss the structure of water aggregates: hexamers and clusters of hexamers. Finally, we discuss the nature of the wetting layer and the formation of ice-like multilayers.

## 37.2 Experimental Aspects

### 37.2.1 Preparation of Water Films

Prerequisite for every surface science experiment that aims at obtaining atomic-level understanding of surface structures is the preparation of a well-defined surface with a minimal degree of contamination by foreign atoms and molecules. The adsorption properties of water are indeed strongly affected by the presence of surface steps, defects, and impurities, which can act as nucleation sites for the formation of one- and two-dimensional water structures.

One of the important roles of steps is to promote the nucleation and growth of ice as well as influence the crystalline structure of the films. For example, a change in the ice structure, cubic as opposed to hexagonal, has been observed in the vicinity of steps on Pt(111), as there is a mismatch in the atomic step height between the metal substrate and the ice bilayer separation [2], as described in more detail later in this chapter. In addition to enhanced nucleation, the higher reactivity of the step edges can promote chemical reactions of water. For instance, on Pt(111) it was found by calculations that, while on the flat surfaces water molecules adsorb intact, on the stepped surface water dissociates forming mixed water/hydroxyl structures through an autocatalytic mechanism promoted by H bonding [3].

Preadsorbed atoms and molecules on the surface can influence the adsorption properties of water, as shown in many coadsorption experiments [4]. Coadsorption effects generally manifest as site blocking, enhanced reactivity, poisoning, structural rearrangements, and solvation, the last one being particular important in regard to water. Most coadsorption studies at the molecular scale have been performed with one of the three species: alkali atoms, oxygen, or carbon monoxide. Alkalis are gener ally known to promote water dissociation on otherwise inactive metal surfaces with the exception of Ru(0001). Therefore, surface science studies on the interaction of water with alkali atoms provide an insight into solvation and charge-transfer effects. The water–oxygen interaction is important to a wide variety of surface phenomena including corrosion, dielectric film growth, selective oxidation catalysis, and mineral surface dissolution.

A common surface impurity in ultrahigh vacuum (UHV) surface science experiments is hydrogen, which arises from the residual gas in the vacuum chamber. Its adsorption is nearly unavoidable on reactive surfaces. Other preadsorbed atoms and molecules on the surface are often related to the adsorption and desorption of molecules from the walls of a UHV chamber, which however can be minimized by careful surface preparation. Other contaminants may arise as a result of segregation of bulk impurities. For instance carbon, a common bulk impurity for Pt-group metals [5], was shown to interact with adsorbed water in STM experiments on Ru(0001) [6]. In the following, we discuss the water adsorption on clean and flat surfaces and omit discussion of coadsorption experiments with other small molecules and atoms.

In a typical experiment, water is first purified by pumping and thawing cycles prior to introduction into the vacuum chamber, which is done via a capillary doser attached to a leak valve or by means of a molecular beam [7, 8]. The later method is preferred because it minimizes adsorption and exchange with the vacuum chamber walls.

### 37.2.2 Experimental Techniques

Most of the available surface science techniques have been employed to study the adsorption properties of water on surfaces. The substantial knowledge about water adsorption we have nowadays is attributed to the complementarities of the different experimental techniques and their indispensable combination with theoretical studies. In this section, we briefly highlight the possibilities and difficulties of the different probes related to water adsorption experiments.

Low-energy electron diffraction (LEED) was extensively employed in early studies of the adsorption of water on single crystals [9]. An important issue of the technique is the potential for electron-beam damage, as electrons are especially effective in inducing dissociation and, to some extent, also desorption. This has resulted in misleading literature reports on commensurate monolayers, especially on Ru(0001). However, when using very low currents and sensitive detectors, LEED is an adequate tool for the determination of the crystalline structure of water overlayers.

X-ray photoelectron spectroscopy (XPS) and X-ray adsorption spectroscopy (XAS) have the advantage of providing direct information about the chemical nature of the adsorbed water molecules, for example, whether water is intact or dissociated, as well as a quantitative measure of the coverage and stoichiometry. A substantial amount of published XPS work reports temperature-programmed X-ray photoelectron spectroscopy (TP-XPS), where the sample is annealed at low rates (0.1–1 K/min) while XPS is recorded. Compared to temperature-programmed desorption (TPD) spectroscopy, TP-XPS enables tracking the nature of the water species as a function of temperature and shows the transitions taking place on the surface at certain temperatures. Like in the case of LEED, XAS and XPS have the potential to produce damage, either directly by photolysis or by means of the secondary electrons ejected from the surface. In contrast, helium atom scattering (HAS) is a nondestructive technique that is also sensitive to the position and ordering of hydrogen atoms and has been successfully used to study the termination of thick ice films [10].

Compared to the surface area averaging techniques LEED, XPS, and HAS, low-temperature STM has been invaluable in the determination of the local atomic structure of water by providing detailed real-space images of monomers, small clusters, and monolayers on many surfaces. A combination of STM with DFT calculations is often required, as the interpretation of STM measurements by itself is not always straightforward. For instance, STM topographic images are difficult to interpret in regard to molecular details such as the orientation of water molecules on the surface and even to differentiate between single $H_2O$ molecules and OH species. Furthermore, the tunneling parameters (tunneling current and bias voltage) need to be carefully chosen, as the tip can easily disturb the water molecules. The strong electric field between the tip and the surface can also interact with the dipole of the water molecules and thus influence their bonding structure. In addition the tunneling electrons can excite, restructure, and dissociate the molecules. Recently, noncontact atomic force microscopy (nc-AFM) as another scanning probe method has been successfully employed to determine the structure of thicker water films [11] and monomers on wide bandgap oxides [12].

Vibrational spectroscopies, including high-resolution electron energy loss spectroscopy (HREELS), infrared (IR), Raman, and sum frequency generation (SFG) spectroscopies have been extensively used to study hydrogen bonding in water clusters and monolayers. The selection rules that determine the intensity of the vibrational modes introduce a new element that needs to be considered in the interpretation. For example, the surface selection rules in IR require that the dynamic dipole has a component perpendicular to the surface. SFG selection rules determine the intensity and shape of the resonance peaks, which depend on the s and p polarization of the three beams involved (IR, incoming and outgoing visible). HREELS has different selection rules depending on whether the scattering is dominated by dipole interaction or by impact. Therefore, the adsorption geometry dictates whether or not the vibrational mode is active and can be detected. These various rules can be used to help determine the orientation of the bonds, by purposely varying the polarization of the photon fields, or through the angular dependence of the incident and scattered electron beams in the case of HREELS. An inherent difficulty in vibrational spectroscopies of water is that the O–H stretching region for isolated $H_2O$ and hydrogen-bonded OH overlap (3400–3600 $cm^{-1}$). Therefore hydrogen-bonded $H_2O$ and OH are sometimes difficult to differentiate. In addition, there have been historically some difficulties leading to an ambiguity in band assignments of water from vibrational spectra [13–15].

## 37.3 Water Adsorption on Metals

### 37.3.1 Water Monomers

#### 37.3.1.1 Theoretical Perspective

A variety of experimental and theoretical methods have been applied to determine the adsorption geometry of water monomers on hexagonal close-packed metal surfaces [4, 9, 16, 17]. The different theoretical approaches include *ab initio* molecular dynamics simulations [18] and DFT [19, 20]. There is a general agreement in all these methods that the water monomer prefers to adsorb on or near a top site, with its molecular plane nearly parallel to the surface, as seen in Figure 37.1 [21]. Michaelides *et al.* found that on Ru, Pt, and Ag the water molecule is slightly displaced laterally from the top site by ~0.3 Å; however, the potential well is quite flat with a change of only ~0.02 eV from the precise atop site [19]. This "flat" energy landscape possibly explains the stability of the small commensurate water clusters with a hexagonal $(\sqrt{3}\times\sqrt{3})R30°$ geometry observed by STM measurements on

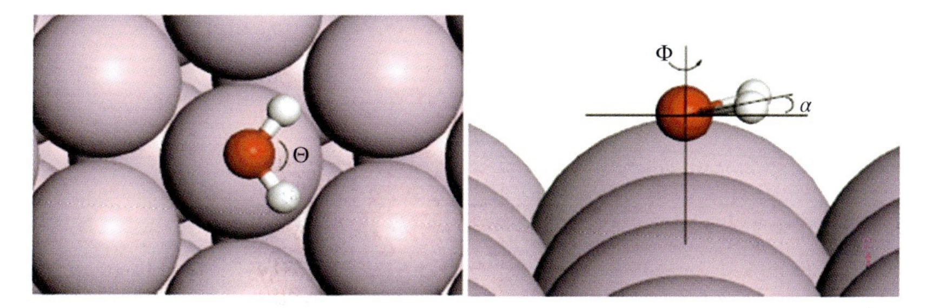


**Figure 37.1** Top and side views of the typical structure of a $H_2O$ monomer adsorbed on a close-packed metal surface. (From Ref. [19] © 2003, American Physical Society (APS).)

close-packed metal surfaces despite the mismatch between the substrate lattice constant and O–O separation of hydrogen-bonded water molecules.

The angle $\alpha$ between the surface and molecular plane varies slightly with the metal, ranging from 6° on Ru(0001) to 15° on Cu(111), with no preference for the azimuthal angle $\Phi$ on any of the hexagonal close-packed surfaces. The $H_2O$ molecule itself deforms only slightly upon adsorption with respect to the gas-phase geometry: The O–H bond length (~0.98 Å) is slightly elongated, and the HOH angle $\Theta$ is expanded by no more than 2° from the calculated gas-phase value of 104° [19]. The calculated bond length between the metal and oxygen varies from 2.25 to 2.36 Å for most of the close-packed metal surfaces, except for Ag and Au with 2.78 and 3.02 Å, which is due to the lower binding energy of water on Ag and Au [19, 20].

The molecular orbitals of the isolated water molecule in the ground electronic configuration are classified according to the $C_{2v}$ symmetry and labeled $1a_1 2a_1 1b_2 3a_1 1b_1$, the first one ($1a_1$) being essentially the 1s core level of O [22]. For a detailed summary of the electronic structure for water adsorbed on single-crystal surfaces and the energy of its molecular orbitals, we refer the reader to Ref. [4]. The two highest occupied orbitals are the $3a_1$ and $1b_1$ orbitals. Because the $3a_1$ orbital is a mixed orbital with some O and H character, it resides approximately 2 eV below the $1b_1$ [21, 23]. The $3a_1$ and the $1b_1$ O "lone-pair" orbital are perpendicular to each other. Consequently, in a parallel adsorption geometry, the interaction of the $H_2O$ molecule with the substrate is determined mainly by hybridization of metal and the $1b_1$ orbitals. Vice versa, an upright water molecule interacts mainly through the $3a_1$ orbital. The adsorption of flat-lying $H_2O$ monomers on 4d metal surfaces leads to an interaction between the $H_2O$ $1b_1$ orbital and the top of the metal d band, resulting in the formation of a set of bonding and antibonding states. The ability of the metal to stabilize this interaction by depopulating the antibonding states determines the water adsorption energies [24, 25]. This stabilization mechanism is viable for Ru and Rh, less favorable for Pd, and infeasible for Ag. This interpretation is largely equivalent to an alternative interpretation of the bonding in terms of an enhancement of the overlap between orbitals as a consequence of a reduction in Pauli repulsion between the $1b_1$ orbital and axial metal d orbitals [26, 27], resulting in a closer approach of the water molecule to the metal surface.

The calculated binding energy of a water monomer on hexagonal close-packed metal surfaces varies between 0.1 and 0.4 eV for the different metals, reflecting the strength of the oxygen–metal bond (see Table 37.1) [19, 20]. The minor discrepancy among the adsorption energies shown in Table 37.1 is related to the parameters

**Table 37.1** Adsorption energies in electron volts of water monomers on close-packed metal surfaces from two different DFT calculations.

| Ru(0001) | Rh(111) | Pd(111) | Pt(111) | Cu(111) | Ag(111) | Au(111) | References |
|---|---|---|---|---|---|---|---|
| 0.38 | 0.42 | 0.33 | 0.35 | 0.24 | 0.18 | 0.13 | [19] |
| 0.409 | 0.408 | 0.304 | 0.291 | — | — | 0.105 | [20] |

used such as unit cell size and energy cutoffs. In most DFT methods, the exchange-correlation functionals do not treat well van der Waals (vdW) forces. However, the relatively weak interaction of the water molecules with the metal surfaces implies that vdW forces should play an important role. For example, Carrasco *et al.* revealed the role of vdW dispersion forces in water–metal bonding by considering different nonlocal vdW density functionals. An enhancement of the adsorption energies (typically > 110 meV/$H_2O$) with respect to the widely used PBE functional was found for Ru(0001), resulting in an adsorption energy of 0.541 eV [28]. The adsorption energy of water on hexagonal close-packed metals is comparable to the hydrogen-bond strength of water $E_{HB}^{water} \approx 0.24$ eV [29]. This equivalence ensures that stable water structures must optimize both the water–metal and the water–water interactions to find stable adsorption structures at the metal interface.

A systematic DFT study comparing the adsorption energies of water monomers on different low-index and stepped surfaces was performed by Tang and Chen on a Cu surface [30]. The calculations identified the top site as the most stable configuration for all the low-index and stepped surfaces. The binding energy varied from 0.19 to 0.44 eV, with the lowest value for the (111) surface followed by (100) < (221) < (211) < (110) < (210). The binding energy correlated linearly with the coordination number of the surface Cu atom and the work function.

#### 37.3.1.2 Experimental Studies

As a result of the weak binding energy of water monomers, isolated water molecules can easily diffuse on hexagonal close-packed surfaces, leading to the formation of water clusters. Therefore, low temperatures are required to restrict the mobility of water and isolate single molecules for sufficiently long times. Morgenstern and Rieder were the first to image water monomers using a low-temperature STM operated at 16 K [31, 32]. In their experiment, the molecules showed as protrusions on Cu(111), with an apparent height of ~80 pm. A similar topography was also observed on Pd(111) [33] and Ru(0001) [34].

The identification of the adsorption site with STM can be achieved by resolving the surface lattice in the surroundings of the molecule. However, one has to take into account the strong interaction between the tip and the adsorbed water. Hence, Fomin *et al.* scanned an area near the molecule at low gap to resolve the atoms while retracting the tip by ~2 Å when scanning near and over the water molecule to circumvent this problem [35]. By extrapolating the lattice from the top and bottom parts of the image, they concluded that the adsorption site of water is atop the Pd atoms, in agreement with theoretical predictions. Likewise, Kumagai used O atoms as reference to triangulate the position of $H_2O$ on Cu(110), where water was also found to adsorb on top sites [36].

In principle, vibrational spectroscopy can also distinguish the water monomer, as its bands will be distinct from other hydrogen-bonded species. However, optical vibrational spectroscopy of monomers is not trivial because the sensitivity is often poor [37]. In addition, owing to the flat adsorption geometry at low coverage, the OD stretch and scissors of the dipole are forbidden and therefore weak in the IR spectra.

### 37.3.2
### Diffusion of Water Monomers and the Formation of Dimers

At temperatures around 40 K, water monomers diffuse on Pd(111) surfaces at a rate of approximately 0.1 Å/min. At this rate, the molecules can be tracked with STM over sequential images, making it possible to observe the aggregation of monomers into dimers. The formation of dimers on Pd(111) is shown in Figure 37.2 [33]. Surprisingly, the mobility of dimers, trimers, and tetramers on Pd(111) was found to be higher than that of the monomer by several orders of magnitude [33]. A first attempt to explain this result was the assumption that in the dimer only one molecules can occupy the most stable binding geometry (on top of Pd atom) if the O–O distance is different from the Pd–Pd distance. However, with this model it is difficult to account for the more than three orders of magnitude in diffusion rate found experimentally.

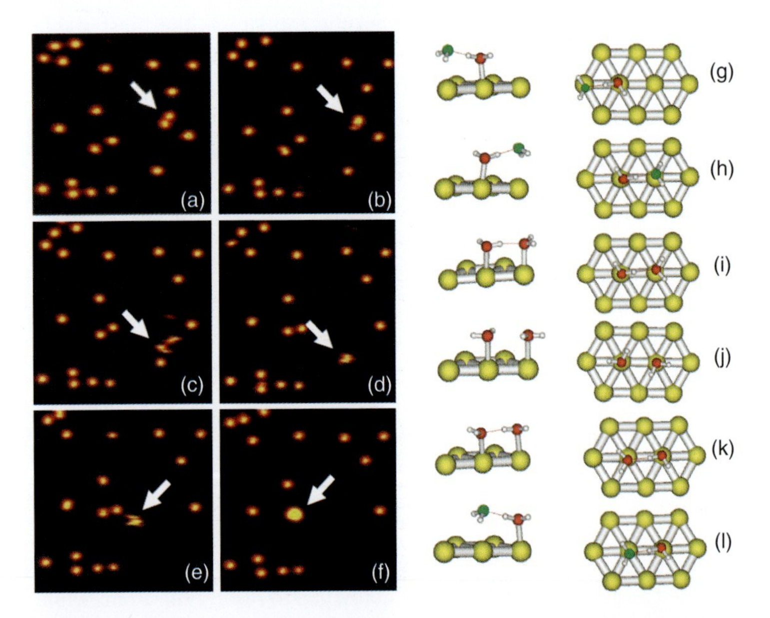


**Figure 37.2** Diffusion mechanism of water monomers on Pd(111) from an experimental perspective (left) and theoretical perspective (right). (a–f) Sequence of STM images (180 Å × 180 Å) showing water molecules (bright dots) adsorbed on Pd(111) at 40 K. Two monomers in (a) join to form a dimer in (b). The dimer diffuses rapidly so that the tip only scans over it for one line before moving to a nearest neighbor site, thus producing a streak in (c). The dimer encounters a third monomer and forms a trimer (d), which diffuses in (e), approaching a pair of nearby monomers. The arrow in (f) points to the pentamer formed by the collision. (g–l) Mechanism for water dimer diffusion (top and side view). Step (g)–(h) involves a nearly free rotation of the dimer; step (h)–(i) is the wagging motion of the dimer, which brings both water molecules to a similar height above the surface from where they can undergo donor–acceptor tunneling interchange (i–k). From step (k) to (l), the dimer restores its equilibrium geometry having translated one lattice spacing. (From: left: Ref. [33] reprinted with permission from AAAS and right: Ref. [38]. Copyright (2004) by APS.)

This indicates that the dimer diffusion may be governed by a different mechanism than thermal diffusion.

Ranea *et al.* [38] proposed a new mechanism for the water dimer diffusion. They used DFT to calculate the adsorption geometries and considered the ability of H bonds to rearrange through quantum tunneling. The starting adsorption geometry of the dimer was calculated to be asymmetric with a surface bound donor molecule and an acceptor molecule adsorbed farther away from the surface, so that it forms no direct bonds with the surface. This allows the dimer to rotate nearly freely around the axis centered at the donor molecule, as shown in Figure 37.2g,h. Another energy configuration minimum is one where both molecules are bound to the Pd atoms, as in Figure 37.2i. This minimum is not as deep as the first one in Figure 37.2g, and requires the overcoming of a barrier of 0.11 eV. In this metastable position, the water molecules can exchange their roles as donor and acceptor through tunneling of the H atoms. Once the two molecules have interchanged their respective roles as donor and acceptor of the H bond, the dimer restores its asymmetric equilibrium structure. The overall result is a net displacement of the dimer by one lattice spacing. Depending on the temperature and the barrier, diffusion can be much faster via a donor–acceptor interchange tunneling mechanism than mere thermal diffusion, explaining the faster diffusion for the dimer than the monomer.

Kumagai *et al.* studied the hydrogen-bond rearrangement within a water dimer and its coupling to molecular vibrations by monitoring the interchange events in real time on Cu(110), also using STM [39]. A large isotope effect was found between two $H_2O$ and $D_2O$, experimentally confirming the mechanism by Ranea which suggested that quantum tunneling of H atoms is involved in the process. In addition, the excitation of the donor–substrate stretch mode was found to effectively assist the interchange tunneling, highlighting the sensitivity of the interchange dynamics to the displacement of the oxygen atoms. Experimental evidence for the rotation of the acceptor molecule in the dimer was provided by Motobayashi *et al.* [40] on Pt(111) and Mugarza *et al.* [6] on Ru(0001). Both imaged the water monomers as a round-shaped protrusion, and the dimers looked like a "flower-like" protrusion with sixfold symmetric lobes in STM. They explained the peculiar shape of the water dimer by the model of a "time-averaged" shape visible as a six-lobed structure produced by the rotating molecule.

Although it is known that thermal energy activates the jump of the molecules from one site to another, the experiments to measure this energy do not indicate which modes are involved in the energy transfer because, upon heating, the energy is apportioned among the various modes according to a Boltzmann distribution. STM has the unique advantage of making possible the excitation of specific vibrational or electronic modes of the water molecule to determine fundamental mechanisms of diffusion and reaction. In this manner, Fomin *et al.* [35] showed that, when the tunneling electrons have an energy equal to or above that of the quantum of the scissoring mode (200 meV for $H_2O$ and 150 meV for $D_2O$), the diffusion rate of water on Pd(111) increases substantially, as shown in Figure 37.3. Similarly, on Cu(111) the experimentally determined threshold energy of 220 mV corresponds to the energy

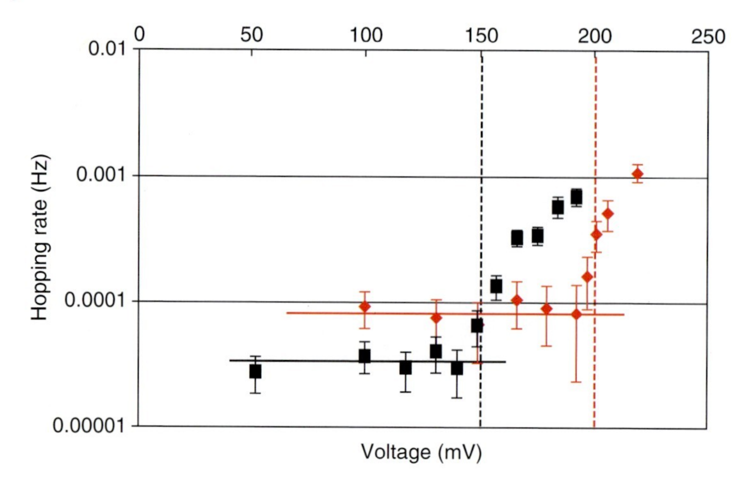


**Figure 37.3** Semilog plot of the hopping rates for $H_2O$ and $D_2O$ molecules versus tunneling voltage on a Pd(111) sample at 40 K with 40 pA tunnel current. Hopping rates rise dramatically for bias voltages higher than 200 mV for $H_2O$ and 150 mV for $D_2O$. Diamond symbols (♦) correspond to $H_2O$ data and squares (■) to $D_2O$. The horizontal lines emphasize the independence of mobility versus voltage below the threshold voltages. The vertical lines represent the energies of the scissor vibration modes of $H_2O$ (200 meV) and $D_2O$ (150 meV). (Reprinted from Ref. [35] with permission from Elsevier.)

of the scissor mode and a second peak at $430 \pm 20$ mV, which is within the range of the O–H stretching frequency [31]. On Ag(111) [41] and Pd(111) [35], it is the excitation of the HOH scissor mode that leads to diffusion. On Ru(0001), diffusion is enhanced by excitation of the OH stretch mode at 450 meV. Higher energy pulses above 1 V lead to dissociation of water molecules on Ru(0001) [6].

## 37.3.3
## Small Water Clusters

### 37.3.3.1 The Water Hexamer

Water adsorbs on close-packed transition metal surfaces with a binding energy that is comparable to the strength of a hydrogen bond between the molecules. Therefore, the structure of water on metal surfaces is dominated by two competing forces: one leading to maximization of water–metal bonding through the oxygen lone-pair, and the other leading to the maximization of the hydrogen bonding among the molecules. Starting with the water monomer as the building block, which prefers to adopt a flat configuration (see Figure 37.1), we can build on the surface various clusters from nearly flat water molecules.

The cyclic water hexamer cluster, which can be considered the smallest piece of ice, has been imaged by several groups on Pt-group metals [42] and on other metal substrates [31, 43]. In the hexamer, each water molecule is located near the preferred atop adsorption site, each accepting and donating one hydrogen bond [43, 44]. On Pd and Ru, the water molecules are stabilized by ~0.3 eV when they form a complete hexamer as compared to the dimer, making hexamer rings the most energetically favorable units on a surface [20, 44]. Depending on the substrate, the hexamers either have a planar structure with all molecules at the same height or a buckled structure with molecules at two distinct heights (see Figure 37.4). Generally, planar hexamers are favored on reactive surfaces, for example, Ru(0001), to which water molecules bond relatively strongly, whereas on noble metals, for example, Cu and Ag [43], buckled hexamers are favored. The vertical displacement between adjacent $H_2O$ molecules in the buckled hexamer is 0.76 Å on Cu and 0.67 Å on Ag.

#### 37.3.3.2 Water Clusters and the Two-Dimensional Water Rules

Since each molecule in an aggregate of flat-lying water molecules can donate two H bonds but accept only one, there is an imbalance between donor and acceptor molecules, and larger networks of flat molecules cannot be formed. Instead, water molecules that accept more than one H bond have to be incorporated into the clusters. If a water molecule accepts two H bonds from other molecules bound to the metal surface, it will not interact significantly with the metal because of steric hindrance and also because of competition for the oxygen lone-pair electrons that are involved both in accepting H bonds and in water–metal bonding [42]. The plane of these molecules is nearly vertical to the surface, with their hydrogen pointing toward the surface ($H_{down}$) or toward vacuum ($H_{up}$) (see Figure 37.5). The difference in stability of $H_{up}$ and $H_{down}$ structures at a given surface generally varies by less than 0.1 eV [45, 46]. Similarly, the energy barrier for $H_{up}/H_{down}$ flipping in the first monolayer is very low [46]. Introducing one double acceptor water molecule destabilizes the hexamer by 0.03 eV/$H_2O$ on Ru [37], while on Pd this energy change is smaller (0.01 eV/$H_2O$) [42]. DFT calculations showed that peripheral double acceptor molecules on Ru and Pd have a greater stability compared to internal double acceptor molecules in a cluster.

STM experiments of small clusters on Pd and Ru have shown that the maximization of both H bonding and O–metal bonding, combined with the greater stability of peripheral $H_{up}$ molecules compared to internal $H_{up}$ molecules, leads to a peculiar growth pattern of water in the form of elongated stripes of honeycomb structures 1–3 cells wide [42, 44]. The growth of more extensive structures requires the

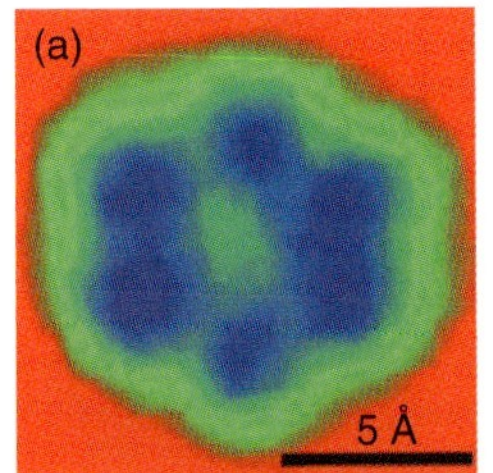

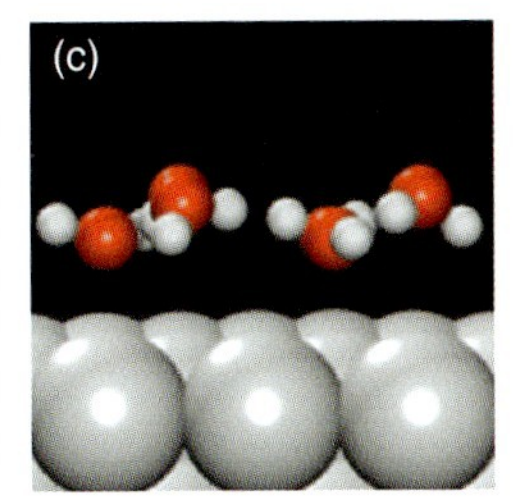


**Figure 37.4** (a) High-resolution STM images of a $H_2O$ hexamer on Cu(111). (b)–(c) Top and side views of the corresponding theoretical structures. (Reprinted by permission from Macmillan Publishers Ltd: [17], copyright (2012).)

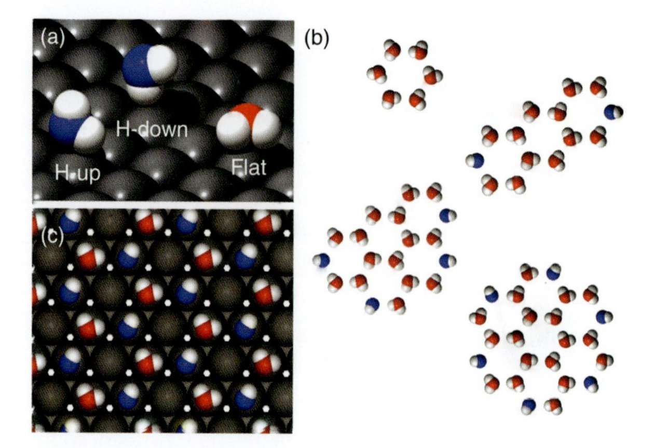


**Figure 37.5** (a) Schematic showing the three common orientations of water molecules on a metal surface: $H_{up}$, $H_{down}$, and flat. (b) Schematic drawings showing water clusters of side-sharing hexagons. The clusters obey the so-called "2D ice rules." These rules require maximization of O–metal bonding and H bonding to adjacent molecules. Ideally, a molecule will donate one or two H bonds while accepting one, the maximum H bonding possible in two dimensions (red colored "flat" molecules parallel to the surface). Beyond the hexamer (top left), each additional hexamer must have at least one "defect" molecule of double-acceptor nature with its plane nearly vertical to the surface (blue colored molecules). (c) Schematic drawing of an extended film.

incorporation of additional double-acceptor, vertically oriented "defect" molecules and is hence only favored for extended layers.

The fact that each water molecule prefers to lie approximately parallel to the surface, maximizing the O–metal bonding and forming H bonds such that O–O–O angles are about 120°, can be considered as a "two-dimensional" ice rule, in similarity to the traditional H-bonding rules for bulk ice [42, 44]. The traditional ice rules by Bernal and Fowler [48] and Pauling [49] for bulk ice imply that a hexagonal network of oxygen atoms are hydrogen-bonded together to create rings of six molecules, each one maintaining a tetrahedral geometry and each O–O axis containing only one hydrogen. To preserve the tetrahedral bonding geometry of water, the oxygen atoms buckle to form "puckered" rings, with each oxygen vertically displaced from its nearest neighbor by ~0.97 Å. Doering and Madey [1] modified these rules to account for the preference of water to bind to a metal surface via its oxygen lone pair. The result is a structure in which water forms two-dimensional islands of ice with a bilayer structure, half of the water molecules binding directly to the metal through the oxygen and the other half hydrogen-bonding to those below. This model predicts ice structures in which each molecule in the upper half of the bilayer has one unsatisfied hydrogen bond, leaving OH dangling "H-up" toward the vacuum (see Figure 37.6). As discussed above, most of the water structures derived from recent DFT calculations and STM experiments on hexagonal close-packed metal surfaces consist of hydrogen-bonded topologies different from the ice bilayer. One example is the cyclic hexamer, and more will follow in the discussion on the water monolayer.

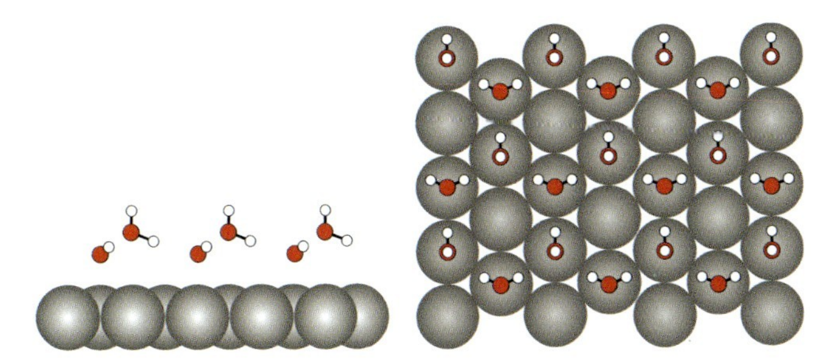

**Figure 37.6** Schematics of the originally proposed ice-like bilayer structure, in registry with the close-packed surface.

#### 37.3.3.3 One-Dimensional Water Chains on Cu(110)

On hexagonal close-packed metal surfaces, such as the (111) planes of face-centered cubic (fcc) and the (0001) planes of hexagonal close-packed (hcp) crystals, cyclic water hexamers form as described in the previous sections, where each water molecule is located near the favored atop adsorption site and accepts and donates just a single H bond. Although hexagonal structures are favored, water can also form other structural motifs, such as pentagons, heptagons, and others. These non-hexagonal motifs have been observed only in small clusters, as well as in grain boundaries and defects. On surfaces that are non-hexagonal, however, the question arises whether water can form other cyclic rings in extended domains. This has been recently proposed to be the case on Cu(110).

On Cu(110) water forms approximately 1-nm-wide chains, which have a zig-zag structure and are oriented along the [001] direction [50–52]. Unlike in early proposed models [51], Carrasco *et al.* used STM, IR spectroscopy, and DFT to conclude that the water chains are built from a face-sharing arrangement of water pentagons[50], with a backbone of molecules adsorbed flat atop Cu atoms, with the ring being completed by a tilted molecule in a double acceptor configuration (see Figure 37.7). The pentagon structure is favored over others because it maximizes the water–metal bonding while maintaining a strong hydrogen-bonding network. In addition, DFT calculations showed that the earlier proposed hexagonal structures are unstable and would lead to structures considerably wider than the images found in the STM studies. Another characteristic of the chains observed in the STM images is that they repel each other along the [110] direction in order to maximize the spacing between them. DFT calculations explained the repulsion with a decreased binding energy as they come closer, following a $1/d$ dependence, consistent with a electrostatic repulsion. At higher coverage, water forms a two-dimensional network with a $(7 \times 8)$ reconstruction with half of the water in a $H_{down}$ and half in a $H_{up}$ configuration [53].

DFT calculations predicted also the structure and relative stability of 1D chains on several other (110) metal surfaces, that is, Ni, Pd, and Ag, which have lattice constants from 3.5 to 4.2 Å [50]. A strong correlation is observed between the relative stability of hexagon versus pentagon chains and the metal lattice parameter, with pentagons favored on substrates with small lattice constants. Ni(110) and Cu(110)

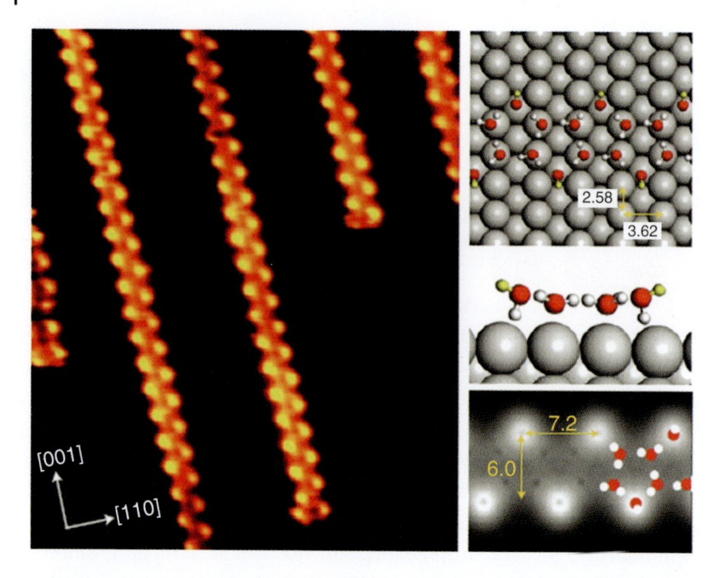


**Figure 37.7** STM image (120 × 140 Å) of water chains growing along the [001] direction along with the proposed equilibrium geometry identified by DFT and the corresponding simulated STM image. Distances are in Å. (Reprinted by permission from Macmillan Publishers Ltd: [50], copyright (2009))

were identified as substrates where pentagons are stable. On Ag(110), there is a preference for hexagons while Pd(110) is a borderline case.

At higher temperatures, water partially dissociates into OH and H. Three distinct phases of water/hydroxyl structures on Cu(110) [52, 54, 55] were observed as the temperature decreased and the water/hydroxyl ratio increased. It starts with pure OH dimers, then extended $1H_2O$:1OH chains from, which at higher coverage are replaced by islands with a distorted, two-dimensional hexagonal c(2 × 2) $2H_2O$:1OH network [54]. The water/hydroxyl chains are rotated compared to the intact ones, along the [1–10] direction. DFT calculations confirm that none of the water and water/hydroxyl structures formed on Cu(110) obeys the usual ice rules; instead, they optimize the water and hydroxyl binding sites rather than simply the overall number of H bonds. Hydroxyl binds in the Cu bridge site in the 1D chain structures, but is displaced to the atop site in the 2D network in order to accommodate water in its preferred atop binding geometry.

#### 37.3.3.4 **Partially Dissociated Water Structures on Ru(0001)**

One-dimensional water structures are also found in the case of partially dissociated water layers. Ru(0001) is a well-studied model substrate, as it allows the growth of mixed $H_2O$:OH phases [56–59]. On the hexagonal Pt-group metal surfaces, besides Ru only on the Rh(111) have hydroxyl structures reported, but only through beam damage [60]. Partial dissociation of water into OH and H species is an activated process on Ru(0001), with the dissociation barrier being near that for desorption [23, 59]. Therefore, water adsorbs nondissociatively at low temperatures in a metastable state. The temperature above which partial dissociation has been observed covers a broad range between 104 K [57] and 150 K [58], while complete dissociation to atomic oxygen was observed above 190 K [57]. The interdependence of time and temperature in advance of the dissociation process explains the wide range of dissociation temperatures reported in the literature [61]. As mentioned earlier, partial dissociation may also be assisted via electron or X-ray irradiation [56, 57, 59, 61].

Initially, DFT calculations suggested that the partially dissociated layers ($OH + H_2O + H$) form monolayers, with the water molecules and hydroxyl fragments hydrogen-bonded in a hexagonal network and with the hydrogen atoms bound to the metal [62]. Weissenrieder *et al.* [57] gave the first experimental evidence for partially dissociated water on Ru(0001) by means of photoelectron spectroscopy. They found an $OH/H_2O$ ratio of 3 : 5 for the partially dissociated layer, indicating a nonstoichiometric $OH/H_2O$ phase, consistent with a partially dissociated layer formed of $(OH)_3(H_2O)_5$ units. Experimental evidence from LEED supported the formation of partially dissociated layer but showed that the layer has a domain structure with narrow dimensions in one direction [37]. STM measurements and DFT calculations by Tatarkhanov *et al.* showed that the mixed $H_2O$–OH structures consist of long, narrow stripes aligned with the three crystallographic directions perpendicular to the close-packed atomic rows of the Ru(0001) substrate [56]. The internal structure of the stripes is a honeycomb network of H-bonded water and hydroxyl species. The width of the $H_2O$/OH stripes varied from 2.5 to 6 lattice constants corresponding to $H_2O$/OH ratios from 4 : 0 to 5 : 3 resulting in average to a 3 : 1 ratio (see Figure 37.8). From DFT simulations, it was found that water prefers to lay flat, bonding via O to Ru. In addition, OH is found to prefer locations in the interior of the stripe.

### 37.3.4
### Wetting Layer

In the 1980s, Doering and Madey performed water adsorption experiments on the hcp Ru(0001) surface, where the distance between neighboring Ru atoms is within 4% of the nearest-neighbor water–water distance in ice Ih [1]. The observed $(\sqrt{3} \times \sqrt{3})R30°$ pattern in LEED led to a the widely used bilayer model, which closely resembles the hexagonal water layers of the (001) basal plane of ice Ih. Soon after, the $\sqrt{3}$ LEED patterns observed on other metal surfaces were similarly interpreted, making this original bilayer the canonic model of the ice monolayer [4]. The nearest-neighbor distance in transition metals and its mismatch with the spacing of water in bulk ice are shown in Table 37.2.

As explained previously, substantial progress has been made in the fundamental understanding of water on metal surfaces, made possible by the application of LEED, XPS, STM, and DFT methods. There is now a consensus that the wetting layer structure deviates substantially from the original ice-like bilayer model, a deviation

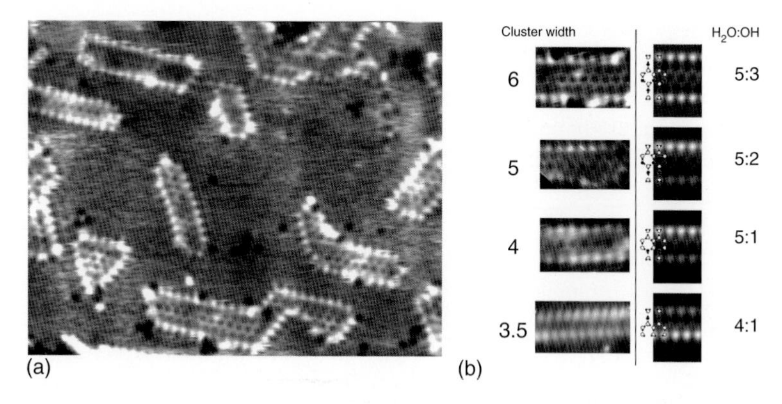


**Figure 37.8** (a) STM image (200 × 160 Å$^2$) of structures formed by dosing 3–4 ML $H_2O$ at 45 K and subsequent annealing to 180 K. The internal structure of the mixed $H_2O$–OH phase can be resolved and consists of elongated honeycomb structures with a high contrast periphery. The dark spots are due to impurity atoms, most likely C and O. The H atoms also present on the surface cannot be resolved in this image due to their weak contrast and rapid mobility. (b) Close-up images (left) showing various types of mixed $H_2O$–OH honeycomb stripes extracted from larger scale STM images. The numbers on the left denote the width of the stripe in units of Ru–Ru atomic distances. STM simulations (right) for the models that best fit each cluster. Drawings of the $H_2O$ and OH species forming the stripe are superimposed in the left part of the calculated images with white and dark circles, respectively. The H atoms appear as small dark circles in all cases. (Reprinted with permission from [56]. Copyright 2008, AIP Publishing LLC.)

**Table 37.2** The lateral expansion required to form a commensurate ($\sqrt{3} \times \sqrt{3}$)R30° hexagonal ice layer on the close-packed face of the metal can be quantified by the disregistry, defined in a simplified manner as $m = (\sqrt{3}a - a_{ice})/a_{ice}$, where $a$ is the lattice constant of the substrate and $a_{ice} = 4.5$ Å [64] is the bulk lattice constant of ice Ih.

| Metal | Ni | Cu | Rh | Ru | Pd | Pt | Ag |
|---|---|---|---|---|---|---|---|
| Lattice spacing, $a$ (Å) | 2.49 | 2.56 | 2.69 | 2.71 | 2.75 | 2.77 | 2.89 |
| Misfit, $m$ (%) | −4.2 | −1.5 | 3.5 | 4.3 | 5.8 | 6.6 | 11.2 |

resulting from the competing forces of water–metal interaction and hydrogen bonding among molecules.

In this section, we will review the two most studied surfaces with respect to wetting layers, that is, Pt(111) and Ru(0001). On both surfaces, recent experiments revealed a non-ice-like wetting layer. The one system for which there is the most compelling evidence for an ice bilayer structure so far is an alloy of platinum and tin, whose surface was specifically tailored with a $\sqrt{3}$ periodicity with a corrugation that facilitated bilayer adsorption [65] by allowing only alternating molecules to chemisorb and stabilizing the $H_{down}$ water structure by reducing the metal–hydrogen repulsion, as compared to a flat surface.

#### 37.3.4.1 First Water Layer on Pt(111)

Water adsorbs molecularly on Pt(111) and desorbs intact near 160 K with zero-order kinetics [4, 9]. Mixed OH/water monolayers have only been obtained by either coadsorption of water and oxygen [66] or by electron damage [67]. Evidence that water dissociation is not favored on Pt(111) is the observation that $O_2$ and $H_2$ react to form water at temperatures below 150 K [68]. The fact that water does not dissociate upon heating on this surface makes possible the use of higher temperatures during growth, a prerequisite to grow well-ordered, thermodynamically stable layers.

Instead of the $(\sqrt{3}\times\sqrt{3})$R30° ice bilayer structure of the wetting layer reported in early LEED measurements [69], where water is strained to reach commensurability with the hexagonal close-packed surface, HAS [70] and low-energy electron diffraction experiments [71, 72] revealed a $(\sqrt{37}\times\sqrt{37})$R25.3° superstructure that changes to a $(\sqrt{39}\times\sqrt{39})$R16.1° one with increasing coverage. Glebov *et al.* [70] suggested a model for the wetting layer slightly modified form the conventional ice bilayer by rotating the ice-like sheets of water molecules and compressing them to fit the structures observed by HAS, as shown in Figure 37.9. In the $\sqrt{39}$ reconstruction, the monolayer is compressed by about 3% relative to bulk ice [71]. Meng *et al.* [20, 73] calculated the structures for the $\sqrt{37}$ and $\sqrt{39}$ superstructures by DFT and obtained binding energies of 0.597 and 0.617 eV per water molecule, which are considerably more stable than the $\sqrt{3}$ structure. Contrary to Glebov *et al.*'s model calculations, later experiments including XAS [26] suggest that water binds with the OH group toward the surface ($H_{down}$) with no dangling OH bonds, giving rise to a hydrophobic water monolayer.

STM experiments by Morgenstern *et al.* [74, 75] showed that water on Pt(111) initially decorates step edges and at higher coverage forms 2D islands on terraces. Three different structures were observed, depending on the coverage and surface temperature. One phase was ascribed to a rotated bilayer yielding a moiré pattern and two superstructure domains of different orientation. However, the molecular structure remained unsolved in these experiments. Recently, independent STM experiments by Nie *et al.* [76] and Standop *et al.* [77] of water deposited at 140 K revealed triangular regions where the water contrast is lower, indicating a lower height, about ~0.3 Å

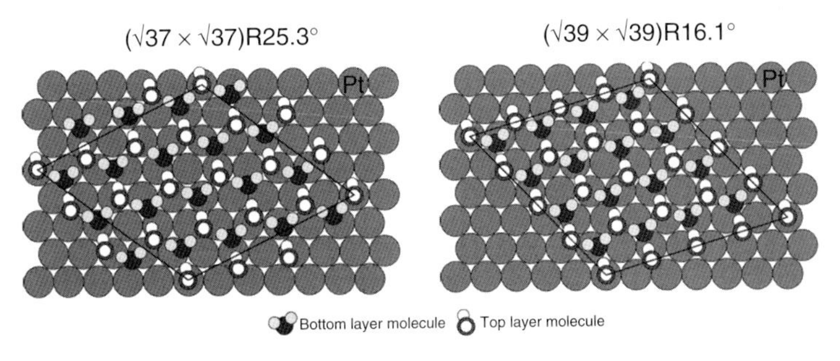


**Figure 37.9** Model showing the $(\sqrt{37}\times\sqrt{37})$R25.3° and $(\sqrt{39}\times\sqrt{39})$R16.1° water superstructures on Pt(111) in a $H_{up}$ configuration. Later experiments and calculations suggest a $H_{down}$ arrangement. (Reprinted with permission from [70]. Copyright 1997, AIP Publishing LLC.)

below the surrounding regions. This difference is small compared to the apparent height of a water layer and must therefore be associated with the orientation of molecules within the monolayer. The dark triangles are interpreted as interstitial defects, wherein a hexagonal ring, rotated relative to the surrounding ones, is surrounded by pentagons and heptagons (see Figure 37.10). The area of the triangles contains flat-lying $H_2O$ molecules, with their O atoms bound to the metal atoms. In ordered patches, the periodicity and rotation of the triangular features are precisely that of the $(\sqrt{37} \times \sqrt{37})R25.3°$ phase. Similarly, the $(\sqrt{39} \times \sqrt{39})R16.1°$ was explained as made by five-, six-, and seven-membered rings of water molecules.

#### 37.3.4.2 First Water Layer on Ru(0001)

The interaction of water with the Ru(0001) surface has been the center of an extended debate. This debate originates in part from the sensitivity of the water adlayer on Ru(0001) to electron, photon, and thermally induced water dissociation reactions on this reactive surface, as discussed in Section 37.3.3.4.

In contrast with the bilayer model, on Ru(0001) the O atoms of the water molecules are nearly coplanar in the two-dimensional layer, with only a small buckling of 0.15 Å, as found by Held and Menzel in 1994 by LEED [78]. This is inconsistent with the buckled bilayer model in which the vertical displacement between O atoms in adjacent molecules is substantially larger (0.96 Å in ice). This

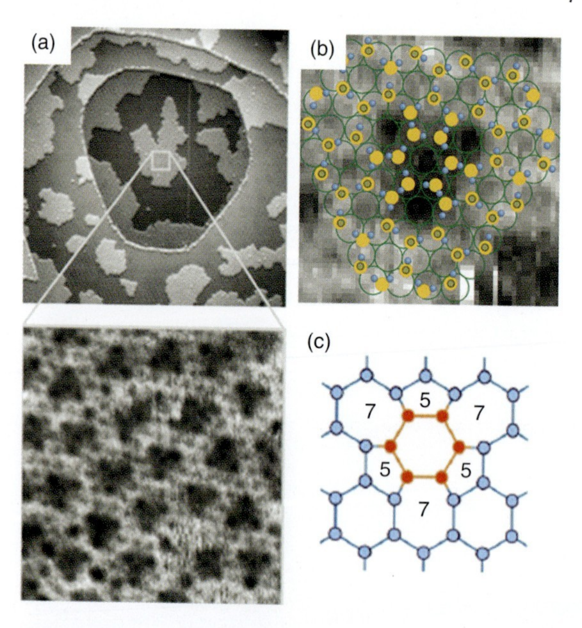


**Figure 37.10** (a) STM images of submonolayer amounts of water on Pt(111) at 140 K. The magnified images show dark triangles, which can be interpreted as interstitial defects (8 nm × 8 nm). (b) Overlay of the molecular model on STM images of a single triangle. (c) Schematic showing a di-interstitial defect consisting of hexagons, pentagons, and heptagons in a hexagonal lattice. (Reprinted with permission from [76]. © (2010) APS.)

result motivated a theoretical analysis of the energetics of water adsorption by Feibelman, who concluded that on Ru the stability of the quasi-planar monolayer could be explained by the dissociation of one-half of the water molecules into H and OH to maximize both hydrogen-bonding and oxygen–metal interactions [62]. This partially dissociated model is also supported by other calculations [23]. Dissociative water adsorption on Ru(0001) was unexpected since water adsorbs molecularly on most close-packed surfaces, including Ni(111), Cu(111), Pd(111), Rh(111), and Pt(111) [4, 60]. Dissociation was supported by the presence of two peaks in the water O 1s photoemission spectrum at 531.3 and 533.3 eV [9], corresponding to adsorbed water and hydroxyl species. Soon after, Andersson *et al.* [59] showed that, while the partially dissociated layer is more stable on Ru(0001), its formation is kinetically hindered by the competition with desorption, both occurring with similar rates around 140–150 K.

Because of the dual contribution to dissociation from phonons during annealing or from electrons, the nature of the film of water remained unresolved. LEED, which is insensitive to H-atom positions, indicates O-atom ordering [78], whereas broad reflection-absorption IR spectroscopy (RAIRS) bands [37] as well as low reflectivity and broad peaks in He scattering [79] imply a disordered phase. Theoretical efforts provided hints, notably that a H-bonded honeycomb network with alternating chains of flat-lying molecules and molecules with dangling H bonds has a particularly low energy [37, 79]. Alternatively, a high-order commensurate structure involving a rotated water bilayer had been suggested based on He atom scattering [80]. Eventually, high-resolution STM measurements supported by DFT calculations resolved the controversy by providing a consistent picture of the structure of the metastable, intact wetting layer on Ru(0001) [81]. This layer consists of hexagonal water units of two types rotated by 30° relative to each other. One of the orientations is in registry with the substrate, with its molecules lying flat and their O atoms forming bonds to the metal atoms lying directly below. The molecules in the rotated hexagons have dangling H bonds, are weakly bound to the substrate, and lie correspondingly higher. The two types of hexagons are connected by pentagons and heptagons. This bonding motif, although nonperiodic, is of similar nature to the wetting structure recently reported on Pt(111) but very different from the conventional ice bilayer. A similar structure with two types of water hexagons, rotated by 30° relative to each other and connected by heptagons and pentagons, is found on Pd(111) [81].

### 37.3.5 Multilayer Adsorption

Knowing the structure of the first layer of water molecules on a surface is a starting point for a general understanding of water structures in contact with a material but not the endpoint. In many cases, DFT calculations indicate that on most metals the first water layer locates its H atoms either in that layer or between it and the surface, instead of dangling into the vacuum ready to form hydrogen bonds with a second layer of molecules [82]. In that sense, it appears as if the first layer might have hydrophobic properties. The hydrophobicity of the wetting layer is a result of maximizing the bonding to the metal, on one hand, and hydrogen bonding within the layers, on the other. Hydrophobicity or hydrophilicity will probably depend on whether the second and higher layers will reconstruct the first layer to adapt to a bulk ice structure, perhaps inducing a change of orientation of some of its molecules [83]. Indeed, recent experiments have shown that the structure and morphology of thin ice films are quite complex.

Multilayers of water provide an ideal model system to study many processes at the ice/gas interface, for example, relevant for processes in atmospheric surface science or astrophysics. Also, phenomena such as the solvation dynamics of excess electrons, which is associated with the reactivity of small molecules and their transport into the ice film, can be studied on thin supported ice films using surface science experiments.

An interesting aspect of the surface of ice films is the proton ordering, as it is expected to significantly influence many ice properties, for example, its interaction with adsorbates. Bulk ice Ih is comprised of orientationally disordered water molecules, giving rise to positional disorder of the hydrogen atoms in the hydrogen-bonded network of the lattice ("proton disordered"). The molecules in the top half of an ice Ih surface bilayer are threefold coordinated, with no second layer above them to form H bonds. Each threefold-coordinated molecule contributes to the surface either a dangling H (but not forming an H bond) or dangling O lone-pair electrons. The extension of the bulk H disorder to the surface would naturally suggest a quasi-random pattern of molecules with dangling H atoms and O lone-pair electrons [84]. However, many ordered proton configurations on the ice surface have been proposed in recent calculations [85–87]. Most prominent is the striped phase with alternating rows of dangling H and dangling O lone-pair atoms, as suggested by Fletcher [87]. Experimentally, the assignment of the exact proton structure is an ambitious task, as the protons have a small footprint in otherwise powerful techniques such as LEED, He scattering, and STM. SFG, X-ray absorption spectroscopy, and work function signals are sensitive to proton arrangements, and could provide a more complete picture.

Cu(111), Ag(111), and Au(111) are non-wetting surfaces, where water is expected to form 3D ice clusters instead of continuous ice films. On Ru(0001), Pd(111), and Pt(111), on the other hand, water forms a wetting layer, making these surfaces a convenient template for the growth of homogenous ice films. The growth of ice films was observed with a variety of surface science techniques. Real-space observations at the molecular scale of extended ice structures with STM at high resolution are, in general, difficult and have been achieved only recently. The reason is that water molecules are easily disturbed by the scanning process due to the large dipole moment and the low energy of the hydrogen bond. Furthermore, molecular vibrations excited from inelastically scattered tunneling electrons starting from approximately 200 mV, which is close to the H-bonding energy, might trigger molecular reorientations as discussed earlier. Finally, the insulating character of ice puts a severe limitation on the thickness of the multilayer structure that can be imaged by

STM. In the following, we will summarize recent progress on the understanding of water multilayer on metals on the example of Pt(111) and Cu(111) surfaces.

#### 37.3.5.1 Thin Ice Films on Pt(111)

By rare-gas physisorption [88] and STM experiments [2], it was found that a water monolayer wets Pt(111) at all temperatures between 20 and 155 K to form a continuous film, completely covering the surface before a second layer is deposited. A detailed description of the wetting layer structure can be found in Section 37.3.4.1. The fully coordinated water monolayer in most metals has no dangling H bonds or lone-pair electrons, giving rise to a hydrophobic monolayer on which water diffusion is facile [76]. Additional water leads to the formation of amorphous water layers at temperatures below 120 K, which grow layer by layer [88]. The low surface energy of the hydrophobic monolayer relative to the crystalline ice surface favors the formation of non-wetting 3D ice crystallites, which grow at temperatures above 135 K. The equilibrium structure of crystalline ice on Pt(111) has the lattice constant of bulk ice, is incommensurate with the substrate, and is rotated 30° relative to the close-packed direction on the Pt(111) [71, 89]. Similar ice films were obtained for $D_2O$ [90].

Recently, Thürmer and Bartelt showed that the local structure of ice films with as many as 30 layers can be imaged by AFM and even with STM, when negative sample biases of less than −6 V and sub-picoampere tunneling currents are used [2, 11, 91]. These scanning probe measurements made possible the determination of the thickness-dependent morphology of ice films. Consistent with TPD measurements, water deposited on Pt(111) below 120 K was found to form amorphous films, whereas metastable cubic ice appeared between 120 and 150 K. The STM measurements revealed also that surface diffusion was significant at temperatures between 115 and 140 K, allowing for the ice-film morphology to relax toward equilibrium [92].

At low temperatures, ice can also exist in a cubic structure in addition to the hexagonal one [84]. The cubic phase of water ice is metastable relative to the Ih structure, with otherwise the same density and the same hydrogen-bonding arrangement and therefore also the same coordination. It differs from hexagonal one only in the stacking sequence of the bilayer. In the equilibrium structure of hexagonal ice Ih, the stacking alternates between bilayers, as shown by the "black" and "blue" lines in Figure 37.11, whereas in the metastable cubic ice the bilayers are stacked in the same way. Thürmer and coworkers observed both the cubic and hexagonal phase in the crystalline ice film on Pt(111) [2, 11], and concluded that cubic ice emerges from screw dislocations caused by the mismatch in the atomic Pt-step height and the ice-bilayer separation.

#### 37.3.5.2 Thin Ice Films on Cu(111)

In this section, we illustrate the multilayer adsorption of ice on Cu(111) as an example of a non-wetting film on close-packed metal surfaces. Mehlhorn and Morgenstern described in detail the structural transformation from amorphous solid water to crystalline ice depending on the temperature [93]. The structure changes from monomer-decorated double bilayers with different superstructures

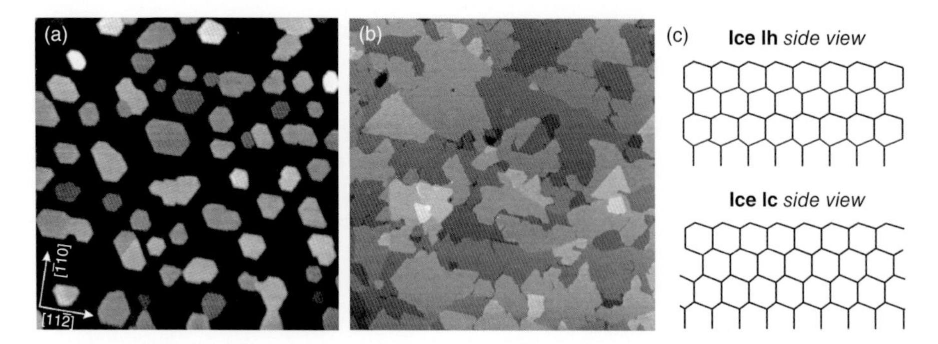


**Figure 37.11** Surface topography of a crystalline ice-multilayer film on Pt(111). (a) 500 nm× 500 nm STM image, showing individual approximately 3-nm-high ice crystals grown at 140 K. (b) STM image (750 nm× 750 nm) of a continuous 4-nm-thick ice film grown onto Pt(111) at 140 K. (c) Schematic side view of an ice Ih and Ic ice film, revealing the different stacking. (Reprinted with permission from (a) [91] and (b) [2]. © (2008) by APS.)

of additional molecules, to pyramidal islands, and finally to nanocrystallites of different heights close to the desorption temperature of 149 K (Figure 37.12). All these structures can be explained by hydrogen bonding within the ice network. None of the structures, however, shows the termination of a classical ice bilayer, with three molecules in each hexagonal ring with dangling H atoms ($H_{up}$) and the

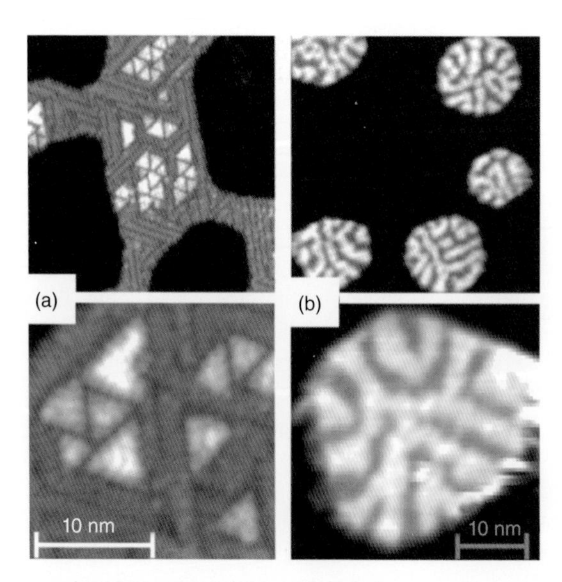


**Figure 37.12** Surface topography of ice clusters on Cu(111) after annealing 1.1 BL $D_2O$ at 145 K and flash to 149 K. (a) Triangular pyramids of water and (b) ice crystallites formed upon annealing near the desorption temperature. (Reprinted with permission from [93]. © (2008) by APS.)

other three bound to the metal or water layer underneath. Instead, there are always additional molecules on top of the $H_{up}$ molecules in the bilayer, as expected for a continuation of the crystalline ice, giving rise to superstructures. This is evidence that a bilayer termination on Cu(111) is not as energetically favorable as often assumed for the hexagonal close-packed metals.

## 37.4
## Water Adsorption on Nonmetallic Surfaces

In this section we give a brief outline of the water adsorption on nonmetallic surfaces, such as metal oxides, salts, and graphene. As shown in previous reviews, the interaction of water with metal oxides [4, 94, 95] and salts [4, 96, 97] has received considerable attention over the last few years.

### 37.4.1
### Metal Oxides

Metal oxides constitute a class of materials with diverse material properties ranging from metals to semiconductors and insulators. Their surfaces play a crucial role in environmental chemistry, in the passivation against corrosion, and as catalysts for partial oxidation reactions [98]. Water adsorption on metal oxides has received considerable attention due to their ability to induce water dissociation. Therefore, the focus is predominantly not only on the water adsorption structure but more on the dissociation pathways.

There are basically two thermal reaction pathways by which water can dissociate on a solid surface [4]. In the first, water dissociates on the surface, but recombination of the fragments to liberate water is favored:

$$H_2O_{(a)} \rightarrow OH_{(a)} + H_{(a)} \rightarrow H_2O_{(g)}.$$

In the second case, water dissociates, but thermodynamics favors other pathways over recombination, the most common being surface oxidation (irreversible dissociation):

$$H_2O_{(a)} \rightarrow OH_{(a)} + H_{(a)} \rightarrow O_{(a)} + H_{2(g)}.$$

The reversible water dissociation is typically observed on oxide surfaces, while irreversible water dissociation is typically observed on metal and semiconductor surfaces.

Whether the water adsorbs intact or dissociates on metal oxides depends often on the surface termination of the oxide. An example is FeO. On FeO(111), water adsorbs molecularly [99, 100], while on $Fe_3O_4$(111) dissociative adsorption is observed [101–103]. A comparison of the FeO(111) and $Fe_3O_4$(111) surface structures shows that the dissociation reaction is directly related to Fe atoms exposed on the latter surface [99, 102]. A table containing a selection of solid oxide surfaces indicating whether water adsorbs molecularly versus dissociative on the surface can be found in Ref. [4].

#### 37.4.1.1 MgO

MgO is one of the most intensively studied metal oxides for water adsorption. Despite substantial efforts in both experimental and theoretical studies [104–110], the possibility of spontaneous dissociation of water on the MgO(001) surface has been a controversial issue. Some theoretical works predicted that dissociation can only occur on defect sites [111, 112]. More recently, a consensus is emerging that a mixed ($H_2O + OH$) monolayer is the most stable on defect-free MgO(001) [47, 104, 113]. An atomic-scale STM study of the adsorption of water on MgO thin films on Ag(100) showed that dissociated water molecules are rarely observed on MgO, indicating that the spontaneous dissociation of isolated water molecules on the defect-free MgO(100) terrace is not thermodynamically favorable [114]. Instead, the water monomers adsorb on the $Mg^{2+}$ sites with the oxygen atom on top of the magnesium cation, and with one hydrogen atom asymmetrically pointing toward the neighboring surface oxygen atom. At monolayer coverage, Heidberg *et al.* observed a c(4 × 2) and a c(2 × 4) by LEED at 150 K [115]. In contrast, Xu and Goodman found a p(3 × 2)/p(2 × 3) LEED after heating at 185 K [116], consistent with neuron scattering data by Demirdjian *et al.* at temperatures between 200 and 270 K [117]. Later, He atom diffraction and LEED measurements by Ferry *et al.* showed that the c(4 × 2) pattern found at temperatures below 180 K turns into a p(3 × 2) at temperatures between 180 and 210 K for the same coverage [118–120]. Recent DFT calculations confirm the existence of the two stable surface structures: a c(4 × 2) structure containing 10 water molecules per unit cell, stable at low temperature; and a p(3 × 2) structure containing six water molecules per unit cell at higher temperature [47]. Both structures contain four hydroxyl groups from the dissociation of two water molecules per surface cell (see Figure 37.13).

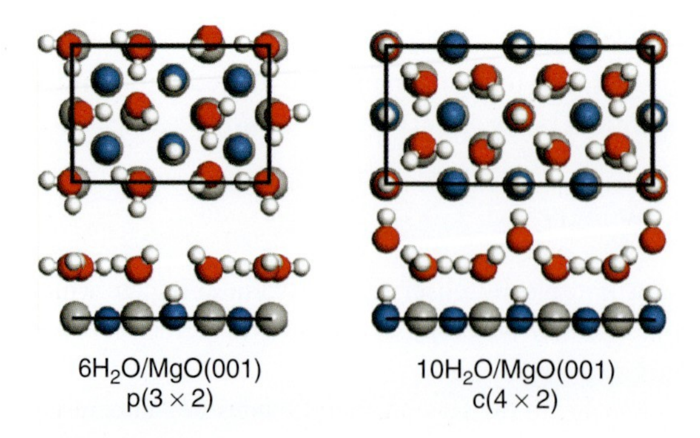


**Figure 37.13** Atomic structures of the ordered water monolayers on the MgO(001) surface. Density functional theory-predicted stable surface structures: a c(4 × 2) structure containing 10 water molecules per unit cell stable at low temperature and a p(3 × 2) structure containing 6 water molecules per unit cell stable at higher temperature. (Reprinted with permission from [47]. Copyright (2011) American Chemical Society.)

In contrast to these low-temperature UHV studies, under ambient conditions of pressure and temperature Newberg *et al.* showed, using XPS, that on a MgO(100) film grown on Ag(100) water dissociates to form a two monolayer-thick $Mg(OH)_2$ film. The dissociation of water provides an OH group bound to the Mg cation, while the remaining H forms another OH with the surface O [121].

### 37.4.2 Salts

The interaction of water with salts is of particular interest to atmospheric and environmental chemistry, because micrometer- and submicrometer-sized crystals of sodium chloride are a major constituent of atmospheric aerosols [122]. These aerosols play a fundamental role as condensation nuclei in cloud droplet formation, and their surface can host a number of chemical reactions between atmospheric components. Mechanistic pathways involving ionic dissociation in the presence of water is the most important first step in the heterogeneous chemistry of these solid electrolytes.

In this section, we will focus on the adsorption of water on sodium chloride, which is often considered as a prototype insulator with large bandgaps and can be prepared reproducibly by cleavage with large terraces and a low defect density. On salt surfaces, not only the adsorption geometry is an important research topic but also the mechanism by which water dissolves salts by pulling apart the ions in a salt crystal.

The cleavage faces of alkali halides are the nonpolar (001) faces. On these faces, the water molecules lie fairly flat against the surface in the most stable adsorption geometry [123–126]. The O atom is close to a Na site and the hydrogen atoms are directed at neighboring Cl ions. In a recent STM study on NaCl thin films on Au(111), isolated water monomers could be imaged with submolecular resolution [127]. However, compared to bulk NaCl, long-range dispersion forces from the Au substrate stabilize a "standing" adsorption geometry of water on the thin film with the molecular plane perpendicular to the surface and aligned with the $Na^+$–$Cl^-$ direction.

DFT calculations estimate adsorption energies of 0.3–0.6 eV for the water monomer on NaCl [123–126]. More recent calculations based on quantum chemistry methods gave an adsorption energy for the monomer on NaCl(001) of 517 meV after substrate relaxation [128]. On the other hand, the experimental values for the adsorption energies range from 600 to 680 meV. These higher values compared to theory might indicate that the measurements reflect adsorption of water at defective sites of NaCl(001) instead of terraces [129, 130].

The atomic structure of water layers at monolayer coverage on NaCl is controversial in UHV studies performed at low temperatures. On the basis of HAS experiments, Bruch *et al.* suggested that ordered two-dimensional water layers have a $(1 \times 1)$ periodicity, with the water molecule bound to the Na atom at temperatures of 148 K [129]. Fölsch *et al.* and Malaske *et al.* observed a stable $c(4 \times 2)$ structure by LEED in the same temperature range [131, 132]. Toennies *et al.* revisited the system and found by HAS that, initially after adsorption, water forms a $(1 \times 1)$ structure, which transforms to a $c(4 \times 2)$ structure after the LEED experiment [133]. This might indicate that the transition from $(1 \times 1)$ to $c(4 \times 2)$ structure is driven by the electron irradiation during the LEED experiments, so that some water molecules desorb, lowering the water coverage and that the water molecules can reorganize on the surface. Both structures have been modeled with DFT by a number of groups [125, 134], with Cabrera *et al.* concluding that the $(1 \times 1)$ and $c(4 \times 2)$ structures were essentially degenerate.

The dissolution of salt in UHV experiments has been studied with ultraviolet photoelectron spectroscopy (UPS) and other techniques [135]. It was found that NaCl starts to dissociate in a UHV environment at ~90 K in the presence of water and becomes significant at around 115 K.

At ambient conditions, water films on salts have been studied by investigating the step morphology and contact potential using scanning polarization force microscopy [136–138]. In these studies, two adsorption regimes have been identified. Below a critical relative humidity, water adsorbs predominantly at defects and steps, where the solvation of ions is observed by their enhanced mobility. Above the characteristic relative humidity, large-scale modification of the step morphology takes place, leading to irreversible changes in the ion distribution.

### 37.4.3 Graphene

Graphene is a two-dimensional one-atom-thick honeycomb lattice of $sp^2$-bonded carbon atoms. Because of its unique properties such as the high mobility of the charge carriers, it is a very promising material for future applications in electronics. Pristine graphene is a semimetal but with no electronic bandgap [139]. A bandgap can be opened by either incorporating heteroatoms or defects into the graphene layer or adsorbing polar molecules on top. In this section we discuss how water molecules adsorb on the surface and how they influence the electronic properties of graphene. This is relevant for the use of graphene in future electronic applications, as water is almost always present in the environment of any device.

Carbon surfaces are generally hydrophobic on a macroscopic scale, and it is thus expected that graphene will interact weakly with water, with no adsorption except at the lowest temperature. In spite of the weak binding of water on graphene, the large dipole moment of water can still influence the graphene electronic structure. Indeed, transport measurements revealed that adsorbed water molecules change the local carrier concentration in graphene, acting as an electron acceptor, which manifests as a change in resistance [140]. In addition, Yavari *et al.* found that the adsorption of water opens a tunable bandgap of up to 0.2 eV depending on the relative humidity [141]. Triggered by these *macroscopic* experiments, several theoretical and experimental studies have been performed to elucidate the adsorption properties of water on graphene from a molecular perspective.

#### 37.4.3.1 Adsorption of Water Monomers

Studying the adsorption properties of water monomers on graphene is a challenge for both experiment and theory [63]. The reason is that dispersion forces and hydrogen bonding are interactions not well described by DFT. Experimentally also, observation of weakly bound monomers is a difficult task because water prefers to form clusters.

The adsorption energy of water on graphene from quantum Monte Carlo and random-phase approximations yield values below 100 meV [63]. The adsorption geometry is also very different from that on a metal, with a slight preference for a structure with two hydrogens oriented toward the surface compared to the two-leg structure (see Figure 37.14). First-principles calculations using DFT revealed that the adsorption energy is primarily determined by the orientation of the water molecule and to a lesser degree by the position of the molecule [143]. The energy differences were very small, however, with a variation of 5–6 meV with respect to the orientation, and about 1–2 meV when changing the position. Over defective sites, water molecules can dissociate on graphene, forming C–H and C–OH bonds, which has been demonstrated by both theoretical calculations [144] and vibrational spectroscopy experiments [145]. Experimentally, the dissociation was observed only for water adsorbed at room temperature, while water adsorbed at low temperatures (~100 K) desorbed intact.

#### 37.4.3.2 Adsorption of Water on Metal-Supported Graphene

While water adsorption has been studied experimentally on graphene supported on metals and on SiC, here we will focus mainly on experiments performed on metal supports. Kimmel *et al.* demonstrated that water adsorbs on graphene supported on

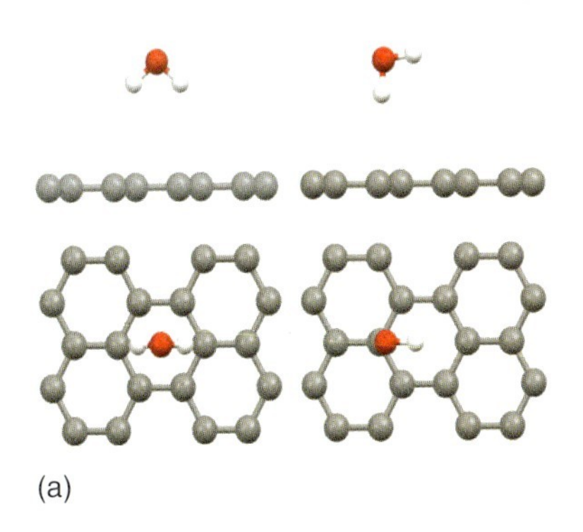

(a)

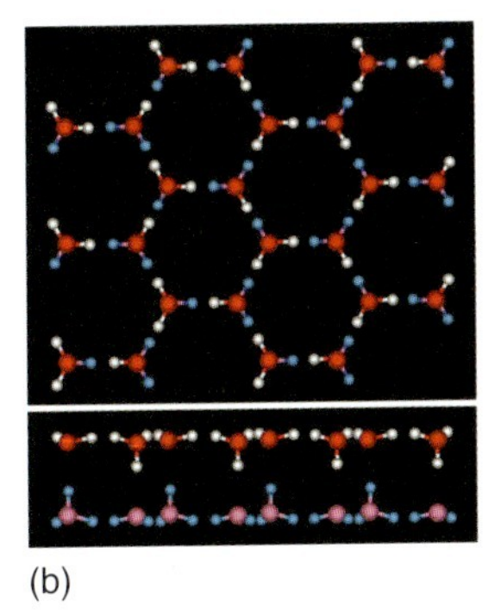

(b)

**Figure 37.14** (a) Water monomers on graphene. Left: The two-leg structure shown from the side (top) and from above (bottom). Right: The one-leg structure shown from the side (top) and from above (bottom). (Reproduced with permission from Ref. [63] © 2011, American Physical Society.) (b) Calculated two-layer ice structure using MD simulations. The white (blue) spheres represent hydrogen atoms in top (bottom) layer. Red (pink) spheres represent oxygen atoms in the top (bottom) layer. The side view showing the two flat layers of molecules with hydrogen bonds connecting the layers. (Reproduced with permission from Ref. [142] © 2009, American Chemical Society.)

Pt(111) at temperatures between ~100 and 130 K [142]. LEED and rare-gas adsorption/desorption experiments have shown that the water layer on graphene/Pt(111) consists of two flat hexagonal sheets of molecules in which the hexagons in each sheet are stacked directly on top of each other (Figure 37.14). This is unlike hexagonal ice, which consists of stacks of puckered hexagonal "bilayers." Such two-layer ice has been predicted for water confined between hydrophobic walls but not previously observed experimentally.

Information on the reactivity of defects and grain boundaries on metal-supported graphene has been provided by STM experiments on Ru(0001) and Cu(111) substrates [146]. These experiments revealed that on Gr/Ru(0001) water splits the epitaxial graphene along line defects, generating numerous fragments at temperatures as low as 90 K, followed by water intercalation under the graphene. On Gr/Cu(111), however, the water-induced splitting of graphene is far less effective, indicating that the substrate plays a key role in modifying the chemical properties of epitaxial graphene. One of the differences between Cu and Ru related to the adsorption of water is that on clean Ru(0001) water can dissociate in contrast to Cu(111).

#### 37.4.3.3 Intercalated Water under Graphene Layers

Only one-atom-thick graphene is believed to be impermeable to all gases and liquids [147]. In spite of this, water can still intercalate under the graphene flakes, starting from defects and edges. The intercalated water is trapped, allowing microscopic studies of its properties at ambient conditions, which would otherwise be very challenging due to the highly dynamic nature of water at room temperature. An interesting question, when water is confined between two surfaces, one hydrophilic and the other hydrophobic, is what determines its structure and diffusivity. Xu *et al.* used AFM to study water trapped between graphene and mica at high humidity [148]. They found that water adlayers grow epitaxially in a layer-by-layer manner between graphene and mica, with a height in agreement with the interlayer spacing found in ice. Only films thicker than two layers showed a liquid-like behavior. The boundaries of the intercalated water film are polygonal, with sides forming angles of approximately 120° and aligned with the compact lattice directions of the mica surface, supporting the notion that the intercalated water forms an ice-like layer [149]. On the other hand, intercalation of water under exfoliated graphene on a hydrophilic, amorphous $SiO_2$ substrate does not show crystalline features under similar humidity conditions. Kim *et al.* also studied the intercalation pathways of water during the growth of intercalated water between graphene and mica, by exposing samples to high relative humidity (50% RH) for a prolonged time [150]. They observed highly anisotropic growth of water stripes forming patterns oriented along the C–C bond zig-zag direction of the graphene (Figure 37.15). These results suggest that, while the hydrophilic mica substrate determines the structure of water near its surface, graphene guides its diffusion.

Intercalated water islands have been also used to experimentally address the influence of water on the electronic structure of graphene [140]. DFT calculations revealed that there should be no doping effects for single water molecules on

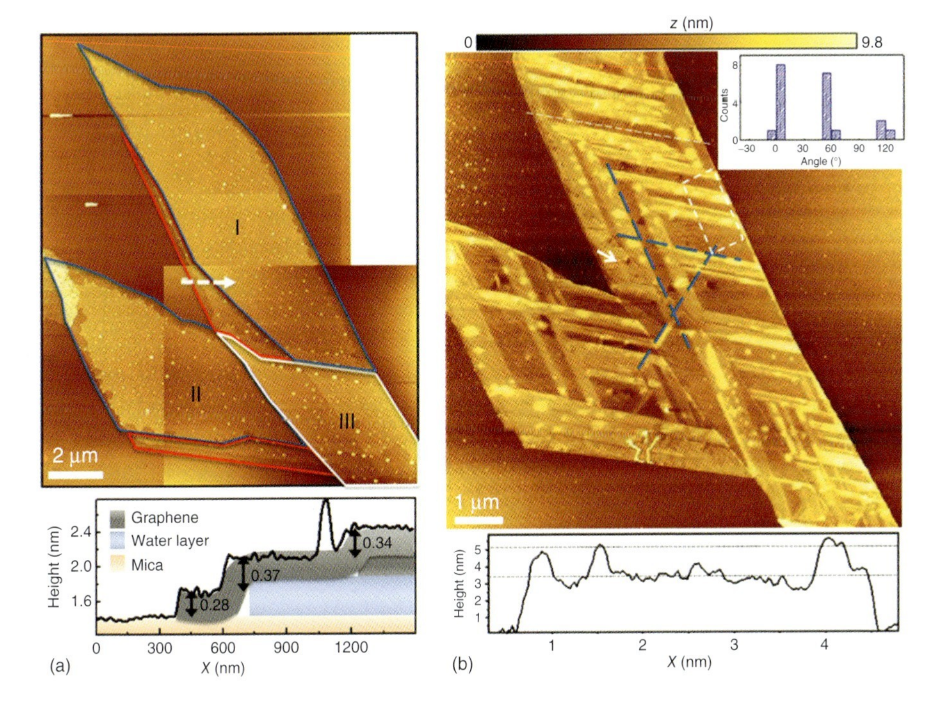


**Figure 37.15** (a) Water intercalated between graphene and mica. AFM topographic image of graphene flakes deposited on mica at 30–40% RH using the mechanical exfoliation method with monolayer ice intercalated. Region III contains a six-layer graphene overlapping the bilayer graphene of regions I and II. Red, blue, and white lines indicate the flake contours of monolayer, bilayer, and few-layer graphene, respectively. (b) Multilayers water stripes formed by diffusion of additional water between graphene and mica. AFM topographic images obtained after exposure to high RH (>50%) for one additional week. (Reprinted by permission from Macmillan Publishers Ltd: [150], Copyright (2013).)

free-standing graphene [143, 151], in seeming contradiction to transport experiments. The reason for the lack of change in the density of states (DOS) close to the Fermi level upon adsorption of the molecule is because the HOMO and LUMO of $H_2O$ are both more than 2 eV away from the Fermi level. This absence of impurity levels close to the Dirac point shows that single water molecules on perfect free-standing graphene sheets do not cause doping. The $H_2O$-induced doping is thus not understood at present. An electrostatic field-mediated mechanism based on the enhanced dipole moments of water clusters offers one possibility [151, 152]. Cao *et al.* have experimentally demonstrated that individual nanometer-sized water clusters, trapped between graphene and an Au(111) substrate, induce strong electron doping in graphene [153]. The observation that the doping effects are highly localized at the sites of the water clusters, together with the strong correlation found between cluster size and the amount of doping, supports the proposed electrostatic field-mediated doping mechanism.

## 37.5 Conclusions

The bilayer model, based of a hexagonal ice structure, was for a long time the "standard model" for water adsorption on hexagonal close-packed metal surfaces. However, recent high-resolution STM imaging and DFT studies have revealed that the wetting layer structure does not reflect the ice bilayer structure, a result that is rationalized by the tendency of the molecules to maximize the bonding to the metal on one hand and hydrogen-bonding within the layers on the other. It was also found that the wetting layer structures are not exclusively built from hexagons but can also include pentagons or combinations of heptagons and pentagons. The $\sqrt{3}$ diffraction patterns observed on earlier studies on many metals are often associated with partially dissociated OH–$H_2O$ overlayers. It seems there is no general model for the adsorption structure of waters, which is not surprising given the very rich phase diagram of water. The discussion thus far has largely focused on the first water layer in contact with the metals. However, there is increasing interest in understanding multilayer water adsorption, and recent progress has provided interesting insights into ice nucleation on metal surfaces.

### Further Reading

For comprehensive overviews of the literature about water adsorption on metal surfaces, the interested reader is directed to Refs [4, 16] and for other complementary mini-reviews on some specific aspects of water at surfaces to Refs [17, 82, 154]. For a comprehensive review of the structure and properties of ice, the book by Petrenko and Whitworth is recommended [84].

### References


1. Doering, D.L. and Madey, T.E. (1982) *Surf. Sci.*, **123**, 305.
2. Thürmer, K. and Bartelt, N.C. (2008) *Phys. Rev. B*, **77**, 195425.
3. Donadio, D., Ghiringhelli, L.M., and Delle Site, L. (2012) *J. Am. Chem. Soc.*, **134**, 19217.
4. Henderson, M.A. (2002) *Surf. Sci. Rep.*, **46**, 5.
5. Musket, R.G., McLean, W., Colmenares, C.A., Makowiecki, D.M., and Siekhaus, W.J. (1982) *Appl. Surf. Sci.*, **10**, 143.
6. Mugarza, A., Shimizu, T.K., Ogletree, D.F., and Salmeron, M. (2009) *Surf. Sci.*, **603**, 2030.
7. Bozack, M.J., Muehlhoff, L., Russell, J.N., Choyke, W.J., and Yates, J.T. (1987) *J. Vac. Sci. Technol., A*, **5**, 1.
8. Huffstetler, R.D. and Leavitt, A.J. (2001) *J. Vac. Sci. Technol., A*, **19**, 1030.
9. Thiel, P.A. and Madey, T.E. (1987) *Surf. Sci. Rep.*, **7**, 211.
10. Braun, J., Glebov, A., Graham, A.P., Menzel, A., and Toennies, J.P. (1998) *Phys. Rev. Lett.*, **80**, 2638.
11. Thürmer, K. and Nie, S. (2013) *Proc. Natl. Acad. Sci. U.S.A.*, **110**, 11757.
12. Torbrügge, S., Custance, O., Morita, S., and Reichling, M. (2012) *J. Phys. Condens. Matter*, **24**, 084010.
13. Nakamura, M. and Ito, M. (2005) *Chem. Phys. Lett.*, **404**, 346.
14. Andersson, S., Nyberg, C., and Tengstål, C.G. (1984) *Chem. Phys. Lett.*, **104**, 305.

15. Ogasawara, H., Yoshinobu, J., and Kawai, M. (1994) *Chem. Phys. Lett.*, **231**, 188.
16. Hodgson, A. and Haq, S. (2009) *Surf. Sci. Rep.*, **64**, 381.
17. Carrasco, J., Hodgson, A., and Michaelides, A. (2012) *Nat. Mater.*, **11**, 667.
18. Izvekov, S. and Voth, G.A. (2001) *J. Chem. Phys.*, **115**, 7196.
19. Michaelides, A., Ranea, V.A., de Andres, P.L., and King, D.A. (2003) *Phys. Rev. Lett.*, **90**, 216102.
20. Meng, S., Wang, E.G., and Gao, S.W. (2004) *Phys. Rev. B*, **69**, 195404.
21. Michaelides, A. (2006) *Appl. Phys. A: Mater. Sci. Process.*, **85**, 415.
22. Siegbahn, H., Asplund, L., and Kelfve, P. (1975) *Chem. Phys. Lett.*, **35**, 330.
23. Michaelides, A., Alavi, A., and King, D.A. (2003) *J. Am. Chem. Soc.*, **125**, 2746.
24. Taylor, C.D. and Neurock, M. (2005) *Curr. Opin. Solid State Mater. Sci.*, **9**, 49.
25. Carrasco, J., Michaelides, A., and Scheffler, M. (2009) *J. Chem. Phys.*, **130**, 184707.
26. Ogasawara, H., Brena, B., Nordlund, D., Nyberg, M., Pelmenschikov, A., Pettersson, L.G.M., and Nilsson, A. (2002) *Phys. Rev. Lett.*, **89**, 276102.
27. Cavalleri, M., Ogasawara, H., Pettersson, L.G.M., and Nilsson, A. (2002) *Chem. Phys. Lett.*, **364**, 363.
28. Carrasco, J., Klimeš, J., and Michaelides, A. (2013) *J. Chem. Phys.*, **138**, 024708.
29. Clay, C. and Hodgson, A. (2005) *Curr. Opin. Solid State Mater. Sci.*, **9**, 11.
30. Tang, Q.-L. and Chen, Z.-X. (2007) *Surf. Sci.*, **601**, 954.
31. Morgenstern, K. and Rieder, K.-H. (2002) *J. Chem. Phys.*, **116**, 5746.
32. Morgenstern, K. and Rieder, K.-H. (2002) *Chem. Phys. Lett.*, **358**, 250.
33. Mitsui, T., Rose, M.K., Fomin, E., Ogletree, D.F., and Salmeron, M. (2002) *Science*, **297**, 1850.
34. Shimizu, T.K., Mugarza, A., Cerda, J.I., Heyde, M., Qi, Y.B., Schwarz, U.D., Ogletree, D.F., and Salmeron, M. (2008) *J. Phys. Chem. C*, **112**, 7445.
35. Fomin, E., Tatarkhanov, M., Mitsui, T., Rose, M., Ogletree, D.F., and Salmeron, M. (2006) *Surf. Sci.*, **600**, 542.
36. Kumagai, T., Shiotari, A., Okuyama, H., Hatta, S., Aruga, T., Hamada, I., Frederiksen, T., and Ueba, H. (2012) *Nat. Mater.*, **11**, 167.
37. Haq, S., Clay, C., Darling, G.R., Zimbitas, G., and Hodgson, A. (2006) *Phys. Rev. B*, **73**, 115414.
38. Ranea, V.A., Michaelides, A., Ramírez, R., de Andres, P.L., Vergés, J.A., and King, D.A. (2004) *Phys. Rev. Lett.*, **92**, 136104.
39. Kumagai, T., Kaizu, M., Hatta, S., Okuyama, H., Aruga, T., Hamada, I., and Morikawa, Y. (2008) *Phys. Rev. Lett.*, **100**, 166101.
40. Motobayashi, K., Matsumoto, C., Kim, Y., and Kawai, M. (2008) *Surf. Sci.*, **602**, 3136.
41. Morgenstern, K., Gawronski, H., Mehlhorn, M., and Rieder, K.-H. (2004) *J. Mod. Opt.*, **51**, 2813.
42. Tatarkhanov, M., Ogletree, D.F., Rose, F., Mitsui, T., Fomin, E., Maier, S., Rose, M., Cerdá, J.I., and Salmeron, M. (2009) *J. Am. Chem. Soc.*, **131**, 18425.
43. Michaelides, A. and Morgenstern, K. (2007) *Nat. Mater.*, **6**, 597.
44. Cerda, J., Michaelides, A., Bocquet, M.L., Feibelman, P.J., Mitsui, T., Rose, M., Fomin, E., and Salmeron, M. (2004) *Phys. Rev. Lett.*, **93**, 116101.
45. Michaelides, A., Alavi, A., and King, D.A. (2004) *Phys. Rev. B*, **69**, 113404.
46. Meng, S., Xu, L.F., Wang, E.G., and Gao, S. (2002) *Phys. Rev. Lett.*, **89**, 176104.
47. Wlodarczyk, R., Sierka, M., Kwapien, K., Sauer, J., Carrasco, E., Aumer, A., Gomes, J.F., Sterrer, M., and Freund, H.J. (2011) *J. Phys. Chem. C*, **115**, 6764.
48. Bernal, J.D. and Fowler, R.H. (1933) *J. Chem. Phys.*, **1**, 515.
49. Pauling, L. (1935) *J. Am. Chem. Soc.*, **57**, 2680.
50. Carrasco, J., Michaelides, A., Forster, M., Haq, S., Raval, R., and Hodgson, A. (2009) *Nat. Mater.*, **8**, 427.
51. Yamada, T., Tamamori, S., Okuyama, H., and Aruga, T. (2006) *Phys. Rev. Lett.*, **96**, 036105.

52. Lee, J., Sorescu, D.C., Jordan, K.D., and Yates, J.T. (2008) *J. Phys. Chem. C*, **112**, 17672.
53. Schiros, T., Haq, S., Ogasawara, H., Takahashi, O., Öström, H., Andersson, K., Pettersson, L.G.M., Hodgson, A., and Nilsson, A. (2006) *Chem. Phys. Lett.*, **429**, 415.
54. Forster, M., Raval, R., Carrasco, J., Michaelides, A., and Hodgson, A. (2012) *Chem. Sci.*, **3**, 93.
55. Shi, Y., Choi, B.Y., and Salmeron, M. (2013) *J. Phys. Chem. C*, **117**, 17119.
56. Tatarkhanov, M., Fomin, E., Salmeron, M., Andersson, K., Ogasawara, H., Pettersson, L.G.M., Nilsson, A., and Cerda, J.I. (2008) *J. Chem. Phys.*, **129**, 154109.
57. Weissenrieder, J., Mikkelsen, A., Andersen, J.N., Feibelman, P.J., and Held, G. (2004) *Phys. Rev. Lett.*, **93**, 196102.
58. Clay, C., Haq, S., and Hodgson, A. (2004) *Chem. Phys. Lett.*, **388**, 89.
59. Andersson, K., Nikitin, A., Pettersson, L.G.M., Nilsson, A., and Ogasawara, H. (2004) *Phys. Rev. Lett.*, **93**, 196101.
60. Shavorskiy, A., Gladys, M.J., and Held, G. (2008) *Phys. Chem. Chem. Phys.*, **10**, 6150.
61. Faradzhev, N.S., Kostov, K.L., Feulner, P., Madey, T.E., and Menzel, D. (2005) *Chem. Phys. Lett.*, **415**, 165.
62. Feibelman, P.J. (2002) *Science*, **295**, 99.
63. Ma, J., Michaelides, A., Alfè, D., Schimka, L., Kresse, G., and Wang, E. (2011) *Phys. Rev. B*, **84**, 033402.
64. Feibelman, P.J. (2008) *Phys. Chem. Chem. Phys.*, **2008**, 4688.
65. McBride, F., Darling, G.R., Pussi, K., and Hodgson, A. (2011) *Phys. Rev. Lett.*, **106**, 226101.
66. Clay, C., Haq, S., and Hodgson, A. (2004) *Phys. Rev. Lett.*, **92**, 046102.
67. Harnett, J., Haq, S., and Hodgson, A. (2003) *Surf. Sci.*, **528**, 15.
68. Sachs, C., Hildebrand, M., Völkening, S., Wintterlin, J., and Ertl, G. (2001) *Science*, **293**, 1635.
69. Firment, L.E. and Somorjai, G.A. (1976) *Surf. Sci.*, **55**, 413.
70. Glebov, A.L., Graham, A.P., Menzel, A., and Toennies, J.P. (1997) *J. Chem. Phys.*, **106**, 9382.
71. Haq, S., Harnett, J., and Hodgson, A. (2002) *Surf. Sci.*, **505**, 171.
72. Zimbitas, G., Haq, S., and Hodgson, A. (2005) *J. Chem. Phys.*, **123**, 174701.
73. Meng, S. (2005) *Surf. Sci.*, **575**, 300.
74. Morgenstern, M., Muller, J., Michely, T., and Comsa, G. (1997) *Z. Phys. Chem.-Int. J. Res. Phys. Chem. Chem. Phys.*, **198**, 43.
75. Morgenstern, M., Michely, T., and Comsa, G. (1996) *Phys. Rev. Lett.*, **77**, 703.
76. Nie, S., Feibelman, P.J., Bartelt, N.C., and Thürmer, K. (2010) *Phys. Rev. Lett.*, **105**, 026102.
77. Standop, S., Redinger, A., Morgenstern, M., Michely, T., and Busse, C. (2010) *Phys. Rev. B*, **82**, 161412.
78. Held, G. and Menzel, D. (1994) *Surf. Sci.*, **316**, 92.
79. Gallagher, M., Omer, A., Darling, G.R., and Hodgson, A. (2009) *Faraday Discuss.*, **141**, 231.
80. Traeger, F., Langenberg, D., Gao, Y.K., and Woll, C. (2007) *Phys. Rev. B*, **76**, 033410.
81. Maier, S., Stass, I., Mitsui, T., Feibelman, P.J., Thürmer, K., and Salmeron, M. (2012) *Phys. Rev. B*, **85**, 155434.
82. Feibelman, P.J. (2010) *Phys. Today*, **63**, 34.
83. Salmeron, M., Bluhm, H., Tatarkhanov, M., Ketteler, G., Shimizu, T.K., Mugarza, A., Deng, X., Herranz, T., Yamamoto, S., and Nilsson, A. (2009) *Faraday Discuss.*, **141**, 221.
84. Petrenko, V.F. and Whitworth, R.W. (1999) *Physics of Ice*, Oxford University Press, New York.
85. Buch, V., Groenzin, H., Li, I., Shultz, M.J., and Tosatti, E. (2008) *Proc. Natl. Acad. Sci. U.S.A.*, **105**, 5969.
86. Pan, D., Liu, L.-M., Tribello, G.A., Slater, B., Michaelides, A., and Wang, E. (2008) *Phys. Rev. Lett.*, **101**, 155703.
87. Fletcher, N.H. (1992) *Philos. Mag. B*, **66**, 109.
88. Kimmel, G.A., Petrik, N.G., Dohnalek, Z., and Kay, B.D. (2005) *Phys. Rev. Lett.*, **95**, 166102.
89. Glebov, A., Graham, A.P., Menzel, A., Toennies, J.P., and Senet, P. (2000) *J. Chem. Phys.*, **112**, 11011.

90. Kimmel, G.A., Petrik, N.G., Dohnalek, Z., and Kay, B.D. (2007) *J. Chem. Phys.*, **126**, 114702.
91. Thürmer, K. and Bartelt, N.C. (2008) *Phys. Rev. Lett.*, **100**, 186101.
92. Nie, S., Bartelt, N.C., and Thürmer, K. (2009) *Phys. Rev. Lett.*, **102**, 136101.
93. Mehlhorn, M. and Morgenstern, K. (2007) *Phys. Rev. Lett.*, **99**, 246101.
94. Al-Abadleh, H.A. and Grassian, V.H. (2003) *Surf. Sci. Rep.*, **52**, 63.
95. Brown, G.E., Henrich, V.E., Casey, W.H., Clark, D.L., Eggleston, C., Felmy, A., Goodman, D.W., Grätzel, M., Maciel, G., McCarthy, M.I., Nealson, K.H., Sverjensky, D.A., Toney, M.F., and Zachara, J.M. (1998) *Chem. Rev.*, **99**, 77.
96. Ewing, G.E. (2006) *Chem. Rev.*, **106**, 1511.
97. Verdaguer, A., Sacha, G.M., Bluhm, H., and Salmeron, M. (2006) *Chem. Rev.*, **106**, 1478.
98. Henrich, V.E. and Cox, P.A. (1996) *The Surface Science of Metal Oxides*, Cambridge University Press.
99. Joseph, Y., Kuhrs, C., Ranke, W., Ritter, M., and Weiss, W. (1999) *Chem. Phys. Lett.*, **314**, 195.
100. Cappus, D., Haßel, M., Neuhaus, E., Heber, M., Rohr, F., and Freund, H.J. (1995) *Surf. Sci.*, **337**, 268.
101. Cutting, R.S., Muryn, C.A., Vaughan, D.J., and Thornton, G. (2008) *Surf. Sci.*, **602**, 1155.
102. Joseph, Y., Ranke, W., and Weiss, W. (2000) *J. Phys. Chem. B*, **104**, 3224.
103. Kendelewicz, T., Liu, P., Doyle, C.S., Brown, G.E. Jr.,, Nelson, E.J., and Chambers, S.A. (2000) *Surf. Sci.*, **453**, 32.
104. Giordano, L., Goniakowski, J., and Suzanne, J. (1998) *Phys. Rev. Lett.*, **81**, 1271.
105. Johnson, M.A., Stefanovich, E.V., Truong, T.N., Günster, J., and Goodman, D.W. (1999) *J. Phys. Chem. B*, **103**, 3391.
106. Kim, Y.D., Stultz, J., and Goodman, D.W. (2002) *J. Phys. Chem. B*, **106**, 1515.
107. Savio, L., Celasco, E., Vattuone, L., and Rocca, M. (2004) *J. Phys. Chem. B*, **108**, 7771.
108. Wu, M.-C., Estrada, C.A., and Goodman, D.W. (1991) *Phys. Rev. Lett.*, **67**, 2910.
109. Yu, Y., Guo, Q., Liu, S., Wang, E., and Møller, P.J. (2003) *Phys. Rev. B*, **68**, 115414.
110. Carrasco, J., Illas, F., and Lopez, N. (2008) *Phys. Rev. Lett.*, **100**, 016101.
111. Langel, W. and Parrinello, M. (1994) *Phys. Rev. Lett.*, **73**, 504.
112. Scamehorn, C.A., Harrison, N.M., and McCarthy, M.I. (1994) *J. Chem. Phys.*, **101**, 1547.
113. Wang, Y. and Truong, T.N. (2004) *J. Phys. Chem. B*, **108**, 3289.
114. Shin, H.-J., Jung, J., Motobayashi, K., Yanagisawa, S., Morikawa, Y., Kim, Y., and Kawai, M. (2010) *Nat. Mater.*, **9**, 442.
115. Heidberg, J., Redlich, B., and Wetter, D. (1995) *Ber. Bunsen-Ges. Phys. Chem. Chem. Phys.*, **99**, 1333.
116. Xu, C. and Goodman, D.W. (1997) *Chem. Phys. Lett.*, **265**, 341.
117. Demirdjian, B., Ferry, D., Suzanne, J., Hoang, P.N.M., Picaud, S., and Girardet, C. (2001) *Surf. Sci.*, **494**, 206.
118. Ferry, D., Glebov, A., Senz, V., Suzanne, J., Toennies, J.P., and Weiss, H. (1996) *J. Chem. Phys.*, **105**, 1697.
119. Ferry, D., Glebov, A., Senz, V., Suzanne, J., Toennies, J.P., and Weiss, H. (1997) *Surf. Sci.*, **377**, 634.
120. Ferry, D., Picaud, S., Hoang, P.N.M., Girardet, C., Giordano, L., Demirdjian, B., and Suzanne, J. (1998) *Surf. Sci.*, **409**, 101.
121. Newberg, J.T., Starr, D.E., Yamamoto, S., Kaya, S., Kendelewicz, T., Mysak, E.R., Porsgaard, S., Salmeron, M.B., Brown, G.E. Jr.,, Nilsson, A., and Bluhm, H. (2011) *Surf. Sci.*, **605**, 89.
122. Knipping, E.M., Lakin, M.J., Foster, K.L., Jungwirth, P., Tobias, D.J., Gerber, R.B., Dabdub, D., and Finlayson-Pitts, B.J. (2000) *Science*, **288**, 301–306.
123. Cabrera-Sanfelix, P., Arnau, A., Darling, G.R., and Sanchez-Portal, D. (2006) *J. Phys. Chem. B*, **110**, 24559.
124. Park, J.M., Cho, J.-H., and Kim, K.S. (2004) *Phys. Rev. B*, **69**, 233403.
125. Yang, Y., Meng, S., and Wang, E.G. (2006) *Phys. Rev. B*, **74**, 245409.

126. Meyer, H., Entel, P., and Hafner, J. (2001) *Surf. Sci.*, **488**, 177.
127. Guo, J., Meng, X., Chen, J., Peng, J., Sheng, J., Li, X.-Z., Xu, L., Shi, J.-R., Wang, E., and Jiang, Y. (2014) *Nat. Mater.*, **13**, 184.
128. Li, B., Michaelides, A., and Scheffler, M. (2008) *Surf. Sci.*, **602**, L135.
129. Bruch, L.W., Glebov, A.L., Toennies, J.P., and Weiss, H. (1995) *J. Chem. Phys.*, **103**, 5109.
130. Fölsch, S. and Henzler, M. (1991) *Surf. Sci.*, **247**, 269.
131. Fölsch, S., Stock, S., and Henzler, M. (1992) *Surf. Sci.*, **264**, 65.
132. Malaske, U., Pfnür, H., Bässler, M., Weiss, M., and Umbach, E. (1996) *Phys. Rev. B*, **53**, 13115.
133. Toennies, J.P., Traeger, F., Vogt, J., and Weiss, H. (2004) *J. Chem. Phys.*, **120**, 11347.
134. Cabrera-Sanfelix, P., Arnau, A., Darling, G.R., and Sanchez-Portal, D. (2007) *J. Chem. Phys.*, **126**, 214707.
135. Borodin, A., Höfft, O., Kahnert, U., Kempter, V., Poddey, A., and Blöchl, P.E. (2004) *J. Chem. Phys.*, **121**, 9671.
136. Luna, M., Rieutord, F., Melman, N.A., Dai, Q., and Salmeron, M. (1998) *J. Phys. Chem. A*, **102**, 6793.
137. Verdaguer, A., Sacha, G.M., Luna, M., Ogletree, D.F., and Salmeron, M. (2005) *J. Chem. Phys.*, **123**, 124703.
138. Dai, Q., Hu, J., and Salmeron, M. (1997) *J. Phys. Chem. B*, **101**, 1994.
139. Geim, A.K. and Novoselov, K.S. (2007) *Nat. Mater.*, **6**, 183.
140. Schedin, F., Geim, A.K., Morozov, S.V., Hill, E.W., Blake, P., Katsnelson, M.I., and Novoselov, K.S. (2007) *Nat. Mater.*, **6**, 652.
141. Yavari, F., Kritzinger, C., Gaire, C., Song, L., Gulapalli, H., Borca-Tasciuc, T., Ajayan, P.M., and Koratkar, N. (2010) *Small*, **6**, 2535.
142. Kimmel, G.A., Matthiesen, J., Baer, M., Mundy, C.J., Petrik, N.G., Smith, R.S., Dohnálek, Z., and Kay, B.D. (2009) *J. Am. Chem. Soc.*, **131**, 12838.
143. Leenaerts, O., Partoens, B., and Peeters, F.M. (2008) *Phys. Rev. B*, **77**, 125416.
144. Kostov, M.K., Santiso, E.E., George, A.M., Gubbins, K.E., and Nardelli, M.B. (2005) *Phys. Rev. Lett.*, **95**, 136105.
145. Politano, A., Marino, A.R., Formoso, V., and Chiarello, G. (2011) *AIP Adv.*, **1**, 042130.
146. Feng, X., Maier, S., and Salmeron, M. (2012) *J. Am. Chem. Soc.*, **134**, 5662.
147. Bunch, J.S., Verbridge, S.S., Alden, J.S., van der Zande, A.M., Parpia, J.M., Craighead, H.G., and McEuen, P.L. (2008) *Nano Lett.*, **8**, 2458.
148. Xu, K., Cao, P., and Heath, J.R. (2010) *Science*, **329**, 1188.
149. Pi, U.H., Jeong, M.S., Kim, J.H., Yu, H.Y., Park, C.W., Lee, H., and Choi, S.Y. (2005) *Surf. Sci.*, **583**, 88.
150. Kim, J.-S., Choi, J.S., Lee, M.J., Park, B.H., Bukhvalov, D., Son, Y.-W., Yoon, D., Cheong, H., Yun, J.-N., Jung, Y., Park, J.Y., and Salmeron, M. (2013) *Sci. Rep.*, **3**, 2309.
151. Wehling, T.O., Katsnelson, M.I., and Lichtenstein, A.I. (2009) *Chem. Phys. Lett.*, **476**, 125.
152. Wehling, T.O., Lichtenstein, A.I., and Katsnelson, M.I. (2008) *Appl. Phys. Lett.*, **93**, 202110.
153. Cao, P., Varghese, J.O., Xu, K., and Heath, J.R. (2012) *Nano Lett.*, **12**, 1459.
154. Schiros, T., Andersson, K.J., Pettersson, L.G.M., Nilsson, A., and Ogasawara, H. (2010) *J. Electron. Spectrosc. Relat. Phenom.*, **177**, 85.

# 38
# Adsorption of (Small) Molecules on Metals

*Reinhard Denecke and Hans-Peter Steinrück*

## 38.1
## Introduction

Most pure metallic surfaces are not inert, and thus do not stay clean (meaning uncovered by adsorbates) under realistic conditions. Exceptions would be noble metals, but even there adsorption has been observed at ambient or higher pressures [1]. In most cases, the adsorption and/or reaction of small molecules and the resulting formation of ultrathin layers of different nature and chemical composition modify the properties of metallic surfaces, such as reflectivity, reactivity, and their corrosion. Therefore, the fundamental understanding of adsorption, which typically is the first elementary step of successive surface reactions or the formation of adsorbate layers, is very important for a general understanding of surface-mediated processes and possible improvement of materials properties. In particular, optimizing heterogeneous catalysts can benefit greatly from a detailed understanding of the processes on a molecular or even atomic level.

In general, adsorption on a surface can take place both from gas phase and liquid phase. Here we focus on adsorption from gas phase. We address an idealized description of adsorption properties, which are obtained under nonequilibrium conditions, where simultaneous desorption can be neglected. This is typically achieved when working at sufficiently low sample temperatures, as will be the case in most of the examples discussed in this chapter. Nevertheless, equilibrium experiments in the presence of a gas phase provide additional information, which can, for example, be obtained by recording adsorption isotherms. However, this should not be the main focus in this chapter, because this has already been described in Chapter 31 of this Volume.

Information about the adsorption process and its resulting adsorbate layer can be obtained in different ways. Common to them is mainly the surface sensitivity of the applied probes, which in most cases is necessary to obtain the surface-related information without being overruled by the much stronger – because much more abundant – bulk information.

In the following, we will address different aspects of the adsorption of small molecules on metal surfaces. After some definitions in Section 38.2, we first discuss

*Surface and Interface Science: Solid-Gas Interfaces I,* First Edition. Edited by Klaus Wandelt.

the process of adsorption in Section 38.3. Section 38.4 then deals with the geometric structure, and finally in Section 38.5 we discuss the electronic structure, bonding, and molecular orientation of adsorbates. In each section, different aspects will be addressed using appropriate example systems. It is important to note here that the aim of this chapter is not to provide a review of all relevant results for a particular adsorption system, for example, CO/Pt(111), but rather use specific adsorption systems to illustrate the level of insight that can be obtained. Thus, a specific adsorption system can be addressed at several instances, with a different focus. For the sake of clarity, we limit our presentation to a small number of molecules, which will be discussed in more or less detail. These include CO, $N_2$, $O_2$, ethylene ($C_2H_4$), and benzene ($C_6H_6$).

## 38.2 Definitions

In this introductory part, we will review some necessary definitions to be able to describe adsorption phenomena in a comprehensive way. Some points have already been introduced throughout the book, so we will only briefly repeat them here.

### 38.2.1 Physisorption and Chemisorption

Adsorption of molecules on surfaces implies some interaction between the electronic systems of both the adsorbate and the substrate. Two generally different scenarios can be distinguished. In case of physisorption, the interaction solely relies on van der Waals forces, which result from mutually induced dipoles in the adsorbate and substrate. It is usually found in the adsorption of inert adsorbates such as light noble gases (see Chapters 31 and 33 in this Volume) and small saturated hydrocarbons, or in condensation processes where thicker layers can be obtained. The other case, called *chemisorption*, describes interactions that change the electronic structure of the constituents beyond polarization effects. These interactions lead to the formation of chemical bonds in one way or another. In most cases, chemisorption is restricted to the first adsorbate layer, as in subsequent layers interactions between the adsorbate molecules dominate, which are in most cases weak. As will be described later, such weakly bound (physisorbed) states can also serve as precursor states for the following chemisorption.

As far as the adsorption process of the intact gas phase species (i.e., molecular adsorption) is concerned, there is no fundamental difference between physisorption and chemisorption. For small molecules, physisorption naturally yields much smaller binding energies; for large organic molecules, however, even physisorption might lead to a stable adsorption state at room temperature or above. We want to stress that there is no strict boundary value for the strength of the interaction (in terms of a certain molecular binding energy value) to separate the two regimes "chemisorption" and "physisorption." In fact, the only valid distinction is through analysis of the electronic, geometric, or vibrational properties. Differential binding energy shifts of orbitals involved in the chemical bond, shift of vibrational bands relative to the gas phase, or changes of intramolecular bond lengths are typical indicators of chemisorption. Regarding possible reactions of adsorbed molecules, only chemisorption is able to change the electronic or geometric structure of the molecules in a way that allows dissociation and, thus, further reactions (see also Chapter 31 in this Volume).

### 38.2.2 Coverage

An important parameter describing both the adsorption process as well as the resulting adsorption state is the amount of adsorbed molecules on a surface. This value is usually called *coverage* and abbreviated with $\theta$. There are two possible definitions of coverage both of which are used in the literature and which should not be mixed up. The first one takes the saturation coverage as the reference value, which is fixed as 1. All other coverages are given relative to this saturation coverage, so that we will refer to this definition as *relative coverage*. It is usually given in fractions, for example, $\theta = 0.5$ monolayers. The second definition takes the number of atoms in the substrate surface layer as reference, that is, 1 ML is one adsorbate species (atom or molecule) per substrate atom. Thus, only if there is an adsorbate species adsorbed on each surface atom will the coverage be 1 ML. While the value of 1 ML can be achieved, for example, for atomic hydrogen, the majority of other (most molecular) adsorbates are larger than metal substrate atoms, and thus even at saturation coverage, values of well below 1 ML are found when using this definition. Since the number of substrate surface atoms per unit area for a specific surface orientation of a certain substrate is known, the second definition describes the absolute number of adsorbate molecules and is thus called *absolute coverage*. In order to distinguish between the two definitions, the unit ML (or MLE – monolayer equivalent) is used for absolute coverage, to refer to the MLE to the number of substrate atoms (note that some authors only use the ratio without the unit "monolayer"; in this case 1 ML is denoted as $\theta = 1.0$); the *relative coverage* is usually given in "monolayers."

It is immediately obvious that both definitions have their advantages and disadvantages. In the case of *relative coverage*, the value does not tell anything about the number of adsorbed molecules, which could be rather different if the molecules have different sizes, for example, CO versus benzene. A rather severe problem arises if the saturation coverage cannot be reached easily or depends on the preparation process (i.e., pressure or temperature). In fact, this seems to be the rule then the exception. Regarding physisorption cases or adsorption of metals on metals, the relative coverage has the advantage that the number of layers and the coverage yield the same integer values. On the other hand, the *absolute coverage* correctly gives an impression about the number of molecules adsorbed on the surface, but it leads to strange values if multilayer growth is concerned, in particular for large molecules where the number of molecules per substrate atom is very small. Since we will be dealing mainly with chemisorption systems, throughout this chapter we will use only the absolute coverage definition, and will give values in ML.

### 38.2.3
### Adsorption Site

If we consider a low-index single-crystal surface, we face a highly symmetric situation. While adding adsorbate molecules or atoms, they would naturally first occupy high-symmetry positions on the surfaces. Such positions could be directly on top of a substrate surface atom, called *on-top* sites, or could be straddled in between two surface atoms in form of a bridge, that is, *bridge* sites. While these two sites are possible on all ordered surfaces, there are differences for higher coordination depending on the actual surface structure. On a hexagonal surface structure, a threefold hollow site would be the next high-symmetry site. Depending on the stacking, *face-centered cubic* (fcc) and *hexagonal close-packed* (hcp) threefold hollows sites are distinguished. On a rectangular surface structure, a *fourfold hollow* site can be found. Figure 38.1 schematically shows these high-symmetry adsorption sites. As long as the interactions between molecules are small, adsorbates in most cases occupy these highly symmetric sites (there are, however, some exceptions). In very dense structures, however, deviations from these positions are observed [2].

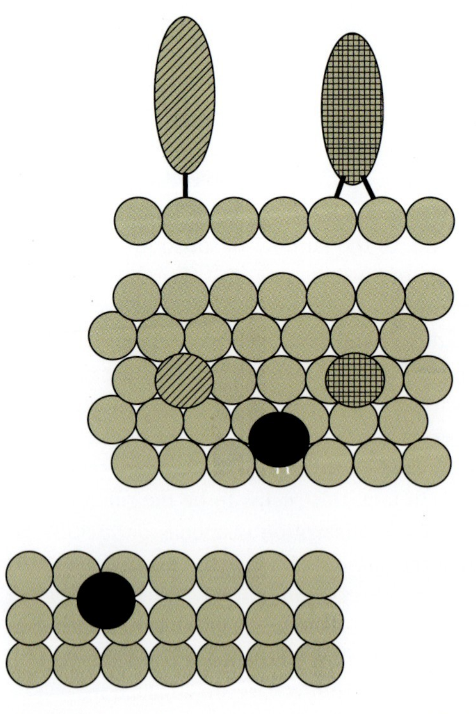

**Figure 38.1** Highly symmetric adsorption sites. ● On-top site, ● bridge site, ● hollow site (threefold-hollow on hexagonal and fourfold-hollow on quadratic or rectangular surfaces).

### 38.2.4
### Adsorption Rate, Sticking Coefficient, and Impingement Rate

The kinetics of an adsorption process is described by the adsorption rate. This rate is defined as the increase in coverage per time interval, that is, $d\theta/dt$. Here we consider only the case where the sample temperature is low enough so that no thermal desorption occurs (note that this might not hold very close to the saturation coverage at a given temperature). A negligible desorption rate means that the equilibrium between the gas phase and the adsorbate phase is on the adsorbate side. If this is true, then only two parameters define the adsorption rate: the number of particles hitting the surface in a unit time, called *impingement rate*, $Z$, and the probability with which these particles adsorb on the surface, called the *sticking coefficient*, $S(\theta)$, which typically depends on coverage $\theta$.

The sticking coefficient $S$ under a given experimental situation is defined as the ratio of the adsorption rate and the impingement rate. In other words, it is the fraction of molecules finally adsorbing on the surface, normalized to the number of molecules impinging on the surface. As such, it can be obtained by analyzing the number of adsorbed atoms or molecules (i.e., the coverage) resulting from a given number of molecules that impinged on the surface, after a certain exposure. The sticking coefficient generally depends on the coverage of a surface. Thus, one has to differentiate between the sticking coefficient for atoms or molecules that hit a clean surface and the sticking coefficient for an already covered surface. While the former is denoted as $S(\theta = 0) \equiv S_0$ and represents the situation on the surface in the absence of adsorbate–adsorbate interactions, the latter is denoted as $S(\theta)$. One should note that measuring the initial sticking coefficient $S_0$ can be very challenging because the determined signal goes to zero as the coverage approaches zero. Thus, $S_0$ typically is understood as the value extrapolated to zero coverage (i.e., in the zero coverage limit) from measurements of $S(\theta)$ at the smallest experimentally accessible coverages. This can sometimes result in rather large error bars for certain systems.

In general, the sticking coefficients depend on a variety of parameters as already addresse in Chapter 31 in this volume. These include the properties of the impinging gas molecules as well as those of the substrate. The relevant gas properties include kinetic energy, rotational energy, vibrational energy, alignment effects (of nonspherical molecules), and incidence angle as well as possible electronic excitations. The relevant substrate properties are chemical composition, surface temperature, surface structure, defect density, surface coverage, and other parameters such as the presence of coadsorbates. Depending on the existence of an activation barrier, one differentiates between nonactivated adsorption (typically with comparatively large sticking coefficients up to unity) and activated adsorption (typically with small to vanishing sticking coefficients). In simple cases, the latter can be described with one single activation barrier $E_A$ that can be overcome with sufficient kinetic energy.

While the coverage dependence of the sticking coefficient can be rather complex, typically two limiting cases are discussed. The first and simplest case implies a uniform adsorption (only one type of adsorption sites with a unique heat of adsorption) and no adsorbate–adsorbate interactions and is denoted as *Langmuir behavior*:

here adsorption takes place only if the site directly reached by impingement from the gas phase is vacant. This would mean that the number of these sites decreases linearly with increasing coverage, that is, $S(\theta) = S_0(1 - \theta)$. The second case is the so-called precursor behavior. In addition to direct chemisorption, this model includes weakly bound (physisorbed) precursor states, either on top of an already occupied area (extrinsic precursor) or above an empty chemisorption state (intrinsic precursor). The molecules thus can follow another adsorption route besides direct adsorption. In these precursor states, the molecules have a certain residence time (which depends on the surface temperature and depth of the potential well), during which they either can diffuse on the surface to find an empty chemisorption site or desorb to the gas phase. Therefore, a higher sticking coefficient can be observed, in particular at higher adsorbate coverage, as compared to the Langmuir behavior. A description of the interplay between these indirect adsorption channels with the direct adsorption channel is provided by the so-called Kisliuk model [3].

The second parameter to describe the adsorption process is the impingement rate of the molecules, which is the flux of molecules per unit time and unit area. With respect to experiment, two different scenarios have to be distinguished. (i) For adsorption from a background gas at temperature $T$ in the chamber, the impingement rate can be calculated from kinetic theory of gases using the Maxwell–Boltzmann distribution to be proportional to the pressure $p$: $Z = p/\sqrt{2\pi m k_B T}$, with $m$ being the mass of the molecules and $k_B$ the Boltzmann constant. (ii) When using a supersonic molecular beam, the pressure is no longer a good parameter, since the molecular beam is far from being in equilibrium. However, the flux of particles hitting the surface can still be given.

## 38.3 Adsorption as a Process

### 38.3.1 Introduction

When talking about adsorption, at least two aspects need to be considered. The first is adsorption as the process during which a gas atom or molecule hits the surface and transfers sufficient energy to eventually end up in a bound state at the surface. To describe this process, one needs experimental and theoretical methods that are capable of dealing with time-dependent processes, that is, the kinetics of the adsorption process. The second is the state of adsorption, that is, the adsorption complex consisting of the adsorbate plus the surface at a given temperature. This adsorption complex is – like the initial state of the pure, clean surface – accessible by thermodynamics and typically requires time-independent methods able to unravel its electronic, geometric, and vibrational structure. Both aspects are closely related to each other.

A large number of adsorbate systems have been studied in the past. In order to show the various aspects, we will present a couple of examples from different areas in the course of this chapter. We mainly concentrate on small molecules.

Here, examples such as CO or $O_2$, $N_2$, and $H_2$ first come to mind. The latter three examples have in common that dissociation of the molecules can take place to result in a chemisorbed atomic case. For reactive processes involving these molecules, such as CO oxidation and hydrogenation, this dissociation is a necessary prerequisite. The dissociation can take place directly upon adsorption or via a weakly bound precursor state. The dissociation path realized depends not only on the particular adsorbate–substrate combination but also on other parameters, such as the surface geometry, that play a crucial role. For a general overview, Ref. [4] can be recommended.

In order to characterize adsorption in detail, we need to apply a number of specific experimental and theoretical methods in order to investigate the whole process starting from a clean substrate and ending with a stable adsorbate structure. We will concentrate here on methods that allow following adsorption as a function of time, at different experimental conditions, that is, surface coverage and temperature. In the following, we will mainly discuss results from two different methods, both of which allow for the identification of different adsorption sites as a function of adsorbate coverage. We concentrate on a small number of examples (CO, $O_2$, $N_2$, benzene), which nevertheless provide an overview on the level of insight that can nowadays be obtained by state-of-the art methods. The case of hydrogen adsorption and its dissociation is the subject of Chapter 36 in this volume and thus will not be addressed here. In the following, we discuss the spectroscopic identification of adsorbates in their specific state on the surface. Thereafter, we will discuss different methods to follow the coverage dependence of the adsorbate properties.

### 38.3.2 Spectroscopic Identification

In general, different kinds of spectroscopies are available that allow specifically addressing adsorbates, their chemical state, and their adsorption sites. These methods can typically be applied to investigate the adsorption process in a time- or coverage-dependent manner. Until the 1990s, an individual measurement at a certain temperature and/or coverage lasted some minutes to an hour, and thus such measurements were performed in a stepwise (interrupted) mode. Nowadays, modern techniques allow measurements on the timescale of seconds, thus enabling us to follow a surface process such as the adsorption process or thermally induced reactions *in situ*. In this section, we will discuss the potential for spectroscopic identification of adsorbed species, while in the next section we will address coverage-dependent *in situ* measurements during the adsorption process. One powerful method to identify adsorbates and their adsorption sites is IR spectroscopy. It is able to detect the intramolecular vibrations of, for example, CO molecules adsorbed on different adsorption sites on Pt(111) [5]. Figure 38.2 shows the development of the absorption signals for the bridge-bonded CO (a) and the on-top bonded CO (b) for increasing coverages at 125 K. While vibrational spectroscopy is able to detect small changes in the vibrational properties, it faces the potential problem that, because of dipole–dipole interactions between the molecules, the excitation strength of the vibrations could vary with coverage, which

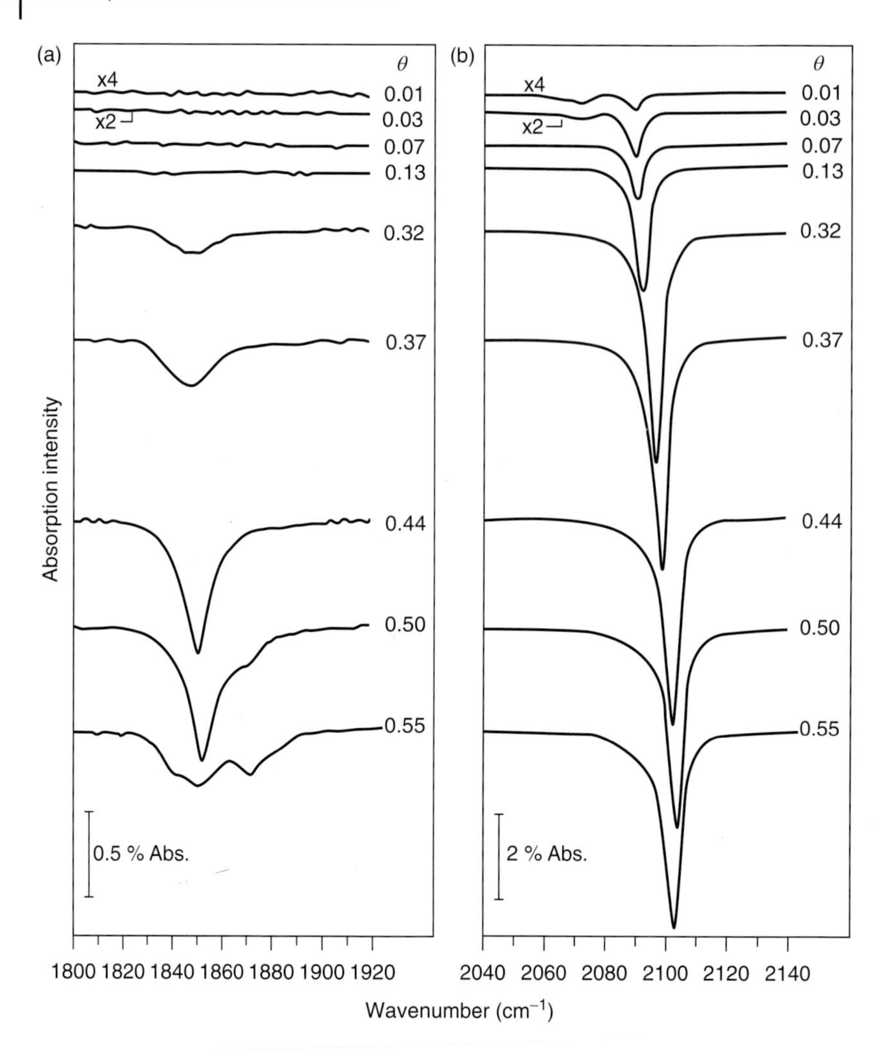


**Figure 38.2** FT-IR spectra measured for varying CO coverages on Pt(111). (a) Energy range showing the CO stretching vibration of the bridge-bonded CO molecules. (b) Energy range covering the on-top CO region. (Reprinted from Ref. [5], Copyright (1987), with permission from Elsevier.)

makes a quantitative analysis of coverages and site occupations sometimes difficult. However, with a suitable calibration this problem can be solved in most cases. In addition, such photon-based methods as IR spectroscopy can also be applied at more realistic pressures (see also Chapter 3.4.1, Volume 1). Then, however, special variants like sum-frequency generation (SFG) or polarization-modulated IR absorption spectroscopy (PMIRAS) are used (see, e.g., [6]).

As a variant of vibrational spectroscopy but with electrons and thus more surface sensitive, high-resolution electron energy loss spectroscopy (HREELS) is also able to detect the different adsorption sites of adsorbates, for example, of CO on Pt(111).

In this case, not only the CO stretching frequencies but also the vibrations of the whole molecule against the surface can be measured [7–11].

A second very powerful method is high-resolution X-ray photoelectron spectroscopy (XPS), which also allows for a quantitative analysis of the adsorbed species, based on the XP signals of the relevant adsorbate core levels (see Chapter 3.2.2, Volume 1). For high kinetic energies of the detected photoelectrons (typically obtained when using a Al K$\alpha$ radiation), the peak intensities are a direct measure of the adsorbate coverage, since the cross sections of core levels are independent of the environment. However, when working with low kinetic energies (as typically done at synchrotron radiation facilities), one has to pay attention to photoelectron diffraction effects, which can lead to pronounced intensity changes for different adsorption sites [12]. This is due to the strong dependence of the backscattering of electrons from neighboring atoms, which is most pronounced at kinetic energies between 50 and 200 eV, but becomes very small above 500 eV. Since most electron analyzers used only integrate the emitted photoelectrons over a very limited angle range, photoelectron diffraction (see Chapter 3.2.2, Volume 1) can cause also coverage-dependent intensity variations. Suitable treatment of such data allows even the derivation of geometric information [12]. Once aware of this potential problem, nevertheless quantitative information can be obtained quite readily. Therefore, XPS is a very powerful tool to investigate adsorbate phases. It was shown already quite early that, for example, the C 1s and O 1s binding energies of CO adsorbed on Pt(111) depend on the type of adsorption site (on-top, bridge) [13]. As a recent example demonstrating the unequivocal identification of adsorption sites, Figure 38.3 shows the C 1s XP spectrum taken at the CO saturation coverage of 0.5 ML at 200 K [14]. Clearly visible are the different binding energies for on-top and bridge site adsorption at binding energy values of 286.9 and 286.0 eV, respectively. At this coverage, both sites are occupied with a 1 : 1 ratio; the fact that the bridge

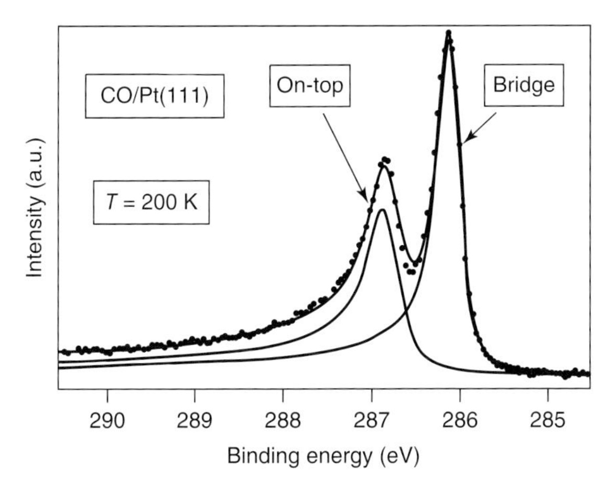


**Figure 38.3** C 1s spectrum of saturated CO layer adsorbed at 200 K on Pt(111). (Reprinted with permission from Ref. [14]. Copyright (2002), AIP Publishing LLC.)

peak has a larger area than the on-top peak is due to the mentioned photoelectron diffraction effects. Note that these effects typically cause a problem only when comparing different adsorption sites but not for the same sites at different coverages; this is due to the fact that backscattering is dominated by the much heavier substrate atoms, with much smaller to negligible contribution from scattering off neighboring molecules. In the example above, one finds a single photoemission line for a certain adsorption site. As we will discuss below, intramolecular vibrational excitation, which can happen during the photoemission process, can lead to extra features due to vibrational fine structure, which are, however, resolved only at sufficiently high energy resolution; examples can be found, for example, for CO on Ni(100) [15] (see also Chapter 3.2.2, Volume 1).

Similar studies both experimentally and theoretically exist for various metal surfaces, including noble metals (see, e.g., [16]). The adsorbate system CO on Cu displays another special type of fine structure in the XPS spectra. Here, some "giant" satellites occur, which are caused by shake-up processes involving certain weak charge transfer between substrate and adsorbate and the specific bonding situation [17–19].

While CO adsorbs vertically on most metal surfaces, there are also reports of flat-lying species. For example, on Cr(110) surfaces CO is reported to adsorb with its axis parallel to the surface. This seems to enable dissociation that can be observed on Cr(110). For higher CO coverages, a vertical species is also found [20]. Interestingly, the same features are also found for 1 ML of Cr deposited on Ru(0001) [21].

Let us next turn to oxygen. This molecule has been studied on a large number of metal surfaces. For an overview of the early literature, we suggest, for example, Ref. [22] (see also Chapter 31 in this Volume). Regarding the adsorption on Pt(111), at low temperatures (30 K) a physisorbed molecular state is observed. It results in a split O 1s core level, as can be seen in Figure 38.4. This paramagnetic splitting is due to the coupling of the spin of the remaining O 1s electron with the spins of the two unpaired electrons, resulting in two different total spin configurations (as observed first in gas-phase electron spectroscopy for chemical analysis, ESCA [23]). Upon heating, two molecular chemisorbed states are found. The proof of the different bonding scheme is given by comparison to valence-band ultraviolet photoelectron spectroscopy (UPS) data. The molecular species are determined to be superoxo- or peroxo-like. Additional heating results in dissociation to atomic oxygen above 150 K [24]. Interestingly, direct adsorption into the peroxo species, which readily dissociates to atomic oxygen, is less likely due to a higher activation barrier as the successive path via the different molecular states, as shown by a combined molecular beam and HREELS study [25]. Thus, preparation of atomic oxygen is more effective when heating a molecular layer adsorbed at low temperatures instead of adsorbing at room temperature. Although generally similar, there are certain differences predicted and observed for oxygen adsorption on other Pt-group metals, for example, Pd and Ni [26].

Another intensely studied molecular adsorbate is $N_2$. On Ni(100), it shows a perpendicular bonding geometry, resulting in two distinguishable N atoms [27]. This is reflected in the XPS and X-ray absorption spectroscopy (XAS) spectra by

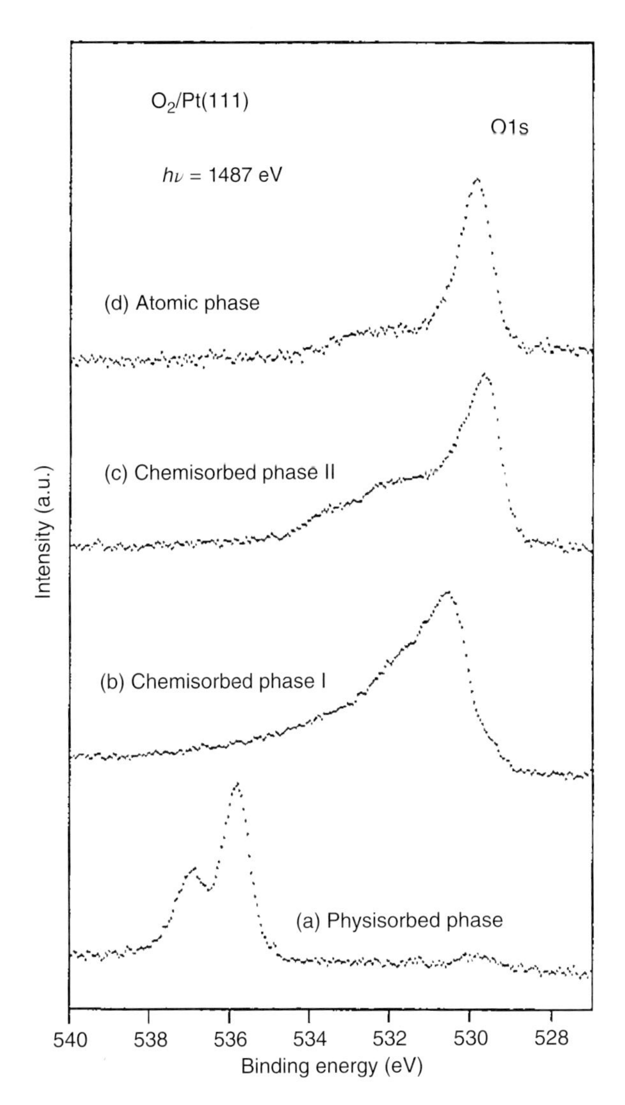


**Figure 38.4** O 1s core level spectra of (a) physisorbed $O_2$ at 25 K, (b) chemisorbed $O_2$ at 90 K, (c) chemisorbed $O_2$ after heating to 138 K, and (d) atomic oxygen (2 × 2) phase after heating to above 150 K. (Reprinted from [24], Copyright (1995), with permission from Elesevier.)

two distinct N 1s binding energies and allows using X-ray emission spectroscopy (XES) specifically on the separate N atoms [28], thus giving very detailed information about the local bonding situation [29]. On Ru(0001), where also a vertical adsorption geometry was found, dissociation of $N_2$ was clearly related to defects [30]. On Fe(111), a flat-lying $N_2$ species could be identified at around 100 K, which readily dissociated for increased temperatures (170 K) into atomic nitrogen [31].

## 38.3.3
## Coverage-Dependent Measurements

In this subsection, we address several examples where the adsorption process is followed in detail by performing coverage-dependent studies.

### 38.3.3.1 CO/Pt(111)

For the first example, we come back to the adsorption of CO on metals. The example CO/Pt(111) has already been introduced, and Figure 38.3 was used as an example to clearly show that different adsorption sites can be unequivocally identified by XPS. When utilizing synchrotron radiation as excitation source, the acquisition of data is feasible on the timescale of seconds. Thus, adsorption processes on this timescale are addressable by XPS, which is termed fast XPS [32]. For a review on this technique, Refs [33, 34] can be recommended. The change in site occupation manifests itself by the varying intensities associated with these lines. Using a peak profile analysis, the areas of the contributions can be determined and plotted versus adsorption time or exposure. The latter value allows also for quantitative comparison of adsorption experiments not using the same gas pressure. In order to have a better control of the adsorption process, a molecular beam can be used as the gas dosing facility. This allows us to "switch on" a nearly constant pressure at time equal to zero, and thus avoids the usual pressure buildup by backfilling a vacuum chamber. Figure 38.5 now shows a typical uptake series for CO/Pt(111) and Figure 38.6 the quantitative analysis of such an uptake experiment at 200 K. Obviously, saturation coverage is reached. If the resulting adsorbate phase is investigated by low-energy electron diffraction (LEED), a well-ordered CO-$(4 \times 2)$ structure is found (see Section 38.4.2). This can be assigned to a surface coverage of 0.5 ML. In addition, the structure contains exactly half of the CO molecules on on-top and half on bridge sites [7, 8], so also the partial coverages can be calibrated (and photoelectron effects can be accounted for). It is quite obvious that the on-top sites are found to be occupied first, followed by the bridge sites only if the on-top sites are almost saturated. Comparing the experimental data for different temperatures (and also desorption data) with results from a multi-site lattice gas model, the resulting adsorption energy difference between both sites is determined to be 95 meV. Here, it is vital to include intermolecular interactions [35]. From Figure 38.6, one would be tempted to derive relative sticking coefficients for the different adsorption sites. However, because of the possibility of site exchange, the partial coverages on on-top and bridge sites cannot be used to derive partial sticking coefficients. O 1s data also show the same details, however with slightly lower sensitivity due to a larger line width.

### 38.3.3.2 O/Ru($10\bar{1}0$)

Another approach to study adsorption uses the core-level information not from the *adsorbate* molecules but from the surface atoms of the *substrate*. This is illustrated by the dissociative adsorption of oxygen on the Ru($10\bar{1}0$) surface. Because of a different electronic structure, surface atoms could be identified that share a different number of adsorbed molecules. A quantitative analysis of the appearance of these different bonding configurations also represents the adsorption process.

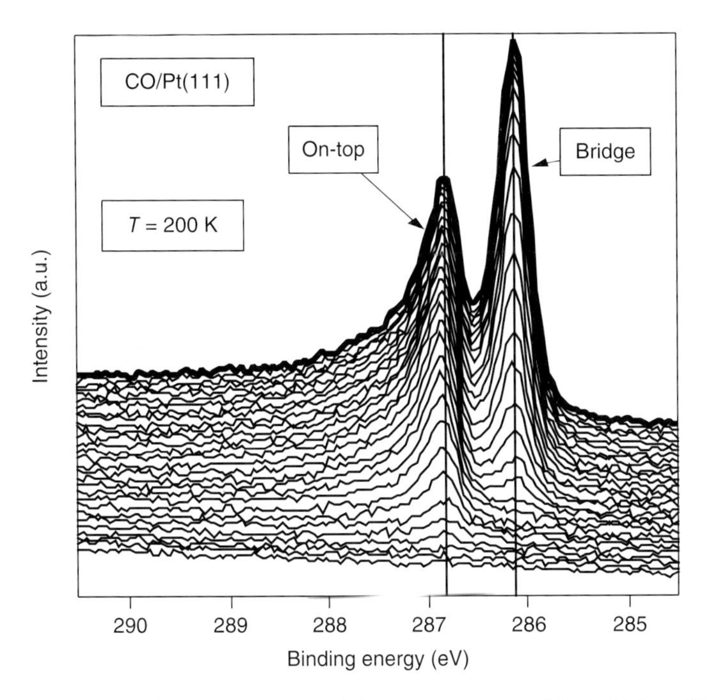


**Figure 38.5** C 1s spectra recorded during CO uptake on Pt(111) at 200 K. Measurements used $sh\nu = 380$ eV; time between spectra 60 s, $p = 1.7 \times 10^{-9}$ mbar, time per spectrum 4.8 s. Bold line: c(4 × 2) LEED structure observed after measurement. (Reprinted with permission from Ref. [14]. Copyright (2002), AIP Publishing LLC.)

The corresponding results are shown in Figure 38.7 [36]. In total, five different Ru species are found. They range from clean surface with two different surface species for the first (light blue) and second (orange) layer besides the bulk signal (at 280.1 eV) to saturated O coverage of 1 ML with Ru atoms in the first layer with two O atoms attached (gray) and Ru in the second layer bonded to one O atom (dark blue). For intermediate coverages, a first layer Ru with a single O bond (red) is observed (see also Chapter 3.2.3, volume 1).

Further characterization of the adsorbate layer can be performed by so-called temperature-programmed X-ray photoelectron spectroscopy (TP-XPS). In that variant, an adsorbate layer prepared at low temperatures is continuously heated while XP spectra are acquired. If the surface coverage of a certain species changes, this quantitative method picks up this change. If molecules dissociate, new species with different electronic structures can be distinguished by a change in core-level binding energy. This can be used to investigate surface reactions such as dissociation but also bimolecular formation reactions. We will come back to that point later (see Section 38.3.4).

#### 38.3.3.3 CO Adsorption on Surface Alloys: $Cu_3Pt(111)$ and PtCo

Staying with small molecules, an additional aspect can be introduced by using specific substrates. While stepped surfaces will be addressed separately, just a few facts

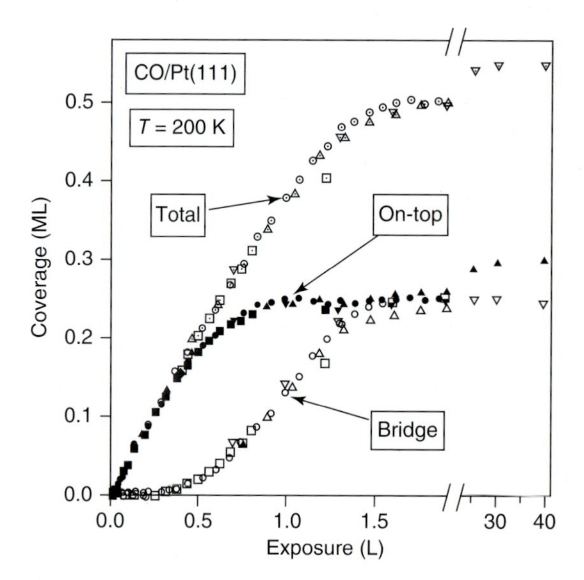


**Figure 38.6** CO coverage as function of exposure on Pt(111) at 200 K. Shown are the total coverage as well as the partial coverages on bridge and on-top sites. Different symbols represent different CO pressures during adsorption. Clearly visible is the saturation coverage of 0.5 ML comprising 0.25 ML CO on-top and 0.25 ML CO in bridge sites. (Reprinted with permission from Ref. [14]. Copyright (2002), AIP Publishing LLC.)

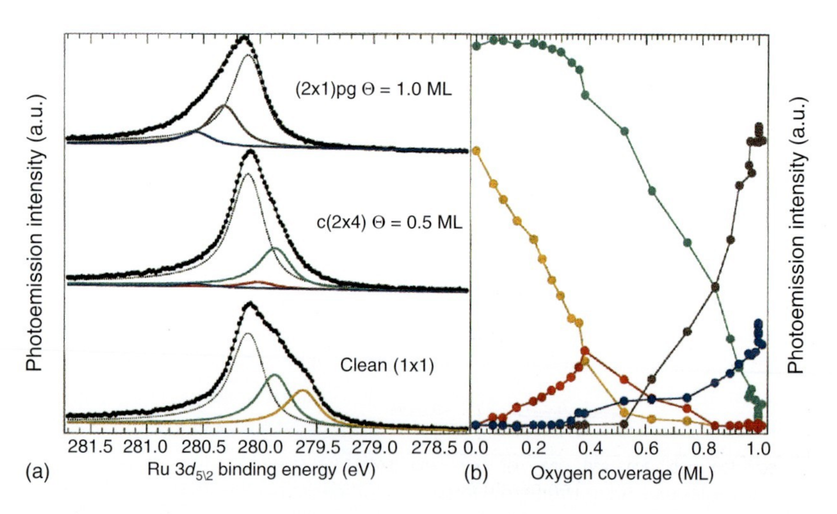


**Figure 38.7** (a) Ru $3d_{5/2}$ data for selected O coverages on Ru(10$\bar{1}$0). The various contributions representing Ru atoms bonded to a different number of O atoms and resulting from a peak fit are included. (b) Variation of the different intensities from the component fit as a function of O coverage. One can clearly follow the change in site occupation. (Reprinted from Ref. [36], Copyright (2000), with permission from Elsevier.)

about adsorption on alloyed surfaces should be mentioned here. Utilizing alloys from mixtures of metals with different adsorption properties towards a certain adsorbate molecule, the question arises as to how the existence and distribution of more than one metallic element in the surface influences the adsorption. A well-studied system is the adsorption of CO on an ordered $Cu_3Pt(111)$ surface where ideally single Pt surface atoms are surrounded only by Cu surface atoms [37] (see also Chapter 31 in this volume). At room temperature, CO adsorbs only on the Pt atoms but with a significantly lower adsorption energy than on a pure Pt(111) surface. This is shown in Figure 38.8. For low CO exposures at 50 K, only adsorption on Pt atoms is observed by temperature-programmed desorption (TPD) data. However, the resulting binding energy is much lower than on Pt(111) where the maximum of CO desorption occurs at around 400 K (see, e.g., [38]). A weakening of the bond has also been observed in HREELS experiments [39]. For high exposures (i.e., coverages), adsorption on Cu and in a physisorbed state is also found. Therefore, obviously the presence of the surrounding Cu atoms changes the electronic structure of the Pt atoms in the $Cu_3Pt(111)$ alloy. This is an example of a ligand effect and has been modeled also in DFT calculations [40].

A typical example of a disordered alloy is the system CO on PtCo. On both pure metals, CO has different binding energies with the bonding to Pt being stronger. Obviously, this preference can also be found in the alloy case, as can be seen from Figure 38.9. Here, an STM (scanning tunneling microscopy) image first shows the clean surface with chemical contrast (Figure 38.9a). The positions of the Co and Pt surface atoms are registered, which are schematically shown in Figure 38.9c.

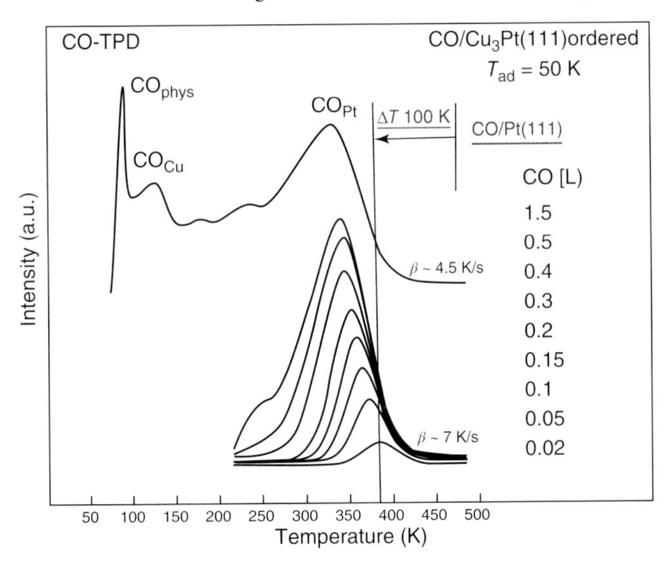


**Figure 38.8** TPD traces of CO adsorbed at 50 K on ordered $Cu_3Pt(111)$. For CO exposures below 0.5 L, only adsorption on Pt atoms is observed. At higher coverages, also weak adsorption on Cu atoms and further physisorption are obvious. The additional vertical lines mark the change in desorption temperature relative to Pt(111). (Reprinted (slightly modified) from Ref. [37], Copyright 1993, with permission from Elsevier.)

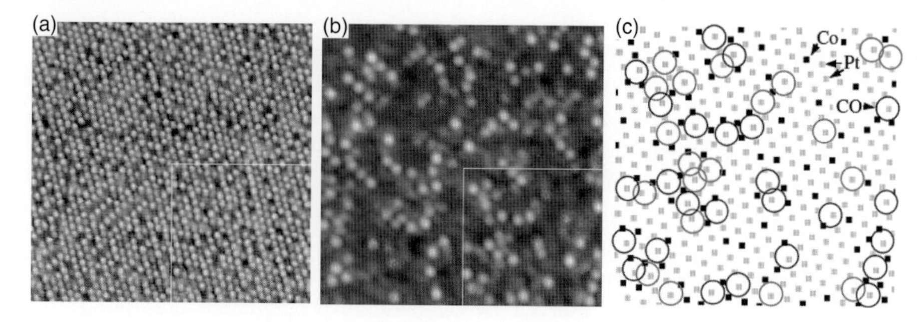


**Figure 38.9** (a) STM image of a clean PtCo surface with chemical contrast ($V = -0.5$ mV; $I = 1.6$ nA). (b) STM image after CO saturation adsorption at room temperature ($V = -0.31$ V; $I = 0.68$ nA). (c) Positions of Co (black) and Pt (gray) atoms from (a) and positions of CO molecules (circles) from (b) within the marked area. (Reprinted with permission from Ref. [41]. Copyright (2001) by the American Physical Society.)

Upon CO adsorption, the CO molecules are imaged on the substrate (Figure 38.9b) and their positions are again marked in (c). From the overlay, one can see that CO molecules are found only at Pt sites, and the probability to find CO increases with the number of Co neighbors. Thus, this is another excellent example of the ligand effect, which is corroborated by DFT calculations [41].

In general, also an ensemble of available surface atoms for bonding could govern the adsorption process, termed the *ensemble effect* [42]. However, because of the close correlation between electronic and geometric properties, a distinction between the simplifying pictures of either ligand (electronic) or ensemble (geometric) effects seems to be outdated [43]. Nevertheless, the terms are still used to describe the specific situations (e.g., [44]).

#### 38.3.3.4 **Benzene/Ni(111)**

While the discussed time-dependent XPS data yield quite detailed results for small molecules such as CO, NO, and others, the situation becomes quickly more complicated for larger molecules. In the case of benzene, the symmetry of the molecule helps obtaining still rather interesting results. On Ni(111), benzene is reported, both from experimental and theoretical work, to adsorb in different adsorption sites depending on coverage [45–49] (see also Section 38.5). An *in situ* XPS study allowed following these site changes continuously [50]. The central finding is that the combined system “benzene molecule and substrate” leads to six indistinguishable C atoms in the case of a threefold hollow site adsorption while two subsets of C atoms form in the case of bridge site adsorption. In Figure 38.10, the two situations are shown. XP spectra distinguish exactly between these situations by different binding energies for the C1 and C2 carbon atoms. Such difference could be supported by DFT calculations, similar to results published for benzene on Pt(111) [51]. With this spectroscopic information, the development of the intensities of the two carbon types can be followed throughout the adsorption process. Figure 38.11 depicts the results for a surface temperature of 200 K. In the top panel, one can clearly see how the development changes abruptly around a benzene coverage

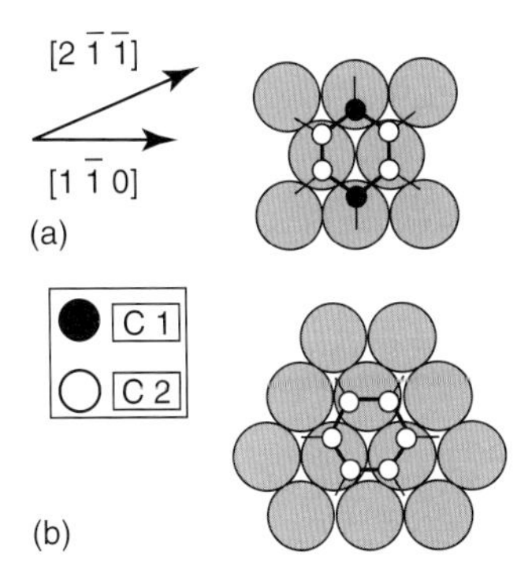


**Figure 38.10** Schematic drawings of the bonding situation of benzene on Ni(111). (a) Bridge configuration with two distinguishable C species (filled and open circles). (b) Hollow site with only a single C species. (Reprinted with permission from Ref. [50]. Copyright (2006) by the American Physical Society.)

of about 0.09 ML. While below this coverage the two carbon signals increase with a fixed ratio of slightly above 0.4–0.5 (see Figure 38.11b), the C1 component vanishes for coverages up to saturation. In this situation, only the C2 component is left. Connecting this to the two adsorption sites, a clear site change is observed (Figure 38.11c). Up to 0.09 ML, only the bridge site is occupied; above that coverage, a continuous change toward the hollow configuration is observed.

This particular system benzene/Ni(111) offers also an interesting isotope effect. Utilizing deuterated benzene, $C_6D_6$, the same site change is observed also for adsorption at 125 K. However, for $C_6H_6$ no such site change occurs at 125 K. Only upon heating to 220 K the site change is observed. A reason for this isotope effect can be found in the different spatial extent of the C-H/D bending mode vibrations due to the different widths of the potential curves and therefore the energetic positions of the respective vibrational levels. The vibrational displacements are significantly larger for $C_6H_6$, thus changing the activation energy for a site change at a coverage (0.09 ML), where already intermolecular interactions play a role [50].

### 38.3.4
### Sticking Coefficient Measurements

The adsorption process and the adsorption rate can be quantitatively described by the sticking coefficient. As already introduced in Section 38.2.4, the adsorption rate depends on the sticking coefficient times the impingement rate $Z$ of the molecules. The sticking coefficient gives the probability of a molecule that hits the surface to finally stay on the surface. From the definition of the adsorption rate, the sticking coefficient can be calculated from the determined slope of the coverage versus time curve ($d\theta/dt$) and from the impingement rate: $S(\theta) = d\theta/dt \times 1/Z$. Therefore, an obvious experiment would determine the time-dependent coverage $\theta(t)$. Such data can be derived either from spectroscopic data, such as, for example, photoelectron spectroscopy data measured *in situ* during exposure of the surface to the impinging molecules, or from TPD data after certain exposure times. Both techniques work nicely. However, the main factor describing the impingement rate is the absolute gas phase pressure on the sample surface. This value cannot be determined accurately enough without appropriate effort.

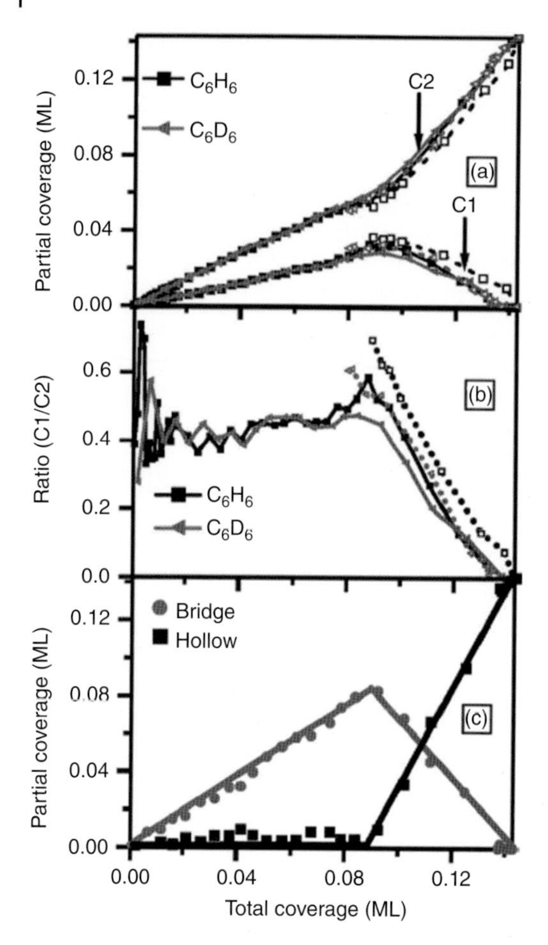


**Figure 38.11** Results of *in situ* adsorption experiments of benzene on Ni(111) at 200 K. (a) Partial coverages of the two C species defined in Figure 38.10 for methane and deuterated methane. (b) Intensity ratio of the C1 and C2 species for $C_6H_6$ and $C_6D_6$. (c) Partial coverage of the bridge and hollow benzene species as a function of total benzene coverage. (Reprinted with permission from Ref. [50]. Copyright (2006) by the American Physical Society.)

Therefore, King and Wells devised a method [52] using only the relative changes of pressures, which can be measured quite accurately, for example, by quadrupole mass spectrometry. The main idea of the method is that the clean substrate surface may act as a sorption pump, which under controlled conditions reduces the partial pressure of the adsorbing molecules in a specific manner. For that, the molecules have to arrive ideally in a directed beam only hitting the surface of interest (molecular beam). The higher the sticking coefficient, the larger the pressure change. If the maximum pressure change is known from some calibration experiment (i.e., the pressure change between "all gas beam molecules are present in the chamber" to "no gas beam molecules are present in the chamber"), the absolute sticking coefficient value can just be calculated from the ratio between the two pressure changes. Maximum pressure change would mean $S = 1$, and no pressure change means $S = 0$. The schematic experiment and the resulting pressure trace are shown in Figure 38.12.

The three-stage supersonic molecular beam contains a first chamber in which the gas expands through a nozzle into vacuum. Here, the molecules reach supersonic velocities. The extraction of the molecules by a skimmer into the second chamber results in a molecular flow of molecules with a narrow energy distribution. The apertures between the second and third chamber can be exchanged so that the size of the molecular beam can be tailored to fit the size of the sample. In the analysis chamber, a rotatable quadrupole mass spectrometer is used for sticking coefficient measurements and scattering experiments. The resulting partial pressure trace shown in Figure 38.12b demonstrates the initial pressure increase if the beam enters the analysis chamber but does not impinge directly on the sample surface. Upon removing the sample flag, the partial pressure is reduced because of the sample acting as an adsorption pump.

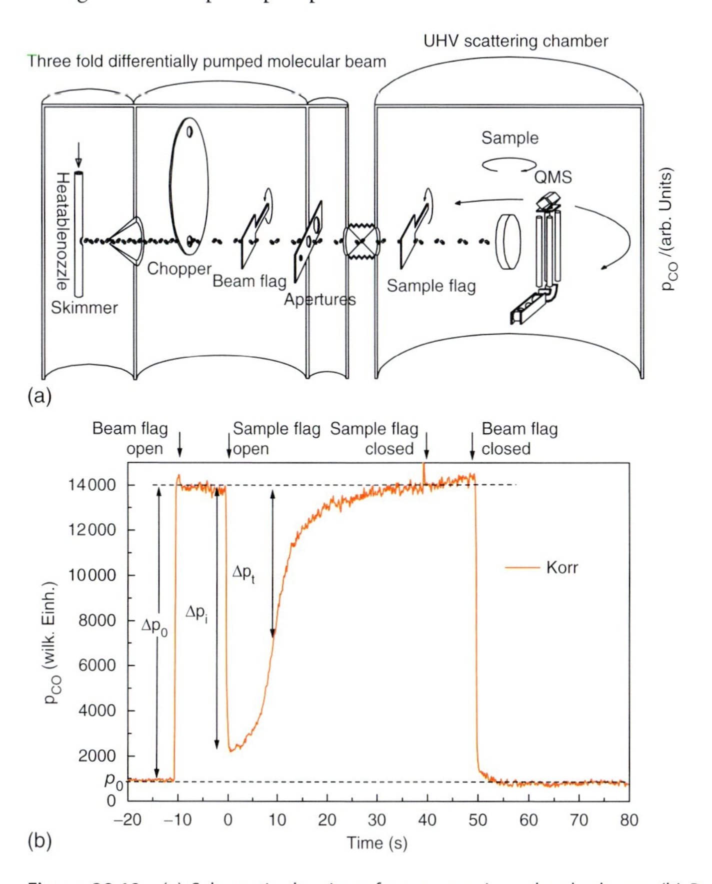


**Figure 38.12** (a) Schematic drawing of a supersonic molecular beam. (b) Pressure trace during an adsorption experiment. In the time interval from −10 to 0 s, the calibration part is run with the sample shielded by the sample flag.

Results of such sticking coefficient measurements are mainly twofold. As a specific parameter, the initial sticking coefficient for negligible coverage (only to be determined by extrapolation) carries information about the adsorption probability. This value depends not only on the particular combination of adsorbate molecule and substrate but also on the kinetic energy of the impinging molecules. In case of a nonactivated adsorption process, a higher kinetic energy leads to a decreasing initial sticking coefficient since the molecules have difficulty in dissipating their high energy. In case of activated adsorption, the adsorption probability increases with increasing kinetic energy. The latter is particularly observed for direct dissociative adsorption, such as for the case of hydrogen adsorption, for example, $H_2$/Cu [53] or $CH_4$/Pt(111) [54, 55].

The other aspect is the shape of the coverage-dependent development of the sticking coefficient. Assuming the so-called Langmuir adsorption, which means that the adsorption energy does not depend on coverage and that only direct adsorption is considered (either the molecule hits an empty site and adsorbs or it desorbs immediately), a linear decrease of the sticking coefficient with increasing coverage is expected. If adsorption via a weakly bound precursor state is also possible, as in the case described in the Kisliuk model [3], then the sticking coefficient stays high until the free adsorption sites get very rare, that is, until close to saturation. By tuning the kinetic energy of the impinging molecules, the two scenarios can be distinguished for certain adsorption systems, as shown in Figure 38.13 for CO on a pseudomorphic monolayer of Cr on Ru(0001). Here, at high kinetic energies, the precursor state is not viable for the molecules and thus only direct adsorption is possible. At low kinetic energies, the molecules can easily adsorb in the precursor state and diffuse on the surface to find the remaining empty adsorption sites. The solid lines in the figure are model calculations using a modified Kisliuk model

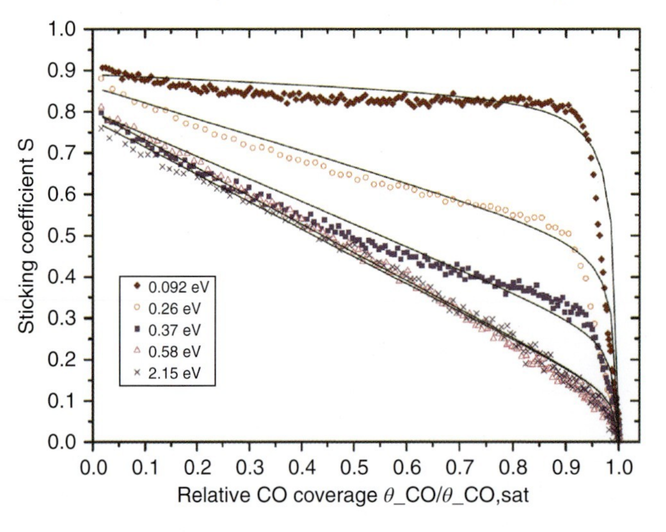


**Figure 38.13** CO sticking coefficient as a function of relative CO coverage on a pseudomorphic Cr monolayer on Ru(0001). For the various traces, the kinetic energy of the CO molecules has been varied by using a supersonic molecular beam.

to include adsorption into a precursor state. With increasing kinetic energy, the sticking coefficient in the precursor state becomes very small, leaving mainly the direct Langmuir adsorption channel [56, 57].

With the method by King and Wells, sticking coefficients well above 1% can be measured. For cases with lower values of $S$, or if additional spectroscopic information is needed, the determination of the exposure-dependent coverage, $\theta(t)$, is helpful. In this case, low sticking coefficients can be determined by sampling the slowly increasing coverage over a longer time interval. This time-dependent coverage can be determined either by spectroscopic techniques such as XPS or by TPD. An example of such an approach using XPS is shown in Figure 38.14.

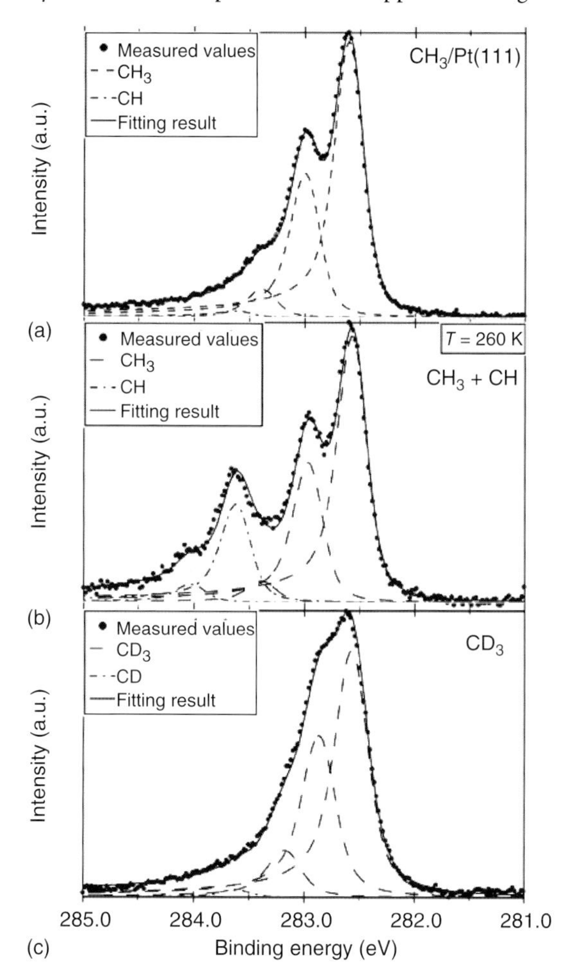


**Figure 38.14** C 1s XP spectra for methane adsorbed dissociatively (as $CH_3$) on a flat Pt(111) (a) at 125 K, (b) at 260 K, and (c) at 125 K using deuterated methane. Dashed lines show model functions emphasizing the vibrationally split components, while the dash-dotted line in (b) shows the contribution of the dissociation product CH. (Reprinted from Ref. [58], Copyright (2004), with permission from Elsevier.)

Methane molecules show physisorption on Pt surfaces only below 70 K [59]. In order to observe chemisorption and thus be able to study surface reactions involving methane, methane needs to dissociate. This can be facilitated by using methane molecules with defined and high kinetic energies from a supersonic molecular beam. The molecules adsorb dissociatively as $CH_3$ and hydrogen. Measuring the C 1s intensity as a function of dosing time allows following the adsorption process *in situ*. During the course of these measurements, spectra like the one shown in Figure 38.14a are obtained [58]. The different peaks and shoulders observed originate from a vibrational splitting due to vibrational excitation of the molecules in the photoemission final state. This can be proven by using deuterated methane, leading to a smaller vibrational energy of the C–H stretching mode and thus to a smaller splitting of the observed components, as shown in Figure 38.14c. Recording the sum of all the C 1s components (which represent something similar to a satellite structure) as a function of time and extrapolating the slope of $\theta(t)$ versus time to zero time (i.e., to zero coverage) gives a measure of the initial sticking coefficient (if the experiment ensures constant adsorption pressure or flux of methane from the molecular beam). Figure 38.15 now displays these values as a function of kinetic energy. The results for Pt(111) are shown in black symbols. From the logarithmic scale of the vertical axis, the activated nature of the dissociation process is obvious. However, the displayed relative sticking coefficient values cannot be taken as absolute numbers because the methane flux is unknown.

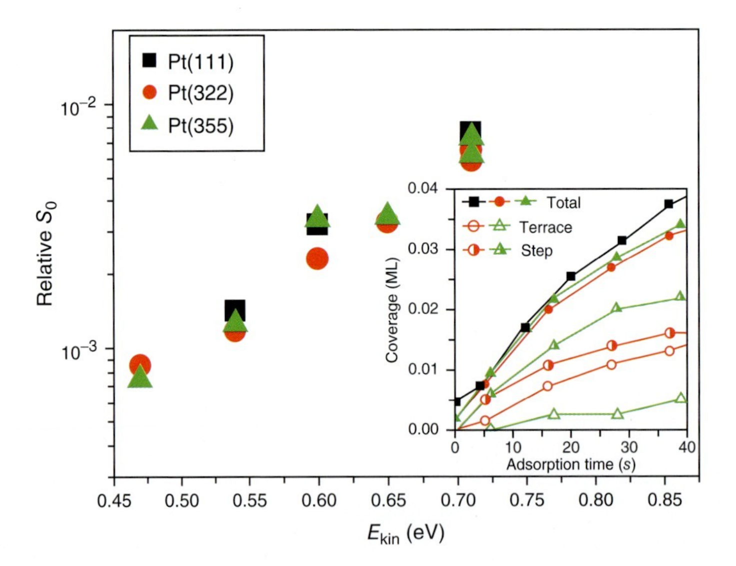


**Figure 38.15** Relative initial sticking coefficient of dissociative methane adsorption at 125 K on Pt(111), Pt(322), and Pt(355) surfaces as a function of kinetic energy. Data have been derived from coverage-dependent XPS experiments. The inset shows the respective time-dependent coverages for the different adsorption species and the total $CH_3$ coverage for the beginning of adsorption. (Reprinted with permission from Ref. [55]. Copyright (2007) American Chemical Society.)

Comparing the data of Pt(111) with those of stepped Pt(355) and Pt(322) surfaces, an interesting finding results [55]. As shown in Figure 38.15 with green and red symbols, the resulting relative sticking coefficients do not vary, although surface defects like steps are usually considered to result in higher surface activity. For a low temperature of 120 K, only the direct adsorption and dissociation channel without further surface reactions is recorded. According to Donald *et al.* [60] and others, methane dissociation at such low temperatures is mainly governed by a tunneling process. This implies that the dissociation process might start already during the approach of the molecule to the surface electronic potential without a bound transition state. Thus, since the steps represent only minor and – especially with increasing distance to the surface (see, e.g., the Tersoff-Hamann model of STM for an analog [61]) – smooth changes of this electron density, no influence of the steps is observed on the primary dissociation step. However, on heating the adsorbed $CH_3$ layer to elevated temperatures, the resulting further dissociation steps to CH (as is shown in Figure 38.19) indeed show a dependence on the presence of steps. The successive reaction steps are found to occur at lower temperatures for the stepped surfaces [55]. With this example, the influence of steps has already been demonstrated. A more detailed account will be given in the next section.

### 38.3.5
### Adsorption on Stepped Surfaces

The small active particles of realistic heterogeneous catalysts exhibit many steps and defects. As we have seen at the end of the last section, the reactivity of steps can differ from that of the flat terraces. Therefore, it is necessary to include the influence of such steps and defects in the model investigations on adsorption, that is, the initial step of a catalytic reaction, described here (see also Chapter 31 in this Volume). One frequently applied approach is to use regularly stepped surfaces in adsorption experiments as described above. An example has already been discussed using the methane adsorption.

Figure 38.16 shows the models of two stepped surfaces, namely Pt(355) and Pt(322) (see also Chapter 4, Volume 2). One can clearly see that in this case (111) terraces containing five atom rows are separated by monatomic steps. While in the case of Pt(355), the step exhibits a (111) facet, Pt(322) has a (100) step facet.

While King and Wells-type measurements allow the determination of the sticking coefficient under certain experimental conditions, they do not provide specific information about the step adsorption, such as adsorption of adsorbates on step or terrace sites. Nevertheless, from the comparison of such experiments for flat and stepped surfaces some information about the influence of the steps can be obtained, as can be seen in Figure 38.17. The coverage-dependent sticking coefficients for $N_2$ on differently oriented W surfaces at 300 K exhibit large differences [62], which reflect different dissociation probabilities on the different local surface geometries.

Spectroscopic techniques are able to not only distinguish different adsorption sites on flat surfaces, as already discussed above, but they can also distinguish between adsorption at step sites and adsorption at terrace sites. As a typical

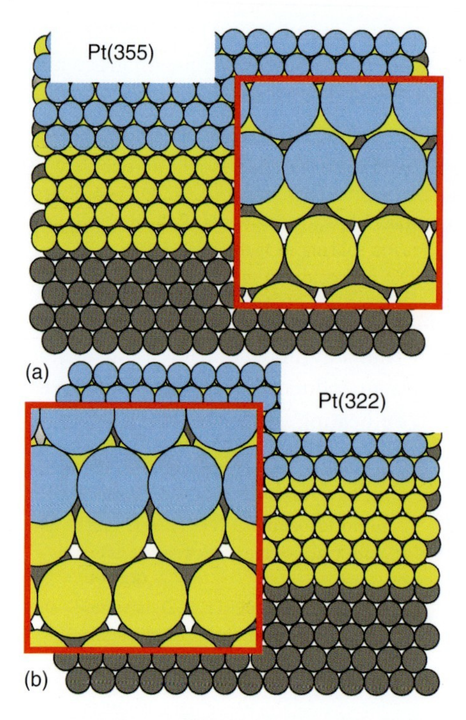


**Figure 38.16** Models of the stepped surfaces (a) Pt(355) and (b) Pt(322). Both are derived from the Pt(111) surface with monatomic steps. The step facets have (111) orientation in case of Pt(355) and (100) orientation in case of Pt(322) (as shown by the insets).

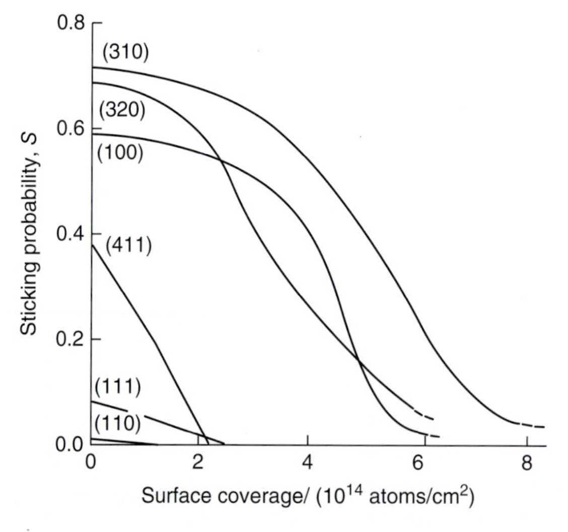


**Figure 38.17** Coverage-dependent sticking coefficients for $N_2$ on different crystallographic planes of W. (Reprinted from Ref. [62], Copyright (1975), with permission from Elsevier, here in the version "Oberflächenphysik des Festkörpers" by M. Henzler, W. Göpel, Teubner 1991.)

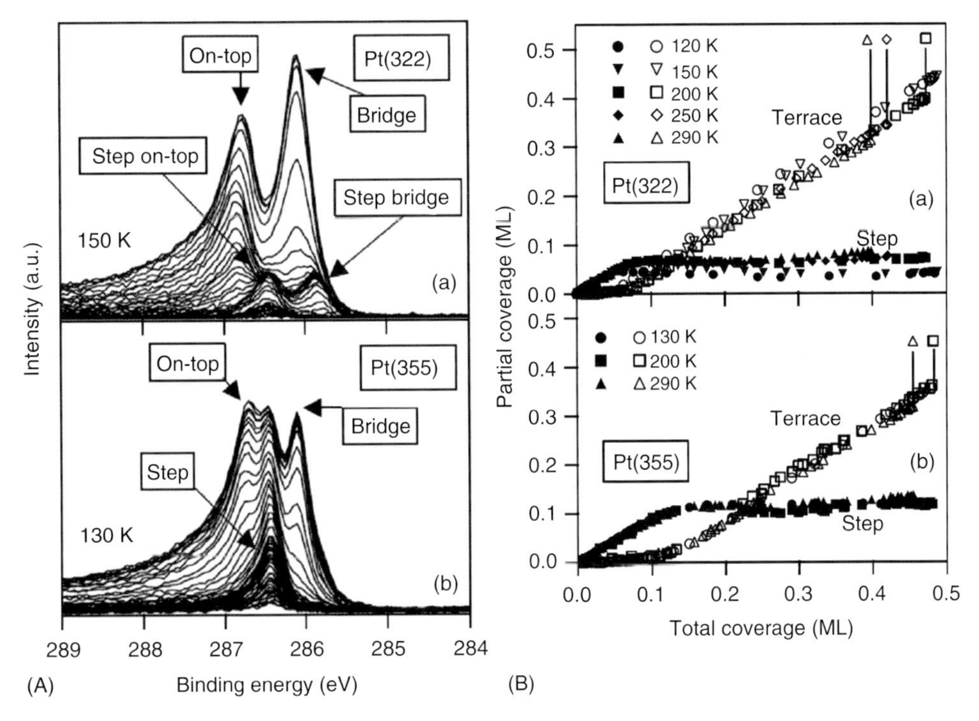


**Figure 38.18** (A) C 1s spectra recorded during CO adsorption on (a) Pt(322) at 150 K and (b) Pt(355) at 130 K. Marked by arrows are the different adsorption sites manifested by different core-level binding energies. (B) Resulting developments of the partial coverages on step and terrace sites versus total coverage for CO on (a) Pt(322) and (b) Pt(355) at various temperatures. (Reprinted from Ref. [63], Copyright (2007), with permission from Elsevier.)

example, in Figure 38.18 we show the spectra of CO adsorption on stepped Pt(355) and Pt(322) surfaces [63]. In addition to the above-mentioned adsorption sites "on-top" and "bridge" on the (111) terraces, there are new peaks associated with the step sites. Interestingly, different site occupation on the steps is observed for the two very similar surfaces. They both exhibit terraces that are five atomic rows wide, but they have different step orientations – (111) for Pt(355) and (100) for Pt(322). From the total-coverage dependence of the partial coverages, one can see that the step sites are occupied first, pointing to a larger adsorption energy on such sites as compared to the flat terraces. Another difference between the two surfaces is the different temperature dependence, which cannot be discussed in detail here [63]. The assignment of the spectral components to the various adsorption sites, especially terrace and step, can be verified by decorating the steps with a row of Ag atoms, which suppresses the step components [64].

A similar observation of the influence of steps on adsorption behavior can also be made for the activated adsorption of methane on Pt surfaces [55]. In the discussion above, we only emphasized the similarity of the measured initial sticking coefficient. Behind such data, the spectroscopic signature of the steps is clearly observed.

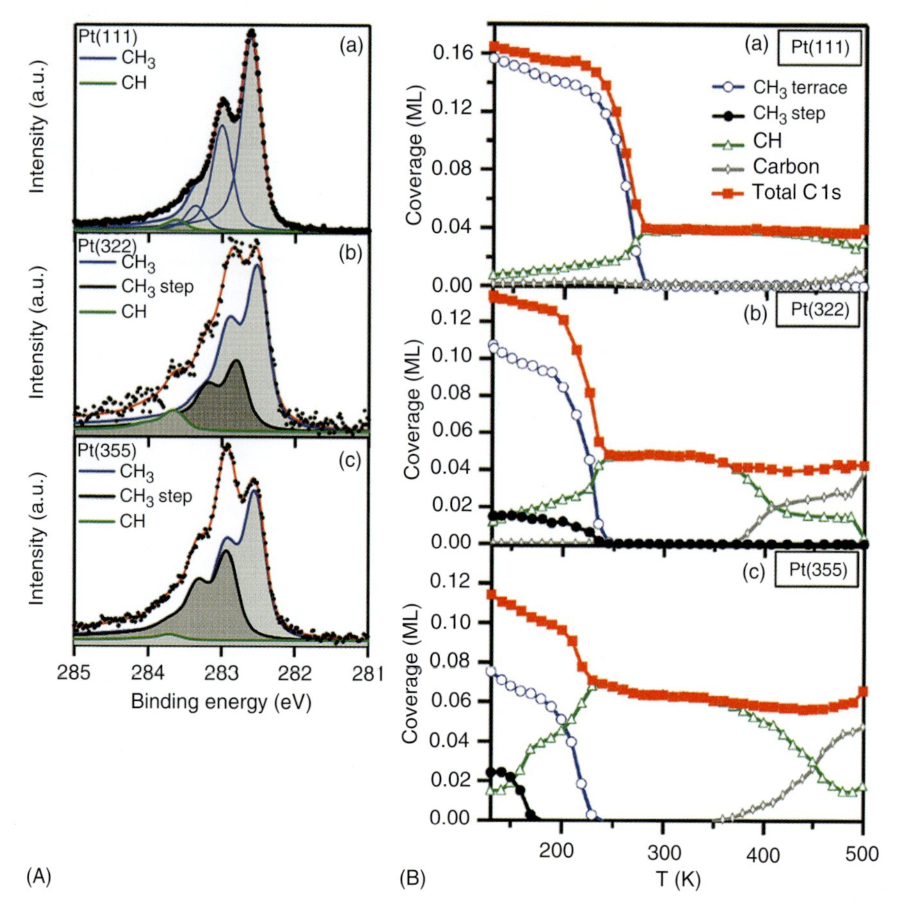


**Figure 38.19** (A) C 1s spectra for saturation coverages following adsorption of methane at 150 K on (a) Pt(111), (b) Pt(322), and (c) Pt(355). Shown are results of a peak-fit, clearly demonstrating the presence of step-related methyl species. (B) Quantitative analysis of data from temperature-programmed XPS of the methyl layers adsorbed at 150 K. Shown are the total C coverage as well as the partial coverages of respective surface species as a function of temperature for (a) Pt(111), (b) Pt(322), and (c) Pt(355). (Reprinted with permission from Ref. [55]. Copyright (2007) American Chemical Society.)

Figure 38.19 shows a comparison between the data obtained for dissociative methane adsorption on Pt(111), Pt(322), and Pt(355) at low temperatures (120 K). In the data analysis, special shifted components have been used to account for the step adsorption. Different relative coverages on the two stepped surfaces are visible. Again, meaningful sticking coefficient data can be derived only by the total methyl coverage obtained from an integral of the C 1s data. However, in order to investigate further dissociation on the surface, a detailed analysis is possible and fruitful. By that, a clear preference of the step sites for dissociation could be found. However, together with the step sites, also the terrace sites show dissociation at lower temperatures. In the end, the stepped surfaces are more reactive toward methane dissociation as compared to the flat surface, as expected.

One aspect in connection to the stepped surfaces is the possibility to manipulate the reactivity of the steps by adsorption/deposition of other atoms. Particularly interesting are either atoms that reduce the reactivity of the steps (like Ag) or that enhance the reactivity (like Cr). On the already mentioned two stepped Pt surfaces, Ag forms rows along the step edges if adsorbed and annealed at room temperature. Upon successive CO adsorption, a clear blocking of the step adsorption sites for CO seems to be obvious from the disappearance of the respective C 1s components, as can be seen in Figure 38.20. However, caution must be exercised because a detailed theoretical investigation has revealed that the electronic structure changes at least partly such that the CO core-level binding energies of CO molecules still adsorbing at the Pt step edge are very similar to those of terrace species. Increasing the coverage such that also terrace sites are occupied finally leads to a reappearance of the step sites in the XP spectra. Again, by involving DFT calculations, it could be clearly shown that, in the presence of CO, the energetically most favorable situation is such that the Ag atoms move partly away from the steps, forming the so-called embedded islands in the upper terrace. As a consequence, step sites are available again for CO adsorption. As a result of the larger total energy gain, this site exchange of the Ag atoms is beneficial [64].

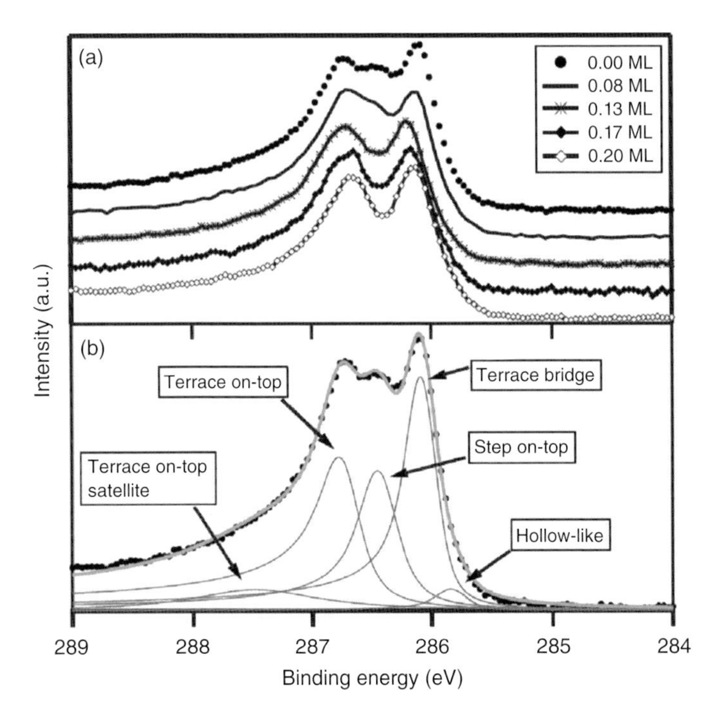


**Figure 38.20** (a) C 1s spectra for CO adsorption on Pt(355) at 200 K for various amounts of predosed Ag (given in monolayers; 0.2 ML corresponds to roughly one atomic row). (b) C 1s spectrum for CO on clean Pt(355). Shown are the different adsorption sites. (Reprinted with permission from Ref. [64]. Copyright (2009), AIP Publishing LLC.)

### 38.3.6
### Coadsorption

Interesting situations in terms of fundamental research as well as technologically relevant systems occur if two or more different types of molecules are coadsorbed together on a surface. Besides the formation of ordered adsorbate structures (many examples can be found throughout volumes 5 and 6), also changes in electronic structure, and thus in the spectroscopic signatures, can be observed.

Starting again from our prototypical system CO on Pt(111), the coadsorption with water molecules is interesting [65]. Starting with a bilayer of water molecules adsorbed on Pt(111) at low temperatures (here 125 K) as discussed in detail in Chapter 37 in this Volume, CO is subsequently adsorbed on this surface. Interestingly, recording the population of on-top and bridge species as described above results in an inverse situation compared to the pure CO adsorption. As shown in the left part of Figure 38.21, now the bridge species seems to be favored over the on-top species. In addition, an additional new species emerges. If the resulting system is heated (right part of Figure 38.21), two aspects need to be mentioned: (i) At 153 K water desorbs. However, this desorption temperature is not the one observed from

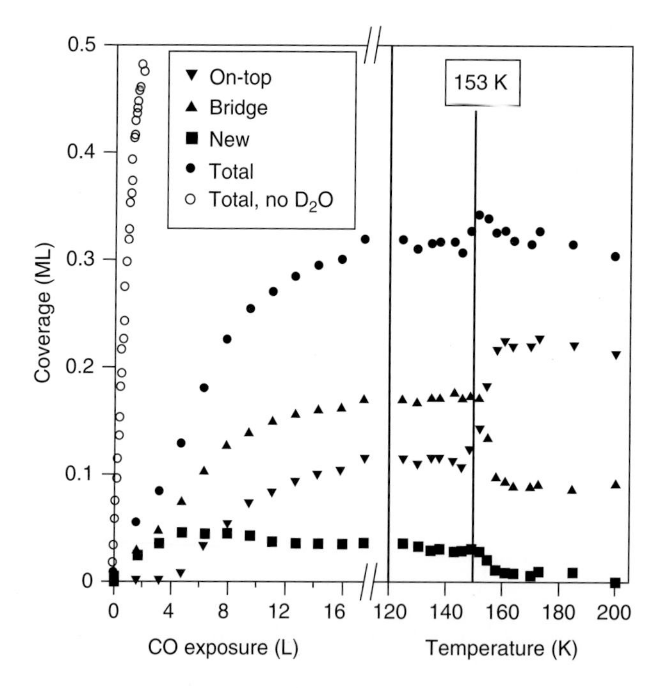


**Figure 38.21** (Left part) Dependence of the CO coverage on different adsorption sites and in total on a Pt(111) surface covered with a bilayer of water, as determined by XPS during CO adsorption at 125 K. For comparison, the resulting CO coverage without water is included. (Right part) Behavior of the coverages during thermal desorption. The vertical line marks the desorption of water. (Reprinted with permission from Ref. [65]. Copyright (2004) American Chemical Society.)

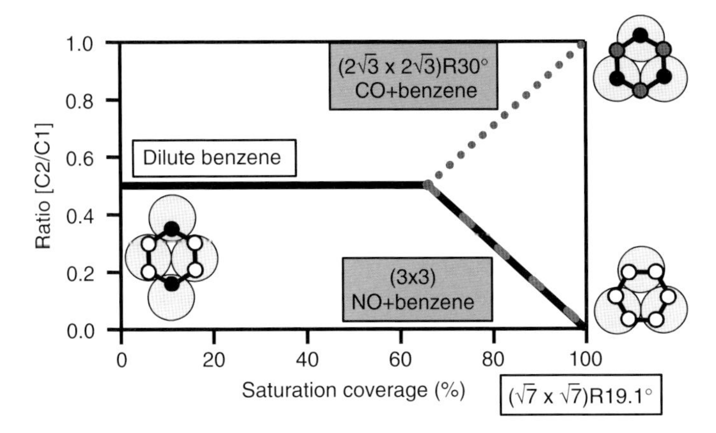


**Figure 38.22** Behavior of the intensity ratio of the two distinguishable C species (C1 black and gray filled circles, C2 white circles) within the benzene molecule as a function of the relative benzene coverage for coadsorption scenarios with CO (dotted line) and NO (dashed line). Included are the proposed adsorption models of the benzene molecules. (Reprinted from Ref. [67].)

a pure water bilayer on Pt(111) (reported to desorb at 168 K [66]) but rather the one for water multilayers (expected at about 155 K [66]); (ii) Upon removal of the water, the site occupation between on-top and bridge sites relaxes back to the usual situation for CO on clean Pt(111). So obviously, water is pushed into multilayers by the coadsorption of CO, which itself occupies now preferentially bridge sites. This observation is interesting in the context of reactions in the presence of water, such as reactions under realistic conditions or in electrochemical applications.

Coming back to the example of benzene adsorption, here also coadsorption has a great influence on the behavior. Figure 38.22 summarizes the observations [67]. Coadsorption with NO leads to the same behavior as for pure benzene, resulting in a hollow site adsorption for the saturated ordered layer with six identical C atoms (see also Figure 38.10), while coadsorption with CO favors the rotated hollow adsorption site with equal number of two kinds of C atoms in the $C_6$ ring. Therefore, the interaction between the coadsorbed molecules can drastically alter the adsorption situation and with this the possible reaction routes on such surfaces.

## 38.4 Adsorbate Structures

### 38.4.1 Introduction

Besides the adsorption process itself, the resulting adsorbate layer also contains important information. For example, using a known ordered structure, coverage values measured with other techniques (e.g., TPD) can be calibrated. In addition, stable adsorbate structures are accessible by equilibrium calculations such as DFT and are therefore an important step in cross-linking experiment and theory. In order to investigate the structure of a monolayer, surface-sensitive techniques are necessary. As already introduced in Chapter 3.2.1 of Volume 1, LEED is a method of choice to determine the surface structure. Photoelectron diffraction would also be possible, especially if no long-range order is present (see Chapter 3.2.2, Volume 1). Another more direct method involves STM, described in Chapter 3.5, Volume 1. Both methods are useful, but their strengths and limitations have to be taken into account. Since the methods have been introduced already, we can directly discuss some examples.

### 38.4.2 Examples

Coming back to the above-discussed example of CO adsorption on Pt(111), we are going to discuss its geometric structure. Figure 38.23 shows a LEED pattern obtained after saturation adsorption at 200 K. Between the basic scattering peaks of the substrate (marked by circles), one observes the diffraction spots due to the c(4 × 2) structure of the CO molecules. This structure is formed by a CO coverage of 0.5 ML, with an equal number of molecules adsorbed in on-top and bridge sites.

Another more complex surface structure is formed by benzene on Ni(111) or Ru(0001). Figure 38.24 shows the $(\sqrt{7} \times \sqrt{7})R19°$ structure of benzene on Ru(0001) as obtained with a kinetic energy of 200 eV, following adsorption at temperatures

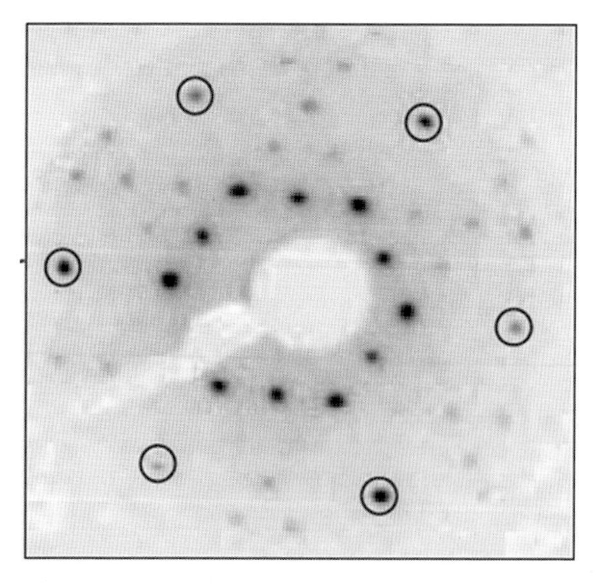

**Figure 38.23** LEED pattern obtained after saturation adsorption of CO on Pt(111) at 200 K. The pattern was obtained with the sample cooled to 100 K and at an electron energy of 94 eV. Substrate peaks are marked by circles. (Reprinted from Ref. [68].)

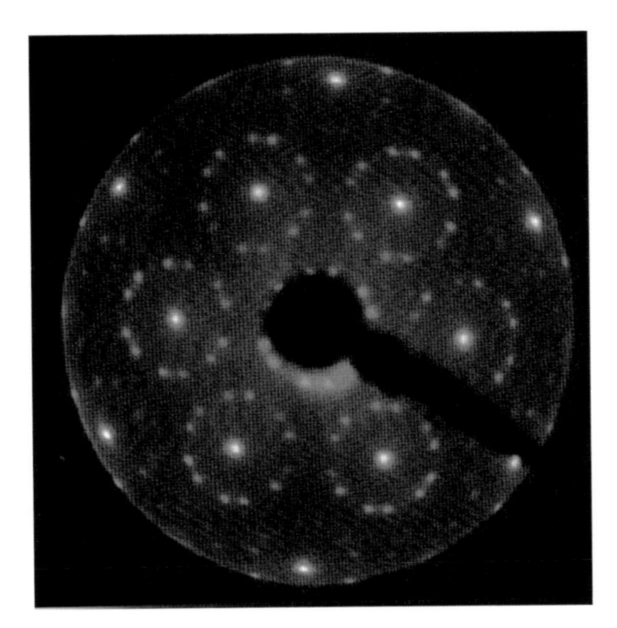

**Figure 38.24** LEED pattern obtained for benzene on Ru(0001) adsorbed at below 270 K. The resulting $(\sqrt{7} \times \sqrt{7})$R19° structure has a benzene coverage of 1/7 (0.143) ML. The pattern has been recorded at a kinetic energy of 200 eV. (Reprinted from Ref. [69], Copyright (2001), with permission from Elsevier.)

below 270 K. The same adsorbate structure is also found on Ni(111). From geometric arguments, a benzene coverage of 1/7 (0.143) ML follows. The situation on Ru(0001) is rather sensitive to the adsorption temperature and resulting coverage, with a total of five different surface structures [69].

In some cases, LEED patterns do not help in determining the surface coverage. One such example is NO on Pt(111). In the range between 100 and 300 K, always a $(2 \times 2)$ superstructure is observed for saturation adsorption. However, the actual NO coverage was found to vary a lot [70]. In this case, only LEED $I$–$V$ experiments, which are sensitive to the actual position of the molecules within the unit cell, can distinguish between the different situations [71]. In addition, high-resolution XPS of both O 1s and Pt 4f levels measured *in situ* during the adsorption process and as a function of surface temperature also revealed that the $(2 \times 2)$ structure is formed by different amounts of NO molecules per unit cell occupying different adsorption sites [70]. Figure 38.25 displays the situation and the corresponding Pt 4f spectra. Upon adsorption of 0.25 ML of NO, still a part of the Pt surface atoms is not connected to adsorbed NO, thus keeping its clean-surface core-level shift (see atoms marked by "S" in Figure 38.25b). The adsorbed NO molecules are found in hollow sites, yielding Pt atoms bonded to a single NO molecule (marked as "$H_1$" in Figure 38.25). For 0.5 ML of NO, the NO on-top position becomes occupied, resulting in the appearance of the respective Pt 4f component "T" in panel (c), while no unperturbed Pt surface atoms ("S") are present anymore. Finally, at 0.75 ML NO

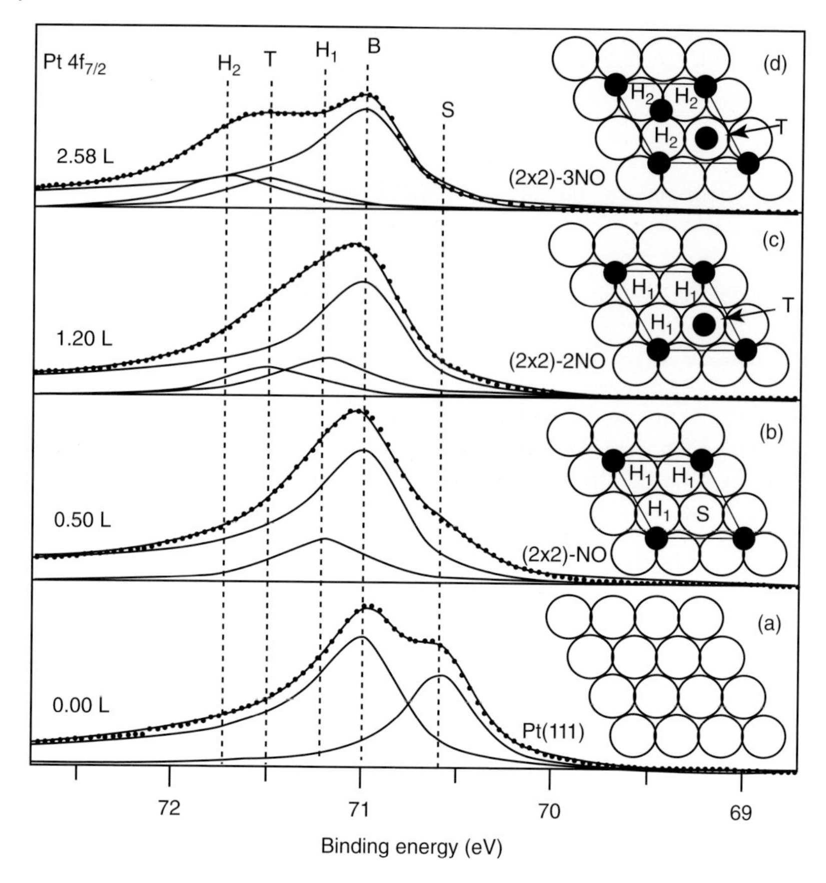


**Figure 38.25** Pt 4f spectra for various exposures of NO on Pt(111). The insets show the proposed surface structure derived from the identification of adsorption sites from the Pt 4f data and O 1s spectra. "S" marks unperturbed Pt surface atoms while "T," "$H_1$," and "$H_2$" mark Pt atoms connected to on-top-bonded, one hollow-bonded, or two hollow-bonded NO molecules, respectively. (Reprinted from Ref. [70], Copyright (2003), with permission from Elsevier.)

coverage, NO molecules bound in adjacent hollow sites yield Pt surface atoms now bonded to two NO molecules (marked as "$H_2$" in panel (d)). So a clear distinction of the surface structures is possible by XPS investigations [70].

Some adsorbate systems have been determined recently using STM. As a word of caution, one has to be careful that a certain STM image showing an ordered surface structure only displays the structure in a very limited surface area. In most cases, additional methods (like LEED) need to be employed to ensure a wide range of reproducibility. On the other hand, an advantage of STM is the possibility to measure *in situ* in the presence of suitable or necessary gas pressures (see, e.g., [2] and comment or [72]).

One of the first accounts of imaging molecules on metal surfaces by STM was the coadsorption system benzene + CO on Rh(111) [73]. While in this particular system the CO molecules could not be observed, the benzene ring could clearly be identified. This can be seen in Figure 38.26. Interestingly, the relation to the Rh substrate was already determined to be such that the benzene molecules seemed to be bonding not in on-top sites. In addition, a clear threefold symmetry signals that not all six carbon atoms have the same electron density. This fits very nicely to recent results by XPS on benzene adsorption on Ni(111). There, also shifts in the C 1s core-level binding energies were observed, being related to certain adsorption sites [50]. The early images were somewhat blurred by today's standards, but imaging molecules without destroying or moving them is still not easy. More recent examples can be found in a review article by Chiang [74]. The most promising way to study adsorption systems in great detail is the combination of STM and DFT simulations of the images, as, for example, in the work of Simic-Milosevic *et al.* [75].

Another example can be again CO on Pt(111). In Figure 38.27, different surface structures as a function of CO coverage are shown. Already in these small areas the limited range of some structures is obvious. For example, the $\sqrt{3} \times \sqrt{3}$ structure is usually not observed in LEED, as the diffraction technique requires a larger ordered area. In this study, only ultrahigh vacuum (UHV) CO adsorption was studied, resulting in maximum CO coverages only slightly above 0.5 ML [76].

Longwitz *et al.* [72] have presented STM images obtained at CO pressures as high as 720 Torr. At this condition, a commensurate structure with a CO coverage of 0.7 ML emerged, showing itself in a typical Moire pattern from interference

**Figure 38.26** STM image of a (3 × 3) surface structure caused by coadsorption of benzene and CO. Only the CO molecules are imaged. The superimposed mesh derived from a LEED study show that the threefold symmetric lobes are not placed on top of a surface atom. $V_{bias} < 0.5$ V. (Reprinted with permission from Ref. [73]. Copyright (1988) by the American Physical Society.)

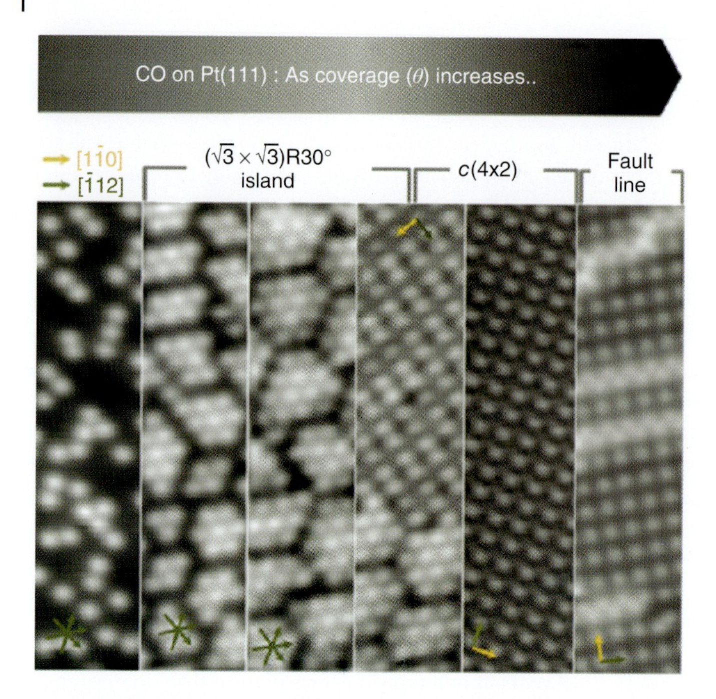


**Figure 38.27** STM images of CO adsorption situations on Pt(111) for increasing CO coverage. High-symmetry directions are marked by colored arrows. (Reprinted with permission from Ref. [76]. Copyright (2013) American Chemical Society.)

between the hexagonal Pt surface and the hexagonal CO structure, being described as $(\sqrt{19} \times \sqrt{19})$ R23.4°-13CO.

## 38.5
## Electronic Structure, Bonding, and Molecular Orientation

In this section, we address the electronic structure and orientation of small adsorbates. We will mostly concentrate on the results obtained by angle-resolved UV photoelectron spectroscopy (ARUPS), closely following review articles by one of the authors [77–79]. The discussed molecules range from CO to the small and highly symmetric hydrocarbon molecules ethylene and benzene, all adsorbed on transition-metal surfaces. The restriction to those molecules is motivated by several reasons: (i) The chosen molecules are considered model systems for the adsorption on single-crystal metal surfaces and often serve as building blocks for larger molecules. A detailed understanding of their adsorption behavior is thus essential when addressing more complicated systems. (ii) All of them are highly symmetric and typically bind in well-defined adsorption geometries on the surface, which permits the investigation of the adsorbate/substrate interaction in great detail. One specifically favorable case is the adsorption on surfaces with twofold symmetry, such as the fcc(110) surfaces, from which detailed information on the molecular orientation and the symmetry of the adsorption complex "adsorbate + substrate" can be derived. (iii) For these molecules, a very detailed and fundamental understanding of the electronic structure, specifically also of lateral interactions in densely packed layers, has been deduced. (iv) They belong to the best studied systems concerning the formation of adsorbate band structures. (v) For some of the molecules, model cluster calculations, slab calculations, force field calculation, or tight-binding band structure calculations are available, which have helped obtaining a deeper understanding.

ARUPS is a well-established and powerful technique to study the electronic structure of adsorbates on solid surfaces [77–93]. It provides information on the electronic states in the outer valence region and thereby on the modifications induced by the physical and/or chemical bond between adsorbate and substrate (see Chapter 3.2, Volume 1 and Chapter 5, Volume 2). By comparing the energetic positions of the electronic levels of the adsorbed species with the corresponding values for the free molecule, direct and detailed information on the nature of the adsorbate/substrate bond and on the chemical identity of the adsorbate can be obtained. Furthermore, from the polarization dependence and the angular dependence of the valence photoemission signals, the orientation of a molecule and the symmetry of the adsorption complex can be deduced.

### 38.5.1
### Electronic Structure, Bonding, and Energetic Considerations

The valence electronic structure of a free or an adsorbed molecule is a more or less unique fingerprint of its chemical state. For adsorbed molecules, one should therefore – at least in principle – be able to identify the chemical identity of an adsorbed species from the ARUPS spectrum and decide whether a molecule binds intact to the surface or decomposes during the adsorption process. From the binding energies of the electronic levels, and in particular from the changes that occur for individual levels upon adsorption, detailed information on the electronic structure of the adsorbate and on the chemical bond between adsorbate and substrate can be deduced.

To illustrate such an analysis, we discuss the changes in the UPS spectra that are observed upon adsorption of a molecule on a surface, using the angle-integrated spectra of gaseous and chemisorbed benzene, as shown in Figure 38.28 [77–79]. The upper spectrum corresponds to the free molecule, with the energetic positions of all valence levels and their assignment indicated. For several peaks, a vibrational fine structure is resolved, which is attributed to vibrational excitations in the ionic final state according to the Franck–Condon principle. Upon adsorption, significant changes are observed, as is evident from the bottom of Figure 38.28, where the UPS spectrum of benzene chemisorbed on Ni(111) is shown. (i) The most evident one is the fact that the spectrum now contains the modified levels of the free molecule plus the signal from the substrate. The substrate-induced intensity is composed of direct emission of the substrate valence bands and secondary electrons due to

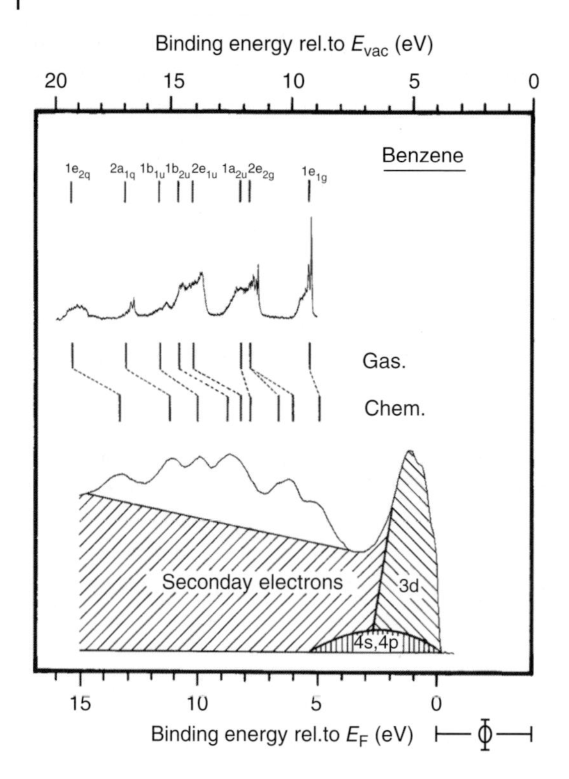


**Figure 38.28** Schematic illustration of the changes in the photoelectron spectra that occur upon chemisorption [77–79]. (Top) UPS spectrum of gaseous benzene and (bottom) UPS spectrum of benzene chemisorbed on Ni(111). Φ: work function of the adsorbate covered surface. (Reproduced from Ref. [77].)

inelastic scattering off the substrate or the adsorbate layer. (ii) The adsorbate levels are significantly broadened because of the initial and final state effects, a common observation for adsorbates on surfaces. The initial state broadening is due to the coupling to the substrate bands [94], and the final state effects result from inelastic processes such as electron–hole pairs and phonon excitations, a reduced lifetime of the excitation due to the coupling to the substrate, and inhomogeneous screening of the ionic final state [95–102]. (iii) In contrast to free atoms and molecules, where the binding energies are referenced to the vacuum level, for adsorbed species they are usually given with respect to the Fermi level. The energy difference between the vacuum level and the Fermi level is the work function, Φ, of the adsorbate-covered substrate. (iv) In comparison to the free molecule, the adsorbate levels undergo a more or less uniform relaxation shift, $E_{Rel}$, toward lower binding energies, which is mainly attributed to the extramolecular screening (final state effect) within the adsorbate layer and by the substrate [84, 95, 103]. (v) For chemisorbed molecules, an additional differential shift of one or several molecular levels toward higher binding energies (i.e., in the opposite direction of the relaxation shift) is observed. This shift reflects the participation of the corresponding orbitals in the chemical bond to the substrate (initial state effect) and is referred to as the *bonding shift*, $E_{Bond}$ [77–79, 84, 95, 102–105].

To a first approximation, the binding energy of electrons in a particular adsorbate molecular level is then given by

$$E_B(\mathrm{ad}) = E_B(\mathrm{gas}) - E_{Rel} + E_{Bond} - \Phi. \tag{38.1}$$

The separation of relaxation shifts and bonding shifts is difficult. In most cases, it is assumed that the relaxation shifts for those orbitals involved in the bond to the substrate and for those unaffected by this interaction are identical. Using this approach, for benzene chemisorbed on metal surfaces, the relaxation shift $E_{Rel}$ is typically between 1.5 and 2.2 eV, and bonding shifts $E_{Bond}$ of 1.0–1.5 eV are observed for the two out-of-plane $\pi$ orbitals ($1a_{2u}$ and $1e_{1g}$), which are made responsible for the bonding interaction to the substrate [106]. The assumption of identical relaxation shifts for all orbitals should, however, be treated with some caution. The change of the intramolecular bond length due to the chemisorptive bond to the substrate can induce differential shifts of molecular orbitals that are not primarily involved in the adsorbate substrate bond. This effect is demonstrated by cluster calculations (see, e.g., Refs [105, 107, 108]), which, indeed, reveal small differential shifts ("chemical shifts") up to several tenths of an electronvolt for those orbitals. One other aspect to be considered when comparing the binding energies $E_B$ of particular adsorbate levels to the corresponding levels of the free molecule is the formation of adsorbate band structures, that is, a dependence of the binding energy $E_B$ on the momentum $\mathbf{k}$ of the emitted electron. Dispersion of electronic states is typically observed for densely packed layers, where the orbitals of neighboring molecules overlap, and the corresponding band width, that is, variation of the binding energy, can be as large as 2 eV for hydrocarbon molecules (see Section 38.5.3), but it is very often neglected in the literature. For a meaningful comparison with the spectrum of the free molecule, an adequate integration over the adsorbate surface Brillouin zone (SBZ) has to be performed.

### 38.5.2 Orientation and Symmetry Selection Rules

The knowledge about the orientation of a molecule with respect to the substrate and the symmetry of the adsorption complex adsorbate + substrate is essential for understanding the bonding interaction of a molecule with a surface. The symmetry of the adsorption complex is in many cases lower than the symmetry of the free molecule. Even for a flat adsorption geometry of a planar molecule, the molecular plane is not a symmetry plane any more because of the presence of the surface. For a single molecule on the surface, that is, in the limit of zero coverage, the interaction of a molecule with the substrate, the so-called adsorbate/substrate interaction, is determined by the electronic structure of both partners, and leads to a well-defined adsorption geometry. For higher adsorbate coverages, lateral interactions within the adsorbate layer have to be taken into account, and the particular arrangement of the molecules is dictated by the interplay between adsorbate/substrate interaction and adsorbate/adsorbate interaction. In many cases, the bonding to the substrate is significantly stronger than intermolecular interactions; nevertheless, since the differences in binding energy between possible adsorption geometries are generally much smaller than the absolute binding energy, the properties of densely packed adsorbate layers can be strongly influenced by sometimes much weaker lateral (steric) interactions, which can lead to changes in adsorption sites, bond lengths, as well as orientation and symmetry, with increasing coverage.

In order to obtain a most detailed understanding of the "vertical" interaction of a molecule with the surface, that is, the adsorbate/substrate interaction, one would ideally investigate the properties of an isolated molecule on this substrate. However, this situation of very low coverages cannot be realized because of the low intensity of the adsorbate-derived levels in ARUPS. As a compromise, studies on dilute adsorbate layers are performed. Typically, coverages from 20% to 50% of the saturated chemisorbed layer are investigated, since measurements at lower coverages are very difficult to assess because of the low signal-to-noise ratio. In order to demonstrate that lateral interactions are indeed negligible or at least are only very small, one then has to verify that no island formation occurs (e.g., by LEED or STM), and/or that no adsorbate band structure is formed (by ARUPS). The measurements on such dilute layers serve as the starting point for DFT calculations or quantum chemical cluster calculations that provide more insight in the nature of the adsorbate/substrate interaction.

To study the influence of adsorbate/adsorbate interactions, typically well-ordered saturated layers are investigated. This guarantees significant lateral interactions but also a well-defined arrangement. Complementary information can be derived from force field calculations that allow the determination of the repulsive energy per molecule in a densely packed layer and finding the minimum energy geometry. Despite the fact that these calculations are often performed for unsupported layers, they nevertheless have proven to be very helpful for the simulation of adsorbate structures.

The electronic structure, orientation, and symmetry of occupied adsorbate valence states can be determined from ARUPS. In the dipole approximation, the differential cross section for the photoemission process is given by Fermi's Golden rule:

$$d\sigma/d\Omega \approx | < \Phi_f|\mu|\Phi_i > |^2 \delta(E_f - E_i - h\omega). \tag{38.2}$$

In the single-particle approximation, the initial state $|\Phi_i>$ refers to the bound electron in a particular orbital, the final state $<\Phi_f|$ to the emitted electron, and $\mu = \mathbf{A_o}.\mathbf{p}$ to the dipole operator; the delta function ensures energy conservation [77–79, 84–91]. In contrast to free molecules, where no specific orientation exists, adsorbed molecules often have a well-defined orientation on the surface. In this situation, all components of the matrix element $<\Phi_f|\mu|\Phi_i>$ are well defined in space. The final state $<\Phi_f|$ is uniquely specified by the kinetic energy $E_{Kin}$ and the momentum **k** of the outgoing electron, as determined by an angle-resolving electron spectrometer. The initial state $|\Phi_i>$ is given by the particular "oriented" orbital in the oriented molecule, and the dipole operator $\mu$ is defined by the polarization of the incoming photon beam. As a consequence, the differential cross section for particular adsorbate levels can be investigated in great detail by polarization-dependent ARUPS measurement.

Principally, information on the orientation of adsorbed molecules could be obtained from the comparison of experimental data with corresponding theoretical calculations of the differential photoionization cross section for the adsorbate levels of interest, at the appropriate photon energy. However, such calculations become very complicated and challenging if the surface and the changes induced by the chemical bond are properly taken into account [109–111]. An alternative and extremely successful approach is the analysis of ARUPS data using the so-called symmetry selection rules [77–79, 83, 84, 112, 113]. Thereby, detailed information on the orientation and symmetry of adsorbed molecules can be deduced.

Symmetry selection rules follow from the fact that, upon photoionization, not only energy but also angular momentum has to be conserved. This leads, for example, to the commonly known dipole selection rule $\Delta l = \pm 1$ for photon absorption or photon emission (fluorescence) in atoms, accounting for the change of the orbital momentum to compensate the spin of the incoming or emitted photon, respectively. In the more generalized form, these rules are derived from the fact that, in order to be nonzero, the dipole matrix element in Equation 38.2

$$M_{\mathrm{if}} = < \Phi_{\mathrm{f}} | \mu | \Phi_{\mathrm{i}} > \quad (38.3)$$

must be totally symmetric, or (for point groups with degenerate representations) at least contain a totally symmetric component [77–79, 83, 84, 91, 112, 113]. The dipole operator $\mu$ transforms as the Cartesian axes $x$, $y$, and $z$ of the point group of the system. The representation of $\mu$ is thus defined by the Cartesian components of the vector potential **A** (or, alternatively the electric field vector **E**) of the incoming radiation. In the following, **E** will be used to denote the polarization of the incoming radiation.

The requirement of a totally symmetric dipole matrix has the following consequence: for a given polarization **E**, excitations from an initial state $\Phi_{\mathrm{i}}$ are allowed only to final states $\Phi_{\mathrm{f}}$ of particular symmetry. In the framework of group theory, one can show that the symmetry of the final states (i.e., of the wave function of the emitted electrons in photoemission), for which $M_{\mathrm{if}} \neq 0$ holds, is given by the direct product of the representations of the initial state and the dipole operator:

$$\Phi_{\mathrm{i}} \times \mathbf{E} = \Phi_{\mathrm{f}}. \quad (38.4)$$

This simply follows from the fact that only the integral of the product of a specific wave function (representation) with itself yields a nonzero value. Based on symmetry selection rules, one can therefore predict whether emission from a specific orbital $\Phi_{\mathrm{i}}$ is allowed or forbidden for a particular detector position and a given polarization. These rules are especially powerful if the polarization vector (i.e., the electric field vector of the incoming light) and/or detector are positioned in a high-symmetry direction, such as the surface normal or mirror planes, of the system. The consequences of these symmetry selection rules for two examples are denoted in Table 38.1 [77, 114] and in Figure 38.29 [79]. Table 38.1 shows the prediction of

**Table 38.1** Symmetry selection rules for perpendicularly oriented CO [114].

| Initial state | Polarization | ⇒ | Final state |
|---|---|---|---|
| $\sigma$ | $\sigma(E_z)$ | ⇒ | $\sigma$ (normal emission) |
| | $\pi(E_x, E_y)$ | ⇒ | $\pi$ (no normal emission) |
| $\pi$ | $\sigma(E_z)$ | ⇒ | $\pi$ (no normal emission) |
| | $\pi(E_x, E_y)$ | ⇒ | $\sigma + \delta$ (normal emission) |

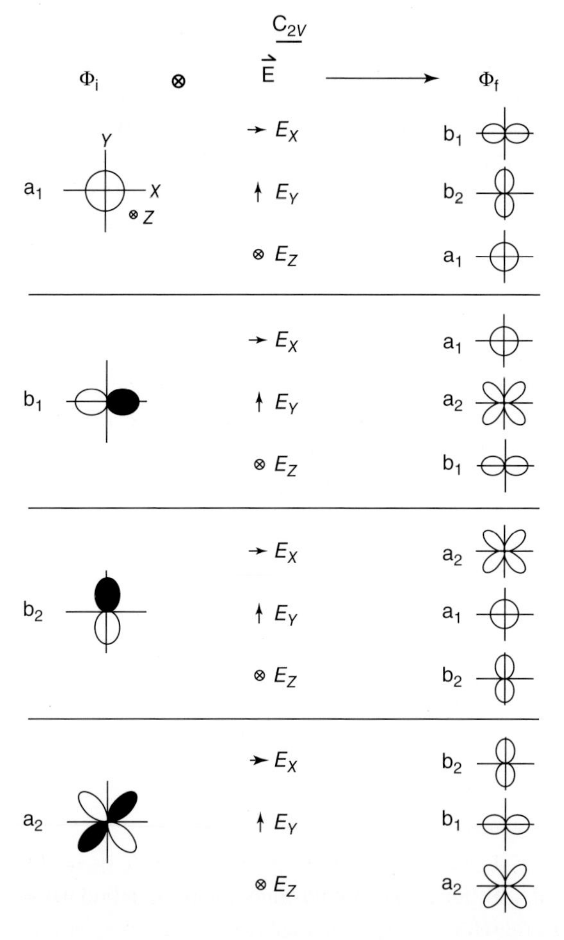


**Figure 38.29** Schematic illustration of the predictions of symmetry selections rules for the point group $C_{2v}$ (Ref. [77–79]). The excitation from a particular initial state $\Phi_i$ (left) with a well-defined polarization **E** (center) leads to a specific final state $\Phi_f$ (right), according to Equation 38.4. $a_1$ final states are allowed at any detector position; $b_1$ ($b_2$) final states are forbidden in normal emission and in the *yz* (*xz*) mirror plane (nodal planes of the final state) but are allowed in the *xz* (*yz*) mirror plane; and $a_2$ final states are forbidden in normal emission and in both mirror planes. (Reproduced from Ref. [77].)

a simple two-atom molecule such as CO or NO, whereas Figure 38.29 schematically illustrates the general behavior for an adsorption complex with $C_{2v}$ symmetry. The application of these rules will be explained when discussing the corresponding examples below.

Symmetry selection rules in many cases enable the unequivocal assignment of peaks in ARUPS spectra to specific orbitals or bands. If, on the other hand, the peaks in a spectrum are already assigned to specific molecular orbitals, symmetry selection rules allow obtaining the orientation and symmetry of the adsorbate. Note that often the peak assignment as well as the determination of the molecular orientation is an iterative process, taking into account the information of all measured molecular levels. One important note of caution is to be made, namely that symmetry selection rules do not allow any predictions about the intensity of a particular final state: even if a final state is allowed by symmetry selection rules, it can show vanishing intensity in the experiment due to cross-sectional effects. This ambiguity is usually overcome by measurements at various photon energies.

In the following, a number of case studies are discussed; for details, see the references given in the headlines.

#### 38.5.2.1 **CO/Ni(100)**

As CO is considered the Drosophila of surface science, we discuss this system as our first example, and demonstrate that on Ni(100) the molecule is adsorbed with its molecular axis along the surface normal [77, 114]. CO belongs to the symmetry class $C_{\infty v}$, and the molecular orbitals are thus classified according to $\sigma$ and $\pi$ symmetry, with respect to the molecular axis. In Figure 38.30, ARUPS of CO on Ni(100) are depicted, which were collected at normal emission ($\theta = 0°$) for different light polarizations. Upon adsorption, the $5\sigma$ Ievel, which is the highest occupied orbital of the free molecule, suffers a differential shift of 2.6 eV to higher binding energies as a result of the strong bonding interaction with the substrate; this shift indicates an adsorption geometry in which the molecules bind to the substrate via the carbon atom. Because of its bonding shift, the $5\sigma$ Ievel is nearly degenerate with the $1\pi$ level; this is evident from the comparison with the gas-phase peaks, which are indicated as vertical bars at the top of the figure. Note that the overall shift to lower binding energy of the UPS peaks for the adsorbate (by ~2.8 eV) is due to relaxation effects that occur on the surface mostly via final state screening of the hole by the metal electrons. The spectra show a pronounced dependence on the emission angle. For the $4\sigma$ level at $E_B = 11.0$ eV, no normal emission is observed if the electric field vector (polarization) is aligned parallel to the surface ($E_x$, $E_y$), which is achieved at normal incidence. On the other hand, strong emission is found when the polarization has a component perpendicular to the surface ($E_z$). From this behavior, one can directly deduce that the molecule is oriented with its molecular axis along the surface normal. In this geometry, normal emission corresponds to a totally symmetric ($\sigma$) final state, and emission from a totally symmetric ($\sigma$) initial state is possible only with a symmetric $\sigma$ component (that is an $E_z$ component) of the electric field vector **E** (see Table 38.1). Since the behavior of the $4\sigma$ and the $5\sigma$ orbitals must be identical, we can assign the peak at $E_B = 7.8$ eV in the upper spectrum to the CO $1\pi$ level, whereas

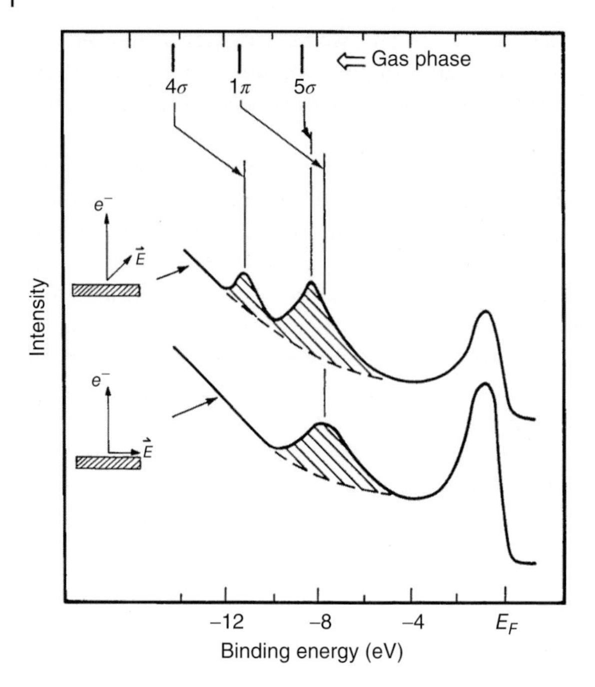


**Figure 38.30** Angle-resolved UPS spectra of CO adsorbed on Ni(100) collected at normal emission with different light polarizations [114]. (a) **E** parallel to the surface. (b) Parallel and perpendicular component of **E**. (Reproduced from Ref. [77].)

the peak in the upper spectrum at $E_B = 8.3$ eV results from the $5\sigma$ orbital. It is important to note the unambiguous assignment of the molecular orientation of CO was possible only based on selection rules. As a further example, the system CO/Pd(111) was discussed in Chapter 3.2, Volume 1 [115].

#### 38.5.2.2 **Ethylene/Ni(110)**

As next example, we discuss the adsorption of ethylene on Ni(111) [77–79, 107, 116] (see also Chapter 40 in this Volume). Ethylene ($C_2H_4$) is a small and symmetric unsaturated hydrocarbon molecule that chemisorbs on transition-metal surfaces. The symmetry of the ethylene molecule is $D_{2h}$. In Figure 38.31, a schematic drawing of the molecule with its symmetry elements is depicted; also shown is an overview of the spatial extension of the ethylene valence orbitals. Upon adsorption, the symmetry is lowered as a result of the asymmetry of the substrate–molecule–vacuum system. The highest possible symmetry is $C_{2v}$; however, the specific adsorbate/substrate interaction can lower the symmetry even further: for $C_2$ symmetry, the only remaining symmetry element is a twofold rotation axis $C_2$, and for $C_s(\sigma_{xz})$ or $C_s(\sigma_{yz})$ symmetry, the remaining symmetry elements are the $\sigma_{xz}$ or the $\sigma_{yz}$ mirror planes, respectively. A lower symmetry leads to modified (relaxed) symmetry selection rules; this is illustrated in Table 38.2, where the correlations between molecular orbitals for various possible point groups are summarized along with the corresponding symmetry selection rules.

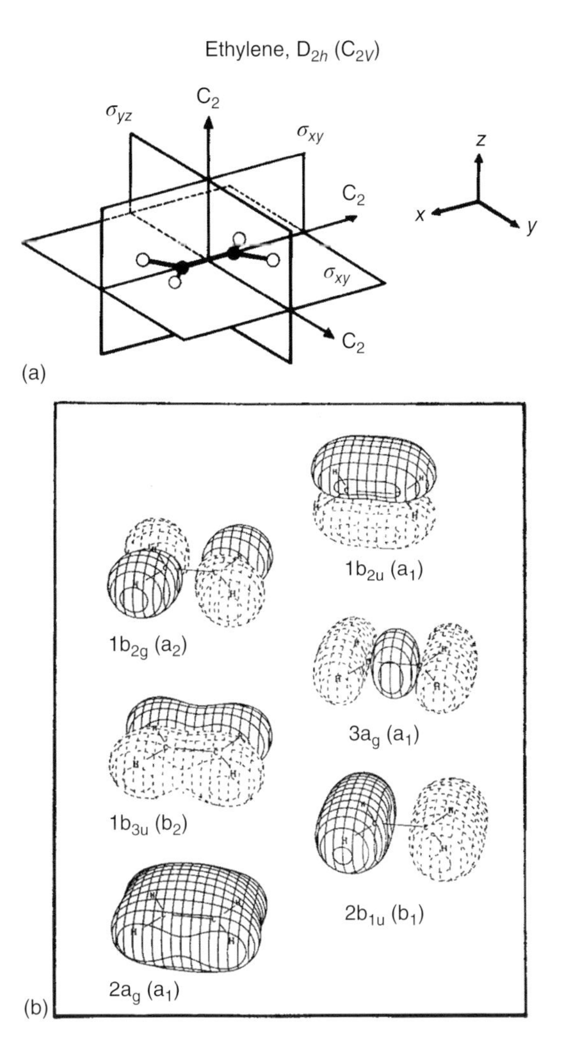


**Figure 38.31** (a) Schematic drawing of ethylene with its symmetry elements. (b) Molecular orbitals of ethylene [116]. (Reproduced from Ref. [79].)

**38.5.2.2.1 Dilute Ethylene Layer** First, we are interested in the adsorbate/substrate interaction; therefore, we discuss ARUPS measurements of a dilute ethylene layer on Ni(110) with a coverage of 0.25 ML, that is, one half of the saturation coverage. The corresponding layer was prepared by the appropriate exposure to ethylene at temperatures below 120 K. A selection of ARUPS spectra is depicted in Figure 38.32. The spectra have been collected in various experimental geometries, with the electric field vector **E** aligned parallel to the close-packed substrate rows ($[1\bar{1}0]$ azimuth) or perpendicular to them ([001] azimuth), and the plane of detection D parallel to **E** (D || **E**, allowed geometry). For normal incidence ($\alpha = 0°$, only $E_x$ or $E_y$), spectra are shown for emission angles $\vartheta$ of 0° (normal emission), 30°, and 60°; for $\alpha = 45°$

**Table 38.2** Symmetry selection rules for several molecular symmetries of ethylene at various experimental geometries.

| $D_{2h}$ | $C_{2v}$ | $E_xD_x$ | $E_yD_y$ | $E_xD_y$ | $E_yD_x$ | $E_zD_x$ | $E_zD_y$ |
|---|---|---|---|---|---|---|---|
| $b_{2u}, a_g$ | $a_1$ | + | + | − | − | N | N |
| $b_{2g}$ | $a_2$ | − | − | + | + | − | − |
| $b_{1u}$ | $b_1$ | N | − | N | − | + | − |
| $b_{3u}$ | $b_2$ | − | N | − | N | − | + |

| $D_{2h}$ | $C_2$ | $E_xD_x$ | $E_yD_y$ | $E_xD_y$ | $E_yD_x$ | $E_zD_x$ | $E_zD_y$ |
|---|---|---|---|---|---|---|---|
| $b_{2u}, a_g$ | a | + | + | + | + | N | N |
| $b_{2g}$ | a | + | + | + | + | N | N |
| $b_{1u}$ | b | N | N | N | N | + | + |
| $b_{3u}$ | b | N | N | N | N | + | + |

| $D_{6h}$ | $C_s(\sigma_{xz})$ | $E_xD_x$ | $E_yD_y$ | $E_xD_y$ | $E_yD_x$ | $E_zD_x$ | $E_zD_y$ |
|---|---|---|---|---|---|---|---|
| $b_{2u}, a_g$ | a′ | N | + | N | − | N | N |
| $b_{2g}$ | a″ | − | N | + | N | − | + |
| $b_{1u}$ | a′ | N | + | N | − | N | N |
| $b_{3u}$ | a″ | − | N | + | N | − | + |

| $D_{6h}$ | $C_s(\sigma_{yz})$ | $E_xD_x$ | $E_yD_y$ | $E_xD_y$ | $E_yD_x$ | $E_zD_x$ | $E_zD_y$ |
|---|---|---|---|---|---|---|---|
| $b_{2u}, a_g$ | a′ | + | N | − | N | N | N |
| $b_{2g}$ | a″ | N | − | N | + | + | − |
| $b_{1u}$ | a″ | N | − | N | + | + | − |
| $b_{3u}$ | a′ | + | N | − | N | N | N |

$E_x$, $E_y$, and $E_z$ denote the Cartesian components of the electric field vector of the incoming light with respect to the surface plane and a flat lying ethylene molecule (see Figure 38.31a); $D_x$ and $D_y$ characterize the alignment of the detection plane. "N" indicates that a band is allowed in normal emission ($\theta'=0°$) and at all other angles, "+" and "−" indicate that emission in the detection plane ($\theta' \neq 0°$) is allowed or forbidden, respectively [107].

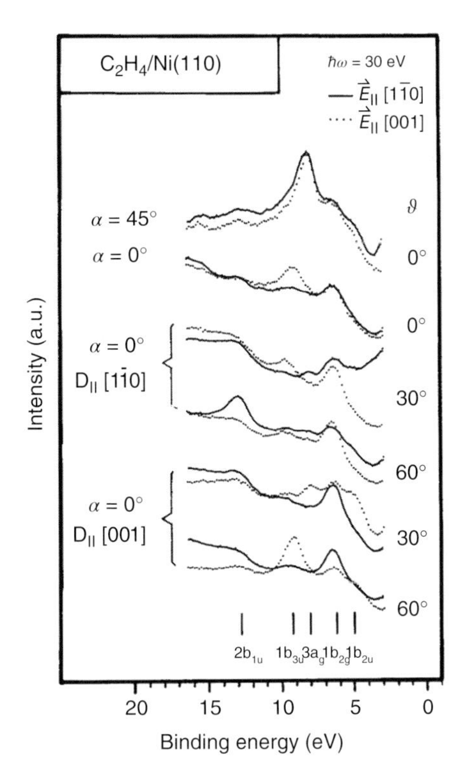


**Figure 38.32** Angle-resolved UPS spectra of the dilute ethylene layer ($\theta = 0.25$) on Ni(110) at different geometries, collected at a photon energy of 30 eV [107, 116]. Orbital positions and assignment are denoted as a bar diagram. D indicates the plane of detection, **E** the orientation of the electric field vector. $\alpha$ and $\vartheta$ are the photon angle of incidence and the electron emission angle, respectively, with respect to the surface normal. (Reproduced from Ref. [78].)

($E_{x+z}$ or $E_{y+z}$), only spectra for $\vartheta = 0°$ are shown. The comparison of the energetic position of the various levels with the corresponding values for the free molecule [118] reveals a bonding shift of 1.1 eV of the $\pi$ orbital ($1b_{2u}$) to higher binding energy due to the chemisorptive bond to the substrate. Note that the binding energies of all levels are independent of the azimuth and polar angle of emission within $\pm 0.1$ eV (see also Figure 38.33a), confirming that lateral interactions are indeed not important; thus the obtained results are representative of the interaction of the individual molecule with the surface.

In the following, we derive the orientation and symmetry of ethylene on Ni(110) from the distinct polarization, as well as the azimuthal and polar angle dependencies of the valence levels by applying symmetry selection rules. The orientation of the molecular plane parallel to the surface is deduced from the behavior of the $3a_g$ and $1b_{2u}$ orbitals at normal emission ($\vartheta = 0°$). In Figure 38.32, both show significant intensity for $\alpha = 45°$ ($E_z \neq 0$), but have vanished for $\alpha = 0°$ ($E_z = 0$). Since both orbitals belong to the totally symmetric $a_1$ (a) representation for $C_{2v}$

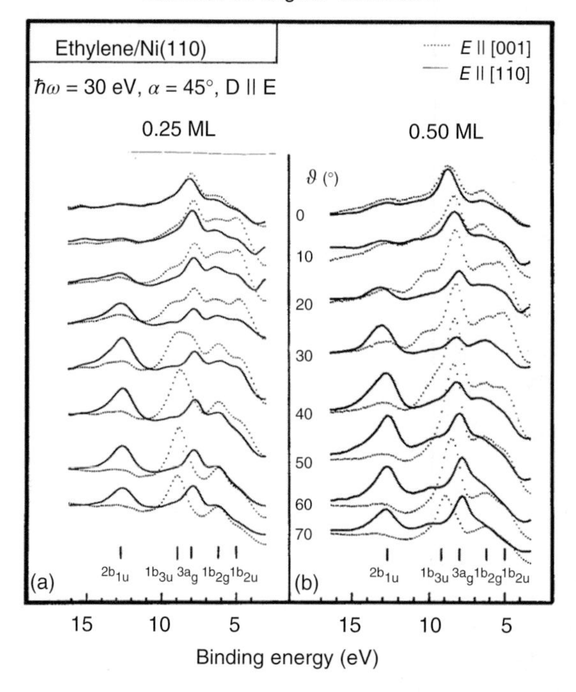


**Figure 38.33** Angle-resolved UPS spectra of ethylene on Ni(110) for increasing polar angles $\vartheta$ at an incidence angle of $\alpha = 45°$ and a photon energy of 30 eV. (a) Dilute layer, $\theta = 0.25$ ML [107]. (b) Saturated c(2 × 4) layer, $\theta = 0.50$ ML [79]. (Reproduced from Ref. [79].)

($C_2$) symmetry, the C–C axis must be parallel to the surface. Information on the azimuthal orientation is derived from the behavior of the $1b_{3u}$ and $2b_{1u}$ orbitals (in Figure 38.33a), which belong to the $b_2$ and $b_1$ representations, respectively, for a flat-lying ethylene molecule with $C_{2v}$ symmetry. The $1b_{3u}$ orbital shows strong emission for **E**||[001] and weak emission for **E**||[1$\bar{1}$0]; the $2b_{1u}$ orbital shows the opposite behavior. By comparison with the predictions from symmetry selection rules for $C_{2v}$ symmetry (see Table 38.2), we conclude that the C–C axis is preferentially aligned with the densely packed substrate rows, that is, the [1$\bar{1}$0] azimuth. While the behavior of the $1b_{2u}$, $3a_g$, $1b_{3u}$, and $2b_{1u}$ orbitals is consistent with $C_{2v}$ symmetry, the nonzero normal emission ($\vartheta = 0°$) from the $1b_{2g}$ orbital ($a_2$ in $C_{2v}$) in all geometries suggests that the symmetry of the adsorption complex is only $C_1$, probably due to a low-symmetry adsorption site and/or a twisting of the adsorbed molecule. In Figure 38.34a, a schematic drawing of the adsorption geometry (with $C_{2v}$ symmetry) of ethylene on Ni(110) is shown.

**38.5.2.2.2 Saturated c(2 × 4) Ethylene Layer** The saturated ethylene layer is well ordered and exhibits a c(2 × 4) LEED pattern with the (0, $\pm 1/2(2n+1)$) order spots missing for normal incidence. The saturation coverage is 0.5 ML, indicating a nonprimitive unit cell containing two ethylene molecules. In Figure 38.33b, the

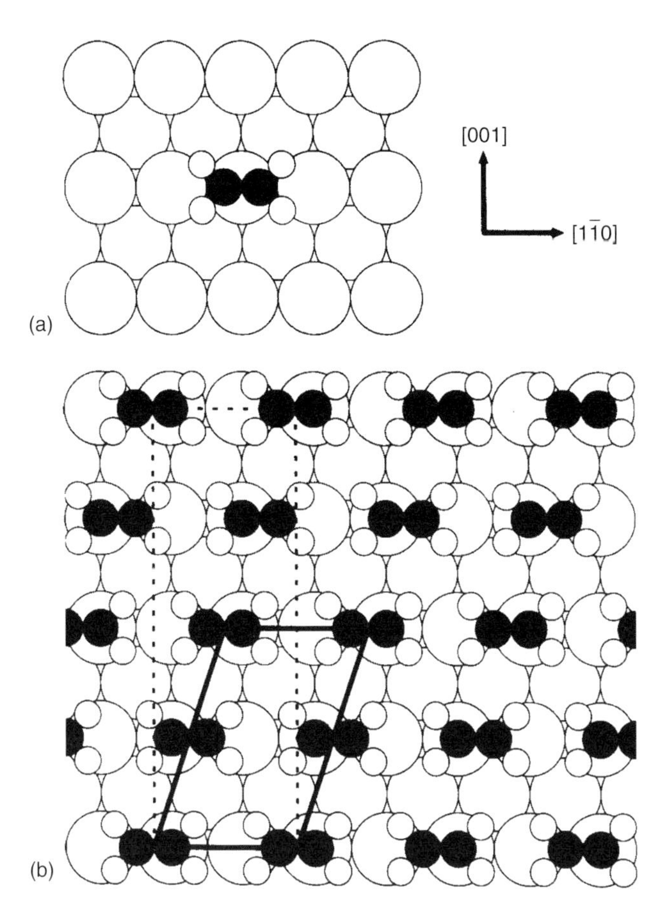


**Figure 38.34** (a) Schematic drawing of the proposed orientation of a single ethylene molecule in the dilute layer on Ni(110) [107], with $C_{2v}$ symmetry. Note that the adsorption site has been chosen arbitrarily. (b) Structural model for the densely packed c(2 × 4) ethylene layer on Ni(110) as determined from LEED, ARUPS, NEXAFS, and force field calculations [79]. (Reproduced from Ref. [79].)

data of the saturated c(2 × 4) ethylene layer are shown. The spectra show essentially the same emission characteristics as those of the dilute layer in Figure 38.33a, which indicates a coverage-independent orientation. However, there is one major difference between the saturated and the dilute layer, namely the significant dispersion (variation of the binding energy with emission angle) of up to 2 eV of various levels for the saturated layer; this dispersion indicates the formation of a two-dimensional (2D) band structure (see Section 38.5.3). From the c(2 × 4) structure and the well-defined azimuthal orientation with the C–C axis aligned parallel to the substrate rows, a compressed real-space structure is deduced, which is schematically depicted in Figure 38.34b. In this structure, the molecules are adsorbed in dense rows along the [1$\bar{1}$0] direction, with every second substrate site being occupied. Neighboring adsorbate rows are shifted relative to each other by half a substrate lattice vector along $[1\bar{1}0]$, and the two molecules in the unit cell are related by a $C_2$ point transformation. Note that, in this structure, translationally equivalent adsorption sites are occupied only in every second row, entailing a nonprimitive oblique unit cell for the c(2 × 4) structure. This structural model was confirmed by near-edge X-ray absorption fine structure (NEXAFS) measurements and by force field calculations [116].

#### 38.5.2.3 Benzene/Ni(110)

Benzene also is a highly symmetric molecule and its adsorption has been studied extensively. As an unsaturated hydrocarbon with a reactive $\pi$ system, it typically chemisorbs on transition-metal surfaces, with few exceptions in a flat adsorption geometry. In the following, we will discuss the behavior of benzene on Ni(110) [77–79, 106, 119]. The free benzene molecule belongs to the $D_{6h}$ point group. A schematic drawing of the molecule is given in Figure 38.35, along with an overview of the spatial extension of its molecular orbitals [120]. Upon adsorption, the symmetry is reduced to $C_{6v}$ or lower depending on the particular interaction with the substrate. In Table 38.3, the correlations between possible point groups are summarized along with the corresponding symmetry selection rules.

**38.5.2.3.1 Dilute Benzene Layer** Also for this example, we start with a dilute layer. In Figure 38.36a, a selected set of ARUPS spectra for the dilute benzene ($C_6D_6$) layer on Ni(110) is shown; its coverage of 0.10 ML corresponds to 40% of the saturation coverage of the chemisorbed layer. This layer was prepared by exposing the Ni(110) surface to 0.7 L $C_6D_6$ at 110 K. From the comparison with the spectrum of the free molecule [118], differential shifts of the two $\pi$ orbitals ($1a_{2u}$, $1e_{1g}$) by ~1.1 eV toward higher binding energies are deduced, which is due to the chemisorptive bonding to the substrate. The different levels show pronounced dependencies on polarization and emission angle but no dispersion within ±0.1 eV, which indicates that lateral interactions are not important. From the data in Figure 38.36a, we conclude that benzene is oriented with its molecular plane parallel to the surface and that two of the C–H bonds are aligned along the [001] azimuths; this arrangement is schematically shown in Figure 38.37a. The parallel orientation of the molecular plane follows from the vanishing normal emission ($\vartheta = 0°$) of the totally symmetric $2a_{1g}$ and $1a_{2u}$ levels for $\alpha = 0°$ (normal incidence, only $E_x$ or $E_y$) as compared to the very strong normal emission for $\alpha = 45°$ ($E_z \neq 0$). The azimuthal orientation follows from the emission behavior of the b-type orbitals for $\alpha = 45°$: the $1b_{1u}$ orbital shows strong emission for **E** || [001], but vanishes for **E** || $[1\bar{1}0]$; for the $1b_{2u}$ orbital the opposite behavior is found, namely strong emission for **E** || $[1\bar{1}0]$ and no emission for **E** || [001]. This behavior indicates that the benzene $\sigma_v$ mirror plane coincides with the [001] substrate azimuth, and the benzene $\sigma_d$ mirror plane coincides with the $[1\bar{1}0]$ azimuth. The analysis of the e-type orbitals reveals a splitting of the $2e_{1u}$ level by 0.3 eV (which is particularly evident at normal emission for $\alpha = 0°$) and of the $1e_{1g}$ level by 0.2 eV; this indicates a lifting of the degeneracy of these levels for the adsorbed molecules and $C_{2v}$ symmetry of the adsorption complex. The double

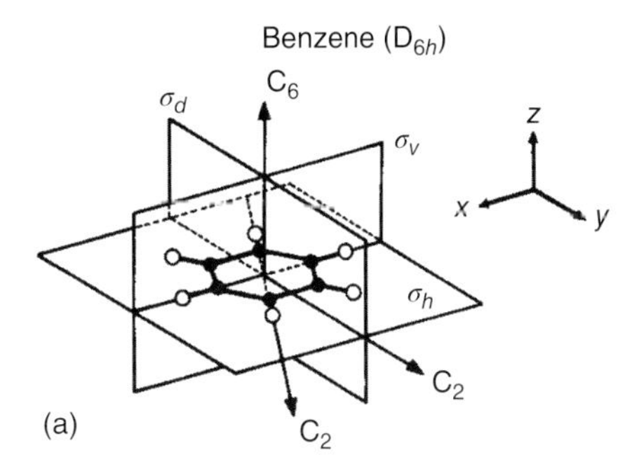


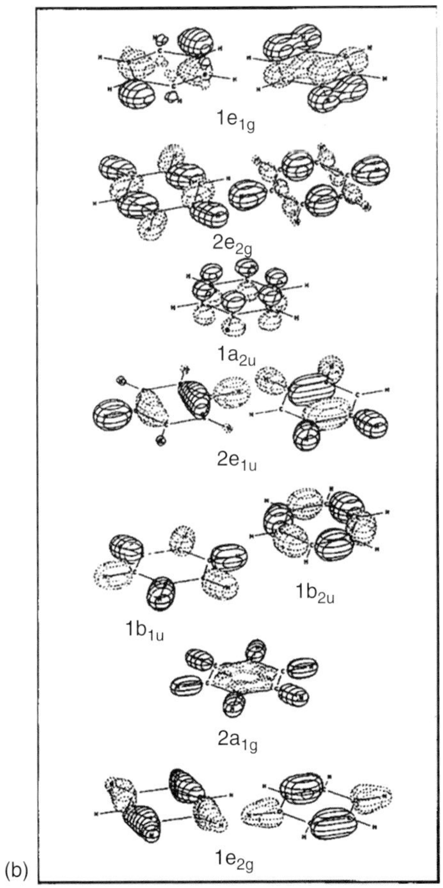


**Figure 38.35** (a) Schematic drawing of benzene with its symmetry elements. Note that only one out of three equivalent $C_2$ rotation axes and $\sigma_d$ and $\sigma_v$ mirror planes are indicated. (b) Molecular orbitals of benzene after Jorgensen and Salem [120]. (Reproduced from [79].)

**Table 38.3** Symmetry selection rules for several molecular symmetries of benzene at various experimental geometries.

| $D_{6h}$ | $C_{6v}$ | | | | | |
|---|---|---|---|---|---|---|
| | | $E_xD_x$ | $E_yD_y$ | $E_xD_y$ | $E_yD_x$ | $E_zD_x$ | $E_zD_y$ |
| $e_{1g}$,$e_{1u}$ | $e_1$ | N | N | N | N | + | + |
| $e_{2g}$,$e_{2u}$ | $e_2$ | + | + | + | + | + | + |
| $a_{1g}$,$a_{2u}$ | $a_1$ | + | + | − | − | N | N |
| $a_{2g}$,$a_{1u}$ | $a_2$ | − | − | + | + | − | − |
| $b_{1u}$ | $b_1$ | + | − | + | − | + | − |
| $b_{2u}$ | $b_2$ | − | + | − | + | − | + |

| $D_{6h}$ | $C_{3v}(\sigma_v)$ | | | | | | |
|---|---|---|---|---|---|---|---|
| | | $E_xD_x$ | $E_yD_y$ | $E_xD_y$ | $E_yD_x$ | $E_zD_x$ | $E_zD_y$ |
| $e_{1g}$,$e_{1u}$ | e | N | N | N | N | + | + |
| $e_{2g}$,$e_{2u}$ | e | N | N | N | N | + | + |
| $a_{1g}$,$a_{2u}$ | $a_1$ | + | + | + | − | N | N |
| $a_{2g}$,$a_{1u}$ | $a_2$ | − | + | + | + | − | − |
| $b_{1u}$ | $a_1$ | + | + | + | − | N | N |
| $b_{2u}$ | $a_2$ | − | + | + | + | − | − |

| $D_{6h}$ | $C_{2v}$ | | | | | | |
|---|---|---|---|---|---|---|---|
| | | $E_xD_x$ | $E_yD_y$ | $E_xD_y$ | $E_yD_x$ | $E_zD_x$ | $E_zD_y$ |
| $e_{1g}$,$e_{1u}$ | $b_1$ | N | − | N | − | + | − |
| | $b_2$ | − | N | − | N | − | + |
| $e_{2g}$,$e_{2u}$ | $a_1$ | + | + | − | − | N | N |
| | $a_2$ | − | − | + | + | − | − |
| $a_{1g}$,$a_{2u}$ | $a_1$ | + | + | − | − | N | N |
| $a_{2g}$,$a_{1u}$ | $a_2$ | − | − | + | + | − | − |
| $b_{1u}$ | $b_1$ | N | − | N | − | + | − |
| $b_{2u}$ | $b_2$ | − | N | − | N | − | + |

| $D_{6h}$ | $C_2$ | | | | | | |
|---|---|---|---|---|---|---|---|
| | | $E_xD_x$ | $E_yD_y$ | $E_xD_y$ | $E_yD_x$ | $E_zD_x$ | $E_zD_y$ |
| $e_{1g}$,$e_{1u}$ | b | N | N | N | N | + | + |
| | b | N | N | N | N | + | + |
| $e_{2g}$,$e_{2u}$ | a | + | + | + | + | N | N |
| | a | + | + | + | + | N | N |

**Table 38.3** *(Continued)*

| $D_{6h}$ | $C_2$ | | | | | | |
|---|---|---|---|---|---|---|---|
| | | $E_xD_x$ | $E_yD_y$ | $E_xD_y$ | $E_yD_x$ | $E_zD_x$ | $E_zD_y$ |
| $a_{1g}$,$a_{2u}$ | a | + | + | + | − | N | N |
| $a_{2g}$,$a_{1u}$ | a | + | + | + | + | N | N |
| $b_{1u}$ | b | N | N | N | N | + | + |
| $b_{2u}$ | b | N | N | N | N | + | + |

$E_x$, $E_y$, and $E_z$ denote the Cartesian components of the electric field vector of the incoming light with respect to the surface plane and a flat lying benzene molecule (see Figure 38.35a); $D_x$ and $D_y$ characterize the alignment of the detection plane. "N" indicates that a band is allowed in normal emission ($\theta'$=0°) and at all other angles, "+" and "−" indicate that emission in the detection plane ($\theta' \neq 0°$) is allowed or forbidden, respectively [119].

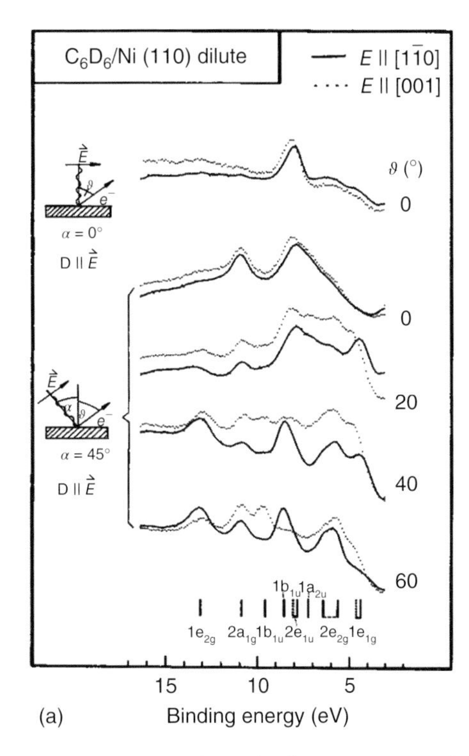


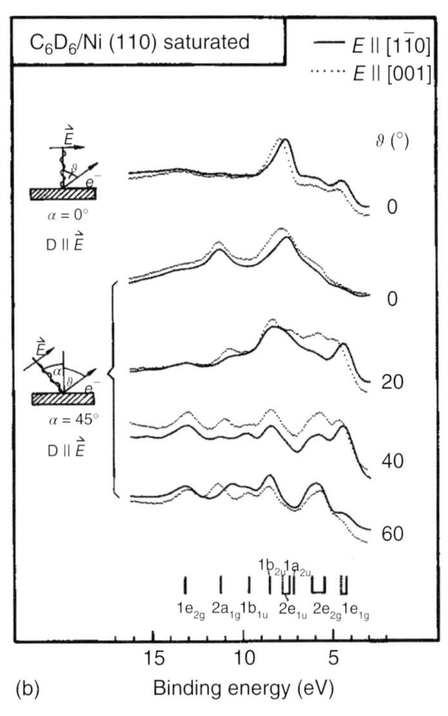


**Figure 38.36** Angle-resolved UPS spectra of benzene on Ni(110) at various different geometries, collected at a photon energy of 30 eV [119]. (a) Dilute layer ($\theta = 0.1$ ML $= 0.4\ \theta_{SAT}$). (b) Saturated c(4 × 2) layer ($\theta = 0.25$ ML). Orbital positions and assignment are indicated as a bar diagram. (Reproduced from Ref. [78].)

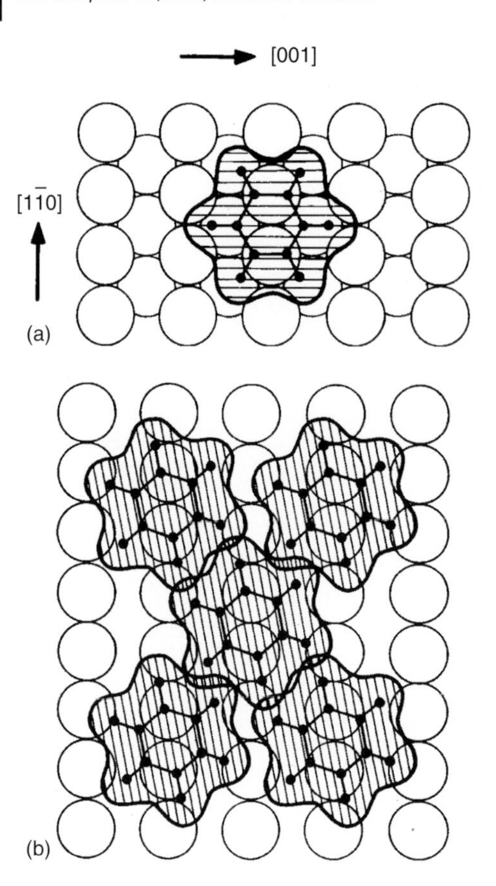


**Figure 38.37** Schematic drawing of the proposed orientation of (a) a single benzene molecule in the dilute layer on Ni(110) and (b) the saturated c(4 × 2) benzene layer on Ni(110). Note that the adsorption site has been chosen arbitrarily [119]. (Reproduced from Ref. [79].)

peak structure of the $2e_{2g}$ level is of different origin, namely a dynamical Jahn–Teller effect [121].

**38.5.2.3.2 Saturated c(4 × 2) Benzene Layer** After addressing the adsorbate/substrate interaction, we now concentrate on the influence of lateral interactions on the molecular orientation. For this purpose, the benzene coverage is increased to its saturation value, namely $\theta_{SAT} = 0.25$ ML. The corresponding layer is well ordered and exhibits a c(4 × 2) LEED pattern. The arrangement of the benzene molecules in this layer is again deduced from the polarization, polar angle, and azimuthal dependencies of the emission from the various orbitals. In the ARUPS spectra in Figure 38.36b, the totally symmetric $2a_{1g}$ and $1a_{2u}$ orbitals show the same behavior as for the dilute layer, that is, strong normal emission for $\alpha = 45°$ and no emission for $\alpha = 0°$, indicating that the planar adsorption geometry is maintained at the higher coverage. However, in contrast to the dilute layer, both the $1b_{1u}$ *and* the

$1b_{2u}$ orbitals are observed in both azimuths $[1\bar{1}0]$ and [001] at $\alpha = 45°$. From this behavior, we conclude that in the saturated layer the molecular mirror planes do not coincide with the substrate mirror planes any more. This lowering in symmetry is attributed to an azimuthal reorientation (rotation) of the benzene molecules in the densely packed saturated layer as a result of strong lateral interactions. The proposed arrangement of the molecules on the surface is schematically illustrated in Figure 38.37b.

In the following, we will try to understand the driving force for the reorientation of the molecules. In Figure 38.38 (a, I–III), the benzene c(4×2) structure on

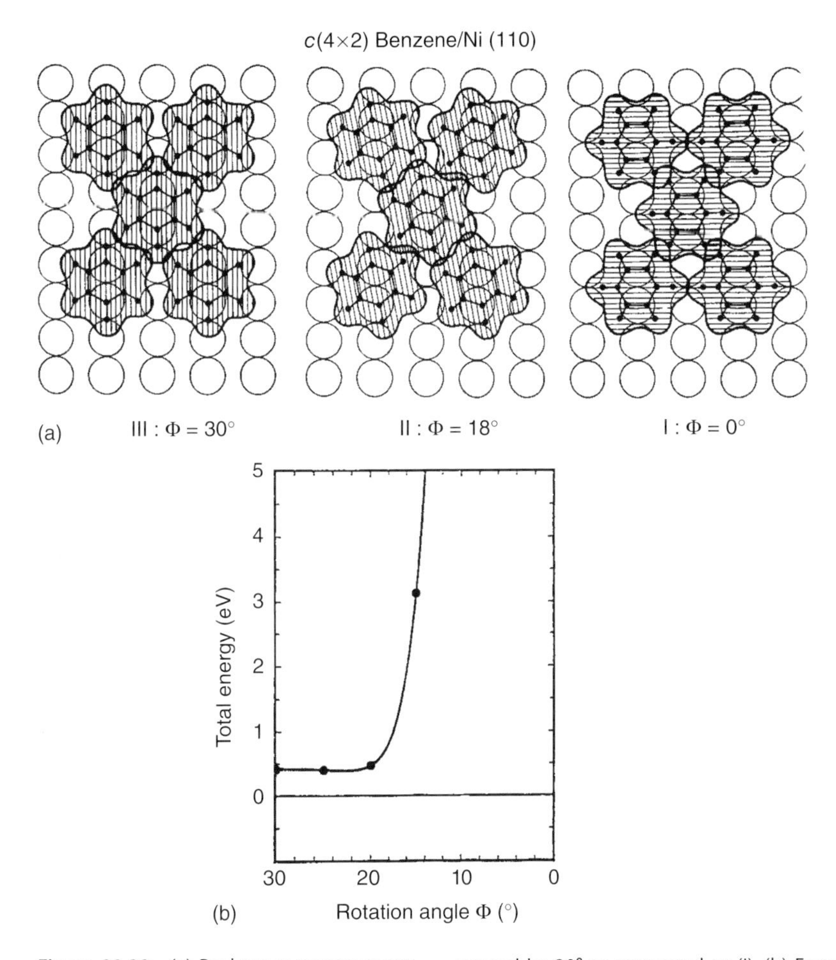


**Figure 38.38** (a) Real-space arrangement for the saturated benzene layer on Ni(110) based on a c(4 × 2) structure [119]. (I) Two hydrogens along [001], as in dilute layer; (II) azimuthally rotated by ~18° as compared to (I); and (III) two hydrogens along $[1\bar{1}0]$ rotated by 30° as compared to (I). (b) Force field calculation of the repulsive energy per molecule for an unsupported benzene layer with the lateral arrangement of the c(4 × 2) layer on Ni(110) [122]. (Reproduced from Ref. [78].)

Ni(110) is depicted for various possible azimuthal orientations of the molecules; the molecules are plotted according to their van der Waals dimension. As the c(4 × 2) layer is densely packed, the van der Waals areas of neighboring molecules have a significant overlap. This overlap is strongest for an azimuthal orientation as for that of the isolated adsorbed molecule (Figure 38.38, I), that is, with two C–H bonds pointing along the [001] directions. Intuitively, this overlap can be reduced by an azimuthal rotation relative to the geometry of the isolated molecule, in order to minimize lateral repulsion and thereby to lower the total energy of the system (Figure 38.38, II and III). These qualitative arguments have been confirmed by force field calculations by Fox and Rösch [122]. They have calculated the repulsive energy per molecule as a function of the rotation angle $\varphi$ for an unsupported free-standing benzene layer (Figure 38.38b). In agreement with the intuitive arguments used above, a very large repulsive energy is observed for the orientation of the isolated molecule ($\varphi = 0°$). When the molecule is azimuthally rotated, the repulsion decreases and reaches a flat minimum at $\varphi \sim 18°$. One should note that, from the calculation, an orientation of benzene with two hydrogens pointing along the $[1\bar{1}0]$ azimuth, that is, rotated by 30° (or equivalently 90°) would also be possible. Such an arrangement, however, would correspond to an alignment of the molecular mirror planes with the substrate's high-symmetry directions (Figure 38.38, III), which is ruled out by the nonzero intensity of the $1b_{1u}$ and $1b_{2u}$ orbitals for both azimuths in Figure 38.36b for $\alpha = 45°$. From these considerations, only the adsorption geometry shown in Figures 38.37b and 38.38, II, which results from the competition between the bonding of the molecule to the substrate (which favors the geometry of the isolated molecule) and repulsive lateral interactions between adjacent molecules, is consistent with the experimental results. Note that the lateral interactions are also reflected in a pronounced dispersion of the $2a_{1g}$ benzene level (0.8 eV) due to the formation of a 2D adsorbate band structure (see Section 38.5.3).

### 38.5.3
### Adsorbate Band Structures

Well-ordered adsorbate layers represent 2D superlattices on a surface with a 2D periodic potential. For small intermolecular distances within the adsorbate lattice (roughly their van der Waals dimensions), the orbital overlap between neighboring molecules is significant and the formation of a 2D adsorbate band structure is expected [77–79, 83, 84, 90, 103, 104]. As a consequence, the binding energy of a specific molecular orbital can depend on the electron momentum of the emitted electron parallel to the surface, $\mathbf{k}_{||}$. Such effects have indeed been observed in the above-mentioned examples, that is, the c(2 × 4) ethylene layer and the c(4 × 2) benzene layer on Ni(110). The adsorbate wave functions then have to be described as two-dimensional Bloch states $\Phi(\mathbf{k}_{||})$, with $\mathbf{k}_{||} = \mathbf{k}_x + \mathbf{k}_y$. The 2D band structure $E_B$ ($\mathbf{k}_{||}$) can be obtained from ARUPS. The magnitude of the electron momentum parallel to the surface, $|\mathbf{k}_{||}|$, is simply obtained from the kinetic energy of the emitted photoelectron $E_{Kin}$ and the polar angle of emission $\vartheta$ via the relationship

$$|\mathbf{k}_{||}| = (2m/(h/2\pi)^2 \cdot E_{Kin})^{1/2} \sin \vartheta. \quad (38.5)$$

This relation is valid if one assumes a free-electron dispersion relation for the final state [84]. The 2D band structure can then directly be plotted as the initial state energy $E_B$ versus $\mathbf{k}_{||}$.

The formation of adsorbate band structures in adsorbate layers had been reported already very early for CO on Ni(100) and Pd(100) by Horn *et al.* [117, 123–125]. Later on, systematic investigations, in selected cases accompanied by model calculations, followed for ordered CO layers on a variety of different substrates (for an overview see Refs [87, 90]).

For larger than diatomic molecules, such as, for example, hydrocarbons, the formation of adsorbate band structures in well-ordered and densely packed layers is often neglected. It typically occurs for those orbitals that show highest overlap with their neighbors (often orbitals with C–H character). It is important to realize that by taking into account only the band structure the electronic properties of such adsorption system can be correctly described. In the following, we will concentrate on those adsorbate systems that have already been discussed in this chapter, namely benzene and ethylene adsorbed on Ni(110). This is not only due to the simplicity of presentation but also due to the fact that these systems represent the best studied systems so far. In contrast to, for example, the fcc(111), fcc(100), and hcp(001) surfaces, investigations on an fcc(110) surface with $C_{2v}$ symmetry have the advantage that the data analysis is not hampered by the existence of symmetry-equivalent domains; therefore, the adsorbate bands can be measured up to $\mathbf{k}_{||}$ values in the second or sometimes even higher Brillouin zones.

#### 38.5.3.1 c(4 × 2) Benzene/Ni(110)

In Figure 38.39a, the adsorbate SBZ of a benzene c(4 × 2) layer on Ni(110) is shown. From the dense packing of this layer, one expects the formation of an adsorbate band structure [77–79, 119]. The ARUPS spectra in Figure 38.36b indeed reveal a significant dispersion of the $2a_{1g}$ level with the emission angle. The corresponding 2D adsorbate band structure has been determined from these spectra and spectra at other photon energies (different symbols) according to Equation 38.5, and is plotted in Figure 38.39b along the $[1\bar{1}0]$ and [001] azimuths. The $|\mathbf{k}_{||}|$ values of the symmetry points in the adsorbate SBZ are also marked in Figure 38.39b. The binding energy at the $\bar{\Gamma}'$ point is 11.3 eV, and the magnitude of dispersion (difference between highest binding energy at the $\bar{\Gamma}'$ point and lowest binding energy at the $\bar{A}'$ point) is 0.8 eV for both azimuths. The observed periodicity of the experimentally determined 2D band structure perfectly reflects that of the c(4 × 2) adsorbate Brillouin zone for both azimuths. Along $[1\bar{1}0]$, the $2a_{1g}$ band closely follows the periodicity $\bar{\Gamma}'\bar{A}'\bar{\Gamma}'\bar{A}'\bar{\Gamma}'$ to the second neighbor adsorbate SBZ, and along [001] the $2a_{1g}$ band exhibits the periodicity $\bar{\Gamma}'\bar{B}'\bar{A}'\bar{B}'\bar{\Gamma}'\bar{B}'$. The binding energy of the $2a_{1g}$ level at the $\bar{A}'$ point can thus be obtained by measuring along either $[1\bar{1}0]$ or [001], with the two values being identical ($10.5 \pm 0.1$ eV), as should be the case. Notably, all other levels show no significant dispersion.

Investigations of densely packed, well-ordered benzene layers have also been performed on hexagonal close-packed transition-metal surfaces, and also

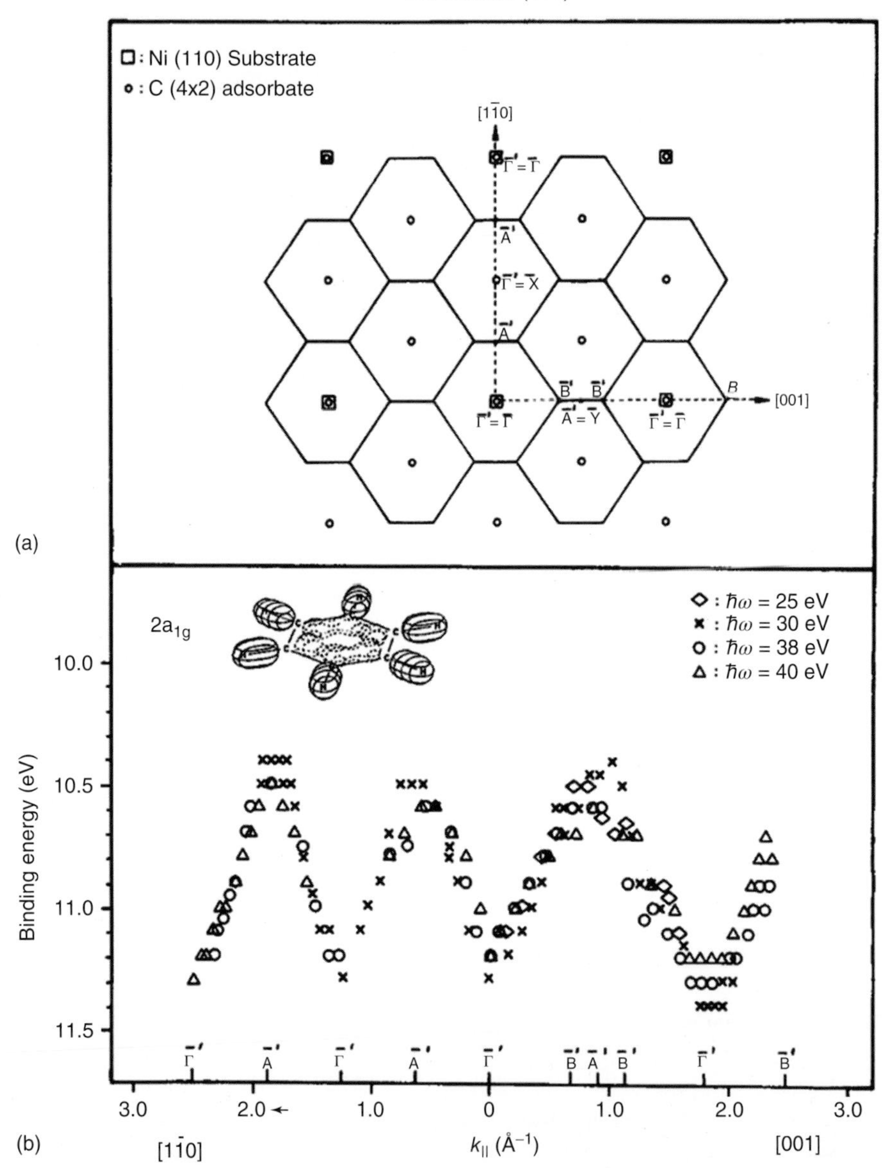


**Figure 38.39** (a) Adsorbate surface Brillouin zone for a c(4 × 2) adsorbate structure. (b) Two-dimensional band structure of the saturated benzene layer on Ni(110) [119]. The insert in (b) shows the $2a_{1g}$ orbital of benzene [120]. (Reproduced from Ref. [78].)

for these systems the formation of 2D band structures was found: for $(\sqrt{7}\times\sqrt{7})R19.1^\circ$ benzene/Ni(111) [126], $(\sqrt{7}\times\sqrt{7})R19.1^\circ$ benzene/Os(001) [127], and $(\sqrt{19}\times\sqrt{19})R23.4^\circ$ benzene/Rh(111) [90], band widths of ~0.4 eV have been observed for the $2a_{1g}$ level. This dispersion is smaller than that observed on Ni(110), which is attributed to larger nearest neighbor distances in these layers. The overall behavior on those surfaces is, however, similar to that on Ni(110). The $2a_{1g}$ band is strongly bonding at the $\overline{\Gamma}'$ point and antibonding at the zone edge (A′ point in Figure 38.39).

In order to obtain a qualitative understanding of the dispersion, we analyze the wave functions of neighboring molecules in a hexagonal layer for the two symmetry points. For that purpose, the $2a_{1g}$ Bloch wave functions of a hexagonal benzene layer are schematically indicated in Figure 38.40 for the $\overline{\Gamma}'$ point and the zone edge. The hexagonal geometry represents the arrangement on Ni(111) and Os(001), and is also a fairly good approximation for the quasi-hexagonal $c(4\times2)$ structure on Ni(110). At the $\overline{\Gamma}'$ point, all wave functions are in phase, which leads to a strongly bonding

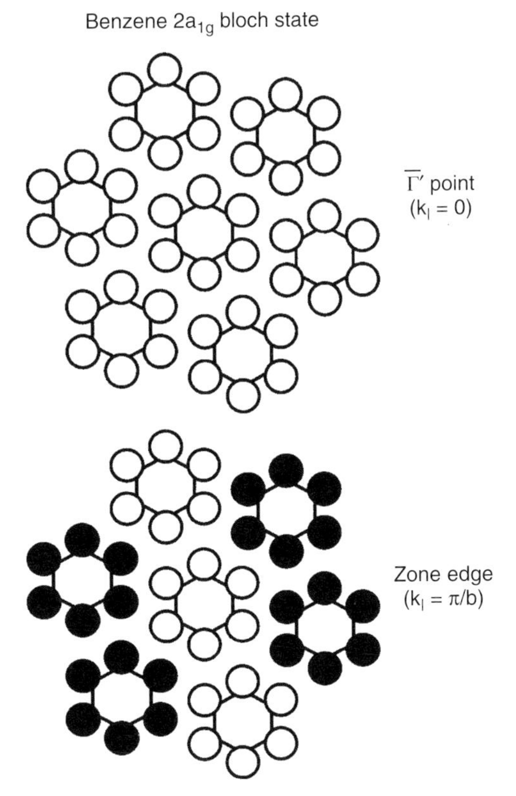


**Figure 38.40** Schematic representation of the $2a_{1g}$ Bloch wave function of benzene molecules in a hexagonal arrangement for the $\overline{\Gamma}'$ point and the zone edge of the first Brillouin zone. (Reproduced from Ref. [79].)

situation for the totally symmetric $2a_{1g}$ orbital. At the zone edge, the wave functions of two neighbors are in phase, while those of the other four neighbors have opposite phase, yielding an overall antibonding situation. From this analysis, one can understand the significant upward dispersion from the $\overline{\Gamma}'$ point to the zone edge, which is found in the experiment (Figure 38.39b).

#### 38.5.3.2 **c(2 × 4) Ethylene/Ni(110)**

The second example addresses the band structure of the saturated c(2 × 4) ethylene layer on Ni(110) [107, 116]. The real space structure of this densely packed layer has already been introduced in the previous section and is shown in Figure 38.33b. Translationally equivalent adsorption sites are occupied only every second row, resulting in a nonprimitive, oblique unit cell of the c(2 × 4) structure. Interestingly, an adsorbate structure with a smaller unit cell containing only one molecule is obtained by neglecting the underlying substrate. The extended adsorbate Brillouin zone as deduced from the c(2 × 4) LEED pattern is illustrated in Figure 38.41a, and the larger adsorbate Brillouin zone for the isolated layer, that is, neglecting the substrate, is depicted in Figure 38.41b. For both structures, the various symmetry points are also indicated.

As pointed out in Section 38.5.2.2, the ARUPS of the saturated c(2 × 4) ethylene layer exhibits significant dispersion for several molecular levels, as is evident from Figure 38.33b. From these data and analogous spectra obtained at other photon energies, the corresponding adsorbate band structure along the $[1\bar{1}0]$ and [001] directions of the substrate can be determined, which is depicted in Figure 38.42. The highest lying $1b_{2u}$-derived band shows essentially no dispersion, which is attributed to its out-of-plane ($\pi$) character, which leads to a negligible overlap of this orbital for neighboring molecules. For all other levels, significant dispersion is found. The $1b_{2g}$ and $2b_{1u}$ bands exhibit strong dispersion along $[1\bar{1}0]$, but no dispersion along [001]. The $3a_g$ and $1b_{3u}$ bands show dispersion for both azimuths. At the $\overline{\Gamma}'$ points ($|\mathbf{k}_{||}| = 0$ and 1.8 Å$^{-1}$ along [001]), these two bands are energetically nearly degenerate, with the binding energy of the $1b_{3u}$ band being even somewhat smaller than that of the $3a_g$ band (8.4 eV vs. 8.6 eV). The magnitude of the dispersion differs for the various bands, with the highest value of 2 eV (!) observed for the $1b_{3u}$ band.

Interestingly, the periodicity of the 2D adsorbate band structure of ethylene on Ni(110) does *not* reflect the symmetry of the adsorbate Brillouin zone based on the c(2 × 4) LEED pattern. Furthermore, the ARUPS spectra do not reveal a splitting (doubling) of any of the adsorbate bands; such a splitting is expected from the existence of two nonequivalent molecules per unit cell. On the other hand, excellent agreement with the experimental data is obtained by a 2D band structure calculation (tight-binding approximation at the extended Hückel level) for an unsupported ethylene layer, that is, with only one adsorbate molecule per unit cell. The corresponding real space unit cell and extended Brillouin zone is shown in Figure 38.41b. In Figure 38.42, this calculated band structure is indicated as solid lines, aligned with respect to the $2b_{1u}$ band of the experimental data at the $\overline{\Gamma}'$ point. For all calculated bands, the measured energy dispersion is mostly well reproduced, with some

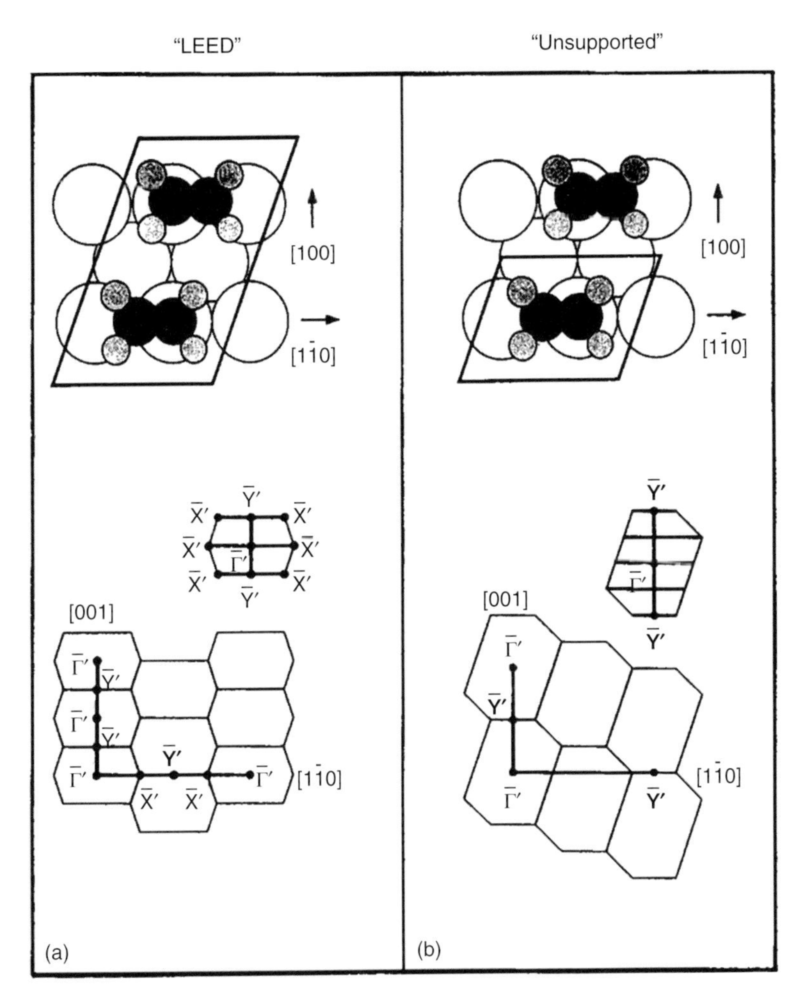


**Figure 38.41** Extended Brillouin zone scheme for the ethylene c(2 × 4) structure on Ni(110) (a) as determined from the c(2 × 4) LEED pattern and (b) for the unsupported ethylene layer. This structure is obtained by simply neglecting the substrate in (a). The experimental $\mathbf{k}_{\|}$ paths along the [1$\bar{1}$0] and [001] azimuths of the Ni substrate are indicated as solid lines. The labels denote the high-symmetry points of the two structures. The real-space unit cells of the corresponding lattices are given at the top of the figure [116]. (Reproduced from Ref. [78].)

small discrepancies for the $3a_g$ band, which indicates the limitation of the approximate band structure method. The only major difference between the experimental spectra and the calculation for the unsupported layer is the fact that the calculated $\pi$-derived band ($1b_{2u}$) lacks the substrate-induced bonding shift of about 0.8 eV to higher binding energies. This is to be expected for a calculation that does not include the substrate. The good agreement between the calculation for the unsupported ethylene layer with the experimental results for the c(2 × 4) layer indicates that the

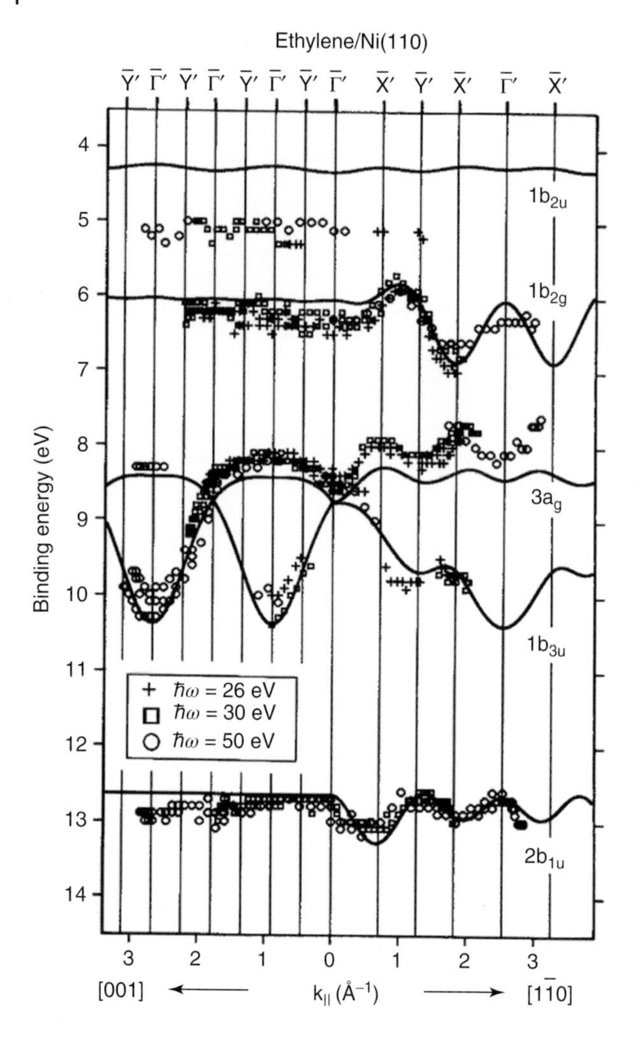


**Figure 38.42** Two-dimensional band structure of the saturated ethylene layer as determined from ARUPS spectra at various photon energies: (+) 26 eV, (□) 30 eV, and (o) 50 eV [116]. The solid lines indicate the calculated band structure for an unsupported layer. The labels correspond to the high-symmetry points of the Brillouin zone as deduced from the c(2 × 4) LEED pattern as displayed in Figure 38.41. (Reproduced from Ref. [78].)

adsorbate band structure of the c(2 × 4)ethylene layer on Ni(110) is dominated by the adsorbate/adsorbate interactions, that is, by the orbital overlap of neighboring molecules. The "vertical" adsorbate/substrate interactions, which are responsible for the interaction with the surface, are essentially decoupled from these lateral interactions and are reflected only in the differential shift of the out-of-plane $\pi$ orbital ($1b_{2u}$), which is mainly responsible for the bonding of the molecule.

Finally, we address the sensitivity of the band structure calculations to the lateral arrangement of the molecules on the surface. It turns out that this sensitivity can

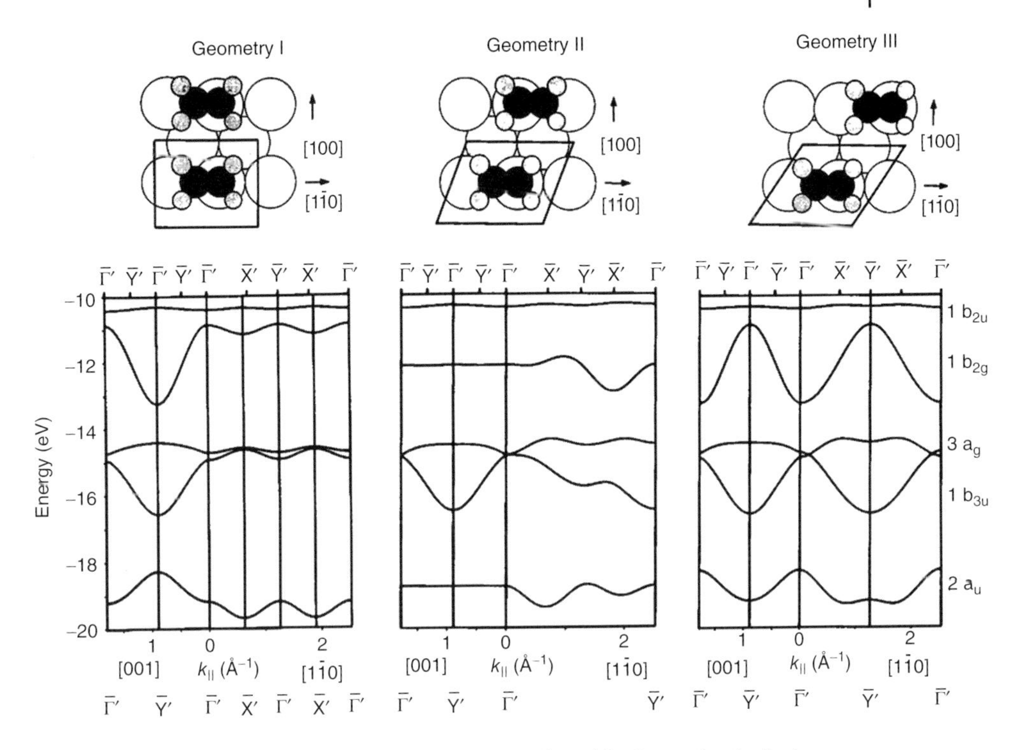


**Figure 38.43** Band structure calculations for the saturated (unsupported) ethylene layer with different lateral arrangements of the molecules [116]. In each case, the relative position of neighboring molecules is shown at the top of the figure with the unit cell of the corresponding lattice. (Reproduced from Ref. [79].)

be used to obtain additional information on the orientation of the molecules on the surface and on the arrangement of neighboring molecules. Figure 38.43 shows the band structure for three different arrangements of ethylene on Ni(110), denoted as geometries I–III. Geometry II corresponds to the geometry derived from ARUPS, NEXAFS, LEED, and force field calculations. The most dramatic differences between the calculations for the different geometries are observed for the $1b_{2g}$ and the $2a_u$ bands, in particular along the [001] azimuth. The $1b_{2g}$ band exhibits a downward dispersion of ~2.4 eV for geometry I, an upward dispersion of ~2.4 eV for geometry III, but is completely flat for geometry II. The latter behavior is observed in the experiment, which is a strong support for the structural model proposed in Figure 38.34b.

## 38.6 Summary

Fundamental aspects of adsorption of small molecules on metals are already quite well understood in great detail from surface science studies on model systems at the specific conditions of the techniques used. This includes the electronic and geometric structure of such systems, as shown in this chapter. In most cases, however, this approach requires vacuum conditions, which are far from the conditions used in technologically interesting catalytic processes. Therefore, extensions of the methods presented here, or sometimes even new methods, are developed to overcome this "pressure gap", or at least to narrow the gap. Some of these techniques are still in the framework of "surface science" and are included in this series. Some others already cross into the chemical technology area and are as such beyond the scope of this series.

## References


1. Kim, J., Samano, E., and Koel, B.E. (2006) Oxygen adsorption and oxidation reactions on Au(211) surfaces: exposures using $O_2$ at high pressures and ozone (O3) in UHV. *Surf. Sci.*, **600**, 4622–4632.
2. Jensen, J.A., Rider, K.B., Salmeron, M., and Somorjai, G.A. (1998) High pressure adsorbate structures studied by scanning tunneling microscopy: CO on Pt(111) in equilibrium with the gas phase. *Phys. Rev. Lett.*, **80**, 1228–1231.
3. Kisliuk, P. (1957) The sticking probabilities of gases chemisorbed on the surfaces of solids. *J. Phys. Chem. Solids*, **3**, 95 and *J. Phys. Chem. Solids* (1958) **5** 1..
4. Darling, G.R. and Holloway, S. (1995) The dissociation of diatomic molecules on surfaces. *Rep. Prog. Phys.*, **58**, 1595.
5. Tüshaus, M., Schweizer, E., Hollins, P., and Bradshaw, A.M. (1987) Yet another vibrational study of the adsorption system Pt(111)-CO. *J. Electron. Spectrosc. Relat. Phenom.*, **44**, 305.
6. Rupprechter, G. (2007) A surface science approach to ambient pressure catalytic reactions. *Catal. Today*, **126**, 3.
7. Froitzheim, H., Hopster, H., Ibach, H., and Lehwald, S. (1977) Adsorption sites of CO on Pt (111). *Appl. Phys.*, **13**, 147–151.
8. Hopster, H. and Ibach, H. (1978) Adsorption of CO on Pt(111) and Pt [6(111) x (111)] studied by high resolution electron energy loss spectroscopy and thermal desorption spectroscopy. *Surf. Sci.*, **77**, 109.
9. Steininger, H., Lehwald, S., and Ibach, H. (1982) On the adsorption of CO on Pt(111). *Surf. Sci.*, **123**, 264.
10. Sheppard, N. (1988) Vibrational spectroscopic studies of the structure of species derived from the chemisorption of hydrocarbons on metal single-crystal surfaces. *Annu. Rev. Phys. Chem.*, **39**, 589.
11. Sheppard, N. and De la Cruz, C. (1998) Vibrational spectra of hydrocarbons adsorbed on metals. Part II. Adsorbed acyclic alkynes and alkanes, cyclic hydrocarbons including aromatics, and surface hydrocarbon groups derived from the decomposition of alkyl halides, etc. *Adv. Catal.*, **42**, 181.
12. Woodruff, D.P. (2007) Adsorbate structure determination using photoelectron diffraction. *Surf. Sci. Rep.*, **62**, 1.
13. Norton, P.R., Goodale, J.W., and Selkirk, E.B. (1979) Adsorption of CO on Pt(111) studied by photoemission, thermal desorption spectroscopy and high resolution dynamic measurements of work function. *Surf. Sci.*, **83**, 189.
14. Kinne, M., Fuhrmann, T., Whelan, C.M., Zhu, J.F., Pantförder, J., Probst, M., Held, G., Denecke, R., and Steinrück, H.-P. (2002) Kinetic parameters of CO adsorbed on Pt(111) studied by in-situ high resolution x-ray photoemission. *J. Chem. Phys.*, **117** (23), 10852.
15. Föhlisch, A., Wassdahl, N., Hasselström, J., Karis, O., Menzel, D., Martensson, N., and Nilsson, A. (1998) Beyond the chemical shift: vibrationally

resolved core-level photoelectron spectra of adsorbed CO. *Phys. Rev. Lett.*, **81**, 1730.

16. Gajdos, M., Eichler, A., and Haffner, J. (2004) CO adsorption on close-packed transition and noble metal surfaces: trends from ab initio calculations. *J. Phys. Condens. Matter*, **16**, 1141.
17. Brundle, C.R. and Wandelt, K. (1977) A LEED, XPS and UPS study of the adsorption of CO on Cu(100). Proceedings of 7th International Vacuum Congress and 3rd International Conference on Solid Surfaces, Wien, Austria, p. 1171.
18. Tillborg, H., Nilsson, A., and Martensson, N. (1993) Shake-up and shake-off structures in core level photoemission spectra from adsorbates. *J. Electron. Spectrosc. Relat. Phenom.*, **62**, 73.
19. (a) K. Schönhammer, O. Gunnarsson (1978) Shape of core level spectra in adsorbates, *Solid State Commun.* **23** (1977) 691; (b) Gunnarsson, O. and Schönhammer, K. (1978) CO on Cu(100) – explanation of the three-peak structure in the x-ray-photoemission-spectroscopy core spectrum. *Phys. Rev. Lett.*, **41**, 1608.
20. Shinn, N.D. and Madey, T.E. (1985) CO chemisorption on Cr(110): evidence for a precursor to dissociation. *J. Chem. Phys.*, **83**, 5928.
21. Engelhardt, M.P., Fuhrmann, T., Held, G., Denecke, R., and Steinrück, H.-P. (2002) Adsorption of CO on ultrathin Cr layers on Ru(0001). *Surf. Sci.*, **512**, 107.
22. Wandelt, K. (1982) Photoemission studies of adsorbed oxygen and oxide layers. *Surf. Sci. Rep.*, **2**, 1.
23. Hedman, P.-F., Heden, C., and Nordling, K.S. (1969) Energy splitting of core electron levels in paramagnetic molecules. *Phys. Lett. A*, **29**, 178.
24. Puglia, C., Nilsson, A., Hernnäs, B., Karis, O., Bennich, P., Mårtensson, N. (1995) Physisorbed, chemisorbed and dissociated $O_2$ on Pt(111) studied by different core-level spectroscopy methods. *Surf. Sci.*, **342**, 119.
25. Nolan, P.D., Lutz, B.R., Tanaka, P.L., Davis, J.E., and Mullins, C.B. (1999) Molecularly chemisorbed intermediates to oxygen adsorption on Pt(111): a molecular beam and electron energy-loss spectroscopy study. *J. Chem. Phys.*, **111**, 3696.
26. Eichler, A., Mittendorfer, F., and Hafner, J. (2000) Precursor-mediated adsorption of oxygen on the (111) surfaces of platinum-group metals. *Phys. Rev. B*, **62**, 4744.
27. Fuggle, J.C., Umbach, E., Menzel, D., Wandelt, K., and Brundle, C.R. (1978) Adsorbate line shapes and multiple lines in XPS: comparison of theory and experiment. *Solid State Comm.*, **27**, 65–69.
28. Sandell, A., Björneholm, O., Nilsson, A., Zdansky, E.O.F., Tillborg, H., Andersen, J.N., and Martensson, N. (1993) Autoionization as a tool for interpretation of x-ray absorption spectra: $N_2$/Ni(100). *Phys. Rev. Lett.*, **70**, 2000.
29. Bennich, P., Wiell, T., Karis, O., Weinelt, M., Wassdahl, N., Nilsson, A., Nyberg, M., Pettersson, L.G.M., Stöhr, J., and Samant, M. (1998) Nature of the surface chemical bond in $N_2$ on Ni(100)studied by x-ray-emission spectroscopy and ab initio calculations. *Phys. Rev. B*, **57**, 9274.
30. Dahl, S., Tornqvist, E., and Chorkendorff, I. (2000) Dissociative adsorption of $N_2$ on Ru(0001): a surface reaction totally dominated by steps. *J. Catal.*, **192**, 381.
31. Grunze, M., Golze, M., Hirschwald, W., Freund, H.-J., Pulm, H., Seip, U., Tsai, M.C., Ertl, G., and Küppers, J. (1984) $\pi$-Bonded $N_2$ on Fe(111): the precursor for dissociation. *Phys. Rev. Lett.*, **53**, 850.
32. Baraldi, A., Barnaba, M., Brena, B., Cocco, D., Comelli, G., Lizzit, S., Paolucci, G., and Rosei, R. (1995) Time resolved core level photoemission experiments with synchrotron radiation. *J. Electron. Spectrosc. Relat. Phenom.*, **76**, 145–149.
33. Baraldi, A., Comelli, G., Lizzit, S., Kiskinova, M., and Paolucci, G. (2003) Real-time x-ray photoelectron spectroscopy of surface reactions. *Surf. Sci. Rep.*, **49**, 169.

34. Papp, C. and Steinrück, H.-P. (2013) In situ high-resolution x-ray photoelectron spectroscopy – fundamental insights in surface reactions. *Surf. Sci. Rep.*, **68**, 446.
35. McEwen, J.-S., Payne, S.H., Kreuzer, H.J., Kinne, M., Denecke, R., and Steinrück, H.-P. (2003) Adsorption and desorption of CO on Pt(111): a comprehensive analysis. *Surf. Sci.*, **545** (1-2), 47.
36. Baraldi, A., Lizzit, S., and Paolucci, G. (2000) Identification of atomic adsorption site by means of high-resolution photoemission surface core-level shift: oxygen on Ru(1010). *Surf. Sci.*, **457**, L354–L360.
37. Schneider, U., Castro, G.R., and Wandelt, K. (1993) Adsorption on ordered Cu,Pt( 111): site selectivity. *Surf. Sci.*, **287–288**, 146.
38. Allers, K.-H., Pfnür, H., Feulner, P., and Menzel, D. (1993) Angular and velocity distributions of CO desorbed from adsorption layers on Ni(100) and Pt(111): examples of non-activated desorption. *Surf. Sci.*, **291**, 167.
39. Becker, C., Pelster, T., Tanemura, M., Breitbach, J., and Wandelt, K. (1999) CO adsorption on $Cu_3Pt(111)$ and $CuPt_3(111)$: a comparative HREELS study. *Surf. Sci.*, **427–428**, 403–407.
40. Hammer, B., Morikawa, Y., and Nørskov, J.K. (1996) CO chemisorption at metal surfaces and overlayers. *Phys. Rev. Lett.*, **76**, 2141–2144.
41. Gauthier, Y., Schmid, M., Padovani, S., Lundgren, E., Buš, V., Kresse, G., Redinger, J., and Varga, P. (2001) Adsorption sites and ligand effect for CO on an alloy surface: a direct view. *Phys. Rev. Lett.*, **87**, 036103.
42. Liu, P. and Norskov, J.K. (2001) Ligand and ensemble effects in adsorption on alloy surfaces. *Phys. Chem. Chem. Phys.*, **3**, 3814.
43. Becker, C. and Wandelt, K. (2007) Tailoring specific adsorption sites by alloying: adsorption of unsaturated organic molecules on alloy surfaces. *Top. Catal.*, **46**, 151.
44. Mancera, L.A., Behm, R.J., and Groß, A. (2013) Structure and local reactivity of PdAg/Pd(111) surface alloys. *Phys. Chem. Chem. Phys.*, **15**, 1497.
45. Schaff, O., Fernandez, V., Hofmann, P., Schindler, K.-M., Theobald, A., Fritzsche, V., Bradshaw, A.M., Davis, R., and Woodruff, D.P. (1996) Coverage-dependent changes in the adsorption geometry of benzene on Ni(111). *Surf. Sci.*, **348**, 89.
46. Steinrück, H.-P., Huber, W., Pache, T., and Menzel, D. (1989) The adsorption of benzene mono- and multilayers on Ni(111) studied by TPD and LEED. *Surf. Sci.*, **218**, 293.
47. Held, G., Bessent, M.P., Titmuss, S., and King, D.A. (1996) Realistic molecular distortions and strong substrate buckling induced by the chemisorption of benzene on Ni(111). *J. Chem. Phys.*, **105**, 11305.
48. Yamagishi, S., Jenkins, S.J., and King, D.A. (2001) Symmetry and site selectivity in molecular chemisorption: benzene on Ni(111). *J. Chem. Phys.*, **114**, 5765.
49. Mittendorfer, F. and Hafner, J. (2001) Density-functional study of the adsorption of benzene on the (111), (100) and (110) surfaces of nickel. *Surf. Sci.*, **472**, 133.
50. Papp, C., Fuhrmann, T., Tränkenschuh, B., Denecke, R., and Steinrück, H.-P. (2006) Site selectivity of benzene adsorption on Ni(111) studied by HR-XPS. *Phys. Rev. B*, **73**, 235426.
51. Zhang, R., Hensley, A.J., McEwen, J.-S., Wickert, S., Darlatt, E., Fischer, K., Schöppke, M., Denecke, R., Streber, R., Lorenz, M., Papp, C., and Steinrück, H.P. (2013) Integrated x-ray photoelectron spectroscopy and DFT characterization of benzene adsorption on Pt(111), Pt(355) and Pt(322) surfaces. *Phys. Chem. Chem. Phys.*, **15**, 20662–20671.
52. King, D.A. and Wells, M.G. (1972) Molecular beam investigation of adsorption kinetics on bulk metal targets: nitrogen on tungsten. *Surf. Sci.*, **29**, 454–482.
53. Anger, G., Winkler, A., and Rendulic, K.D. (1989) Adsorption and desorption kinetics in the systems $H_2/Cu(111)$,

$H_2$(Cu110) and $H_2$/Cu(100). *Surf. Sci.*, **220**, 1.

54. Luntz, A.C. and Bethune, D.S. (1989) Activation of methane dissociation on a Pt(111) surface. *J. Chem. Phys.*, **90**, 1274.
55. Papp, C., Tränkenschuh, B., Streber, R., Fuhrmann, T., Denecke, R., and Steinrück, H.-P. (2007) Adsorption of methane on stepped platinum surfaces. *J. Phys. Chem. C*, **111**, 2177.
56. Denecke, R., Tränken-schuh, B., Engelhardt, M.P., and Steinrück, H.-P. (2003) Adsorption kinetics of CO on Cr/Ru surfaces. *Surf. Sci.*, **532–535**, 173–178.
57. Engelhardt, M.P. (2003) PhD thesis. Bimetallische Cr/Ru- und Cu/Ru-Schichten – Präparation und Adsorptionsverhalten von CO, Friedrich-Alexander-Universität Erlangen-Nürnberg.
58. Fuhrmann, T., Kinne, M., Whelan, C.M., Zhu, J.F., Denecke, R., and Steinrück, H.-P. (2004) Vibrationally resolved in-situ XPS study of activated adsorption of methane on Pt(111). *Chem. Phys. Lett.*, **390** (1-3), 208.
59. Carlsson, A.F. and Madix, R.J. (2000) The dynamics of argon and methane trapping on Pt(111) at 30 and 50 K: energy scaling and coverage dependence. *Surf. Sci.*, **458**, 91–105.
60. Donald, S.B., Navin, J.K., and Harrison, I. (2013) Methane dissociative chemisorption and detailed balance on Pt(111): dynamical constraints and the modest influence of tunneling. *J. Chem. Phys.*, **139**, 214707.
61. Tersoff, J. and Hamann, D.R. (1983) Theory and application for the scanning tunneling microscope. *Phys. Rev. Lett.*, **50**, 1998.
62. Singh-Boparai, S.P., Bowker, M., and King, D.A. (1975) Crystallographic anisotropy in chemisorption: nitrogen on tungsten single crystal planes. *Surf. Sci.*, **53**, 55.
63. Tränkenschuh, B., Papp, C., Fuhrmann, T., Denecke, R., and Steinrück, H.-P. (2007) The dissimilar twins – a comparative, site-selective in-situ study of CO adsorption and desorption on Pt(322) and Pt(355). *Surf. Sci.*, **601**, 1108.
64. Streber, R., Tränkenschuh, B., Schöck, J., Papp, C., Steinrück, H.-P., McEwen, J.-S., Gaspard, P., and Denecke, R. (2009) Interaction between silver nanowires and CO on stepped platinum surfaces. *J. Chem. Phys.*, **131**, 064702.
65. Kinne, M., Fuhrmann, T., Tränkenschuh, B., Zhu, J.F., Denecke, R., and Steinrück, H.-P. (2004) $D_2O$ adsorption and coadsorption with CO studied by in-situ high resolution x-ray photoelectron spectroscopy. *Langmuir*, **20** (5), 1819.
66. Haq, S., Harnett, J., and Hodgson, A. (2002) Growth of thin crystalline ice films on Pt(111). *Surf. Sci.*, **505**, 171.
67. Papp, C. (2007) In-situ investigations of adsorbed hydrocarbons – model systems of heterogeneous catalysis. PhD thesis. Friedrich-Alexander-Universität Erlangen-Nürnberg.
68. Kinne, M. (2004) Kinetische Untersuchungen von Oberflächenreaktionen mittels hochaufgelöster Röntgen-Photoelektronenspektroskopie – Oxidation von CO auf Pt(111) und zugehörige Elementarschritte. PhD thesis. Friedrich-Alexander-Universität Erlangen-Nürnberg.
69. Braun, W., Held, G., Steinrück, H.-P., Stellwag, C., and Menzel, D. (2001) Coverage-dependent changes in the adsorption geometries of ordered benzene layers on Ru(0001). *Surf. Sci.*, **475**, 18.
70. Zhu, J.F., Kinne, M., Fuhrmann, T., Denecke, R., and Steinrück, H.-P. (2003) In-situ high resolution XPS studies on adsorption of NO on Pt(111) surfaces. *Surf. Sci.*, **529**, 384.
71. Matsumoto, M., Tatsumi, N., Fukutani, K., and Okano, T. (2002) Dynamical low-energy electron diffraction analysis of the structure of nitric oxide on Pt(111). *Surf. Sci.*, **513**, 485.
72. Longwitz, S.R., Schnadt, J., Vestergaard, E.K., Vang, R.T., Lægsgaard, E., Stensgaard, I., Brune, H., and Besenbacher, F. (2004) High-coverage structures of carbon monoxide adsorbed on Pt(111) studied by high-pressure scanning tunneling

microscopy. *J. Phys. Chem. B*, **108**, 14497.

73. Ohtani, H., Wilson, R.J., Chiang, S., and Mate, C.M. (1988) Scanning tunneling microscopy observations of benzene molecules on the Rh(111)-(3 x 3) ($C_6H_6$+ 2CO) surface. *Phys. Rev. Lett.*, **60**, 2398.
74. Chiang, S. (2011) Imaging atoms and molecules on surfaces by scanning tunnelling microscopy. *J. Phys. D: Appl. Phys.*, **44**, 464001.
75. Simic-Milosevic, V., Bocquet, M.-L., and Morgenstern, K. (2009) Adsorption orientation and STM imaging of meta-benzyne resolved by scanning tunnelling microscopy and ab initio calculations. *Surf. Sci.*, **603**, 2479.
76. Yang, H.J., Minato, T., Kawai, M., and Kim, Y. (2013) STM investigation of CO ordering on Pt(111): from an isolated molecule to high-coverage superstructures. *J. Phys. Chem. C*, **117**, 16429.
77. Steinrück, H.-P. (1994a) Angle-resolved UV-photoelectron spectroscopy. *Vacuum*, **45**, 715.
78. Steinrück, H.-P. (1994b) Angle-resolved UV photoelectron spectroscopy of ethylene and benzene on nickel. *Appl. Phys. A*, **59**, 517.
79. Steinrück, H.-P. (1996) Angle-resolved photoemission studies of adsorbed hydrocarbons. *J. Phys. Condens. Matter*, **8**, 6465.
80. (a) Cardona, M. and Ley, L. (eds) (1978) *Photoemission in Solids I*, Springer, New York; (b) Cardona, M. and Ley, L. (eds) (1979) *Photoemission in Solids II*, Springer, New York.
81. Feuerbacher, B., Fitton, B., and Willis, R.F. (eds) (1978) *Photoemission and the Electronic Properties of Surfaces*, John Wiley & Sons, Inc., New York.
82. Smith, N.V. and Himpsel, F.J. (1983) in *Handbook on Synchrotron Radiation*, vol. 1B (ed. E.E. Koch), North Holland, Amsterdam.
83. Richardson, N.V. and Bradshaw, A.M. (1981) Symmetry and the Electron Spectroscopy of Surfaces, in *Electron Spectroscopy Theory, Techniques and Applications*, vol. 4 (eds C.R. Brundle and A.D. Baker), Academic Press, New York, p. 153–196.
84. Plummer, E.W. and Eberhardt, W. (1982) Angle-resolved photoemission as a tool for the study of surfaces. *Adv. Chem. Phys.*, **49**, 533.
85. Ertl, G. and Küppers, H. (1985) *Low Energy Electrons and Surface Chemistry*, Wiley-VCH Verlag GmbH & Co. KGaA, Weinheim.
86. Siegbahn, H. and Karlsson, L. (1982) Photoelectron Spectroscopy, in *Handbuch der Physik XXXI* (ed. W. Mehlhorn), Springer, Berlin, p. 215–468.
87. Freund, H.J. and Neumann, M. (1988) Photoemission of molecular adsorbates. *Appl. Phys. A*, **47**, 3.
88. Eberhardt, W. (1992) Angle-Resolved Photoelectron Spectroscopy, in *Synchrotron Radiation Research: Advances in Surface and Interface Science*, Techniques, vol. 1 (ed. R.Z. Bachrach), Plenum Press, New York, p. 139–197.
89. Netzer, F.P. (1990) Determination of structure and orientation of organic molecules on metal surfaces. *Vacuum*, **41**, 49.
90. Freund, H.-J. and Kuhlenbeck, H. (1995) Band-Structure Determination of Adsorbates, in *Applications of Synchrotron Radiation*, Springer Series in Surface Sciences, vol. 35 (ed. W. Eberhardt), Springer, Berlin, p. 9–63.
91. Scheffler, M. and Bradshaw, A.M. (1983) in *The Chemical Physics of Solid Surfaces and Heterogeneous Catalysis*, vol. 2 (eds D.A. King and D.P. Woodruff), Elsevier, Amsterdam, p. 165–257.
92. Woodruff, D.P. and Delchar, T.A. (1994) *Modern Techniques in Surface Science*, Cambridge University Press, Cambridge.
93. Netzer, F.P. and Ramsey, M.G. (1992) Structure and orientation of organic molecules on metal surfaces. *Crit. Rev. Solid State Mater. Sci.*, **17**, 397.
94. Gadzuk, J.W. (1974) Surface molecules and chemisorption: I. Adatom density of states. *Surf. Sci.*, **43**, 44.
95. Menzel, D., (1978) Core Level Photoemission from Adsorption Layers, in

*Photoemission and the electronic properties of surfaces*, (eds B. Feuerbacher, B. Fitton, and R.F. Willis) Wiley & sons, New York, Chapter 13, p. 381–408.

96. Gadzuk, J.W. (1976) Vibrational excitation in photoemission spectroscopy of condensed molecules. *Phys. Rev. B*, **14**, 5458. Vibrational excitation, hole delocalization, and photoelectron line shapes of molecules *Phys. Rev. B*, (1979), **20**, 515.
97. Gadzuk, J.W., Holloway, S., Mariani, C., and Horn, K. (1982) Temperature-dependent photoemission line shapes of physisorbed xenon. *Phys. Rev. Lett.*, **48**, 1288.
98. Norton, P.R., Tapping, R.L., Broida, H.P., Gadzuk, J.W., and Waclawski, B.J. (1978) High resolution photoemission study of condensed layers of nitrogen and carbon monoxide. *Chem. Phys. Lett.*, **53**, 465.
99. Fuggle, J.C. and Menzel, D. (1979) XPS, UPS and XAES studies of the adsorption of nitrogen, oxygen, and nitrogen oxides on W(110) at 300 and 100 K: I. Adsorption of $N_2$, $N_2O$ and $NO_2/N_2O_4$. *Surf. Sci.*, **79**, 1.
100. Höfer, U., Breitschafter, M., and Umbach, E. (1990) Photoemission line shapes of adsorbates: conclusions from the analysis of vibration-resolved N2 spectra. *Phys. Rev. Lett.*, **64**, 3050.
101. Bertolo, M., Hansen, W., and Jacobi, K. (1991) Hopping of the photohole during photoemission from physisorbed N2: The influence of band formation on vibrational excitation. *Phys. Rev. Lett.*, **67**, 1898.
102. Gustafsson, T. and Plummer, E.W., (1977) Valence Photoemission from Adsorbates, in *Photoemission and the electronic properties of surfaces*, (eds B. Feuerbacher, B. Fitton, and R.F. Willis) Wiley & sons, New York, Chapter 12, 353.
103. Hoffmann, R. (1988) A chemical and theoretical way to look at bonding on surfaces. *Rev. Mod. Phys.*, **60**, 601.
104. Holloway, S. and Norskov, J. (1991) *Bonding at Surfaces*, Liverpool University Press, Liverpool.
105. Rösch, N. (1992) in *Cluster Models for Surface and Bulk Phenomena*, NATO Advanced Study Institute, Series B, Physics, vol. 283 (eds G. Pacchioni, P.S. Bagus, and F. Parmigiani), Plenum Press, New York, p. 251.
106. G. Held and H.-P. Steinrück, *Cyclic Hydrocarbons, Landolt-Börnstein "Physics of Covered Solid Surfaces – Adsorbed Layers on Surfaces"*, Editor: H. P. Bonzel, Vol. III/42, Subvolume A4, Chapter 3.8.7. (2005) pp. 300-369.
107. Weinelt, M., Huber, W., Zebisch, P., Steinrück, H.P., Pabst, M., and Rösch, N. (1992) The electronic structure of ethylene on Ni(110): an experimental and theoretical study. *Surf. Sci.*, **271**, 539.
108. Ramsey, M.G., Steinmüller, D., Netzer, F.P., Neuber, M., Ackermann, L., Lauber, J., and Rösch, N. (1992) Cyanogen on Ni(110): an experimental and theoretical study. *Surf. Sci.*, **260**, 163.
109. Davenport, J.W. (1976) Ultraviolet photoionization cross sections for $N_2$ and CO. *Phys. Rev. Lett.*, **36**, 945. *J. Vac. Sci. Technol.*, (1978), **15**, 433..
110. Schichl, A., Menzel, D., and Rösch, N. (1984) Shape resonance energies in the $X\alpha$ scattered-wave cluster model. A study of CO and N2 adsorbed on Ni surfaces. *Chem. Phys. Lett.*, **105**, 285.
111. Dubs, R.L., Smith, M.E., and McKoy, V. (1988) Studies of angle-resolved photoelectron spectra from oriented NiCO: a model for adsorbed CO. *Phys. Rev. B*, **37**, 2812.
112. Nyberg, G.L. and Richardson, N.V. (1979) Symmetry analysis of angle-resolved photoemission: polarization dependence and lateral interactions in chemisorbed benzene. *Surf. Sci.*, **85**, 335.
113. Hofmann, P., Horn, K., and Bradshaw, A.M. (1981) Orientation of adsorbed benzene from angle-resolved photoemission measurements. *Surf. Sci.*, **105**, L260.
114. Smith, R.J., Anderson, J., and Lapeyre, G.J. (1976) Adsorbate orientation using angel-resolved polarization-dependent photoemission. *Phys. Rev. Lett.*, **37**, 1081.

115. Miranda, R., Wandelt, K., Rieger, D., and Schnell, R.D. (1984) Angle-resolved photoemission of CO chemisorption on Pd(111). *Surf. Sci.*, **139**, 430.
116. Weinelt, M., Huber, W., Zebisch, P., Steinrück, H.P., Reichert, B., Birkenheuer, U., and Rösch, N. (1992) Ethylene adsorbed on Ni(110): an experimental and theoretical determination of the two-dimensional band structure. *Phys. Rev. B*, **46**, 1675.
117. Horn, K., Scheffler, M., and Bradshaw, A.M. (1978) Photoemission from physisorbed xenon: evidence for lateral interactions. *Phys. Rev. Lett.*, **41**, 822.
118. (a) D.W. Turner, C. Baker, A.D. Baker and C.R. Brundle (1970) *Molecular Photoelectron Spectroscopy*, John Wiley & Sons, Ltd, London; (b) Stroscio, J.A., Bare, S.R., and Ho, W. (1984) The chemisorption and decomposition of ethylene and acetylene on Ni(110). *Surf. Sci.*, **148**, 499.
119. Huber, W., Weinelt, M., Zebisch, P., and Steinrück, H.P. (1991) Azimuthal reorientation of adsorbed molecules induced by lateral interactions: benzene/Ni(110). *Surf. Sci.*, **253**, 72.
120. Jorgensen, W.L. and Salem, L. (1973) *The Organic Chemists Book of Orbitals*, Academic Press, New York.
121. Eiding, J., Domcke, W., Huber, W., and Steinrück, H.P. (1991) Jahn-Teller effect of the $2e_{2g}$ level of chemisorbed benzene. *Chem. Phys. Lett.*, **180**, 133 (Erratum: *Chem. Phys. Lett.*, (1992), **191**, 203).
122. Fox, T. and Rösch, N. (1991) Modeling of the geometry of densely packed chemisorbed overlayers: small organic molecules on Ni(110) and Pd(110). *Surf. Sci.*, **256**, 159.
123. Horn, K., Bradshaw, A.M., and Jacobi, K. (1978) Angular resolved UV photoemission from ordered layers of CO on a Ni(100) surface. *Surf. Sci.*, **72**, 719.
124. Horn, K., Bradshaw, A.M., Hermann, K., and Batra, I.P. (1979) Adsorbate band formation: the chemisorption of CO on Pd (100). *Solid State Commun.*, **31**, 257.
125. Batra, I.P., Hermann, K., Bradshaw, A.M., and Horn, K. (1979) Theoretical and experimental studies of band formation on CO adlayers. *Phys. Rev. B*, **20**, 801.
126. Huber, W., Zebisch, P., Bornemann, T., and Steinrück, H.P. (1991) *Surf. Sci.*, **258**, 16.
127. Graen, H.H., Neuber, M., Neumann, M., Odörfer, G., and Freund, H.J. (1990) Lateral interaction in ordered hydrocarbon overlayers — C–H band dispersion of adsorbed benzene. *Europhys. Lett.*, **12**, 173.

# 39
# Surface Science Approach to Heterogeneous Catalysis

Günther Rupprechter

## 39.1
## Modeling Catalysts with Surface Science Methodology

"Make things (models) as simple as possible, but not simpler!" is a famous quote from Albert Einstein. This statement may be true for many things in science (and life!) but it applies specifically well to model systems of heterogeneous catalysts. When catalytic processes are modeled by a "surface science" approach, one must be very cautious whether studying just a small and simple fraction of a complex system turns out to be relevant at all. Before discussing all these obstacles (usually called *gaps*), let us consider why we need model studies. Why not examine just "real" technological catalysts?

The characterization and, particularly, the control of the *active sites* of a catalyst is the "holy grail" in heterogeneous catalysis research and a major step toward green chemistry, because this would allow 100% selectivity without undesired or harmful byproducts. However, industrial-grade heterogeneous catalysts that are used in chemical synthesis, environmental technology, and energy conversion [1–6] are typically rather complex materials. In many cases, they are composed of noble/transition-metal or alloy nanoparticles supported on porous (sometimes mixed) oxides [7–9], and in some cases they may be boosted by promoters. The loading of precious metal is often on the order of less than 1 wt%, and promoters may be present in much smaller amounts and are sometimes undetectable. The exact location and distribution of these components is often unknown, as are the "active surface configurations" that are present during the ongoing catalytic reaction, that is, under functioning (operando) conditions [10]. For metals, the shape and thus the distribution of various facets, edges, steps, defects (vacancies, twin boundaries, dislocations, etc.) may change depending on reaction environment [11–19]. The oxide support may have different terminations (metal, oxygen, or mixed) depending on synthesis and gaseous environment [20–25]. Furthermore, both metal and oxide surface structures may not be simple truncations of the well-known bulk structures. Indeed, in many cases they may be very different (relaxation, reconstruction), and for oxides they are largely unknown. The degree of oxide hydroxylation and the presence of defects (grain boundaries, oxygen

*Surface and Interface Science: Solid-Gas Interfaces I*, First Edition. Edited by Klaus Wandelt.

vacancies, polarons, etc.) may significantly affect the catalytic properties [26–28]. With respect to mono- or bimetallic metal nanoparticles [29], the nature of sites at the metal–oxide interface (termed *adlineation sites, phase-boundary sites,* or *three-face boundary sites* [30–32]) may also be largely unknown, despite advanced high-resolution transmission electron microscopy (HRTEM) [6, 33–38] and other techniques. Clearly, even catalysts with seemingly simple composition ("just noble metal and an oxide") exhibit significant complexity.

Consider Pt or Pd nanoparticles supported on a high-surface-area oxide. Pt-$Al_2O_3$ or Pd-$Al_2O_3$ (or supported on $SiO_2$) are used for the isomerization, hydrogenation, or oxidation of gas feeds in synthetic chemistry (e.g., for hydrocarbons) or emission control (e.g., CO oxidation). When HRTEM was applied to catalysts in the late 1970s, the atomic structure of the supported nanoparticles could be directly imaged for the first time [34, 37, 39]. Nevertheless, in case of low metal loading, the nanoparticles are so small that sufficient contrast cannot be achieved.

In Figure 39.1a, TEM is unable to detect the metal nanoparticles of a 2 wt% Pd-$Al_2O_3$ catalyst because the oxide support itself exhibits nanosized crystalline regions (although the presence of Pd in a specific sample area can be proven by energy-dispersive X-ray fluorescence, EDX). The metal loading can be increased until the nanoparticles are detected by TEM, but this catalyst is technically less relevant (Figure 39.1b). For the catalyst shown in Figure 39.1b, the Pd nanoparticles are ~3–4 nm in diameter (some of them marked by arrows), are of cuboctahedral shape, and exhibit mostly (111) and (100) surface facets (see the high-resolution image in Figure 39.1b). In favorable cases, HRTEM also provides information on steps on the nanoparticle surfaces (Figure 39.1c) [42], on metal–support interfaces,

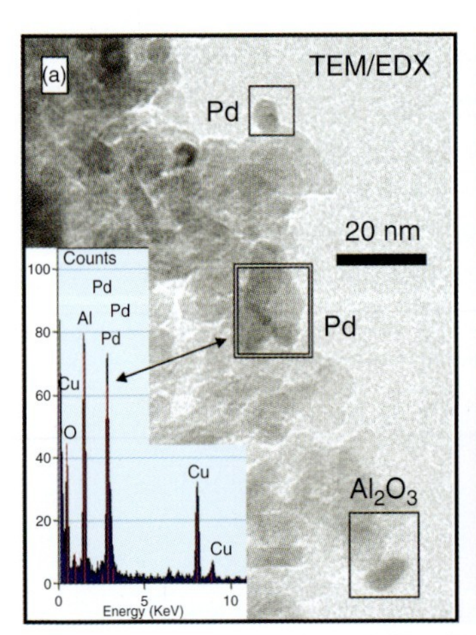


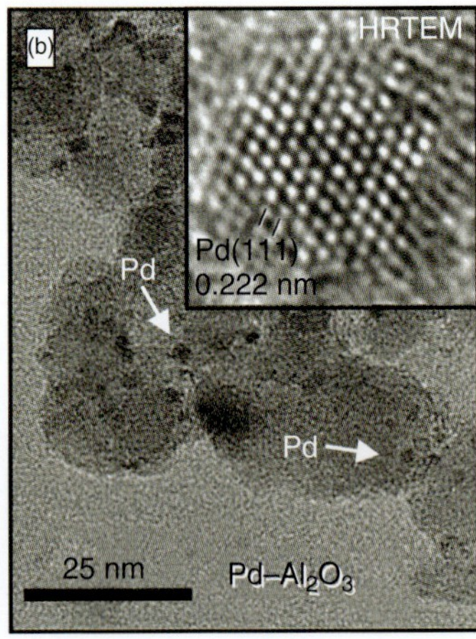


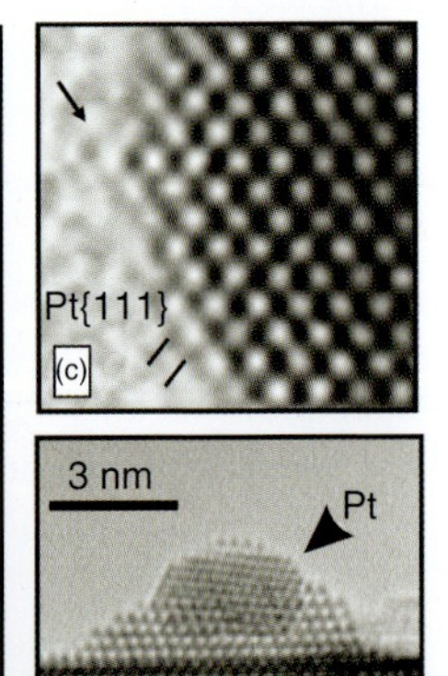


**Figure 39.1** (a) Transmission electron micrograph of a technological (impregnated) 2 wt% Pd-$Al_2O_3$ catalyst. (b) 5 wt% Pd-$Al_2O_3$ catalyst. The inset shows the atomic structure of a cuboctahedral Pd particle, as revealed by high-resolution imaging. (c) HRTEM image of surface steps on a Pt nanoparticle. (d) HRTEM image of a mixed-metal oxide catalyst. (Adapted from Refs [40, 41] with permission from Elsevier.)

or even on partial oxide encapsulation (decoration) of the metal nanoparticles (Figure 39.1d) [43].

Of course, there are also indirect ways of characterizing such small entities, for example, via the adsorption of probe molecules such CO, hydrogen, or ammonia, that is, by titrating the surface atoms via chemisorption [44]. Using this method, the dispersion $D$, that is, the ratio between the number of surface noble metal atoms and the total number of noble metal atoms in a catalyst, could be determined (assuming a specific ratio between adsorbed molecules and surface atoms). For a certain particle shape, for example, hemispheres, the nanoparticle diameter could then be calculated (also assuming selective adsorption on the metal or correcting for adsorption on the support or for hydrogen absorption). Temperature-programmed techniques (TPD, TPR, TPO) can then be used to differentiate between different adsorption sites. Likewise, Fourier transform infrared spectroscopy (FTIR) of adsorbed molecules enables one to identify different adsorption sites on the nanoparticles [45].

Altogether, these methods of (HR)TEM, chemisorption, temperature-programmed methods and infrared represent the most important characterization tools for heterogeneous catalysts, providing information on nanoparticle size, shape, atomic (surface) structure, number and type of adsorption sites, binding strength of adsorbates, and so on. Nowadays, this is typically complemented by various other methods such as physisorption, X-ray diffraction (XRD), EDX, edge X-ray absorption, X-ray absorption spectroscopy (XAS), Raman spectroscopy, X-ray photoelectron spectroscopy (XPS), and so on (see Volume 1 and Chapter 31 in this Volume). Nevertheless, all these methods still integrate over the whole catalyst and thus assume perfect homogeneity (despite HRTEM showing that many inhomogeneities in structure and composition may exist in a catalyst).

This complexity is the driving force to utilize simplified model systems to examine fundamental processes in heterogeneous catalysis [5, 6, 13–16, 19, 32, 46–52], ranging from monometallic low-Miller-index single-crystal surfaces to (promoted) (bi)metallic nanoparticles on mixed-oxide supports. One should note that this was particularly true about 50 years ago when the first surface science model studies were performed. Those days it was nearly impossible to obtain detailed molecular-level information for industrial catalysts, whereas nowadays also technological studies can be performed on a very sophisticated level, challenging the surface science approach.

### 39.1.1
### Overview on Model Catalysts of Various Complexities

In the following, the most important types of model systems for metal–oxide catalysts are presented, in a sequence of low to high complexity (Figure 39.2).

#### 39.1.1.1 Single-Crystal Surfaces

Single-crystal surfaces of noble and transition metals have been most frequently used for fundamental studies [48, 54, 55]. Their strength, that is, providing a homogeneous (typically low-index) surface, is also their weakness. The mere

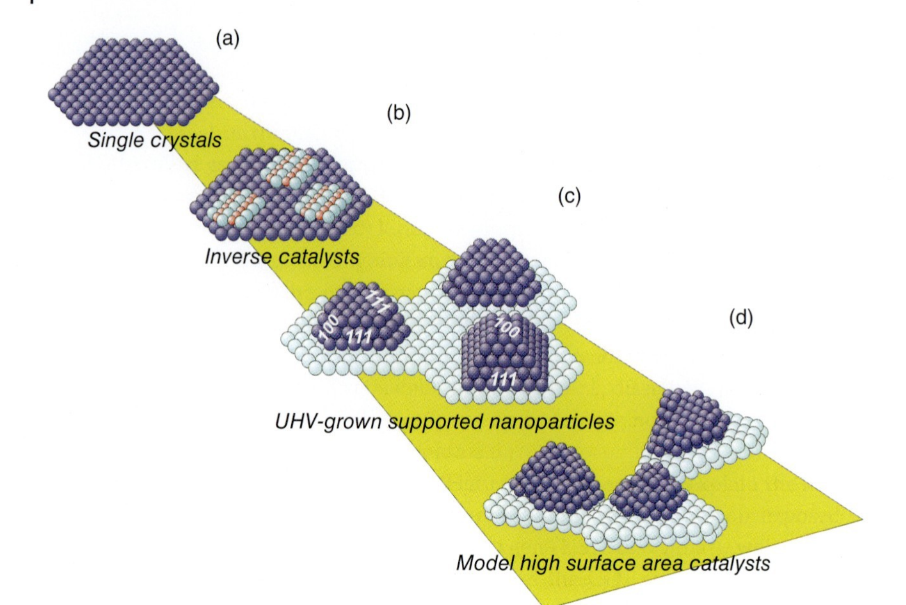


**Figure 39.2** Schematic view of the most common model catalyst. (a) Smooth and stepped noble metal single crystals. (b) Thin oxide films grown on single-crystal substrates (inverse catalysts). (c) Ultrahigh-vacuum-grown metal nanoparticles supported by thin oxide films (grown on single-crystal substrates). (d) Model high-surface-area catalysts. Systems (a)–(c) are prepared under ultrahigh vacuum, which allows control of catalyst morphology and composition (cleanliness). System (d) provides a link between surface science-based and industrial catalysts. (Adapted from Ref. [53] with permission from IOP Publishing.)

presence of a specific metal excludes several complexities of a technological catalyst: (i) macroscopic bulk crystals versus nanocrystals (with implications on atomic and electronic structure [56]), (ii) single termination versus several different communicating surface facets [57], (iii) the absence of an oxide support and of metal–oxide interfaces, and (iv) various other effects including the absence of hydroxylation, spillover, and reverse spillover. Further complications may arise simply from "finite size" effects [58, 59]. Compared to that of a nanoparticle, the volume of a macroscopic single crystal is infinite. When atomic species are dissolved in a metal, such as hydrogen, oxygen, carbon, or other metals like Zn [60–62], nanoparticles will be saturated very quickly with these species whereas they basically vanish in the bulk of single crystals. Consequently, the properties of a metal nanoparticle are not simply a superposition of the properties of its individual surface facets. Overall, this may lead to artificial differences in the adsorption and catalytic properties of single crystals and nanoparticles, and great care has to be taken in the interpretation of such data [58, 63].

Nevertheless, low- and high-index single-crystal surfaces are still very powerful model systems. There structural homogeneity allows the elucidation of structure–property relationships since the results of integrating methods apply to the entire surface. The structural simplicity is also an advantage for comparison with theoretical studies, since modeling surfaces by, for example, density functional theory (DFT) methods is still easier than cluster calculations. Also, microscopic methods such as scanning tunneling microscopy (STM) or photoemission electron microscopy (PEEM) can be more easily applied to smooth single-crystal surfaces. Whereas STM reaches atomic resolution both for the surface structure and adsorbed molecules (Figure 39.3a) [16, 55, 70, 71], the parallel imaging principle of PEEM allows observation of surface processes on a mesoscopic scale [72–74]. Starting from the smooth and densely packed, for example, face-centered cubic (fcc) (111) and (100) surfaces, the surface roughness can be increased by switching to (110), (211), (311), and so on, surfaces, finally approaching stepped and kinked configurations (e.g., 557, etc.). Using this approach, the structure sensitivity of hydrocarbon reactions on Pt surfaces has been demonstrated early on by Somorjai and coworkers [75], and the complexity of CO oxidation was explained by Ertl [76]. The effect of defect or cow-coordinated sites can also be examined by comparing smooth (annealed) and sputtered surfaces.

Apart from the clean (monometallic) single-crystal surfaces, these model systems can be modified in many ways. Very simply, adsorption/desorption experiments can just be repeated several times without intermittent sample cleaning [77]. In case of CO or hydrocarbon dissociation, the surface will be continuously affected by carbonaceous species, often blocking typically threefold hollow sites. Another route is to pre-expose the clean surface to, for example, oxygen or hydrogen or to hydrocarbons at elevated temperature (to create $CH_x$ species). Of course, single crystals can also be modified by ultrathin layers of other metals and, depending on annealing temperature, the overlayers may be present as individual atoms, islands, or continuous thin films but they may also alloy with the substrate.

In summary, despite their simplicity and the limitations mentioned above, studies of single-crystal surfaces provide a wealth of information on the elementary steps of catalytic reactions, which has led to the discovery of relaxation and reconstruction reconstruction, of adsorbate-induced of adsorbate-induced reconstruction, of coadsorption and site-blocking phenomena, of structure sensitivity, of thickness-dependent properties, and so on [14, 48, 54, 55].

#### 39.1.1.2 Inverse Model Catalysts (Oxide Overlayers/Islands on Metal Single Crystals)

In order to examine the metal–oxide interface, thin oxide islands can be grown on metal substrates [31, 32, 65, 78, 79]. A major finding was that, for specific reactions, a rate maximum was observed for around half-monolayer oxide coverage, clearly pointing to active sites at the phase boundary. Such rate enhancement at metal–oxide interfaces has often been termed *strong metal support interaction* (SMSI) but this is an erroneous use of the original term SMSI (which refers to the loss of adsorption capacity and of catalytic activity [80] due to, for example, encapsulation of metal particles by (sub)oxide overlayers [30, 81, 82]). Similarly, when two-dimensional (monolayer or very thin) islands of metal were grown on oxide substrates, this has also been sometimes wrongly called SMSI. Studies of oxide overlayers on metal are still an active field or research, with $CeO_2$/Pt [79],

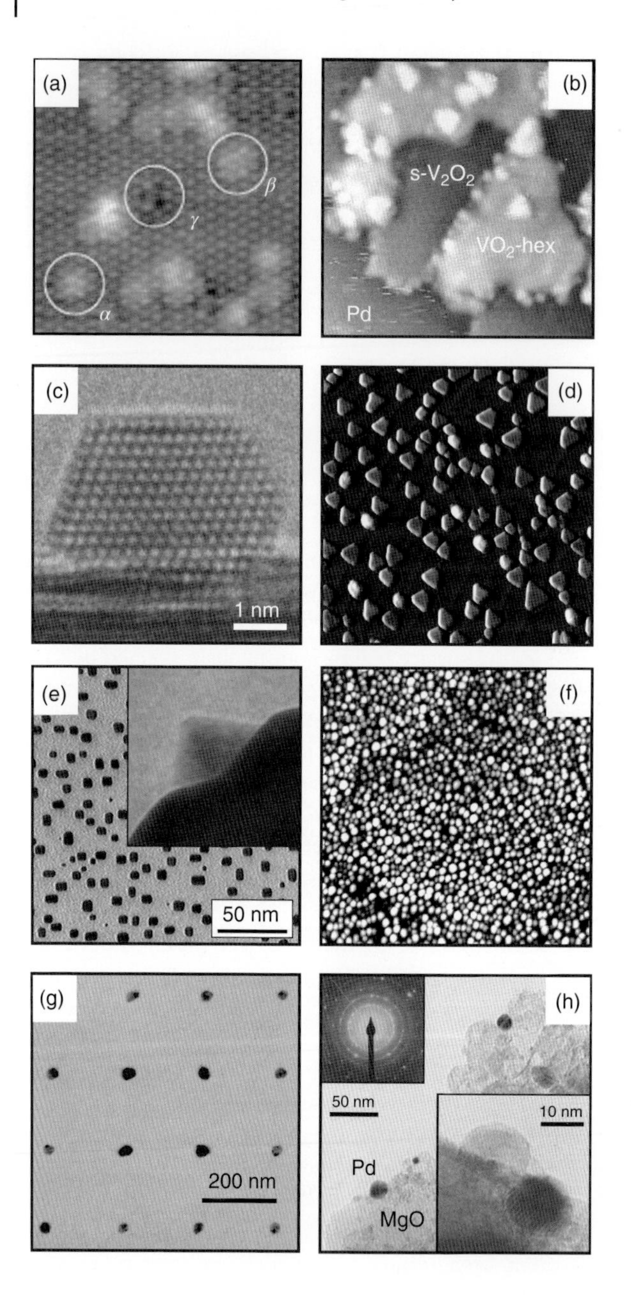
(a)
β
γ
α
(b)
s-V2O2
VO2-hex
Pd
(c)
1 nm
(d)
(e)
50 nm
(f)
(g)
200 nm
(h)
50 nm
10 nm
Pd
MgO

$CeO_2$/Au [83], $Fe_2O_3$/Pt [84, 85] $TiO_x$/Pt [86], and $VO_x$/Pd [78] being among recent examples. One should note, however, that the number of these studies is, as compared to single crystals, rather limited. Apart from the abundance, the inverse catalysts still await extended studies using a wider range of spectroscopic and microscopic methods. Finally, one should not forget that phase boundaries between oxide islands on a planar metal single crystal may still behave differently from those between nanoparticles and an oxide support.

#### 39.1.1.3 **Supported Nanoparticle Model Catalysts**

The most realistic model systems apparently consist of oxide-supported metal nanoparticles and their key ingredient is the model support [5, 24, 51]. Early attempts used oxide single crystals, but their insulating nature strongly limits the application of surface-sensitive methods that mostly rely on charged particles. A major breakthrough was the development of well-ordered ultrathin and atomically flat oxide films as model supports. To date, many technological oxides have been modeled, including $Al_2O_3$ grown on NiAl(110) [5, 16] or $Ni_3Al$(111) [87, 88], $SiO_2$ grown on Mo(112) [89–91], $CeO_2$ grown on Ru(0001) [92], $V_xO_y$, $Fe_3O_4$ grown on Pt(111) [84, 85, 93, 94], MgO grown on Ag(100) [25, 95], $Nb_2O_5$ grown on $Cu_3Au$(100) [96], $ZrO_2$ grown on $Pt_3Zr$ [97, 98], and so on (for reviews see [24, 99, 100, 101]). Typically, these oxide films are a few atomic layers to a few nanometers in thickness with sufficient electrical/thermal conductivity, and are grown on conducting substrates (Figure 39.4a; steps 1 and 2), so that surface science methods can typically be applied without charging problems.

In a subsequent step, mono- or bimetallic nanoparticles are then grown on the oxide via physical vapor deposition and nucleation and growth processes (Figure 39.4a; step 3) [6, 15, 18, 36, 102–105]. Using this approach, the number density of nanoparticles (particles per centimeter square of support) as well as the particle size can be controlled via the substrate temperature and the amount of evaporated metal, respectively (some morphological parameters are indicated at the bottom of Figure 39.4).

This approach is illustrated for Pd-$Al_2O_3$/NiAl(110) model catalysts in more detail (Figure 39.4). The alumina film was grown by oxidation of a NiAl(110) alloy single crystal at $10^{-5}$ mbar $O_2$ at 523 K followed by annealing at 1000 K. The oxide atomic structure was examined by low-temperature (4 K) STM [103, 106] and

**Figure 39.3** Model catalysts most frequently used for investigations of elementary processes of heterogeneous catalysis. (a) Atomically resolved STM image of a (nominally) clean Pd(111) surface (6.0 × 6.6 nm; 4 pm corrugation; subsurface impurities are marked). (Adapted from Ref. [64].) (b) STM image of an inverse model catalyst: vanadium oxide islands on Pd(111). (Adapted from Ref. [65].) (c) Au particle on MgO(100). (Adapted from Ref. [66].) (d) STM image of cuboctahedral palladium nanocrystals grown on $Al_2O_3$-NiAl(110) at 300 K. (Adapted from Ref. [67].) (e) Pyramidal platinum particles (grown on NaCl(100)) supported by $Al_2O_3$ [42] (the inset shows a profile (side) view of a single particle). (f) STM image of palladium nanoparticles grown on $Al_2O_3$-NiAl(110) at 90 K. (Adapted from Ref. [67].) (g) Platinum nanoparticle array grown on $SiO_2$ by electron beam lithography [68]. (h) Impregnated palladium-MgO catalyst, with the inset showing a profile view [69].

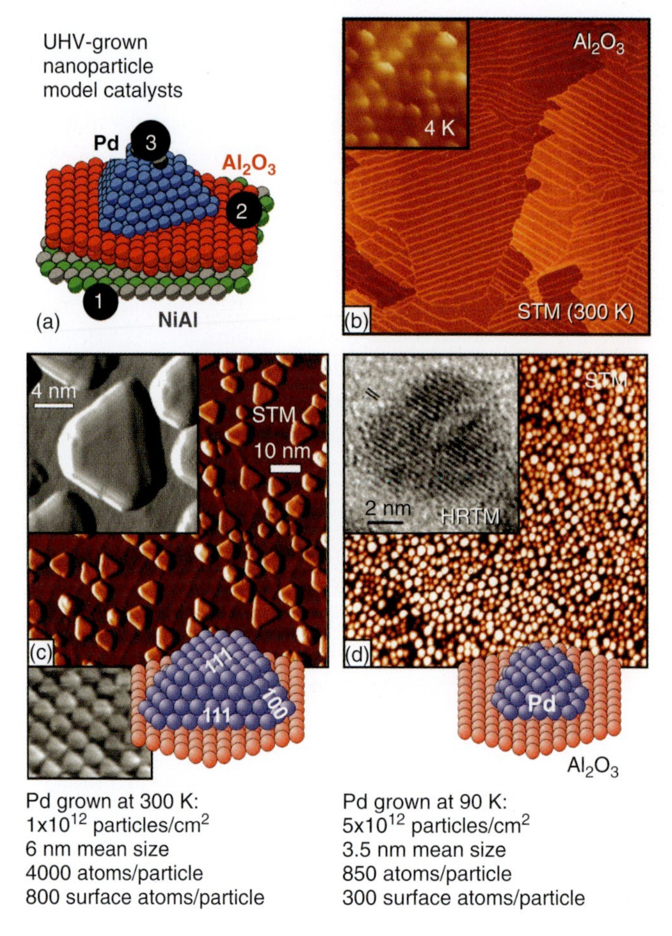


**Figure 39.4** (a) Graphical illustration of the preparation of Pd nanoparticles supported by $Al_2O_3$/NiAl(110). (b) STM image of the thin $Al_2O_3$ support (500 × 500 nm, (adapted from Refs [5, 18, 102])) acquired at 300 K, and atomic resolution obtained at 4 K. (Inset; (adapted from Ref. [103]).) (c) STM image (100 × 100 nm, (adapted from Refs [67, 104, 114])). (d) HRTEM image of Pd nanoparticles, with the insets showing individual particles at higher magnification. Controlling the growth conditions allows the preparation of Pd particles with different morphology/surface structure, for instance well-faceted truncated cuboctahedron (c) or rougher, defective particles (d).

surface XRD [107]. The alumina film was only ~0.5 nm thick and hydroxyl-free, and its exact structure deviated from that of bulk aluminas [102, 107, 108]. Preparation routes to hydroxylate model oxides have also been worked out [109].

Pd nanoparticles were then grown on the thin oxide film by physical vapor deposition (electron beam evaporation of a Pd rod). Figure 39.4c,d shows the STM results for Pd nanoparticles grown on $Al_2O_3$/NiAl(110) at 300 and 90 K substrate temperatures [5, 18, 67, 102]. For the dependence of the island density, of the mean particle diameter, and of average number of atoms per particle on the nominal film thickness and the growth temperatures, see Refs [6, 102]. Similar nucleation studies were performed for Pt, Rh Ir, and others [102, 104, 110].

Upon Pd deposition (nominal thickness of 2 ML (monolayer)) at 300 K, the Pd particles were preferentially located along line defects (nucleation density $\sim 1 \times 10^{12}$ particles/cm$^2$) [67, 104, 110, 111]. At 90 K, the reduced Pd mobility led to a higher nucleation density ($\sim 5 \times 10^{12}$ particles/cm$^2$) and thus to a more homogeneous distribution of particles on alumina terraces. The $Al_2O_3$ temperature during Pd deposition also influenced the particle shape and surface structure. For the given conditions, Pd particles grown at 90 K had a mean size of 3.5 nm and were of rounded (irregular, about hemispherical) shape (~850 atoms/particle; about 300 surface atoms). No atomically resolved STM images could be obtained, but HRTEM suggested a high number of low-coordination sites (defects, steps; see the inset in Figure 39.4d [42, 112]). Pd particles grown at 300 K had a mean size of 6 nm with a hexagonal outline, suggesting a cuboctahedral shape (about 4000 atoms/particle; about 800 surface atoms; Figure 39.4c). Atomically resolved STM images of individual particles (see inset in Figure 39.4c) indicated (111) top facets and (111) and (100) side facets [5, 111, 113, 114], with the (111) facets comprising about 80% of the particle surface (and the remaining ~20% being (100) facets). For a detailed account of particle structural parameters (mean size, number of (surface) atoms) and of surface site statistics (relative distribution of terrace, edge, corner, interface atoms, etc.), refer to Tables 1 and 2 in Ref. [36].

The nanoparticle model catalysts can then be used to model many catalytic processes, ranging from (co)adsorption via diffusion to reactivity studies. Despite their resemblance to technological catalysts, there are still differences to the "real world." Apart from the differences already discussed, one could question whether an ultrathin, sometimes trilayer oxide (O–Me–O on a substrate), can really model a macroscopic powder oxide. There may be differences in the atomic structure (in particular defect structure) and in the degree of hydroxylation (ultrahigh vacuum (UHV)-grown oxides are typically nonhydroxylated). Most importantly, for ultrathin films the electronic structure of the nanoparticles may be additionally affected by the underlying substrate as, for example, shown for Au nanoparticles on very thin MgO layers grown on Ag(100) [25]. The oxide–substrate interface may also provide a "sink" for oxygen or hydrogen atoms, which is not present for technological catalysts [58, 63]. Similarly, when metal atoms become mobile at elevated temperatures, they may react (alloy) with the substrate (and "disappear"), rather than sinter as nanoparticles do in a technological catalyst.

Nevertheless, supported nanoparticle model catalysts are, of course, the most realistic models. Apart from UHV growth, alternative approaches have been used that are more similar to industrial wet-chemical synthesis, such as spin coating [51]. Recently, atomic-layer deposition (ALD) has also been utilized [115, 116]. For some time, electron beam lithography (EBL) was used to prepare ordered arrays of metal nanoparticles [117–120]. However, the obtained particle sizes (>10 nm) were too large, there were too much synthesis residues ($CH_x$) from removing the photoresist, and overall EBL was too time consuming, which eventually hindered its wider application.

#### 39.1.1.4 High-Surface-Area Model Catalysts

Finally, one should mention "high-surface-area model catalysts," that is, technological catalysts prepared by standard wet-chemical routes but using clean (residue-free) precursors and oxides, often with somewhat larger metal loading to obtain higher quality spectroscopic and microscopic results, approaching those of surface science studies. These models thus provide a bridge between UHV models and industrial catalysts, especially when corresponding studies are performed under similar conditions both for model and applied catalysts [121–124].

### 39.1.2 Dynamic Changes of Catalysts

Upto this point, we have mainly considered the (model) catalyst in its "as-prepared" state, that is, after preparation. However, when materials are used for (catalytic) processes at elevated temperature and in various gas atmospheres, their structure and composition may change significantly. In fact, the transformation of technological heterogeneous catalysts from the "as-prepared" to the "active-state" has been extensively studied [15, 32, 34, 35, 39, 47, 125–137].

As illustrated in Figure 39.5, different effects may occur during the activation (e.g., by oxidation/reduction), lifetime (operation), and regeneration (rejuvenation) of oxide-supported metal catalysts.

#### 39.1.2.1 Surface Restructuring

In order to remove carbonaceous synthesis residues or other contaminants, repeated cycles of oxidation and reduction (up to ~973 K) are frequently carried

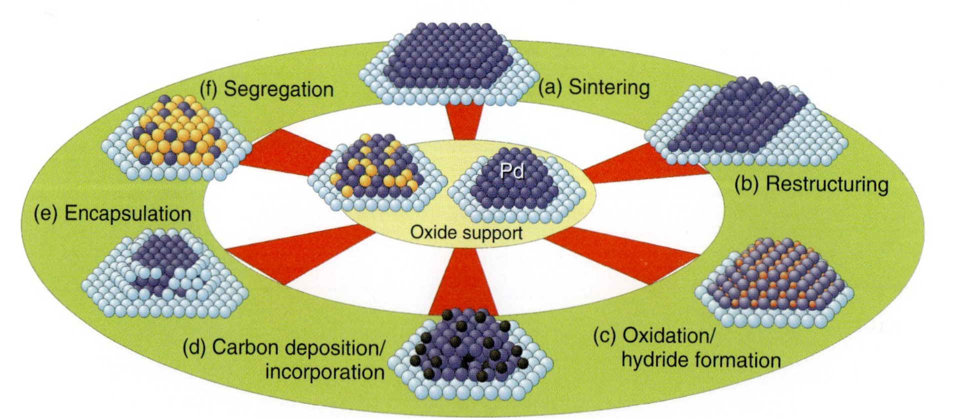


**Figure 39.5** Illustration of some structural and compositional changes of oxide-supported nanoparticles that may occur during the catalysts operation. (a,b) Sintering and restructuring. (c,d) Oxide/hydride formation and coking. (e) Particle encapsulation. (f) Surface segregation of bimetallic particles. Similar changes may occur for single-crystal model catalysts. (Adapted from Ref. [53] with permission from IOP Publishing.)

out. This catalyst (re-)activation may lead to either sintering (Figure 39.5a) or redispersion of metal nanoparticles [138]. Even when the particle size (dispersion) remains unchanged, the particle shape and surface structure typically change (reshaping, restructuring). Such changes depend on the relative surface energies of the metal, metal oxide, and support oxide (wetting and de-wetting behavior) and can produce low-index (smooth and high-coordinated) or high-index (rough and low-coordinated) surfaces (Figure 39.5b) [15]. The changes in surface energy are induced by gas ($O_2$, $H_2$) adsorption, by the removal of contaminants, or by partial reduction of the support oxide (metal-support interaction [30, 80]). When a catalyst has reached a stable activity after activation, it is still subject to mid- or long-term effects, which may reduce its activity/selectivity. For example, rough (low-Miller-index) particle facets may smoothen, leading to less active low-index facets [120, 137, 139]. This transformation may again originate from gas adsorption, weakening the metal–metal bonds (adsorbate-induced restructuring) [140]. Vice versa, exposure to CO roughens, for example, Pt surfaces [141] but also promotes sintering [142].

#### 39.1.2.2 Compositional Changes

Catalysts often alter their performance upon initial exposure to the reactant feed. During the so-called "induction period", a stable activity is reached that may be lower or higher than the initial activity, often accompanied by pronounced compositional changes [143]. For instance, Pd-oxide phases [100, 144] and Pd-hydrides [58, 63, 145, 146] may be formed during oxidation and hydrogenation reactions, respectively (Figure 39.5c). For hydrocarbon reactions, the formation of carbonaceous species or of a metal-carbon phase (Figure 39.5d) may not be detrimental but rather required for high activity [147–154]. In contrast, in many cases activity quickly drops in the first minutes/hours of a reaction, for example, when the most active sites decompose hydrocarbons, leading to self-poisoning by $CH_x$. For hydrocarbon (olefin, diene, or alkyne) hydrogenation or ($CH_4$) reforming, carbon deposition (coking) is a frequent cause of long-term deactivation, with the carbon species successively reducing the accessible noble metal area or even destroying the catalyst (carbon filament growth). Pronounced compositional changes may particularly occur for bimetallic nanoparticles, for example, when – depending on the reaction environment – one constituent segregates on the surface (Figure 39.5f) [155–159], with pronounced effects on catalytic performance. When complex oxides such as perovskites are used as electrodes in electrochemical cells, such compositional changes may not only depend on the gas environment but may even be driven by the applied potential [160].

#### 39.1.2.3 Metal–Support Interaction

For reducible oxide supports, exposure to hydrogen can lead to the formation of (sub)oxides that encapsulate and thus deactivate the nanoparticles (SMSI [80]) (Figure 39.5e). The most prominent SMSI oxides are clearly $TiO_2$ and $CeO_2$ [30, 35, 81, 86]. However, the SMSI effect is not necessarily detrimental for activity. When the encapsulation is partially reversed (e.g., by oxidation), the nanoparticle surface remains partly covered by patches of (sub)oxide, and this in fact increases the metal–support interface, which may lead to improved activity [123, 161, 162]. This state can also be achieved by hydrogen reduction at moderate temperatures, leading to the so-called WMSI state (weak metal support interaction) [30]. Partial reduction of the oxide support may not only lead to oxide overgrowth but may also create bimetallic nanoparticles. For example, when Pd-ZnO [123], Pd-$Ga_2O_3$ [163–165], or Pd-$In_2O_3$ catalysts are reduced or simply exposed to reaction (methanol steam reforming, MSR) conditions, PdZn, $Pd_2Ga$, and PdIn bimetallic (intermetallic, IMC) nanoparticles are formed, respectively, which represent the active phase [29, 62, 166, 167].

### 39.1.3 Functioning Model Catalysts – Mind the Gap!

All the described potential structural and compositional changes that are induced by the reaction environment illustrate the importance of examining (model) catalysts under functioning (*operando* [168]) conditions. Notwithstanding the detailed understanding of catalytic processes gained only from experiments under UHV, the relevance of surface-sensitive studies under UHV for technological catalysis has repeatedly been questioned [36, 169–171] since there is at least 10 orders of magnitude pressure difference between typical UHV surface science investigations and applied catalysis, corresponding to the so-called *pressure gap.*

Initially it has been speculated that at high pressure (>1 bar) the obtained (saturation) coverages likely exceed those accessible under UHV. This may lead to different (co)adsorbate structures (compressed overlayers), but it also has been suggested that adsorbate geometries may be present that are even unknown ("high-pressure species"). If this were true, one would expect a strongly modified surface chemistry, different from that revealed by surface science. However, to date no unknown "high-pressure species" have been discovered by applying "*in situ* methods" that are capable of providing surface-sensitive information at (or near) ambient pressure (such as SFG, PM-IRAS, NAP-XPS, NAP-STM, etc. [10, 172, 173]). All high-pressure configurations were compatible with the well-known species identified previously under UHV.

There may be still differences between adsorbates under UHV and ambient pressure, especially at elevated (reaction) temperatures. In order to obtain high coverages, UHV investigations are often restricted to cryogenic temperature (due to limitations of gas pressure $<10^{-6}$ mbar). However, at low temperature, the adsorbate mobility is also low, which may lead to nonequilibrium adsorbate structures, CO on Pd(111) [174] and Pt(111) [175, 176] being well-known examples. Such structures are then not characteristic of the catalytic situation. Nevertheless, upon annealing or upon gas dosing during cooling, the nonequilibrium structures may convert to equilibrium structures. Furthermore, if the active species is weakly bound, it will appear only at high pressure (high coverage), after all the strongly adsorbing sites on the surface are occupied by more strongly bonded species [177, 178].

As already mentioned, apart from the mere adsorbate structures on a pristine surface, model catalyst surfaces may undergo restructuring at elevated pressures and temperatures (e.g., step formation, surface roughening) [11, 42, 68, 120, 137, 161, 179–181]. The composition of the surface may also depend on the gas pressure; for example, a surface may change from that of a metal with adsorbed oxygen to a surface metal oxide [53, 144, 182–187] or to a metastable (subsurface) oxide that cannot be identified in UHV or by other analyses [53, 188–190].

Furthermore, when catalytic reactions are studied at millibar pressures, the high impingement rates of reactants (being about nine orders of magnitude higher than under UHV conditions) may enable the observation of processes that may seem absent under UHV. For example, the scission of the methanolic C–O bond on Pd(111) was absent under UHV (upon Langmuir exposures) but was observed at elevated pressures by near-ambient pressure X-ray photoelectron spectroscopy (NAP-XPS) [191]. Likewise, upon $CO_2$ exposure, carbonate formation on $Al_2O_3$ thin films did not occur in UHV but could be detected only at millibar gas pressure [122].

It is apparent that such "pressure effects" have a strong impact on the catalytic properties and that *in situ* measurements under elevated pressure are desirable. Concerning the "pressure gap", one should note that the surface science approach still provides very valuable information. In its most active state at high pressure and high temperature, a catalyst typically exhibits only low coverage; otherwise, there would be self-poisoning by the reactants. As a consequence, *in situ* spectroscopy of active catalysts is very difficult because the adsorbate surface concentrations are typically low. Vice versa, when coverages are high(er) and well suited for *in situ* spectroscopy, the catalyst is typically not in its most active state and often mainly populated by spectator species. Differentiation between active and spectator species is very challenging but can, for example, be mastered by concentration modulation infrared spectroscopy [163, 192].

#### 39.1.3.1 Surface-Sensitive Studies on Model Catalysts Under Realistic Conditions

Model catalysis bridging both the material and pressure gaps must thus integrate several approaches. First, surface-sensitive techniques are required that can operate under technologically relevant conditions, that is, at least in the 1–1000 mbar pressure range. In this respect, photon-based techniques such as sum frequency generation (SFG) and polarization-modulation infrared reflection absorption spectroscopy (PM-IRAS) provide surface vibrational spectra of adsorbates from UHV up to atmospheric pressure [36, 52, 193]. Although electron spectroscopies are typically limited to pressures $<10^{-4}$ mbar, recent developments in NAP-XPS allow the determination of complementary chemical information at pressures up to $\sim$1 mbar (applying differential pumping and electrostatic lenses to focus the photoelectrons) [194–197]. Direct structural information under millibar pressure can be provided by high-pressure STM [71, 198, 199]. When one admits millibar pressures to model catalysts, special care must also be taken of the cleanliness of the gas [36]. For example, CO must be cleaned from volatile Ni and Fe carbonyls [195, 200, 201], which would otherwise lead to metal deposition on the model catalyst.

Second, apart from single crystals, model catalysts with increasing complexity should be employed, finally approaching the complex properties of supported metals (Figure 39.2). Nevertheless, the metal nanoparticles should still exhibit well-defined surface facets to allow more reliable data interpretation and a comparison with single-crystal and theoretical results.

A third gap may be related to the *complexity* of the reactions studied. The most frequently examined reactions in surface science-based model catalysis are CO oxidation and ethylene hydrogenation, which, of course, are much simpler than most technologically relevant catalytic reactions. A fourth gap may be a *methodology gap*, as discussed in more detail below.

#### 39.1.3.2 Experimental Setup for Surface Spectroscopy or Surface Microscopy during Catalysis

Figure 39.6 shows an experimental setup capable of performing surface spectroscopy on model catalysts from UHV to ambient pressure. This requires an apparatus that combines a UHV preparation and surface analysis chamber (Figure 39.6a) with a UHV–high pressure reaction cell (Figure 39.6c,d), which allows performing vibrational spectroscopy on the catalyst under working conditions [36, 42, 202, 203]. The (upper) UHV section is equipped for sample preparation and characterization. The model catalyst can then be transferred under UHV to the reaction cell with the help of a manipulator. During this operation, the sample holder is inserted into an arrangement of three differentially pumped, spring-loaded Teflon seals (Figure 39.6d), and the reaction cell is separated from the UHV part [36, 202]. Vibrational spectroscopy can then be performed in the reaction cell, either under UHV or at pressures up to 1 bar (e.g., by PM-IRAS (Figure 39.6b) or SFG (Figure 39.6d)). The high-pressure cell is also interfaced to a gas chromatography (GC) setup for product analysis.

A multipurpose UHV system for performing *in situ* surface (reaction) microscopy on model catalysts by PEEM [72, 73, 204] is shown in Figure 39.7. It again consists of two independently operated chambers connected with each other by a sample transfer line, thus allowing a common reactive gas atmosphere in the $10^{-4}$–$10^{-9}$ mbar range. The microscopy chamber (right side) is equipped with sample preparation facilities, PEEM, MS, and LEED, and the spectroscopy chamber (left side) is equipped with XPS (hemispherical energy analyzer and twin anode). This provides the global (MS, mass spectrometry) as well as the local "laterally resolved" (PEEM) kinetics on heterogeneous model catalysts such as a polycrystalline metal foil or oxide-supported metal particles (micrometer to nanometer range). XPS provides complementary information on the chemical composition of the samples in UHV and under the same reactive gas atmosphere as during PEEM imaging. Both chamber sections are equipped with a high-purity gas supply system and argon ion sputtering.

In PEEM, a magnified image of the sample surface is created by photoelectrons emitted upon UV or X-ray illumination [48, 72]. In the studies described below, a deuterium discharge UV lamp was utilized, with an intensity maximum at approximately 190 nm (photon energy ~6.5 eV; focused to about 1 mm spot on the sample). PEEM images were recorded by a high-speed CCD camera, and magnification was

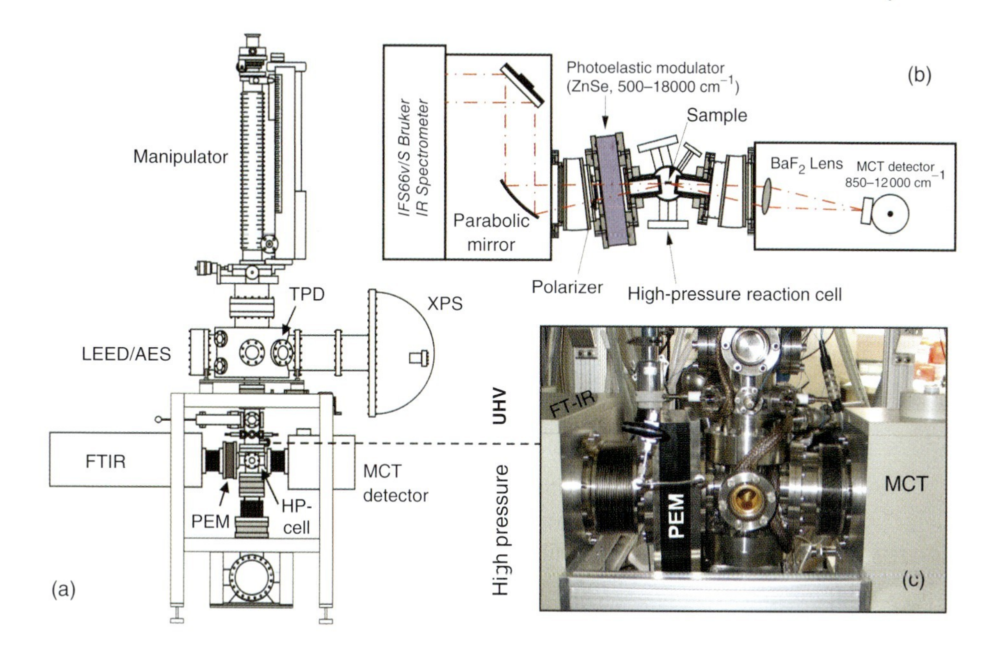


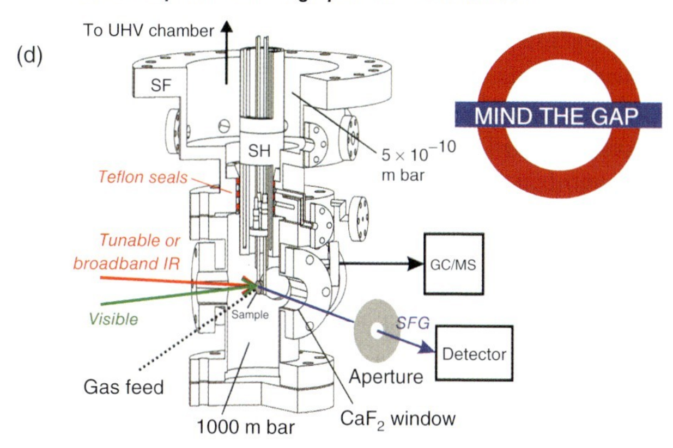


**Figure 39.6** (a) Experimental setup combining a UHV surface analysis chamber with a UHV–high-pressure reaction cell optimized for PM-IRAS spectroscopy. Pre- and post-reaction surface analysis under UHV can be performed by XPS, LEED, AES, and TPD. The optical setup and the high-pressure reaction cell are shown in (b) and (c), respectively. (Adapted from Ref. [113] with permission. Copyright (2007) Elsevier.) (d) A high-pressure reaction cell for *in situ* optical spectroscopy on model catalysts shown in cross section: sample holder (SH) and sealing flange (SF), housing three differentially pumped, spring-loaded Teflon seals. The cell is coupled to an ultrahigh-vacuum (UHV) sample preparation and analysis system (not shown). A single crystal is mounted on the sample holder, which is inserted into the Teflon seals. (Adapted from Ref. [42] with permission. Copyright (2001) The PCCP Owner Societies.)

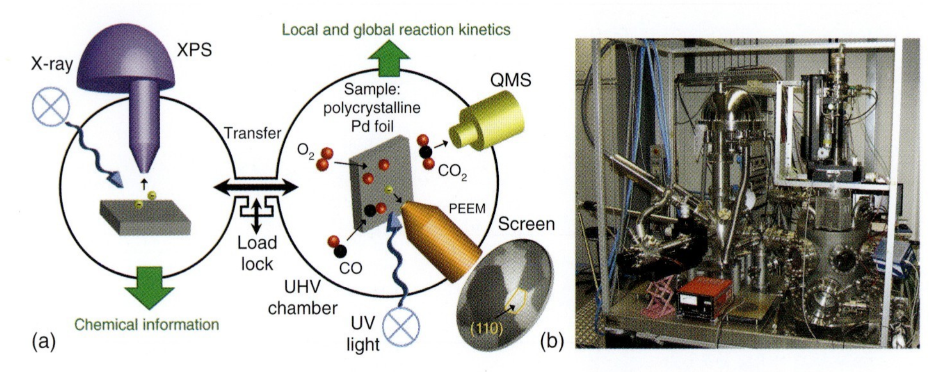


**Figure 39.7** Experimental setup for performing *in situ* surface (reaction) microscopy on model catalysts by PEEM while simultaneously monitoring reaction rates by MS, providing the global (MS) as well as the local "laterally-resolved" (PEEM) kinetics on heterogeneous model catalysts. (a) scheme, (b) fotograph. Complementary information on the chemical composition of samples in UHV and upon gas exposure is provided by XPS. (Adapted from Ref. [204] with permission. Copyright (2013) Springer.)

calibrated by comparison with optical or electron micrographs of the same sample position (see also Chapter 49, volume 6).

## 39.2 Case Studies

In the following, selected case studies are discussed illustrating the benefits of the surface science approach to heterogeneous catalysis. The three subsections focus on (i) surface spectroscopy, (ii) surface microscopy, and (iii) activity/selectivity of specific catalytic reactions. These studies bridge both the materials and pressure gap and are carried out both on model and technological catalysts, which should clearly reveal the relevance of model studies.

### 39.2.1 Understanding Selectivity Using the Surface Science Approach: The PdZn System for Methanol Steam Reforming (MSR)

For many years, the surface science approach to heterogeneous catalysis was mainly directed toward understanding the activity of a catalytic surface for a specific reaction. The prime examples include CO oxidation, hydrocarbon (mostly olefins, in particular ethylene) hydrogenation, CO + NO reaction, and a few others. These studies have been repeatedly reviewed [3, 13, 36, 48, 50, 53, 205] and will not be discussed here. As a major result, the effects of competitive adsorption, site blocking/poisoning, structure sensitivity (with typically higher activity of low-coordinated sites) have been elucidated. Nevertheless, for these types of reactions "selectivity" is typically not an issue – unless one considers unwanted carbon formation (coking) as a side reaction. In applied heterogeneous catalysis, however, "selectivity" is more critical than mere activity because the separation of unwanted by-products is tedious (leading to the complexity gap mentioned above).

Below we will therefore discuss surface science studies that led to a detailed understanding of the selectivity of supported Pd-ZnO catalysts utilized for MSR. MSR is a promising reaction to generate hydrogen (and $CO_2$) from methanol and water [167, 206]. Methanol might serve as a fuel replacement, and is easy to store and handle, enabling on-site or on-board generation of hydrogen, which could then be used for fuel cells. This is advantageous over using compressed or liquefied hydrogen, but only when the undesired side reaction of methanol decomposition (MDC to $H_2$ and CO) can be avoided because CO poisons the fuel cell's (typically Pt-based) anode catalyst.

The technically applied catalyst for MSR is Cu-ZnO, but this system is pyrophoric and also exhibits sintering. Iwasa and coworkers [166, 167] have suggested a series of alternative catalysts such as Pd-ZnO, Pd-$Ga_2O_3$, or Pd-$In_2O_3$. All three systems exhibited high MSR selectivity with improved thermal stability, and it was soon realized that the active phases PdZn, $Pd_2Ga$, and PdIn were formed under reaction conditions via partial reduction of the support oxide surrounding the Pd nanoparticles, followed by metal alloying. The hydrogen required for reduction is formed via methanol dehydrogenation on Pd, but the alloys or, in fact, the IMCs [62, 207] can also be formed via prereduction of the catalyst. IMCs are single-phase compounds consisting of metals, with the crystal structure of an IMC being different from the structure of the constituting metals. Because of a partly covalent or ionic bond character, the structural stability of IMCs hinders surface segregation (of one constituent), which is frequently observed for alloys and a major reason for deactivation [208].

Nevertheless, as in many cases in catalysis, it remained unclear why marginal changes in catalyst synthesis, activation, regeneration, and operation may have so marked effects on MSR selectivity (desired $CO_2 + H_2$ vs. undesired $CO + H_2$). This calls for a thorough surface science investigation using both UHV-based PdZn near-surface alloys as model systems [61, 124, 209–215] and "high-surface-area model catalysts," that is, ZnO-supported, wet chemically prepared PdZn nanoparticles [123, 216]. For this enterprise, several *in situ* techniques have been combined, including vibrational (PM-IRAS and FTIR) spectroscopy, NAP-XPS, and X-ray absorption spectroscopy (XAS) for characterization of active phases present during MSR, their stability, and mechanisms.

To set the stage, let us first consider the technological Pd-ZnO catalyst, which could substitute commercially used Cu/ZnO/$Al_2O_3$ in MSR. One has to confess, however, that even the improved catalytic properties of PdZn can hardly compensate its price (being about 1500 times more expensive than the technologically applied Cu). The catalytic properties of Pd supported on ZnO (Figure 39.8a) are clearly very different from those of Pd on inert supports [167]. Pd-ZnO is highly selective to MSR, yielding $H_2$ and $CO_2$ [123, 167], whereas Pd/inert support is selective to MDC to $H_2$ and CO [19, 191, 217, 218]. The difference is not due to different Pd particle sizes or surface structures but "simply" due to the alloy

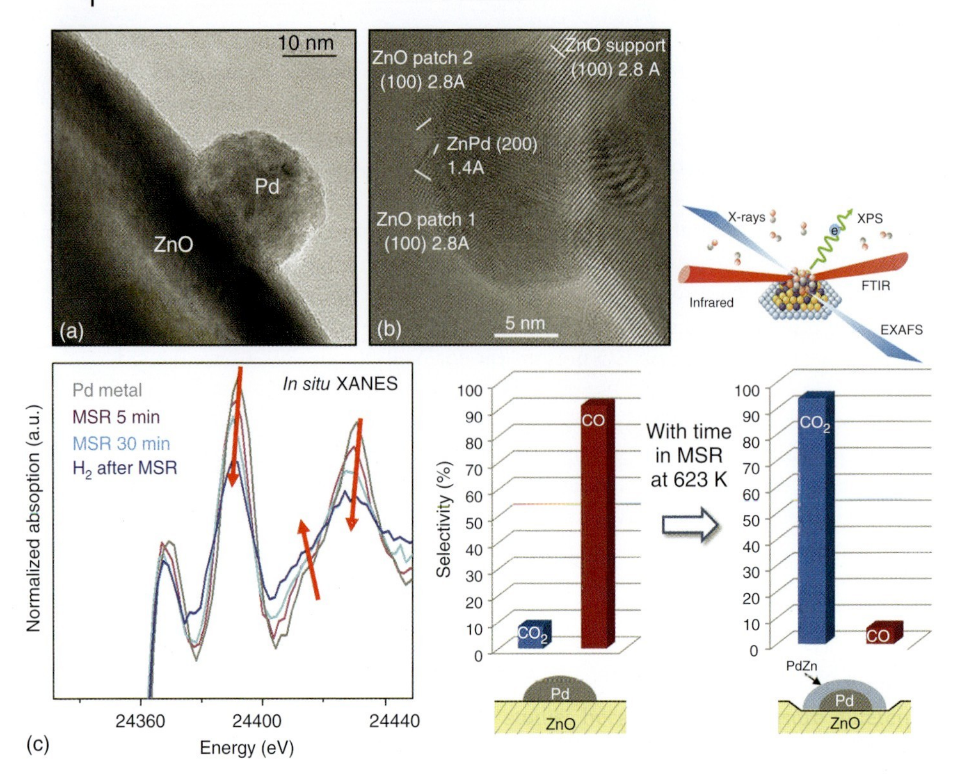


**Figure 39.8** (a) HRTEM images of a Pd-ZnO catalyst (as prepared) and (b) of PdZn/ZnO after alloy formation. The individual PdZn nanoparticle has two ZnO patches on the surface, both exhibiting ZnO(100) lattice fringes of 280 pm. The ZnO support is also clearly visible with the same lattice spacings. The PdZn lattice distances (140 pm) are assigned to the (200) lattice planes. (c) Left: Pd K edge XANES spectra obtained upon exposure of 7.5 wt% Pd/ZnO to MSR conditions at 623 K without pre-reduction ($p_{CH_3OH} = p_{H_2O} = 20$ mbar). The arrows illustrate the changes due to *in situ* formation of PdZn. Right: Selectivity to CO and $CO_2$, which are representative for the selectivity to MDC and MSR, respectively, after 5 and 60 min time-on-stream in MSR at 623 K. (Panels (a,c) adapted with permission from Ref. [123]. Copyright (2011) American Chemical Society.) (Panel (b) adapted with permission from [162]. Copyright (2013) Wiley-VCH.)

formation between Pd and reduced Zn. Nevertheless, the exact atomic details of these processes remain unknown.

The structure and composition of wet-chemically prepared Pd nanoparticles supported on ZnO apparently change under reaction conditions, which affects selectivity [123, 162]. As illustrated in Figure 39.8c, *operando* quick-EXAFS was utilized for real-time examination of the formation of the PdZn alloy in a reactive methanol/water environment (producing the required $H_2$), monitoring the structural and electronic changes via the near-edge region [123]. The amplitude of the oscillations after the edge (characteristic for metallic Pd) strongly decreases upon PdZn alloying (which is also observed upon catalyst reduction in pure $H_2$). As the nanoparticles change from Pd to (surface) PdZn, the selectivity switches from MDC to MSR (Figure 39.8c; as observed by GC and/or MS). The structural and electronic properties of the active phase resemble those of the tetragonal PdZn IMC with a 1 : 1 stoichiometry [207]. The same morphological alterations can also be observed *ex situ*, for example, by performing HRTEM [162, 219] or XRD [167] before and after reaction or reduction (Figure 39.8b). Nevertheless, it is certainly beneficial to monitor the changes of the working catalyst.

The effect of oxidative treatments has been debated as to whether it positively or negatively affects selectivity. According to Quick-EXAFS [123], it leads to the decomposition of the PdZn overlayer and the formation of a Pd overlayer partially covered by ZnO patches, which is MDC selective (Figure 39.9). However, the oxidative treatment also removes carbonaceous contaminants and, upon reaction or reduction, a "fresh" PdZn overlayer is thus produced. After oxidation, even higher activity and selectivity to $H_2$ and $CO_2$ were observed, likely due to the formation of a more homogeneous PdZn alloy (with less remaining Pd patches) or to the formation of a more Zn-rich surface alloy. Accordingly, oxygen treatment initially reduces selectivity but enables the eventual re-establishment of the clean selective phase, explaining the controversy in literature. Aberration-corrected HRTEM [162], which revealed PdZn nanoparticles partially covered by ZnO patches, corroborated this picture (Figure 39.8b). All results clearly indicate the different selectivity of Pd and PdZn. Nevertheless, for the technological system it is difficult to obtain more detailed, that is, atomic and molecular level, information.

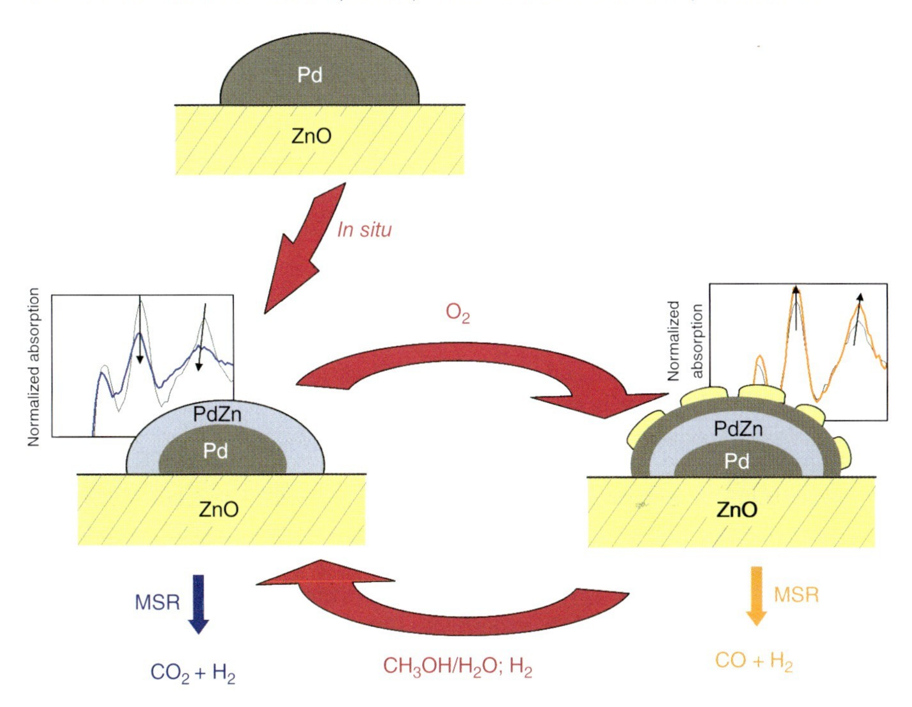


**Figure 39.9** Schematic illustration of the structural changes of Pd-ZnO MSR catalysts. (Adapted with permission from [123]. Copyright (2011) American Chemical Society.)

The PdZn system and the MSR reaction were thus also examined using surface science model catalysts. PdZn 1 : 1 surface alloys were prepared under UHV by Zn evaporation on Pd(111) single crystals followed by annealing (Figure 39.10). Although this seems simple, intense research by a number of research groups was required for developing this model system [61, 209–213, 221, 222], with initially somewhat different results. Key to a (nearly) complete understanding was to combine several surface-sensitive structure and composition methods, such as low-energy electron diffraction (LEED) (and later STM), XPS (and later LEIS (low-energy ion scattering)), as well as probing surface alloying by temperature-programmed desorption (TPD) via CO adsorption (and later electron energy loss spectroscopy (EELS) and infrared reflection absorption spectroscopy (IRAS)). Finally, a very straightforward preparation route was established: namely the deposition of 3 ML Zn on Pd(111) at or below room temperature, followed by annealing at 550 K. This leads to a well-ordered PdZn 1 : 1 surface alloy (see the LEED images in Figure 39.10 and the XPS spectra in Figure 39.11), with up to six layers of thickness (when >3 ML Zn were deposited, the excess Zn evaporated during annealing). The "2 × 2" LEED pattern is misleading because the stable (111) surface of a PdZn 1 : 1 alloy exhibits (2 × 1) periodicity, as indicated by DFT [221]. Because of the threefold symmetry of the hexagonal (111) substrate, the "(2 × 2)" LEED pattern is simply the superposition of the diffraction patterns of three (2 × 1) surface domains with different rotational orientations (rotated by 60°). The surface of the 1 : 1 PdZn/Pd(111) alloy exhibits a p(2 × 1) structure, which has been explained by the surface consisting of alternating rows of Pd and Zn (Figure 39.10) [221]. Using STM, this was later confirmed by Weirum *et al.* [212].

In order to directly monitor the formation of the desired PdZn alloy, XPS is the best suited method, by examining the Pd 3d (Figure 39.11) and Zn 2p regions (not shown). Starting with clean Pd(111) (green trace in Figure 39.11), upon deposition of

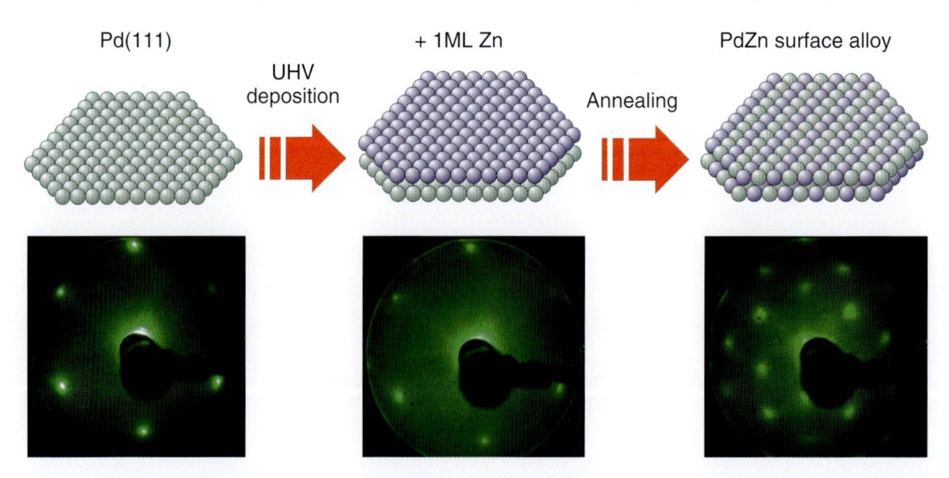


**Figure 39.10** UHV preparation of PdZn 1 : 1 surface alloys by Zn deposition on Pd(111) and subsequent annealing. The bottom shows the LEED pattern (at 100 eV) corresponding to each step. (Adapted with permission from Ref. [220]. Copyright (2014) Springer.)

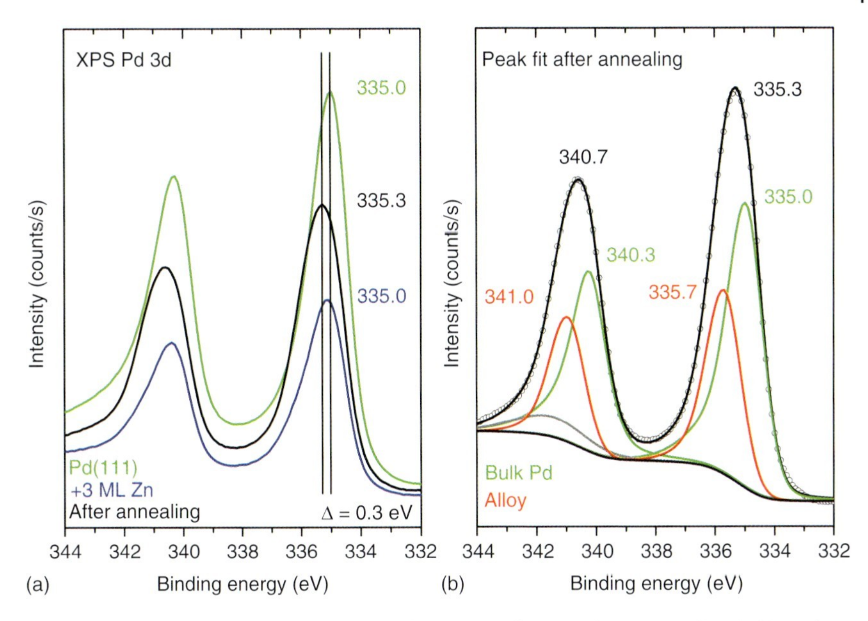


**Figure 39.11** XPS (Pd 3d) studies of alloy formation of Zn overlayers on Pd(111). (a) Evolution of the Pd signal with 3 ML Zn initially deposited on Pd(111) at 300 K. (b) Peak fit of the Pd 3d region after annealing to 550 K (i.e., of black trace in (a)).

3 ML Zn at 90 K the intensity of the Pd 3d signal (blue trace) was reduced to ~40%, with the peak shape slightly asymmetric toward higher binding energy (BE). Nevertheless, since alloying had not yet occurred in the Zn overlayer on Pd, the peak maximum remained at a BE of 335.0 eV, which is characteristic of bulk Pd [195]. Alloying is induced by annealing (5 min) at 550 K (black trace), upon which the Pd signal increased by 28%, accompanied by a +0.3 eV shift of the peak maximum, characteristic of alloying. Contrary to the Pd 3d peaks, the intensity of the Zn $2p_{3/2}$ signal decreased (~15%) upon annealing, and the peak maximum shifted from 1021.5 eV (after deposition) to 1020.9 eV. The observed peak shifts originate from a redistribution of charge from Zn toward Pd upon alloying, as predicted by DFT [61], leading to a charge situation of $Zn^{0.4+}Pd^{0.4-}$ [207] (despite this correct notation, we will continue to use "PdZn" below).

Fitting the Pd 3d alloy peak (Figure 39.11b) reveals that two components are present: one component at 335.7 eV (red) due to Pd atoms in the PdZn surface alloy, and the other at (335.0 eV, green) due to the Pd atoms of the underlying Pd(111) substrate. The third (black) component in the peak fit (341.5 eV) originates from a Pd plasmon excitation [223].

The preparation of the 1 : 1 PdZn model surface alloy seems simple but its final structure and composition strongly depend on the annealing temperature, and to some extent also on the amount of Zn initially deposited. This is illustrated in Figure 39.12a, which plots the XPS intensities of Pd (green), Zn (blue), and PdZn (red) as a function of the annealing temperature. A "stability window" can be identified for the PdZn surface alloy, in between the annealing temperatures of

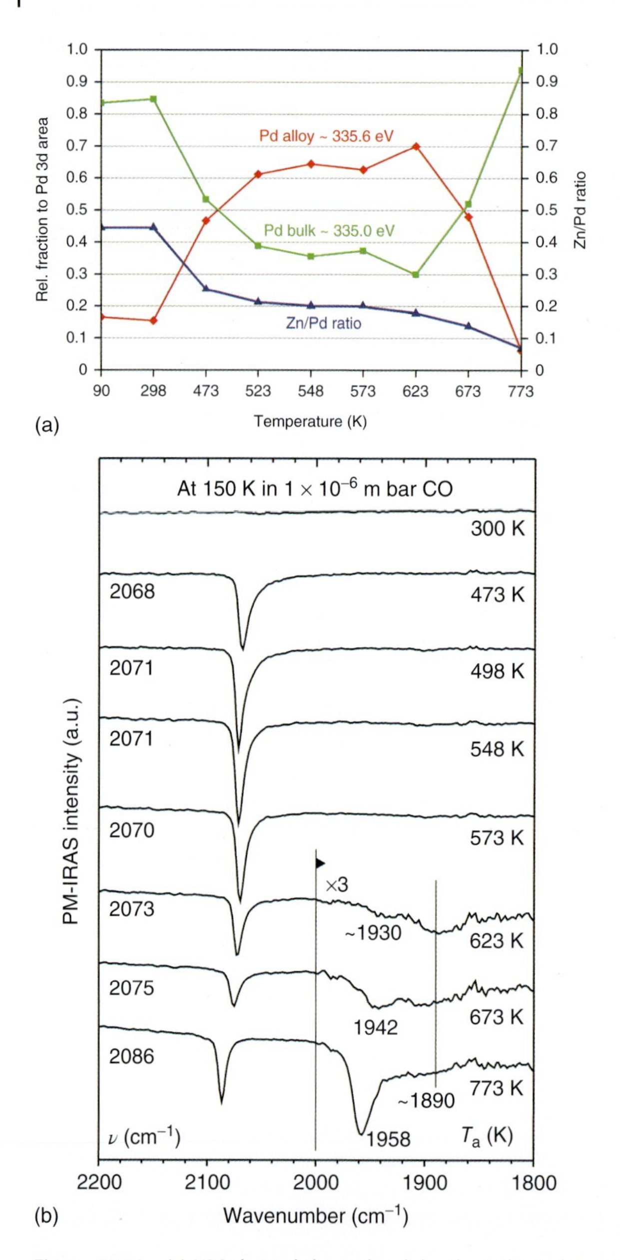


**Figure 39.12** (a) XPS-derived thermal stability window of PdZn surface alloys. (b) Surface alloy formation/decomposition upon annealing derived from (PM-)IRAS. 2 ML of Zn/Pd(111). Spectra were acquired at 150 K in $10^{-6}$ mbar CO, after successive annealing to the indicated temperatures. The IR intensity in the frequency range of multiply coordinated CO (below 2000 $cm^{-1}$) was multiplied by 3 in some spectra. (Adapted from Ref. [214] with permission from the American Chemical Society.)

~450 and 600 K. At lower $T$, alloy formation is not complete, and at higher $T$ the PdZn alloy starts to decompose by Zn dissolution in the Pd bulk (only above 800 K Zn desorbs to the gas phase [214, 224, 225]).

The formation and decomposition of the PdZn surface alloy can also be monitored via IRAS using CO as the probe molecule. Figure 39.12b shows the effect of exposing the model system to CO after annealing at increasing temperatures. Initially, 2 ML Zn were evaporated at 90 K on clean Pd(111) so that spontaneous intermixing between deposited Zn and the Pd substrate could be excluded. Starting from the unalloyed Zn/Pd(111) layered system, the sample was heated up to 300 K and then cooled down in $10^{-6}$ mbar CO to 150 K (90 K $min^{-1}$). After that, an IRAS spectrum was acquired at 150 K in $10^{-6}$ mbar CO. Then, the sample was heated in $10^{-6}$ mbar CO atmosphere to 473 K, cooled once more to 150 K, and another IR spectrum was recorded. This procedure was repeated several times while stepwise increasing the annealing temperature up to 773 K.

No CO adsorption could be observed for annealing temperatures up to 423 K (Figure 39.12b). As CO does not significantly adsorb on Zn under these conditions [226], the absence of CO adsorption clearly shows that the Zn layers are intact and not yet alloyed with palladium. Only after annealing at 473 K, CO adsorption occurs, characterized by an IR band at 2068 $cm^{-1}$ as described in detail in Chapter 3.4.1, Volume 1. At higher annealing temperature, this band gains intensity and shifts to higher wavenumbers as a result of increasing CO coverage. Higher coverage increases the intermolecular dipole interactions between neighboring CO molecules (dynamic dipole–dipole coupling [227]), which induces the frequency shift to higher wavenumbers. Additionally, increasing coverage reduces the amount of back bonding from the metal d electrons into the antibonding $2\pi^*$ orbitals of the CO molecules, weakening the metal–CO bond and again resulting in a blue shift of the internal C–O stretch. The maximum intensity of this band is obtained after heating the sample to 548 K (2071 $cm^{-1}$), which is attributed to CO linearly bound on (surface) Pd atoms in the PdZn alloy. The strong increase of the CO signal between 473 and 573 K can be explained by two facts: (i) the ongoing alloying process between these two temperatures, and (ii) a reconstruction of the PdZn surface layer that leads to higher CO coverage [215]. As discussed below (see Figure 39.17), in the presence of CO, the surface undergoes a reconstruction, with the row structure changing to a zig-zag structure and the maximum CO coverage increasing from $\theta \sim {}^1/_3$ ML to $\theta \sim {}^1/_2$ ML [215]. At this coverage, every Pd atom on the surface is occupied by a CO molecule of on-top geometry.

After annealing at 623 K, the on-top band was clearly weaker, and a signal due to bridge-bonded CO at ~1930 $cm^{-1}$ was detected. (For the correlation of CO adsorption site and frequency, see Chapter 3.4.1, Volume 1.) This indicates the onset of alloy decomposition since bridge-bonded CO requires patches of three or more Pd atoms at the surface, which are not present in the PdZn(111) $2 \times 1$ surface periodicity. As mentioned, alloy degeneration starts by Zn diffusion into the Pd bulk, because Zn desorption from PdZn surfaces was reported to occur above 800 K [214, 224, 225]. Finally, in the spectrum obtained after heating the sample to 773 K, IR bands at 2086 $cm^{-1}$ (on-top CO) and 1958 $cm^{-1}$ (bridge-bonded CO) as well as a shoulder at

1890 $cm^{-1}$ (hollow-bonded CO) could be observed, closely resembling CO on clean Pd(111) under these conditions.

The XPS and IRAS results on the PdZn stability window (473–600 K) thus agree well. TPD is not well suited to examine alloy formation because the surface alloy forms and decomposes during the TPD run. Nevertheless, for preannealed PdZn, TPD indicated a CO saturation coverage of about 0.5 ML and a desorption temperature of ~210 K (corresponding to one CO molecule per Pd surface atom and a BE of ~0.6 eV) [215, 220, 228].

To this point, the PdZn model system had been examined by several groups using a variety of experimental methods (LEED, XPS, TPD, IRAS, EELS, STM), as well as by theory (DFT). So one could provocatively ask: what more could be learned about the PdZn system? The desired (and required) level of increased detail can be best illustrated by comparing measurements by XPS, AES, and LEIS of the PdZn alloy stability window (Figures 39.12 and 39.13). When using the typical laboratory Al and Mg X-ray sources, the kinetic energies of the Pd-related photoelectrons are around 1 keV, which, in turn, corresponds to an inelastic mean free path of ~1 nm. Consequently, the probing depth of XPS encompasses ~2 nm, that is, up to eight atomic layers (see Chapters 3.2.1 and 3.2.3, Volume 1). The Zn Auger electron spectroscopy (AES) measurements (excited by electron bombardment) average over 3+ layers. For most cases, especially the AES "surface sensitivity" should be sufficient, but it is not for the current PdZn system, particularly when considering changes in surface composition between 550 and 650 K. For XPS Pd 3d, there is hardly any difference (Figure 39.12), whereas Zn AES (open dots) indicates a pronounced loss of Zn. Typically, this has been interpreted as a decrease of the Zn concentration via bulk dissolution, occurring homogeneously at the PdZn surface and in deeper layers. Thus, under such conditions, a Zn-depleted, Pd-rich surface was assumed to be present.

However, the LEIS measurements (see Chapter 3.3, Volume 1), monitoring the evolution of the normalized Zn intensities (Figure 39.13, filled dots), indicate a different scenario (the Ne+ ions interact only with the surface monolayer and thus yield utmost surface sensitivity). Zn deposited on the Pd surface starts to diffuse into the (sub)surface layers at around 300 K, at annealing temperatures of 400–550 K a stability plateau with 1 : 1 Pd:Zn elemental composition is observed, and diffusion of Zn from the PdZn surface/subsurface layers into the Pd bulk sets in at ~623 K [124, 213, 214].

However, for the temperature range 550–650 K, LEIS indicates a much smaller change in *surface* composition than AES or XPS; that is, according to LEIS, the PdZn 1 : 1 surface is only slightly changed. Apparently, the Zn loss must rather affect the PdZn layers *below the surface*. As a consequence, up to an annealing temperature of 573 K, a "PdZn multilayer" is present, with a thickness of about four to six layers (see insets in Figures 39.13b and 39.15), whereas upon annealing at 600 K or higher, a "PdZn monolayer" is formed by loss of Zn from the second and deeper layers. DFT had already predicted that the surface 1 : 1 PdZn layer is more stable than the deeper "subsurface" layers [61, 212]. Only when heated to 700 K or higher does the monolayer decompose by Zn bulk dissolution and subsequent Zn desorption (evaporation).

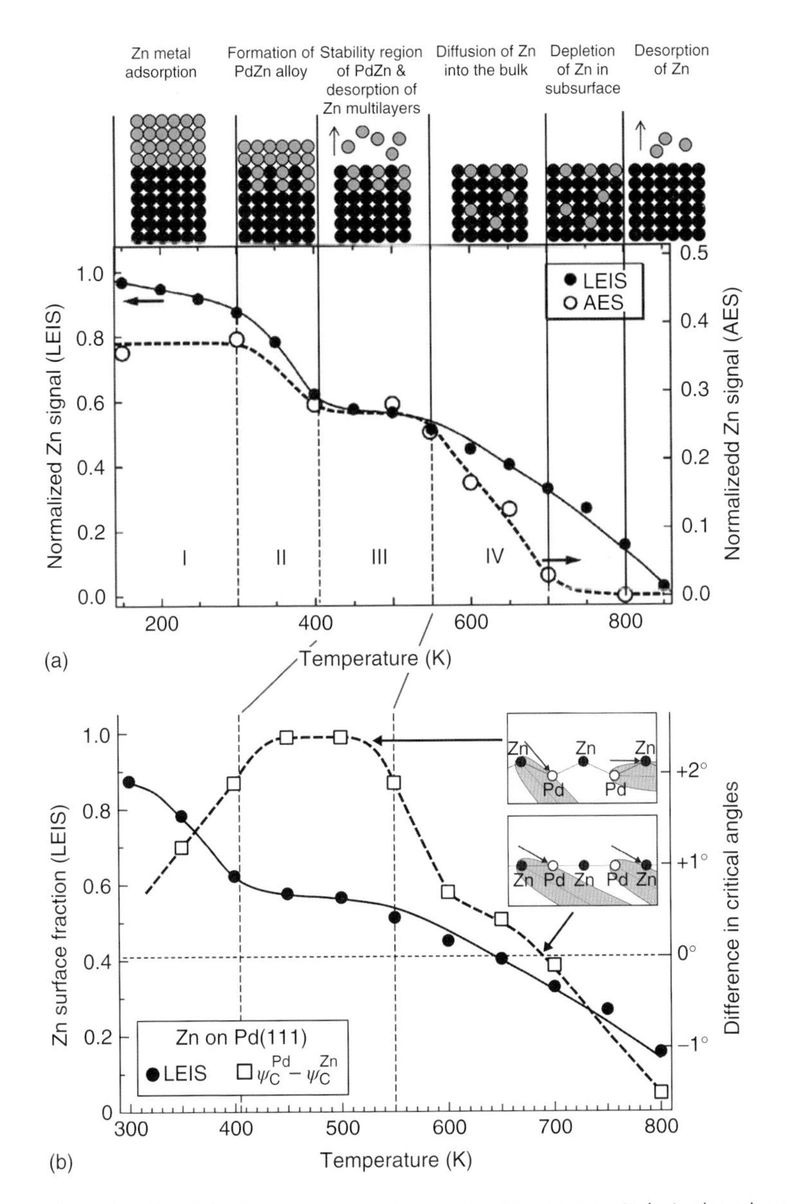


**Figure 39.13** (a) Surface alloy composition derived from LEIS and AES after different annealing temperatures for 2.2 ML of Zn/Pd(111). (Reproduced with permission from Ref. [214]. Copyright (2010) American Chemical Society.) (b) Filled circles: Pd:Zn surface fractions derived from LEIS for Zn films deposited on single-crystal Pd(111). Open squares: difference in critical angles for backscattering of 5 keV Ne ions from Pd and Zn atoms, respectively. In the schematic side views, a corrugated and a noncorrugated (2 × 1) PdZn surface are depicted. Arrows and gray shadow cones indicate the critical angles at which backscattering from Pd and Zn atoms, respectively, sets in. For illustration purposes, the corrugation is largely exaggerated. (Reproduced with permission from Ref. [124]. Copyright (2010) Wiley.)

Note that both the PdZn multilayer and monolayer exhibit the same surface composition (1 : 1 PdZn) and 2 × 1 structure. However, because of the different composition of the second, third, and so on, (subsurface) layers, that is, PdZn versus pure Pd, the coordination of the surface atoms is markedly different. This shows up when angle-dependent LEIS is acquired both on the multilayer and monolayer PdZn, revealing a different surface corrugation, depending on the number of alloy layers, that is, alloy thickness [124, 214, 229]. The PdZn multilayer exhibits corrugation with Zn-out/Pd-in (corrugation about 0.025 nm), whereas the PdZn monolayer exhibits almost no corrugation with rather Zn-in/Pd-out, again in agreement with DFT predictions [61, 230].

Based on this detailed knowledge, the PdZn multilayer and monolayer provide excellent test structures for catalytic activity/selectivity in the MSR reaction. The kinetic tests were performed in an all-glass, UHV-compatible, high-pressure reaction cell (operated as recirculating batch reactor) that was attached to a UHV preparation and characterization chamber [231]. Figure 39.14 shows a reactivity test performed at millibar gas pressure by temperature-programmed reaction of the reactants MeOH and water (with He filled up to 1 bar), using mass spectroscopic detection of the reactants and products [124, 213].

Figure 39.14a (upper panel) shows that the multilayer PdZn 1 : 1 surface alloy converts $CH_3OH$ and water to $CO_2$ and formaldehyde ($CH_2O$) up to a temperature of 573 K, whereas at temperatures above 573 K, $CH_2O$ is consumed (negative formation rate) and converted to CO via dehydrogenation (in all cases $H_2$ is also produced, of course). The different selectivity can be understood based on the previous surface characterization results. A temperature of 573 K is the upper limit of the multilayer stability; above 573 K, progressive transformation to the PdZn monolayer alloy takes place.

This was confirmed by performing the reaction on a monolayer PdZn 1 : 1 surface alloy, prepared either by annealing the multilayer above 573 K before the reaction or by directly preparing a monolayer by limiting the amount of evaporated Zn. The PdZn monolayer exhibited no $CO_2$ selectivity even at reaction temperatures below 573 K (Figure 39.14a, lower panel); it rather behaved very similar to the pure Pd substrate. This is remarkable because the *surface compositions* of the 1 : 1 PdZn monolayer and multilayer are virtually identical. It is important to note that the Zn/Pd ratios obtained by intermediate or postreaction LEIS/AES analysis correspond to the values depicted in Figure 39.13 at the respective temperature; that is, the thermal change of the near-surface composition is hardly affected by exposure to the reaction environment. Based on these findings, it can be suggested that the PdZn multilayer is "$CO_2$ selective" whereas the PdZn monolayer is "CO selective" (Figure 39.14, middle), despite the identical surface stoichiometry of PdZn 1 : 1 for both catalysts.

How can this be explained? The answer lies in the effect of the subsurface composition on the electronic and geometric structure of the topmost catalytically active PdZn layer. In order to examine the electronic structure of the multilayer and monolayer alloys and of potential adsorbed intermediates during the ongoing reaction, NAP-XPS was performed *in situ* (Figure 39.15; Pd 3d, Zn 3d, and valence band regions). The Pd 3d level of the multilayer alloy exhibits its typical maximum

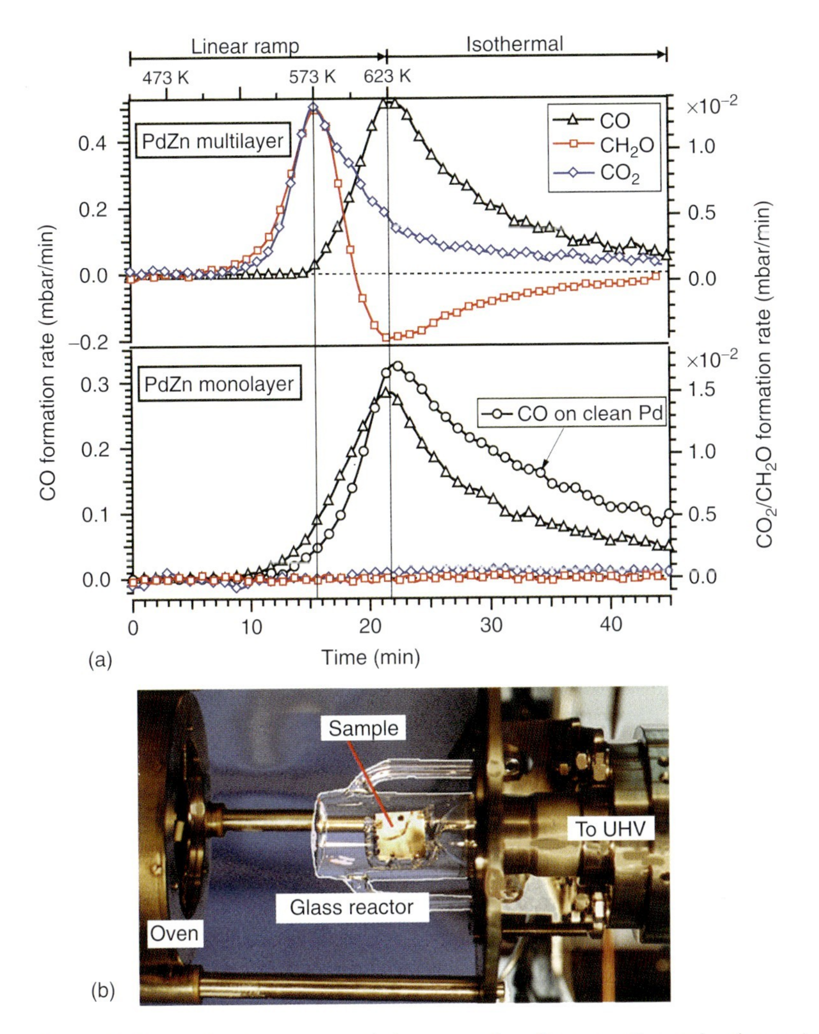


**Figure 39.14** Ambient-pressure catalysis on PdZn model catalysts. (a) Temperature-programmed methanol steam reforming on the multilayer PdZn 1 : 1 alloy grown on Pd foil (upper panel) versus MSR reaction on a "Zn-lean" monolayer PdZn surface and MSR on a clean Pd foil (lower panel). Reaction conditions: 12 mbar methanol, 24 mbar water, 977 mbar He; linear temperature ramp (9.0 K/min) up to 623 K, followed by isothermal reaction for 24 min. The decrease of the CO formation rate in the isothermal region is caused by progressive methanol consumption and carbon poisoning of the catalyst surface. Experiments at 0.12 mbar/0.24 mbar methanol/water (i.e., the same partial pressures as used in the NAP-XPS studies) reveal essentially the same trends. Complete reaction mass balance involving stoichiometric hydrogen formation was verified by mass spectrometry analysis. (b) Photograph of the all-glass high-pressure cell: catalyst sample in the glass frame, Pyrex glass reaction cell, oven, port to UHV. (Reproduced with permission from Ref. [124]. Copyright (2010) Wiley.)

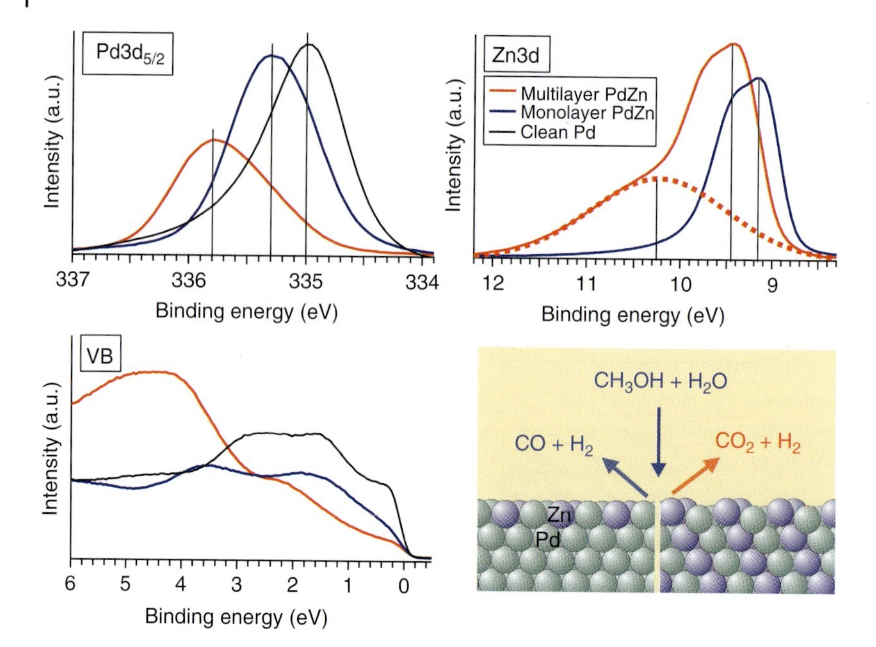


**Figure 39.15** Near-ambient-pressure XPS spectra (Pd 3d, Zn 3d, and valence-band (VB) regions; BESSY II) acquired *in situ* during MSR on the PdZn 1 : 1 multilayer (red traces) and monolayer alloy (blue traces). For comparison, the corresponding spectra for "pure" Pd are included (black traces). The oxidized ZnOH component is highlighted by the dotted red line (top right panel). To obtain equal information depth, the Pd 3d spectra were recorded with 650 eV photon energy and the Zn 3d and VB regions with 120 eV. Reaction conditions: 0.12 mbar methanol, 0.24 mbar water, 553 K. (Reproduced with permission from Ref. [124]. Copyright (2010) John Wiley and Sons.)

at 335.8 eV, whereas for the monolayer surface alloy it is at 335.3 eV. The higher intensity of the monolayer Pd 3d signal and the concomitant decrease of the Zn 3d signal reflect the subsurface Zn-lean character of this surface. The valence band spectra reveal another important difference. Whereas the multilayer alloy shows a "Cu-like" valence band [209], that is, a significantly reduced density of states between ~2 and 0 eV, the monolayer alloy, despite its 1 : 1 surface composition, rather resembles a strongly modified Pd surface with an increased density of states close to the Fermi level and upward shifted Pd 4d features, approaching the valence band spectrum of clean Pd. The *most important* difference is, however, evident from the Zn 3d spectra acquired under MSR reaction conditions (i.e., in the presence of $H_2O$). Whereas the multilayer alloy activates water by forming a Zn–OH species, resulting in the 10.25 eV shoulder in Figure 39.15 (dashed-red peak fit, right panel), this species is not present on the monolayer alloy under identical conditions (blue curve). Consequently, the monolayer alloy does not activate water, and the $CH_2O$ formed from methanol is not converted to $CO_2$ but simply dehydrogenated to CO.

The Pd 3d and valence band results were complemented by corresponding C 1s spectra (Figure 39.16). Signatures of $CH_2O$ or related oxygenates were observed up

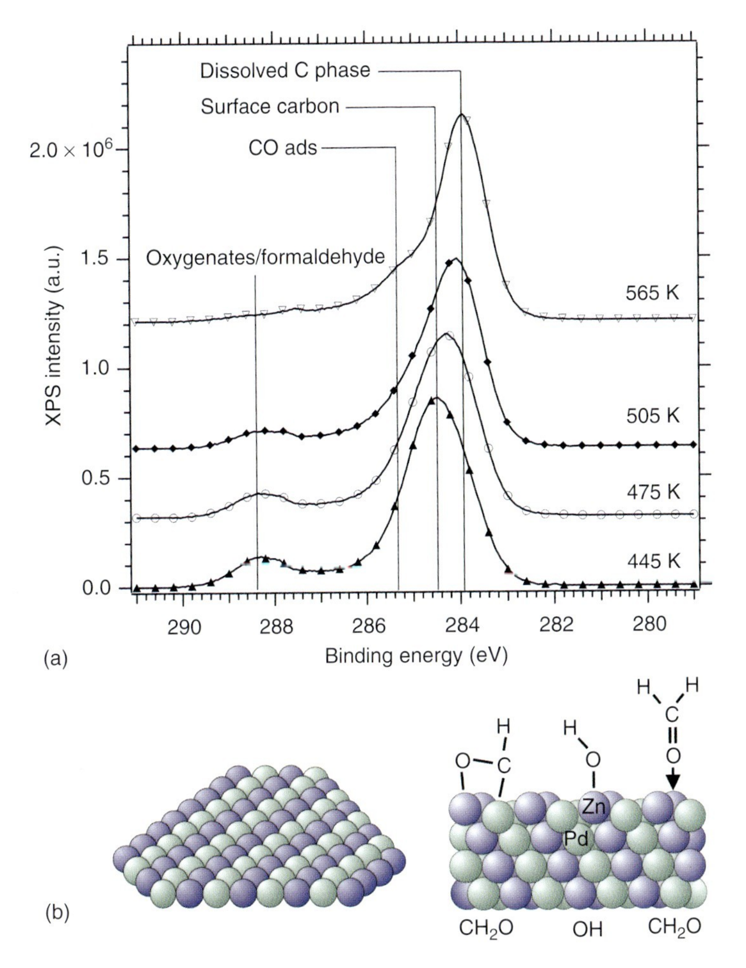


**Figure 39.16** (a) C 1s spectra obtained *in situ* during MSR (0.12 mbar methanol + 0.24 mbar water) on the multilayer PdZn surface alloy (Pd(111) + 3ML Zn annealed at 503 K in vacuum). The peak centered at 284.5 eV is likely due to disordered surface C from methanol C–O bond-breaking. (b) p(2 × 1) surface structure of the 1 : 1 multilayer and monolayer PdZn alloys on Pd(111) and likely surface intermediates on the multilayer PdZn alloy. (Reproduced with permission from Ref. [124]. Copyright (2010) John Wiley and Sons.)

to 505 K, that is, for the multilayer alloy. Around 570 K and higher, the surface $CH_2O$ was replaced by CO, pointing to the transformation of multilayer to monolayer alloy. O 1s data could not be reliably evaluated because of its strong overlap with the Pd 3p signal.

Altogether, the results provide clear evidence that both the geometric and the electronic structure of the multilayer and monolayer alloys are significantly different. The $CO_2$-selective multilayer alloy features a lowered density of states close to the Fermi edge, and the low-lying Pd 4d states as well as surface ensembles of PdZn exhibit a "Zn-up/Pd-down" corrugation, which represents the "bifunctional" active sites both for efficient water activation and for steering of methanol dehydrogenation toward $CH_2O$ (Figure 39.16b). The PdZn monolayer fails to activate water, and thus the only feasible reaction is methanol dehydrogenation.

With respect to potential molecular reaction mechanisms (from $CH_3OH$ toward $CO_2$), there are several scenarios of different formaldehyde species interacting either with OH(ads) or O(ads) in different bonding geometries. For example, dioxomethylene has been proposed as an intermediate, but pinning down the detailed reaction path remains a challenge both for experiment and theory. With respect to MDC, DFT pointed out two main differences between (the currently applied) Cu, PdZn, and Pd [61, 232]. Whereas $CH_2O$ is rather easily decomposed to CO on Pd [150, 191], the barriers on Cu and PdZn are significantly higher, so there is a higher probability of adsorbed formaldehyde reacting with water. Furthermore, the breaking of the methanolic C–O bond is more difficult on Cu and PdZn than on Pd [232], leading to less coking [220].

In summary, the pronounced electronic and structural effects on the catalytic properties of the topmost active PdZn layer are driven by the composition of the second ("subsurface") and deeper layers, due to variations of the Zn coordination of both Pd and Zn surface atoms; that is, catalysis that is more than skin deep! The properties of the active sites are not sufficiently described by considering only the constituting atoms of the active ensemble, but they require the consideration of neighboring sites. This limits the applicability of a "single-site" concept at least for systems with extended electronic structures. One should note, however, that the electronic structure of ultrathin layers on various substrates may clearly deviate from that of thicker (bulk-like layers), which is a classical topic of surface science [16, 25, 95, 98, 233–235].

One can now critically ask whether the picture developed based on ideal model catalysts really applies to technological (powder) catalysts consisting of PdZn nanoparticles supported on ZnO. To this end, we have applied CO adsorption at millibar pressure to "titrate" the accessible surface sites, both for supported PdZn nanoparticles [123, 216] and for single-crystalline PdZn surface alloys (Figure 39.17). In excellent agreement, in both cases solely on-top adsorbed CO is observed, with vibrational frequencies $\sim$2070 $cm^{-1}$. On PdZn, CO exclusively adsorbs on the Pd atoms, whereas adsorption on bridge and hollow sites (typical of Pd metal) is absent. This is due to the "site isolation" of Pd, with bridge and hollow adsorption sites being unstable, likely due to increased Pd–Pd distances in PdZn. When compared to metallic Pd, both for PdZn nanoparticles and surface alloys a $\sim$20 $cm^{-1}$ red shift of on top CO was observed (Figure 39.17b) [123, 215], reflecting a change in electronic properties due to charge transfer from Zn to Pd, as predicted by DFT [61]. This results in increased back-donation from Pd d bands to CO, thereby weakening the internal C–O bond. The charge transfer from Zn to Pd also shifts the core-level binding energies (see the XPS and VB (valence band) spectra in Figure 39.15) [209]. Upon Pd-Zn formation, the Pd $3d_{5/2}$ peak shifts from 335.0 eV in metallic Pd by $\sim$0.6–0.8 eV to higher values (Figure 39.15). Likewise, upon alloy formation, Zn 3d shifts by $\sim$0.3 eV to lower binding energies. Analogous

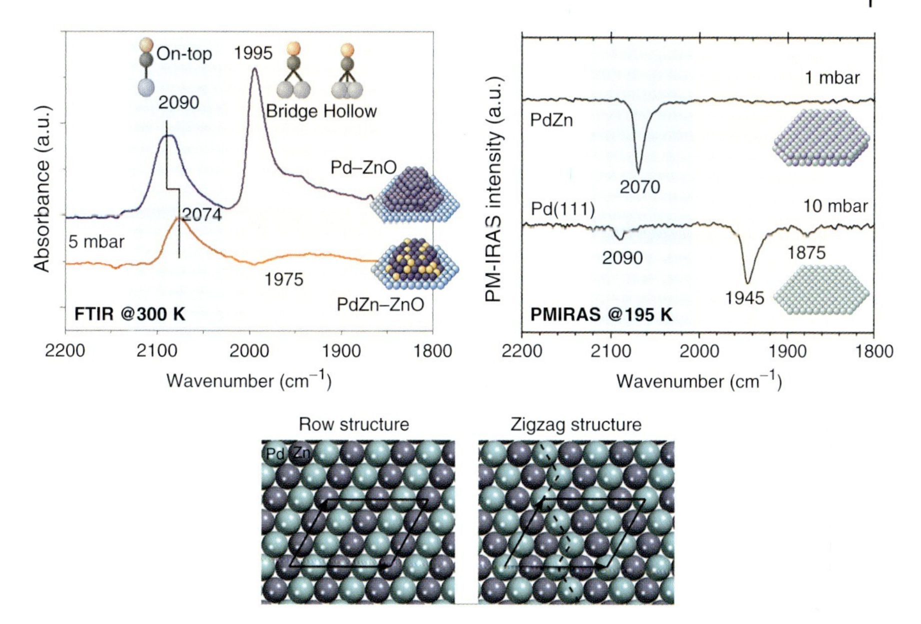


**Figure 39.17** (a) FTIR spectra of CO adsorption (5 mbar, 303 K) on 7.5 wt% Pd nanoparticles on ZnO reduced at 303 K (upper spectrum, representing Pd/ZnO) and at 623 K (lower spectrum, representing PdZn/ZnO). (Reproduced with permission from Ref. [123], Copyright (2011) American Chemical Society.) (b) PM-IRAS spectra of CO adsorption at 195 K on a 4-ML PdZn/Pd(111) model catalyst annealed to 550 K and on Pd(111). (Reproduced with permission from [214], Copyright (2010) American Chemical Society.) (c) Illustration of the suggested geometric surface structures (Pd: cyan; Zn: blue): bulk terminated row structure and reconstructed zigzag geometry. The latter arrangement has higher stability in CO atmosphere. (Reproduced with permission from Ref. [215]. Copyright (2012) American Chemical Society.)

shifts were reported for PdZn/ZnO nanoparticles [167, 236]. The similarity of properties detected by FTIR, PM-IRAS, and XPS confirm the good comparability of model and applied PdZn bimetallic catalysts, which are corroborated by DFT [61]. Combining TPD, PM-IRAS, and DFT, it was further suggested that, upon CO adsorption, the 1 : 1 PdZn(111) surface reconstructs from the row to a zigzag-like structure (Figure 39.17c) [215]. The zigzag arrangement is in agreement with the absence of bridge CO sites and the experimentally observed saturation coverage of $1/2$ ML [220, 228]; a direct experimental proof by, for example, STM has not yet been achieved, though.

### 39.2.2
### Observing Laterally Resolved Kinetics by Surface Microscopy: CO Oxidation on Pt and Pd Surfaces

CO oxidation on noble metals is the prototypical catalytic reaction of model catalysis [76, 177, 205, 237], CO being sometimes called the "drosophila of surface science"

[238]. The interest in this system seems to be renewed by every novel instrument, which is then rapidly applied to the same "old twist" (such as enhanced animation technology induces new releases of "King Kong"). Among the novel methods were, for example, SFG [36, 239], high-pressure STM [198, 240], surface X-ray diffraction (SXRD) [241, 242], and NAP-XPS [195, 196, 243].

Classically, CO oxidation on noble metals is considered to be a Langmuir–Hinshelwood type reaction [48]. Because of asymmetric site blocking, the active surface is covered by adsorbed (dissociated) oxygen and reacts with adsorbed CO. On CO-covered surfaces, however, oxygen dissociation is hindered and thus they are inactive (CO poisoning). Clearly, in cases of oxygen excess and/or conditions when the metal oxide is more stable, the only reaction that can occur is between CO and the (surface) oxide. Inherent activities (or reaction probabilities) on Pd (metal) and $Pd_5O_4$ surface oxide have been reported [144, 244]. Of course, $Pd_5O_4$ can be reduced by CO and thus produce $CO_2$, but its inherent activity is about 5 times lower than that of chemisorbed oxygen. Thus, claims of a higher oxide activity – based on a comparison with the inactive CO-poisoned metallic surface – are simply not correct. For small nanoparticles and excess oxygen, the nanoparticles may indeed be oxidic (when the reaction with CO is not rapid enough to reduce them), but this picture cannot be generalized. For example, Zorn *et al.* [245] have determined the upper limit of the oxidation state of Pd nanoparticles supported on $Al_2O_3$ during CO oxidation (50 mbar CO, 50 mbar $O_2$, 503 K). In the absence of the reductant CO (50 mbar $O_2$, 503 K), the Pd particles were only partially oxidized, whereas in the presence of the reductant CO the surface oxide was rapidly reduced already at room temperature. Thus, under these specific conditions the reduction of the surface oxide is faster than its formation, so that an oxygen-covered metallic surface is present under *in situ* conditions rather than the surface oxide [245].

An effect that has received much less attention than the determination of the active phase is the heterogeneity of catalytic surfaces. In technological catalysis, measurements average over billions of nanoparticles that vary in particle size, shape, surface structure (depending on the crystal habit various fractions of different facets, steps, kinks, defects may be present, see [36]), composition, and so on. To systematically examine these differences, catalysts are prepared with varying mean particle size (preferentially monodisperse, sometimes with shape control [141, 246–251]) and (surface) composition. For catalytic measurements, averaging is still applied, though. In surface science-based model catalysis, heterogeneity is typically addressed by examining several single crystals of different surface terminations (most frequently carried out for (111), (110), and (100) faces [77]). For completeness, it should be mentioned that polycrystalline surfaces and even powders were also examined by photoemission of adsorbed xenon (PAX) [252, 253]. Examining several single-crystal terminations is, however, tedious and time-consuming work, and sometimes the conditions of surface preparation (cleanliness, defect density) and catalytic tests may not be exactly reproducible for different crystals. Furthermore, a relatively large parameter space of reactant gas pressure and reaction temperature must be investigated, so there is no catalytic model system with a complete set of reported reactivity data (not even for CO oxidation).

In the following, we present an alternative and more efficient approach, enabling the examination of the *locally resolved* activity of different crystallographic terminations of a specific metal, as developed by Suchorski *et al.* [72, 254]. Inspired by the pioneering work of Ertl and coworkers [76, 255], PEEM was employed to study the CO oxidation reaction, but not on (homogeneous) single crystals [256] but on (heterogeneous) polycrystalline Pt foil [72, 73, 254].

In a first step, the polycrystalline model catalyst must be characterized. Using UV-PEEM, a magnified image of the sample surface was created by the photoelectrons emitted upon UV illumination (here a deuterium discharge lamp was utilized; intensity maximum at ~190 nm/~6.5 eV). Figure 39.18 shows that the polycrystalline Pt foil consisted of domains with different contrast, typically a few hundred micrometers in diameter, formed after UHV annealing for several hours at 1100 K [72, 73, 257]. In UV-PEEM, the yield of the photoelectrons (contrast) depends on the local work function of the surface, which allows the identification of (clean) grains with different surface orientations (based on the known work function values of clean single crystals [72]). Figure 39.18 (left) displays a PEEM image of a clean polycrystalline Pt foil (field of view ~500 μm), comprised of

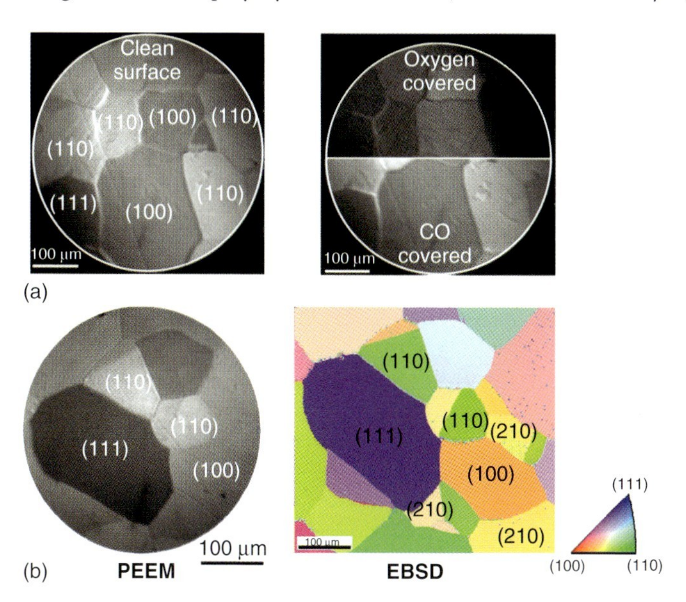


**Figure 39.18** Imaging the surface termination of specific domains on polycrystalline metal foil. (a) Comparison of PEEM images of clean, oxygen-covered, and CO-covered Pt foil. Because adsorbed oxygen increases the work function of the Pt surface, the oxygen-covered surface appears dark in PEEM. (Reproduced with permission from Ref. [257]. Copyright (2011) Elsevier.) (b) On polycrystalline Pd foil different domains are imaged and identified via work function differences in PEEM and via electron backscatter diffraction (EBSD) in an electron microscope. (Reproduced with permission from Ref. [258]. Copyright (2013) American Chemical Society.)

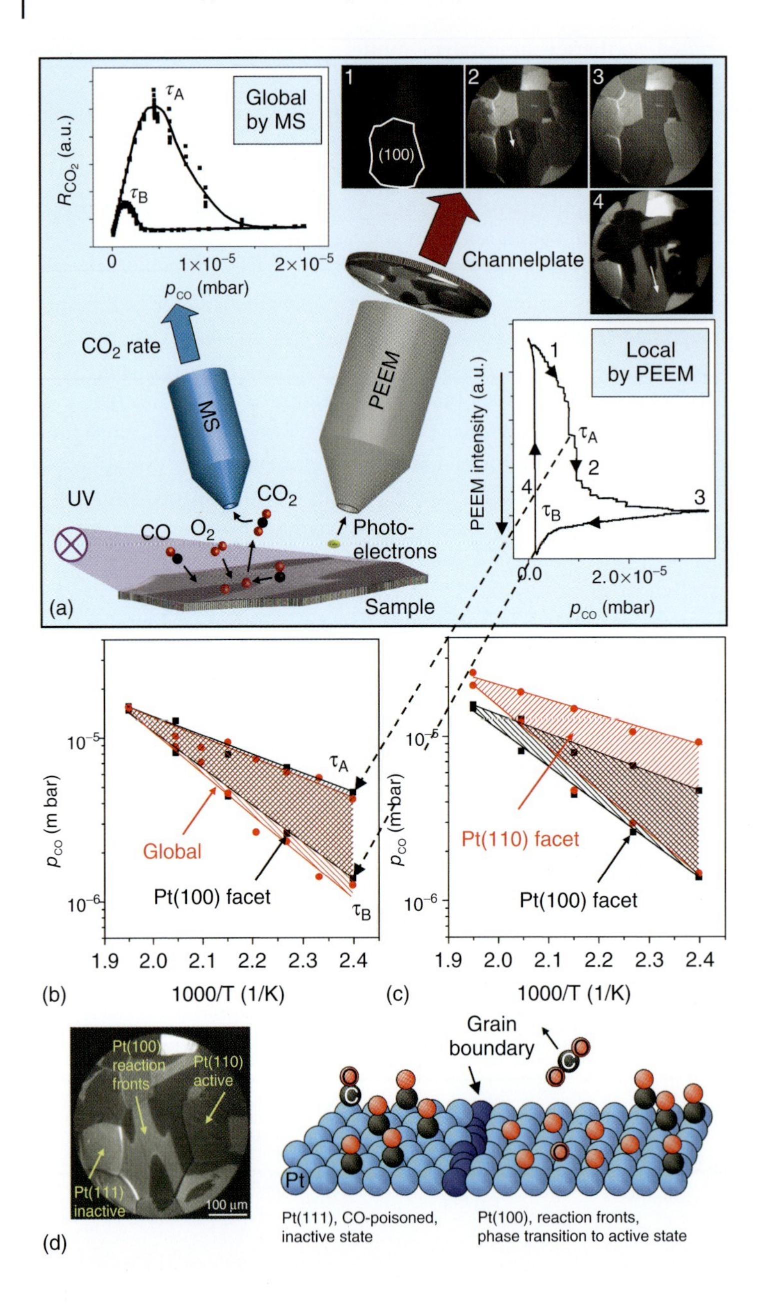

Global by MS
$R_{CO_2}$ (a.u.)
$\tau_A$
$\tau_B$
0
1×10⁻⁵
2×10⁻⁵
$p_{CO}$ (mbar)
1
(100)
2
3
4
Channelplate
$CO_2$ rate
MS
PEEM
Local by PEEM
PEEM intensity (a.u.)
0.0
2.0×10⁻⁵
$p_{CO}$ (mbar)
UV
CO
$O_2$
$CO_2$
Photo-electrons
Sample
(a)
$p_{CO}$ (m bar)
Global
Pt(100) facet
Pt(110) facet
1000/T (1/K)
(b)
(c)
Pt(100) reaction fronts
Pt(110) active
Pt(111) inactive
100 μm
Grain boundary
Pt
Pt(111), CO-poisoned, inactive state
Pt(100), reaction fronts, phase transition to active state
(d)

(111)-, (110)-, and (100)-oriented surface domains, with the surfaces with higher work function appearing as dark and those with lower work function as bright. The resolution of PEEM is ~1 μm, and the exact magnification was calibrated by comparison with optical or electron micrographs of the same sample position.

Apart from the (clean) surface terminations, PEEM also allows monitoring adsorbed molecules and differentiating them (e.g., CO and atomic O), again based on the correlation of the local image intensity with the work function values of the corresponding adsorbate-covered single crystals [72]. Figure 39.18a (right) shows PEEM images of the same polycrystalline Pt foil after gas adsorption. Upon oxygen or CO adsorption, the relative brightness of the domains changes according to the known work function changes upon adsorption of molecules/atoms [72]. On Pt, both O and CO increase the work function, but O-covered surfaces appear the darkest. These adsorbate-induced changes corroborate the crystallographic assignment of the clean domains, but their surface termination was further analyzed by electron backscatter diffraction (EBSD) in an electron microscope (Figure 39.18b).

These work function/contrast changes are the basis of imaging surface processes or catalytic reactions on the micrometer scale (see also Chapter XX, volume Y). During CO oxidation, the CO or oxygen coverage governs the rate of $CO_2$ formation (see below) [259], which means that the local PEEM image intensity (brightness) serves as a descriptor of the local reaction rate, which allows real-time imaging of surface reactions.

The principle of *simultaneously* obtaining global and local reaction kinetics is illustrated in Figure 39.19 [72, 254, 257]. As mentioned, these experiments were performed in a multipurpose UHV system consisting of two connected (but independent) chambers, one for PEEM/LEED/MS and the other for XPS (Figure 39.7). The reactants CO and $O_2$ are dosed via leak valves, and to allow reactant pressures up to $10^{-4}$ mbar, the PEEM intensifier section was differentially pumped and two in-line apertures (diameter 4 and 0.3 mm) were placed along the photoelectron

**Figure 39.19** Local kinetics of CO oxidation on individual crystalline grains of a Pt foil. (a) Principle of the local kinetic measurements by PEEM. Left inset: global $CO_2$ production rate versus CO partial pressure as measured by MS during catalytic CO oxidation at $T = 417$ K and $p_{O_2} = 1.3 \times 10^{-5}$ mbar. Right inset: hysteresis plot of the local PEEM intensity of an individual Pt(100) facet versus CO partial pressure, as measured *in situ* by PEEM at the same reaction conditions. The selected domain is indicated in frame (1). The numbers along the hysteresis curve indicate the corresponding PEEM frames: (1) oxygen-covered active surface; (2) kinetic transition at $\tau_A$; (3) CO-covered inactive surface; and (4) kinetic transition at $\tau_B$; (1) oxygen-covered, active surface, catalytic cycle is going to be closed. (b) Corresponding kinetic phase diagram for a Pt(100) facet (black squares) in comparison to the global (MS-measured) diagram (red dots). Hatched areas correspond to regions of bistability, regions below the hatched areas represent the active steady state, and regions above the hatched areas indicate the inactive steady state. (c) The same as in (b), but in comparison to a local diagram for a Pt(110) facet (red dots). (d) The scheme on the right is a graphical representation of the PEEM image on its left. (Panels (a–c) reproduced with permission from Ref. [257]. Copyright (2011) Elsevier.) (Panels (d,e) reproduced with permission from [72]. Copyright (2010) Wiley.)

trajectory. For better time resolution, PEEM images were recorded by a high-speed CCD camera.

The global kinetics of CO oxidation is determined by a mass spectrometer (located near the sample) in the typical averaging mode of MS-based measurements (e.g., for background gas, molecular beams, or TPD), providing the global $CO_2$ rate produced by the entire polycrystalline sample (see left inset in Figure 39.19a). *In parallel,* the local (laterally resolved) kinetics on the individual grains is *simultaneously* determined by imaging the ongoing surface reaction by PEEM (see right inset in Figure 39.19a). On Pt, the oxygen-covered surface is active and governs the rate of $CO_2$ formation, so that the dark PEEM areas are the active regions. The oxidation of CO on platinum seems to be a simple surface reaction but it exhibits several complexities, such as hysteresis (bistability), oscillations, dissipative structures, and chaotic behavior [76, 256]. (Here we focus on bistability because oscillations are extensively described in Chapter 49, volume 6.)

The global (overall) reaction kinetics (Figure 39.19a) serves to explain the kinetic behavior of Pt foil at a reaction temperature of 417 K. The Pt foil was exposed to a constant partial background pressure of oxygen ($1.3 \times 10^{-5}$ mbar), while the CO partial pressure ($p_{CO}$) was increased/cycled, and the $CO_2$ production rate ($R_{CO_2}$) was monitored by MS. Upon exposing the oxygen-covered Pt surface to increasing CO pressure, $R_{CO_2}$ increased (high activity) until a kinetic transition point at pressure $\tau_A$ was reached. When $p_{CO}$ exceeded $\tau_A$, the Pt surface switched from oxygen-covered to CO-covered, marked by the loss of catalytic activity due to CO self-poisoning (oxygen cannot adsorb on the CO-covered surface). When $p_{CO}$ was decreased, the catalyst remained in the poisoned (low-reactivity) steady state until a second transition pressure $\tau_B$ was reached (at which the surface switched back from CO-covered to oxygen-covered). Accordingly, at $p_{CO}^{\tau_B} < p_{CO} < p_{CO}^{\tau_A}$, the catalyst can adopt one of two steady states, depending on the sample history. The resulting hysteresis is characteristic for a bistable reaction behavior, which originates from the asymmetric inhibition of dissociative oxygen adsorption by CO. $O_2$ needs two neighboring adsorption sites per molecule for dissociation and cannot adsorb on a densely packed, CO-covered Pt surface, whereas on the oxygen-covered Pt surface CO easily adsorbs and reacts with $CO_2$ [256].

At constant $p_{O_2}$ but different reaction temperatures, the values of the transition pressures $\tau_A$ and $\tau_B$ of the hysteresis are shifted (because the CO poisoning is temperature dependent). Plotting the kinetic results, that is, $\tau_A$ and $\tau_B$ pressures versus the reciprocal temperature, then yields a so-called kinetic phase diagram (Figure 39.19b). As mentioned, such a measurement provides the global *spatially averaged* reaction kinetics as typical for MS reaction studies, the averaging being over a catalyst area in the centimeter square range.

How do we address *spatially resolved* kinetics? The corresponding mass spectrometric route would be using a micrometer-sized nozzle ("sniffer") positioned on (or scanned over) specific surface spots [260]. However, the resolution is limited, it is not a parallel measurement, and also the exact nozzle position relative to the surface structure underneath can be determined only indirectly. In contrast, PEEM enables parallel (i.e., not scanning) imaging and surface characterization at the same time.

As mentioned, the contrast changes (caused by work function changes induced by oxygen or CO adsorption on the clean Pt foil) and their intensity analysis provide the basis for "watching" the ongoing $CO + O_2$ reaction. The resolution of the locally resolved reaction kinetics is given by the PEEM resolution of ~1 μm (i.e., every image pixel is like a micrometer-sized mass spectrometer).

When CO oxidation is performed under conditions as described previously (resulting in the hysteresis plot of Figure 39.19a), PEEM microscopy provides a high-speed "movie" (500 frames per second) of the ongoing reaction (and the reaction is simultaneously monitored by MS) [72, 73]. Figure 39.19a (inset) displays selected PEEM video frames acquired during the hysteresis cycle (note that several grains are in the field of view). When the CO pressure is increased, for example, the (100) facet becomes brighter when switching to a CO-covered state (frames 2 and 3 in Figure 39.19a). When the CO pressure is reduced, the (100) domain switches back to dark, that is, O-covered (frame 4 in Figure 39.19a; one reaction front is marked by a white arrow). The image sequence is also displayed in Figure 39.19d.

To quantitatively analyze the locally resolved kinetics, the (local) PEEM intensity was integrated for a specific, selected domain (region of interest), and plotted as a function of $p_{CO}$, which again identified the corresponding transition pressures $\tau_A$ and $\tau_B$. As an example, Figure 39.19a (right inset) shows the locally measured hysteresis curve for the (100) domain marked in frame 1. Kinetic transitions appear as pronounced drop/increase in the local PEEM intensity hysteresis curve. Other (100) domains on the foil behaved identically because of the identical gas pressure and reaction temperature all over the sample. Analogous measurements of a series of local hysteresis curves at different reaction temperatures then allowed the construction of a *local* domain-specific phase diagram for (100) domains (Figure 39.19b,c) [257].

As a parallel imaging technique, PEEM provides this type of information for the entire field of view at any reaction time. Therefore this approach was applied to (110)- and (111)-oriented domains as well, with the results again summarized in the domain-specific phase diagrams of Figures 39.19c and 39.20b [72, 73, 257]. Despite numerous previous studies on single crystals, the kinetic phase diagrams for Pt(100) and Pt(110) obtained from the PEEM foil measurements were indeed novel.

It is also remarkable that the (111) facet, the most popular (and most frequently studied) in surface science [77], is the least active for the current reaction! Pt(111) gets deactivated at the lowest CO pressure and also needs the lowest CO pressure to reactivate. Consider an imaginary point (A) in the facet-resolved kinetic phase diagram of Figure 39.20b [72, 73]. At a CO pressure that is in the "gap" of the bistability ranges, the (111) facet would already be CO-poisoned ($p_{CO} > \tau_A(111)$), whereas the (100) and (110) facets would be active and would not have reached their maximum activity yet (just below $\tau_A(100)$ and $\tau_A(110)$). Consequently, under specific reaction conditions, some facets of a large catalytic particle would be active while others may be inactive.

As mentioned, one key advantage of the PEEM approach is that at a given time the reactant gas pressure and temperature are identical for all surface terminations (domains), allowing direct and simultaneous comparison of their

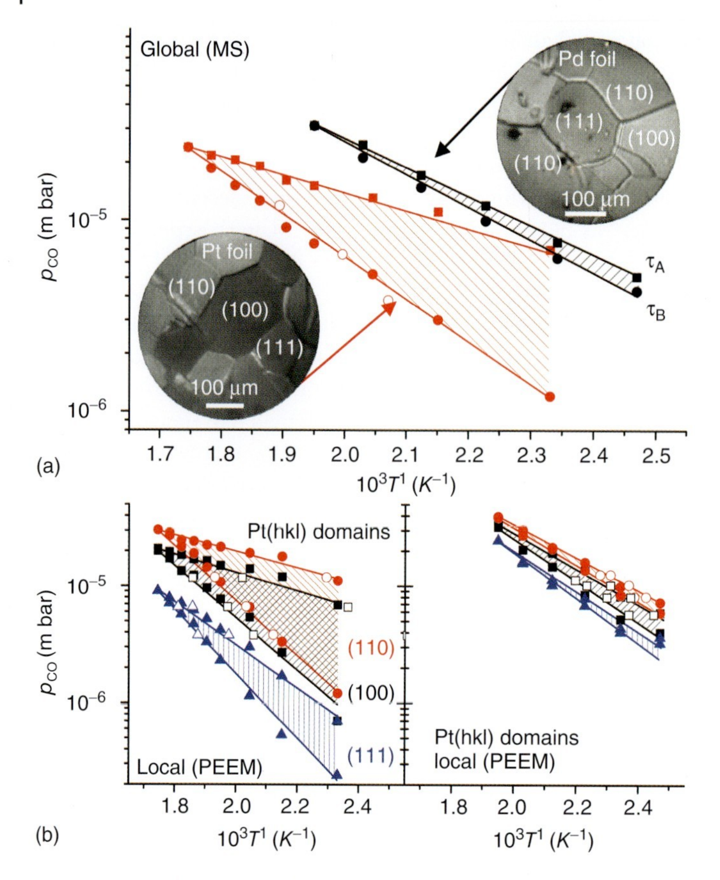


**Figure 39.20** Comparison of the CO oxidation performance of palladium and platinum. (a) Global kinetic phase diagrams (by MS) at constant oxygen pressure ($p_{O_2} = 1.3 \times 10^{-5}$ mbar) of polycrystalline Pt (filled red squares and circles) and Pd (black squares and circles). Open circles are ignition points for Pt. (b) Corresponding local kinetic phase diagrams for individual Pt(*hkl*) domains (left) and Pd(*hkl*) (right), obtained by local PEEM intensity analysis. Open symbols are the local ignition and extinction points. At an imaginary point A, the (111) domains are already inactive, whereas the (100) and (110) domains are still in a high-reactivity regime. (Reproduced with permission from Ref. [73]. Copyright (2012) John Wiley and Sons.)

catalytic properties [72]. The same is true for sample preparation/cleaning/modification (e.g., creating defects by sputtering [258] or surface oxidation [204]) because *all* domains are treated identical. If one were to carry out similar measurements on separate single crystals, this level of "identical modifications" can hardly be achieved.

Nevertheless, one can question whether a (111) domain on the foil is really representative of a Pt(111) single-crystal surface. On the foil, there are certainly grain boundaries, and one expects more defects in general. Thus, in order to critically assess the validity of the locally resolved kinetics, the phase diagram of

(111)-oriented domains on Pt foil (Figure 39.20b) was compared with that of a Pt(111) single crystal (measured by MS [79]). Despite the fact that the measurements were performed in different UHV chambers, the deviations were rather small; for example, in the temperature interval 489–493 K the $\tau_B$ pressure for (111) single crystal and (111) domain on Pt foil were $3.1 \times 10^{-6}$ and $4.8 \times 10^{-6}$ mbar, respectively (the difference likely due to the different positions of the pressure gauges relative to the sample). The good similarity with single crystal data was further supported by the observation of elliptical patterns (reaction fronts) on the (110) facets of Pt foil [261]. The anisotropy of the Pt(110) surface leads to different front propagation velocities along the main crystallographic directions, thus producing elliptical patterns. The relative front propagation velocities along [110] and [001] directions are in quantitative agreement with single-crystal data for Pt(110), as described in detail in Ref. [261].

A further question may be related to the effect of the grain boundaries, because higher activity is typically reported for defect sites [258]. The global reaction kinetics of Pt foil was thus modeled by summing up the weighted contributions (weighted average of the transition pressures) of the different domains, based on the domain area distribution of annealed foil of (100) : (110) : (111) = 2 : 2 : 1 [72]. Indeed, the global phase diagram constructed from the superposition of domain-specific local phase diagrams of the individual (100), (110), and (111) terminations (based on local PEEM intensity analysis) perfectly coincides with the experimental global phase diagram obtained by MS [72]. This indicates that the PEEM imaging/analysis monitored all relevant surface catalytic processes and that the facets, rather than the domain boundaries, governed the reaction rate. Otherwise, there should have been a significant deviation between the global MS rates (accounting for all active surface sites) and the weighted sum of local PEEM rates (including only "visible" domain sites in the field of view). Alternatively, one can also use the area-integrated PEEM intensities to construct the phase diagrams, and also these coincide well with the global MS data.

The good agreement points to a quasi-independent behavior of the individual domains (despite being surrounded by other different domains) and to a minor role of the domain boundaries. This is further supported by the observation that the propagating reaction fronts were confined *within* the grain boundaries (i.e., reaction fronts did not propagate across domain boundaries; see Figure 39.19a, frame 4). Subsequent atomic force microscopy (AFM) studies have finally shown that the domain boundaries are in fact rather "cracks," physically separating the different domains [257]. The independent reaction behavior of the individual domains can thus be rationalized by considering potential mechanisms of reactive coupling (synchronization) between the facets. Apparently, the "cracks" (domain boundaries) effectively hinder diffusion coupling, and the total reactant pressure ($10^{-5}$ mbar) is also too low to synchronize the facets via gas-phase coupling.

How relevant are the current model results for supported metal nanoparticles, since the adjacent planes of a small individual nanoparticle may behave differently from (separated) single-crystal surfaces of the same termination [57]? As shown by field ion microscopy (FIM) of CO oxidation on the facets of Pt field-emitter tips (facet size about 5 nm) [262, 263], for nanoparticles the kinetic transitions occur *coherently* for adjacent facets: that is, the kinetic transitions on (111), (100), and (110) facets occurred simultaneously. A kinetic phase diagram could thus be created only for the whole tip (because the nanofacets behaved identically). This was also observed for Pd nanoparticles supported on alumina using molecular beam methods [264]. Diffusion coupling was responsible for synchronization of local transitions on such small length scales. At pressures higher than $10^{-4}$ mbar, gas-phase coupling provides the synchronization [265, 266], and at atmospheric pressure, heat transfer contributes significantly or even dominates spatial coupling [262, 264]. For the current PEEM measurements, the formation of surface oxide can be excluded. Surface oxides are formed only at orders of magnitude higher oxygen pressure, as indicated by high local brightness in PEEM [204, 255]. Furthermore, nanoparticles, in particular when the reaction is run in oxygen excess, are clearly more prone to oxidation [141, 245, 267] than metal foils [144].

Apart from merely comparing the locally resolved, "facet-specific" activities, a more detailed microscopic picture can be obtained from a thorough analysis of the PEEM images and by using complementary techniques. On individual (100)-type grains of polycrystalline Pt foil, reaction-induced surface morphology changes (hills and valleys) were identified by optical differential interference contrast microscopy and AFM [257]. On the apparently isotropic Pt(100) surface, PEEM observations reveal a high degree of propagation anisotropy for the reaction front propagation both for oxygen and CO fronts. Whereas anisotropy vanishes for oxygen fronts at temperatures above 465 K, it is maintained for CO fronts at all temperatures studied (417–513 K), which was explained by a change in the front propagation mechanism [257]. On individual [113]-oriented domains, the observed front propagation velocities and the degree of their anisotropy agreed well with earlier observations on Pt(110) single crystals (again demonstrating the validity of the foil results) [261].

#### 39.2.2.1 The "Methodology Gap"

Apart from the well-known *materials* and *pressure* (and maybe *complexity*) gaps, there may also be a *methodology gap* between surface science and applied catalysis. For CO oxidation, surface science experiments (using MS, molecular beams, etc.) frequently keep the temperature constant and vary the reactant pressure/impingement rate [36, 72, 79, 144, 199, 205, 268, 269] (leading, e.g., to the hysteresis plots discussed above), whereas for technological catalysts usually the gas feed ("flow") composition is kept constant and the temperature is raised until conversion increases [245, 270, 271], leading to the well-known ignition curves [272, 273].

The motivation for such measurements lies in the so-called cold start problem. Although modern three-way catalytic converters (TWCs) efficiently remove pollutants from automobile exhausts, most (80%) of the emissions are emitted when the TWC is still cold, that is, before the catalytic converter warms up to its "ignition" temperature (at which the reaction rate rapidly increases). Clearly, there is strong interest in lowering this critical temperature, for example, by adding materials that are active around room temperature such as nanoscale gold [274, 275] or cobalt oxide [276, 277] or their combination [278].

The critical ignition temperature was originally defined as the temperature at which the heat generated by the exothermic reaction exceeds the dissipated heat, so that the catalyst heats up and no external heating is required anymore ("light-off"). For CO oxidation on model catalysts in UHV, the heat generated by the reaction can be neglected and the catalytic ignition problem is thus reduced to the "pure kinetics," that is, the temperature-triggered kinetic transition from the low-rate (CO poisoned) steady state to a high-rate (O-covered) steady state [279–281].

It is thus worthwhile to apply the locally resolved-kinetics PEEM approach also to the catalytic ignition in CO oxidation on micrometer-sized (*hkl*) domains of polycrystalline Pt *and* Pd foil, since both metals are used in TWCs [73]. This allows a direct comparison of the ignition properties of low-Miller-index Pt(*hkl*) and Pd(*hkl*) domains under identical reaction conditions, yielding the inherent reaction behavior. The combined MS/PEEM approach yields the *global* (Figure 39.21a) and *local* (Figure 39.21b) ignition kinetics for Pd foil, for example, by scanning the temperature from 372 to 493 K for constant/isobaric gas feed ($p_{CO} = 5.8 \times 10^{-6}$ mbar and $p_{O_2} = 1.3 \times 10^{-5}$ mbar). Upon increasing the temperature, the global $CO_2$ rate suddenly increases, indicating the transition $\tau_B^*$ from the state of low catalytic activity (CO-poisoned surface; video frame 1 in Figure 39.21a, dark contrast; the asterisk indicates the isobaric measurement) to the state of high catalytic activity at which

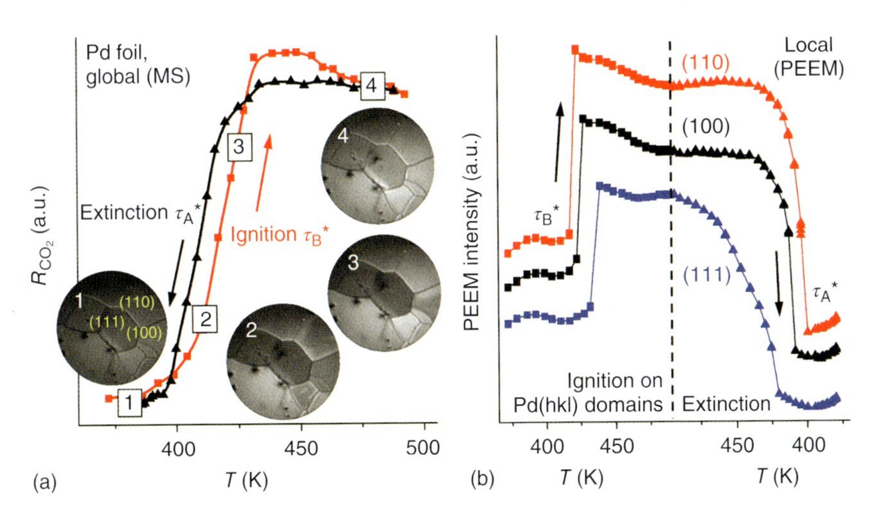


**Figure 39.21** (a) Ignition (red squares) and extinction curves (black triangles) on polycrystalline Pd foil, that is, the $CO_2$ production rate measured globally by MS on cyclic variation of the sample temperature (rate: 0.5 K/s) at constant $p_{CO} = 5.8 \times 10^{-6}$ mbar and $p_{O2} = 1.3 \times 10^{-5}$ mbar. The simultaneously recorded PEEM video sequences illustrate the ignition process: Frame (1) inactive CO-covered surface; (2) ignition sets in on (110) domains; (3) ignition continues on (100) domains; and (4) oxygen-covered active surface. (b) Laterally resolved ignition/extinction measurements: local PEEM intensity for individual (110), (100), and (111) domains during the same cyclic temperature scan as in (a). The vertical dashed line indicates the switch from heating to cooling. (Adapted with permission from Ref. [73]. Copyright (2012) John Wiley and Sons.)

the surface becomes oxygen-covered (frame 4; bright contrast). Note that, because of the different work functions, the contrast on Pd is inverted compared to that on Pt.

Analogous to the MS signal in the overall $CO_2$ reaction rate, the jumps in the *local* PEEM intensity represent the local kinetic transitions on the individual grains (Figure 39.21b). These transitions do not occur simultaneously on the different orientations but show a pronounced structure sensitivity with critical temperatures of 417 K for Pd(110), 423 K for Pd(100), and 432 K for Pd(111). A similar observation was made for reaction extinction, that is, for the transition $\tau_A^*$ from the high reactivity (O-covered) to the low reactivity (CO-covered) state upon cooling the sample. The extinction curve of the *global* $CO_2$ production rate appears to be "smoothened out" (black curve in Figure 39.21a), whereas *local* extinction on the individual grains occurs rather sharp and independent from each other (Figure 39.21b). This clearly demonstrates the limits when an averaging technique (here MS) is applied to a heterogeneous sample.

Similar as for Pt foil, the kinetic transitions of CO oxidation on Pd foil were also studied by varying the $CO/O_2$ pressure ratio at constant temperature ($p_{O_2} = 1.3 \times 10^{-5}$ mbar and $T = 449$ K; right inset of Figure 39.22a). The global $CO_2$ formation rate exhibits a hysteresis upon cyclic variation of the CO partial pressure, as is evident from the gap between the kinetic transition $\tau_A$ (high to low reactivity state) and the reverse transition $\tau_B$. In between, the system is bistable, that is, it can be either in the high or in the low reactivity steady state, depending on the sample history. The $\tau_A$ and $\tau_B$ points are again temperature-dependent, thus a (global) kinetic phase diagram can be constructed for the Pd foil, summarizing the kinetic transitions (Figure 39.22a).

For comparison, the kinetic transition points ($\tau_A^*/\tau_B^*$) extracted from global isobaric ignition/extinction experiments (shown in the left inset of Figure 39.22a for constant $p_{CO} = 5.8 \times 10^{-6}$ mbar and $p_{O_2} = 1.3 \times 10^{-5}$ mbar) are also plotted in the isothermally obtained diagram. The *isobaric* variation of the reaction temperature results in kinetic transitions that are *quantitatively* matching the kinetic phase diagram obtained by the *isothermal* variation of the CO pressure [73]. In this figure, the *isothermal surface science world* and the *isobaric applied catalysis world* meet and mingle!

Figure 39.22b displays the corresponding *local* PEEM analysis of the same transitions for an individual Pd(100) domain. The left inset again shows the ignition/extinction experiment, and the right inset presents the transitions obtained via variation of the CO pressure at constant $T$ and $p_{O_2}$. Again, the local isothermal and isobaric data are equivalent.

The specific properties of Pt and Pd for CO oxidation are evident from comparing the kinetic phase diagrams [73]. The *global* reactivity of Pt and Pd foil is compared in Figure 39.20a, whereas Figure 39.20b displays the *local* kinetic transitions of the individual Pt(*hkl*) and Pd(*hkl*) domains. The main differences between Pt and Pd are (i) the global and the respective local kinetic phase diagrams of Pd foil are shifted to higher CO partial pressure, and (ii) the bistability range is much narrower for Pd than for Pt foil. This means that for Pd the transition $\tau_A$ from the high to the low reactivity state occurs at higher CO partial pressure than for Pt,

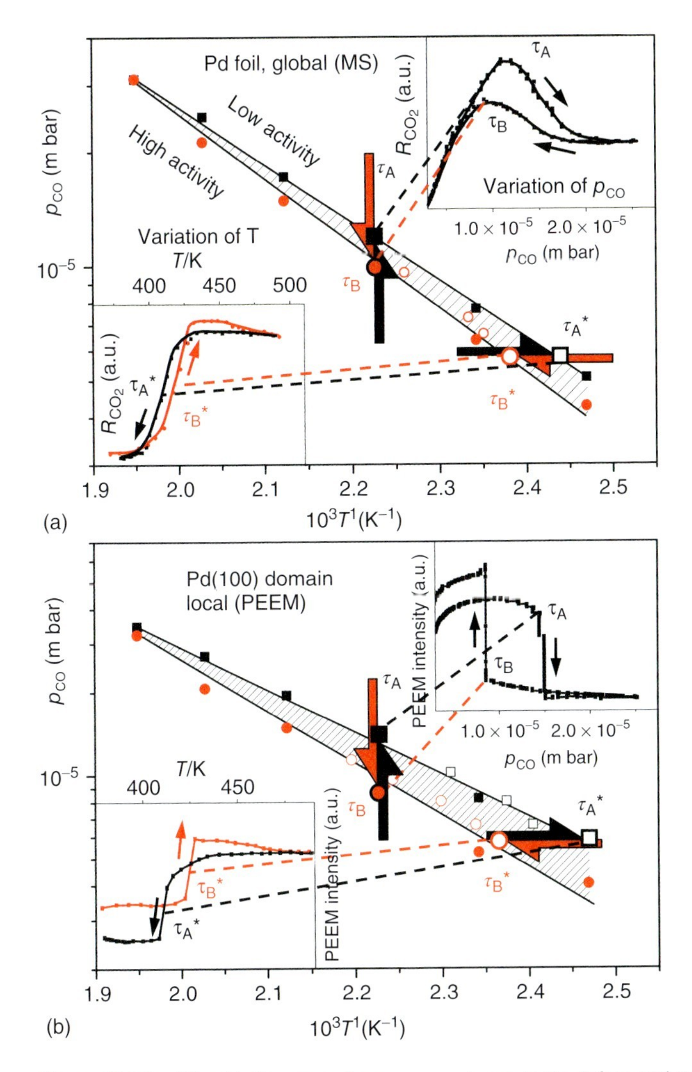


**Figure 39.22** CO oxidation on polycrystalline Pd foil. (a) Global (MS) kinetic phase diagram and (b) local (PEEM) kinetic phase diagram of an individual Pd(100) domain on the Pd foil. Note the agreement of the transition points $\tau_A^*$ and $\tau_B^*$ obtained by varying $T$ (via the ignition/extinction curves shown in the left insets) with those obtained via cyclic variation of $p_{CO}$ (via the poisoning/reactivation curves in the right insets). The dashed regions indicate the range of bistability. (Reproduced with permission from Ref. [73]. Copyright (2012) John Wiley and Sons.)

and that the reverse transition $\tau_B$ also occurs at a higher CO/oxygen ratio than for Pt. In other words, under the current conditions, Pd is the better (more CO-tolerant) catalyst than Pt because *more* CO is needed to poison the Pd surface and a *lower* oxygen/CO ratio is sufficient to "reactivate" the Pd surface. Besides that, the bistability regime of Pd disappears at a lower temperature than in the case of Pt, namely at $T_{Pd} = 513$ K in contrast to $T_{Pt} = 573$ K, so already at lower temperature Pd cannot be poisoned by CO anymore.

In order to rationalize the experimental observations, DFT was applied to calculate the adsorption energies of CO and oxygen on clean Pd(*hkl*) and Pt(*hkl*) domains, as well as the reaction barriers [73]. The bistability regions for Pt(111), Pt(100), Pt(110), and Pd(111) were simulated by a micro-kinetic model based on the conventional Langmuir–Hinshelwood mechanism, both simulating ignition/extinction and hysteresis curves. In all cases, the kinetic simulations confirmed the experiment: the kinetic phase diagram of Pd is located at higher CO pressure, and the bistability region of Pd is considerably narrower than that of Pt. Also the DFT-derived order of the local kinetic phase diagrams corresponded with the experimentals results: Pt(111) is easiest to poison, Pt(100) is in the middle, and Pt(110) is still active at the highest CO pressures [73].

Both experiments and theory showed that Pt(*hkl*) domains are deactivated at lower CO pressure than Pd(*hkl*) (at given oxygen pressure and temperature). This is mainly due to the adsorption energies of oxygen that are higher on Pd(*hkl*) than on Pt(*hkl*); that is, oxygen is more strongly bound to Pd. Thus, CO poisoning of the Pd surface occurs at higher CO pressure. The Pd foil thus also "reactivates" already at a considerably higher CO pressure than the Pt foil does. The disappearance of the bistability regime at the so-called cusp point, which occurs at lower temperature for Pd than for Pt, can be explained by the generally lower desorption temperature of CO on Pd(*hkl*).

Further efforts to bring the PEEM approach closer to "real-world catalysis" included studying the effect of defects (artificially created on the polycrystalline Pd foil by STM-controlled $Ar^+$ sputtering) [258]. On the defect-rich foil, the independent reaction behavior of the individual (single-crystal-like) Pd(*hkl*) domains vanished. It changed to a correlated reaction behavior, that is, the reaction fronts propagated unhindered across the grain boundaries. This is due to the fact that (i) the specific microstructure of the different terminations is destroyed by sputtering (leading to more or less the same surface structure), and (ii) the cracks between the domains were filled by sputtered metal. However, the defect-rich surface showed a significantly higher CO tolerance, leading to a shift of both the global (MS-measured) and the local (PEEM-measured) kinetic diagrams toward higher CO pressure [258]. Recently, locally resolved PEEM imaging of reaction kinetics was also applied to polycrystalline Pd powder pressed onto different substrates such as Pt foil or $Al_2O_3$ or $ZrO_2$ thin film oxides [282], which again brings model catalysis closer to applied catalysis.

### 39.2.3
### Kinetic Studies on Nanoparticle Model Catalysts – Ethylene and 1,3-Butadiene Hydrogenation

In the proceeding sections, the focus has been on surface spectroscopy and surface microscopy, complemented by kinetic studies, of course. In the following, studies are presented that are rather centered around the catalytic properties of model catalysts [36, 201, 283–286], determined by atmospheric pressure reactions performed in UHV-compatible high-pressure cells (batch mode with gas recirculation; see Figure 39.6) [202] with quantitative online GC analysis of the reactants and products. Clearly, these studies were also accompanied to some extent by surface characterization and spectroscopy.

"Ethylene" (ethene $C_2H_4$) hydrogenation to ethane ($C_2H_6$), being a classical surface science reaction [18, 58, 77, 101, 279–285], was employed as a first test reaction for Pd-$Al_2O_3$/NiAl(110) model catalysts (conditions: 50 mbar $C_2H_4$, 215 mbar $H_2$, 770 mbar He) [18, 36, 58, 201, 294]. Figure 39.23b displays a conversion versus time plot for Pd particles (mean size 3.5 nm) at various reaction temperatures, with the straight lines indicating a constant reaction rate (turnover frequency, TOF; defined

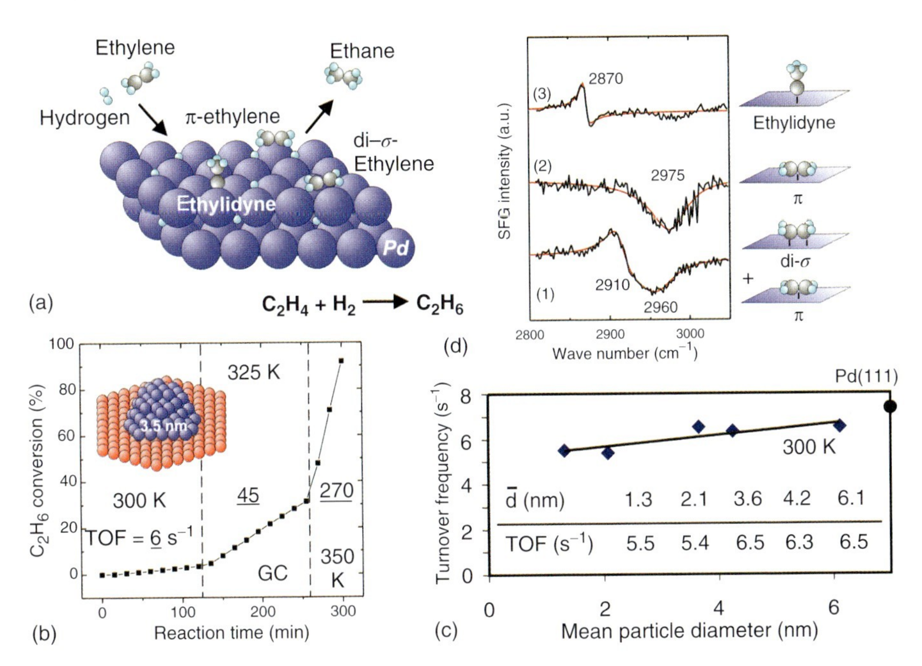


**Figure 39.23** (a) Schematic illustration of the various adsorbed ethene species and their adsorption positions on Pd(111). (b) Ethene hydrogenation on a Pd/$Al_2O_3$/NiAl(110) model catalyst with a mean palladium particle diameter of 3.5 nm. The reaction was carried out with 50 mbar of $C_2H_4$, 215 mbar of $H_2$, and 770 mbar of He at temperatures in the range 300–350 K. Because the high-pressure SFG cell (see Figure 39.6) was used as a recirculation batch reactor, the conversion increased with time. Turnover frequencies for the various temperatures are indicated. (c) Ethene hydrogenation activity of various Pd/$Al_2O_3$/NiAl(110) model catalysts (mean palladium particle diameter of 1–6 nm), illustrating the structure insensitivity of this reaction. The TOF characterizing the reaction on Pd(111) is marked by the circle. (d) SFG spectra of $C_2H_4$ species on Pd(111). Exposures were 2.5 L of $C_2H_4$ at 100 K (1), 1 L of $H_2$, followed by 2.5 L of $C_2H_4$ at 100 K (2). Spectrum (3) was acquired at 300 K after annealing of the sample in $5 \times 10^{-7}$ mbar of $C_2H_4$ at temperatures from 100 to 300 K. (Panels (b,c) adapted from Refs [18, 201] with permission. Copyright (2003, 2002) Elsevier.) ((d) Adapted from Ref. [58] with permission. Copyright (2005) Elsevier.)

as the number of product molecules reacted per Pd surface atom per second) at a given temperature. Figure 39.23c compares the steady-state TOFs at 300 K for various (mean) Pd particle sizes. As expected for a structure-insensitive reaction, the steady-state TOF at 300 K (~6 $s^{-1}$) was independent of the particle size (1.3–6.1 nm; size determined by STM [102]) and similar to the TOF of Pd(111). The nanoparticles were stable with "time-on-stream", and the steady-state TOFs, reaction orders (ethylene: −0.3; hydrogen: 1), and the activation energy (about 50–60 kJ/mol) were very similar to those reported for wet-chemically synthesized (powder) catalysts [295, 296], demonstrating that Pd-$Al_2O_3$/NiAl(110) model catalysts closely mimicked the properties of technological catalysts (as intended by the model catalyst approach!).

For mechanistic studies, *in situ* surface spectroscopy is required. The reaction presumably proceeds via the stepwise hydrogenation of ethylene, as proposed already by Horiuti and Polanyi [297]. The various ethylene adsorbate configurations have been determined by different techniques, including LEED surface crystallography, TDS (thermal desorption spectroscopy), UPS (ultraviolet photoelectron spectroscopy), HREELS (high-resolution electron energy loss spectroscopy), XPS, IRAS, NEXAFS (near-edge X-ray absorption fine structure), DFT, and so on [18, 289, 298–309]. Figure 39.23a shows a schematic of $C_2H_4$ adsorption and hydrogenation (see also Chapter 40 in this Volume).

Following an "easy" approach, ethene adsorption has been investigated on single-crystal Pd(111) by SFG spectroscopy [18, 58, 201, 310]. At 100–200 K, ethene adsorbed in a di-σ-configuration with a characteristic peak at 2910 $cm^{-1}$ ($\nu_S(CH_2)$; Figure 39.23d, trace (1)). The (gas-phase like) carbon–carbon double bond of $C_2H_4$ is broken and the carbon atoms attain nearly $sp^3$ hybridization. Two $\sigma$ bonds are formed with the underlying palladium surface, yielding di-σ-bonded ethene at bridge sites. The second weak peak at ~2960 $cm^{-1}$ can be attributed to the $\nu_S(CH_2)$ of π-bonded ethane adsorbed on top of a single Pd atom. Now the C–C bond is parallel to the surface, with the distance between the two $sp^2$ carbon atoms being almost unchanged with respect to the gas-phase molecule [298–300].

When this layer was annealed to room temperature, nearly all of the ethene desorbed, and only a small amount dehydrogenated to ethylidyne [104, 307]. Only when ethene was adsorbed at room temperature [58, 311], a signal for ethylidyne ($M{\equiv}C{-}CH_3$) could be observed at 2870 $cm^{-1}$ ($\nu_S(CH_3)$; Figure 39.23d, trace (3)). This means that a hydrogen shift had occurred from one carbon atom to the other to form ethylidene ($=CH{-}CH_3$) [312], followed by further dehydrogenation to form ethylidyne $\equiv C{-}CH_3$, which is oriented upright and located at an fcc threefold hollow metal site [58, 298–300, 303, 306, 307, 311, 313]. Ethylidyne is then stable at temperatures up to ~400 K, and at higher temperatures dehydrogenation continues, producing $CH_x$ species and finally graphitic precursors. A Pd(111) surface covered only with π-bonded ethene ($\nu_S(CH_2)$ at 2975 $cm^{-1}$) can be produced by first adsorbing hydrogen or oxygen (which block the threefold hollow sites), followed by ethene adsorption at 100 K (Figure 39.23d, trace (2)) [306, 314].

SFG spectra acquired during ambient-pressure $C_2H_4$ hydrogenation on Pd(111) (under conditions of high activity) did not detect any resonances, suggesting that both di-σ-bonded ethene and ethylidyne were either absent or present in very small amounts ([18, 310]). In light of the high activity, π-bonded ethene may thus be the reactive species because π-bonded ethene produces only a small SFG signal in the C–H stretching frequency range, as a consequence of the C–H bonds being nearly parallel to the metal surface [309] (according to the surface-dipole selection rule for metal surfaces, dynamic dipoles parallel to the surface plane are cancelled by image dipoles inside the metal [227]). Furthermore, under active conditions, the surface concentration of π-bonded $C_2H_4$ may also be low. PM-IRAS experiments should be able to detect the deformation modes of π-bonded ethene at about 1000 $cm^{-1}$. However, for an active state (TOF of about 15 $s^{-1}$ at a $C_2H_4/H_2$ molar ratio of 1 : 4) also PM-IRAS did not detect surface species [36, 294]. Only for a lower activity (TOF of 4 $s^{-1}$ and a $C_2H_4$:$H_2$ molar ratio of 1 : 1) a single peak at 1339 $cm^{-1}$ was detected, characteristic of ethylidyne ($\delta CH_3$ of $\equiv C-CH_3$) [104, 311, 315]. $C_2H_4$ dehydrogenation to ethylidyne is apparently suppressed when a fourfold excess of hydrogen is used (H adsorbed in hollow sites seems to block the sites required for ethylidyne formation).

Thus, π-bonded ethene seems to be the most likely intermediate, but a contribution of di-σ-bonded ethene cannot be excluded. Ethylidyne, on the other hand, is only a spectator in $C_2H_4$ hydrogenation. This is corroborated by the observation that, once an ethylidyne layer had been formed, it could not be hydrogenated away by increasing the hydrogen pressure. When $C_2H_3$ is further decomposed, it leads to catalyst deactivation by $CH_x$ poisoning. *In situ* spectra of supported palladium nanoparticles under ambient conditions did also not provide evidence of adsorbed ethene species.

Finally, a more complex ambient pressure hydrogenation reaction involving *selectivity* is discussed, namely 1,3-butadiene (BD) hydrogenation [284, 285]. Butadiene hydrogenation produces 1-butene, *trans*-2-butene, *cis*-2-butene, and *n*-butane (see Figure 39.24c), with 1-butene being the industrially most interesting product. To examine size effects on activity and selectivity, Pd-$Al_2O_3$ model catalysts with mean particle diameters ranging from 2 to 8 nm were again employed. The nanoparticles exhibited various shapes, including hemispherical to (111)-oriented truncated cuboctahedra, as determined by STM studies [18, 102, 111]. This is a tremendous advantage because the relative abundance of specific surface sites (such as terrace, edge, interface atoms, etc.) can be accurately determined (see the surface site statistics in Tables 1 and 2 of Ref. [36]). Once the *exact number* and *type of available surface sites* are known, more accurate TOFs can be obtained.

Figure 39.24a,b show the product distribution versus reaction time for Pd particles of 2.1 and 6.1 nm mean diameter, respectively. Particle size effects on rate and selectivity are described in detail in [285]; here the focus is on the particle size dependence of the *initial* activity (for a reaction time of ~60 min). Figure 39.24c shows the initial TOF as a function of mean Pd particle diameter. For this plot, the rate (number of product molecules produced per second) was normalized (divided) by the total number of Pd surface atoms (i.e., averaged over the entire surface of all Pd nanoparticles), yielding a pronounced particle size dependence, indicating

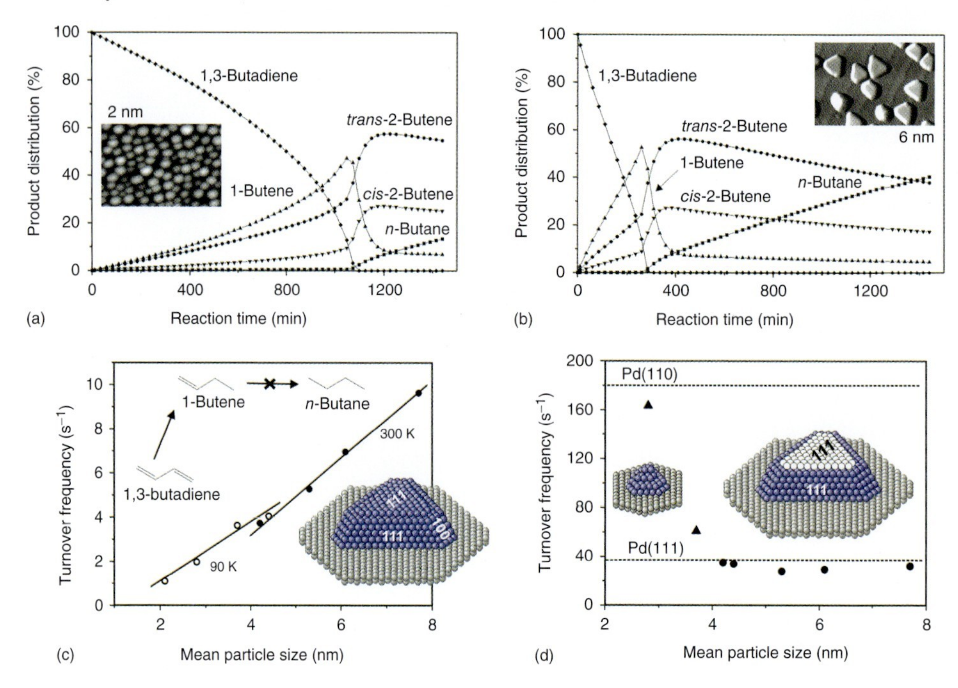


**Figure 39.24** 1,3-Butadiene hydrogenation on different $Pd/Al_2O_3/NiAl(110)$ model catalysts: product distribution versus reaction time at 373 K ($p_{1,3\text{-butadiene}}$: 5 mbar; $p_{H2}$: 10 mbar; Ar added up to 1 bar). (a) Mean Pd particle size 2.1 nm, (b) 6.1 nm. For both catalysts, the number of surface Pd atoms is $\sim 8 \times 10^{14}\ cm^{-2}$. The TOF for different $Pd\text{-}Al_2O_3/NiAl(110)$ catalysts as a function of mean particle size, (c) normalized by the total number of Pd surface atoms, and (d) normalized by using the number of Pd atoms on incomplete (111) facets, using a truncated cubo-octahedron as structural model (shown as inset). The TOF values for Pd(111) and Pd(110) under identical reaction conditions are included (dashed lines). (Adapted from Ref. [284] with permission. Copyright (2006) The Royal Society of Chemistry.)

that larger Pd particles are more active than smaller ones, in agreement with earlier reports [316–323].

However, when normalizing to the total number of surface atoms, any information on the different sites is lost. It is well known from single-crystal studies that Pd(110) is 5 times more active than Pd(111) [283, 324]. Therefore, the absolute activity of a catalyst (i.e., the total number of butadiene molecules converted) should also be correlated with the total number of the various specific surface sites of the Pd nanoparticles (such as (111) or (100) facets, step and edge sites, interface sites). A good correlation between rate and surface sites was obtained only for Pd atoms within incomplete (111) surface facets [285]. An incomplete (111) facet is shown in white in the model in Figure 39.24d (such incomplete surface facets were observed by HRTEM [42] and STM [104]). Accordingly, Figure 39.24d displays the particle size dependence of the reaction rate (filled dots), normalized to the number of Pd atoms in incomplete (111) facets. In this plot, the TOF is clearly particle size-*independent*, at least for Pd particles >4 nm. This suggests that the Pd atoms in the (111) facets are the active sites, which is supported when the normalized rate on incomplete (111) nanoparticle facets is compared to the rate of BD hydrogenation measured under identical reaction conditions on a Pd(111) single crystal [283]. This study is a prime example of how the "materials gap" between heterogeneous catalysis on metal nanoparticles and surface science on single crystals can be bridged on a *quantitative* basis. For small Pd particles (mean size ~2–3 nm), normalization was more difficult because these particles no longer exhibit well-developed facets (a ~2 nm particle is shown in Figure 39.24d; these data points are marked as ▲). Clearly, the TOF suggests that the catalytic activity of small Pd particles approaches that of Pd(110).

The purely kinetic measurements were extended by complementary spectroscopic studies, but performed on Pd(111) (Figure 39.25) [286]. Contrary to the behavior on Pd nanoparticles [284, 285], the Pd(111) single crystal exhibits a lower selectivity toward butene formation, with *n*-butane formed already in the first minutes of the

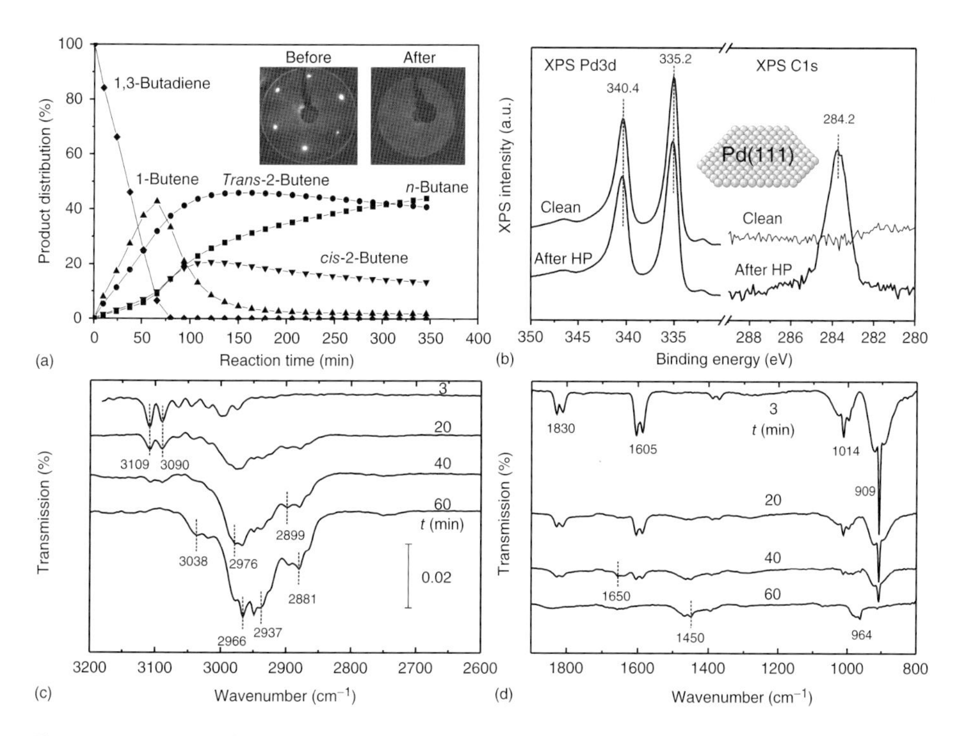


**Figure 39.25** (a) Product distribution versus reaction time for 1,3-butadiene hydrogenation on Pd(111) at 300 K ($p_{1,3\text{-butadiene}}$: 5 mbar; $p_{H2}$: 10 mbar; Ar added up to 1 bar). (b) Post-reaction XPS analysis in the Pd 3d and C 1s region (after 60 min high-pressure reaction). (c,d) IRAS measurements during the catalytic reaction at 300 K ($p_{1,3\text{-butadiene}}$: 3.3 mbar; $p_{H2}$: 6.7 mbar) after 3, 20, 40, and 60 min reaction time. (Adapted from Ref. [286] with permission. Copyright (2007) Elsevier.)

reaction, as evident from GC and/or IRAS. The hydrogenation process is accompanied by re-adsorption of butenes, mainly 1-butene, producing *n*-butane through hydrogenation and *trans/cis*-2-butene through isomerization. With decreasing 1-butene concentration, hydrogenation of 2-butenes sets in. This was corroborated by IRAS experiments, which detected only the gas-phase species (Figure 39.25c,d). No surface species could be observed, probably due to the flat adsorption geometry of the different molecules on the surface. An absence of surface signals was also reported in IRAS studies of butadiene hydrogenation on Pt(111) and $Pt_3Sn(111)$ [325]. Adsorption studies of BD on Pd(110) by HREELS and NEXAFS have also shown that the molecule adsorbs parallel on the surface [326].

Postreaction analysis by XPS in the C 1s region (Figure 39.25b) revealed a band at 284.2 eV, corresponding to adsorbed butadiene and/or carbonaceous deposits [286]. Quantification of this peak revealed a total carbon coverage of 0.3 ML. Nevertheless, deactivation due to carbon deposition was a minor effect, as indicated by the kinetics of the subsequent butene hydrogenation reaction. Temperature-dependent XPS spectra acquired after butadiene adsorption at 100 K in UHV indicated a high stability of the diene molecule on Pd(111), with hardly any desorption and/or decomposition up to 500 K. This strong interaction of the diene molecule with the Pd surface must be responsible for the blocking effects on hydrogen adsorption and butene re-adsorption, as reported previously for Pd nanoparticles supported on $Al_2O_3$ [284, 285]. However, care must still be taken when comparing UHV results with the atmospheric pressure kinetic studies because of the pressure difference and of the absence of hydrogen in the UHV measurements (very likely, adsorbed H would modify the diene adsorption properties). A high-temperature treatment of the butadiene-covered Pd(111) surface indicated decomposition processes (dehydrogenation) above 500 K and (partial) carbon dissolution into the Pd bulk above 700 K. For mechanistic insights and for studies on PdSn and PtSn alloys, refer to [283, 325, 327–330] and references therein, as well as the following chapter in this Volume.

## 39.3 Synopsis

Model catalysis has come a long way. A vast variety of technological catalysts can already be modeled, and there is an impressive array of surface characterization methods in our hands and increasingly more complex reactions are being tackled. The development of new analytical *in situ* methods will continue, but especially time-resolved (TR) measurements will become more and more important. TR ultrafast (ps, fs) laser spectroscopy is well established [331–334], but TR-XPS is rather at its beginning [335, 336]. Nevertheless, this chapter could provide only a small insight, even just a glimpse, into the impressive body of work reported by many research groups all over the world, as reflected by the references.

The surface science approach to heterogeneous catalysis has turned out to be very successful, and the final frontier has not yet been reached (Figure 39.26). The voyages

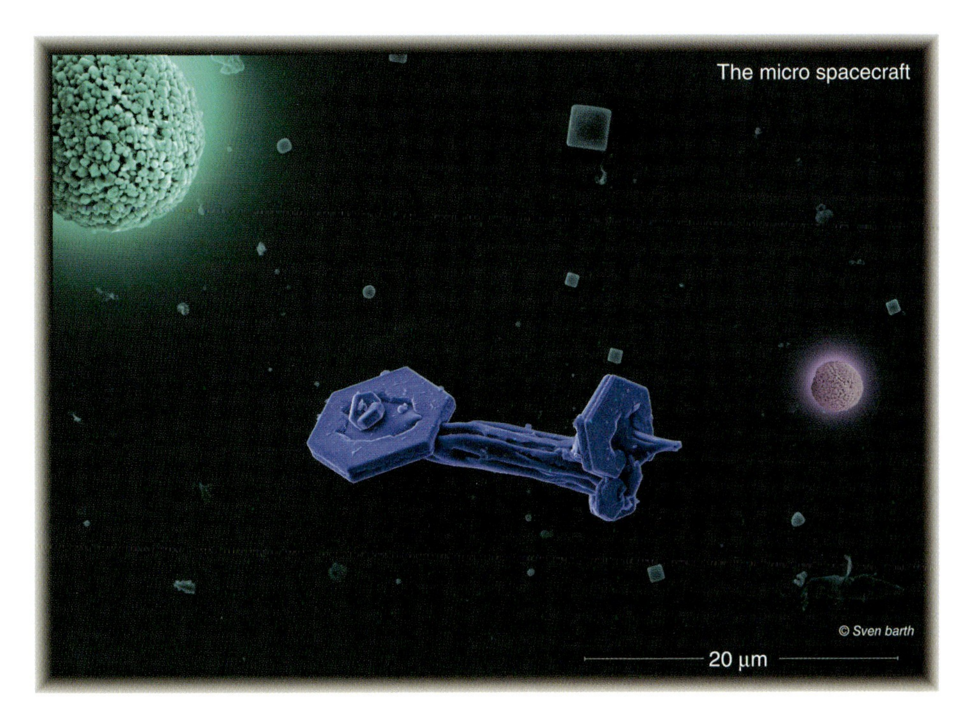


**Figure 39.26** SEM image of agglomerated ZnS crystals. (Image courtesy of S. Barth; Technische Universität Wien.)

of the Enterprise *Model Catalysis* are to be continued, as is its mission: to explore new model systems, to seek out new techniques and new reaction networks, and to boldly go where no one has gone before.


## Acknowledgments

I am very grateful for the contributions of previous and current coworkers and collaborators whose names appear in the list of references. PhDs: T. Dellwig, H. Unterhalt, M. Morkel, M. Borasio, F. Höbel, K. Zorn, D. Vogel, C. Spiel, A. Haghofer, H. Holzapfel, A. Wolfbeisser, L. Hao. Postdocs: L. Hu, P. Galletto, O. Rodriguez de la Fuente, V.V. Kaichev, J. Silvestre-Albero, A. Bandara, N. Barrabes, A. Bukhtiyarov. Habilitanden: K. Föttinger, C. Weilach, C. Rameshan.

I am particularly indebted to the following colleagues: H.-J. Freund (FHI Berlin), G.A. Somorjai (University of California, Berkeley), R. Schlögl (FHI Berlin), Y. Suchorski (TU Wein), K. Hayek[†], B. Klötzer and S. Penner (University of Innsbruck), J. Libuda (University Erlangen-Nuremberg), M. Bäumer (University Bremen), V.I. Bukhtiyarov (Boreskov Institute of Catalysis Novosibirsk), J. van


[†] Passed away.

Bokhoven (ETH Zurich and PSI), K.M. Neyman (ICREA and University Barcelona), H. Grönbeck (Chalmers University Gothenburg), D. Lennon (University Glasgow), C.R. Henry (University Marseille), J.J Calvino, and J.A. Perez-Omil (University of Cadiz).

## References

1. Ertl, G., Knözinger, H., Schüth, F., and Weitkamp, J. (2008) *Handbook of Heterogeneous Catalysis*, VCH-Verlag, Weinheim.
2. Somorjai, G.A. (1994) *Introduction to Surface Chemistry and Catalysis*, John Wiley & Sons, Inc., New York.
3. Chorkendorff, I. and Niemantsverdriet, J.W. (2003) *Concepts of Modern Catalysis and Kinetics*, Wiley-VCH Verlag Gmbh, Weinheim.
4. Rioux, R. (2010) *Model Systems in Catalysis – Single Crystals to Supported Enzyme Mimics*, Springer, New York.
5. Freund, H.-J., Bäumer, M., and Kuhlenbeck, H. (2000) Catalysis and surface science what do we learn. *Adv. Catal.*, **45**, 333.
6. Rupprechter, G. (2010) in *Model Systems in Catalysis - Single Crystals to Supported Enzyme Mimics* (ed R.M. Rioux), Springer, New York, pp. 319–344.
7. Stacchiola, D.J., Senanayake, S.D., Liu, P., and Rodriguez, J.A. (2012) Fundamental studies of well-defined surfaces of mixed-metal oxides: special properties of MOx/$TiO_2$(110) {M = V, Ru, Ce, or W}. *Chem. Rev.*, **113**, 4373–4390.
8. Wachs, I.E. and Routray, K. (2012) Catalysis science of bulk mixed oxides. *ACS Catal.*, **2**, 1235–1246.
9. Rupprechter, G. and Penner, S. (2010) in *Model Systems in Catalysis – Single Crystals to Supported Enzyme Mimics* (ed R.M. Rioux), Springer, New York, pp. 367–394.
10. Knözinger and Gates, B.C. (2006/2007) *Adv. Catal.*, **50–52**.
11. Somorjai, G.A. (1991) The flexible surface. Correlation between reactivity and restructuring ability. *Langmuir*, **7**, 3176.
12. Somorjai, G.A. and Rupprechter, G. (1998) The flexible surface. *J. Chem. Educ.*, **75**, 161.
13. Goodman, D.W. (1995) Model studies in catalysis using surface science probes. *Chem. Rev.*, **95**, 523.
14. Campbell, C.T. (1997) Ultrathin metal films and particles on oxide surfaces: structural, electronic and chemisorptive properties. *Surf. Sci. Rep.*, **27**, 1–111.
15. Henry, C.R. (1998) Surface studies of supported model catalysts. *Surf. Sci. Rep.*, **31**, 235.
16. Freund, H.-J. (1997) Adsorption of gases on complex solid surfaces. *Angew. Chem. Int. Ed. Engl.*, **36**, 452.
17. Freund, H.-J. (2002) Clusters and islands on oxides: from catalysis via electronics and magnetism to optics. *Surf. Sci.*, **500**, 271.
18. Freund, H.-J., Bäumer, M., Libuda, J., Risse, T., Rupprechter, G., and Shaikhutdinov, S. (2003) Preparation and characterization of model catalysts: from ultrahigh vacuum to in situ conditions at the atomic dimension. *J. Catal.*, **216**, 223–235.
19. Bäumer, M., Libuda, J., Neyman, K.M., Rösch, N., Rupprechter, G., and Freund, H.-J. (2007) Adsorption and reaction of methanol on supported palladium catalysts: microscopic-level studies from ultrahigh vacuum to ambient pressure conditions. *Phys. Chem. Chem. Phys.*, **9**, 3541–3558.
20. Henrich, V.E. and Cox, P.A. (1994) *The Surface Science of Metal Oxides*, Cambridge University Press, Cambridge.
21. Diebold, U. (2003) The surface science of titanium dioxide. *Surf. Sci. Rep.*, **48**, 53.
22. Freund, H.J. and Goodman, D.W. (2008) Ultrathin oxide films, in *Handbook of Heterogeneous Catalysis* (eds G. Ertl, H. Knözinger, F. Schüth, and J. Weitkamp), Wiley-VCH Verlag Gmbh, Weinheim.

23. Wöll, C. (2007) The chemistry and physics of zinc oxide surfaces. *Prog. Surf. Sci.*, **82**, 55–120.
24. Kuhlenbeck, H. and Freund, H.-J. (2006) in *Landolt-Börnstein: Physics of Covered Solid Surfaces, III-42* (ed H.P. Bonzel), Springer, pp. 332–403.
25. Pacchioni, G. and Freund, H.-J. (2013) Electron transfer at oxide surfaces. The MgO paradigm: from defects to ultrathin films. *Chem. Rev.*, **113**, 4035.
26. Paier, J., Penschke, C., and Sauer, J. (2013) Oxygen defects and surface chemistry of ceria: quantum chemical studies compared to experiment. *Chem. Rev.*, **113**, 3949–3985.
27. Esch, F. *et al.* (2005) Electron localization determines defect formation on ceria substrates. *Science*, **309**, 752.
28. Barrabes, N., Föttinger, K., Llorca, J., Dafinov, A., Medina, F., Sa, J., Hardacre, C., and Rupprechter, G. (2010) Pretreatment effect on Pt/$CeO_2$ catalyst in the selective hydrodechlorination of trichloroethylene. *J. Phys. Chem. C*, **114**, 17675–17682.
29. Föttinger, K. and Rupprechter, G. (2014) In situ spectroscopy of complex surface reactions on supported Pd-Zn, Pd-Ga and Pd(Pt)-Cu nanoparticles. *Acc. Chem. Res.*, **47**, 3071–3079.
30. Hayek, K., Kramer, R., and Paal, Z. (1997) Metal-support boundary sites in catalysis. *Appl. Catal., A*, **162**, 1.
31. Levin, M.E., Williams, K.J., Salmeron, M., Bell, A.T., and Somorjai, G.A. (1988) Alumina and titania overlayers on rhodium: A comparison of the chemisorption catalytic properties. *Surf. Sci.*, **195**, 341–351.
32. Hayek, K., Fuchs, M., Klötzer, B., Reichl, W., and Rupprechter, G. (2000) Studies of metal-support interactions with "real" and "inverted" model systems: reactions of CO and small hydrocarbons with hydrogen on noble metals in contact with oxides. *Top. Catal.*, **13**, 55.
33. Datye, A.K., Hansen, P.L., and Helveg, S. (2006) Atomic-scale imaging of supported metal nanocluster catalysts in the working state. *Adv. Catal.*, **50**, 77–95.
34. Datye, A.K. and Smith, D.J. (1992) The study of heterogeneous catalysts by high-resolution transmission electron-microscopy. *Catal. Rev. Sci. Eng.*, **34**, 129.
35. Bernal, S., Botana, F.J., Calvino, J.J., and Pérez-Omil, J.A. (1995) *Catal. Today*, **23**, 219.
36. Rupprechter, G. (2007) Sum frequency generation and polarization-modulation infrared reflection absorption spectroscopy of functioning model catalysts from ultrahigh vacuum to ambient pressure. *Adv. Catal.*, **51**, 133–263.
37. Henry, C.R. (2005) Morphology of supported nanoparticles. *Prog. Surf. Sci.*, **80**, 92.
38. Gai, P.L., Boyes, E.D., Helveg, S., Hansen, P.L., Giorgio, S., and Henry, C.R. (2007) Atomic-resolution environmental transmission electron microscopy (ETEM) for probing gas-solid reactions in heterogeneous catalysis. *MRS Bull.*, **32**, 1–7.
39. José-Yacamán, M., Díaz, G., and Gómez, A. (1995) Electron microscopy of the catalysts: the present, the future and the hopes. *Catal. Today*, **23**, 161.
40. Rupprechter, G. and Weilach, C. (2007) Mind the gap! Spectrosocpy of catalytically active phases. *Nano Today*, **2**, 20–29.
41. Blanco, G., Calvino, J.J., Cauqui, M.A., Corchado, P., López-Cartes, C., Colliex, C., Pérez-Omil, J.A., and Stephan, O. (1999) Nanostructural evolution under reducing conditions of a Pt/CeTbOx CATALYST: a new alternative system as a twc component. *Chem. Mater.*, **11**, 3610–3619.
42. Rupprechter, G. (2001) Surface vibrational spectroscopy from ultrahigh vacuum to atmospheric pressure: adsorption and reactions on single crystals and nanoparticle model catalysts monitored by sum frequency generation spectroscopy. *Phys. Chem. Chem. Phys.*, **3**, 4621.
43. Fuchs, M., Jenewein, B., Penner, S., Hayek, K., Rupprechter, G., Wang, D., Schlögl, R., Calvino, J.J., and Bernal, S. (2005) Interaction of Pt and Rh nanoparticles with ceria and silica

supports: ring opening of methylcyclobutane and CO hydrogenation after reduction at 373–723 K. *Appl. Catal., A*, **294**, 279.

44. Gruber, H.L. (1962) Chemisorption studies on supported platinum. *J. Phys. Chem.*, **66**, 48.
45. Eischens, R.P., Francis, S.A., and Pliskin, W.A. (1956) The effect of surface coverage on the spectra of chemisorbed CO. *J. Phys. Chem.*, **60**, 194.
46. Ertl, G. (1990) Advances. *Adv. Catal.*, **37**, 1.
47. Poppa, H. (1993) Model studies in catalysis on UHV deposited clusters. *Catal. Rev. Sci. Eng.*, **35**, 359.
48. Ertl, G. (1994) Reactions at well-defined surfaces. *Surf. Sci.*, **299/300**, 742.
49. King, D.A. (1994) Chemisorption on metals: a personal review. *Surf. Sci.*, **299/300**, 678.
50. Somorjai, G.A. (1996) Modern surface science and surface technologies: an introduction. *Chem. Rev.*, **96**, 1223.
51. Gunter, P.L.J., Niemantsverdriet, J.W.H., Ribeiro, F.H., and Somorjai, G.A. (1997) Modelling supported catalysts in surface science. *Catal. Rev. Sci. Eng.*, **39**, 77.
52. Somorjai, G.A. and McCrea, K.R. (2000) Sum frequency generation: surface vibration spectroscopy studies of catalytic reactions on metal single-crystal surfaces. *Adv. Catal.*, **45**, 385.
53. Rupprechter, G. and Weilach, C. (2008) Vibrational studies of surface-gas interactions at ambient pressure. *J. Phys. Condens. Matter*, **20**, 184020.
54. Somorjai, G.A. (1994) The surface science of heterogeneous catalysis. *Surf. Sci.*, **299–300**, 849.
55. Goodman, D.W. (1994) Catalysis: from single crystals to the "real world". *Surf. Sci.*, **299/300**, 837.
56. Freund, H.-J. (1999) Introductory lecture: oxide surfaces. *Faraday Discuss.*, **114**, 1.
57. Zhdanov, V.P. and Kasemo, B. (1997) Knetics of rapid heterogeneous reactions on the nanometer scale. *J. Catal.*, **170**, 377.
58. Morkel, M., Rupprechter, G., and Freund, H.-J. (2005) SS finite size CO-H Pd particles C2H4. *Surf. Sci. Lett.*, **588**, L209.
59. Bowker, M., Cookson, L., Bhantoo, J., Carley, A., Hayden, E., Gilbert, L., Morgan, C., Counsell, J., and Yaseneva, P. (2011) The decarbonylation of acetaldehyde on Pd crystals and on supported catalysts. *Appl. Catal., A*, **391**, 394–399.
60. Yudanov, I.V., Neyman, K.M., and Rösch, N. (2004) Subsurface H C N O in Pd clustern. *Phys. Chem. Chem. Phys.*, **6**, 116.
61. Neyman, K.M., Lim, K.H., Chen, Z.-X., Moskaleva, L.V., Bayer, A., Reindl, A., Borgmann, D., Denecke, R., Steinrück, H.-P., and Rösch, N. (2007) Microscopic models of PdZn alloy catalysts: structure and reactivity in methanol decomposition. *Phys. Chem. Chem. Phys.*, **9**, 3470–3482.
62. Armbrüster, M., Behrens, M., Cinquini, F., Föttinger, K., Grin, Y., Haghofer, A., Klötzer, B., Knop-Gericke, A., Lorenz, H., Ota, A., Penner, S., Prinz, J., Rameshan, C., Révay, Z., Rosenthal, D., Rupprechter, G., Sautet, P., Schlögl, R., Shao, L., Szentmiklósi, L., Teschner, D., Torres, D., Wagner, R., Widmer, R., and Wowsnick, G. (2012) How to control the selectivity of palladium-based catalysts in hydrogenation reactions: the role of sub-surface chemistry. *ChemCatChem*, **4**, 1048–1063.
63. Rupprechter, G., Morkel, M., Freund, H.-J., and Hirschl, R. (2004) Sum frequency generation and density-functional studies of CO-H interaction on Pd(111). *Surf. Sci.*, **554**, 43.
64. Rose, M.K., Borg, A., Mitsui, T., Ogletree, D.F., and Salmeron, M. (2001) Subsurface impurities in Pd(111) studied by scanning tunneling microscopy. *J. Chem. Phys.*, **115**, 10927.
65. Surnev, S., Kresse, G., Ramsey, M.G., and Netzer, F.P. (2001) Novel interface-mediated metastable oxide phases: vanadium oxides on Pd(111). *Phys. Rev. Lett.*, **87**, 086102.
66. Giorgio, S., Henry, C.R., Chapon, C., Nihoul, G., and Penisson, J.M. (1991) Electron-beam-induced transformations

of gold particles epitaxially grown on MgO microcubes. *Ultramicroscopy*, **38**, 1.

67. Heemeier, M., Stempel, S., Shaikhutdinov, S., Libuda, J., Bäumer, M., Oldman, R.J., Jackson, S.D., and Freund, H.-J. (2003) On the thermal stability of metal particles supported on a thin alumina film. *Surf. Sci.*, **523**, 103.
68. Eppler, A., Rupprechter, G., Anderson, E.A., and Somorjai, G.A. (2000) The thermal and chemical stability and adhesion strength of Pt nanoparticle arrays supported on silica studied by TEM and AFM. *J. Phys. Chem. B*, **104**, 7286.
69. Bertarione, S., Scarano, D., Zecchina, A., Johanek, V., Hoffmann, J., Schauermann, S., Frank, M., Libuda, J., Rupprechter, G., and Freund, H.-J. (2004) Surface reactivity of polycrystalline MgO supported Pd nanoparticles as compared to thin film model catalysts: infrared study of CO adsorption. Part I. *J. Phys. Chem. B*, **108**, 3603.
70. Besenbacher, F. (1996) Scanning tunneling microscopy studies of metal surfaces. *Rep. Prog. Phys.*, **59**, 1737.
71. Lauritsen, J.V. and Besenbacher, F. (2006) Model catalyst surfaces investigated by scanning tunneling microscopy. *Adv. Catal.*, **50**, 97–147.
72. Suchorski, Y., Spiel, C., Vogel, D., Drachsel, W., Schlögl, R., and Rupprechter, G. (2010) Local reaction kinetics by imaging: CO oxidation on polycrystalline platinum. *ChemPhysChem*, **11**, 3231–3235.
73. Vogel, D., Spiel, C., Suchorski, Y., Trinchero, A., Schlögl, R., Grönbeck, H., and Rupprechter, G. (2012) Local light-off in catalytic CO oxidation on low-index Pt and Pd surfaces: a combined PEEM, MS and DFT study. *Angew. Chem. Int. Ed.*, **51**, 10041–10044.
74. Suchorski, Y. and Rupprechter, G. *Surf. Sci. doi:10.1016/j.susc.2015.05.021.*
75. Davis, S.M., Zaera, F., and Somorjai, G.A. (1982) Surface structure and temperature dependence of light alkane skeletal rearrangement reactions catalyzed over platinum-single crystal surfaces. *J. Am. Chem. Soc.*, **104**, 7453.
76. Ertl, G. (2008) Reactions at surfaces: from atoms to complexity nobel prize lecture. *Angew. Chem. Int. Ed.*, **47**, 3524.
77. Rupprechter, G. and Somorjai, G.A. (2006) in *Landolt-Börnstein: Physics of Covered Solid Surfaces* (ed H.P. Bonzel), Springer, pp. 243–330.
78. Surnev, S., Ramsey, M.G., and Netzer, F.P. (2003) Vanadium oxide surface studies. *Prog. Surf. Sci.*, **73**, 117.
79. Suchorski, Y., Wrobel, R., Becker, S., and Weiss, H. (2008) CO oxidation on a CeOx/Pt(111) inverse model catalyst surface: catalytic promotion and tuning of kinetic phase diagrams. *J. Phys. Chem. C*, **112**, 20012–20017.
80. Tauster, S.J., Fung, S.C., and Garten, R.L. (1978) Strong metal-support interactions – group-8 noble-metals supported on $TiO_2$. *J. Am. Chem. Soc.*, **100**, 170.
81. Braunschweig, E., Logan, A.D., Datye, A.K., and Smith, D.J. (1989) Reversibility of strong metal-support interactions on $RhTiO_2$. *J. Catal.*, **118**, 227.
82. Bernal, S., Calvino, J.J., Cauqui, M.A., Gatica, J.M., Larese, C., Perez-Omil, J.A., and Pintado, J.M. (1999) Some recent results on metal/support interaction effects in $NM/CeO_2$ (NM: noble metal) catalysts. *Catal. Today*, **50**, 175.
83. Rodriguez, J.A., Ma, S., Liu, P., Hrbek, J., Evans, J., and Pérez, M. (2007) Activity of CeOx and TiOx nanoparticles grown on Au(111) in the water-gas shift reaction. *Science*, **318**, 1757–1760.
84. Ying-Na, S., Livia, G., Jacek, G., Mikolaj, L., Zhi-Hui, Q., Claudine, N., Shamil, S., Gianfranco, P., and Hans-Joachim, F. (2010) The interplay between structure and CO oxidation catalysis on metal-supported ultrathin oxide films. *Angew. Chem. Int. Ed.*, **49** (26), 4418–4421.
85. Earl, M.D., Ke, Z., Yi, C., Helmut, K., Shamil, S., and Hans-Joachim, F. (2015) Growth of $Fe_3O_4$(001) thin films on Pt(100): Tuning surface termination with an Fe buffer layer. *Surf. Sci.*, **636**, 42–46.

86. Dulub, O., Hebenstreit, W., and Diebold, U. (2000) Imaging cluster surfaces with atomic resolution: the strong metal-support interaction state of Pt supported on $TiO_2(110)$. *Phys. Rev. Lett.*, **84**, 3646.
87. Degen, S., Becker, C., and Wandelt, K. (2004) Thin alumina films on Ni3Al(111): **a** template for nanostructured Pd cluster growth. *Faraday Discuss.*, **125**, 343–356.
88. Hamm, G., Barth, C., Becker, C., Wandelt, K., and Henry, C.R. (2006) Surface structure of an ultrathin alumina film on Ni3Al(111): a dynamic scanning force microscopy study. *Phys. Rev. Lett.*, **97**, 126106.
89. Schroeder, T., Adelt, M., Richter, B., Naschitzki, N., Bäumer, M., and Freund, H.-J. (2000) Epitaxial growth of $SiO_2$ on Mo(112). *Surf. Rev. Lett.*, **7**, 7.
90. Chen, M.S., Santra, A.K., and Goodman, D.W. (2004) The structure of thin $SiO_2$ films grown on Mo(112). *Phys. Rev. B*, **69**, 155404.
91. Lu, J.-L., Kaya, S., Weissenrieder, J., Todorova, T.K., Sierka, M., Sauer, J., Gao, H.-J., Shaikhutdinov, S., and Freund, H.-J. (2006) Formation of one-dimensional crystalline silica on a metal substrate. *Surf. Sci. Lett.*, **600**, L164–L168.
92. Lu, J.-L., Gao, H.-J., Shaikhutdinov, S., and Freund, H.-J. (2006) Morphology and defect structure of the $CeO_2(111)$ films grown on Ru(0001) as studied by scanning tunneling microscopy. *Surf. Sci.*, **600**, 5004–5010.
93. Weiss, W. and Schlögl, R. (2000) An integrated surface science approach towards metal oxide catalysis. *Top. Catal.*, **13**, 75.
94. Lemire, C., Meyer, R., Heinrich, V.E., Shaikhutdinov, S., and Freund, H.-J. (2004) The surface structure of $Fe_3O_4(111)$ films as studied by CO adsorption. *Surf. Sci.*, **572**, 103.
95. Sterrer, M., Risse, T., Pozzoni, U.M., Giordano, L., Heyde, M., Rust, H.P., Pacchioni, G., and Freund, H.-J. (2007) Control of the charge state of metal atoms on thin MgO films. *Phys. Rev. Lett.*, **98**, 096107.
96. Starr, D.E., Mendes, F.M.T., Middeke, J., Blum, R.-P., Niehus, H., Lahav, D., Guimond, S., Uhl, A., Klüner, T., Schmal, M., Kuhlenbeck, H., Shaikhutdinov, S., and Freund, H.-J. (2005) Preparation and characterization of well-ordered, thin niobia films on a metal substrate. *Surf. Sci.*, **599**, 14.
97. Antlanger, M., Mayr-Schmölzer, W., Pavelec, J., Mittendorfer, F., Redinger, J., Varga, P., Diebold, U., and Schmid, M. (2012) Pt3Zr(0001): a substrate for growing well-ordered ultrathin zirconia films by oxidation. *Phys. Rev. B*, **86**, 035451 (035459 pp).
98. Li, H., Choi, J.J., Mayr-Schmölzer, W., Weilach, C., Rameshan, C., Mittendorfer, F., Redinger, J., Schmid, M., and Rupprechter, G. (2015) The growth of an ultrathin zirconia film on Pt3Zr examined by-HR-XPS, TPD, STM and DFT. *J. Phys. Chem. C*, **119**, 2462–2470 *doi:10.1021/jp5100846.*
99. Sterrer, M. and Freund, H.-J. (2012) in *Textbook on Surface and Interface Science* (ed K. Wandelt), p. 096107.
100. Lundgren, E., Mikkelsen, A., Andersen, J.N., Kresse, G., Schmid, M., and Varga, P. (2006) Surface oxides on close-packed surfaces of late transition metals. *J. Phys. Condens. Matter*, **18**, 481.
101. Helmut, K., Shamil, S. and Hans-Joachim, F. (2013) Well-ordered transition metal oxide layers in model catalysis – a series of case studies. *Chem. Rev.*, **113** (6), 3986–4034.
102. Bäumer, M. and Freund, H.-J. (1999) Metal deposits on well-ordered oxide films. *Prog. Surf. Sci.*, **61**, 127.
103. Kulawik, M., Nilius, N., Rust, H.-P., and Freund, H.-J. (2003) Atomic structure of antiphase domain boundaries of a thin $Al_2O_3$ film on NiAl(110). *Phys. Rev. Lett.*, **91**, 256101.
104. Frank, M. and Bäumer, M. (2000) *Phys. Chem. Chem. Phys.*, **2**, 3723.
105. Heemeier, M., Carlsson, A.F., Naschitzki, M., Schmal, M., Bäumer, M., and Freund, H.-J. (2002) Preparation and characterization of a model bimetallic catalyst: Co-Pd nanoparticles supported on $Al_2O_3$. *Angew. Chem. Int. Ed.*, **41**, 4073.

106. Schmid, M., Shishkin, M., Kresse, G., Napetschnig, E., Varga, P., Kulawik, M., Nilius, N., Rust, H.-P., and Freund, H.-J. (2006) Oxygen-deficient line defects in an ultra-thin aluminum oxide film. *Phys. Rev. Lett.*, **97**, 046101.
107. Stierle, A., Renner, F., Streitel, R., Dosch, H., Drube, W., and Cowie, B.C. (2004) X-ray diffraction study of the ultrathin $Al_2O_3$ layer on NiAl (110). *Science*, **303**, 1652.
108. Kresse, G., Schmid, M., Napetschnig, E., Shishkin, M., Köhler, L., and Varga, P. (2005) Structure of the ultrathin aluminum oxide film on NiAl(110). *Science*, **308**, 1440.
109. Libuda, J., Frank, M., Sandell, A., Andersson, S., Brühwiler, P.A., Bäumer, M., Martensson, N., and Freund, H.-J. (1997) Interaction of rhodium with hydroxylated alumina model substrates. *Surf. Sci.*, **384**, 106.
110. Bäumer, M., Libuda, L., Sandell, A., Freund, H.-J., Graw, G., Bertrams, T., and Neddermeyer, H. (1995) The growth and properties of Pd and Pt on $Al_2O_3$/NiAl(110). *Ber Bunsenges Phys. Chem.*, **99**, 1381.
111. Højrup Hansen, K., Worren, T., Stempel, S., Lægsgaard, E., Bäumer, M., Freund, H.-J., Besenbacher, F., and Stensgaard, I. (1999) Palladium nanocrystals on $Al_2O_3$: structure and adhesion energy. *Phys. Rev. Lett.*, **83**, 4120.
112. Unterhalt, H., Rupprechter, G., and Freund, H.-J. (2002) Vibrational sum frequency spectroscopy on Pd(111) and supported Pd nanoparticles: CO adsorption from ultrahigh vacuum to atmospheric pressure. *J. Phys. Chem. B*, **106**, 356.
113. Rupprechter, G. (2007) A surface science approach to ambient pressure catalytic reactions. *Catal. Today*, **126**, 3.
114. Napetschnig, E., Schmid, M., and Varga, P. (2007) Pd, Co and Co-Pd clusters on the ordered alumina film on NiAl(110): contact angle, surface structure and composition. *Surf. Sci.*, **601**, 3233.
115. Stair, P.C. (2012) Synthesis of supported catalysts by atomic layer deposition. *Topics in Catalysis*, **55** (1–2), p. 93–98.
116. Wegener, S.L., Marks, T.J., and Stair, P.C. (2012) Design Strategies for the Molecular Level Synthesis of Supported Catalysts. *Accounts of Chemical Research*, **45** (2), pp. 206–214.
117. Jacobs, P.W., Wind, S.J., Ribeiro, F.H., and Somorjai, G.A. (1997) Nanometer size platinum particle arrays. *Surf. Sci.*, **372**, L249.
118. Eppler, A., Rupprechter, G., Guczi, L., and Somorjai, G.A. (1997) Model catalysts fabricated using EBL and PLD. *J. Phys. Chem. B*, **101**, 9973.
119. Baldelli, S., Eppler, A.S., Anderson, E., Shen, Y.R., and Somorjai, G.A. (2000) SFG on model catalysts fabricated using EBL. *J. Chem. Phys.*, **113**, 5432.
120. Rupprechter, G. and Freund, H.-J. (2001) Adsorbate-induced restructuring and pressure-dependent adsorption on metal nanoparticles studied by electron microscopy and sum frequency generation spectroscopy. *Top. Catal.*, **14**, 3–14.
121. Föttinger, K., Schlögl, R., and Rupprechter, G. (2008) The mechanism of carbonate formation on Pd-Al2O3 catalysts. *Chem. Commun.*, 320–322.
122. Weilach, C., Spiel, C., Föttinger, K., and Rupprechter, G. (2011) Carbonate formation on Al2O3 thin film model catalyst supports. *Surf. Sci.*, **605**, 1503–1509.
123. Föttinger, K., van Bokhoven, J.A., Nachtegaal, M., and Rupprechter, G. (2011) Dynamic structure of a working methanol steam reforming catalyst: in situ quick-EXAFS on Pd/ZnO nanoparticles. *J. Phys. Chem. Lett.*, **2**, 428–433.
124. Rameshan, C., Stadlmayr, W., Weilach, C., Penner, S., Lorenz, H., Hävecker, M., Blume, R., Rocha, T., Teschner, D., Knop-Gericke, A., Schlögl, R., Memmel, N., Zemlyanov, D., Rupprechter, G., and Klötzer, B. (2010) Subsurface-controlled $CO_2$-selectivity of PdZn near-surface alloys in H2 generation by methanol steam reforming. *Angew. Chem. Int. Ed.*, **49**, 3224–3227.

125. Wang, T., Lee, C., and Schmidt, L.D. (1985) Shape and orientation of supported Pt particles. *Surf. Sci.*, **163**, 181.
126. Gillet, M. and Renou, A. (1979) Crystal habits of Pd and Pt particles: Nature of surface sites. *Surf. Sci.*, **90**, 91.
127. Jefferson, D.A. and Harris, P.J.F. (1988) Direct imaging of an adsorbed layer by high-resolution electron microscopy. *Nature*, **332**, 617.
128. Hayek, K. (1989) Electron microscopy and structure -activity correlations of oriented thin film catalysts. *J. Mol. Catal.*, **51**, 347.
129. Burkhardt, J. and Schmidt, L.D. (1989) Comparison of microstructures in oxidation and reduction of Rh and Ir particles on $SiO_2$ and $Al_2O_3$. *J. Catal.*, **116**, 240.
130. Giorgio, S., Henry, C.R., and Chapon, C. (1990) Structure and morphology of small palladium particles (2–6 nm) supported on MgO micro-cubes. *J. Cryst. Growth*, **100**, 254.
131. Schmidt, L.D. and Krause, K.R. (1992) *Catal. Today*, **12**, 1035.
132. Logan, A.D., Sharoudi, K.S., and Datye, A.K. (1991) Oxidative restructuring of rhodium metal surfaces: correlations between single crystals and small metal particles. *J. Phys. Chem.*, **95**, 5568.
133. Henry, C.R., Chapon, C., Duriez, C., and Giorgio, S. (1991) Growth and morphology of Pd particles epitaxially deposited on a MgO (100) surface. *Surf. Sci.*, **253**, 177.
134. Giorgio, S., Henry, C.R., Chapon, C., and Roucau, C. (1994) A high-resolution TEM study of the annealing of Pd particles supported on Mgo. *J. Catal.*, **148**, 534–539.
135. Rupprechter, G., Seeber, G., Hayek, K., and Hofmeister, H. (1994) Epitaxial noble metal particles upon oxidation and reduction. A model system for supported metal catalysts. *Phys. Status Solidi A*, **146**, 449.
136. Rupprechter, G., Hayek, K., Rendón, L., and José-Yacamán, M. (1995) Epitaxially grown catalyst particles of platinum, rhodium, iridium, palladium and rhenium studied by electron microscopy. *Thin Solid Films*, **260**, 148.
137. Rupprechter, G., Hayek, K., and Hofmeister, H. (1998) Electron microscopy of thin-film model catalysts: activation of alumina-supported rhodium nanoparticles. *J. Catal*, **173**, 409.
138. Sushumna, I. and Ruckenstein, E. (1987) Redispersion of Pt/alumina via film formation. *J. Catal.*, **108**, 77.
139. Ramachandran, A.S., Anderson, S.L., and Datye, A.K. (1993) The effect of adsorbate and pretreatment on the shape and reactivity of silica-supported platinum particles. *Ultramicroscopy*, **51**, 282.
140. Van Hove, M.A. and Somorjai, G.A. (1994) Adsorption and adsorbate-induced restructuring: a LEED perspective. *Surf. Sci.*, **299/300**, 487.
141. Haghofer, A., Sonström, P., Fenske, D., Föttinger, K., Schwarz, S., Bernardi, J., Al-Shamery, K., Bäumer, M., and Rupprechter, G. (2010) Colloidally prepared Pt-nanowires versus impregnated Pt nanoparticles: comparison of adsorption and reaction properties. *Langmuir*, **26**, 16330–16338.
142. Parkinson, G.S., Novotny, Z., Argentero, G., Schmid, M., Pavelec, J., Kosak, R., Blaha, P., and Diebold, U. (2013) Carbon monoxide-induced adatom sintering in a Pd-Fe3O4 model catalyst. *Nat. Mater.*, **12**, 724.
143. Demoulin, O., Rupprechter, G., Seunier, I., Clef, B.L., Navez, M., and Ruiz, P. (2005) Investigation of parameters influencing the activation of a Pd/$\gamma$-alumina catalyst during methane combustion. *J. Phys. Chem. B*, **109**, 20454.
144. Gabasch, H., Knop-Gericke, A., Schlögl, R., Borasio, M., Weilach, C., Rupprechter, G., Penner, S., Jenewein, B., Hayek, K., and Klötzer, B. (2007) Comparison of the reactivity of different Pd-O species in CO oxidation. *Phys. Chem. Chem. Phys.*, **9**, 533.
145. Palczewska, W. (1975) Catalytic reactivity of hydrogen on palladium and nickel hydride phases. *Adv. Catal.*, **24**, 245.
146. Ceyer, S.T. (2001) H beneath the surface. *Acc. Chem. Res.*, **34**, 737.
147. McNamara, J., Jackson, S., and Lennon, D. (2003) Butane dehydrogenation over

Pt/alumina: activation, deactivation and the generation of selectivity. *Catal. Today*, **81**, 583–587.

148. Kennedy, D.R., Webb, G., Jackson, S.D., and Lennon, D. (2004) Propyne hydrogenation over alumina-supported palladium and platinum catalysts. *Appl. Catal., A*, **259**, 109–120.
149. Bowker, M., Morgan, C., Perkins, N., Holroyd, R.P., Fourre, E., Grillo, F., and MacDowall, A. (2005) $C_2H_4$ Pd110 Pd as carbon sponge. *J. Phys. Chem. B*, **109**, 2377.
150. Borasio, M., Rodríguez de la Fuente, O., Rupprechter, G., and Freund, H.-J. (2005) In situ studies of methanol decomposition and oxidation on Pd(111) by PM-IRAS and XPS spectroscopy. *J. Phys. Chem. B Lett.*, **109**, 17791.
151. Gabasch, H., Hayek, K., Klötzer, B., Knop-Gericke, A., and Schlögl, R. (2006) Carbon incorporation in Pd(111) by adsorption and dehydrogenation of ethene. *J. Phys. Chem. B*, **110**, 4947.
152. Teschner, D., Vass, E., Hävecker, M., Zafeiratos, S., Schnörch, P., Sauer, H., Knop-Gericke, A., Schlögl, R., Chamam, M., Wootsch, A., Canning, A.S., Gamman, J.J., Jackson, S.D., McGregor, J., and Gladden, L.F. (2006) Alkyne hydrogenation over Pd catalysts: a new paradigm. *J. Catal.*, **242**, 26–37.
153. Bowker, M., Counsell, J., El-Abiary, K., Gilbert, L., Morgan, C., Nagarajan, S., and Gopinath, C.S. (2010) Carbon dissolution and segregation in Pd(110). *J. Phys. Chem. C*, **114**, 5060–5067.
154. Bowker, M., Holroyd, R., Perkins, N., Bhantoo, J., Counsell, J., Carley, A., and Morgan, C. (2007) Acetaldehyde adsorption and catalytic decomposition on Pd(110) and the dissolution of carbon. *Surf. Sci.*, **601**, 3651–3660.
155. Park, J.Y., Zhang, Y., Grass, M., Zhang, T., and Somorjai, G.A. (2008) Tuning of catalytic CO oxidation by changing composition of Rh-Pt bimetallic nanoparticles. *Nano Lett.*, **8**, 673–677.
156. Bloxham, L.H., Haq, S., Jugnet, Y., Bertolini, J.C., and Raval, R. (2004) Trans-1,2-dichloroethene on Cu50Pd50(110) alloy surface: dynamical changes in the adsorption, reaction, and surface segregation. *J. Catal.*, **227**, 33–43.
157. Saint-Lager, M.C., Jugnet, Y., Dolle, P., Piccolo, L., Baudoing-Savois, R., Bertolini, J.C., Bailly, A., Robach, O., Walker, C., and Ferrer, S. (2005) Pd8Ni92(110) surface structure from surface X-ray diffraction. Surface evolution under hydrogen and butadiene reactants at elevated pressure XRD Pd8Ni92 110 structure under H and butadiene. *Surf. Sci.*, **587** (3), 229–235.
158. Caballero, G.E.R. and Balbuena, P.B. (2006) Surface segregation phenomena in Pt-Pd nanoparticles: dependence on nanocluster size. *Mol. Simul.*, **32**, 297–303.
159. Wolfbeisser, A., Klötzer, B., Mayr, L., Rameshan, R., Zemlyanov, D., Bernardi, J., Föttinger, K., and Rupprechter, G. (2015) Surface modification during methane decomposition on Cu- promoted Ni-ZrO2 catalysts. *Catal. Sci. Technol.*, **5**, 967–978.
160. Opitz, A.K., Nenning, A., Rameshan, C., Rameshan, R., Blume, R., Hävecker, M., Knop-Gericke, A., Rupprechter, G., Fleig, J., and Klötzer, B. (2015) Improving electrochemical water-splitting kinetics by polarization-driven formation of near-surface $Fe^0$: an in-situ XPS study on perovskite-type electrodes. *Angew. Chem. Int. Ed.*, **54**, 2628–2632.
161. Rupprechter, G., Seeber, G., Goller, H., and Hayek, K. (1999) Structure-activity correlations on $Rh/Al_2O_3$ and $Rh/TiO_2$ thin film model catalysts after oxidation and reduction. *J. Catal.*, **186**, 201.
162. Friedrich, M., Penner, S., Heggen, M., and Armbrüster, M. (2013) Teamwork ZnPd and ZnO results in high CO2-selectivity in methanol steam reforming. *Angew. Chem. Int. Ed.*, **52**, 4389–9432.
163. Haghofer, A., Ferri, D., Föttinger, K., and Rupprechter, G. (2012) Who is doing the job? Unraveling the role of Ga2O3in methanol steam reforming on Pd2Ga/Ga2O3. *ACS Catal.*, **2**, 2305–2315.
164. Haghofer, A., Föttinger, K., Girgsdies, F., Teschner, D., Knop-Gericke, A., Schlögl, R., and Rupprechter, G. (2012)

In situ study of the formation and stability of supported $Pd_2Ga$ methanol steam reforming catalysts. *J. Catal.*, **286**, 13–21.
165. Haghofer, A., Föttinger, K., Nachtegaal, M., Armbrüster, M., and Rupprechter, G. (2012) Microstructural changes of supported intermetallic nanoparticles under reductive and oxidative conditions: an in situ X-ray absorption study of $Pd/Ga_2O_3$. *J. Phys. Chem. C*, **116**, 21816–21827.
166. Iwasa, N., Mayanagi, T., Ogawa, N., Sakata, K., and Takezawa, N. (1998) New catalytic functions of Pd–Zn, Pd–Ga, Pd–In, Pt–Zn, Pt–Ga and Pt–In alloys in the conversions of methanol. *Catal. Lett.*, **54**, 119.
167. Iwasa, N. and Takezawa, N. (2003) New supported Pd and Pt alloy catalysts for steam reforming and dehydrogenation of methanol. *Top. Catal.*, **22**, 215–224.
168. Banares, M.A. (2005) Operando methodology: combination of in situ spectroscopy and simultaneous activity measurements under catalytic reaction conditions. *Catal.*, **100**, 74.
169. Thomas, J.M. and Somorjai, G.A. (1999) *Topics in Catalysis Special Issue on "In situ Characterization of Catalysts"*.
170. Schlögl, R. and Zecchina, A. (2001) *Topics in Catalysis Special Issue on "In situ Characterization of Catalysts"*.
171. Somorjai, G.A. (1996) Surface Science at high pressure. *Z. Phys. Chem.*, **197**, 1.
172. Stierle, A. and Molenbroek, A.M. (eds) (2007) Special Issue on "Novel in situ probes for nanocatalysis". *MRS Bull.*, **32**.
173. Lundgren, E. and Over, H. (2008) Special issue: in situ gas–surface interactions: approaching realistic conditions. *J. Phys. Condens. Matter.*, **20**, 180302.
174. Morkel, M., Unterhalt, H., Salmeron, M., Rupprechter, G., and Freund, H.-J. (2003) SS CO Pd(111) non-equilibrium coadsorption. *Surf. Sci.*, **532-535**, 103.
175. Kruse Vestergaard, E., Thostrup, P., An, T., Hammer, B., Lægsgaard, E., and Besenbacher, F. (2002) Comment on "high pressure adsorbate structures studied by scanning tunneling microscopy: CO on Pt(111) in equilibrium with the gas phase". *Phys. Rev. Lett.*, **88**, 259601.
176. Longwitz, S., Schnadt, J., Kruse Vestergaard, E., Vang, R.T., Lægsgaard, E., Stensgaard, I., Brune, H., and Besenbacher, F. (2004) High-Coverage structures of carbon monoxide adsorbed on Pt(111) studied by high-pressure scanning tunneling microscopy. *J. Phys. Chem. B*, **108**, 14497.
177. Somorjai, G.A. and Rupprechter, G. (1999) Molecular studies of catalytic reactions on crystal surfaces at high pressures and high temperatures by infrared-visible Sum Frequency Generation (SFG) surface vibrational spectroscopy. *J. Phys. Chem. B*, **103**, 1623.
178. Suchorski, Y., Drachsel, W., and Rupprechter, G. (2008) High-field versus high-pressure: weakly adsorbed CO species on Pt(111). *Ultramicroscopy*, **109**, 430–435.
179. Scheibe, A., Günther, S., and Imbihl, R. (2003) Selectivity changes due to restructuring of the Pt(533) surface in the $NH_3 + O_2$ reaction. *Catal. Lett.*, **86**, 33–37.
180. Somorjai, G.A. (2004) On the move. *Nature*, **430**, 730.
181. Berkó, A., Ménesi, G., and Solymosi, F. (1996) Scanning tunneling microscopy study of the CO-induced structural changes of rh crystallites supported by $TiO_2(110)$. *J. Phys. Chem.*, **100**, 17732–17734.
182. Stampfl, C. and Scheffler, M. (1999) Density functional theory study of the catalytic oxidation of CO over transition metal surfaces. *Surf. Sci.*, **433–435**, 119–126.
183. Leisenberger, F.P., Koller, G., Sock, M., Surnev, S., Ramsey, M.G., Netzer, F.P., Klötzer, B., and Hayek, K. (2000) Surface and subsurface O on Pd111. *Surf. Sci.*, **445**, 380.
184. Over, H., Kim, Y.D., Seitsonen, A.P., Wendt, S., Lundgren, E., Schmid, M., Varga, P., Morgante, A., and Ertl, G. (2000) Atomic scale structure and

catalytic reactivity of the $RuO_2(110)$ surface. *Science*, **287**, 1474–1476.

185. Lundgren, E., Kresse, G., Klein, C., Borg, M., Andersen, J.N., De Santis, M., Gauthier, Y., Konvicka, C., Schmid, M., and Varga, P. (2002) A two-dimensional oxide on Pd(111). *Phys. Rev. Lett.*, **88**, 246103.
186. Reuter, K., Frenkel, D., and Scheffler, M. (2004) The steady-state of heterogeneous catalysis, studied by first-principles statistical mechanics. *Phys. Rev. Lett.*, **93**, 116105.
187. Gabasch, H., Unterberger, W., Hayek, K., Klötzer, B., Kleimenov, E., Teschner, D., Zafeiratos, S., Hävecker, M., Knop-Gericke, A., Schlögl, R., Han, J.Y., Ribeiro, F.H., Aszalos-Kiss, B., Curtin, T., and Zemlyanov, D. (2006) In situ XPS study of Pd(111) oxidation at elevated pressure, part 2: Palladium oxidation in the 10(-1) mbar range. *Surf. Sci.*, **600**, 2980–2989.
188. Knop-Gericke, A., Hävecker, M., Schedel-Niedrig, T., and Schlögl, R. (2001) Characterisation of active phases of a copper catalyst for methanol oxidation under reaction conditions: an in situ X-ray absorption spectroscopy study in the soft energy range. *Top. Catal.*, **15**, 27.
189. Ketteler, G., Ogletree, D.F., Bluhm, H., Liu, H., Hebenstreit, E.L.D., and Salmeron, M. (2005) In situ spectroscopic study of the oxidation and reduction of Pd(111). *J. Am. Chem. Soc.*, **127**, 18269.
190. Gabasch, H., Hayek, K., Klötzer, B., Unterberger, W., Kleimenov, E., Teschner, D., Zafeiratos, S., Hävecker, M., Knop-Gericke, A., Schlögl, R., Aszalos-Kiss, B., and Zemlyanov, D. (2007) Methane oxidation on Pd(111): in situ XPS identification of active phase. *J. Phys. Chem. C*, **111**, 7957–7962.
191. Morkel, M., Kaichev, V.V., Rupprechter, G., Freund, H.-J., Prosvirin, I.P., and Bukhtiyarov, V.I. (2004) Methanol dehydrogenation and formation of carbonaceous overlayers on Pd(111) studied by high-pressure SFG and XPS spectroscopy. *J. Phys. Chem. B*, **108**, 12955.
192. Urakawa, A., Bürgi, T., and Baiker, A. (2008) Sensitivity enhancement and dynamic behavior analysis by modulation excitation spectroscopy: principle and application in heterogeneous catalysis. *Chem. Eng. Sci.*, **63**, 4902–4909.
193. Föttinger, K., Weilach, C., and Rupprechter, G. (2012) Sum Frequency Generation (SFG) and Infrared Reflection Absorption Spectroscopy (IRAS), in *Characterisation of Solid Materials: From Structure to Surface Reactivity* (eds J. Vedrine and M. Che), Wiley-VCH Verlag GmbH, Weinheim.
194. Ogletree, D., Bluhm, H., Lebedev, G., Fadley, C., Hussain, Z., and Salmeron, M. (2002) A differentially pumped electrostatic lens system for photoemission studies in the millibar range. *Rev. Sci. Instrum.*, **73**, 3872–3877.
195. Kaichev, V.V., Prosvirin, I.P., Bukhtiyarov, V.I., Unterhalt, H., Rupprechter, G., and Freund, H.-J. (2003) High-pressure studies of CO adsorption on Pd(111) by X-ray photoelectron spectroscopy and sum-frequency generation. *J. Phys. Chem. B*, **107**, 3522.
196. Hävecker, M., Mayer, R.W., Knop-Gericke, A., Bluhm, H., Kleimenov, E., Liskowski, A., Su, D.S., Follath, R., Requejo, F.G., Ogletree, D.F., Salmeron, M., Lopez-Sanchez, J.A., Bartley, J.K., Hutchings, G.J., and Schlögl, R. (2003) In situ investigation of the nature of the active surface of a vanadyl pyrophosphate catalyst during *n*-butane oxidation to maleic anhydride. *J. Phys. Chem. B*, **107**, 4587.
197. Bukhtiyarov, V.I., Kaichev, V.V., and Prosvirin, I.P. (2005) X-ray photoelectron spectroscopy as a tool for in-situ study of the mechanisms of heterogeneous catalytic reactions. *Top. Catal.*, **32**, 3.
198. Jensen, J.A., Rider, K.B., Salmeron, M., and Somorjai, G.A. (1998) High pressure adsorbate structures studied by scanning tunneling microscopy: CO on Pt(111) in equilibrium with the gas phase. *Phys. Rev. Lett.*, **80**, 1228.
199. Hendriksen, B.L.M., Bobaru, S.C., and Frenken, J.W.M. (2005) Bistability and

oscillations in CO oxidation studied with scanning tunneling microscopy inside a reactor. *Catal. Today*, **105**, 234.

200. Küppers, J., Nitschke, F., Wandelt, K., Ertl, G., and Brundle, C.R. (1979) Search for strongly adsorbed CO at Cu single crystal surfaces using UPS. *J. Chem. Soc., Faraday Trans.*, **175**, 984–986.
201. Rupprechter, G., Unterhalt, H., Morkel, M., Galletto, P., Hu, L., and Freund, H.-J. (2002) *Surf. Sci.*, **502-503**, 109.
202. Rupprechter, G., Dellwig, T., Unterhalt, H., and Freund, H.-J. (2001) CO adsorption on Ni(100) and Pt(111) studied by infrared-visible sum frequency generation spectroscopy: design and application of an SFG-compatible UV-high-pressure reaction cell. *Top. Catal.*, **15**, 19.
203. Kung, K.Y., Chen, P., Wei, F., Rupprechter, G., Shen, Y.R., and Somorjai, G.A. (2001) Ultrahigh vacuum high-pressure reaction system for 2-infrared 1-visible sum frequency generation studies. *Rev. Sci. Instrum.*, **72**, 1806.
204. Vogel, D., Budinska, z., Spiel, C., Schlögl, R., Suchorski, Y., and Rupprechter, G. (2013) Silicon oxide surface segregation in CO oxidation on Pd: an in situ PEEM, MS and XPS study. *Catal. Lett.*, **143**, 235–240.
205. Libuda, J. and Freund, H.-J. (2005) Molecular beam experiments on model catalysts. *Surf. Sci. Rep.*, **57**, 157.
206. Holladay, J.D., Wang, Y., and Jones, E. (2004) Review of developments in portable hydrogen production using microreactor technology. *Chem. Rev.*, **104**, 4767–4790.
207. Armbrüster, M., Behrens, M., Föttinger, K., Friedrich, M., Gaudry, E., Matam, S., and Sharma, H. (2013) The intermetallic compound ZnPd and its role in methanol steam reforming. *Catal. Rev.*, **55**, 289–367.
208. Zafeiratos, S., Piccinin, S., and Teschner, D. (2012) Alloys in catalysis: phase separation and surface segregation phenomena in response to the reactive environment. *Catal. Sci. Technol.*, **2**, 1787–1801.
209. Bayer, A., Flechtner, K., Denecke, R., Steinrück, H.-P., Neyman, K.M., and Rösch, N. (2006) Electronic properties of thin Zn layers on Pd(111)during growth and alloying. *Surf. Sci.*, **600**, 78–94.
210. Jeroro, E., Lebarbier, V., Datye, A., Wang, Y., and Vohs, J.M. (2007) Interaction of CO with surface PdZn alloys. *Surf. Sci.*, **601**, 5546–5554.
211. Jeroro, E. and Vohs, J.M. (2008) Zn modification of the reactivity of Pd(111) towards methanol and formaldehyde. *J. Am. Chem. Soc.*, **130**, 10199.
212. Weirum, G., Kratzer, M., Koch, H.P., Tamtögl, A., Killmann, J., Bako, I., Winkler, A., Surnev, S., Netzer, F.P., and Schennach, R. (2009) Growth and desorption kinetics of ultrathin Zn layers on Pd(111). *J. Phys. Chem. C*, **113**, 9788–9796.
213. Rameshan, C., Weilach, C., Stadlmayr, W., Penner, S., Lorenz, H., Hävecker, M., Blume, R., Rocha, T., Teschner, D., Knop-Gericke, A., Schlögl, R., Zemlyanov, D., Memmel, N., Rupprechter, G., and Klötzer, B. (2010) Steam reforming of methanol on PdZn near-surface alloys on Pd(111) and Pd foil studied by in-situ XPS, LEIS and PM-IRAS. *J. Catal.*, **276**, 101–113.
214. Stadlmayr, W., Rameshan, C., Weilach, C., Lorenz, H., Hävecker, M., Blume, R., Rocha, T., Teschner, D., Knop-Gericke, A., Zemlyanov, D., Penner, S., Schlögl, R., Rupprechter, G., Klötzer, B., and Memmel, N. (2010) Temperature-induced modifications of PdZn layers on Pd(111). *J. Phys. Chem. C*, **114**, 10850–10856.
215. Weilach, C., Kozlov, S.M., Holzapfel, H., Föttinger, K., Neyman, K.M., and Rupprechter, G. (2012) Geometric arrangement of components in bimetallic PdZn/Pd(111) surfaces modified by CO adsorption: a combined DFT, PM IRAS, and TPD study'. *J. Phys. Chem. C*, **116**, 18768–18778.
216. Conant, T., Karim, A.M., Lebarbier, V., Wang, Y., Girgsdies, F., Schlögl, R., and Datye, A. (2008) Stability of bimetallic Pd-Zn catalysts for the steam reforming of methanol. *J. Catal.*, **257**, 64–70.

217. Schauermann, S., Hoffmann, J., Johánek, V., Hartmann, J., Libuda, J., and Freund, H.-J. (2002) Catalytic activity and poisoning of specific sites on supported metal nanoparticles. *Angew. Chem. Int. Ed.*, **41**, 2532–2535.
218. Bertarione, S., Scarano, D., Zecchina, A., Johanek, V., Hoffmann, J., Schauermann, S., Frank, M., Libuda, J., Rupprechter, G., and Freund, H.-J. (2004) Surface reactivity of polycrystalline MgO supported Pd nanoparticles as compared to thin film model catalysts: Infrared study of CH3OH adsorption. Part II. *J. Catal.*, **223**, 64.
219. Penner, S., Jenewein, B., Gabasch, H., Klötzer, B., Wang, D., Knop-Gericke, A., Schlögl, R., and Hayek, K. (2006) Growth and structural stability of well-ordered PdZn nanoparticles. *J. Catal.*, **241**, 14.
220. Holzapfel, H.H., Wolfbeisser, A., Rameshan, C., Weilach, C., and Rupprechter, G. (2014) PdZn surface alloys as models of methanol steam reforming catalysts: molecular studies by LEED, XPS, TPD and PM-IRAS. *Top. Catal.*, **57**, 1218–1228.
221. Chen, Z.-X., Neyman, K.M., Gordienko, A.B., and Rösch, N. (2003) Surface structure and stability of PdZn and PtZn alloys: density-functional slab model studies. *Phys. Rev. B*, **68**, 075417.
222. Gabasch, H., Penner, S., Jenewein, B., Klötzer, B., Knop-Gericke, A., Schlögl, R., and Hayek, K. (2006) Zn adsorption on Pd(111): ZnO and PdZn alloy formation. *J. Phys. Chem. B*, **110**, 11391.
223. Zemlyanov, D., Aszalos-Kiss, B., Kleimenov, E., Teschner, D., Zafeiratos, S., Hävecker, M., Knop-Gericke, A., Schlögl, R., Gabasch, H., Unterberger, W., Hayek, K., and Klötzer, B. (2006) In situ XPS study of Pd(111) oxidation. part 1: 2D oxide formation in 10(-3) mbar O-2. *Surf. Sci.*, **600**, 983–994.
224. Kratzer, M., Tamtögl, A., Killmann, J., Schennach, R., and Winkler, A. (2009) Preparation and calibration of ultrathin Zn layers on Pd(1 1 1). *Appl. Surf. Sci.*, **255**, 5755–5759.
225. Koch, H.P., Bako, I., Weirum, G., Kratzer, M., and Schennach, R. (2010) A theoretical study of Zn adsorption and desorption on a Pd(111) substrate. *Surf. Sci.*, **604**, 926–931.
226. Rodriguez, J.A. (1996) Physical and chemical properties of bimetallic surfaces. *Surf. Sci. Rep.*, **24**, 225–287.
227. Hoffmann, F.M. (1983) Infrared reflection-absorption spectroscopy of adsorbed molecules. *Surf. Sci. Rep.*, **3**, 107–192.
228. Tamtögl, A., Kratzer, M., Killman, J., and Winkler, A. (2008) Adsorption/desorption of H2 and CO on Zn-modified Pd(111). *J. Chem. Phys.*, **129**, 224706.
229. Stadlmayr, W., Penner, S., Memmel, N., and Klötzer, B. (2009) Growth, thermal stability and structure of ultrathin Zn layers on Pd(111). *Surf. Sci.*, **603**, 251.
230. Koch, H.P., Bako, I., and Schennach, R. (2010) Adsorption of small molecules on a (2 × 1) PdZn surface alloy on Pd(1 1 1). *Surf. Sci.*, **604**, 596–608.
231. Reichl, W., Rosina, G., Rupprechter, G., Zimmermann, C., and Hayek, K. (2000) Ultrahigh vacuum compatible all-glass high pressure reaction cell for accurate and reproducible measurement of small reaction rates. *Rev. Sci. Instrum.*, **71**, 1495.
232. Lim, K.H., Chen, Z.X., Neyman, K.M., and Rösch, N. (2006) Comparative theoretical study of formaldehyde decomposition on PdZn, Cu, and Pd surfaces. *J. Phys. Chem. B*, **110**, 14890–14897.
233. Zonnevylle, M.C., Rhodin, T., Wandelt, K., Konrad, B., and Weser, T. (1990) Valence band structure of monolayer alloy films of copper and gold on Ru(0001). *Vacuum*, **41**, 449–452.
234. Kalki, K., Pennemann, B., Schröder, U., Heichler, W., and Wandelt, K. (1991) Properties of noble metal and binary alloy monolayer films on Ru(001). *Appl. Surf. Sci.*, **48/49**, 59–68.
235. Willerding, B., Grüne, M., Linden, R.-J., Boishin, G., Wandelt, K., Postnikov, A., and Braun, J. (1999) Photoelectron spectra of thin alloy films: measurements and relativistic calculations for

the system c(2×2) CuMn on Cu(100). *Surf. Sci.*, **433–435**, 886–889.
236. Zsoldos, Z., Sarkany, A., and Guczi, L. (1994) XPS evidence of alloying in Pd/ZnO catalysts. *J. Catal.*, **145**, 235–238.
237. Campbell, C.T., Ertl, G., Kuipers, H., and Segner, J. (1980) A molecular beam study of the catalytic oxidation of CO on a Pt(111) surface. *J. Chem. Phys.*, **73**, 5862.
238. Rupprechter, G., Dellwig, T., Unterhalt, H., and Freund, H.-J. (2001) High-pressure carbon monoxide adsorption on Pt(111) revisited: a sum frequency generation study. *J. Phys. Chem. B*, **105**, 3797.
239. Su, X., Cremer, P.S., Shen, Y.R., and Somorjai, G.A. (1997) High pressure CO oxidation on Pt(111) monitored with Infrared-visible Sum Frequency Generation (SFG). *J. Am. Chem. Soc.*, **119**, 3994.
240. Hendriksen, B.L.M. and Frenken, J.W.M. (2002) CO oxidation on Pt(110): scanning tunneling microscopy inside a high-pressure flow reactor. *Phys. Rev. Lett.*, **89**, 046101.
241. Ackermann, M.D., Pedersen, T.M., Hendriksen, B.L.M., Robach, O., Bobaru, S.C., Popa, I., Quiros, C., Kim, H., Hammer, B., Ferrer, S., and Frenken, J.W.M. (2005) Structure and reactivity of surface oxides on Pt(110) during catalytic CO oxidation. *Phys. Rev. Lett.*, **95**, 255505.
242. Lundgren, E., Gustafson, J., Mikkelsen, A., Andersen, J.N., Stierle, A., Dosch, H., Todorova, M., Rogal, J., Reuter, K., and Scheffler, M. (2004) Pd 100 oxid SXRD Kinetic hindrance during the initial oxidation of Pd(100) at ambient pressures. *Phys. Rev. Lett.*, **92**, 046101.
243. Pantförder, J., Pöllmann, S., Zhu, J.F., Borgmann, D., Denecke, R., and Steinrück, H.-P. (2005) New setup for in situ x-ray photoelectron spectroscopy from ultrahigh vacuum to 1 mbar. *Rev. Sci. Instrum.*, **76**, 014102.
244. Schalow, T., Brandt, B., Laurin, M., Schauermann, S., Libuda, J., and Freund, H.-J. (2006) CO oxidation on partially oxidized Pd nanoparticles. *J. Catal.*, **242**, 58–70.
245. Zorn, K., Giorgio, S., Halwax, E., Henry, C.R., Grönbeck, H., and Rupprechter, G. (2010) CO oxidation on technological Pd/$Al_2O_3$ catalysts: oxidation state and activity. *J. Phys. Chem. C*, submitted.
246. Lee, H., Habas, S.E., Kweskin, S., Butcher, D., Somorjai, G.A., and Yang, P. (2006) Morphological control of catalytically active platinum nanocrystals13. *Angew. Chem. Int. Ed.*, **45**, 7824–7828.
247. Burda, C., Chen, X., Narayanan, R., and El-Sayed, M.A. (2005) Chemistry and properties of nanocrystals of different shapes. *Chem. Rev.*, **105**, 1025–1102.
248. Xia, Y., Yang, P., Sun, Y., Wu, Y., Mayers, B., Gates, B., Yin, Y., Kim, F., and Yan, H. (2003) One-dimensional nanostructures: synthesis, characterization, and applications. *Adv. Mater.*, **15**, 353–389.
249. Borchert, H., Jürgens, B., Zielasek, V., Rupprechter, G., Giorgio, S., Henry, C.R., and Bäumer, M. (2007) Pd nanoparticles with highly defined structure on MgO as model catalysts: an FTIR study of the interaction with CO, O2 and H2 under ambient conditions. *J. Catal.*, **247**, 145–154.
250. Tsung, C.K., Kuhn, J.N., Huang, W.Y., Aliaga, C., Hung, L.I., Somorjai, G.A., and Yang, P. (2009) Sub-10 nm platinum nanocrystals with size and shape control: catalytic study for ethylene and pyrrole hydrogenation. *J. Am. Chem. Soc.*, **131**, 5816–5822.
251. Semagina, N. and Kiwi-Minsker, L. (2009) Recent advances in the liquid-phase synthesis of metal nanostructures with controlled shape and size for catalysis. *Catal. Rev. Sci. Eng.*, **51**, 147–217.
252. Hulse, J., Küppers, J., Wandelt, K., and Ertl, G. (1980) UV-photoelectron spectroscopy from xenon adsorbed on heterogeneous metal surfaces. *Appl. Surf. Sci.*, **6**, 453–463.
253. Milun, M., Pervan, P., and Wandelt, K. (1989) Interaction of oxygen with a polycrystalline Palladium surface over a wide temperature range. *Surf. Sci.*, **218**, 363–388.

254. Suchorski, Y. and Rupprechter, G. Local reaction kinetics by imaging. *Surf. Sci.* *doi:10.1016/j.susc.2015.05.021.*
255. Lauterbach, J., Hass, G., Rotermund, H.H., and Ertl, G. (1993) The formation of subsurface oxygen on Pt(100). *Surf. Sci.*, **294**, 116–130.
256. Imbihl, R. and Ertl, G. (1995) Oscilla tory kinetics in heterogeneous catalysis. *Chem. Rev.*, **95**, 697–733.
257. Vogel, D., Spiel, C., Suchorski, Y., Urich, A., Schlögl, R., and Rupprechter, G. (2011) Mapping local reaction kinetics by PEEM: CO oxidation on individual (100)-type grains of Pt foil. *Surf. Sci.*, **605**, 1999–2005.
258. Vogel, D., Spiel, C., Schmid, M., Stöger-Pollach, M., Schlögl, R., Suchorski, Y., and Rupprechter, G. (2013) The role of defects in the local reaction kinetics of CO oxidation on low-index Pd surfaces. *J. Phys. Chem. C*, **117**, 12054–12060.
259. Ertl, G. (1991) Oscillatory kinetics and spatio-temporal self-organization in reaction at solid surfaces. *Science*, **254**, 1750–1755.
260. Johansson, M., Lytken, O., and Chorkendorff, I. (2007) The sticking probability of hydrogen on Ni, Pd and Pt at a hydrogen pressure of 1 bar NOZZLE. *Top. Catal.*, **46**, 175–187.
261. Spiel, C., Vogel, D., Suchorski, Y., Drachsel, W., Schlögl, R., and Rupprechter, G. (2011) Catalytic CO oxidation on individual (110) domains of a polycrystalline Pt foil: local reaction kinetics by PEEM. *Catal. Lett.*, **141**, 625–632.
262. Suchorski, Y., Imbihl, R., and Medvedev, V.K. (1998) Compatibility of field emitter studies of oscillating surface reactions with single crystal measurements: catalytic CO oxidation on Pt. *Surf. Sci.*, **401**, 392–399.
263. Spiel, C., Vogel, D., Schlögl, R., Rupprechter, G., and Suchorski*, Y. Spatially coupled catalytic ignition of CO oxidation on Pt: mesoscopic versus nano-scale, in press. *doi:10.1016/j.ultramic.2015.05.012.*
264. Johánek, V., Laurin, M., Grant, A.W., Kasemo, B., Henry, C.R., and Libuda, J. (2004) Chemical bistability on catalyst nanoparticles. *Science*, **304**, 1639.
265. Yamamoto, S.Y., Surko, C.M., and Maple, M.B. (1995) Spatial coupling in heterogeneous catalysis. *J. Chem. Phys.*, **103**, 8209–8216.
266. Slinko, M.M., Ukharskii, A.A., and Jaeger, N.I. (2001) Global and nonlocal coupling in oscillating heterogeneous catalytic reactions: the oxidation of CO on zeolite supported palladium. *Phys. Chem. Chem. Phys.*, **3**, 1015–1021.
267. Schalow, T., Brandt, B., Starr, D.E., Laurin, M., Shaikhutdivov, S.K., Schauermann, S., Libuda, J., and Freund, H.-J. (2006) Size-dependent oxidation mechanism of supported Pd nanoparticles. *Angew. Chem. Int. Ed.*, **45**, 3693.
268. Suchorski, Y., Wrobel, R., Becker, S., Strzelczyk, B., Drachsel, W., and Weiss, H. (2007) Ceria nanoformations in CO Oxidation on Pt(111): promotional effects and reversible redox behaviour. *Surf. Sci.*, **601**, 4843–4848.
269. Ozensoy, E. and Vovk, E.I. (2013) In-situ vibrational spectroscopic studies on model catalyst surfaces at elevated pressures. *Top. Catal.*, **15-17**, 1569–1592.
270. Carlsson, P.-A., Österlund, L., Thormählen, P., Palmqvist, A., Fridell, E., Jansson, J., Zhdanov, V.P., and Skoglundh, M. (2004) *J. Catal.*, **226**, 422–434.
271. Carlsson, P.-A., Zhdanov, V.P., and Skoglundh, M. (2006) Self-sustained kinetic oscillations in CO oxidation over silica-supported Pt. *Phys. Chem. Chem. Phys.*, **8**, 2703–2706.
272. Rinnemo, M., Kulginov, D., Johannson, S., Wong, K.L., Zhdanov, V.P., and Kasemo, B. (1997) Catalytic ignition in the CO–$O_2$ reaction on platinum: experiment and simulations. *Surf. Sci.*, **376**, 297–309.
273. Eigenberger, G. (1978) Kinetic instabilities in heterogeneously catalyzed reactions—I: rate multiplicity with langmuir-type kinetics. *Chem. Eng. Sci.*, **33**, 1255–1261.
274. Haruta, M. (2002) Catalysis of gold nanoparticles deposited on metal oxides. *Cat. Tech.*, **3**, 102.

275. Meyer, R., Lemire, C., Shaikhutdinov, S.K., and Freund, H.-J. (2004) Surface chemistry of catalysis by gold. *Gold Bull.*, **37**, 72–124.
276. Xie, X., Li, Y., Liu, Z.-Q., Haruta, M., and Shen, W. (2009) Low-temperature oxidation of CO catalysed by Co3O4 nanorods. *Nature*, **458**, 746–749.
277. Meyer, W., Biedermann, K., Gubo, M., Hammer, L., and Heinz, K. (2008) Surface structure of polar Co3O4(111) films grown epitaxially on Ir(100)-(1×1). *J. Phys. Condens. Matter*, **20**, 265011.
278. Cunningham, D.A.H., Kobayashi, T., Kamijo, N., and Haruta, M. (1994) Influence of dry operating conditions: observation of oscillations and low temperature CO oxidation over Co3O4 and Au/Co3O4 catalysts. *Catal. Lett.*, **25**, 257–264.
279. Zhdanov, V.P. and Kasemo, B. (1994) Kinetic phase transitions in simple reactions on solid surfaces. *Surf. Sci. Rep.*, **20**, 113–189.
280. Bowker, M., Jones, I.Z., Bennet, R.A., Esch, F., Baraldi, A., Lizzit, S., and Comelli, G. (1998) Shedding light on catalytic ignition: coverage changes during CO oxidation on Pd(110). *Catal. Lett.*, **51**, 187–190.
281. Bowker, M., Jones, I.Z., Bennett, R.A., Esch, F., Baraldi, A., Lizzit, S., and Comelli, G. (1998) Shedding light on catalytic ignition: coverage changes during CO oxidation on Pd(110). *Catal. Lett.*, **51**, 187–190.
282. Datler, M., Bespalov, I., Rupprechter, G., and Suchorski, Y. (2015) Analyzing the reaction kinetics for individual catalytically active components: CO oxidation on a Pd powder supported by Pt foil. *Catal. Lett., doi:10.1007/s10562-015-1486-7.*
283. Silvestre-Albero, J., Rupprechter, G., and Freund, H.-J. (2005) Atmospheric pressure studies of selective 1,3-butadiene hydrogenation on Pd single crystals: effect of CO addition. *J. Catal.*, **235**, 52.
284. Silvestre-Albero, J., Rupprechter, G., and Freund, H.-J. (2006) From Pd nanoparticles to single crystals: 1,3-butadiene hydrogenation on well-defined model catalysts. *Chem. Commun.*, 80.
285. Silvestre-Albero, J., Rupprechter, G., and Freund, H.-J. (2006) Atmospheric pressure studies of selective 1,3-butadiene hydrogenation on well-defined $Pd/Al_2O_3$/NiAl(110) model catalysts: effect of Pd particle size. *J. Catal.*, **240**, 58.
286. Silvestre-Albero, J., Borasio, M., Rupprechter, G., and Freund, H.-J. (2007) Combined UHV and ambient pressure studies of 1,3-Butadiene adsorption and reaction on Pd(111) by GC, IRAS and XPS. *Catal. Commun.*, **8**, 292–298.
287. McCrea, K.R. and Somorjai, G.A. (2000) SFG-surface vibrational spectroscopy studies of structure sensitivity and insensitivity in catalytic reactions: cyclohexene dehydrogenation and ethylene hydrogenation on Pt(111) and Pt(100) crystal surfaces. *J. Mol. Catal. A*, **163**, 43.
288. Sheppard, N. and De La Cruz, C. (1996) Vibrational spectra of hydrocarbons adsorbed on metals. Part I. Introductroy principles, ethylene, and the higher acyclic alkenes. *Adv. Catal.*, **41**, 1.
289. Sheppard, N. and De La Cruz, C. (1998) Vibrational spectra of hydrocarbons adsorbed on metals. Part II. Adsorbed acyclic alkynes and alkanes, cyclic hydrocarbons including aromatics, and surface hydrocarbon groups derived from the decomposition of alkyl halides, etc. *Adv. Catal.*, **42**, 181.
290. Shaikhutdinov, S., Heemeier, M., Bäumer, M., Lear, T., Lennon, D., Oldman, R.J., Jackson, S.D., and Freund, H.-J. (2001) Structure–reactivity relationships on supported metal model catalysts: adsorption and reaction of ethene and hydrogen on $Pd/Al_2O_3$/NiAl(110). *J. Catal.*, **200**, 330.
291. Doyle, A., Shaikhutdinov, S., Jackson, S.D., and Freund, H.-J. (2003) Hydrogenation on metal surfaces: why are nanoparticles more active than single crystals? *Angew. Chem. Int. Ed.*, **42**, 5240.
292. Ibach, H. (1994) Electron energy loss spectroscopy: the vibration

spectroscopy of surfaces. *Surf. Sci.*, **299/300**, 116.
293. Steiniger, H., Ibach, H., and Lehwald, S. (1992) Surface reactions of ethylene and oxygen on Pt(111), c2h4 hydr. *Surf. Sci.*, **117**, 685–688.
294. Borasio, M. (2006) Polarization modulation infrared reflection absorption spectroscopy on Pd model catalysts at elevated pressure. PhD thesis. Free University Berlin.
295. Beebe, T. and Yates, J.T. (1986) *J. Am. Chem. Soc.*, **108**, 663.
296. Davis, R.J. and Boudart, M. (1991) Hydrogenation of alkenes on supported PdAu clusters. *Catal. Sci. Technol.*, **1**, Proceedings, TO-CAT-1 (Tokyo, 1990), Kodansha, Ltd., Tokyo, 129–134.
297. Horiuti, I. and Polanyi, M. (1934) Exchange reactions of hydrogen on metallic catalysts. *Trans. Faraday Soc.*, **30**, 1164.
298. Kesmodel, L.L. and Gates, J.A. (1981) Ethylene adsorption and reaction on Pd(111): an angle-dependent eels analysis. *Surf. Sci.*, **111**, L747.
299. Gates, J.A. and Kesmodel, L.L. (1983) Thermal evolution of acetylene and ethylene on Pd(111). *Surf. Sci.*, **124**, 68–86.
300. Lloyd, D.R. and Netzer, F.P. (1983) The electronic structure of the high temperature phase of ethylene adsorbed on Pd(111). *Surf. Sci.*, **129**, L249.
301. Stuve, E.M. and Madix, R.J. (1985) Bonding and dehydrogenation of ethylene on palladium metal: vibrational spectra and temperature programmed reaction studies on Pd(100), TPD HREELS C2H4 Pd(100). *J. Phys. Chem.*, **89**, 105.
302. Camplin, J., Eve, J., and McCash, E. (1997) *Surf. Rev. Lett.*, **4**, 1371.
303. Sandell, A., Beutler, A., Jaworowski, A., Wiklund, M., Heister, K., Nyholm, R., and Andersen, J.N. (1998) Adsorption of acetylene and hydrogen on Pd(111): formation of a well-ordered ethylidyne overlayer. *Surf. Sci.*, **415**, 411.
304. Ogasawara, H., Ichihara, S., Okuyama, H., Domen, K., and Kawai, M. (2001) Orientation of unsaturted hydrocarbons on Pd(110). *J. Electron. Spectrosc. Relat. Phenom.*, **114**, 339.
305. Sekitani, T., Takaoka, T., Fujisawa, M., and Nishijima, M. (1992) Interaction of ethylene with the hydrogen-preadsorbed palladium (110) surface: hydrogenation and hydrogen-deuterium-exchange reactions. *J. Phys. Chem.*, **96**, 8462.
306. Stacchiola, D., Azad, S., Burkholder, L., and Tysoe, W.T. (2001) An investigation of the reaction pathway for ethylene hydrogenation on Pd (111). *J. Phys. Chem. B*, **105**, 11233.
307. Sock, M., Eichler, A., Surnev, S., Andersen, J.N., Klötzer, B., Hayek, K., Ramsey, M.G., and Netzer, F.P. (2003) High-resolution electron spectroscopy of different adsorption states of ethylene on Pd(1 1 1). *Surf. Sci.*, **545**, 122.
308. Neurock, M. and van Santen, R.A. (2000) A first principles analysis of C-H bond formation in ethylene hydrogenation. *J. Phys. Chem. B*, **104**, 11127–11145.
309. Ge, Q. and Neurock, M. (2002) Correlation of adsorption energy with surface structure: ethylene adsorption on Pd surfaces. *Chem. Phys. Lett.*, **358**, 377–382.
310. Rupprechter, G. (2004) Surface vibrational spectroscopy on noble metal catalysts from ultrahigh vacuum to atmospheric pressure. *Annu. Rep. Prog. Chem. Sect. C*, **100**, 237–311.
311. Kaltchev, M., Thompson, A.W., and Tysoe, W.T. (1997) Reflection-absorption infrared spectroscopy of ethylene on palladium(111) at high pressure. *Surf. Sci.*, **391**, 145.
312. Cremer, P.S., Stanners, C., Niemantsverdriet, J.W., Shen, Y.R., and Somorjai, G.A. (1995) The conversion of di-sigma bonded ethylene to ethylidine on Pt (111) by sfg. *Surf. Sci.*, **328**, 111.
313. Stacchiola, D., Kaltchev, G., Wu, G., and Tysoe, W.T. (2000) The adsorption and structure of carbon monoxide on ethylidyne-covered Pd(111). *Surf. Sci.*, **470**, L32.
314. Stacchiola, D., Burkholder, L., and Tysoe, W.T. (2002) Ethylene adsorption on Pd(111) studied using infrared

reflection–absorption spectroscopy. *Surf. Sci.*, **511**, 215.

**315.** Frank, M., Bäumer, M., Kühnemuth, R., and Freund, H.-J. (2001) Adsorption and reaction of ethene on oxide-suported Pd, Rh and Ir particles. *J. Vac. Sci. Technol., A*, **19**, 1497.

**316.** Tardy, B., Noupa, C., Leclercq, C., Bertolini, J.C., Hoareau, A., Treilleux, M., Faure, J.P., and Nihoul, G. (1991) Catalytic hydrogenation of 1,3-butadiene on Pd particles evaporated on carbonaceous supports: particle size effect. *J. Catal.*, **129**, 1–11.

**317.** Arnold, H., Döbert, F., and Gaube, J. (1997) in *Handbook of Heterogeneous Catalysis* (eds G. Ertl, H. Knözinger, and J. Weitkamp), Wiley-VCH Verlag GmbH, Weinheim, p. 2165.

**318.** Hammerschaimb, H.U. and Spinner, J.B. (1988) U.S. Patent 4774375.

**319.** Furlong, B., Hightower, J., Chan, T., Sarkany, A., and Guczi, L. (1994) 1,3-Butadiene selective hydrogenation over Pd/alumina and CuPd/alumina catalysts. *Appl. Catal., A*, **117**, 41.

**320.** Bertolini, J.C., Delichere, P., Khanra, B., Massardier, J., Noupa, C., and Tardy, B. (1990) Electronic properties of supported Pd aggregates in relation with their reactivity for 1,3-butadiene hydrogenation. *Catal. Lett.*, **6**, 215.

**321.** Bond, G.C., Webb, G., Wells, P.B., and Winterbottom, J.M. (1965) The hydrogenation of alkadienes. Part I. The hydrogenation of buta-1,3-diene catalysed by the Noble Group VIII metals. *J. Chem. Soc.*, 3218.

**322.** Boitiaux, J.P., Cosyns, J., and Vasudevan, S. (1983) Hydrogenation of highly unsaturated-hydrocarbons over highly dispersed palladium catalyst. 1. Behavior of small metal particles. *Appl. Catal.*, **6**, 41–51.

**323.** Molnár, A., Sárkány, A., and Varga, M. (2001) Hydrogenation of carbon-carbon multiple bonds: chemo-, regio- and stereo-selectivity. *J. Mol. Catal. A: Chem.*, **173**, 185.

**324.** Massardier, J., Bertolini, J.C., and Renouprez, A. (1988) Proceedings 9th International Congress on Catalysis, Calgary, Alberta, vol. 3, p. 1222.

**325.** Jugnet, Y., Sedrati, R., and Bertolini, J.C. (2005) Selective hydrogenation of 1,3-butadiene on Pt3Sn(111) alloys: comparison to Pt(111). *J. Catal.*, **229**, 252–258.

**326.** Katano, S., Ichihara, S., Ogasawara, H., Kato, H.S., Komeda, T., Kawai, M., and Domen, K. (2002) Adsorption structure of 1,3-butadiene on Pd(110). *Surf. Sci.*, **502–503**, 164.

**327.** Zhao, H. and Koel, B.E. (2005) Hydrogenation of 1,3-butadiene on two ordered Sn/Pt(111) surface alloys. *J. Catal.*, **234**, 24–32.

**328.** Valcarcel, A., Clotet, A., Ricart, J.M., Delbecq, F., and Sautet, P. (2004) Comparative DFT study of the adsorption of 1,3-butadiene, 1-butene and 2-cis/trans-butenes on the Pt(1 1 1) and Pd(1 1 1) surfaces. *Surf. Sci.*, **549**, 121.

**329.** Breinlich, C., Haubrich, J., Becker, C., Valcárcel, A., Delbecq, F., and Wandelt, K. (2007) Hydrogenation of 1,3-butadiene on Pd(111) and PdSn/Pd(111) surface alloys under UHV conditions. *J. Catal.*, **251**, 123–130.

**330.** Vigné, F., Haubrich, J., Loffreda, D., Sautet, P., and Delbecq, F. (2010) Highly selective hydrogenation of butadiene on Pt/Sn alloy elucidated by first-principles calculations. *J. Catal.*, **275**, 129–139.

**331.** Richter, L.T., Petrallimallow, T.P., and Stephenson, J.C. (1998) Vibrationally resolved sum-frequency generation with broad-bandwidth infrared pulses. *Opt. Lett.*, **23**, 1594.

**332.** Hess, C., Wolf, M., and Bonn, M. (2000) Direct observation of vibrational energy delocalization on surfaces: CO onRu(001). *Phys. Rev. Lett.*, **85**, 4341.

**333.** Symonds, J.P.R., Arnolds, H., Zhang, V.L., Fukutani, K., and King, D.A. (2004) Broadband femtosecond sum-frequency spectroscopy of CO on Ru1010 in the frequency and time domains. *J. Chem. Phys.*, **120**, 7158.

**334.** Roeterdink, W.G., Aarts, J.F.M., Kleyn, A.W., and Bonn, M. (2004) Broadband sum frequency generation spectroscopy to study surface reaction kinetics: a temperature-programmed study of CO oxidation on Pt(111). *J. Phys. Chem. B*, **108**, 14491.

335. Fadley, C.S. (2010) X-ray photoelectron spectroscopy: progress and perspectives. *J. Electron. Spectrosc. Relat. Phenom.*, **178–179**, 2–32.
336. Whelan, C.M., Neubauer, R., Borgmann, D., Denecke, R., and Steinrück, H.-P. (2001) A fast x-ray photoelectron spectroscopy study of the adsorption and temperature-dependent decomposition of propene on Ni(100). *J. Chem. Phys.*, **115**, 8133–8140.

# 40
# Adsorption of Unsaturated and Multifunctional Molecules: Bonding and Reactivity

*Jan Haubrich and Klaus Wandelt*

## 40.1
## Introduction

The adsorption and reactivity of molecules on surfaces has been a long-standing subject in scientific research since it is a central puzzle piece to a fundamental understanding and control of a variety of important processes such as heterogeneous catalysis, photocatalysis, semiconductor and film growth, and electrochemistry as well as environmental problems such as pollutant degradation on soils and the interaction of chemical compounds with plant surfaces or skin. The types of surfaces targeted range correspondingly from mono- or multimetallic compounds, stoichiometric or reduced oxides, pure and doped semiconductors to the sufaces of skin and plant leaves.

As usual in fundamental research, the primary goals in surface and interface chemistry studies are the development of an in-depth understanding and the derivation of guidelines to tailor and control new or existing processes. Ideally, all the details of the reactions can be identified unambiguously and incorporated into analytic and predictive models. The first critical step in understanding any interface chemical problem is the adsorption and bonding of molecules at the interface, be it a reactant on a catalyst particle or a signal molecule on biological surfaces. The adsorption is, in most cases, the initial mechanistic step before entering a reaction channel or inducing other functions. Hence, the bonding on surfaces is a key question to be addressed from a fundamental perspective and a prerequisite for a thorough understanding of any surface process.

However, most real surfaces are simply too complex to be studied in exhaustive detail in a finite time even with the most advanced experimental or theoretical tools available today. Complexity can arise, for example, from the simultaneous presence of various facets on small particles, from "imperfectness" of the surfaces, due, for instance, to steps, kinks, surface or subsurface defects, from the dynamical changes induced by a gaseous or liquid phase in contact with the interface; from the presence of contaminants or promoters, from the influence of an external potential or light; or from the size of the reactant or spectator molecules (see Chapter 39 in this Volume).

*Surface and Interface Science: Solid-Gas Interfaces I,* First Edition. Edited by Klaus Wandelt.

Many of these factors can be avoided or selectively eliminated by approximating a real problem through a well-defined, controlled model of the same process, giving rise to the "model catalysis" or "model surface science" approach. Model surfaces, molecules, and reaction systems can be studied in great detail and their complexity subsequently increased by incorporating further factors present in the real-world problem. This approach allows a systematic and stepwise evolution of the understanding of each contributing factor.

In this chapter, we focus on model studies on a selected problem from heterogeneous catalysis: the hydrogenation of double bonds in "small" hydrocarbons. The hydrogenation of unsaturated hydrocarbons on metal catalysts is of particular interest in view of its importance of processes in, among others, the refining or pharmaceutical industry [1–3]. On the reactant side, we have selected molecules with increasing complexity to guide the reader toward the current state of knowledge on the molecule–catalyst interaction and hydrogenation of these compounds. One particular aspect in our discussion will be the nature of the adsorption bond of the reactants, products, and surface species, which has multiple implications for governing the reactivity. Although often useful for a qualitative prediction of reactivities, it is, however, only one aspect in understanding the role of catalysts.

Since the adsorption of hydrogen on metal surfaces has been very widely studied [4–9] and the present level of understanding has already been discussed in Chapter 36 in this volume, we will focus our attention primarily on the hydrocarbon aspect of the reactivity. Among the probe molecules most suitable for these studies is ethene, the simplest of all alkenes. The longer chained propene and the butene isomers serve as examples for higher olefins, while 1,3-butadiene introduces the complexity of a chemically identical second function. This symmetry is broken up in the α,β-unsaturated aldehydes, which still have two adjacent conjugated double-bonds, yet one of them is an aldehydic type whereas the other one is an olefinic carbon–carbon bond. While the pure unsaturated carbon–carbon bonds usually form only di-σ type (bidentate, $\mu_2\eta^2$) and or π type (monodentate, $\mu_1\eta^2$) coordinations to the substrates, the aldehydic carbon–oxygen double bonds can also show monodentate on-top interactions ($\mu_1\eta^1$) with a coordination through one of the oxygen lone pairs. This variety of coordination modes leads to increasingly complex adsorption behavior and reactivity patterns.

In applied catalysis, transition-metal catalysts based on Ni ("Raney Nickel"), Pd ("Lindlar's catalyst"), or Pt ("Adams's catalyst") are widely used for hydrogenation reactions [2, 3, 10, 11]. Their activity, selectivity, and stability can be modified by promoters such as Sn or Pb, which can lead to the formation of bimetallic phases or alloys. In order to study the process on the real supported catalysts, single-crystal surfaces of transition-metal catalysts have proven to be suitable models. On mono and bimetallic model catalysts, the behavior of the reactants can be mimicked under the right conditions while the degree of complexity is sufficiently reduced to make such systems tractable with today's experimental and theoretical tools. Particularly, Pt(111) and $Pt_xSn(111)$-based model catalysts with well-characterized properties and structural ensembles have enabled many groups to advance our present understanding of the hydrogenation and interaction mechanisms of double bonds on solid surfaces and are an ideal example of case studies, although we will also draw several parallels to other transition metals where useful.

The hydrogenation of olefins in heterogeneous catalysis has been reviewed in many books and articles in the past [1–3, 10, 12–16]. It usually follows a Horiuti–Polanyi [17] mechanism, in which hydrogen dissociates upon adsorption and attacks the unsaturated reactant bond that is activated by the catalyst. Alternatively, this process can be viewed as an insertion of the olefin into a metal–hydride bond. The Horiuti–Polanyi [17] mechanism corresponds to a Langmuir–Hinshelwood mechanism, one of the two limiting cases in surface chemistry. The other limit is the Eley–Rideal mechanism, which describes a reaction in which one of the two reactants is not bonded to the catalyst surface prior to reaction.

A key step for the reaction mechanism is the initial bonding and activation of the reactants by the catalysts. In the past decades, numerous concepts have been developed for the bonding and reactivity of molecules and the effects of the catalyst material, most of which are of analytical nature. On the catalyst side, among the first mechanistic concepts is the idea of Taylor that the unsaturated character of an active site can control the chemical reactivity at the atomic level [18]. Later, Boudart extended this idea by dividing surface reactions into structure-sensitive or -insensitive processes, establishing a connection between the structure and activity of a catalyst [19]. Alloying metal catalysts with promoters was first studied in detail by Sachtler, who strictly distinguished *ensemble* (geometric/template) or *ligand* (electronic) effects [20]. With the advancement of surface science, it is, however, becoming clearer that both effects are usually closely linked [21–23].

On the molecular side, the Dewar–Chatt–Duncanson [24–26] model for olefin bonding and the Blyholder [27] model of CO adsorption are amongst the first detailed theoretical descriptions of the interaction of molecules with metal atoms or extended surfaces. Both show the importance of the interplay between electron donation from molecules to electronic states at the bonding site on the surface and the back-donation into antibonding states of $\pi^*$ symmetry of the C=O or C=C double bonds. Although originally derived in organometallic chemistry for the bonding of ethene in metal-π-complexes of certain compounds, the Dewar–Chatt–Duncanson [24–26] model has been found to be valid in surface chemistry as well. It formulates a connection between the size of the bond strength between the olefin and the metal site and the reduction of the carbon–carbon bond order, which leads to elongated C=C distances, an increase of the hybridization from $sp^2$ toward $sp^3$, and an increasing perturbation of vibrational frequencies such as that of the $\nu$(C=C) stretching mode. Analogously, the general back-bonding in the case of the Blyholder model leads to perturbations of the C–O bond length and stretching vibration.

The frontier orbital model by Hoffmann [28] and the d-band model by Hammer and Nørskov [23, 29] have proven to be very useful for understanding the bonding of reactants with transition-metal surfaces from their electronic properties. The latter model has been particularly successful in giving simple explanations for the bonding and reactivity of a variety of small molecules such as $O_2$, $N_2$, CO, or $CH_4$ [23, 29, 30] (see Chapter 38).

Recently, Nørskov and coworkers successfully estimated the transition state energetics of elementary steps from reaction energetics obtained by electronic structure theory [21, 30] based on the observation that activation energies of (rate-limiting) processes can be estimated from simple thermodynamic descriptors such as total reaction enthalpies using a linear Brønsted–Evans–Polanyi (BEP) correlation [31, 32]. Since the reaction energetics are obtained from bonding energies of reactants and intermediates on transition metals, the use of the so-called linear scaling relations also links the activation barriers to the electronic structure as approximated with the "d-band model" [21, 30]. This represented another big step toward a predictive model that correlates catalytic reactivity with simple chemical and physical properties.

Many cases have already been identified in which reactivity can essentially be described as a sole function of the adsorption energies of the reactants [21, 30], an approach that may be widely applicable. However, also cases with poor electronic structure–reactivity correlations have been reported, in particular for larger reactants. For example, activation barriers for dehydrogenation of hydrocarbons and the correlation of bonding energies of several hydrocarbons and intermediates on mono- and bimetallic transition-metal surfaces to their respective d-band centers have proven to be difficult [33]. Although roughly linear correlations with d-band positions were obtained, the scattering around the linear fits was found to be quite wide [33].

Simultaneously, it was found that the adsorption energies in the case of larger multifunctional molecules can behave fundamentally differently from the true interaction strength with the catalyst [34–38]. Since the interaction strength with the catalyst is commonly considered a good measure of chemical activation, the distinction between adsorption and interaction strength is crucial (see Sections 40.3.1.5 and 40.3.4.4).

Thus, a thorough understanding of the molecule–surface interaction of larger molecules showing more complex coordination modes will be necessary to employ it in the framework of the BEP or "linear scaling relationships." In the following sections, we will guide the reader through a review of the advances obtained from model catalytic studies up to the present level of understanding and the many question yet to be resolved.

## 40.2
## The Choice of Simple, Well-Defined Model Catalysts

Well-characterized substrates enable detailed and unambiguous studies of interfacial processes. To mimic bimetallic catalyst particles by well-defined systems, either single-crystal bulk alloys can be employed or a monometallic single crystal can be alloyed with a second component, which leads to thick or thin alloy surfaces. In the case of the Pt–Sn model system, which we choose to focus on in this chapter, $Pt_3Sn$ bulk alloys are available with different surface terminations as well as two ultrathin Pt–Sn surface alloys of single monolayer thickness (Figure 40.1).

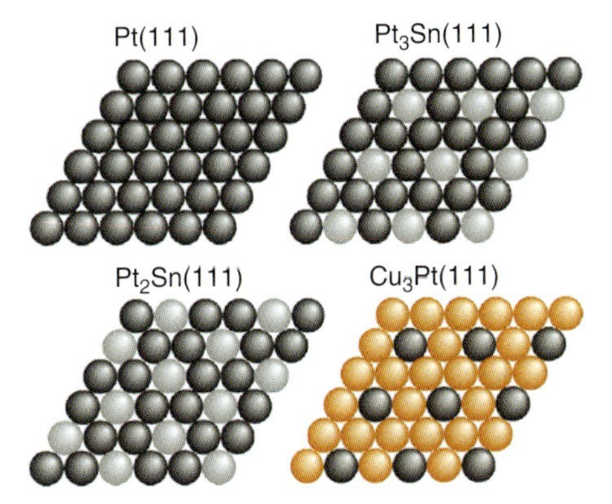


**Figure 40.1** Well-defined mono- and bimetallic model catalyst surfaces with a (111) symmetry that can be used to study electronic and geometric (or related ligand and ensemble) effects in surface chemistry. (Reproduced from Ref. [36].)

Surface alloys can be easily and reproducibly synthesized by physical vapor deposition on a Pt(111) surface. The obtained superstructures depend on the amount of tin deposited and the temperature the sample is heated to after deposition. Surface alloys have been widely studied, and much is known about their geometric as well as electronic properties.

Below 320 K, tin grows in a Stranski–Krastanov mode on a Pt(111) substrate and forms superlattices with $(\sqrt{3}\times\sqrt{3})$R30° and c(4 × 2) periodicity without alloying into the substrate [39]. From low-energy electron diffraction (LEED) investigations, two surface alloys were reported upon annealing the surfaces with different tin coverages to 1000 K: [39] for small coverages of 0.6 ML, a $Pt_3Sn$/Pt(111) surface alloy with a p(2 × 2) periodicity ($\Theta_{Sn} = 0.25$) was observed, whereas for coverages above $\Theta_{Sn} = 0.6$ tin atoms per platinum atom, the formation of a $Pt_2Sn$/Pt(111) surface alloy with a $(\sqrt{3}\times\sqrt{3})$R30° structure ($\Theta_{Sn} = 0.33$) occurred [39]. For the sake of simplicity, we will refer to those alloy surfaces as $Pt_2Sn$(111) and $Pt_3Sn$(111) in further discussions. At exposures between these two limits, the LEED patterns showed a mixture of both structures [39]. Also, scanning tunneling microscopy (STM) experiments carried out by Batzill *et al.* showed both superstructures to be present in different domains after the preparation of the sample [40].

Our own experiments using LEED and Auger electron spectroscopy (AES) to construct a schematic surface phase diagram showed that also at low tin coverages a $(\sqrt{3}\times\sqrt{3})$R30° structure can be formed, but requiring lower annealing temperatures (500–600 K) [35] as compared to Paffett's work [39]. At these low tin coverages, exclusively this alloy is obtained. With increasing Sn deposition, the temperature range for this surface alloy increases, and also a p(2 × 2) structure becomes accessible which is observed around 1000 K and exhibits only a small formation temperature interval of ~75 K. Between both temperature ranges, the LEED images confirmed the coexistence of both surface structures, as reported by Paffett and coworkers [39]. Above 1075 K, the superstructure spots in the LEED images become increasingly diffuse and disappear. Theoretical studies by Teraoka *et al.*

predicted that the $Pt_2Sn(111)/Pt(111)$ surface alloy was thermodynamically more stable than the $Pt_3Sn(111)/Pt(111)$ surface alloy [41]. In the case of monolayer alloys, the computed order–disorder transition temperature for ordering to a $(\sqrt{3} \times \sqrt{3})R30°$ structure is found to be higher than that for forming a $p(2 \times 2)$ surface structure.

$Pt_2Sn(111)$ and $Pt_3Sn(111)$ surface alloys also exist as possible terminations on the $Pt_3Sn$ bulk crystal, which was studied by Atrei *et al.* using I/V-LEED [42, 43]. Both surfaces show significant buckling: the Sn atoms protruded beyond the platinum by 23 and 30 pm on the $Pt_2Sn(111)$-$(\sqrt{3} \times \sqrt{3})R30°$ and $Pt_3Sn(111)$-$p(2 \times 2)$ surfaces, respectively [42]. On the (111)-surface of the $Pt_3Sn$ bulk alloy, the buckling was found to be reduced to 11 pm. Low-energy alkali ion scattering (LEIS) experiments conducted by Overbury *et al.* showed a buckling of 22 pm on both surface alloys [44, 45]. This buckling has important implications for the interaction of molecules with the surface, as we will discuss for selected probe molecules below. Overbury and coworkers also showed that both surface alloys prepared on Pt(111) are single layer alloys, suggesting that the excess tin diffuses into the Pt bulk.

At high tin coverages of 4–5 ML, Galeotti *et al.* obtained contradictory results in X-ray photoelectron diffraction (XPD) experiments, where (see below) a second $p(2 \times 2)$ structure was formed already around 400 K [46]. Their analysis indicates that a thicker "bulk" alloy is formed under these conditions. Indeed, the LEED experiments at high tin exposures performed by Haubrich *et al.* showed a weak diffraction pattern corresponding to a $p(2 \times 2)$ structure around 500 K [35, 37, 38, 47]. However, the low transition temperature at which this $p(2 \times 2)$ structure appeared indicated that this was a different alloy than the one formed around 1000 K.

The influence of the electropositive tin atoms on the adsorption properties of these surface alloys consists of a combination of linked effects. They can be categorized as geometric and electronic effects or "ensemble" and "ligand" effects as defined by Sachtler, who first developed this concept [48]. A strong modification of the electronic properties of the surface Pt sites [40] as well as a site blocking effect for adsorption and reaction of hydrocarbons exerted by tin atoms is known [38, 39, 49, 50]. This results from a weaker interaction of tin with carbon atoms in molecules compared to the bonding to Pt sites.

Alloying a larger atom into the matrix of smaller Pt atoms (atomic diameters of Pt and Sn are 277 and 301 pm, respectively [39]) has been shown to lead to a widening of the d band, which, for conservation of the d-band population, will undergo an energetic downshift. The charge density at the Fermi edge of the surface alloys was, in fact, found to be reduced in STM experiments [40], in agreement with the damping and lowering of the d-band center energy obtained from theoretical simulations of the density of states (DOS) [50] and measured with ultraviolet photoelectron spectroscopy (UPS) [51]. The alloying with tin shifts the d-band center from $-1.93$ eV for Pt(111) to $-2.09$ eV for the $Pt_3Sn$ alloy surface and to $-2.12$ eV for $Pt_2Sn(111)$, and reduces the overlap of the Pt d orbitals forming the d band [50]. Because of the interaction of the Pt d band with the Sn sp bands, this also gives rise to an electron transfer from Sn to Pt: At the density functional theory–generalized gradient approximation (DFT-GGA) level, tin loses 0.42 electrons and Pt gains $0.13e^-$ on the $(\sqrt{3} \times \sqrt{3})R30°$

surface [50]. On the p(2 × 2) structure, the electron density of the Sn atoms decreases by $0.53e^-$, while the Pt atoms gain only $0.10e^-$. Therefore, the Pt atoms on the surface alloys are more negatively charged than on pure platinum. Confirmation was obtained from the photoelectron studies by Rodriguez *et al.*, who concluded from their analysis of the Pt–Sn bonds in the $Pt_2Sn(111)$ surface alloy a Sn(5s,5p) to Pt(6s,6p) charge transfer and an increased electrophilic character of Sn [52].

## 40.3 Selected Case Studies

### 40.3.1 The Simplest Linear Mono-Olefins: Ethene

#### 40.3.1.1 TPD, TPRS, and Reaction Studies

Ethene ($C_2H_4$) is the simplest probe molecule with an olefinic C=C double bond, and is used for developing a fundamental understanding of its bonding to surfaces and its reactivity in many processes such as hydrogenation, decomposition, oxidation, or epoxidation. It has been intensely studied on Pt-based single-crystal model catalysts, especially over the past five decades, and a broad base of results is available today, which turns this molecule into a prime candidate for case studies.

The hydrogenation of ethene to ethane is the simplest model reaction for hydrogenation processes of olefinic C–C bonds. The role metals can play in this reaction was discovered by Senderens and Sabatier in 1897 [54], and has since been studied under well-defined conditions on Pt single crystals in ultrahigh vacuum (UHV) as well as under elevated pressures with supported catalysts [55–62]. Despite such research efforts, different opinions still exist on the exact mechanism of the hydrogenation. This is particularly complicated due to the effects exerted by the support and the role of potential carbonaceous deposits such as ethylidyne, which can act as spectator species or poison catalysts, or even assume the role of hydrogen transfer agents or intermediates [55–59].

The mechanisms proposed originally by Horiuti and Polanyi [17] starts with ethylene bonding in the di-σ(CC) configuration. Hydrogen atoms are added stepwise from the surface, involving the formation of ethyl ($-C_2H_5$) intermediates, which was confirmed by Zaera with TPRS (Temperature Programmed Reaction Spectroscopy) and IR reflection absorption spectroscopy (IRAS) [57]. Moreover, Zaera found that the formation of the ethyl intermediates was rate-limiting under UHV conditions. This mechanism, however, has often been questioned in the past decades [59, 62, 63]. Other mechanisms, including concerted steps [10], direct reactions of ethene with molecular hydrogen [64], or a different active ethene configuration [57, 59, 62, 63], have been proposed.

Studies at elevated pressures of up to 1 bar (760 Torr) ethene employing sum-frequency generation (SFG), which identified both the di-σ(CC) as well as the π(CC) structures from the C–H stretching vibration patterns, suggested that the π form was likely the actual surface species entering the hydrogenation pathway [62]. This might indicate a pressure gap between the observation of H–D exchange [57, 65], hydrogenation, and self-hydrogenation [55, 57, 62] under UHV and in high pressures environments.

Also, molecular beam experiments by Ofner and Zaera [55] produced further evidence that weakly bonded ethene such as $\pi$(CC) structures or even a closely related $\pi$-like precursor state to chemisorption could be essential for the hydrogenation activity.

Hence, one key question for a proper understanding of the mechanism and modeling the reaction concerns the form of adsorbed ethene that undergoes hydrogenation, that is, the bonding of the molecule at the starting point of the pathway. The active species is not necessarily the most stable or abundant adsorption form, although the more strongly adsorbed di-$\sigma$ mode in case of olefins is often argued to be better activated than the weakly bonded ones. The precursor participating in the reaction can also be a metastable minority species, which may be hard to detect using surface science techniques if their lifetime is too short. Clearly, understanding the bonding of the reactants is a prerequisite to developing simple mechanistic insights for any surface-mediated reaction.

Ethene is usually found to chemisorb in the di-$\sigma$(CC) or $\pi$(CC) configuration to different surface sites, while also weakly bonded (physisorbed) multilayers and adsorption-precursor states have been reported [49, 55, 57, 59, 65–71]. The $\pi$-coordinated ethene binds to the metal atoms in a coordination geometry that can best be described as an $sp^2$ hybridization in the valence bond (VB) picture, and the di-$\sigma$ structures are thought to bond with $sp^3$-hybridized C atoms, commonly in twofold bridging sites [66–70, 72, 73]. However, also bonding in threefold and fourfold hollow sites has been reported (see Ref. [66] and references therein). The adsorption energy can vary by an order of magnitude depending on the chemical character of the metal atoms, the use of dopants, or the oxidation of the surface [66, 67, 69, 70].

While Ni and Pt show a preference for the di-$\sigma$ form, on Pd the $\pi$ complex is formed in larger proportions [66]. On Cu, with a significantly smaller lattice constant and different electronic properties, ethylene basically physisorbs only in a flat, unperturbed geometry with $\pi$ coordination to the surface [66]. Analogous to the bonding of CO [27], the di-$\sigma$ and $\pi$ complexes may be considered to differ in the degree of back-bonding from filled d orbitals in the metal into the vacant $\pi^*$ antibonding molecular orbital (MO) of the ethylene molecule according to the Dewar–Chad–Duncason model [24–26, 66]. This back-donation is substantial for the di-$\sigma$ configurations (as far as still applicable to complexes with bond order near unity, showing only bonds of $\sigma$ symmetry), but for the much weaker interacting $\pi$ complexes it plays a much smaller role and in many instances can be negligible as judged, for example, from the molecular vibrational frequencies [66].

Thermal desorption spectroscopy (TDS) and TPRS studies on Pt and bimetallic Pt alloys have provided further insights into the adsorption process and decomposition of ethene, providing a foundation for a correlation with spectroscopic and theoretical data [49, 57, 65, 67, 74–76]. A few case studies shall be highlighted here in order to introduce the interaction of ethene with the chosen transition-metal surfaces.

On Pt(111), ethene adsorbs in several different chemical states. A multilayer, similar to condensation, forms upon adsorption of large doses below 37 K [67]. For doses below monolayer coverages, it can reversibly and weakly adsorb below 52 K [59, 67], while at higher temperatures the molecules also chemisorb irreversibly to the surface [75]. The adsorption state formed at 52 K is similar to a physisorbed or precursor-mediated state that grows at low coverages for temperatures above 200 K, as evidenced by studies of the adsorption kinetics using the dynamic molecular beam method [55], IRAS [59], and HREELS (high-resolution electron energy loss spectroscopy) measurements [36, 74, 76–78]. Only ~20% of an adsorbed ethylene monolayer desorbs upon annealing, a part of which is hydrogenated to ethane in the process [57]. The majority of the molecules populating this irreversibly adsorbed state undergo decomposition to ethylidyne and further to carbon-rich products at elevated temperature [74, 77–83] since the activation energy for desorption exceeds that for the initial decomposition step (see further Ref. [57]).

A comparison of the thermal desorption spectra of ethylene monolayers on Pt(111) and the two $Pt_3Sn(111)$ and $Pt_2Sn(111)$ surface alloys is shown in Figure 40.2 [75]. The desorption of reversibly chemisorbed ethylene on Pt(111) exhibits first-order kinetics and a maximum rate that is measured around $\sim 280 \pm 10$ K [65, 75, 76], with an additional smaller shoulder amounting to roughly 25% of the peak area developing at ~240–250 K for medium doses [57, 75, 76]. The desorption maximum shifts to lower temperatures with increasing tin content in the surface, that is, to ~230–240 K on $Pt_3Sn(111)$ and ~185 K on $Pt_2Sn(111)$ [49, 75, 76]. The peak areas and, hence, the monolayer coverages of ethane appear to be very similar on all three surfaces [49, 75, 76]. The shift to lower temperatures indicates a significant weakening of the bonding by the alloying effects. Using Redhead analysis [84] assuming a frequency prefactor of $10^{13}$ Hz, desorption activation energies of 70–75 kJ/mol are obtained on Pt(111), which decrease stepwise by ~20% due to alloying with Sn to ~59 kJ/mol on $Pt_3Sn(111)$ and ~47 kJ/mol on $Pt_2Sn(111)$ [49, 65, 75, 76].

The decomposition of ethylene on Pt(111) starts above 240 K [57, 65, 74–76, 79–83, 85] and proceeds through a series of elementary steps that include dehydrogenation and rehydrogenation mechanisms at lower temperatures, while at elevated temperature also C–C dissociation occurs. It leads to desorption of molecular $H_2$ in a series of rather well-defined peaks (Figure 40.3a,b) related to the formation of ethylidyne ($\equiv C{-}CH_3$) [79–83, 85, 86] between ~240 and 365 K and hydrogen-poorer carbonaceous $C_xH_y$ surface species at higher temperatures [57, 65, 75, 76]. A desorption-limited recombinative $H_2$ peak of second order is measured at ~300 K close to the desorption temperature found for pure dissociative $H_2$ adsorption on Pt(111) at lower hydrogen coverages [51]. This fraction of hydrogen is attributed to a C–H dissociation during the transformation of ethylene to ethylidyne, which corresponds to a loss of 25% of the C–H bonds. The Pt catalyst is found to lower the bond activation energy from a value of $D_{300}(C{-}H) = 111.2$ kcal/mol [87] strongly [75]. Indirect proof of the formation of surface-adsorbed hydrogen atoms in this temperature range is obtained in the form of a small fraction of self-hydrogenation of ethene to ethane, which is found to evolve at ~280 K during

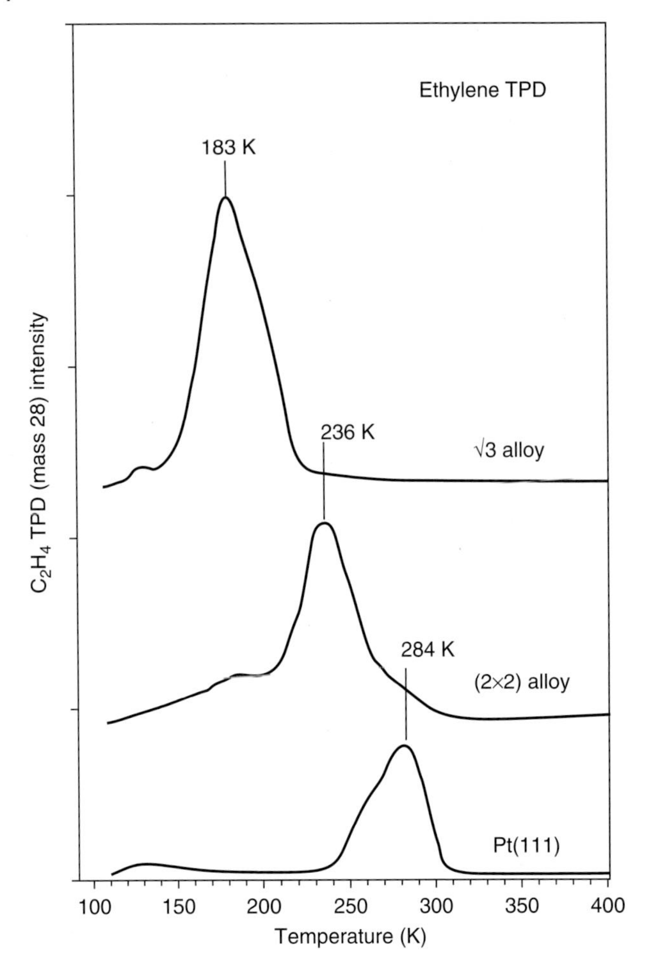


**Figure 40.2** TPD spectra of molecular ethylene desorption from Pt(111) and the $Pt_3Sn(111)$-p(2 × 2) and $Pt_2Sn(111)$-($\sqrt{3} \times \sqrt{3}$)R30° surface alloys after adsorption of a monolayer at 100 K. (Image reproduced from Ref. [75].)

the TPRS experiments [57, 76] as well as extensive H–D studies performed by Salmeron and Somorjai [65].

Between 490 and 500 K [57, 65, 75, 76], a second desorption state is measured, which leads to a surface species with a stoichiometry of ~$C_2H_1$ [75]. Further $H_2$ desorption occurs in a weaker and broader peak around 640 K [57, 75]. This decomposition process is almost entirely suppressed on the Pt–Sn surface alloys. Only small $H_2$ desorption signals can be discerned typically at the same temperatures as observed on Pt(111), which suggests that the alloying process left tiny patches of pure Pt or other defects, still exhibiting sufficiently low reaction barriers for the necessary transformation steps to ethylidyne [75, 76]. This is likely due to a combination of lowering the desorption activation energy and increasing the barriers of

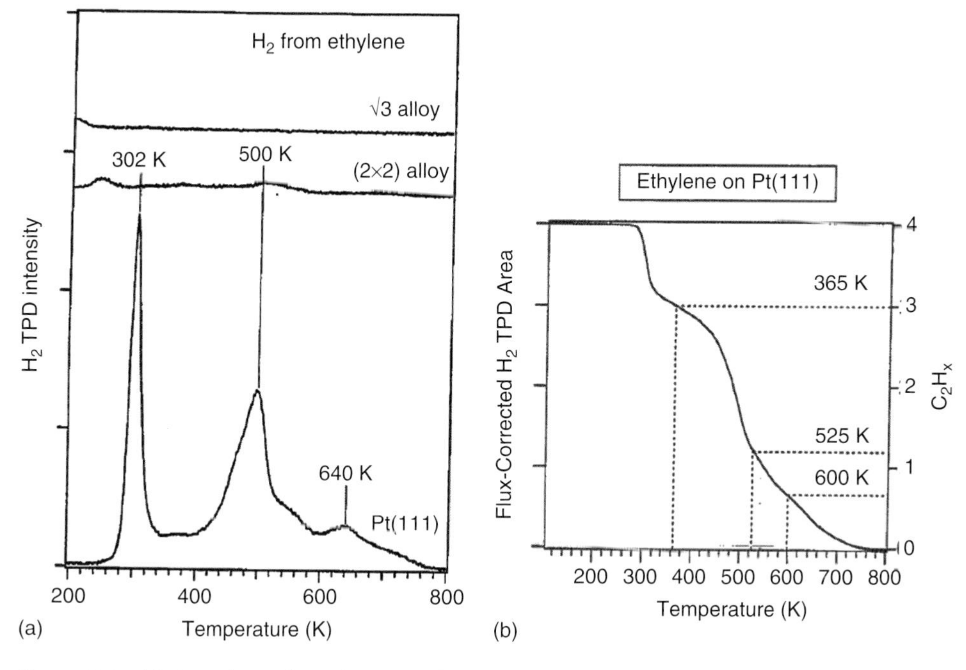


**Figure 40.3** (a) $H_2$ evolution from decomposition of ethylene monolayers adsorbed at 100 K on Pt(111) and the $Pt_3Sn(111)$-p(2 × 2) and $Pt_2Sn(111)$-($\sqrt{3} \times \sqrt{3}$)R30° surface alloys. (b) Stoichiometry of the resulting hydrocarbon surface species computed from the $H_2$ desorption peak areas after correction for kinetic energy of the molecules assuming that only negligible amounts of hydrogen are left after annealing to 800 K. (Images reproduced from Ref. [75]; heating rate $\beta = 4$ K/s.)

the elementary steps of the decomposition due to the changed surface ensemble by alloying, although a qualitative alternation or even elimination of a corresponding reaction pathway cannot be entirely excluded. Clearly, an ensemble of two directly adjacent threefold hollow sites of pure Pt atoms allows a dehydrogenation and decomposition pathway with surmountable reaction barriers, while eliminating this reaction site by incorporation of Sn changes the pathway selectivity dramatically toward molecular desorption.

#### 40.3.1.2 Ethene Bonding on Pt(111), $Pt_3Sn(111)$, and $Pt_2Sn(111)$: Spectroscopy and Microscopy

Insights into the bonding of the different adsorbed ethene species on Pt-based model catalysts have been derived from a number of experimental techniques including XANES/NEXAFS (X-ray Adsorption Near Edge Structure)/(near-edge X-ray absorption fine structure), UPS, vibrational spectroscopies such as IRAS and HREELS, STM, as well as from theoretical studies. Although the majority of the studies have been carried out in vacuum, the fundamental findings are usually transferable to bonding under higher pressures.

Initial studies performed by Koestener *et al.* using NEXAFS to investigate the bonding of ethene on Pt(111) at 90 K and the formation of ethylidyne at 300 K were interpreted as adsorption in a pure π configuration residing flat on the surface, which after injection of thermal energy turned into upright ethylidyne [88]. Surprisingly, these authors found C–C bond distances that corresponded nominally to single bonds for both the adsorbate geometry (1.48 Å) and ethylidyne (1.45 Å). In the light of newer results showing the di-σ(CC) configuration being dominant at 90 K, it appears that these authors did in fact characterize the latter structure.

The formation of π(CC) ethene at 52 K on Pt(111) and its conversion to ethylidyne upon annealing to 300 K, leading to concomitant appearance of di-σ(CC) ethene, were studied with UPS by Hugenschmidt *et al.* [67] (Figure 40.4). These authors found only marginal changes of the MOs involved in the surface bonding from gas-phase or multilayer ethene, consistent with a minimal distortion of the molecule in this configuration. The latter, interpreted in the VB picture rather than the MO scheme, indicates that no rehybridization from the $sp^2$ hybrid orbitals in flat ethene has occurred. These findings were later also confirmed by further UPS [89] and NEXAFS studies [90] by Cassuto *et al.*, which suggested a flat adsorption geometry at 52 K with only a negligible elongation of the C=C bond, thus retaining its C=C bond order [90]. The latter is also indicative of a negligible back-donation contribution to the π* MO.

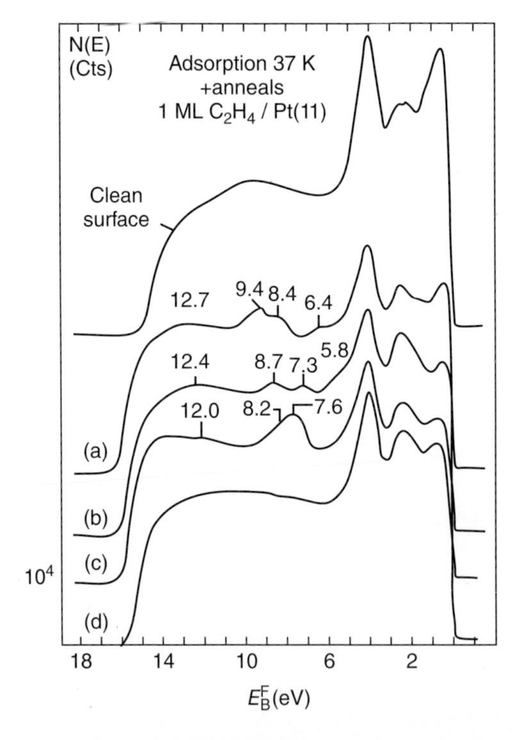


**Figure 40.4** UPS data recorded during annealing experiments of 1 ML ethene on Pt(111) to 37 K (a), 52 K (b), 300 K (c), and 500 K (d). (Taken from Hugenschmidt *et al.* [67].)

Alloying effects on ethene chemistry due to Sn have been studied by Paffett *et al.*, who analyzed photoelectron and vibrational HREEL spectra at low temperatures and concluded that significant changes in the electronic properties of the Pt atoms at the bonding sites lead to weakening of the adsorption compared to Pt(111) and contribute to the suppression of the dehydrogenation reactions [49]. However, these authors considered the electronic effects as secondary, whereas the changes in the atomic ensembles on the two alloy surfaces were more crucial, likely leading to much increased activation barriers for C–H bond breaking.

Besides photoelectron-based techniques, especially vibrational spectroscopies have proven useful to characterize the interaction of ethene with Pt(111) and the Pt–Sn surface alloys. Several studies by Ibach and coworkers using HREELS were performed in the late 1970s and early 1980s in order to characterize the adsorption and decomposition of ethene on Pt(111) [74, 77, 78]. The vibrational fingerprints and particularly the identification of $\nu$(C–C) or $\nu$(Pt–C) stretching modes and associated shifts have been used as indirect information to derive the bonding modes and strengths of hydrocarbon molecules and intermediates on surfaces. Between 140 and 260 K, ethylene was observed to form di-$\sigma$(CC) structures before undergoing transformation to ethylidyne and other carbonaceous intermediates [74, 77, 78].

Complementary IRAS studies performed by Kubota *et al.* [59] and Zaera [57] detected evidence of $\pi$-ethene formation at low and high pressures. On a Pt(111) crystal exposed to $10^{-3}$ Torr ethene at 112 K, IR spectroscopy showed vibrational signals corresponding to the formation of weakly $\pi$-bonded molecules on a surface saturated with di-$\sigma$(CC) ethene. Using the Clausius–Clapeyron equation and integrated IR peak areas assumed to be linearly proportional to the coverage, the heat of adsorption $\Delta H_{ads}$ of $\pi$-ethene under these conditions was estimated at $40 \pm 10$ kJ/mol [59]. This value is lower than most experimental values found for adsorption in a di-$\sigma$ configuration, which range between 46 and 77 kJ/mol [65, 86, 91].

Under conditions of even higher pressures and temperatures, namely in a mixture of 35 Torr $C_2H_4$ and 100 Torr $H_2$ at 295 K, only very small amounts of $\pi$-ethene were measured on Pt(111) besides large fractions of di-$\sigma$ bonded molecules and ethylidyne [62]. These studies were reported by Cremer *et al.* using SFG measurements of the $\nu$(C–H) stretching region. Although only 4% of the detected ethene species were estimated to be bonded in $\pi$ configurations, they were found to be the key species undergoing hydrogenation.

STM recently produced spectacular evidence for the coexistence of $\pi$- and di-$\sigma$ ethene on Pt(111) at lower temperatures. Okada *et al.* imaged both forms using low-temperature STM after cooling down from an adsorption temperature of 50 to 4.7 K (Figure 40.5) [92]. The form with the greater apparent height was attributed to $\pi$-bonded ethylene at on-top Pt sites, whereas the di-$\sigma$ bonded ethylene occupied symmetric bridging sites. Under the influence of slight potential changes from the tip ($\Delta U \approx 0.1$ V), these authors were able to interconvert the two configurations by STM manipulations, indicating that such a process might be facile with the thermal energy available closer to room temperature. Similar observations were made on Pd(110).

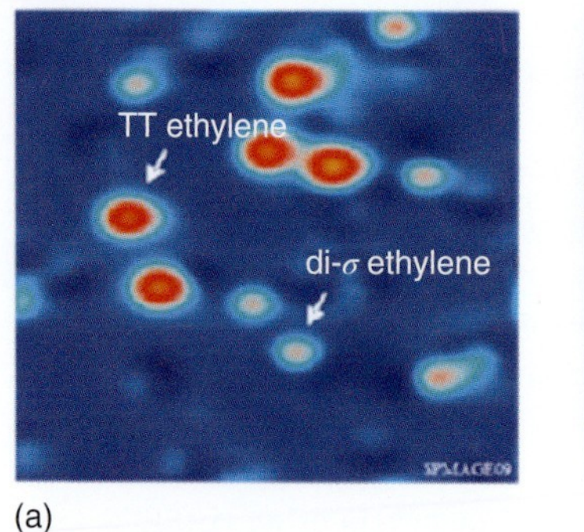


(a)

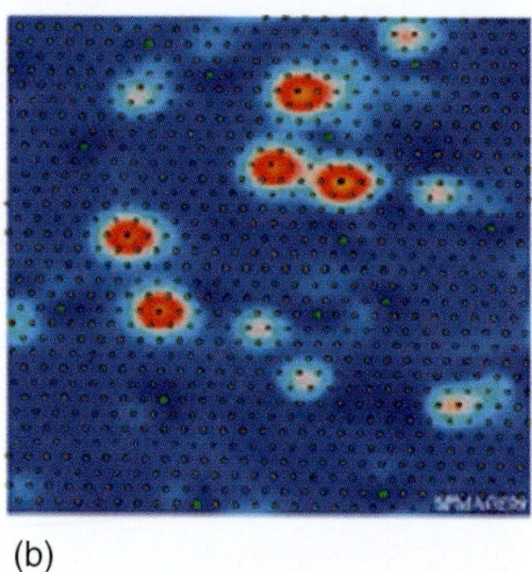

(b)

**Figure 40.5** (a) STM images obtained at 4.7 K showing two forms of ethylene on the Pt(111) surface following exposure at a surface temperature of 50 K to $C_2D_4$ ($5 \times 5\,nm^2$, $V_{Sample} = 0.1$ V, 1.0 nA). (b) Same image as in part (a) but with a superimposed grid of Pt atom positions based on the known adsorption sites for acetylene, which are visible in both images as faint spots. Pt atoms next to the acetylene molecules are highlighted in green. (Images from Ref. [92].)

#### 40.3.1.3 Theoretical Studies of Ethene Adsorption on Pt-Based Surfaces

A number of detailed theoretical studies on the adsorption and reactivity of ethene on Pt(111), $Pt_3Sn$, and $Pt_2Sn$ surfaces have been published in recent years, shedding more light on the formation and properties of different ethene surface complexes [36, 93–101].

Early studies consisted of extended Hückel (EH) and tight binding (TB) calculations of the two coordination modes of ethene and several reaction intermediates including ethylidyne on Pt(111) [98–101]. These studies provided clear support for the preference of a di-σ-bonded species, and developed a good understanding of the orbitals and electronic effects involved in the bonding. Furthering these insights, newer theoretical studies relying largely on DFT have been reported since the millennium.

DFT studies by Miura *et al.* (BLYP functional) found di-σ ethene in symmetric bridge sites to be most stable with an adsorption energy ($E_{ads}$) of 95 kJ/mol on the (111) surface of a $Pt_7$ cluster (see Figure 40.6c). A π configuration with the ethene C–C axis oriented along a Pt–Pt–Pt direction exhibited only $E_{ads} = 25$ kJ/mol (see Figure 40.6b) [94]. The di-σ form also remained more competitive on hydrogen-covered surfaces. Shen *et al.* obtained values of 116 and 71 kJ/mol for the di-σ and π forms, respectively, on a larger $Pt_{19}$ cluster [93].

Several intermediary adsorption sites over threefold hollow sites were considered by Miura *et al.*, of which only one "triangular" structure, with two σ-type coordinations from one carbon to two adjacent Pt atoms and a bond from the remaining carbon to the third Pt atom of the site, was stable with an energy of 53 kJ/mol (see Figure 40.6e,f) [94]. However, no vibrational analysis was carried out for the structures to ensure that real local minima had been obtained. Further, the cluster surface was frozen and not allowed to relax during the formation of the adsorption complex.

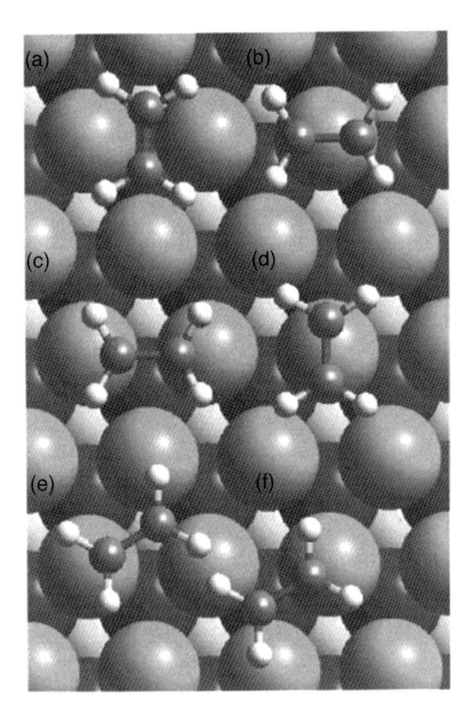


**Figure 40.6** Adsorption configurations considered by Watson *et al.* in slab calculations for ethene on Pt(111). π-Ethene across bridge (a, $E_{ads} = -9$ kJ/mol) and atop sites (b, C–C axis along Pt–Pt–Pt direction, $E_{ads} = -86$ kJ/mol), di-σ in symmetric bridge site (d, $E_{ads} = -128$ kJ/mol), π-ethene atop site with C–C axis orthogonal to unit cell axis (c, "atop hollow," $E_{ads} = -85$ kJ/mol), triangular structure in fcc and hcp hollow sites (e, $E_{ads} = -75$ kJ/mol and f, $E_{ads} = -71$ kJ/mol). Unfortunately, no vibrational analysis was performed for the obtained structures.(Image taken from Ref. [95].)

Besides adsorption, the hydrogenation pathways with stepwise hydrogen addition as proposed by Horiuti and Polanyi were studied by the authors. They found that the activation energy for the second H addition to an ethyl intermediate, which is formed from both ethene adsorption configurations in a first step, might be rate-limiting with a value of ~111 kJ/mol. While the first step leading to ethyl intermediates was more favorable for di-σ starting structures on the clean surface Pt(111), the authors concluded that coadsorbed species such as hydrogen and carbon deposits will adjust the energy balance in a way that also the hydrogenation pathway starting from the π form will play a role under catalytic conditions.

Although very useful, such studies on cluster models must be treated with caution as long as the cluster size is small compared to extended surface models, as observed by Watson and coworkers [95]. Their DFT slab calculations using the PW91 functional of the generalized gradient approximation (GGA) flavor focused on similar possible adsorption configurations on Pt(111) (Figure 40.6) and found adsorption energies of 128 and 86 kJ/mol for the di-σ and π configurations, respectively.

Importantly, Watson *et al.* analyzed the surface relaxations (Figure 40.7) at the bonding site and considered the additional gain in adsorption energy resulting from the outward movement of the Pt atoms [95], effectively increasing the bond strengths and interactions between the carbon and the metal atoms. In the di-σ configuration, a displacement by 0.235 Å out of the surface plane in the relaxed

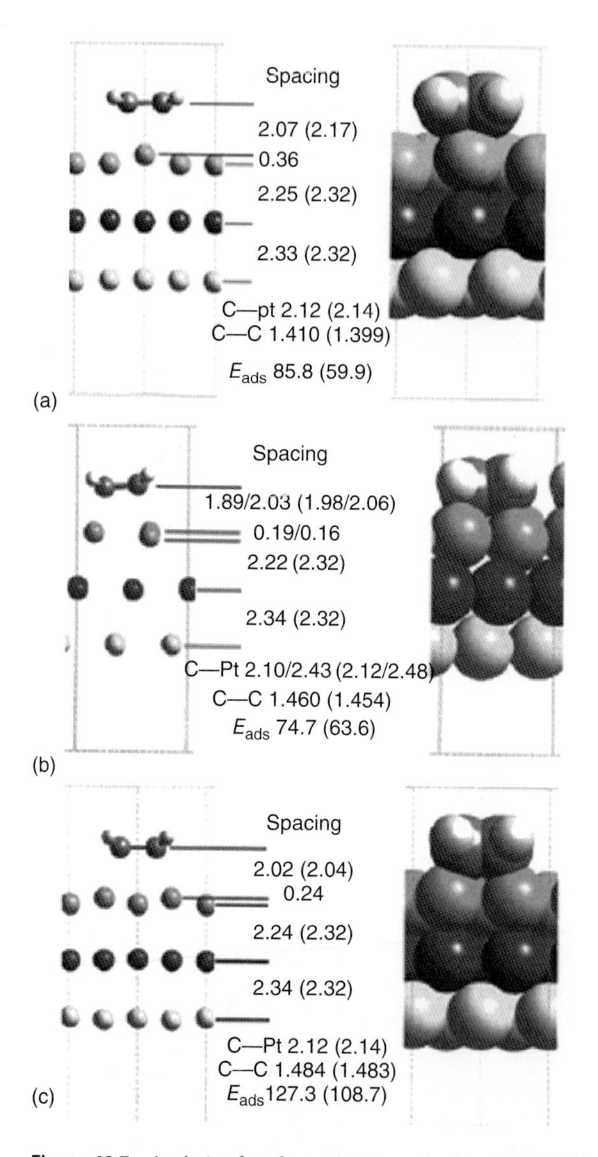


**Figure 40.7** Analysis of surface relaxations (in Å) for ethene on fully relaxed Pt(111) slab models with three-layer thickness. (a) Atop site (see Figure 40.4b), (b) fcc hollow site, and (c) bridge site. Values in parentheses show the corresponding distances with fixed surfaces, respectively. (Reproduced from Ref. [95].)

adsorption complex amounted to a fraction of ~19 kJ/mol of the adsorption energy for the two bonds, while for the π configuration an even larger contribution of 26 kJ/mol was found because of an almost 60% larger relaxation of the single Pt atom of 0.356 Å [95]. These findings highlight the well-known critical role of surface relaxation on the bonding.

Mittendorfer *et al.* expanded the understanding of the adsorption of various unsaturated hydrocarbons including ethene, 1-butene, and 1,3-butadiene on Pt(111) and Pd(111) slab models by computing "ab initio thermodynamic phase diagrams" from the Gibbs free energy of various coverages and superstructures as a function of temperature and pressure (Figure 40.8) [102]. This enabled a ranking of the relative stability of the adsorbates as a function of pressure and temperature. Moreover, these authors analyzed several key properties affected by the adsorption on slab models (DFT-PW91), that is, geometric, electronic (work function), energetic, and vibrational parameters, in order to gain new insights into the adsorption strength of the various probe molecules selected in their work. Consistent with Watson *et al.* [95] or Miura *et al.* [94], they concluded that the mono-olefins bind preferentially in a di-σ(CC) configuration above a bridge site.

The first theoretical study (to the knowledge of the authors) transiting to Pt–Sn alloy surfaces was reported in 2004 by Fearon and Watson [97], who computed the di-σ ethene configurations on Pt(111) and ultrathin $Pt_3Sn(111)$ and $Pt_2Sn(111)$ surfaces modeled over four-layer-thick Pt(111) bulk slabs. The calculations were carried out using DFT-PW91, and the surfaces were kept frozen during the optimizations. At constant coverages, the effect of tin on the electronic properties was found to be small, but the adsorption energies were strongly reduced. On Pt(111), adsorption energies of 109 and 86 kJ/mol were obtained for p(2 × 2) and ($\sqrt{3} \times \sqrt{3}$)R30° ethene

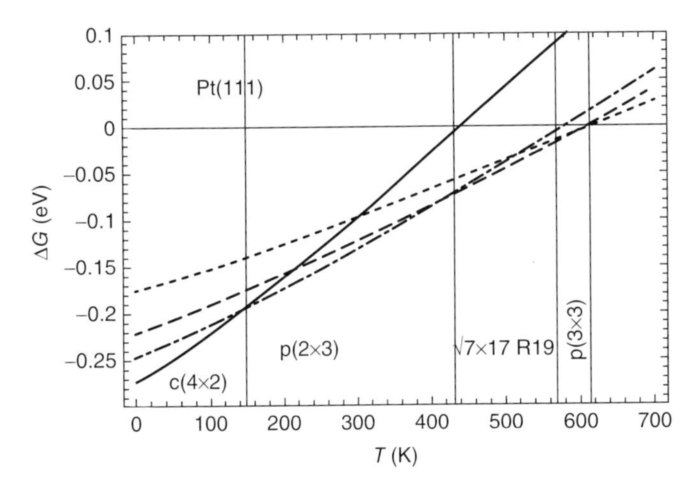


**Figure 40.8** Computed phase diagram of the Gibbs free energy of several ethene adsorption superstructures on Pt(111) as a function of temperature. While a c(4 × 2) superstructure is thermodynamically most stable at low temperatures, transitions to phases with different ordering occur upon temperature increase. (Taken from Ref. [102].)

superstructures [97]. As can be expected, the stronger lateral interactions at higher coverages already lead to a sizable reduction in adsorption energies. In a $Pt_3Sn(111)$-p($2\times2$) unit cell, the alloying reduces $E_{ads}$ of the di-$\sigma$ structure by 25–84 kJ/mol. On the $Pt_2Sn(111)$-($\sqrt{3}\times\sqrt{3}$)R30° surfaces, the reduction was slightly larger with 26 kJ/mol, giving a total adsorption energy of 60 kJ/mol.

Much insight has been gleaned on the bonding of ethene; also, the pathways for its hydrogenation and dehydrogenation have been investigated with first-principles approaches on the various Pd [103–105] and Pt [106–109] surface terminations with the emergence of more powerful computer architectures in the past two decades. Since C–H bond formation and dissociation mechanisms are closely related and can occur in hydrogenation and dehydrogenation processes, a number of research efforts were directed at exhaustive studies of intermediates and elementary reaction steps that hydrocarbon species can undergo. We will briefly review a selection of particularly insightful contributions to highlight the thermal chemistry of ethene decomposition and hydrogenation reactions.

Watwe *et al.* studied the stabilities of a large set of C2-surface intermediates including ethyl, vinyl, vinylidene, and ethylidine on Pt(111) that can be encountered, for example, in de- and hydrogenation of ethane, ethene, or acetylene, using DFT computation on small cluster models [106]. Particularly, molecular relaxation of the C–H groups was found to be important for the stability, contributing as much as 100 kJ/mol through distance and angular distortions. The bond energies of these fragments, calculated by VB or average bond energies (from gas-phase enthalpies), were correlated with the energetic position of the d-band center according to the model of Nørskov and Hammer [110]. Generally the d band was computed to be sharper and higher in energy for small coordination numbers, indicating weaker interactions [106].

The search for the intermediates in ethane, ethene, and acetylene transformations was taken a step further by Chen *et al.*, who recently published the first exhaustive study of the reaction energetics and activation barriers connecting all these stationary points on Pt(111) and Pt(211) [107]. Besides confirming the high stability of ethylidyne moieties on Pt(111) due to high activation barriers for subsequent decomposition reactions, these authors found that the barrier for C–H dissociation or hydrogenation steps was substantially smaller (roughly 20–130 kJ/mol) compared to that for internal isomerization mechanisms (1,2-H shifts, e.g., from $CH_2$–CH to $CH_3$–C species; ~190 kJ/mol), which are not surface-mediated and very similar to the gas-phase processes [107]. Also, the activation of various species for C–C bond cleavage exhibited significantly larger barriers (~85–215 kJ/mol), indicating that de- and rehydrogenation steps should occur at much lower temperatures than C–C bond cleavage in pure hydrocarbons, consistent with the experiments mentioned before. Interestingly, the correlation of activation barriers to reaction enthalpies, giving BEP relationships [31, 32], showed much better linear dependence for C–C–C bond dissociations, while for the elementary steps involving hydrogen the spread was large. Thus, for C–H mechanisms the BEP correlations require individual corrections before they can be used for predictions, as was also found earlier in a related study on various mono- and bimetallic transition-metal model catalysts [33].

The decomposition of ethylidyne was also studied by Rösch and coworkers in a series of computational works starting with Pd(111) and Pt(111) and extending toward other transition metals such as Rh and Ni [107–109]. They confirmed the results by Chen and Vlachos [107]. On Pt(111), they studied three basic mechanisms for the conversion of ethene to ethylidyne: (M1) via vinyl; (M2) via vinyl, vinylidene; and (M3) via ethyl, ethylidene to ethylidyne [108]. The two former mechanisms involving transformations to vinyl intermediates were found to be competitive. Importantly, these authors also treated coverage effects on the reaction energetics and kinetics, which primarily increased the activation barrier by up to 30 kJ/mol.

In comparative studies, this group established that the typical barriers of the hydrogenation–dehydrogenation reactions on Pt(111) with 19–92 kJ/mol were slightly lower than the corresponding elementary barriers of 25–120 kJ/mol on Pd(111) [107–109]. Generally, dehydrogenation reactions were concluded to be thermodynamically and kinetically easier on Ni(111) and Rh(111) than on Pd(111) and Pt(111) [109]. By exploring the bond energies and stabilities of the intermediates in their chosen pathway (Figure 40.9), they suggested that $C_2$ decomposes exothermically on Pd(111), Pt(111), and Rh(111), to atomic carbon, whereas on Ni(111) the formation of $C_2$, a precursor to graphene and coke, is preferred. The binding energies of species with high hydrogen content such as ethylene and vinyl were largest on Pt(111), while those with a lower hydrogen content exhibited the largest binding energies on Rh(111) and Ni(111).

Further similarities in the reactivity of ethene on other transition-metal surfaces to that on Pt(111) are obvious when exploring the literature [69, 70, 104]. A particularly instructive work employing DFT analysis of reaction pathways on Pd(111) to explore the uncertainty in the choice of the hydrogenation precursor was reported by Neurock and coworkers, and we briefly want to discuss it here. These authors performed an investigation of the Horiuti–Polanyi mechanism [17] for hydrogenation of ethene on Pd(111) cluster and slab models and found two regimes with different rate-limiting steps at low and at high coverage [104]. In the former regime, di-σ

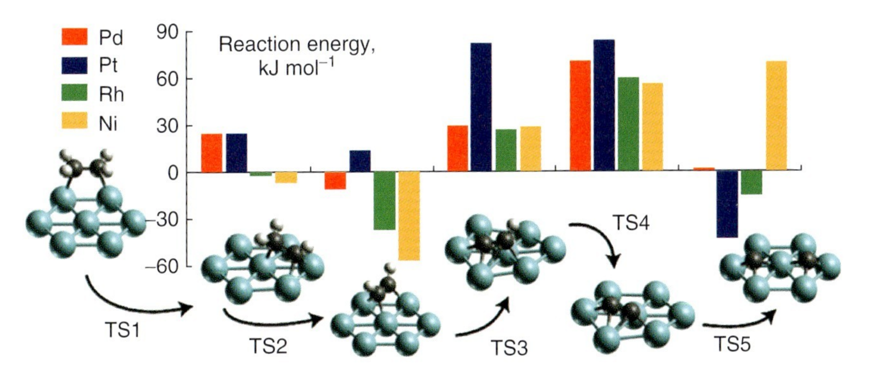


**Figure 40.9** A considered decomposition pathway for a comparative study of ethene decomposition on various transition-metal (111) facets: transformation of di-σ ethene via $C_2H_3$ (vinyl), $C_2H_2$ (di-σ/π acetylene), $C_2H$ (ethynyl), a carbon dimer to two coadsorbed carbon atoms. (Image taken from Ref. [109].)

ethene is the precursor state undergoing attack by hydrogen atoms, and π-ethene is found to convert into this adsorption configuration [104]. The apparent activation energy, calculated with respect to gas-phase ethene, was 26 kJ/mol, whereas the underlying elementary barriers of the formation of surface ethyl groups (first step) and their hydrogenation to ethane (second step) were computed at 72 and 71 kJ/mol with respect to adsorbed reactants. At high coverages, a different mechanism was concluded to be effective: because of a weakening of the M–H and M–C bonds, the activation barrier for ethyl formation from π-ethene was lowered to a value of only 36 kJ/mol, which was considerably lower than that obtained for a di-σ form under these conditions (82 kJ/mol) [104].

#### 40.3.1.4 **Assignment of di-σ and π-Ethene: HREELS Experiments and DFT Simulations**

In the following, we present two investigations that we performed to study the bonding of ethene on five different model surfaces. One goal of these works was to track the changes on the population of the π and di-σ configurations induced by alloying and ensemble effects on surfaces with strongly varying Pt content. For simple on-top adsorption, the effects are primarily of electronic nature and lead to a transition from predominantly di-σ(CC)-coordinated species on Pt(111) to π(CC)-bonded species on $Pt_2Sn(111)$ or $Pt_3Cu(111)$ surface alloys for monolayer coverages at ~100 K.

Moreover, our second goal was to further the understand the correlation between vibrational characteristics of surface species and their interaction with the surfaces, that is, the relationship between frequencies and shifts of vibrations that are directly or indirectly sensitive to the strength of the adsorption. The idea for these studies arose from the "seeming" inconsistencies that we found when analyzing this correlation between adsorption energies of given configurations of the multifunctional molecules crotonaldehyde and prenal, which are strongly decreased by alloying the Pt(111) model surfaces with Sn, and related molecular vibration frequencies. Contrary to our initial expectation, this decrease in adsorption energies did not lead to significant shifts or frequency changes of, for example, the molecule–surface stretching vibrations that ought to be most sensitive to the bonding parameters. However, these aldehydes form a variety of mixed adsorption phases with numerous different configurations – which will be discussed in Section 40.3.4 – and show a large number of vibrations and couplings, which render them too complex to serve as an easy model. Ethene, for which a closer analysis quickly confirmed similar "peculiarities," serves as a much simpler case whose understanding may then be extended to the bonding of large molecules, the indirect and direct roles exerted by alloying effects, and the correlation of vibrational properties to the details of the surface bonding. Therefore, we chose to begin with this molecule as an educational example.

Initially, we established TPRS reference data and recorded enhanced HREELS spectra for the adsorption of ethene on the Pt(111) surface. Since our temperature-programmed desorption (TPD) experiments, performed to ascertain the temperature points at which vibrational data could be most useful to identify monolayer adsorption species and decomposition intermediates, are in excellent agreement with previous literature [49, 57, 65, 75], we will only briefly point out the results that are important for further discussion. In brief, we determined the -dosages for monolayer saturation of the three surfaces, that is, the maximal doses before a multilayer desorption peak occurred, and confirmed the temperatures of the evolution of ethene reported in the previous studies. The self-hydrogenation and decomposition products ethane and $H_2$ were consistently observed only on nonalloyed Pt(111) model catalysts. Using Redhead analysis [84] and assuming first-order kinetics and a pre-exponential factor of $10^{13}$ $s^{-1}$, the activation energies of desorption from Pt(111) decreased from 74 to 71 kJ/mol for exposures equivalent to submonolayer to saturation coverage of the surface.

On the surface alloys, the desorption from ethene monolayers is shifted to lower temperature with increasing Sn content and desorption activation energies of 59 and 47 kJ/mol are obtained on $Pt_3Sn(111)$ and $Pt_2Sn(111)$, suggesting that the bonding to the surface is weakened by the alloying effects.

The trend of lowering adsorption energies with increasing Sn content of the surface alloys is reproduced by DFT computations of the stabilities of ethene adsorption structures. Since the details of the theoretical treatment we chose are important for judging the error bar and significance of the results used for our further discussions, we will briefly summarize the key parameters of our modeling of ethene, prenal, or crotonaldehyde adsorption, amongst other probe molecules [35–38].

We used the periodic DFT approach implemented in the Vienna *ab initio* Simulation Package (VASP) [111, 112], which is a plane-wave-based implementation that allows easy control of the basis set quality through an energy cutoff for the plane wave expansion. The electron–core interactions are described by the projector-augmented wave (PAW) method, which is a frozen core method that utilizes the exact valence wave functions instead of pseudo-potentials [113].

We chose the Perdew–Wang 91 functional at the GGA level [114], which performs well for structural, electronic, and energetic parameters on metal surfaces. Although hybrid functionals often give better results in gas-phase calculations and for semiconductors than the ones of GGA flavor (e.g., for "reproducing" bandgaps with standard DFT, despite it being only a ground state and not a true excited states method), they usually offer only modest improvements for metallic systems at a much increased computational cost [115]. Thus, the standard GGA functionals represent the best compromise for an efficient computation without sacrificing significant accuracy. More important in this context is the failure of all of these functionals to correctly capture the van der Waals contributions to the interactions [116]. Not only for larger molecules but also for ethene this can lead to a sizable underestimation, especially if the adsorption energies are inherently weak such as for atop $\pi$(CC) configurations on the Pt–Sn and Cu surfaces. Unfortunately, correction schemes for this have been implemented in VASP only lately for routine use and are still under further development [117].

Our surface models consist of slabs in a supercell with sufficient vacuum space above and below. The lateral dimensions of the supercell, that is, the model surface unit cell vectors, allow tuning over the "theoretical coverage" $\Theta$. Depending on the surface superstructure and desired coverage $(2\times2)$, $(3\times2)$, $(3\times3)$, and

$(\sqrt{3}\times\sqrt{3})R30°$, supercells were computed with one adsorbed molecule per unit cell ($\theta = 1/4$, 1/6, 1/9, or 1/12 ML, respectively).

Idealized (111) surfaces were constructed from four-layer metal slabs on which the adsorption can take place on only one side. For modeling the ultrathin alloy surfaces $Pt_2Sn$/Pt(111), $Pt_3Sn$/Pt(111), and $Cu_3Pt$/Pt(111), the first layer was replaced by a corresponding bimetallic composition. For the sake of simplicity, we will simplify this notation to $Pt_2Sn$(111), $Pt_3Sn$(111), and $Cu_3Pt$(111), implying surface alloys unless otherwise mentioned.

Each slab is separated from its periodic image in the $z$-direction by a vacuum corresponding in size to five metal layers (~11.5 Å). Since the molecules are adsorbed on one side of the slab only, the unit cell has a net dipole moment, resulting in "artificial" electrostatic interactions between the periodic slabs, which, for example, affect the total energy. Therefore, a dipole correction has been applied to the energy and calculated potential [118]. This correction does not exceed 30 meV for Pt and 60 meV for the alloys.

For geometry optimizations of all structures, all degrees of freedom of the adsorbed molecule and the upper two layers of the surface were included. The two lower layers of Pt below are frozen at the bulk structure parameters ($d_{NN} = 2.82$ Å, optimized from Pt bulk calculations). On the surface alloys, lattice parameters of the surface unit cells are imposed by the underlying Pt bulk, and the larger Sn atoms in the first layer have been found to evade this compression by an outward displacement of 23 and 30 pm on $Pt_2Sn$(111) and $Pt_3Sn$(111), respectively, which induces a Pt–Sn distance of 2.86 Å [42]. The adsorption energy is defined as the difference between the energy of the whole adsorption system and that of the pure slab and a gas-phase molecule.

The calculated adsorption energies for ethene (Table 40.1) confirm that the di-σ(CC) is generally the more stable form at coverages between 1/4 and 1/9 ML, ranging from −100 kJ/mol on Pt(111) to −66 and −57 kJ/mol on $Pt_3Sn$(111) and $Pt_2Sn$(111). The π(CC) structure adsorbs with sizably smaller values, namely −65, −41, and −46 kJ/mol, respectively, which indicates a decreasing gap between the relative stabilities of both configurations with increasing dilution of the Pt content of the surface. While on Pt(111) the difference amounts to 35 kJ/mol, it decreases to only 11 kJ/mol on $Pt_2Sn$(111). Therefore, with increasing Pt content an increase

**Table 40.1** Calculated adsorption energies of diσ ethene and π-bonded ethene on the Pt(111), $Pt_3Sn$/Pt(111), and $Pt_2Sn$/Pt(111) surfaces.

| | $E_{ads}$(di-σ) (kJ/mol) | | | $E_{ads}$(π) (kJ/mol) | | | $\Delta E_{ads}$ (kJ/mol) | | |
|---|---|---|---|---|---|---|---|---|---|
| | This work | Ref. [106] | Ref. [119] | This work | Ref. [106] | Ref. [119] | This work | Ref. [106] | Ref. [119] |
| Pt(111) | −100 | −117 | −101.4 | −65 | −73 | −65.4 | 35 | 44 | 36.0 |
| $Pt_3Sn$/Pt(111) | −66 | −86[a)] | — | −41 | −52[a)] | — | 25 | 34 | — |
| $Pt_2Sn$/Pt(111) | −57 | — | — | −46 | — | — | 11 | — | — |

a) $Pt_3Sn$(111) bulk alloy.

of the population of the $\pi$ state versus the di-$\sigma$ form on the surface can be expected, which is indeed found in our HREELS experiments.

The computed adsorption energies are in surprisingly good agreement with experimentally measured activation energies of desorption except on Pt(111), where only a partial desorption was measured besides the competing decomposition of a large part of the ethene monolayer. This indicates a lowering of the surface coverage by a phase transition or by desorption of weaker structures, such as, for example, the $\pi$ species, that indeed would agree much better. One has to keep in mind, however, that the DFT results correspond to adsorption at zero kelvin, without zero-point vibrational corrections or entropic contributions, whereas "desorption activation energies" are deduced from TPD that typically correspond to a kinetically controlled out-of-equilibrium process.

A comparison of our theoretical results with former studies shows good agreement as well. Not surprisingly, our DFT adsorption energies on Pt(111) correspond very well with the values found by Hirschl *et al.* using the same DFT package [119]. These values are somewhat smaller than values reported by Watwe *et al.* [106], who used only three-layer-thick metal slabs with lesser flexibility of surface relaxation.

By comparing the experimental and calculated HREELS spectra, it is possible to identify the formed ethene adsorption complexes and completely assign the measured loss peaks to their molecular vibrational modes. The HREELS experiments were performed with a primary energy of 3.0 eV in specular geometry at an angle of 60° off the surface normal, which leads to scattering predominantly by dipolar interactions and only small contributions from impact scattering. This expectation was confirmed by test measurements performed by tuning the analyzer 3° and 6° off the specular direction, which showed the elimination of the counts from the strongly forward-focused, dipolar-scattered beam and allowed the detection of weak signals reported in the literature from excitation of symmetry-allowed vibrations by impact scattering [74, 120].

On Pt(111), the HREEL spectra recorded for a saturation coverage of ethene remain virtually unchanged upon successive annealing steps up to ~215 K and recooling to 100 K (Figure 40.10). Taking the TPRS results into account and assuming that this thermal activation procedure induces only irreversible chemical transformations which are "frozen" upon recooling, this shows that in this temperature range ethene remains stable as molecular entity on the surface. All vibrational spectra of surface species in this temperature range can be assigned to a combination of the signals of di-$\sigma$(CC) and $\pi$(CC) ethene adsorption complexes, which were simulated with DFT (see the bar spectra in Figure 40.10 and Table 40.2).

For the detailed analysis, we simulated the HREELS spectra of all possible adsorption configurations by calculation of the harmonic frequencies at the $\Gamma$ point from diagonalization of a Hessian matrix [121], which was set up by perturbation of the optimized geometry along the three Cartesian coordinate axes by finite displacements of $\pm 0.025$ Å. The harmonic frequencies were obtained as the eigenvalues along with the normal modes as eigenvectors.

Only the atoms of the molecule were perturbed while the surface atoms were frozen. This is a valid approximation due to the (reduced) mass of the oscillators,

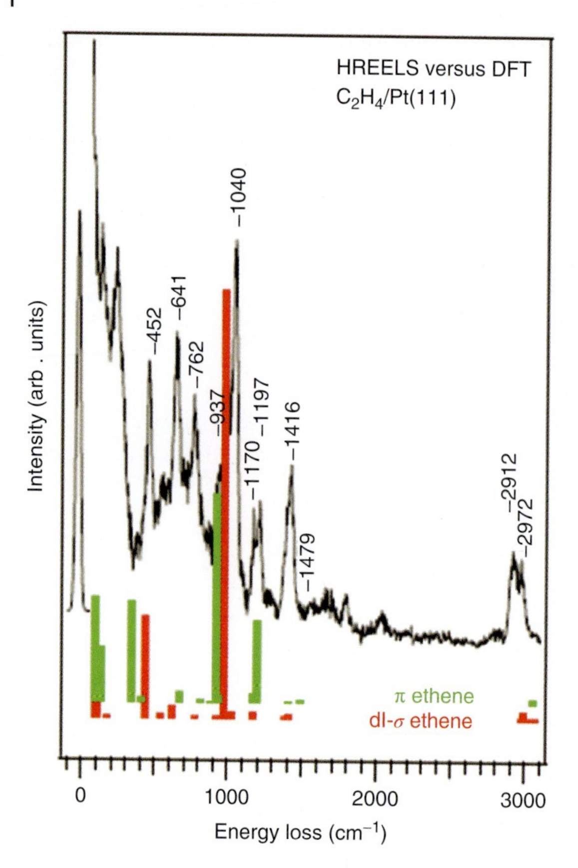


**Figure 40.10** Comparison of the experimental HREEL spectrum for a saturation coverage of ethene on Pt(111) with the calculated HREEL spectra of the corresponding di-σ- and π-bonded ethene species. (Image taken from Ref. [76].)

the Pt surface atoms being an order of magnitude heavier than the atoms of the molecule. The neglect of the surface phonons and the coupling between the latter and low-lying adsorbate vibrations lead to sizable errors only for modes below ~200 $cm^{-1}$ since the Pt metal and alloy phonons are at much lower energies. Hence, mostly frozen rotations and translational modes are affected by this. Test calculations have been performed with the surface atoms of the first layer included in the vibrational analysis, which led to shifts at the order of ~30 $cm^{-1}$ for very low lying normal modes and also to changes of the computed loss intensities.

In the C–H stretching frequency range between ~2500 and 3500 $cm^{-1}$, the strong anharmonicity of almost all vibrational modes leads to large errors and renders this range of the spectrum much less useful than the fingerprint region between ~300 and 1500 $cm^{-1}$, where the error is usually negligible for our purposes (typically much smaller than the experimental resolution of ~30 $cm^{-1}$).

**Table 40.2** Assignment of ethene and ethylidyne vibrational modes on the Pt(111) surface.

| | Normal mode | Di$\sigma$ ethene | | | $\pi$ Ethene | | Ethylidyne | | | |
|---|---|---|---|---|---|---|---|---|---|---|
| | | DFT | Experiment | Ref. [79] | DFT | Experiment | Normal mode | DFT | Experiment | Ref. [79] |
| $\nu_1$ | $\nu_{as}(CH_2)$ | 3078 | | | 3152 | | $\nu_{as}(CH_3)$ | 3016 | 2939 | 2952 |
| $\nu_2$ | $\nu_{as}(CH_2)$ | 3054 | 2972 | 3000 | 3129 | | $\nu_{as}(CH_3)$ | 3013 | | |
| $\nu_3$ | $\nu_s(CH_2)$ | 2993 | 2912 | 2919 | 3057 | | $\nu_s(CH_3)$ | 2941 | 2923 | 2888 |
| $\nu_4$ | $\nu_s(CH_2)$ | 2985 | | | 3051 | | $\delta(CH_3)$ | 1401 | 1406 | 1420 |
| $\nu_5$ | $\delta_s$ $(CH_2)$, $[\nu(CC)]$ | 1416 | 1416 | 1427 | 1477 | 1479 | $\delta(CH_3)$ | 1398 | | |
| $\nu_6$ | $\delta_{as}(CH_2)$ | 1387 | 1387 | | 1411 | | $u(CH_3)$ | 1319 | 1337 | 1347 |
| $\nu_7$ | $\rho_{as}$ $(CH_2)$ | 1177 | 1170 | | 1202 | 1197 | $\nu(CC)$ | 1099 | 1115 | 1129 |
| $\nu_8$ | $\nu(CC)$, $[\delta_s$ $(CH_2)]$ | 1036 | 1040 | 1048 | 1178 | | $\gamma(CH_3)$ | 956 | 966 | 984 |
| $\nu_9$ | $\omega_{as}$ $(CH_2)$, $[\omega_{as}(PtC)]$ | 1037 | | | 964 | | $\gamma(CH_3)$ | 955 | | |
| $\nu_{10}$ | $\omega_s$ $(CH_2)$, $\nu(CC)$ | 974 | 987 | 983 | 948 | 937 | $\nu_{as}(Pt_3C)$ | 465 | 469 | 597 |
| $\nu_{11}$ | $\rho_s$ $(CH_2)$ | 922 | 935 | 661 | 900 | | $\nu_{as}(Pt_3C)$ | 460 | | |
| $\nu_{12}$ | $\tau(CH_2)$ | 787 | 762 | 790 | 802 | | $\nu_s(Pt_3C)$ | 405 | 421 | 427 |
| $\nu_{13}$ | fR | 635 | 641 | | 677 | | fR | 162 | 251 | 307 |
| $\nu_{14}$ | $\nu_{as}(PtC)$, $[\omega_{as}(CH_2)]$ | 535 | | | 401 | | fT | 152 | 166 | |
| $\nu_{15}$ | $\nu_s(PtC)$ | 435 | 452 | | 325 | 354 | fT | 113 | | |
| $\nu_{16}$ | fT | 176 | 234 | | 100 | | | | | |
| $\nu_{17}$ | fR | 137 | 129 | | 75 | | | | | |
| $\nu_{18}$ | fT | 123 | | | 10 | | | | | |

Reproduced from Ref. [76].

The intensities of the HREEL spectra are calculated considering only the dipolar scattering mechanism by the formula given in Ref. [120]. The loss intensities are proportional to the square of the dynamic dipole moments and to a weighing function that depends on experimental parameters (primary energy, scattering geometry) and the inverse of the energy.

The energetic positions of the experimental loss peaks are reproduced well by the calculated spectra, especially in the intermediate frequency region (300–2000 $cm^{-1}$) above the surface phonons and below the C–H stretching region. The analysis of the experiments using DFT allows the verification of the presence of the proposed $\pi$-bonded ethene species unambiguously, since both the measured and calculated spectra show strong dipolar loss signals at characteristically different frequencies. Thirteen of the experimentally detected loss peaks are assigned to the corresponding normal modes of the di-$\sigma$(CC) configuration, whereas four further losses can be traced back to the $\pi$ species.

The most intensive loss features due to the di-$\sigma$ structure stem from the $\delta_s(CH_2)$ mode at ~1416 $cm^{-1}$ (which is strongly coupled to the $\nu(CC)$ mode leading to the signal at ~1040 $cm^{-1}$), the $\omega_s(CH_2)$ mode (987 $cm^{-1}$), a frustrated rotation (641 $cm^{-1}$), and finally the symmetric $\nu_s(PtC)$ molecule–surface stretching vibration (452 $cm^{-1}$). These vibrational assignments agree well with the earlier studies of the di-$\sigma$ species by Steininger *et al.* [79].

The main contribution of the π species is the characteristic double peaks at 1170 and 1197 $cm^{-1}$. In this region, the di-σ species shows only a weak loss signal at 1177 $cm^{-1}$, but an intense loss is found in the calculated spectrum of the π species at 1202 $cm^{-1}$. Similarly, the experimental peak at 937 $cm^{-1}$ can be assigned to an intense peak calculated at 948 $cm^{-1}$.

Above ~250 K, the irreversibly adsorbed ethene monolayer on Pt(111) undergoes decomposition to ethylidyne, as can be seen from the HREEL spectrum recorded after a short annealing step to 269 K (not shown further; see Ref. [76]). New loss peaks identifying these ethylidyne moieties bonded in threefold hollow sites appear, and the intensities of the previously present loss peaks of the adsorption complexes decrease. The new peaks grow and become better defined by further annealing treatment, and at ~400 K an HREEL spectrum corresponding almost solely to ethylidyne species (e.g., at 1406, 1377, 1115, 966, 469, and 421 $cm^{-1}$) is detected. After heating to 518 K, ethylidyne quickly undergoes further transformations and only two broad loss features (1026, 792 $cm^{-1}$) are present, which can be assigned to a methylidene (CH) surface species and vanish at even higher temperatures.

Likewise, on the two Pt–Sn surface alloys, the formed surface species can be identified by their vibrational properties, which reveal an increase in the relative contribution of the π-ethene species with increasing Sn fraction. This indicates a perturbation of the Pt adsorption sites, which for an on-top configuration is mostly of an indirect electronic origin ("ligand effect" according to Sachtler [48]). However, also a small geometric effect is imposed by the neighboring Sn atoms, which is reflected by the bonding distances of the Pt atoms to the adjacent metal atoms in the matrix, which indicates a release of compression by increase in Pt–Sn bond lengths as compared to the pure Pt–Pt bonds [42].

The HREEL spectra of the ethene monolayers (Figures 40.11 and 40.12), which are fully reversibly bound in case of the alloy surfaces, are very similar to the ones recorded on Pt(111). Small changes of the intensities of corresponding vibrations due to the alloying are measured and computed, but the frequencies of normal modes are surprisingly similar, especially for those sensitive to the surface bond strength given the weakening of the adsorption energies. In particular, the measured position of the $\nu_s$(Pt–C) stretching vibration (~450 $cm^{-1}$) of di-σ ethene is virtually constant on all three surfaces, which implies a very local chemical character of the adsorption bond. Analogous to the concept of group frequencies in analytic chemistry, the vibrational character of, for example, an aldehydic group with its characteristic $\nu$(C=O) stretching mode is only weakly affected by its chemical environment and the potential well determining its frequency; that is, the bonding appears to be more or less similar for all aldehydes. The same apparent contradiction was recognized previously in the experiments and calculations for prenal and crotonaldehyde adsorption, which will be discussed in detail in Section 40.3.4.

The HREEL spectra of an ethene monolayer on $Pt_3Sn(111)$ (Figure 40.11; annealed to 230 K) show a relatively dominant double peak feature at ~1165 and ~1222 $cm^{-1}$, which can be explained only by a large fraction of the π species coexisting with the di-σ form.

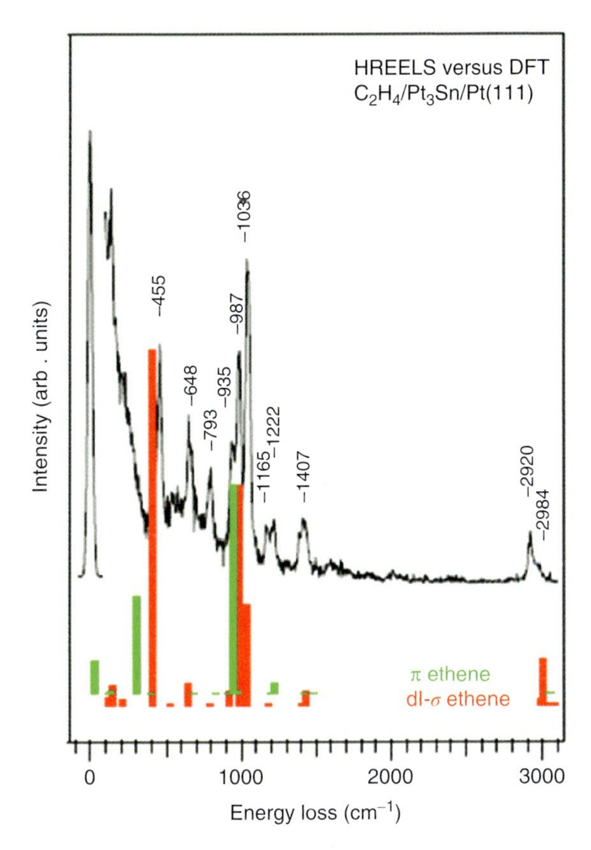


**Figure 40.11** Comparison of the experimental HREEL spectrum for a saturation coverage of ethene on the $Pt_3Sn(111)$ surface alloy with the calculated HREEL spectra of the corresponding di-σ- and π-bonded ethene species. (Image taken from Ref. [76].)

Similar features are measured on the $Pt_2Sn(111)$ surface, also suggesting a mixture of σ- and π-bonded ethene. On this Sn-richer surface, the vibrations characteristic of the on-top π configuration are even more pronounced than on $Pt_3Sn(111)$. A larger abundance of the π-bonded species is suggested by the relatively strong peaks at 1182 and 1498 $cm^{-1}$. The latter peak corresponds to the $\nu$(C=C) stretching vibration, which, contrary to the simple surface selection rules, is detectable since it shows a component of movement normal to the surface (also a very small, albeit changing, charge transfer depending on the perturbation is to be expected). Interestingly, the partial desorption of ethene measured in TPD at 163 K can be traced back with HREELS: it appears to coincide with the vanishing of the peak at 1498 $cm^{-1}$ and the weakening of the peaks at 966 and 1212 $cm^{-1}$ (Figure 40.12), which can tentatively be attributed to the desorption of a π-bonded ethene species.

Although the computed vibrational frequencies allow an unambiguous interpretation, they do not clearly reproduce the loss intensities quantitatively (Figures 40.10–40.12). This may be due to a number of reasons that are inherent in

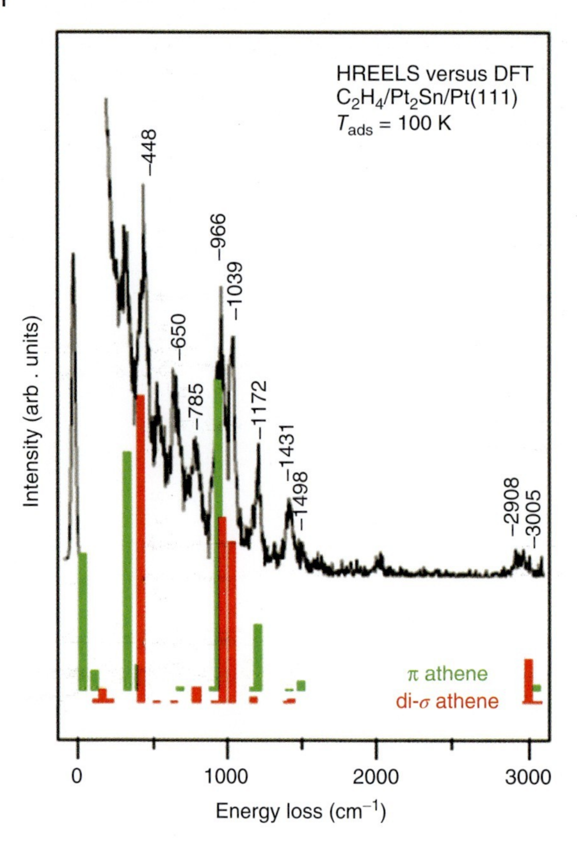


**Figure 40.12** Comparison of the experimental HREEL spectrum for saturation coverage of ethene on $Pt_2Sn/Pt(111)$ with the calculated theoretical HREEL spectra of the corresponding di-σ- and π-bonded ethene species. (Image taken from Ref. [76].)

the theoretical approach as well as in the nature of the experiments and system of study. For example, the dynamic dipole moments obtained in the DFT calculations are strongly affected by the choice of the theoretical method and the model. Especially, the number of bulk layers, the choice of surface coverage, disorder effects by "random" orientations along the high-symmetry axis of the unit cells and adsorption sites (e.g., the sixfold rotation of a molecule bonded on a Pt(111) on-top site), or mixing of different configurations play a role. However, due to the enormous computational cost, the mixing of configurations and disorder effects were not treated in the periodic model used, which replicates the same configurations and orientations of molecules and dipole moments in adjacent supercell images for "ideal" lateral coupling.

In contrast, the experiments show mixed phases, which raises the question as to the population of the states present on the surface. For the sake of simplicity, we assumed all species to be present in equal amounts, although the calculations show sizable differences in their adsorption energies. In an ideal world with unlimited computational resources, one should aim to compute a phase diagram of the total free energies by varying the (relative) coverages of each configuration on the surface, and, thus, explicitly treat (dis)ordering and lateral effects on energies, vibrational frequencies, and intensities. We found that particularly this drawback of the model leads to sizable errors for α,β-unsaturated aldehydes, where at least five configurations are detected on Pt(111) at low temperatures in UHV; when fewer structure were stable, as was the case on the Pt–Sn surfaces, the agreement was found to be much better.

Another important source of error arises from the HREELS method, which does not purely excite through the long-range dipolar scattering mechanism mediated by the oscillating electric fields of the primary electrons, although in reflecting geometries this is generally assumed to be the most dominant contribution for vibrations with large dynamic dipole moments [120]. While our theoretical analysis was restricted to dipolar scattering, experimentally, also impact scattering and, for certain chemical systems, even resonant scattering may contribute sizably.

#### 40.3.1.5 Discrepancy Between Vibrational Properties and Adsorption Energies: A Theoretical Analysis of the Bonding

At first glance, it is unexpected that the vibrational frequencies of corresponding normal modes of the two ethene adsorption configurations, di-σ and $\pi$, on the three Pt-based surface are affected very little by the alloying with Sn, especially in the light of the significant change of their activation energies of desorption determined from TPD or the adsorption energies calculated by DFT [36, 76]. Similar peculiarities had been analyzed previously for crotonaldehyde and prenal on the same surfaces [34–38].

The similarity of the vibrational properties infers a correspondingly similar C–C bond order, which can be estimated using the $\pi\sigma$ parameter introduced by Stuve and Madix [73]. This descriptor (Table 40.3) gives almost identical hybridizations for each of the two types of complexes on the three surfaces: ~0.5 for the π ($sp^2$) and ~1 for di-σ ethene ($sp^3$). Consequently, the bond order of the C–C bond of the di-σ adsorbed ethene is close to unity (such as for the reference case of $sp^3$-hybridized carbon in gas-phase ethane), whereas that of the π species is close to 1.5. Both configurations show a certain degree of rehybrization due to the new surface bonding, with the change for the di-σ configuration being naturally the largest.

**Table 40.3** $\pi\sigma$ parameter introduced by Stuve and Madix [73] for π and di-σ ethene on the three surfaces.

| | di-σ ethene | $\pi$ Ethene |
|---|---|---|
| Pt(111) | 0.97 | 0.58 |
| $Pt_3Sn$/Pt(111) | 0.99 | 0.47 |
| $Pt_2Sn$/Pt(111) | 0.94 | 0.51 |

For the evaluation the calculated energies of the vibration modes were used.

The geometric parameters of the corresponding structures of each configuration obtained from DFT are very similar on the Pt(111) and the two Pt–Sn surfaces. Notably, the *r*(C–C) and *r*(C–Pt) bond distances are almost unaffected by the presence of Sn. For the di-$\sigma$ structure, for example, a C–C bond of 149 pm and symmetric C–Pt bond lengths of ~213 pm are obtained consistently on Pt(111) $Pt_3Sn(111)$, and $Pt_2Sn(111)$ at low coverages (i.e., 1/9, 1/12, and 1/9 ML, respectively). In line with the identical bond orders and similar vibrational frequencies, this implies that also the back-bonding from the surface into the $\pi^*$ orbital of ethene is very similar in all cases, and hence not responsible for the changes in the adsorption energies.

This leaves us with the puzzle of understanding the relationship between the vibrational properties and the activation energies of desorption, and, hence, the proper description of the interaction between the molecule and the surface. Apparently, the frequencies of molecule–surface modes or the indirect shifts of other affected vibrations are not, or at least not solely, governed by the adsorption energy. Also, the similar perturbation of the geometries, that is, the C–C bond elongation (bond order) or carbon rehybridization, cannot be deduced from the changing adsorption energies.

In the simple picture of a harmonic oscillator consisting of two finite masses, the vibrational frequencies are determined by the force constant $k$ of the connecting spring, which represents the chemical bonding ($\nu \sim \sqrt{k}$). The bond energy or strength in case of a diatomic molecule is related to the vibrational frequency through the shape and depth of the potential well of the interaction (in case of the harmonic oscillator, the depth and shape are given by the force constant $k$, whereas the real Morse potential is determined by two formally independent parameters, the width a and the bond energy $D_e$). Small anharmonic corrections do not invalidate this picture [122]. Thus, the vibrational frequencies can be used as a measure of the interaction between the bonding groups.

The measure commonly used in the literature to describe the bond strength to a surface is the adsorption energy, which is usually expected to directly correlate with the vibrational properties of the adsorption complex. In an "ideal case" of a monohaptic configuration, the surface–molecule stretching vibration should provide reliable information on the shape of the potential well of the bond with the surface atom. For multihaptic adsorption configurations of larger molecules with several non-symmetry-identical bonds, the correlation of specific vibrational frequencies to an average bond strength must be considered with caution but can give nonetheless simple and useful insights. It may, however, be misleading if one part of the molecule binds significantly stronger than the other, in which case a more detailed "per bond" or "per functional group" analysis is more meaningful for understanding both the vibrational properties and the adsorption. An example for molecules of the latter kind could be long-chain mono-alkenes, for which the contribution of the covalent bonding through the C=C moiety can be easily exceeded by van der Waals bonding of the remaining saturated chain.

Now, since apparently the adsorption energy alone does not allow understanding the vibrational properties of ethene or other larger adsorbed molecules, the question arises as to which potential determines the vibrational properties and gives a measure of the interaction with the surface and what factors enter the balance that ultimately results in the adsorption energy.

In order to gain further insights into this puzzle, a series of model studies were performed π-ethene on Pt(111), the two Pt–Sn surface alloys, and two more surfaces that allow further modification of the interaction with the surface sites: the bulk $Cu_3Pt(111)$ alloy and pure Cu(111). The Pt content in the surface layer of these models ranges from 100% for Pt(111), to 75% and 66% for $Pt_3Sn(111)$ and $Pt_2Sn(111)$, to 25% for $Cu_3Pt(111)$, and 0% for pure Cu(111). Previous HREEL studies had shown that on the $Cu_3Pt(111)$ bulk alloy, π-ethene bonded to the Pt atoms isolated in the Cu matrix [123, 124] is being formed [125], while on Cu(111) a very weakly perturbed and adsorbed ethene species is being formed [126]. The adsorption energy on $Cu_3Pt(111)$ is significantly stronger compared to that on Cu(111), as judged from TPD and in the range of energies measured on $Pt_2Sn(111)$ [125]. In the theoretical analysis, the adsorption is modeled on a $Cu_3Pt(111)$ *monolayer alloy* on Cu(111), which was justified *a posteriori* by the excellent agreement between the experimental and theoretical data. On the Pt and Pt–Sn surfaces, the assignment by DFT was employed to identify the key vibrational modes of the π species in the mixed phases with the di-σ forms from the experiments.

The analysis of the vibrations of the π configurations in the HREEL spectra recorded for saturation coverages (Figure 40.13a) and the corresponding computed DFT spectra (Figure 40.13b) clearly shows that on all Pt-containing surfaces only minor frequency differences occur, whereas the spectra recorded or computed on Cu(111) differ significantly.

Further discussion is focused on key normal modes for the surface interaction such as the easily discernible symmetric $\nu_s$(Pt–C) stretching vibration. This peak is clearly visible on $Pt_2Sn(111)$ at 309 $cm^{-1}$ (39 meV), where the π species is more abundant than on Pt(111) and $Pt_3Sn(111)$. On $Cu_3Pt(111)$, a slight shift to a lower frequency of 290 $cm^{-1}$ (36 meV) suggests a weakening of the Pt–C bond. Finally, on Cu(111) only the antisymmetric metal–carbon stretch $\nu_{as}$(Cu–C) vibration can be identified in the measured spectrum at 250 $cm^{-1}$ (31 meV). This value is much lower than the corresponding $\nu_{as}$(Pt–C) values of 442 $cm^{-1}$ (51 meV) and 395 $cm^{-1}$ (49 meV) on $Pt_2Sn(111)$ and $Cu_3Pt(111)$, respectively.

The ν(C–C) stretch and $\delta_s(CH_2)$ deformation modes are coupled, and lead to two peaks that are clearly visible on $Cu_3Pt(111)$ at 1202 $cm^{-1}$ (149 meV) and 1492 $cm^{-1}$ (185 meV). These vibrations are a good indicator of the hybridization state of ethene since they appear at ~1039 $cm^{-1}$ (129 meV) and 1431 $cm^{-1}$ (176 meV) for di-σ ethene on Pt(111). Unfortunately, the exact positions of the corresponding vibrations of π-bonded ethene on Pt(111) and the Pt–Sn surface alloys are obscured, but they should be expected at around ~1200 $cm^{-1}$ (150 meV) and ~1490 $cm^{-1}$ (185 meV) for these systems, also according to DFT. On Cu(111), the related peaks are clearly visible at ~1290 $cm^{-1}$ (160 meV) and 1540 $cm^{-1}$ (191 meV), much closer to their respective gas-phase or multilayer positions.

The values deduced from the experimental and theoretical data suggest that the frequencies of the vibrations on the Pt-containing surfaces vary only slightly with respect to those of ethene adsorbed on Cu(111). Not very surprisingly,

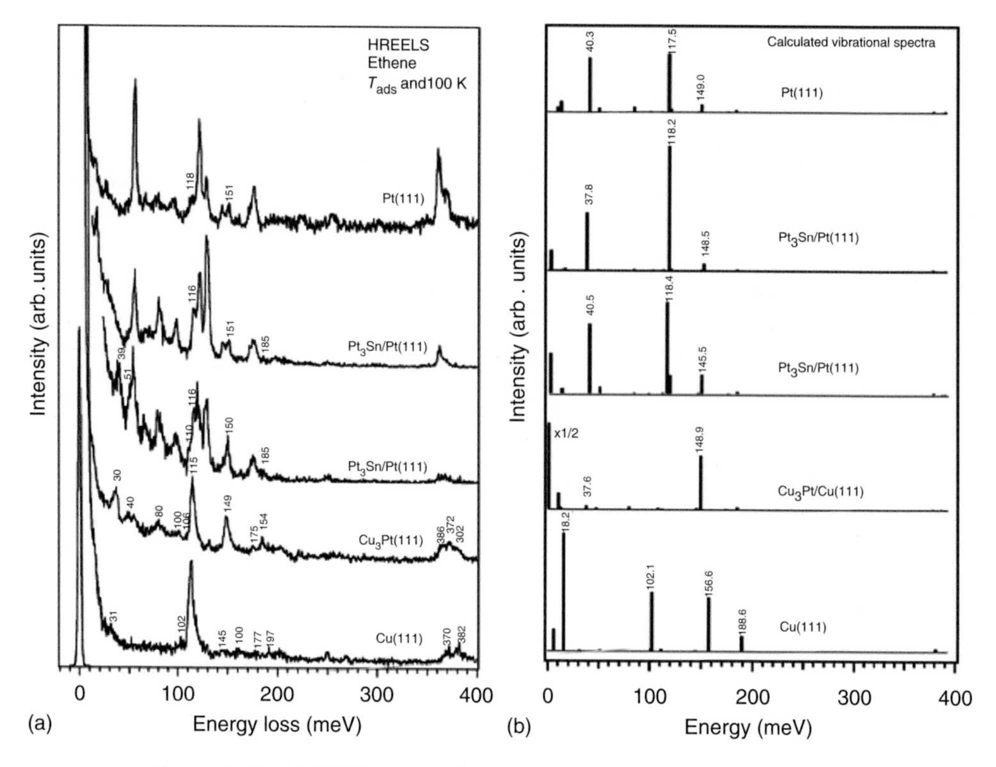


**Figure 40.13** (a) HREEL spectra of ethene monolayers adsorbed at 100 K on the five chosen model surfaces with decreasing Pt content (top to bottom). The spectra are recorded at 100 K in specular geometry with a primary energy of 3.0 eV. Note that except on the $Cu_3Pt(111)$ surface and Cu(111), always a mixed phase of di-$\sigma$ and $\pi$ configurations is formed. (b) Corresponding calculated vibrational HREEL spectra of $\pi$(C=C) adsorbed ethene. (Images taken from Ref. [36].)

also the geometry parameters of $\pi$-ethene on $Cu_3Pt(111)$ turn out to be almost identical with those on the other surfaces exhibiting Pt adsorption sites, whereas on Cu(111) the structure resembles that of gas-phase ethene [36]. The C–C bond length, for example, for $\pi$-ethene adsorbed at an on-top platinum site does not depend significantly on the surface stoichiometry, varying only minimally between 140 and 141 pm. The value of ~141 pm is significantly larger than the value for adsorption atop Cu (136 pm; gas-phase ethene 133 pm). Hence, the back-bonding on $Cu_3Pt(111)$ must be close to that on Pt(111) or the Pt–Sn surface alloys and not responsible for the adsorption energy decrease. This shows that the surface–molecule bonding is a very local phenomenon with respect to the surface site, much as group bonding in gas-phase molecules is (e.g., an aldehyde group retains the properties of an aldehyde and is only slightly affected by the chemical environment, i.e., bonding is also local with respect to the rest of the molecule).

A closer inspection of the geometries of the adsorbed π complexes reveals that in contrast to the molecule, the metal atoms of the surface site behave quite differently depending on the surface composition: the degree of outward relaxation of the bonding Pt atoms varies strongly (Figure 40.14). On Pt(111), the $\pi$(CC) configuration leads to an outward relaxation of the Pt atom by 28 pm (for the two Pt atoms of the di-σ: 18 pm), which increases to 43 pm (28 pm) and 68 pm (42 pm) in the $P_3Sn$ and $Pt_2Sn$ surface alloys. In contrast, the buckling on $Cu_3Pt(111)$ is limited to 32 pm, while on Cu(111) the surface atom below the molecule is pulled up by 25 pm despite the lack of van der Waals contribution in the functional kernel. Obviously, this effect is particularly pronounced on the Pt–Sn surface alloys, which is also observed for other configurations of the olefins or aldehydes such as prenal and crotonaldehyde [36, 38, 50, 116, 125, 127–129].

The energetic effects of the surface relaxation and the perturbation of the molecular species upon formation of the adsorption structure can be analyzed in more detail by a simple approach that allows decomposition of the calculated heat of adsorption into several components (Figure 40.15):

$$E_{int} = E_{ads} - E_{def,surf} - E_{def,mol}. \quad (40.1)$$

The adsorption $E_{ads}$ is split into three terms: the molecular deformation ($E_{def,mol}$) and surface deformation energies ($E_{def,surf}$), which account for the geometry changes of the adsorbed molecule and surface layers due to the respective rehybridization and surface relaxation upon adsorption, and the interaction energy ($E_{int}$) gained upon forming the adsorption complex from both perturbed entities. The molecular deformation energy ($E_{def,mol}$) is obtained as the energy difference between the fully relaxed gas-phase molecule and the isolated molecule frozen in the exact geometry of the adsorption structure. Likewise, the surface deformation energy ($E_{def,surf}$) is calculated by comparing the energies of the fully relaxed clean surface and the isolated, perturbed surface in the adsorption complex. These two energies have to be "invested" during the adsorption. In return, the interaction energy ($E_{int}$) is gained by the formation of the fully relaxed molecule–surface bond. Thus, the overall,

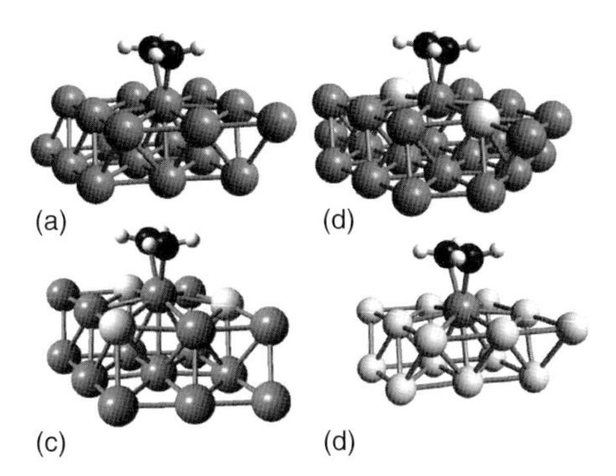


**Figure 40.14** Images of the optimized adsorption complexes of π-ethene on Pt(111) (a) and the $Pt_3Sn(111)$ (b), $Pt_2Sn(111)$ (c), and $Cu_3Pt(111)$ (d) surface alloys. (Reproduced from Ref. [36].)

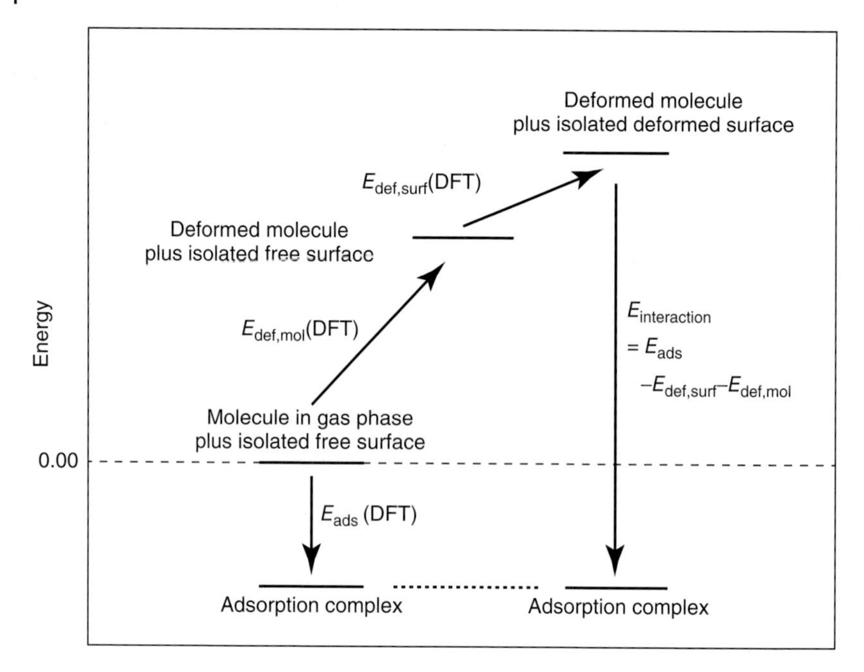


**Figure 40.15** Energy decomposition scheme for analysis of contributions to the surface interaction. (Taken from Ref. [38].)

adsorption energy is a compromise between the distortion energies of the molecule and surface, and the resulting stabilization due to the interaction.

Through this procedure it is possible to estimate the strength of the molecule–surface bond, that is, its interaction energy. This approach is strictly valid for bonding without large charge transfers between the interacting entities or excitation effects, but still useful in cases with sufficiently small charge transfers such as expected for olefins, aldehydes, and most other hydrocarbons on Ni, Pd, Pt, or Cu surfaces. Moreover, in the present case of ethene, the comparison of results from the energy analysis is also simplified by the observation that the charge transfer from $\pi$ states to unoccupied metal d states and the corresponding d–$\pi^*$ back-donation are very similar on the three Pt–Sn surfaces and, hence, leads to similar increases for $E_{int}$. In case of large charge transfers accompanying ionic interactions or covalent interactions with significant ionic contributions due to strongly differing electronic properties, additional terms for charge transfer or electron excitation must be taken into account. Several analysis schemes including charge-transfer considerations such as, for example, the "natural orbitals for chemical valence (NOCV) and the extended transition state (ETS)" method by the group of Ziegler have been developed for isolated molecules from localized orbital calculations in the past in order to study bonding, antibonding, geometric, and excitation contributions to chemical bonds (see Ref. [130] and references therein for a more thorough discussion of this and various schemes). DFT studies by these authors on the bonding of ethene in gas-phase nickel(II) and palladium(II) complexes with diimine ("Brookhart") and salicylaldiminato ("Grubbs") ligands, which can be compared to the "Zeise-Salz" with a π coordination of $C_2H_4$, showed that the donation and back-donation add about −35 and −20 to 30 kJ/mol, respectively, to the total ("orbital") interaction energy. This indicates that our calculated interaction energies could be underestimated consistently by ~50 kJ/mol (note that only ~$0.3e^-$ is donated in each direction) [131]. However, this leaves qualitative trends unaffected, and a relative comparison of this group of model systems is still valid.

Consistent with the implications from the vibrational properties, the highest value for $E_{int}$ is found on Pt(111) (Table 40.4). The $E_{int}$ value for $Pt_3Sn(111)$ is only less than ~2% lower, thus explaining the similar hybridization and vibrational properties of the adsorbed molecules on these surfaces. Even on the other two Pt-containing alloys, the interaction energies are barely 14% lower than on Pt(111). This is even more surprising considering that, at the same time, the heat of adsorption $E_{ads}$ of π-ethene is reduced by more than 40% compared to that on Pt(111). The values of the interaction energies $E_{int}$ on the surface alloys are, however, still considerably higher than the value found for ethene on Cu(111) (41 kJ/mol), which is only 32% of the that found on Pt(111).

This indicates that the nature of the adsorption on the surface is quite different in case of a Cu surface site in contrast to a Pt surface atom. Although this by itself is hardly surprising, the finding that the interaction energies of Pt-bonded π-ethene are only affected very slightly by changes of the "matrix" surrounding the surface Pt atoms (≤14% here) ought to be noted, especially considering that the surface Pt fraction decreases simultaneously to only 25%. Hence, also the interaction with a surface site, that is, the adsorption, is to first order, indeed, a very local phenomenon.

The trend of the interaction energies $E_{int}$ is reflected by the vibrational properties and explains the large difference in the observed HREELS spectra between Cu(111) and the Pt-based model surfaces, as well as the close similarity of the spectra recorded on Pt(111) and the two Pt–Sn surface alloys. Apparently, the vibrational frequencies, especially those of the directly sensitive molecule–stretching modes,

**Table 40.4** Energetic contributions (in kJ/mol) for π-bonded ethene on Pt(111), the Pt-containing surface alloys, the $Cu_3Pt(111)$ surface alloy, and pure Cu(111).

| Surface supercell | | π(CC) ethene | | | | Diσ(CC) ethene | | | |
|---|---|---|---|---|---|---|---|---|---|
| | $E_d$ | $E_{ads}$ | $E_{surf}$ | $E_{mol}$ | $E_{int}$ | $E_{ads}$ | $E_{surf}$ | $E_{mol}$ | $E_{int}$ |
| Pt(111)-3×3 | −1.84 | −76 | 19 | 34 | −129 | −106 | 15 | 134 | −255 |
| $Pt_3Sn$/Pt(111)-2$\sqrt{3}$ | −1.98 | −46 | 45 | 36 | −126 | −66 | 35 | 142 | −243 |
| $Pt_2Sn$/Pt(111)-3×3 | −2.13 | −41 | 39 | 33 | −112 | −57 | 50 | 144 | −250 |
| $Cu_3Pt$/Cu(111)-2×2 | −2.10 | −46 | 27 | 39 | −112 | — | — | — | — |
| Cu(111)-3×3 | −2.44 | −24 | 11 | 6 | −41 | — | — | — | — |

The energy of the d-band centers of gravity $E_d$ (in eV) of the bonding Pt surface sites have been calculated with respect to the Fermi level and with increased accuracy using denser 5×5×1 $k$-point meshes and a higher cut-off threshold of 500 eV for the basis set expansion. Good linear correlations between $E_{int}$ and $E_d$ ($R^2 = 0.89$) as well as $E_{ads}$ and $E_d$ ($R^2 = 0.84$) are found.

correlate much better with the interaction energies and seem to probe the true interaction potential to the surface. Both are at odds with the adsorption energy, or "adsorption strength," which in turn is probed by TPD. This observation raises many questions that will need to be investigated in the future, for example, as to the roles of adsorption strength and interaction strength for chemical activation by a catalyst, and how tuning the latter relative to the former can affect activities or selectivities, for example, between desired reaction pathways, and combustion, coking, or poisoning.

The similarity of the interaction energies can be traced back to a similarity in the electronic properties of the adsorption complexes [36]. The DOS projected on the carbon and the bonding metal atoms indicates that the rehybridization of ethene orbitals (in fact bands in a periodic model) and surface bands leads to new states at very similar energetic positions on the surfaces with Pt sites (Figure 40.16): they are only weakly dependent on the exact composition.

On Pt(111), an interaction between the C $2p_x$, $2p_y$, and the Pt $5d_{z^2}$, Pt $5d_{xz}$, and Pt $5d_{yz}$ orbitals (Figure 40.16a, −7.7, −6.9 eV, and several states between −5.5 and −4 eV) is recognized in the DOS analysis. The C $2p_z$ and the Pt $5d_{z^2}$ states show bonding interactions, which are dominant at −6.9, −4.3, and −3.6 eV. The nature of these states can be visualized by the partial charge densities of the bonding ensemble in small energy intervals (~±0.3 eV) around each of these DOS features (Figure 40.17). The charge densities related to the DOS features closest to the Fermi level (−3.6 and −4.6 eV) correspond to overlaps between the C $2p_z$ orbitals (or $\pi$(C=C)-orbital, respectively) and Pt 5d states (mostly the $5d_{z^2}$ orbital; Figure 40.17c,d). Because of the distortion and lowering of symmetry also, the C $2p_x$ and $2p_y$ and the H 1s can contribute to the states formed around −4.6 eV. At −6.9 eV below the Fermi energy (Figure 40.17b), an overlap between the Pt $5d_{z^2}$ orbital with the C $2p_x$, $2p_y$, and C $2p_z$ orbitals leads to another bonding orbital. Finally, around −7.7 eV (Figure 40.17a), a charge distribution is found that resembles the Pt $5d_{xz}$ orbital overlapping with the C–H $\sigma$ bonds and shows a nodal plane through the Pt and both C nuclei. The partial charge densities of the corresponding structures on $Pt_3Sn(111)$, $Pt_2Sn(111)$, and $Cu_3Pt(111)$ are very similar to that on Pt(111), both energy- and shape-wise, in agreement with the interaction energy analysis and the vibrational characterization.

In contrast, on Cu(111) hardly any changes to the DOS of the bonding metal atom are observed compared to surface atoms on clean Cu(111), and only slight interactions between the C $2p_z$ and the Cu $3d_{z^2}$ states are computed at −4.4 and −4.1 eV below the Fermi energy (Figure 40.16e). The partial charge density of the bands between −4.4 and −4.2 eV (Figure 40.17e) shows mainly the C–C $\pi$ bond formed by the C $2p_z$ orbitals of ethene, which is slightly perturbed as a result of a weak overlap with the Cu $3d_{z^2}$ orbital, thus supporting the conclusion that ethene is only weakly interacting with this surface. In contrast to the Pt-containing surfaces, no further rehybridization of copper d and carbon p states is found at lower energies, which reflects the trend toward a much lower interaction energy $E_{int}$.

In all cases of this analysis, the importance of the molecular and surface deformations for the final value of $E_{ads}$ has to be stressed, even for the comparatively

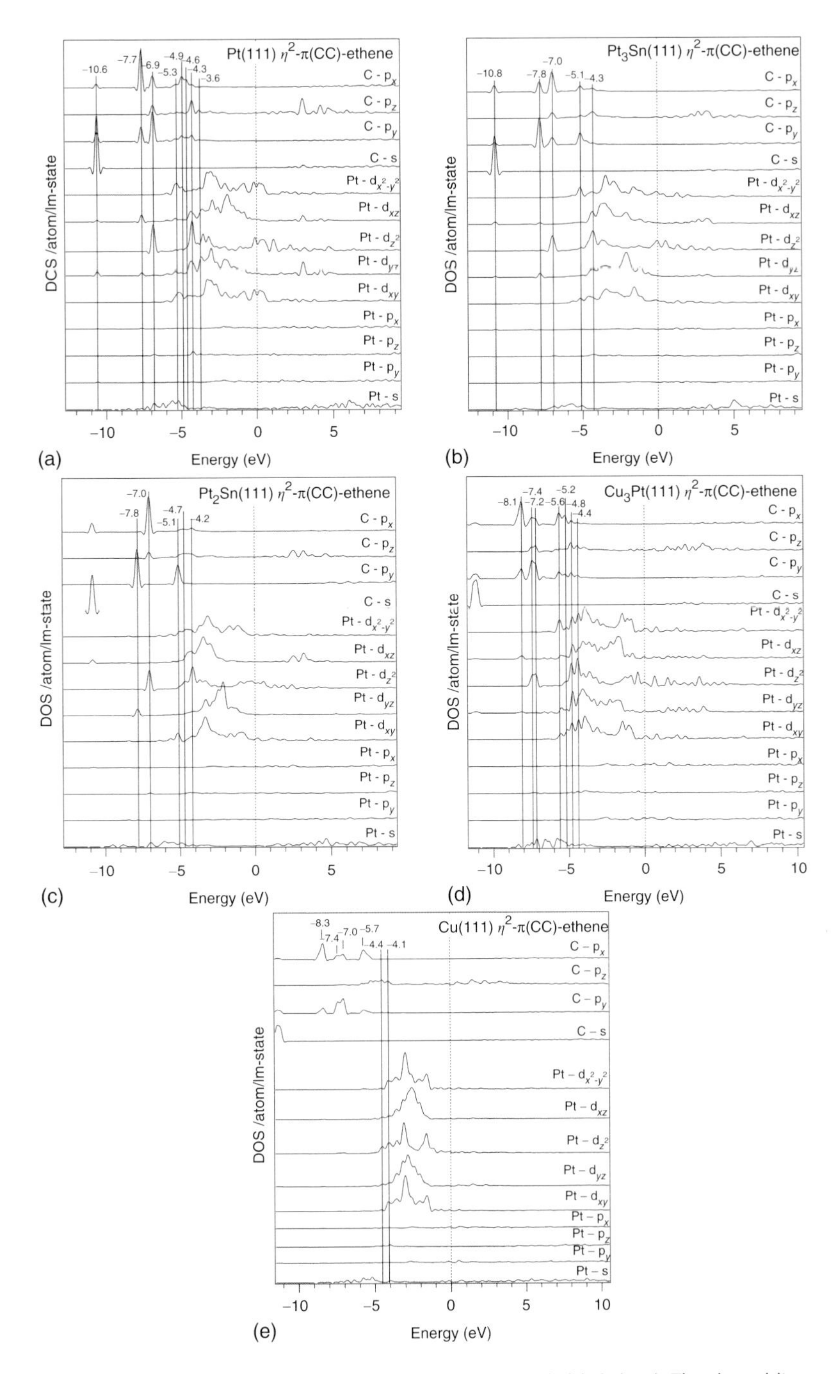


**Figure 40.16** Projected densities of states of the bonding metal and carbon atoms on the investigated surfaces: $\pi$-ethene on (a) Pt(111), (b) $Pt_3Sn(111)$, (c) $Pt_2Sn(111)$, (d) $Cu_3Pt(111)$, (e) Cu(111). The dotted lines represent the Fermi energy, and the thicker lines mark obvious hybridization between Pt and C states. (Images reporiduced from Ref. [36].)

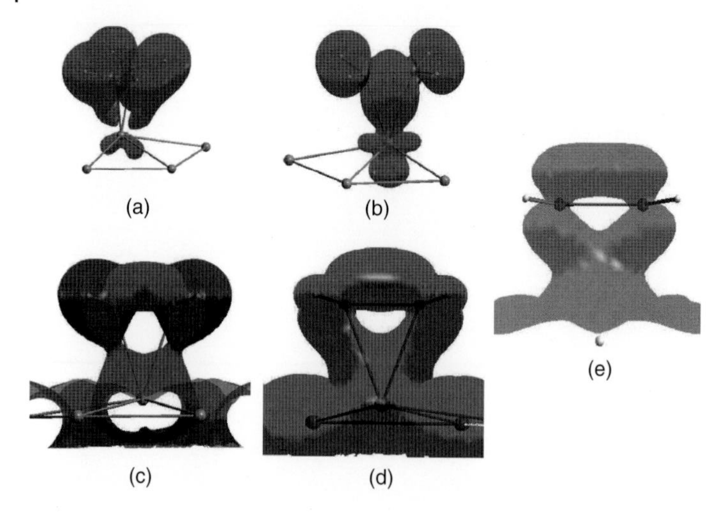


**Figure 40.17** (a–d) Schematic partial charge density representations of the π-ethene on Pt(111). The energy intervals used to plot the partial charge densities were −8.0 to −7.4 eV (a), −7.2 to −6.6 eV (b), −4.9 to −4.0 eV (c), and −3.9 to −3.3 eV (d). (e) Partial charge density for-π ethene in the energy interval from −3.9 to −4.5 eV on Cu(111). (The absolute isocontour values have been varied between the images for better visualization.)

weakly bonded π-ethene. In particular, the strong outward relaxations of the respective bonding Pt atom, which can easily be seen in the structure models (Figs. 40.7 and 40.14), are very important in this context. The latter result in energy costs of ∼19–45 kJ/mol on all Pt-containing surfaces and are correlated with the degree of outward relaxation, which are largest on the two Pt–Sn alloy surfaces.

Although these values are large and already represent a sizable fraction of the reduction of the interaction energies to the respective adsorption energies, the second relaxation component was found to be equally important for the π configurations: the molecular deformation energies range from 32 to 39 kJ/mol on the Pt and Pt-alloy surfaces, whereas it amounts to only 6 kJ/mol on Cu(111). Thus, in the cases of strong chemisorption, it roughly matched the contribution by the surface relaxation. For configurations such as the di-σ form or larger multifunctional molecules including crotonaldehyde or prenal, the molecular deformation term was found to be much more important by a factor of 2–4 [36, 38, 127]. The structural distortions of the molecular backbone and the associated molecular deformation energies are therefore important factors, which, together with the surface-bond formation, govern the vibrational shifts of indirectly perturbed normal modes such as the ν(C=C) stretching modes.

In the early days of surface science, surfaces were considered to be very rigid. However, this picture had to be abandoned with the arrival of new powerful tools such as I/V-LEED in the 1960s and 1970s, and the relaxation of surfaces and molecules upon adsorption became an intense field of study. The LEED-I/V studies of ethylidyne on Pt(111) by Somorjai, Kesmodel, Van Hove, and coworkers, for example, provided unprecedented structural details on surface complexes [83, 85, 132–137].

Thus, it is not surprising that similar analyses of the surface bonding of unsaturated molecules had already been performed by other groups in the past [67, 95, 138–143]. However, to the best of our knowledge, only in the study of benzene adsorption performed by Morin *et al.* [140] were the surface and the molecular relaxation considered as contributions to the surface interaction. Also for benzene on Pt(111), the role of the surface distortion was found to be moderate in comparison to the cost of shaping the benzene molecule for optimal interaction [140]. The distortion of the molecule and the strength of its interaction with the surface were found to be correlated with the molecular HOMO–LUMO (highest occupied molecular orbital - lowest unoccupied molecular orbital) gap [140].

Other groups selectively focused on either of the relaxation components, such as Watson *et al.* [95] who recognized the importance of surface deformations for the heat of adsorption of di-$\sigma$ ethene on Pt(111). In contrast, Hugenschmidt *et al.* performed UPS studies of ethene adsorption on bare and cesium-containing Pt surfaces with particular attention to the molecular deformation [67]. Noteworthy in this context are also several studies by Brizuela *et al.*, who used semiempirical Hückel calculations to investigate the bonding behavior of cyclopentane, cyclopentene, and cyclopentadienyl on Pt(111) [141–143]. These authors studied the effect of the perturbation of the surface on the electronic band structure and the formation of optimal adsorbate–surface orbitals for bonding and back-bonding.

Clearly, the energy costs related to the relaxation of the surface as well as the molecule upon adsorption play important roles for the interaction with a surface and, thus, affect the overall balance: the adsorption strength. They vary depending on a number of factors, including the properties of the surface material, the molecule/adsorbate species, coordination, coverage, and others such as the likely external potentials in electrochemistry.

Our analysis of the example of $\pi$-ethene adsorbed on Pt(111), $Pt_3Sn(111)$, $Pt_2Sn(111)$, $Cu_3Pt(111)$, and Cu(111) reveals that the vibrational spectra and interaction properties are closely linked. While vibrational spectroscopy of the final adsorption complex probes the "true" interaction strength ($E_{int}$), other methods such as TPD provide the "net" adsorption energy ($E_{ads}$). While the adsorption strength can vary, the underlying interaction strength between the perturbed molecule and the relaxed surface is affected very little, resulting in almost unchanged frequencies, for example, for the molecule–surface stretching modes. Interestingly in this case, the geometry and the rehybridization of the adsorbed molecule vary only little as a function of the Pt content in the surface layer, and the corresponding electronic states due to the bonding to the Pt atoms are also very similar. The changes of the energy required for adjusting the surface bonding site are key to the changing adsorption energies in this case study. These findings are in line with observations for di-$\sigma$ ethene [36], propene [116], crotonaldehyde [34, 37, 38], prenal [35, 38], cyclopentene [125, 128], cyclohexene [129], butenes, and butadiene [127] on Pt(111) and the Pt–Sn surface alloys and appear to be of a more general nature for mid-sized mono- and multifunctional molecules.

## 40.3.2
## Examples of Higher Mono-Olefines: Propene and Butenes

### 40.3.2.1 Propene

Numerous studies similar to those of ethene have been completed in the past decades on higher olefins. We want to limit our discussion by just pointing out the key findings for a small number of them, focusing on propene and the four butene isomers since these already illustrate the point of generality of the findings obtained for ethene. Most studies on Pt-based model surfaces focus on Pt(111) [62, 65, 75, 102, 144–156], and only a few authors have studied its behavior on Pt–Sn surfaces [116, 157].

On Pt(111), TPRS studies have found a decomposition of chemisorbed propene in several steps leading to desorption of $H_2$ at temperatures above 296 K up to almost 800 K (Figure 40.18) [65, 144, 145, 147, 157]. Again, the lowest desorption state, measured at 296 K for a full monolayer, follows a recombinative kinetics of adsorbed H atoms and is hence desorption limited, while the higher temperature evolution signals of $H_2$ at 330, 432 K, and around 624 K are reaction limited [157]. This suggests a partial dehydrogenation of propene below 296 K just as in the case of ethene, already pointing to the formation of substoichiometric species such as, for example, an alkylidyne. Since propene, unlike ethene, possesses vinylic as well as allylic C–H bonds, the similarity also suggests that the first $H_2$ evolution peak may be correlated with transformations of the former hydrogens.

Besides the dehydrogenation channel, also a reversible chemisorption of a fraction of propene is measured on Pt(111) (Figure 40.18) [65, 144, 145, 157]. About two-thirds of the saturation uptake of the molecules at an exposure temperature of 100 K desorb again on annealing at ~272 K, while the remainder undergoes decomposition. Neglecting small temperature shifts and assuming first order kinetics, an activation energy of desorption of 73 kJ/mol (17.4 kcal/mol) has been estimated at low coverages ($T_{des} = 284$ K) with the typical frequency factor of $10^{13}$ Hz [157]. In comparison, only very small amounts of propene decompose on the $Pt_3Sn(111)$ surface alloy, while no decomposition occurs on the tin-richer $Pt_2Sn(111)$ surface (Figure 40.18). This again shows the importance of threefold Pt hollow sites for the decomposition pathway.

In contrast, neither the initial sticking coefficient of the molecule nor its saturation coverage upon adsorption at 100 K was found to be affected by the second alloy component ("promoter") despite a much weaker chemisorption bonding in comparison to pure Pt(111). Besides the desorption from a physisorbed state evident at ~120 K on all surfaces, substantial shifts of the monolayer desorption temperatures, and thus much weaker desorption activation energies have been observed by these authors on the Pt–Sn surface alloys (Figure 40.19). The molecular desorption of propene occurs at temperatures as low as ~232 K on $Pt_3Sn(111)$ and 180 K on $Pt_2Sn(111)$, resulting in Redhead [84] estimates of $E_{des} = 62$ kJ/mol (14.8 kcal/mol) and 49 kJ/mol (11.7 kcal/mol), respectively [157].

Vibrational spectroscopy studies by Avery *et al.* and Cremer *et al.* shed additional light onto the bonding and reactivity of propene on Pt(111). Consistently,

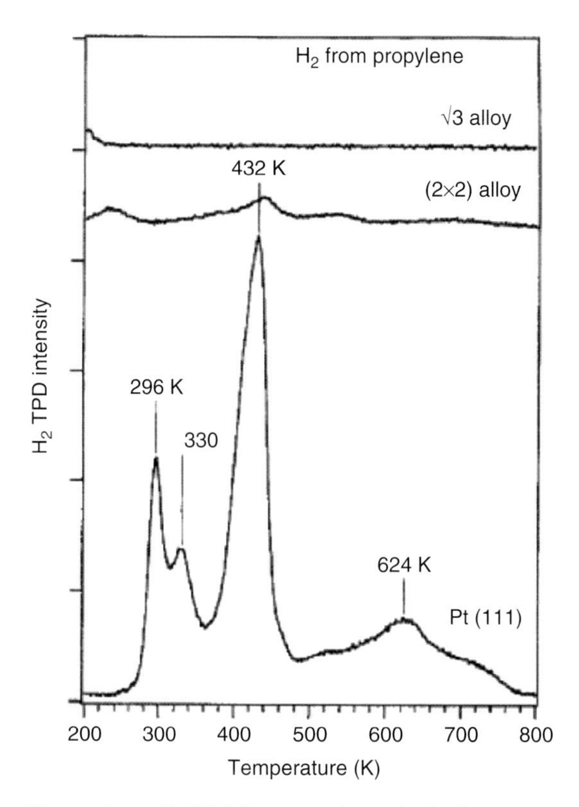


**Figure 40.18** $H_2$ TPRS from propylene adsorbed on Pt(111) and the two Pt–Sn surface alloys $Pt_3Sn(111)$-p(2 × 2) and $Pt_2Sn(111)$-($\sqrt{3} \times \sqrt{3}$)R30°. (Reproduced from Ref. [157]; heating rate $\beta = 4$ K/s.)

HREELS [147] (Figure 40.20) and SFG [62] show that, under UHV conditions and at temperatures below ~230 K, propene also adsorbs predominantly in a di-$\sigma$ configuration. Besides strong similarities of characteristically perturbed modes including the C–H stretching modes, HREELS measurements in specular as well as off-specular geometry detected the $\nu$(C=C) stretching vibration at 1050 cm$^{-1}$, almost identical to the value for ethene [147].

Upon annealing to 300 K, a transformation to a very stable propylidyne ($Pt_3$–C–$CH_2$–$CH_3$) species was deduced with HREELS, which was correlated to the $H_2$ desorption peak at ~296 K [147]. Above 400 K, the adsorbed species became increasingly hydrogen-deficient, and, finally, above ~425 K, only mixtures of fragments consisting largely of CH groups bonded to several metal atoms were suggested to be present.

In contrast, at near-atmospheric pressures (40 Torr of $C_3H_6$ and 723 Torr of $H_2$) and room temperature, a shift of the dominant population toward the molecular $\pi$-adsorption complex occurs, as observed with SFG in *in situ* experiments [62]. By monitoring surface intermediates under reaction conditions, a dominant hydrogenation pathway was identified from a variety of possibilities: a stepwise

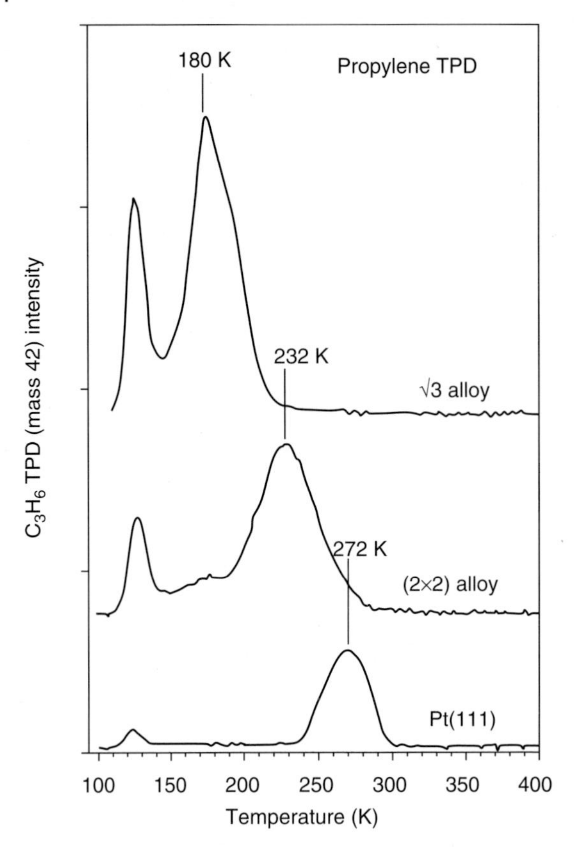


**Figure 40.19** Desorption of molecular propene adsorbed at 100 K on Pt(111) and two Pt–Sn surface alloys. (Reproduced from Ref. [157]; heating rate $\beta = 4$ K/s.)

hydrogen addition of π-bonded propylene through a 2-propyl intermediate to propane (Figure 40.21). Under UHV conditions, propylene was observed to decompose to propylidyne and then to vinylmethylidyne in the absence of hydrogen. Hence, the adsorption behavior of propene follows that of ethene rather closely in UHV as well as under conditions of increased pressures and temperatures.

Theoretical studies support the assignments derived from the vibrational data, and showed that also for this larger olefin the di-σ configuration is generally more stable than the corresponding $\pi$(CC) on Pt(111) and the two Pt–Sn surface alloys. Valcarcel and coworkers established a consistent description of two adsorption configurations of propene on Pt(111) slab and cluster models, proposing that the molecular adsorption occurs in a di-σ coordination with the C=C double bond parallel to the surface on a bridge site, which results in a binding energy of ~40–50 kJ/mol (~0.4–0.5 eV) [146].

Additional insights into the adsorption on not only Pt(111) but also three Pt–Sn surfaces including the $Pt_3Sn(111)$ termination and $PtSn_2(111)$ surface were obtained in a very detailed study performed recently by Nykanen and

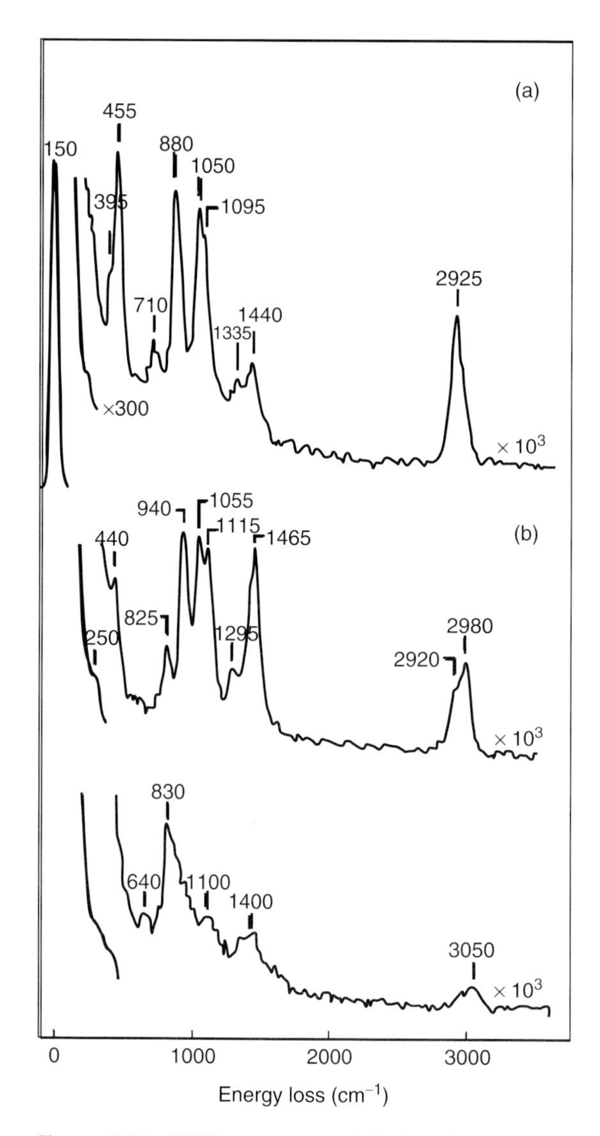


**Figure 40.20** HREEL spectra recorded after adsorption of a monolayer of propene on Pt(111) and annealing to (a) 170 K, (b) 300 K, and (c) 425 K followed by recooling for data acquisition. (Reproduced from [147].)

coworkers [116]. Importantly, these authors also addressed the lack of van der Waals contribution of most standard GGA functionals and investigated the difference between adsorption strength and interaction strength for propene and propane.

Propane adsorption was found to be mediated by dispersion interactions, which led to a lack of site preference, but the bonding of propene is highly covalent and the coordination and site geometries follow those of ethene very closely [116]. On

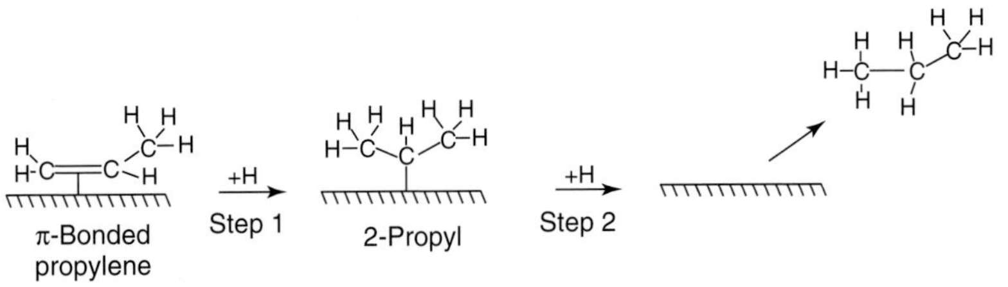

**Figure 40.21** Proposed steps of the propene hydrogenation mechanism involving a surface intermediate that was identified with *in situ* SFG spectroscopy at near-ambient pressure of a mixture of $H_2$ and $C_3H_6$ on Pt(111). (From Ref. [62].)

surfaces exhibiting Pt bridge sites, such as the considered Pt(111) and the $Pt_3Sn$ terminations, the molecule prefers a di-$\sigma$ site over a $\pi$ site. The computed adsorption energies of the di$\sigma$(CC) configurations vary between 73 kJ/mol (−0.75 eV) and 81 kJ/mol (−0.83 eV) on Pt(111), of which ~30 kJ/mol is due to the dispersion forces (Table 40.5). In contrast, the adsorption on the $PtSn_2$(111) surface is weak and clearly dominated again by van der Waals interactions.

Alloying Pt(111) with Sn was reported to weaken the adsorption due to geometric and relaxation effects, but the electronic effects were found to be small [116]. The adsorption energies quickly weaken to only about −40 kJ/mol (0.41 to −0.54 eV) on the $Pt_3Sn$(111) bulk and surface alloys.

However, a closer analysis of the bonding by these authors established that, also in the case of covalently bonded propene, the interaction strength was reduced by less than 10% by alloying (Table 40.6) [116]. Key to this finding were again the changing relaxation energy costs arising from the perturbation of the surface and the molecule upon formation of the adsorbate complex. Moreover, an analysis of the relationship between the computed interaction energies $E_{int}$ and the energetic position of the d-band center of gravity for the considered systems revealed a linear correlation (Figure 40.22). Finally, these authors investigated the implications of the alloying with Sn for the selective propane dehydrogenation to propene.

**Table 40.5** Propene adsorption energies on various $Pt_xSn_y$(111) with two different functionals.

| XC functional | RPBE | | vdW DF | |
|---|---|---|---|---|
| coverage[a] | 1/4 | 1/8 | 1/4 | 1/8 |
| Pt(111) | −0.42 | −0.51 | −0.75 | −0.83 |
| $Pt_3Sn$/Pt(111) | 0.05 | −0.07 | −0.41 | −0.48 |
| $Pt_3Sn$(111) | −0.06 | −0.24 | −0.44 | −0.54 |
| $PtSn_2$(111) | 0.03 | −0.02 | −0.29 | −0.23 |

a) Coverages are given in monolayer and adsorption energies in electronvolts.
Revised Perdew–Burke–Enzerhoff/RPBE and the corrected functional within the Grimme-D2 scheme/vdW-DF = van der Waals-Density functional.
From Ref. [116].

**Table 40.6** Energy analysis showing the interaction energy between a deformed molecule and perturbed surface with the respective geometry parameter fond in the adsorbate complex, and the corresponding energy factors invested in the relaxation of the two components. The coverage is 0.13 ML.

| XC functional | RPBE | | | vdW-DF | | |
|---|---|---|---|---|---|---|
| | $E_{int}$ | $E_{def,surf}$ | $E_{def,mol}$ | $E_{int}$ | $E_{def,surf}$ | $E_{def,mol}$ |
| Pt(111) | −2.42 | 0.20 | 1.74 | −2.60 | 0.19 | 1.58 |
| $Pt_3Sn$/Pt(111) | −2.29 | 0.40 | 1.82 | −2.52 | 0.38 | 1.70 |
| $Pt_3Sn(111)$ | 2.25 | 0.2 | 1.74 | 2.46 | 0.26 | 1.66 |

Taken from Ref. [116].

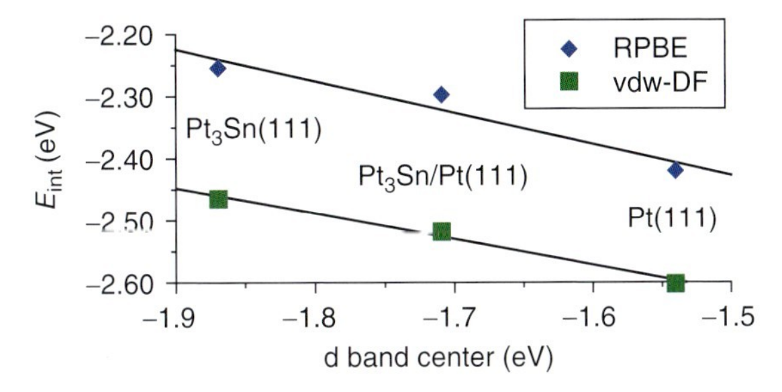


**Figure 40.22** Plot showing the linear relation between interaction energies ($E_{int}$) and the energy of the d-band center of gravity for the RPBE and vdW-DF functional used by Nykanen *et al.* (Reproduced from Ref. [116].)

#### 40.3.2.2 **Butene Isomers**

The adsorption and reaction behavior of the four butene isomers, *n*- or 1-butene, *E*-2-butene (often also called *trans*-butene), *Z*-2-butene (*cis*-butene), and *iso*-butene (2-methyl-propene) (Figure 40.23), is largely an extension of those of the smaller ethene and propene olefins. All these olefins have a similar chemisorption bond strength and reactivity on Pt(111) [65, 75, 147, 148, 157, 158], and TPRS shows that the molecules adsorb intact below ~200 K under UHV conditions. NEXAFS studies showing the disappearance of the $\sigma^*$(C=C) resonance indicate that a strongly chemisorbed state is formed, which corresponds to a di-σ(CC) coordination for all butene isomers [158]. Above 200 K, all isomers undergo a decomposition process that proceeds via highly stable alkylidyne species for all except the 2-butenes [65, 75, 144, 147, 148, 157].

HREEL studies on Pt(111) resulted in well-defined spectra at temperatures between 90 K and the onset of the initial dehydrogenation process, evidencing that also the butene isomers prefer a di-σ(CC) coordination in the undissociated state [147, 148]. The vibrational data coincide with the characteristics of other di-σ adsorbed olefins; for example, the $\nu$(C=C) stretching modes of C–C bond interacting covalently with the surface of 1-butene, the 2-butenes, and *iso*-butene are observed at 1055 $cm^{-1}$ [147], 1040 $cm^{-1}$ [148], and ~1000 $cm^{-1}$ [147], respectively. This has been confirmed by reflection absorption IR spectroscopy (RAIRS) model

1-Butene | *E*-2-Butene | *Z*-2-Butene | *iso*-Butene

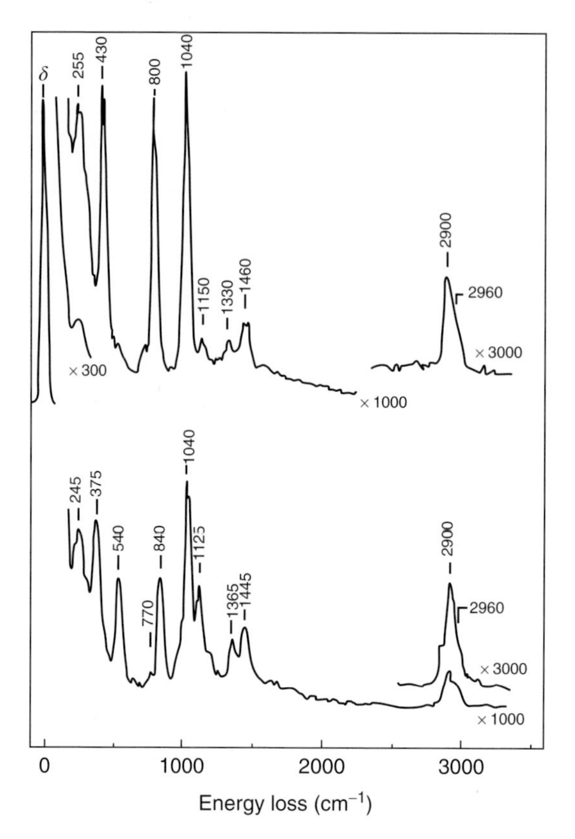


**Figure 40.23** HREEL spectra of *Z*-2-butene and *E*-2-butene monolayers at 170 K on Pt(111). (Reproduced from Ref. [148])

studies, which showed that also the adsorbed butenes will adopt more weakly bound π complexes at close to monolayer saturation or elevated pressures, while di-σ bonding is preferred initially at lower coverages [155]. The results of the vibrational analysis detailed by Avery *et al.* are surprisingly analogous to the findings on ethene and propene and reveal that the substituent effects on the di-σ(CC) interaction mechanism are relatively minor for this group of probe molecules [147, 148].

Although the C=C stretching frequencies of the *Z*- and *E*-2-butenes appear at identical positions in the adsorbed form at low temperatures, their vibrational spectra differ significantly from each other, showing that chemisorption occurs into two distinct di-σ(CC) adsorption structures without isomerization (Figure 40.23) [148]. A cis–trans equilibration has been observed with RAIRS and

TPRS using isotopically labled isomers above ~230 K but generally favors selectivity toward the cis form of olefins [153–155], which was attributed to the relative adsorption energies based on the experimental evidence [155]. The thermodynamic preference for the (*s*)-cis (or *Z*-) configurations of the surface complexes has recently been confirmed by DFT [152]. Because of differences in the twisting of the structures around the central C–C bond leading to a much larger molecule deformation energy for the trans forms, especially in the di-σ configurations, the energetics from the gas phase are inverted and the cis forms become slightly more stable. Since the π species are those that participate directly in hydrogenation and isomerization reactions, the stabilization of the adsorbed π(CC)-cis form is concluded to govern the preferential isomerization of *E*- to *Z*-2-butene on Pt(111) surfaces [152, 155]. The preference is even enhanced by the presence of hydrogen on the surface, which weakens the interaction of the *E* isomers with the surface more than the *Z* forms (Figure 40.24) [152, 155].

Unfortunately, so far no experimental vibrational studies on the Pt–Sn model catalysts have been published, to the best of our knowledge, and the best information available on the adsorption and reactivity of butenes comes from several extensive TPRS and LEED studies. TPRS shows a substantial shift of the monolayer desorption states of chemisorbed molecules to lower temperatures for all four

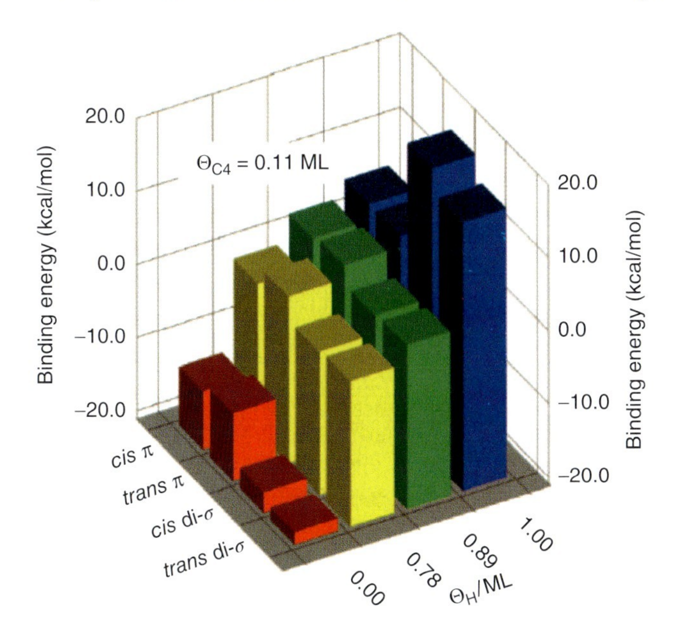


**Figure 40.24** Binding energies for *Z* (*cis*) and *E* (*trans*)-2-butene in their di-σ- and π-bonded forms on a Pt(111)-3 × 3 supercell ($\Theta_{but}$ = 1/9 ML with respect to Pt atoms) with increasing amounts of coadsorbed atomic hydrogen ($0 \leq \Theta_H \leq 1$ ML). The alkene adsorption is disfavored with the presence of hydrogen: the most stable species on the clean surface is the di-σ(CC)-*trans* butene, but at high H coverages it is the π(CC)-*cis* configuration. (From Ref. [152].)

butene isomers and near-quantitative suppression of the decomposition pathways due to the alloying with Sn.

On Pt(111), the monolayer desorption state for the four butene isomers is found with relative variations of less than 10 K between ~290 and ~260 K for low and high initial coverages (Table 40.7; examples of 1-butene shown in Figure 40.25) [65, 75, 147, 148, 157], which coincides closely with the desorption temperatures of propene and ethene. This translates into desorption activation energies with a narrow range of ~73 to 71 kJ/mol according to Redhead analysis ($\nu = 10^{13}\ s^{-1}$). The similarity of the desorption temperatures for ethene, propene, and the butenes indicates no major "inductive effect" on their covalent Pt–C bond strength. The chemisorption bond strength of these di-σ coordinated species is essentially the same [157]. Considering butenes as alkyl-substituted ethene derivates reveals only a very small trend toward decreasing adsorption energy as the amount of methyl substitution increases [157].

Upon alloying the Pt(111) model surfaces with Sn, the chemisorption bonds of the butenes are substantially weakened (Table 40.7; Figure 40.25). On the $Pt_3Sn(111)$ surface alloy, the desorption maximum of the reversibly bonded molecules occurs between 240 and 268 K for low initial coverages, whereas this temperature range is further decreased to 180 and 222 K on the Pt-poorer ultrathin $Pt_2Sn(111)$ alloy [75, 157]. Hence the desorption activation energies are decreased by ~10–15% on the former surface and by ~30% on the latter. The relatively large effect on the adsorption energy when comparing the (2 × 2) and $\sqrt{3}$ alloys was concluded to be due to the elimination of pure Pt threefold sites [75], which, as discussed before, induces electronic changes to the Pt atoms of the binding sites and indirectly affects the bonding by increased repulsive interactions with the protruding Sn atoms.

Neither the initial sticking coefficients nor the monolayer saturation coverages for these small alkenes show any significant effect of alloyed Sn on the two Sn/Pt(111) surface alloys when compared with the clean Pt(111) surface at 100 K [157]. The monolayer saturation coverages on all surfaces are between 60% and 100% higher for 1-butene compared to the corresponding values obtained for the *Z*-2-butene or *iso*-butene [75].

On the Pt–Sn surface alloys, the decomposition of the butene monolayers is strongly suppressed as in the case for the smaller olefins, which is directly linked

**Table 40.7** Temperatures of the desorption maxima of reversibly adsorbed butene molecules at low surface coverages (linear heating rate $\beta = 4$ K/s) and corresponding desorption activation energies obtained from Redhead analysis using a frequency prefactor of $\nu = 10^{13}\ s^{-1}$ and assuming first order kinetics.

| | Desorption maximum *T* at low coverage (desorption activation energy) | | |
|---|---|---|---|
| **Butene isomer** | **Pt(111)** | **$Pt_3Sn$/Pt(111)-p(2 × 2)** | **$Pt_2Sn$/Pt(111)-(3 × 3)R30°** |
| 1-Butene [75] | 288 K (73 kJ/mol) | 268 K (67 kJ/mol) | 222 K (57 kJ/mol) |
| *E*-2-Butene [154] | 230 K (59 kJ/mol) | N/A | N/A |
| *Z*-2-Butene [75] | 270 K (71 kJ/mol) | 254 K (65 kJ/mol) | 200 K (50 kJ/mol) |
| *Iso*-butene [75, 157] | 280 K (71 kJ/mol) | 240 K (62 kJ/mol) | 180 K (46 kJ/mol), shoulders up to 260 K |

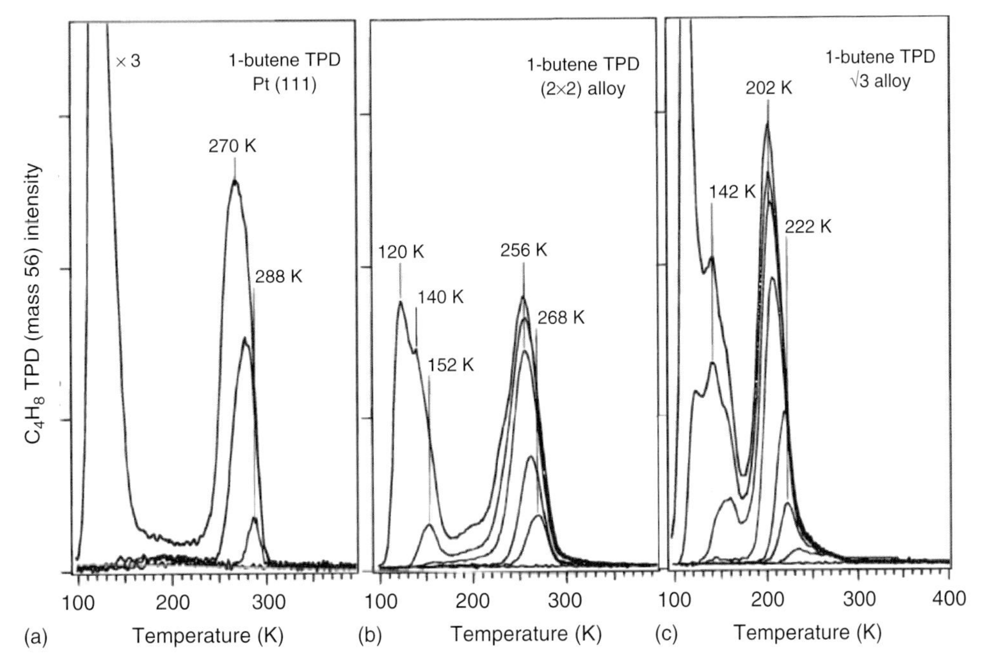


**Figure 40.25** Desorption spectra of increasing initial coverages of 1-butene on (a) Pt(111), (b) $Pt_3Sn$/Pt(111)-p(2 × 2) and (c) $Pt_2Sn$/Pt(111)-(3 × 3)R30° surface alloys (heating rate $\beta = 4$ K/s; Taken from Ref. [75].)

to the weakened chemisorption bonding and lowered desorption barriers of the competing desorption pathways [75, 157]. The decomposition of the butenes occurs almost exclusively on Pt(111) and starts at elevated temperatures above ~200 K. About 47–63% of the chemisorbed monolayer dehydrogenates to form the corresponding butylidynes already below room temperature [75].

In contrast, less than 7% of the monolayers of the butene isomers decomposed on the Pt–Sn surface alloys, as judged from the weak $H_2$ peaks in the range 402–408 K and from AES [75, 157]. This is consistent with the importance of adjacent "Pt-only" threefold hollow sites for the decomposition of the smaller olefins.

Here we concentrate on 1-butene on Pt(111) as an example to illustrate the complexity of the decomposition for the four butene isomers (Figure 40.26) [75]. Several $H_2$ desorption states are recorded with TPRS, suggesting the occurrence of a series of different elementary dehydrogenation steps, some of which occur only at higher coverages. The $H_2$ desorption states at ~296 K agree well with the temperature range for recombinative desorption of H surface atoms [51] and are therefore likely desorption limited, while the remaining (at least four) states up to 620 K must be reaction limited [75]. The number of separate elementary steps suggests that several surface intermediates are formed in sequential or parallel pathways, some of which may be stable enough for spectroscopic characterization either *in situ* or with the quenching/recooling technique.

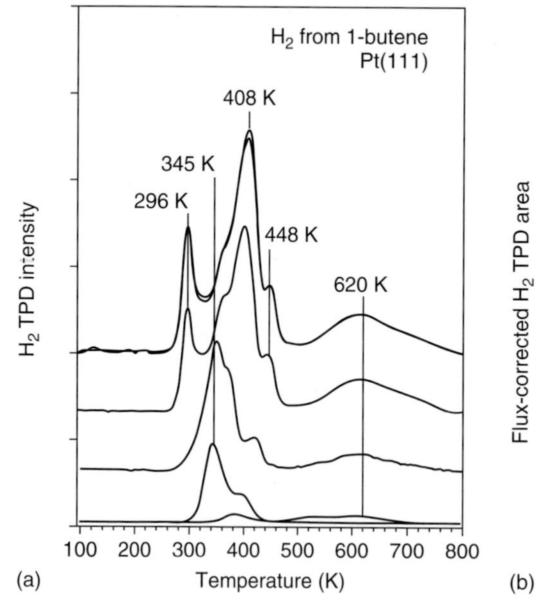


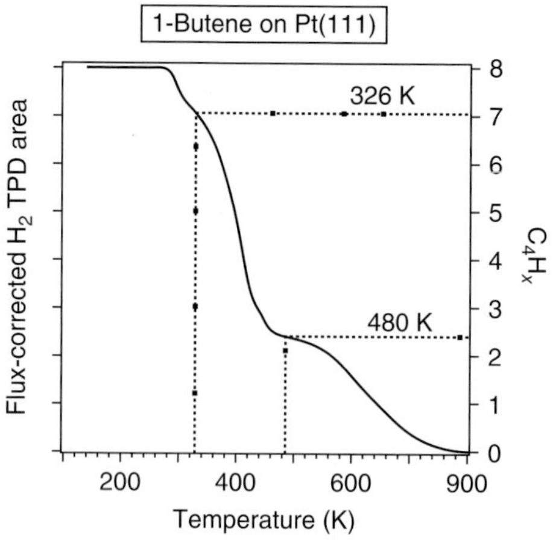


**Figure 40.26** (a) Comparison of the $H_2$ desorption spectra from 1-butene decomposition on Pt(111), the (2 × 2) and ×3 alloy surfaces. (b) Peak areas relative to full stoichiometry of the initial monolayer or flux and fragmentation pattern analysis from TPRS can reveal the average C:H ratio of the surface layer at specific temperatures during the experimental run (heating rate $\beta = 4$ K/s). (Taken from Ref. [75].)

Indeed, several HREELS and RAIRS studies have been reported that shed light on the thermal reactivity of butenes on Pt(111) surfaces above the "static" molecular regime. The vibrational studies show unequivocally that the isomers with terminal double bonds enter different pathways than the 2-butene isomers [148, 155]: both 1-butene and *iso*-butene are found to form alkylidynes at 270–300 K, namely *n*-butylidyne ($Pt_3C–(CH_2)_2–CH_3$ and *i*-butylidyne ($Pt_3C–CH–(CH_3)_2$) [147, 155], whereas *E*- and *Z*-2-butene likely form a very different 1,3-dimethyl acetylenic species (see Ref. [155] and references therein) instead. This provides indirect evidence that, under these reaction conditions on Pt(111), no 2,1-double bond migration or vice versa can occur. The transformation of 1-butene to 1-butylidyne proceeds via 1-butyl intermediate, which has been detected with IR spectroscopy in the temperature range between ~210 and 270 K [155].

Near 450 K, the thermal desorption data indicate average C:H compositions of approximately $C_4H_2$ for the species derived from 1-butene and $C_4H_4$ from *iso*-butene [147]. The HREEL spectra from all the straight-chain butenes including the *E*- and *Z*-2-butenes and even 1,3-butadiene match each other closely at this temperature, which suggests a merging of the reaction pathways. Above ~500 K, also the C–C bond breaking mechanisms become accessible through the thermal activation, and further dehydrogenation steps take place until hydrogen-efficient carbonaceous species remain on the surface [147, 148].

Experimental model studies have also been reported on the fundamental mechanism of the C=C double bond hydrogenation of the butene isomers on Pt(111) surfaces [153–155]. Employing surface IR spectroscopy and TPRS, Lee *et al.* chose the strategy of studying specific surface intermediates formed selectively via dissociation of carbon–halogen bonds in iodo [153] and bromohydrocarbons [155].

Thermal activation of adsorbed halo-alkanes, for example, on transition-metal surfaces leads to the formation of surface alkyl species [69, 70, 159, 160], but also other types of species including metallacycles [161–164] and alkenyls [164–166] can be produced via this route. Since carbon–halogen bond breaking is very facile and occurs already below 200 K [155], it also allows access to more reactive intermediates under well-defined conditions and presents a means to study specific reaction pathways from selected starting points. The alkyl moieties can be further activated by thermal energy to undergo β-hydride eliminations to yield adsorbed alkenes [70, 159, 167], which has been shown for a variety of $C_4$ hydrocarbons using isotopic labeling with deuterium [153–155].

Importantly, Lee *et al.* concluded that β-hydride eliminations from butyl intermediates to adsorbed butenes, and subsequent insertion of the C=C double bond into metal-hydrogen bonds (adsorbed atomic hydrogen), can explain the hydrogenation, dehydrogenation, and H–D exchange products that desorb from the surface [153, 155]. Generally, a tendency for the $C_4$ species toward dehydrogenation starting from the inner carbons (2° and 3° carbon centers) and hydrogen addition to unsaturated, terminal carbons was observed. Consistent with the previous discussion, coadsorption of hydrogen or deuterium led to a direct competition between hydrogenation and dehydrogenation pathways. The validity of the Horiuti–Polanyi mechanism [17] and the nonparticipation of allylic intermediates in the catalytic hydrogenation processes of butene isomers were, thus, also confirmed by these experiments [155].

Theoretical studies on butene adsorption and reactivity, on the other hand, have mostly been carried out in the greater framework of 1,3-butadiene hydrogenation. Hence, we will only give a very brief introduction to these works here and discuss them in more detail in the next section. Besides the study of Delbecq *et al.*, who computed the relative adsorption strengths as well as the deformation and interaction energies of the di-σ(CC) and π(CC)-structures on clean and hydrogen covered Pt(111) [152], two further studies provide insights into the adsorption of butene isomers.

Just like in the case of ethene, Mittendorfer and coworkers carried out an *ab initio* thermodynamic assessment of Gibbs free adsorption enthalpy of 1-butene on Pt(111) and Pd(111) as a function of surface ordering, pressure, and temperature. Using standard DFT [102], they obtained a wide interval for the stability of di-σ(CC) coordinated forms. A weaker adsorption of the olefins on Pd(111) was proposed to be important for the higher hydrogenation activity of the latter metal.

This question was taken up in a DFT study by Valcarcel *et al.*, who showed that the most stable adsorption structure for the butene isomers on a clean Pt(111) slab (with a vacuum region above) is indeed the di-σ-mode [151]. In agreement with a weaker adsorption strength, also the relaxed surface geometry is closer to the gas-phase values on Pd than on Pt, which leads to different spectroscopic fingerprints despite identical binding modes. Thus, also for the butene mono-olefins the adsorption mode and the competition between the two coordination possibilities closely follow the trend outlined by ethene and propene.

### 40.3.3
### A Bifunctional Molecule with Chemically Similar Double Bonds: 1,3-Butadiene

Butadiene is the simplest example of a conjugated polyene and a suitable probe for the adsorption and reaction behavior of more complex multifunctional molecules. It opens simple access routes to study the variety of known reactions of dienes and polyenes in heterogeneous catalytic processes such as the selective hydrogenation to mono-olefines using transition metals (Figure 40.27). Therefore, this molecule is also, from an applied perspective, important to understand.

The selective hydrogenation of 1,3-butadiene to butenes is an efficient route to upgrade the $C_4$ fractions produced by naphtha steam cracking and to avoid polymerization of dienes [168]. Its interaction with various transition and noble metals has been widely studied, mostly on supported catalysts [2, 10, 54, 155, 168–184]. Activity and selectivity for this process depend strongly on the nature of the metal catalyst. An ideal catalytic process would convert butadiene selectively to one of the butene isomers, then lead to deeper hydrogenation toward butane while also suppressing side reactions such as coking. However, only mixtures of butene isomers and butane are produced over platinum catalysts [168, 170–172]. Also, Pd is a widely used catalyst for this reaction. It shows a lower overall activity compared to Pt, but usually achieves better selectivities and lower butane formation rates [172, 179–181]. The higher selectivity of Pd has been attributed to different factors: while some have proposed that the difference stems from the stronger adsorption of 1,3-butadiene on the Pd catalysts relative to that of the butenes (whereas their bonding is more similar on Pt) [101, 102, 168, 171, 172, 180, 185, 186], recent DFT studies have found that the reason is not solely this competition. The theoretical analysis

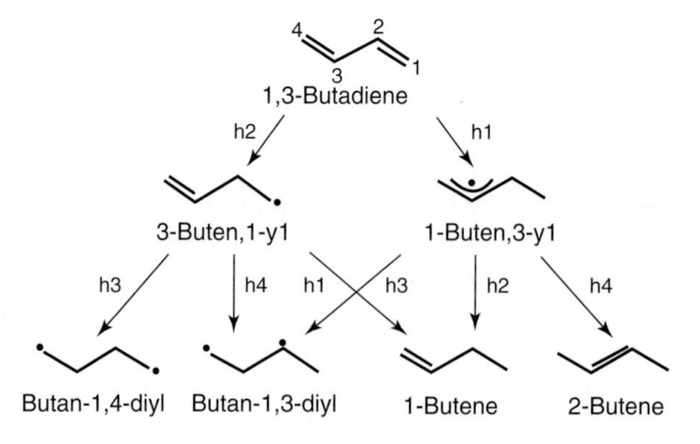


**Figure 40.27** Hydrogenation scheme of butadiene. The diradicals formed after addition of the second H atom can be further hydrogenated to butane. (Reproduced from Ref. [127].)

attributed the improvement in selectivity mainly to the stability of specific reaction intermediates and the related height of activation barriers [127, 151, 187].

The performance of Pt catalysts can be improved by the addition of a promoter such as Sn [171, 188] or Ge [185]. Pt–Sn bimetallic catalysts, supported as well as single-crystal-derived, have been reported to be effective in the selective hydrogenation of alkadienes [171]. The addition of Sn resulted in decreased activity but in increased selectivity for unsaturated hydrocarbon products and enhanced catalyst lifetime due to the reduction of coke formation.

Similarly, sintered Pt–Ge intermetallic powders, tested under conditions of elevated pressures and temperatures (298–523 K, 6.6 kPa alkene and up to 19.9 kPa $H_2$), show high selectivity toward butenes and disfavor butane production [185]. The authors of this study concluded from *ex situ* X-ray photoelectron spectroscopy (XPS) analysis that an electron transfer from Ge to $Pt_5$d orbitals, analogous to the Sn case, as well as geometrical changes of the Pt active sites, account for the changed reactivity. The rate-determining step in butadiene hydrogenation was determined to be the adsorption kinetics, highlighting the critical role of the competition between adsorption strengths of the reactants and the products, also for this system.

Model catalytic hydrogenation studies of 1,3-butadiene on Pt(111) single-crystal surfaces under UHV conditions unfortunately only found a complete dehydrogenation to $H_2$ and surface carbon species, as reported for the mono-olefins [173, 189]. The onset of the $H_2$ signals appears at ~340–370 K for low and high coverages of chemisorbed molecules, which is within the typical desorption-limited $H_2$ evolution range [189]. Two further $H_2$ desorption peaks were identified at 398 and 585 K, and confirmed that the decomposition process is a multistep mechanism.

Alloying Sn into the Pt(111) surface also completely inhibited the decomposition of this hydrocarbon as judged from the lack of carbon deposits observed with AES and the suppression of $H_2$ evolution [189]. In contrast to the butene isomers, butadiene adsorbs reversibly only on the p(2 × 2) and ($\sqrt{3} \times \sqrt{3}$)R30° Pt–Sn surface alloys with similar monolayer coverages as on Pt(111) (~0.15–0.17 ML). Desorption from the chemisorbed monolayers was measured with rate maxima at 341 and 292 K for low coverages, giving Redhead desorption activation energies (first-order desorption, $\nu = 10^{13}$ Hz, $\beta = 3.6$ K/s) of 88 and 75 kJ/mol [189]. These values are considerably larger than the corresponding desorption parameters of the butene isomers and, as will be shown below, imply that the second C=C bond plays a significant role in the surface interaction.

Moreover, because of the competition of desorption and decomposition pathways, the Redhead energies can be used as lower limits for activation energies for dissociation of the vinylic C–H bonds on the alloyed model catalysts [189]: despite a stronger chemisorption of 1,3-butadiene, butadiene appears to be less prone to decomposition since it has no allylic β-CH bonds as present, for example, in 1-butene.

Adsorption of 1,3-butadiene on Pt–Sn surface alloys precovered with hydrogen adatoms from a hydrogen atom source showed the opening of another reaction channel: a hydrogenation to butenes with 100% selectivity (Figure 40.28) [173]. The reaction showed rate-limited behavior in TPRS experiments with product desorption peaks between 280 and 350 K, translating into activation barriers estimated at

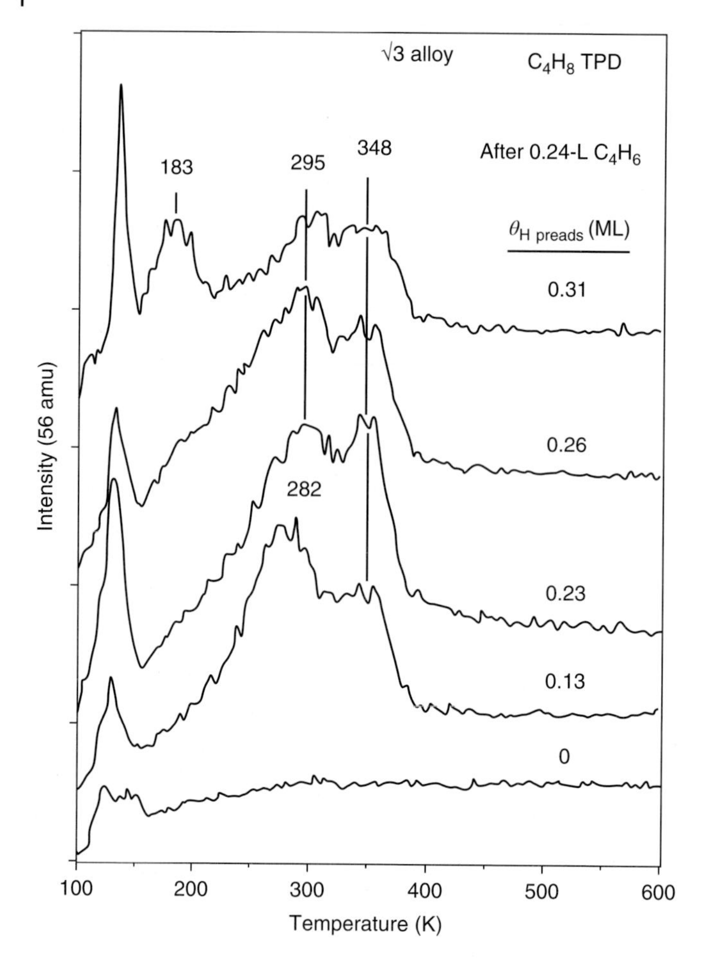


**Figure 40.28** TPR spectra of butene ($C_4H_8$) evolution after dosing 0.24 L of 1,3-butadiene on clean and H-precovered $(\sqrt{3}\times\sqrt{3})R30°$ Pt–Sn surface alloys at 100 K. (Reproduced from Ref. [173].)

~91 and 72 kJ/mol on the p(2 × 2) and $(\sqrt{3}\times\sqrt{3})R30°$ surface alloys, respectively. No further hydrogenation to butane was detected. At very high exposures of hydrogen, additional butene desorption signals were detected below 200 K.

The conversion was highest on the p(2 × 2) alloy and reached 100% when high hydrogen coverages were employed [173]. The preadsorbed hydrogen, however, competed strongly with the chemisorption of 1,3-butadiene under UHV conditions. Hydrogen acted as a site-blocker on both alloys and suppressed the hydrocarbon adsorption above coverages of 0.49 ML on the p(2 × 2) (Figure 40.29) and 0.34 ML on the $(\sqrt{3}\times\sqrt{3})R30°$ surface alloy, which was concluded to be indicative of the role of pure threefold Pt sites as binding sites of the diene.

Building on TPD/TPRS experiments, Avery *et al.* employed HREELS also to investigate the surface structures formed from 1,3-butadiene adsorption and

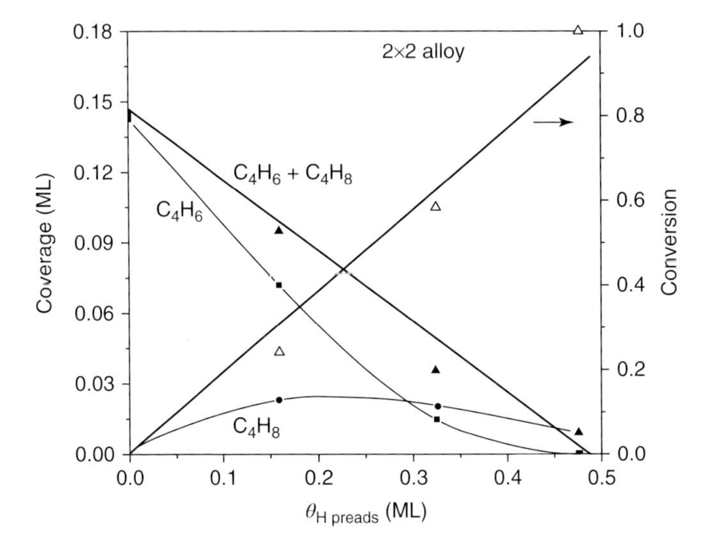


**Figure 40.29** Influence of hydrogen atom precoverage on the amount of 1,3-butadiene adsorption, desorption, and hydrogenation on the p(2 × 2) $Pt_3Sn(111)$ surface alloy. (Reproduced from Ref. [173].)

reaction on Pt(111) (Figure 40.30) [148]. Prior to the onset of $H_2$ desorption, at low-temperatures of 170 K the vibrational spectra showed nondissociatively adsorbed butadiene species. The HREEL spectra changed characteristically after annealing to 300 K, which was taken as an indication for a configuration change of the surface species without overall hydrogen loss. The interpretation of the spectra allowed a reasonable suggestion for a terminally $\eta^2$-1,2-di-σ(CC) butadiene configuration at 170 K, in which only one half of the molecule interacted directly with the surface. After annealing, a $\eta^4$-1,2,3,4-tetra-$\sigma$ configuration of higher hapticity was implied in which both C=C functions of the molecule developed direct bonds with surface Pt atoms. Annealing these molecules to even higher temperatures was found to lead to similar decomposition pathways and mixtures of analogous surface products as in the case of 1-butene or the 2-butenes as judged from very similar HREELS results.

Experimental confirmation for the formation of 1,2-di-σ(CC) and 1,2,3,4-tetra-$\sigma$ butadiene configurations on Pt(111) has also been reported in later RAIRS and PM-IRRAS (Polarization-Modulated Infrared Reflection Absorption Spectroscopy) studies [155, 171]. RAIR spectra suggested a preferential *trans*-butadiene configuration at medium coverages based on a methylene deformation vibration ($\delta(CH_2) = 1371\,cm^{-1}$) as well as the rocking and wagging modes for C–H ($1023\,cm^{-1}$) and methylene (995, 922, 905, and $890\,cm^{-1}$) groups [171].

NEXAFS studies, on the other hand, performed by a number of groups are not entirely consistent with this picture [168, 172]. Bertolini and coworkers studied the changes in the 1,3-butadiene bonding on Pd(111) and Pt(111) at 95 and 300 K [172]. At 95 K, the molecules were weakly bonded and showed little distortion effects on both surfaces. The C K-edges showed intense $\pi^*$ transitions at grazing incidence polarization of the incident beam, which were strongly damped at normal

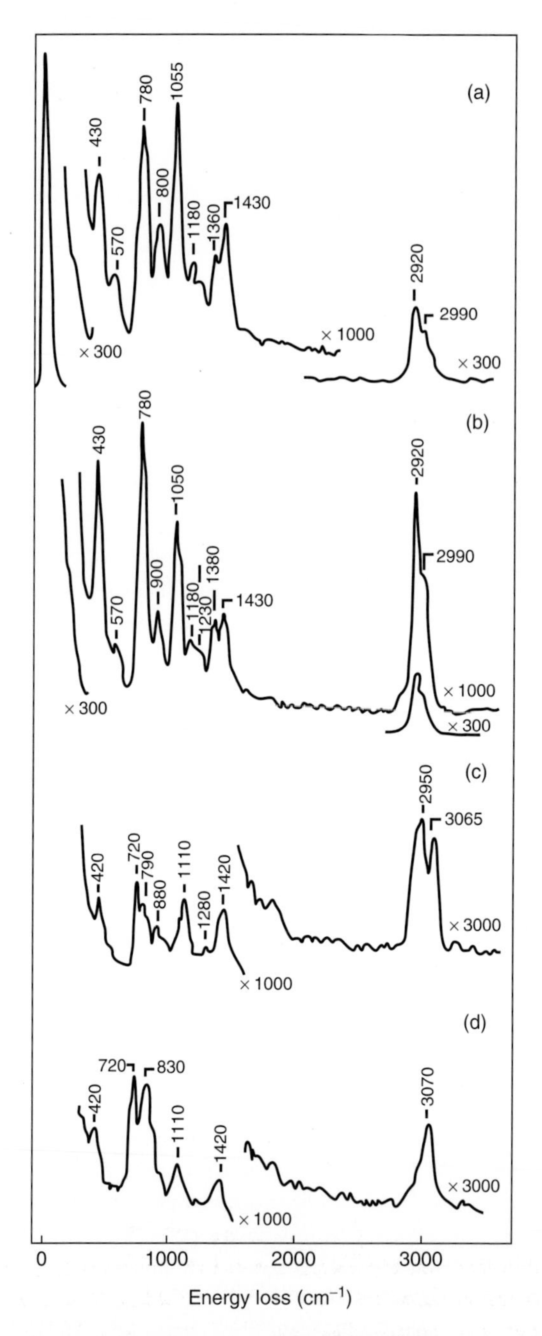


**Figure 40.30** Thermal evolution of the HREEL spectra of a 1,3-butadiene monolayer on Pt(111): (a) 170 K, (b) 300 K, (c) 385 K, and (d) 450 K. Spectra have been recorded after recooling. (Image reproduced from Ref. [148].)

incidence. Simultaneously, $\sigma^*_{C-C}$ resonances were measured that followed the opposite trend, which indicated a planar orientation of the $\pi$ systems to the surface (perpendicular $\pi$ orbitals) and led to the conclusion that 1,3-butadiene binds in a rather undistorted $\pi$ configuration under these conditions. Independent studies performed at the same time by Tourillon and coworkers on Pt(111), Pd(111), Pd(110), and $Pd_{50}Cu_{50}$(111) samples combining NEXAFS with HREELS and UPS also confirmed a weakly bonded 1,3-butadiene species on all surfaces, which was consistent with $\pi$ bonding [168].

At 300 K, however, the adsorption of butadiene is changed dramatically and depends strongly on the nature of the substrate: on Pd substrates, a di-$\pi$ bonded form is concluded from the observation of the $\pi_1^*-\pi_2^*$ splitting due to the intact conjugated $\pi$ system, whereas on Pt(111) this feature was absent while also the $\sigma^*_{C=C}$ resonance at ~300–305 eV was weakened, from which a terminal $\eta^2$-1,4-di-$\sigma$ mode, a metallacycle with a "free" C=C bond, was proposed [168, 172]. In the molecule, half of the carbon atoms rehybridize from $sp^2$ to $sp^3$, as indicated by HREELS [172].

The results were interpreted in the light of activities and selectivities for butadiene hydrogenation: better selectivities to butenes and higher activities compared to Pt(111) were measured on the Pd-based systems, which were less prone to side reactions such as isomerizations [168, 172]. The reason was ascribed, on one hand, to the different binding configurations, namely the higher reactivity of the $\eta^4$-di-$\pi$(CC)-species for hydrogenation, which is related to the findings for smaller alkenes discussed in the previous sections [168]. On the other hand, especially for understanding the trends within the range of tested Pd orientations and Pd–Cu alloy, also adsorption energetics had to be considered. Not only did the activity decrease from Pt(111) to Pd(111) despite higher adsorption strength and molecular distortion in the chemisorbed state of the diene reactant, but it also decreased in the sequence $Pd_{50}Cu_{50}(111) > Pd(110) > Pd(111)$. The molecular distortion and adsorption strengths for butadiene, which were deduced from the $\pi_1^*-\pi_2^*$ splitting in the NEXAFS spectra, followed the opposite trend [168].

These observations were contradictory to the inverse relationship between the activation energies and reactant adsorption strength suggested by Cosyns [169], but could be understood from the added competition between the different surface species. Because of the bifunctional nature of 1,3-butadiene, selectivity is introduced into the hydrogenation process. Moreover, the relationship between adsorption strength and activity of a single reactant becomes more complex by the competition with intermediately formed species, namely the butenes.

Prior theoretical computations of the adsorption of ethylene and 1,3-butadiene on the Pt(111), Pt(110), and Pd(111) facettes of small metal clusters using the EH method, which also found a preference for $\pi$ coordination on Pd and di-$\sigma$ bonding on Pt surfaces, provided insight into this interplay [101]. The higher adsorption energy of butadiene was proposed to be the driving force for the facile displacement and desorption of weakly bonded butene from the Pd(111) face after hydrogenation and explains the higher selectivity compared to Pt(111). Consistently, on Pt(111), the adsorption energies were similar and consequently ought to yield a mixture of products [101].

Also, Tourillon *et al.* used the argument that the competition between (re)adsorption of the mono-olefins and the binding of the diene on the catalyst governed the selectivities. These authors pointed out two potentially important factors that had previously been neglected: changes in the number of active species on the respective catalysts, and the impact of hydrogen adsorption and diffusion dynamics. The characterization of the adsorption structures and strengths allowed for a simple qualitative understanding of the activities and selectivities of the Pt and Pd catalysts. This analysis was found to be useful in a number of later studies [171, 190]. Thus, for example, activities were successfully correlated with adsorption energies of 1,3-butadiene in the low-temperature hydrogenation over several bimetallic Pt catalysts alloyed with another transition metal such as Co, Ni, or Cu and supported over $\gamma$-$Al_2O_3$. This correlation allowed extrapolation of the activity of novel catalysts [190].

In a similar alloying study focusing on hydrogenation activities on Pt(111) and the p(2 × 2) and ($\sqrt{3} \times \sqrt{3}$)-R30° terminations of a $Pt_3Sn(111)$ bulk alloy, Jugnet *et al.* measured substantially improved hydrogenation selectivity to butenes after alloying and simultaneously decreased activity by an order of magnitude compared to Pt(111) [171]. A dependence of the product distribution and activities on the adsorption mode and strength of 1,3-butadiene and the competing butene isomers was invoked in this case also. On the Pt–Sn alloys, changes in the adsorption modes are necessitated by the different surface atom ensembles, eliminating adsorption configurations possible on pure Pt(111), which was concluded to be one reason for the different catalytic performances.

On Pt(111), a zero-order kinetics with respect to 1,3-butadiene consumption was observed, which indicated a surface fully saturated in reactive butadiene species. This in turn implied a weaker di-$\sigma$ bonded form as the most probable candidate to undergo hydrogenation, and also an unfavorable competition with butene adsorption, thus leading to poorer selectivity and deep hydrogenation to butane [171].

The saturation of the surface with reactants suggested another role that heteroatoms in the alloy surfaces play for the control of selectivity besides their modification of the electronic structure of the active sites. Because of a dilution effect by spatially separating the adsorption sites, repulsive interactions between surface species are relieved and different adsorption states can become energetically favorable on the alloy surfaces [171]. Hence, again changes in adsorption configurations are induced that can lead to different reaction pathways and products.

The analysis performed by Jugnet *et al.* was partially based of two DFT studies, the works of Mittendorfer *et al.* and Valcarcel *et al.*, who both calculated a number of stable adsorption structures of 1,3-butadiene on Pt(111) (Figure 40.31) [102, 151].

Mittendorfer found that the most favorable adsorption structure in his set of stable minima on Pt(111) was a $\eta^4$-1,2,3,4-tetra-$\sigma$ coordination ($E_{ads} = -150$ kJ/mol at 0.14 ML coverage, Figure 40.31f) with butadiene in a trans configuration (or also called (*s*)-*trans* to point out the conjugated single-bond character), independent of

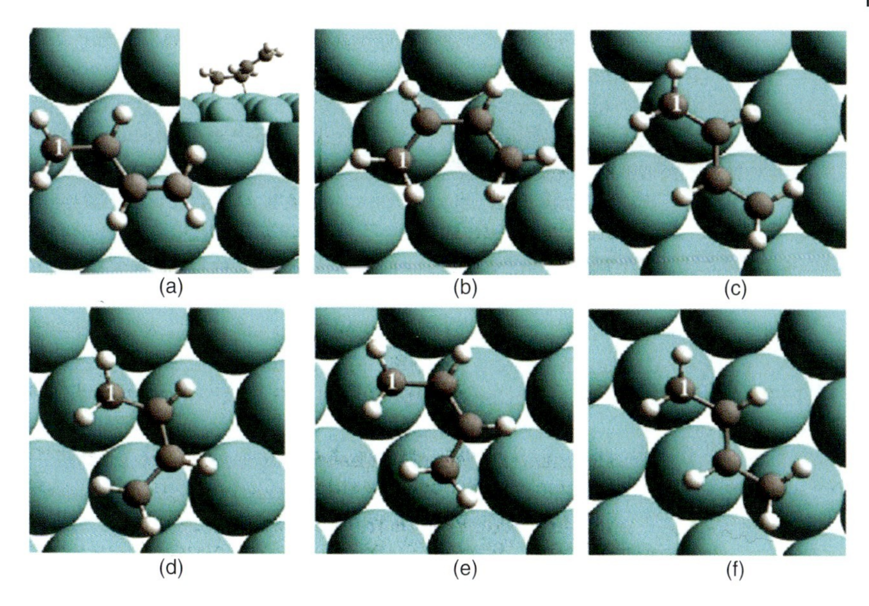


**Figure 40.31** Adsorption modes of 1,3-butadiene on Pt(111) and Pd(111): $\eta^2$-di-σ(CC)-*trans* (a), $\eta^4$-di-π-*cis* (b), $\eta^4$-di-π-*trans* (c), $\eta^4$-1,2-di-σ-3,4-π-*cis* (d), $\eta^4$-1,4-di-σ-2,3-π-*cis* (e), and $\eta^4$-1,2,3,4-tetra-σ-*trans* (f). Carbon atom number 1 is displayed for clarity. (From Ref. [151].)

the chosen surface coverage [102]. Although this result is different from the findings on Pt(111) and Pd(111) discussed above, it was supported by a comparison of the vibrational frequencies of this structure with experimental HREELS data available from Avery and Sheppard [148] By this analysis, two of the three most probable adsorption modes, that is, the $\eta^2$-di-σ(CC) configurations, were ruled out on Pt(111) despite their quite high adsorption energies, whereas a good agreement for the $\eta^4$-tetra-σ structure was obtained. Another adsorption structure, the $\eta^4$-1,2-di-σ(CC)-3,4-π(CC) mode of (*s*)-*cis*-butadiene, in which each C=C bond formed a di-σ or a π bonding shape around two Pt atoms of a threefold hollow site, could not be differentiated by its vibrational frequencies. At higher coverages such as one-fourth the theoretical ML, it was computed to become energetically competitive (~70 kJ/mol) with the $\eta^4$-tetra-σ form. Also these authors, having computed much lower adsorption energies for 1-butene, attributed the higher selectivity of Pd to the competition with butadiene bonding [102].

Valcarcel *et al.* extended the number of tested configurations (Table 40.8) on Pt(111) and Pd(111), and computed also HREEL intensities besides vibrational frequencies for the identification of the experimentally present adsorbate complexes (Figure 40.32) [151]. As mentioned in the previous section, also the bonding of 1-butene and *cis*/*trans*-2-butenes was studied.

Their first finding was that the adsorption on both metals was quite similar: identical most stable adsorption structures, the previously reported $\eta^4$-1,2,3,4-tetra-σ-*trans* and $\eta^4$-1,2-di-σ(CC)-3,4-π(CC), were obtained with comparable adsorption energies in the range of −140 to −160 kJ/mol in the larger supercells.

**Table 40.8** Adsorption energies (kJ/mol) per molecule and per Pt atom (in parenthesis) of stable structures of 1,3-butadiene on Pt(111) as a function of coverage.

| 0 | 1/3 | 1/4 | 1/6 | 1/9 |
|---|---|---|---|---|
| 1,3-Butadiene | | | | |
| Di-$\pi$-*cis* | | | | −114(−13) |
| Di-$\pi$-*trans* | | | −110(−18) | −122(−14) |
| di-$\sigma$ | −45(−15) | −78(−20) | −85(−14) | −89(−10) |
| 1,2-di-$\sigma$-3,4-$\pi$ | | −75(−19) | −122(−20) | −140(−16) |
| 1,4-di-$\sigma$-2,3-$\pi$ | | | −137(−23) | −150(−17) |
| 1,2,3,4-Tetra-$\sigma$ | | | −156(−26) | −160(−18) |
| l-Butene | | | | |
| $\pi$ | | | | −66(−7) |
| di-$\sigma$ | −45(−15) | −83(−21) | −91(−15) | −93(−10) |
| 2-*cis*-Butene | | | | |
| $\pi$ | | | | −58(−6) |
| di-$\sigma$ | | | | −80(−9) |
| 2-*trans*-Butene | | | | |
| $\pi$ | | | | −48(−5) |
| di-$\sigma$ | | | | −79(−10) |

Reproduced from Ref. [151].

Also a "(pseudo-)metallacycle-type mode," a 1,4-di-$\sigma$(CC)-2,3-$\pi$(CC) form related to the structure proposed by Bertolini, Tourillon, and coworkers from NEXAFS [168, 172], was computed to be a local minimum, which was only slightly disfavored by ~10 kJ/mol. Comparison of published [148] and computed HREEL spectra confirmed that the 1,2,3,4-tetra-$\sigma$ adsorption is the most probable adsorption structure under low-temperature and low-pressure conditions and did not exclude the 1,4-metallacycle as a minority species (Figure 40.32) [151].

The major difference in the adsorption on the Pd(111) were the markedly smaller molecular distortions upon adsorption, which despite similar adsorption modes, resulted in different vibrational frequencies on the two surfaces [151]. An energy decomposition analogous to the one discussed for ethene but neglecting the contribution of the substrate relaxation allowed a crude correlation of the molecular deformation costs to the degree of distortion and the perturbation of the vibrational normal modes. Indeed, it is important to remember that the internal distortions of the geometry and bonding of the adsorbate species are one of the two determining factors for the perturbation of the frequencies from the gas phase to those in the surface complex, with the indirect effects by the new surface bonds being the other.

Contrary to expectation, the $\eta^4$-di-$\pi$ configurations expected from HREELS on Pd(111) [148] were obtained as stable forms only at low coverages. Just like the $\eta^2$-$\pi$(CC) forms of the butenes, they were less stable than most of the other forms. Surprisingly, the adsorption energies of 1-butene on these metals were also quite close, −67 kJ/mol versus −79 kJ/mol on Pt(111) and Pd(111), respectively. Therefore, these authors concluded that the different selectivities for the hydrogenation

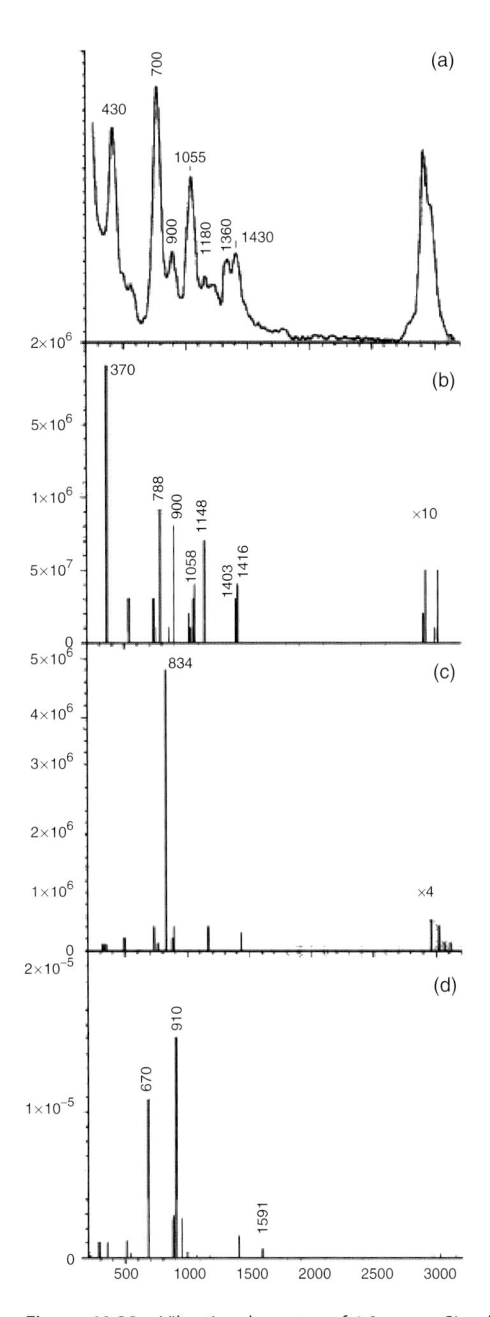


**Figure 40.32** Vibrational spectra of 1,3-butadiene adsorption modes on Pt(111). Experimental EELS spectrum reproduced from Ref. [148] (a), 1,2,3,4-tetra-$\sigma$ (b), 1,4-di-σ-2,3-$\pi$ "metallacycle" (c), and di-σ (d). Simulated spectra have been calculated for a coverage of 1/6-ML; frequencies are given in $cm^{-1}$. Notations ×10 and ×4 account for multiplication factors with respect to the di-σ spectrum (d).

reactions of butadiene on Pt(111) and Pd(111) cannot be satisfactorily explained by the competition between diene and alkene adsorption [151].

Taking up the problem of selectivity, Valcarcel *et al.* explored the role of intermediates and transition states for the Horiuti–Polanyi mechanism [17] in a subsequent study using DFT with a generalized gradient functional [187]. The structures of various mono- and dihydrogenated butadiene intermediates on both metal surfaces were calculated, which showed that radical species were generally better stabilized by Pt than by Pd.

The key to selectivity was found to be the relative stability of the butan-1,3-diyl intermediate (Figure 40.33c) [187]. Its energy with respect to the diradical in the gas phase is −365 kJ/mol on Pt(111) and −326 kJ/mol on Pd(111), rendering it less stable and more reactive on the Pd surface. The differences in the stabilization arise from the interaction energy between the diradical and the surface (−434 kJ/mol on Pt and −380 kJ/mol on Pd), while the distortion energy of the species and the surface is within 9 kJ/mol on both metal surfaces as was shown with the energy decomposition analysis.

Key to the pathway leading primarily to the formation of butane is the butan-1,3-diyl intermediate, which determines the overall selectivity of 1,3-butadiene hydrogenation [187]. In this pathway, 1,3-butadiene transforms in a two dihydrogenation sequence via a 2-buten-1-yl intermediate (Figure 40.33b) and the butan-1,3-diyl species to butane. On Pt(111), the better stabilization of the butan-1,3-diyl intermediate favors this transformation slightly over a competing pathway via 2-buten-1-yl to 1-butene, resulting in lowest activation barriers of 99 kJ/mol versus 106 kJ/mol for both processes, respectively. On Pd(111), in contrast, the stabilization of the butan-1,3-diyl intermediate is smaller and the pathway toward this radical is not accessible, whereas the pathway to 1-butene production is more favorable by 39 kJ/mol.

Thus, taking into account all possible sources of error, Valcarcel *et al.* concluded that their calculations predicted a full selectivity to butene on Pd(111) in agreement with the experiments [187]. Moreover, the pathway to 1-butene was found consistently to have lower activation energies than that to 2-butene on both metals, in agreement with the observed distribution of products. For the Pt catalyst, the two pathways are roughly equally likely, which explains its poor selectivity to butenes and the formation of butane from the beginning of the reaction. The formed butenes will be displaced into the gas phase as long as butadiene is present in the gas phase, since the adsorption of butadiene is strongly favored, but over time the selectivity can decrease further.

Hence, these authors provided a detailed understanding of the hydrogenation mechanism on Pt(111) and Pd(111) model catalysts and showed two key factors governing the selectivity of the butadiene hydrogenation: kinetics, that is, energetics of intermediates and transition states, as well as the adsorption competition between reactants and products.

Vigné *et al.* built further on this result and explored the role of the promoter Sn for selective hydrogenation of 1,3-butadiene with DFT calculations of adsorption

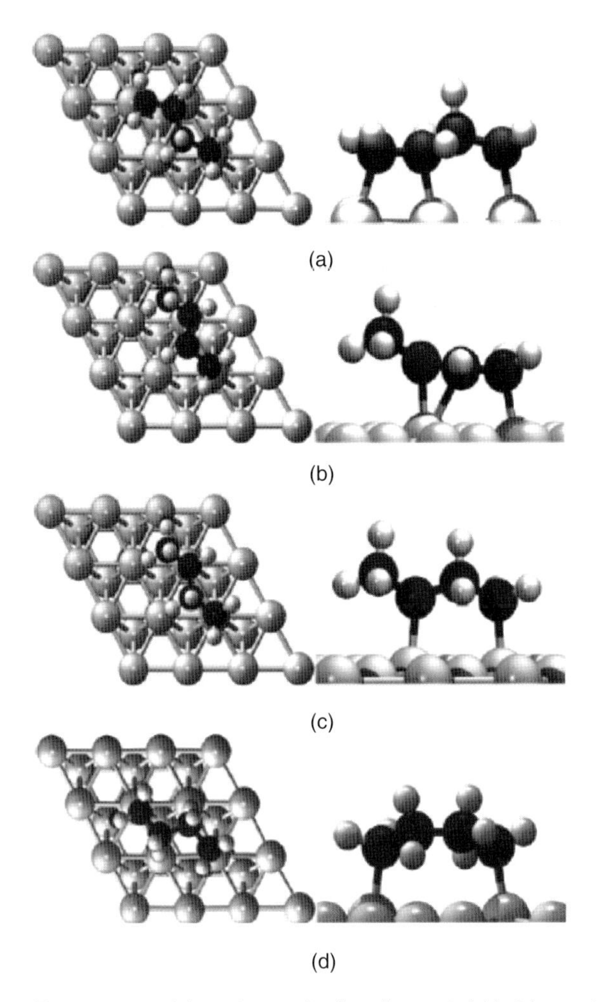


**Figure 40.33** Adsorption modes for 1-buten-4-yl (a), 2-buten-1-yl (b), butan-1,3-diyl (c), and butan-1,4-diyl (d) on Pd(111) and Pt(111) metal surfaces. (Reproduced from Ref. [187].)

structures, pathways, intermediates, and transition state on the $Pt_2Sn$/Pt(111)-($\sqrt{3} \times \sqrt{3}$)R30° surface alloy [127]. Generally, in line with the expectations based on other studies, the adsorption energies of 1,3-butadiene, the 1- and 2-butenes, and all the intermediate species are reduced by the alloying compared to pure platinum. The Pt–C bonds are somewhat destabilized, and Pt–Sn bonds are unstable for these alkenes and diene [50, 191]. With respect to the reference model Pt(111), the most stable adsorption mode of butadiene at low coverages (1/9 ML) changes from the 1,2,3,4-tetra-$\sigma$-*trans* to the 1,4-di-σ-2,3-π-*cis* "metallacycle" form ($E_{ads} = -74$ kJ/mol at the GGA level) followed closely in energy by a 1,2-π-3,4-π-*trans* mode form ($E_{ads} = -63$ kJ/mol) [127]. As the necessary binding site of four Pt atoms is eliminated due to the ensemble effect with Sn, the 1,2,3,4-tetra-$\sigma$-*trans* simply cannot be realized with four Pt ligands on the $Pt_2Sn$(111)

surface alloy. Consistently, also the adsorption energy of 1-butene is computed at a smaller value of −49 kJ/mol. The comparison of the calculated adsorption energies (thermodynamic) of butadiene and 1-butene therefore yielded a surprisingly good agreement with the activation energies of desorption deduced from the TPD experiments (75 and 54 kJ/mol [173, 189], respectively) even when neglecting the frequency factor dependence.

The reduction in the adsorption energy of these alkenes on the Pt/Sn surfaces was found to be consistent with the correlation to the energetic lowering of the d-band center of gravity from 2.39 to 2.57 eV below Fermi energy upon alloying [50, 191], which results in a smaller orbital interaction. In addition, also the electron transfer from Sn to surface Pt atoms present on the alloy surfaces was proposed to cause a weakening of the adsorption due to increasing Pauli repulsions [127]. In conjunction with the elevated position of the Sn atoms [42], the larger repulsive interactions of the surface species with substrate atoms led to higher structure deformations and related energy penalties, of which the dominant contribution could be traced back to the relaxation of the substrate [127]. Since the increase of the energy penalties was larger for di-σ coordinated structures than for the π structures, the adsorption energy of species involving π bonding was affected less by the weakening. This in turn changed the relative stabilities of the di-σ and π forms on the alloy surface to favor the latter. Similar findings had also been reported by these authors for unsaturated aldehydes on Pt–Sn surface alloys, and they will be discussed therefore in more detail in the next sections [35, 37, 38, 192].

Since the cis isomer is more stable than the trans form on the surface alloy, contrary to the gas phase where the *trans*-1,3-butadiene is preferred by 17 kJ/mol, energetically feasible conversion mechanisms are needed to reach the most stable adsorption form. A multistep trans–cis conversion mechanism from the 1,2-π-3,4-π-*trans* structure to the most stable 1,4-di-σ-2,3-π-*cis* form was identified on the surface with activation barriers below 40 kJ/mol, which should be a thermally facile processes above room temperature [127].

Concerning the high selectivity for 1,3-butadiene hydrogenation on this Pt–Sn surface alloy, the authors found by *ab initio* thermodynamic analysis that, under typical reaction conditions (360 K, 15 Torr), only butadiene is adsorbed, while butene is expected to desorb immediately before any further attack by hydrogen atoms can take place [127].

Similar reaction pathways and intermediates as on pure Pt(111) were found, but the energetics had been changed dramatically by inclusion of the promoter [127]. Assuming a $\eta^4$ coordination of butadiene as starting point, two groups of pathways were identified that either led via a 3-buten-1-yl intermediate to 1-butene, butan-1,3-diyl, and butan-1,4-diyl as the first mono-hydrogenated products (Figure 40.34a), or via a 1-buten-3-yl intermediate to 2-butene as well as 1-butene and butan-1,3-diyl (Figure 40.34b). The attack of a hydrogen atom on the terminal carbon (C1) to form the 1-buten-3-yl intermediate in the second group was found to have a much lower activation barrier (71 kJ/mol) than an attack on a central carbon (C2; 136 kJ/mol) to form a 3-buten-1-yl species. The computed value is quite close to the experimental activation energy of 72 kJ/mol determined by Zhao *et al.* with TPRS, indicating that this is the rate-limiting elementary step [173].

Since higher barriers result in much lower reaction rates, the first group of pathways was essentially ruled out [127]. Among the four subsequent mechanisms that the 1-buten-3-yl intermediate can undergo, further hydrogenation to 1-butene shows by far the lowest barriers (overall 33 kJ/mol). All other pathways exhibit larger activation barriers, particularly if leading to metallacycles, which is key to explaining the high selectivity to 1-butene compared to Pt(111). The butane-1,3-diyl intermediate is energetically strongly destabilized by the presence of Sn and, besides the adsorption competition with new butadiene reactants, prevents butane formation. The second most favorable subsequent process yields 2-butene in two elementary steps with a maximum barrier of 50 kJ/mol. Although 2-butene is found in the experiments, the difference to the barrier heights of the 1-butene formation was too large to explain the sizable amount formed. A possible explanation could be the isomerization from 1-butene to 2-butene as observed experimentally, likely through a process of dehydrogenation–rehydrogenation steps depending on the hydrogen coverage.

In comparison to the reaction energetics in Pt(111), the conclusion on the role of the promoter tin on the hydrogenation reaction consisted of two parts: First, tin exerts a function as a site blocker due to its much weaker interaction to the surface species, eliminating adsorption modes and increasing the relaxation of the remaining structures such that transition states in unselective pathways are strongly penalized. Second, the geometric effects of Sn are linked to electronic effects that in conjunction weaken the surface bonding, deciding the competition between butadiene and butenes and weakening the coordination of the C=C double bonds so that the energetics in the selective pathway to butenes are favored.

This study greatly adds to the understanding of a model reaction of a simple multifunctional molecule. In the next section, we will focus on a little more complex probe molecule, in which both functional double bonds are chemically different, which greatly increases the complexity and number of options to be investigated.

### 40.3.4 Bifunctional Molecules with Chemically Different Double Bonds: Prenal, Acrolein, Crotonaldehyde

#### 40.3.4.1 Overview of Recent Experimental Studies

In this last section, we will discuss real catalytic and model catalytic studies of the hydrogenation of α,β-unsaturated aldehydes on mono- and bimetallic catalysts, also with a focus on Pt and the Pt–Sn system. The α,β-unsaturated aldehydes have conjugated C=C and C=O functions, which break up the equivalence of the double bonds in 1,3-butadiene. This can greatly enhance the complexity of any involved process, although it is still a rather simple multifunctional molecule with only two different chemical functions.

First we will review the studies of the past decades in this area, before narrowing down the discussion on prenal (3-methyl-2-butenal, see Figure 40.35). It has been

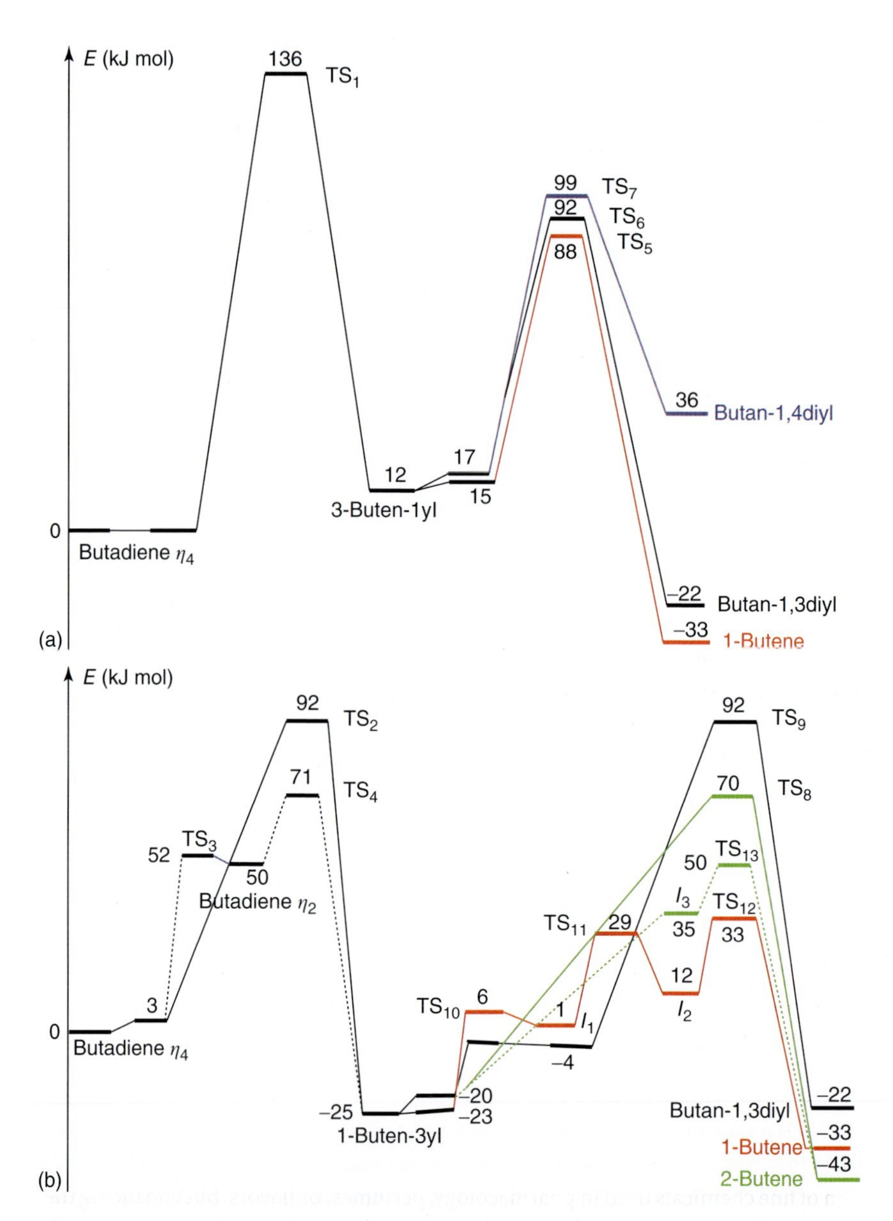


**Figure 40.34** Butadiene hydrogenation energy profiles (kJ/mol) on $Pt_2Sn/Pt(111)$-$(\sqrt{3} \times \sqrt{3})R30°$ via 3-buten-1-yl (a) and 1-buten-3-yl intermediates (b). The energy reference is the energy of butadiene coadsorbed with two hydrogen atoms at infinite distance. (Reproduced from Ref. [127].)

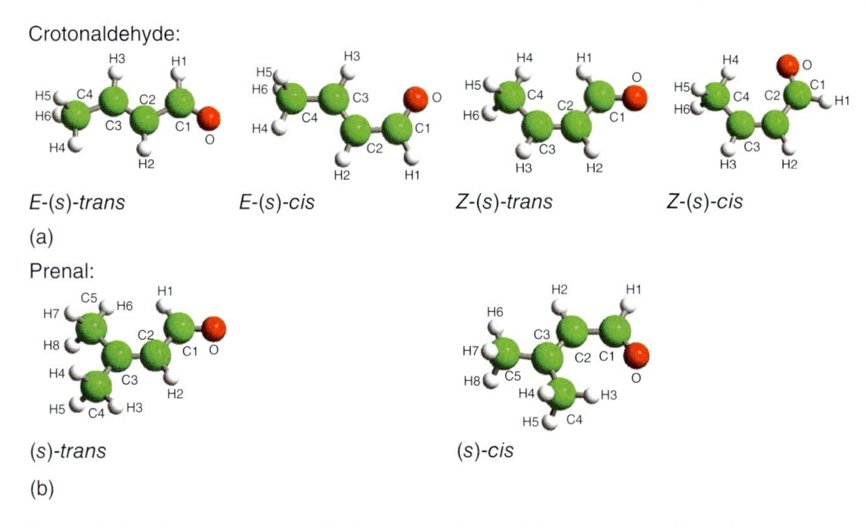


**Figure 40.35** Isomers of crotonaldehyde (a) and prenal (b) in the gas phase. The isomers of acrolein are identical to those of prenal, except that the two terminal $CH_3$ substituents are replaced by hydrogen.

intensively studied in the past just like acrolein (2-propenal) or crotonaldehyde (2-butenal, Figure 40.35), and exists in two energetically close isomers in the gas phase or on surfaces: the (*s*)-*trans* and the (*s*)-*cis* form differing in the steric configuration around the central single bond, which does not show free rotation due to the conjugated π system. Crotonaldehyde possesses not only this central cis/trans isometry but also an *E*/*Z* isomery around the C=C bond since the β-carbon is only singly methyl-substituted. This enhances the configuration space to four isomers that are energetically similar: *E*-(*s*)-*trans*, *Z*-(*s*)-*trans*, *E*-(*s*)-*cis*, and *Z*-(*s*)-*cis*. This renders croton-aldehyde more difficult to study and less suitable as a probe molecule.

While more information is available on the hydrogenation pathways of acrolein on Pt(111), the adsorption and behavior of prenal has been studied on a wider range of surfaces including the Pt–Sn surface alloys. Thus, we will highlight key findings for prenal, and refer to acrolein for the question of selectivity in the hydrogenation process of these molecules.

The study of α,β-unsaturated aldehydes in surface chemistry and catalysis is interesting from a fundamental point of view, but it has also technical and economic relevance. The selective hydrogenation of α,β-unsaturated aldehydes to α,β-unsaturated alcohols (Figure 40.36) is an important catalytic process for the industrial production of fine chemicals used in pharmacology, perfumes, or flavors, but enhancing the product yields remains a major challenge [1, 3, 193].

The difficulties in the control of the selectivity in this reaction arise from the thermodynamic preference to hydrogenate the C=C double bond [194, 195]. The less favorable hydrogenation of the C=O bond has to be performed at low temperatures under kinetic control. Currently, the easiest method to obtain the unsaturated alcohols in high yield (>90%) in the laboratory is by reduction using metal hydrides such

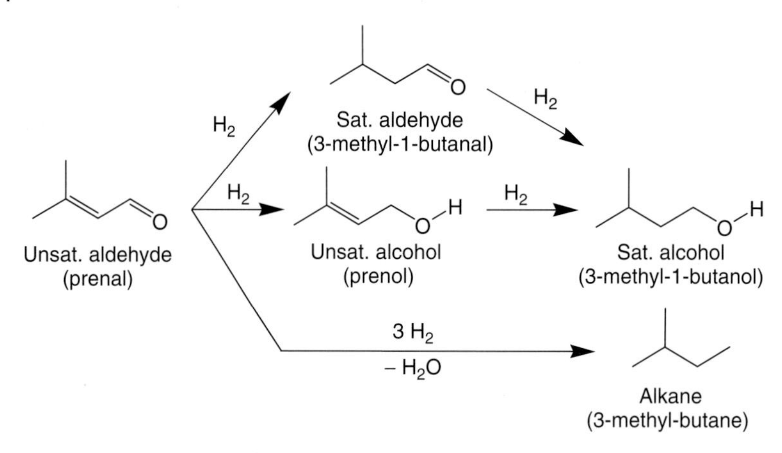


**Figure 40.36** Hydrogenation reactions of α,β-unsaturated aldehydes to unsaturated alcohols, saturated aldehydes, saturated alcohols, or alkanes: the case of prenal. (From Ref. [35].)

as the derivates of $NaBH_4$ or $LiAlH_4$ [196], but this method is hardly feasible on an industrial scale. This highlights the need for the design of sophisticated catalysts with high selectivity toward unsaturated alcohols and sufficient activity that also must be stable against coking and poisoning.

Experimentally and theoretically, the investigation of the hydrogenation and adsorption of α,β-unsaturated aldehydes has proven to be a difficult task. The multifunctionality of the molecules with their two double bonds allows a huge variety of different interactions with a substrate. The coordination types range from pure $\eta^2$-di-σ and $\eta^2$-$\pi$ interactions of either the C=C or C=O double bonds to low coordinated $\eta^1$-top interactions with the aldehydic oxygen, to di-σ(C3,O) metallacycles bonded via the terminal atoms, and to $\eta^3$ and $\eta^4$ adsorption structures, which again can consist of combinations of di-σ and π interactions. The number of conceivable binding geometries will be multiplied when promoters such as Fe or Sn modify the reaction sites on the substrate. This renders these complex systems prime candidates to combine several approaches for a more thorough analysis.

Reactivity studies on supported catalysts have shown that Pd, Ni, Cu, or Rh are much less selective toward the unsaturated alcohols than Pt, Co, Ru, or Ir, which, nonetheless, also do not show an inherently high selectivity [188, 193, 195, 197–218]. The selectivity is greatly influenced by the nature of the support, the size of the catalyst particles, and the reaction conditions. Especially, promoters that can form electropositive species on the surface were suggested as a promising route to tune the performance [200–202].

On Pt(111) and $Pt_{80}Fe_{20}(111)$ single crystals, the selectivity of the hydrogenation of crotonaldehyde and prenal toward the unsaturated alcohol was found to depend mainly on kinetic parameters. It has been increased by alloying with Fe, although the main product on both surfaces remained the saturated aldehyde [197].

Studies on $TiO_2$-supported Pt catalysts by Coloma and coworkers showed that crotonaldehyde can displace CO preadsorbed on the top and bridge sites of the catalyst surface [212]. Unfortunately, their fourier-transform infrared spectroscopy (FT-IR) experiments did not allow the determination of the adsorption site or mode of crotonaldehyde, but an indirect influence on the adsorption properties was observed due to the metal–support interaction that modified the catalyst particles and their the reaction sites.

More information on the bonding of acrolein on Pt(111) is available from NEXAFS and FT-IR studies performed by Bournel *et al.* [219, 220], who proposed that the molecules were physisorbed at 95 K in flat orientations. Complementary RAIRS measurements by de Jesús and coworkers contradicted the physisorption argument. These authors also detected flat geometries for acrolein and crotonaldehyde on Pt(111), but both were sizably chemisorbed. The flat bonding was thought to involve both double bonds and to be the origin of the low selectivity on pure platinum catalysts [221, 222]. The measured shifts of the vibrational bands in their experiments suggested that acrolein was adsorbed mainly via the C=O group (e.g., a di-σ(C=O) form), whereas crotonaldehyde showed a strong di-σ(CC) interaction to the surface. After providing thermal activation energy, the decarbonylation of the adsorbed crotonaldehyde to propylene and CO, desorbing subsequently around 340 and 440 K, was observed. This was accompanied by the desorption of $H_2$ at 280 and 440 K, suggesting the dehydrogenation of unidentified surface species.

The bonding of the aldehydes can vary with pressure and temperature, and phase transitions between mixed phases of different configurations can occur. Depending on the nature of the species that undergoes hydrogenation, like for the π-bonded mono-alkenes, this can lead to a pressure gap. *In situ* SFG vibrational spectroscopy recorded during kinetic measurements of the hydrogenation of acrolein, crotonaldehyde, and prenal over Pt(111) at 1 Torr of aldehyde and 100 Torr of hydrogen in the temperature range 295–415 K showed that acrolein is bonded in flat orientations [223]. Although the structures involved coordination by the C=O bond, the C=C bond participated as well: a mixed phase of $\eta^2$-di-σ(CC)-*trans*, $\eta^2$-di-σ(CC)-*cis*, and highly coordinated $\eta^3$ or $\eta^4$ species was proposed. The SFG results of crotonaldehyde and prenal on Pt(111) implied $\eta^2$ surface intermediates in both cases. For prenal, even higher coordinated species became detectable as the temperature was raised to 415 K, in agreement with its enhanced C=O hydrogenation. The kinetics showed little structure sensitivity except that Pt(100) led to higher decarbonylation while Pt(111) had a slightly higher selectivity toward crotyl alcohol.

The structural dependence of the reactivity and the active sites was also analyzed by Birchem *et al.*, who compared the hydrogenation of prenal on a stepped Pt(553) surface to that on Pt(111) in thermal desorption/reaction and high-pressure hydrogenation experiments [198, 199]. Based on kinetic models, the selectivities toward the different products were attributed to the formation of different surface species formed on the available surface sites: steps were identified as reaction sites for the formation of the saturated aldehyde, and flat terraces for the formation of the saturated alcohols. The structural effects of steps also led to an increase of the activity due to the higher reactivity of low-coordinated step atoms for hydrogen dissociation.

Comparison of the hydrogenation reactions of acrolein, crotonaldehyde, and prenal on $SiO_2$- and $Al_2O_3$-supported Pt catalysts by Marinelli *et al.* showed that the selectivity was improved by increased steric demands created by the substitution of the C=C bond with methyl groups, especially when Fe was added as a promoter [200, 201]. Their results suggested that the bonding to the catalysts occurred predominantly through the C=O group, which is thereby activated, whereas the C=C moiety remained unperturbed.

Similar to Fe, also Co exhibits a positive effect on the selectivity [202]. *In situ* studies on silica-supported bimetallic Pt–Co catalysts showed that cobalt is octahedrally coordinated and completely reduced in alloyed particles. Subsequent hydrogenation experiments showed an increased selectivity toward the crotyl alcohol, which was ascribed to the formation of $Pt^{\delta-}$–$Co^{\delta+}$ sites by electron transfer. The aldehydic C=O group bonds to these $Co^{\delta+}$ sites, thereby getting activated for attack of the hydrogen atoms.

Birchem *et al.* also observed a shift in the selectivity toward the saturated alcohol by alloying the Pt(553) surface with Sn, which was found to be located near the step edges, and also inducing a local reconstruction of the surface before growing islands on the terraces at high coverages [198, 199]. The increased selectivity was ascribed to electronic effects due to partially oxidized tin moieties.

The enhancement of the selectivity by alloying with Sn was also analyzed by Santori and coworkers, who found that the preparation procedure of the Pt–Sn catalysts was critical for their performance [188]. The nature and properties of the obtained active phase strongly influenced the level of selectivity to unsaturated alcohols in the crotonaldehyde hydrogenation. Especially organo-bimetallic catalysts led to high selectivities toward crotyl alcohol (80% at 5% conversion).

In contrast, Jerdev *et al.* found a similar selectivity on Pt(111), $Pt_3Sn(111)$, and $Pt_2Sn(111)$ when probing the hydrogenation reaction of crotonaldehyde with XPS, LEED, and gas chromatography [195]. Since the primary product was still the saturated aldehyde, they concluded that the selectivity of the hydrogenation could not be solely due to the Sn–Pt alloy formation. Although the activity of the catalyst was nearly doubled upon alloying, the activity did not depend significantly on the surface alloy formed (the activity was only 10% higher in the case of the tin-richer $(\sqrt{3} \times \sqrt{3})R30°$ versus the $p(2 \times 2)$ surface structure). The reaction kinetics implied that the $H_2$ partial pressure was involved in the rate-limiting elementary step for the formation of all products except the alkane (butane). Approximate activation energies for butane and butanal formation amounted to 46.8 and 35.9 kJ/mol, respectively.

Also on Pt–Sn surfaces, the role of $\eta^1$ coordination types for the bonding and selective hydrogenation was found to be important. High-resolution XPS investigations of Janin *et al.*, supported by DFT, pointed to a $\eta^3$ coordination of crotonaldehyde with an interaction through the C=C bond and a bond from the aldehydic oxygen on Pt(111) [224, 225]. Alloying with Sn resulted in a strong decrease of the calculated adsorption energies accompanied by a change of the preferred adsorption structure. The authors interpreted the changes in the XPS spectra as a transition to a mixture of a $\eta^3$ adsorption structure with O–Sn interaction and a vertical form bonded only by an oxygen lone pair ($\eta^1$-top-OSn). On the Sn-richer $Pt_2Sn(111)$ surface, the $\eta^1$ geometry was more abundant than on the $Pt_3Sn(111)$.

Contrary to the behavior on pure platinum surfaces, crotonaldehyde was proposed to bind in a $\eta^1$-top geometry already on pure Rh catalysts supported on partially reduced $TiO_2$ [211]. *Ex situ* XPS experiments of the prepared Rh catalysts pointed to the formation of 20% oxidized $Rh^{\delta+}$ interface sites at the support edge (BE of the Rh $3d_{5/2}$ signal shifted from ~307 eV in the metallic state to ~308.5 eV), which were considered to be governing the activation of the C=O group through the preferential coordination. However, diffuse reflectance fourier transform infrared spectroscopy (DRIFT) spectra recorded under reaction conditions also identified other crotonaldehyde species strongly adsorbed through the C=C bond and weakly coordinated through both the C=C and C=O bonds. The vapor-phase hydrogenation on these catalysts led mainly to the butanal, but the selectivity toward crotyl alcohol was increased (to almost 10%) on more reduced supports, presumably by larger strong metal–support interactions (SMSIs).

Alloying Rh catalysts with Sn was found by Nishiyama *et al.* to further increase the trend toward $\eta^1$ coordination with acrolein [206]. On silica-supported Rh–Sn bimetallic catalysts, two absorption bands of carbonyl groups were observed at 1670 and 1720 $cm^{-1}$, which increased with the Sn:Rh ratio. Apparently, acrolein bonded through the oxygen of the C=O moiety to surface Sn sites. The absorption bands of the activated carbonyl group at 1670 $cm^{-1}$ readily disappeared on contact with $H_2$, unlike the gas-phase band at 1720 $cm^{-1}$.

The same phenomenon was observed for acrolein with $Co/SiO_2$ catalysts, where the dominant reaction involved propan-1-ol formation via propanal [207], and for crotonaldehyde with Cu catalysts on different carbon supports [203, 205] and on $Ni/TiO_2$ and $Pt/TiO_2$ catalysts [204]. Analogous mixtures of flat and vertical $\eta^1$ adsorption modes and related frequency red shifts of 50–70 $cm^{-1}$ of $\nu$(C=O) vibrations were reported on these titania-supported Ni and Pt catalysts reduced at a higher temperature of 773 K, which showed large SMSI effects [204]. After reduction at lower temperatures (non-SMSI state), only species strongly adsorbed through the C=C bond or weakly coordinated through both the C=C and the C=O bonds were inferred. The $\eta^1$ species was discovered to be present when crotyl alcohol was formed, and were proposed to be bonded at the interfacial sites in contrast to the flat structures.

In a related work, Iwasa *et al.* measured the influence of the support on Pd-based monometallic and bimetallic catalysts [210]. Gas-phase hydrogenation in a flow reactor showed that Pd/ZnO catalysts were more selective toward the unsaturated aldehyde compared to $Pd/Al_2O_3$ or $Pd/SiO_2$ catalysts, but the major product was still the saturated aldehyde.

On various carbon-supported Cu catalysts, the turnover frequencies were found to correlate with the number of $Cu^{1+}$ surface sites. DRIFT spectra recorded under reaction conditions revealed evidence for a mono-hydrogenated intermediate, which fitted well into a Langmuir–Hinshelwood model for the reaction kinetics [203, 205].

The hydrogenation of a series of α,β-unsaturated aldehydes with increasing methyl group substitution of the C=C double bond – acrolein, crotonaldehyde, 2-methylbut-2-enal, and 3-methylbut-2-enal – has been studied over $Cu/Al_2O_3$

and sulfur-modified $Cu/Al_2O_3$ catalysts [208, 209]. In the absence of modification, the reaction products were found to be mainly the corresponding saturated alcohol and the aldehyde, whereas pretreatment with sulfur compounds significantly enhanced the rate of unsaturated alcohol formation, particularly for acrolein and crotonaldehyde. The origin of the enhancement was also proposed to be electronic effects: the formation of modified $Cu^0$ and $Cu^+$ surface sites. Unlike $Cu/Al_2O_3$ [208, 209], Pd–$Cu/SiO_2$ catalysts did not favor crotyl alcohol formation from crotonaldehyde hydrogenation over thiophene-modified catalysts because the Cu component was preferentially poisoned, leaving active Pd sites that favored C=C bond hydrogenation [226].

Thus, the evidence generally indicates that the bonding of the molecule to the surface plays an important role in determining the selectivity and activity. However, for a deeper understanding, also theoretical studies proved essential. Numerous theoretical studies also combined with experiment added useful insight into the bonding and reactivity of α,β-unsaturated aldehydes. While concentrating in the following section on the example of prenal, we also review other relevant theoretical works on crotonaldehyde or acrolein [34, 35, 37, 38, 50, 119, 121, 227–232].

#### 40.3.4.2 Prenal Adsorption on a Pt(111) Surface: HREELS and DFT Studies

First, a preparation procedure for a saturated monolayer of molecularly adsorbed prenal was derived from the thermal reaction behavior. TPRS experiments of prenal adsorbed on Pt(111) show the desorption of a large fraction of CO and $H_2$ besides that of a small amount of molecular prenal. At low doses, the desorption of intact prenal is completely suppressed; it is irreversibly and strongly chemisorbed on the platinum surface. Several desorption states of prenal are observed when subsequently increasing the exposures above the monolayer dose: a state at 199 K, then also a weak low-temperature shoulder at ~180 K, and ultimately also the desorption from the physisorbed multilayer at 163 K. A Redhead analysis [56] with a pre-exponential factor of $\nu = 10^{13}\ s^{-1}$ resulted in desorption activation energies of 51, 45, and 42 kJ/mol for these desorption states, respectively.

The decomposition process of irreversibly chemisorbed prenal consists of several steps, as evidenced by the sequence of $H_2$ (295, 400, and 473 K) and CO (414 K) peaks. The first $H_2$ desorption state and the CO evolution are desorption-limited, whereas the other $H_2$ states are reaction-limited. All states saturated prior to the onset of intact prenal desorption at 199 K. Only the first $H_2$ desorption showed large temperature shifts with varying coverage, consistent with a second-order kinetics expected for recombinative $H_2$ formation from adsorbed H atoms. From these TPRS results, one can already expect that a variety of different prenal species are present on the surface, showing different adsorption energies and behavior.

HREEL spectra of the irreversibly chemisorbed prenal moieties were recorded following annealing after a multilayer dose to the Pt(111) sample to 205 K to desorb all other prenal species. Spectra recorded after a subsequent annealing to 250 K showed no sign of decomposition reactions and were very similar to those at 205 K.

In order to analyze the complex patterns in experimental HREEL spectra, a large number of conceivable coordination types for both prenal isomers were computed and their HREELS fingerprints simulated. From the set of initial structures, 10 stable structures ($-24 \geq E_{ads} \geq -59$ kJ/mol at $\Theta = 1/9$ ML) were obtained. The stable structures can be grouped into five strongly bonded species ($E_{ads} \leq -47$ kJ/mol) and five weakly stable forms. The latter forms belong to the $\eta^1$-top, the $\eta^2$-di-σ(CO), and the $\eta^2$-$\pi$(CC) types, which exhibit simulated vibrational patterns that are inconsistent with the observed HREEL spectra between 169 and 300 K and may be excluded in agreement with their lower adsorption energies. The set of strongly bonded prenal forms consists of the $\eta^2$-di-σ(CC)-(*s*)-*trans*, $\eta^2$-di-σ(CC)-(*s*)-*cis* and the flat $\eta^3$-di-σ(CC)-σ(O)-(*s*)-*cis*, $\eta^4$-di-σ(CC)-di-σ(CO)-(*s*)-*trans*, and $\eta^4$-$\pi$(CC)-di-σ(CO)-(*s*)-*trans* structures (Figure 40.37). They are similar to those of crotonaldehyde [50, 227, 37] and acrolein [50, 121, 227, 230] in Pt(111). However, the additional methyl group leads to a weakening of the adsorption energies by ~25% for prenal as compared to crotonaldehyde.

A coverage increase leads to a destabilization of several structures, particularly of the $\eta^3$ and $\eta^4$ forms, which are slightly more stable in relative terms at lower coverages. The adsorption energies of $\eta^2$-di-σ(CC) structures decreased from about −50 kJ/mol at 1/9 ML to about −45 kJ/mol at 1/6 ML coverage, whereas those of the $\eta^3$-(*s*)-*cis* and $\eta^4$ structures decreased from −50 to −38 and −47 to −37 kJ/mol, respectively. Even more pronounced was the decrease for the $\eta^4$-(*s*)-*trans* form, being almost 50% from −59 to −31 kJ/mol [37, 38].

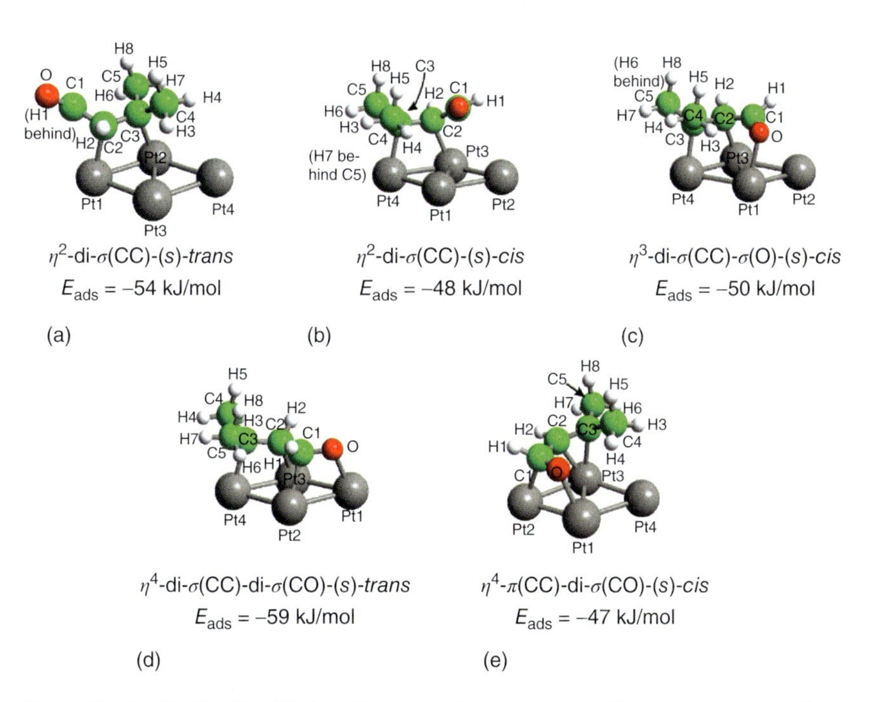


**Figure 40.37** The five identified, stable prenal adsorption modes on Pt(111) (structures obtained from the DFT optimizations at a theoretical coverage of 1/9 ML): (a) $\eta^2$-di-σ(CC)-(s)-trans, (b) $\eta^2$-di-σ(CC)-(s)-cis, (c) $\eta^3$-di-σ(CC)-σ(O)-(s)-cis, (d) $\eta^4$-di-σ(CC)-di-σ(CO)-(s)-trans, (e) $\eta^4$-$\pi$(CC)-di-σ(CO)-(s)-cis.

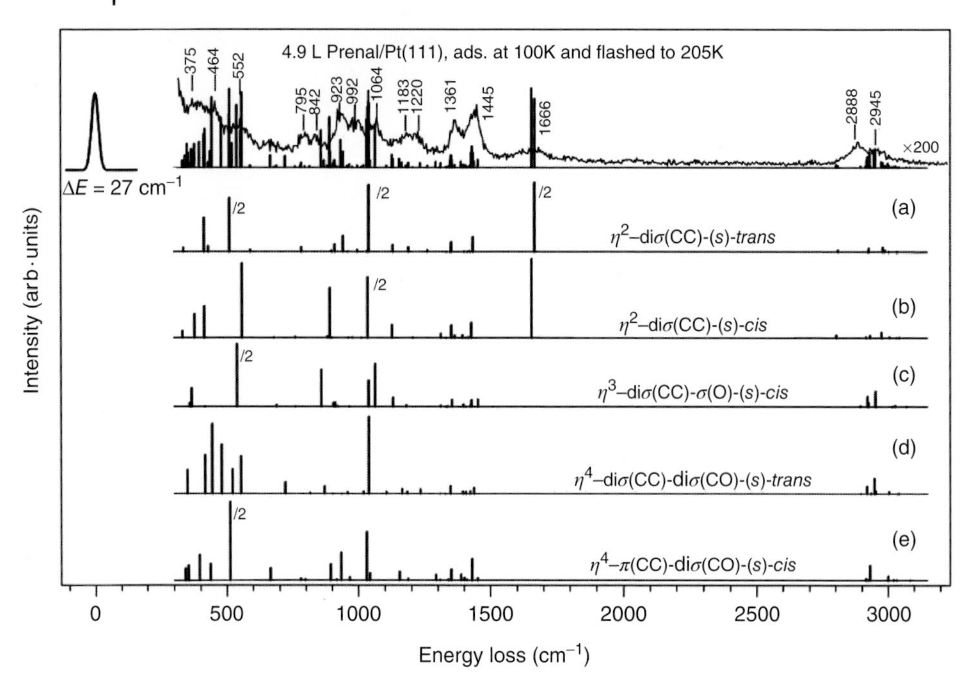


**Figure 40.38** Comparison of the HREELS spectra of 4.9 L prenal/Pt(111) after heating to 205 K with the computed spectra of the five identified most stable adsorption structures (theoretical coverage 1/9 ML). (From Ref. [35].)

The comparison of the simulated HREEL spectra of the five most stable prenal structures with the experiment recorded after annealing to 205 K suggests that a complex mixed phase of these forms is present on the Pt(111) surface under UHV conditions (Figure 40.38).

Although the calculated absolute intensities differ considerably from the measurements, especially for the vibrations involving a strong dipole moment variation, several strong points can be derived for the existence of the five species. Briefly, a strong loss peak at 842 $cm^{-1}$ identifies the $\eta^3$-di-σ(CC)-σ(O)-(*s*)-*cis* structure in the mixed phase. It is the only one sufficiently active in this region with its out-of-plane deformation mode $\gamma_s$(C1H1-C2H2) computed at 860 $cm^{-1}$. Another clear indicator of the presence of the $\eta^2$ and $\eta^3$ structures is measured at 1064 $cm^{-1}$: this band is assigned to the $\nu$(C1–C2) normal modes of the $\eta^2$-(*s*)-*trans* (1039 $cm^{-1}$), the $\eta^2$-(*s*)-*cis* (1036 $cm^{-1}$), and the $\eta^3$-(*s*)-*cis* (1063 $cm^{-1}$) structures. Two more adsorption structures, that is, the $\eta^4$-di-σ(CC)-di-σ(CO)-(*s*)-*trans* and $\eta^4$-$\pi$(CC)-di-σ(CO)-(*s*)-*cis* (Table 40.4) configurations, are compatible with the experimental data. Although it is difficult to identify each of these structures by a single loss peak, they can be considered to be present on the surface because, consistent with the other three adsorption modes, they are also energetically similarly stable and their computed HREEL spectra improve the overall fit. No significant coverage dependence of the vibrational fingerprints has been found by comparison of calculations of prenal in the (3 × 3) unit cell to smaller (2 × 2) and (3 × 2) unit cells, consistent with the absence of detectable vibrational shifts in experimental HREEL spectra for varying submonolayer doses.

Generally, the fit between the experimental and DFT data was worst on Pt(111), which has to do with a number of approximations already mentioned in Section 40.3.1.4. The more complex the surface mixture, that is, the more the species that contribute, the worse the agreement, whereas in simpler cases such as prenal on $Pt_3Sn(111)$ or ethene on Pt(111), the fit is much better [35, 37].

Anharmonicity is a source of systematic error that renders an interpretation of $\nu$(C–H) stretching modes around 3000 $cm^{-1}$ quite useless, it but does not affect modes in the fingerprint region below ~1500 $cm^{-1}$ much [35, 37]. Contributions from impact scattering, which appear in addition to the usually stronger dipole scattering losses in the experiments, are also neglected.

However, besides anharmonicity, impact scattering, and the systematic errors due to the computational parameters (especially the GGA functional and lack of dispersion), the periodic model introduces significant problems in the case of disordered phases [35, 37]. Both the frequencies and the intensities are affected by lateral interactions and coupling, a fact that is not described by the periodic supercell approach with only one species per supercell. Its images are of course always oriented identically, and no configuration, site, or orientation disorder is reproduced. In contrast, LEED experiments performed in a similar manner as the HREELS measurements showed no superstructure spots, which is indicative of a disordered adsorbed phase as one could expect for these molecules.

A probative example of this problem is the rather weak signal around 1660 $cm^{-1}$ on Pt(111) (Figure 40.38), which corresponds to the computed intense $\nu$(C1=O) stretching mode of the $\eta^2$-di-σ(CC)-(*s*)-*trans* (1666 $cm^{-1}$) and the $\eta^2$-di-σ(CC)-(*s*)-*cis* form (1655 $cm^{-1}$). Also several other calculated vibrations of these two $\eta^2$ configurations are intense in regions in which much weaker but very broad bands are detected experimentally. A coupling-induced splitting between the various configurations in different orientations would lead to a dispersion of the intensity into wider but weaker peaks, consistent with the experiments.

Finally, the more complex the adsorbate phase, the greater the difficulties to determine relative coverages of the different species on the surface for a comparison with the experiments since the coverage and disorder dependence of the adsorption energies in a mixed phase (similar to mixing enthalpies) pose a very challenging problem. Hence, for the sake of simplicity, we decided to assume identical coverages for all contributing species despite their different adsorption energies. This crude approximation still allowed the five most strongly adsorbed prenal adsorption modes to be identified in the experimental HREEL spectra.

The adsorption structures and energies were in line with previous EH cluster [228] and periodic DFT calculations [119, 227] and followed the general trend observed for acrolein and crotonaldehyde [121, 227, 228, 230].

Delbecq *et al.* also obtained several flat adsorption modes with molecule–surface interactions through the C=C and C=O bonds ($\eta^3$ and $\eta^4$) for prenal at low coverages, while at higher coverages the molecule switched to a less coordinated geometry

[227, 228]. This can be realized by interacting with the substrate via the C=C double bond only ($\eta^2$ forms) or even solely by the aldehydic oxygen atom ($\eta^1$-top).

Hirschl *et al.* computed similar structures for prenal on a $Pt_{80}Fe_{20}(111)$ bulk alloy surface, concluding that the aldehyde also assumes flat $\eta^2$-di-σ(CC), $\eta^3$, and $\eta^4$ adsorption geometries on this system [119]. A promoter effect was observed in the form of a strong stabilization of the $\eta^3$ and $\eta^4$ adsorption structures, which interacted with the electropositive Fe and induced changes in the segregation profile. The contribution of the C=O interaction with the Fe centers is in line with the studies by Beccat *et al.* [197] and Marinelli and coworkers [200, 201]. In contrast to the Pt/Sn, Rh, Rh/Sn, Co, and Cu catalysts mentioned above, all vertical top adsorption geometries turned out to be uncompetitive to the flat ones in their DFT calculations at low coverages.

Crotonaldehyde was found to behave quite similarly to prenal on Pt(111) [37, 227]. The analysis of HREELS spectra of the chemisorbed monolayer species with DFT simulations also indicated the formation of $\eta^2$-di-σ(CC)-*E*-(*s*)-*trans* ($E_{ads} = -76$ kJ/mol), $\eta^2$-di-σ(CC)-*E*-(*s*)-*cis* (-74), $\eta^3$-di-σ(CC)-σ(O)-*E*-(*s*)-*cis* (-77), $\eta^4$-di-σ(CC)-di-σ(CO)-*E*-(*s*)-*trans* (-80), and $\eta^4$-π(CC)-di-σ(CO)-*E*-(*s*)-*cis* (-69) adsorption structures, which, according to TPRS measurements also decompose above ~300 K. The adsorption energies of the five most stable crotonaldehyde forms at a low coverage of 1/9 ML are ~20–30% stronger than the corresponding prenal forms [37, 38] but ~20% below the values computed for acrolein in similar supercells [227].

In another combined HREELS and DFT investigation, Loffreda *et al.* showed that also acrolein forms a mixture of adsorption modes on Pt(111) at low temperatures [121]. This was one of the first successful attempts to simulate HREEL spectra for such complex systems. Consistent with the previous total energy calculations [227], different adsorption modes were identified as a function of the acrolein coverage: at high coverages, a phase of $\eta^2$-di-σ(CC)-(*s*)-*trans* and $\eta^2$-di-σ(CC)-(*s*)-*cis* acrolein, and at low coverages, a mixture of $\eta^3$ and $\eta^4$ adsorption modes. For all observed adsorption structures, a full vibrational analysis was performed and an interpretation of the measured loss signals presented.

The total energy results of acrolein on Pt(111) were improved subsequently with an *ab initio* thermodynamic analysis, allowing the calculation of the adsorption free energies of the configurations at more realistic temperature and pressure conditions (Figure 40.39) [230]. Similar to the results under model catalytic conditions, the thermodynamic analysis showed a transition from a mixed phase of mainly $\eta^3$-di-σ(CC)-(*s*)-*cis* and $\eta^4$-di-σ(CC)-di-σ(CO)-(*s*)-*trans* forms at low coverages (1/12–1/9 ML) to $\eta^2$-di-σ(CC) adsorption modes near saturation (1/4 ML). The transition was explained from the Gibbs free adsorption energy diagrams, but the energy differences were so subtle that only a qualitative trend was proposed, considering the calculations' error bars. Under the conditions used in Ref. [200] (350 K, 33 mbar), all of the $\eta^2$, $\eta^3$, and $\eta^4$ adsorption modes were competitive with the surface free adsorption energies of ~2 meV/Å$^2$. Thus, the behavior observed for acrolein at low-temperature/low-pressure (UHV) conditions with HREELS and DFT, as well as in the *ab initio* thermodynamic analysis, matches well with the SFG

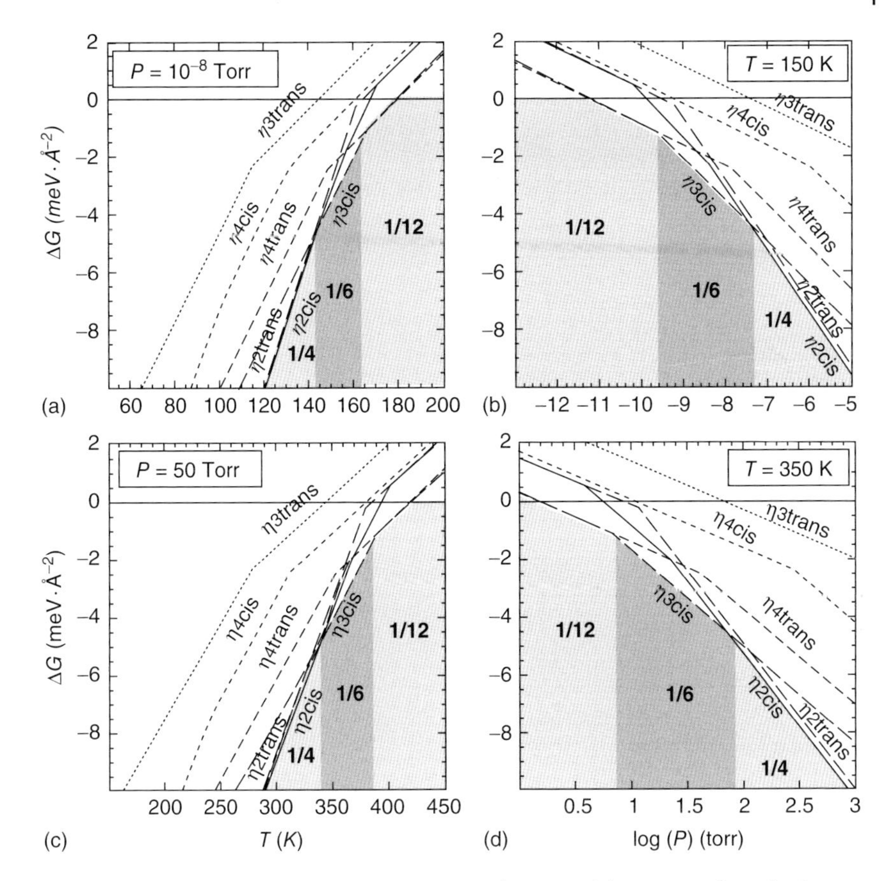


**Figure 40.39** Minimal Gibbs free adsorption energy curves $\Delta G_{ADS}(T, P)$ for the various adsorption modes as a function of the temperature for a fixed pressure (a,c) and as a function of the pressure for a fixed temperature (b,d). Shaded regions mark the stable phases. (Reproduced from Ref. [230].)

studies by Kliewer *et al.* [223], implying that over a large range of conditions there is no significant "pressure gap" for this molecule.

The detection of different coordination modes on the surface has important implications for the catalytic reactivity, since they may be conceived to lead to several competitive reaction channels yielding various products. This may play an important role for the selectivity in the hydrogenation process; however, without further studies one cannot exclude that only the energetics and kinetics of the elementary steps of the hydrogenation process itself determine the overall reactivity.

With the identified structures as starting points, Loffreda *et al.* subsequently performed two systematic studies of the reaction pathways in acrolein hydrogenation with DFT studies (Figure 40.40) [231, 232]. Analyzing the activation energies for the various possible monohydrogenation pathways of the C=C and C=O bonds, a kinetic model was formed that allowed the reproduction of the low selectivity of the hydrogenation toward propenol on Pt(111) [231]. The desorption step of the

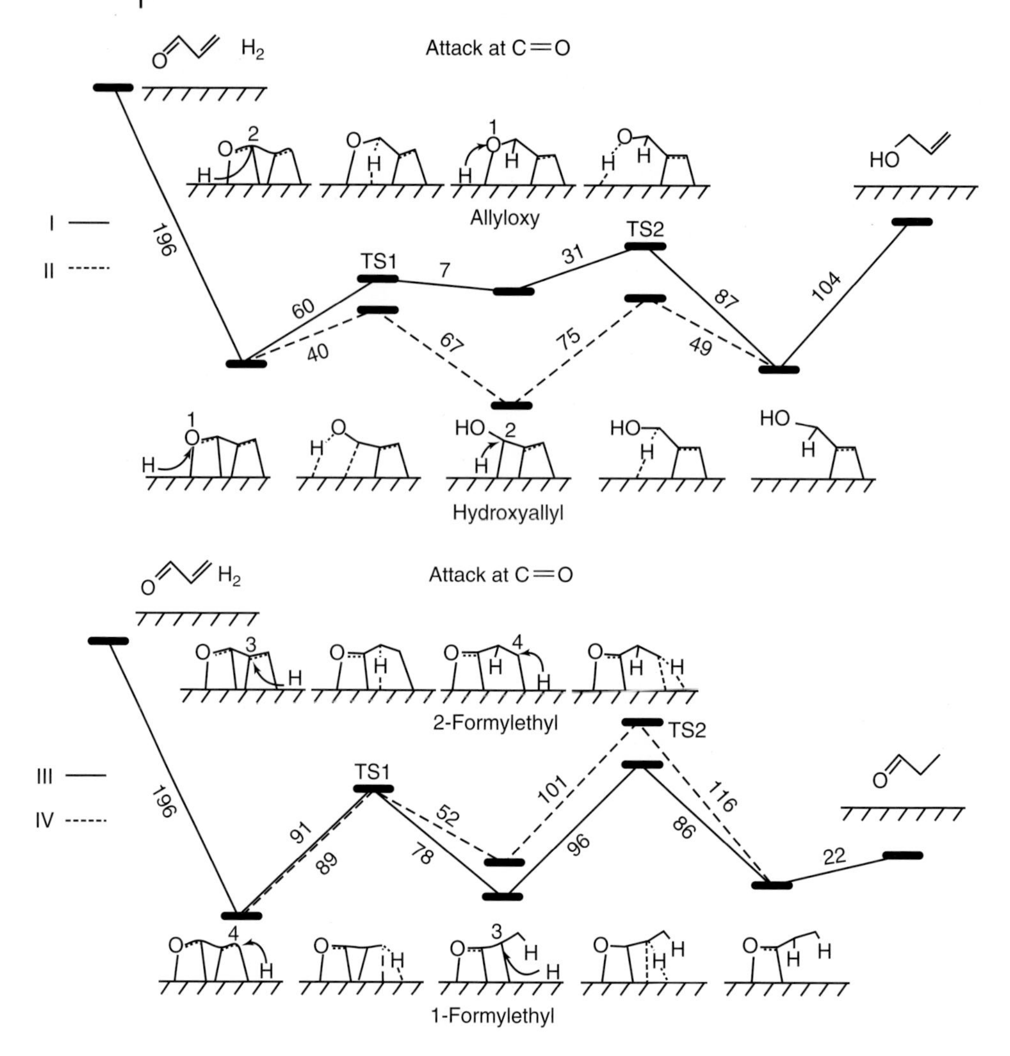


**Figure 40.40** Energy profiles of the hydrogenation pathways for acrolein on Pt(111): attack on the C=O bond (pathways I and II) and on the C=C bond (pathways III and IV). The energies (kJ/mol) are relative to the reactants in the gas phase. The adsorption energies of acrolein and two hydrogen atoms are represented in the initial coadsorption state. The transition states are labeled TS1 and TS2 for the first and the second hydrogenation, respectively [231].

product molecules turned out to be crucial for the overall selectivity. Although the formation of the unsaturated alcohol was favored by reaction energetics, its higher adsorption energy led to an accumulation on the surface. The desorption steps of propanal and propenol were both nonactivated, but the desorption energies were largely different: 22 kJ/mol for propanal and 104 kJ/mol for propenol. The saturated aldehyde, propanal, despite not being the most abundant reaction product on the catalyst, achieved a high yield due to its facile desorption.

The reaction barriers for hydrogenation of the C=O bond (~2–20 kJ/mol for the attack on an O atom and ~40–70 kJ/mol for the attack on a C atom) were smaller than those for hydrogenation of the C=C bond on Pt(111), which is contrary to common intuition [232]. This arises from the existence of metastable precursor states for the O–H bond formation where H adatoms can coadsorb closer to a partially decoordinated molecule. The attack by the hydrogen on the aldehydic oxygen follows a new mechanism where the C=O moiety is not directly bonded to the Pt surface, which can be best described as an intermediate type between the two limiting cases, the Langmuir–Hinshelwood and Rideal–Eley mechanisms. The proximity of both reactants in these precursor states provides early transition states on the potential energy surface, contrary to the initial state of the attack at the C=C bond. The attack at the C=C group follows a Langmuir–Hinshelwood mechanism instead, with high activation barriers (~85 kJ/mol) and late transition states according to the Hammond–Leffler postulate.

Thus, the selectivity in the case of acrolein hydrogenation on Pt(111) has been successfully understood from a combination of reaction barriers for the various elementary steps in the competing pathways and the following desorption of the products.

Recently, also the decomposition of the irreversibly chemisorbed aldehydes on Pt(111) at elevated temperatures was studied by Haubrich *et al.* [229] For prenal, a minimum-energy reaction pathway was found that connected several spectroscopically observed reaction intermediates such as the $\eta^2$-isobutenyl and $\eta^1$-isobutylidyne species. The latter was determined to be stable and comparable to ethylidyne or propylidyne. Importantly, the DFT analysis showed that each detectable intermediate was followed by an elementary step with a higher activation barrier, rendering them experimentally accessible by the quenching techniques. The sequence of elementary steps was in agreement with the observed desorption states of $H_2$ and CO and connected the $\eta^3$-di-$\sigma$(CC)-$\sigma$(O)-(*s*)-*cis* species to $\eta^1$-isobutylidyne. Initially, the aldehydic hydrogen was lost to the surface, and desorbed recombinatively (295 K). The energetically next-best step was found to be the decarbonylation leading to CO desorption (414 K), which was followed by several subsequent dehydrogenation and rehydrogenation steps prior to the detection of $\eta^1$-isobutylidyne at ~440 K. Annealing the sample to higher temperatures resulted in less well-resolved HREEL spectra and suggested a further decomposition to other smaller hydrocarbon species that could not be assigned unambiguously.

#### 40.3.4.3 **Prenal Adsorption Structures on $Pt_3Sn(111)$ and $Pt_2Sn(111)$ Alloy Films**

On the two Pt–Sn surface alloys, TPD experiments show essentially only reversible prenal adsorption without any decomposition aside from minimal amounts due to defects. On $Pt_3Sn(111)$, a prenal desorption occurs for exposures below a multilayer around 161 K. For smaller exposures, only a weak desorption at ~184 K is detected, which is assigned to the prenal monolayer (Redhead: $E_{a,des} = 47$ kJ/mol). Likewise, on $Pt_2Sn(111)$, a weak desorption signal from the monolayer was measured at 216 K (Redhead: $E_{a,des} = 56$ kJ/mol). Thus, spectra recorded after annealing a multilayer to 170 K on $Pt_3Sn(111)$, and 200 K on $Pt_2Sn(111)$ can be expected to correspond to the most stable monolayer species.

Also for the analysis of the HREEL spectra on the $Pt_3Sn(111)$ and $Pt_2Sn(111)$ surface alloys, numerous conceivable prenal adsorption structures have been computed. Theoretical dipolar HREEL spectra have been simulated for all stable local minima.

The optimizations for prenal on $Pt_3Sn(111)$ (29 initial geometries considered) resulted in six flat $\eta^2$, $\eta^3$, and $\eta^4$ configurations with weak adsorption energies ($E_{ads} > -20$ kJ/mol) and two stable vertical geometries of $\eta^1$-top-(*s*)-*trans*-OSn type ($E_{ads} = -39$ and $-31$ kJ/mol), in which prenal interacts directly with the Sn atoms (Figure 40.41). Surprisingly, the $\eta^1$-top-(*s*)-*trans*-OSn(1) form is even more strongly bonded to the PtSn surface than the equivalent $\eta^1$-top-(*s*)-*trans* structure on Pt(111) (−32 kJ/mol). Among the flat structures, $\eta^3$-di-σ(CC)-σ(O)-(*s*)-*cis*-OSn is the most stable one (−20 kJ/mol), which indicates a weakening of the bonding of almost 66% compared to the corresponding configuration on a pure Pt(111) site.

In the HREEL spectra recorded on $Pt_3Sn(111)$ after annealing a prenal multilayer to 170 K, the two most stable adsorption complexes ($\eta^1$-top-(*s*)-*trans*-OSn type) were identified, whereas the fingerprints of the weaker bonded flat adsorption modes failed to reproduce the data (Figure 40.42). The ν(C1=O) and ν(C2=C3) vibrations of the two $\eta^1$ complexes lead in particular to a broad and intensive loss signal between 1590 and 1650 $cm^{-1}$. Although this frequency region was also specific for the $\eta^2$-di-σ(CC)-(*s*)-*trans* bonding ($E_{ads} = -10$ kJ/mol; the *cis* forms was unstable), the simulated spectrum of the latter structure differed considerably from the experiments.

The loss intensities were much weaker on the surface alloy than on Pt(111), consistent with the lower desorption yield in TPRS. In terms of vibrational frequencies, the spectra recorded after annealing to 170 K differed very little from those recorded for a multilayer, which seems to imply a weakly physisorbed state. Since also most

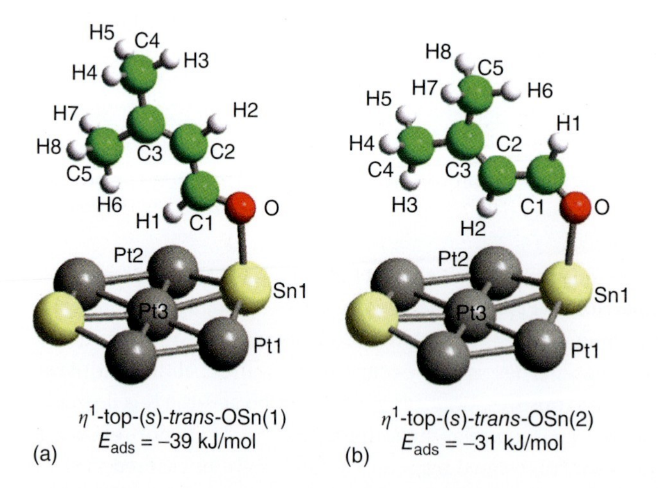


**Figure 40.41** The $\eta^1$-top-OSn adsorption structures of prenal on $Pt_3Sn(111)$: (a) $\eta^1$-top-(s)-trans-OSn(1), (b) $\eta^1$-top-(s)-trans-OSn(2) (obtained from the DFT optimizations at a theoretical coverage of 1/12 ML; taken from Ref. [35].)

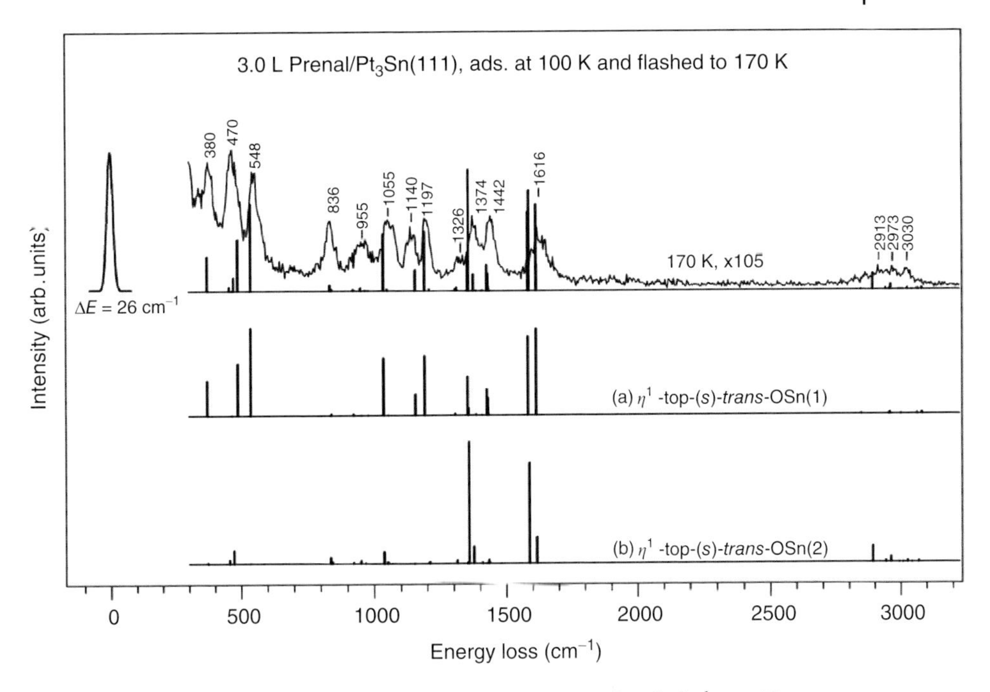


**Figure 40.42** Comparison of the theoretical spectra of the two identified $\eta^1$-top-OSn adsorption structures (theoretical coverage 1/12 ML) and the experimental HREELS spectrum (3.0 L of prenal annealed to 170 K) on the p(2 × 2)-$Pt_3Sn(111)$ surface alloy. (From Ref. [35].)

of the computed vibrations of the two $\eta^1$-top-(*s*)-*trans*-OSn forms differ little from the values in a prenal multilayer due to the molecule's rather unperturbed state, the characteristic intensity patterns related to their specific orientations were critical for their identification and the differentiation from physisorption.

On the $Pt_2Sn(111)$ surface alloy, out of 36 tested prenal adsorption structures only a small number of flat structures ($\eta^2$-di-σ(CC), $\eta^3$, and $\eta^4$ coordination) was stable, all of them being very weakly chemisorbed ($E_{ads} > -20$ kJ/mol). Similar to the $Pt_3Sn(111)$ case, the vertical $\eta^1$-top form is the most stable one on the Sn atoms ($E_{ads} = -33$ kJ/mol; Figure 40.43). The adsorption energies are strongly decreased compared to Pt(111). For the six flat structures that remained stable, an average decrease of ∼30–45 kJ/mol (∼50–80%) was computed with respect to Pt(111) and of ∼10 kJ/mol in comparison to $Pt_3Sn(111)$.

The HREEL spectra recorded for the chemisorbed monolayer on the $Pt_2Sn(111)$ after annealing to 200 K were also found to be very similar to those of the multilayer (Figure 40.44). The experiments were best interpreted with the relatively strongly adsorbed $\eta^1$-(*s*)-*trans*-top-OSn form. The measured bands at 1371 and 1444 $cm^{-1}$ as well as at 1141 and 1202 $cm^{-1}$ were reproduced fairly well by DFT spectra of this structure and the appearance of a loss signal at 1625 $cm^{-1}$ indicated the presence of an $\eta^1$ form. However, two features were measured at 718 and 956 $cm^{-1}$ that were not predicted by the DFT spectra of the $\eta^1$ structures. The spectra of the

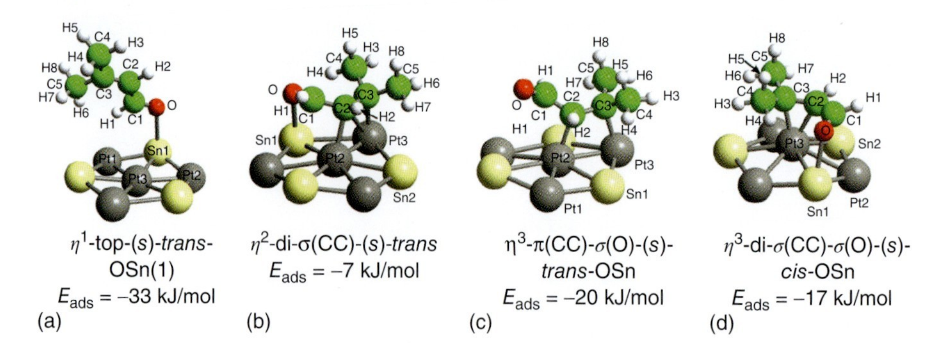


**Figure 40.43** The three most stable configurations of prenal on the $Pt_2Sn(111)$ surface alloy (a) $\eta^1$-top-(s)-trans-OSn(1), (b) $\eta^2$-di-σ(CC)-(s)-trans, (c) $\eta^3$-π(CC)-σ(O)-(s)-trans-OSn, and (d) $\eta^3$-di-σ(CC)-σ(O)-(s)-cis-OSn (DFT optimizations at a theoretical coverage of 1/9 ML) (From Ref. [35].)

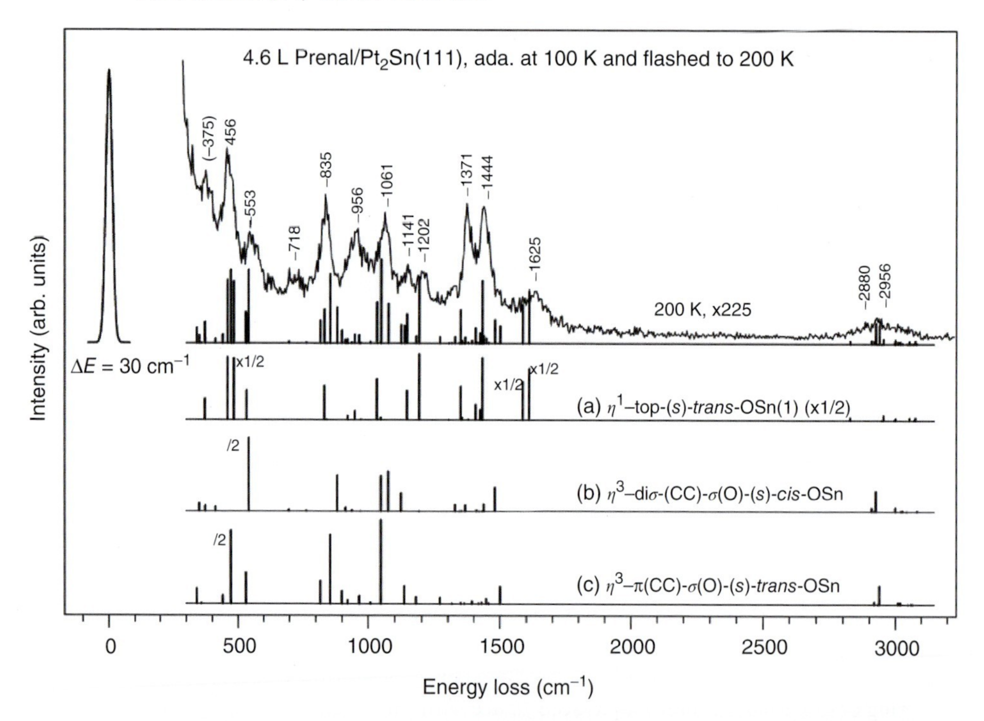


**Figure 40.44** Comparison of the theoretical spectra of the three most stable adsorption structures (theoretical coverage 1/9 ML) and the experimental HREELS spectrum (4.6 L of prenal annealed to 200 K) on the $(\sqrt{3} \times \sqrt{3})R30°$-$Pt_2Sn(111)$ surface alloy. (From Ref. [35].)

more weakly bonded flat adsorption structures such as $\eta^3$-di-σ(CC)-σ(O)-(*s*)-*cis*-OSn ($E_{ads}$ = −17 kJ/mol) form did not improve the fit with the HREEL data and even showed several wrong signals. Therefore, the best correspondence with the experimental data on the $Pt_2Sn(111)$ surface was obtained with the $\eta^1$-(*s*)-*trans*-top-OSn form, in agreement with the predictions from the total energy calculations.

In contrast to the behavior of prenal on the $Pt_3Sn(111)$ and $Pt_2Sn(111)$ surface alloys, crotonaldehyde formed $\eta^1$-*E*-(*s*)-*trans*-top-OSn structures and adsorption modes that were of the same types as those on Pt(111) despite the presence of the Sn atoms. While for prenal neither the experimental data nor the theoretical computations performed at higher coverages (1/6 and 1/4 ML) indicated any coverage dependence of the mixed adsorbate phases, crotonaldehyde showed very distinct phase transitions. At high coverages (≥1/6 ML in the corresponding theoretical slab models), $\eta^2$-di-σ(CC)-*E*-(*s*)-*trans*, $\eta^2$-di-σ(CC)-*E*-(*s*)-*cis*, and two energetically competitive $\eta^1$-top-*E*-(*s*)-*trans*-OSn structures were identified on both surface alloys, whereas at lower coverages (1/9 ML) $\eta^2$-diσ(CC)-*E*-(*s*)-*trans*, the $\eta^3$-diσ(CC)-σ(O)-*E*-(*s*)-*cis*-OSn, and the $\eta^4$-π(CC)-diσ(CO)-*E*-(*s*)-*cis*-OSn forms were detected. On $Pt_3Sn$, also the $\eta^2$-diσ(CC)-*E*-(*s*)-*cis* structure, which is not stable on $Pt_2Sn$, was observed.

#### 40.3.4.4 General Observations: Alloying Effects, Adsorption and Interaction Energies

The crucial question for a deeper understanding of the selective hydrogenation of the α,β-unsaturated aldehydes is how alloying changes the interaction and hence activation of the reactants with the catalyst. With the exception of the $\eta^2$-di-σ(CC) modes, almost all $\eta^1$, $\eta^2$, $\eta^3$, and $\eta^4$ structures on the two surface alloys formed an O–Sn bond, which appears to be required in order to achieve a reasonable adsorption energy. None of the typical $\eta^3$ and $\eta^4$ forms including an interaction of the aldehydic oxygen with Pt was stable [35, 37, 38].

The transition to the $\eta^1$-top-OSn configurations for prenal implies that only the C=O bond is activated by the interaction with the surface alloys. Since for crotonaldehyde and acrolein also the $\eta^3$ form with an interaction of the aldehydic oxygen with Sn and the $\eta^2$-di-σ(CC) forms are competitive on the surface alloys, the C=C bond also may be attacked by hydrogen atoms from the surface. Hence, for prenal a higher hydrogenation selectivity to the unsaturated alcohol was expected [35, 37, 38, 50], analogous to the conclusion of a favorable substituent influence drawn by Marinelli *et al.* for Pt–Fe catalysts [200, 201].

The TPD/TPRS experiments [37, 35] showed that the adsorption on the Pt–Sn surface alloys is considerably weakened since only low-temperature desorption states are measured and the decomposition at higher temperatures is suppressed [35, 37, 38]. This is consistent with the reduced adsorption energies obtained from DFT, and was related to an increase in the electron transfer from Sn to Pt and a lowering of the d-band center [50] in accordance with the d-band model [23]. Since similar structures of all coordination types were obtained on all model catalyst surfaces, no evidence for a true ensemble effect in the sense of Sachtler [48], that is, a site blocking that eliminates certain configurations, was observed [35, 37, 38]. However, structures with higher hapticities were found to be destabilized preferentially.

In order to rationalize the decrease in the adsorption energies, the interaction strengths of the molecules to the surface were analyzed according to Equation 40.1 discussed in Section 40.3.1 [38]. Also here, the simple energy decomposition scheme was assumed to be applicable since the electron transfer (donation/back-donation) between the molecule and the surface was expected to be small [50, 227].

The energy cost for displacing the metal atoms at the adsorption site for an optimal interaction with the molecules was small relative to the distortion energy of the molecule on all three surfaces [38]. Importantly, only small differences in the interaction energies $E_{int}$ between crotonaldehyde and prenal were obtained, while the relaxation penalties $E_{def,mol}$ and $E_{def, surf}$ varied more and were concluded to be responsible for the weaker adsorption energy of prenal compared to that of crotonaldehyde. Neither the interaction energies nor the relaxation costs exhibited simple linear trends as a function of the Sn fraction. Both depend strongly on the bonding configuration.

For the low-hapticity $\eta^1$ types, the relaxations of surface and molecule were almost negligible. This was consistent with the similarity of the optimized geometries to the gas-phase structures and the weak perturbation of the metal atoms [35, 37, 38]. The $\eta^2$, $\eta^3$, and $\eta^4$ configurations, in contrast, showed considerable surface relaxations and, hence, surface deformation costs. Because of the additional Pauli repulsions from the second methyl substituent [191], which resulted in further relaxations of the metal atoms, these were on average larger by ~5 kJ/mol for prenal. The repulsive effects were especially pronounced on the alloys, where the Sn atoms protrude out of the surface planes.

The molecular distortion energies were the largest components for the flat coordination types, and varied significantly with changing hapticity and bonding mechanism. They were larger for structures interacting via a di-$\sigma$ mechanisms than for those exhibiting $\pi$ bonding. For the $\eta^2$ and $\eta^3$ structures, the distortion consistently required ~140–170 kJ/mol, which was concluded to indicate that the di-$\sigma$(CC) bonding was almost solely responsible for the relaxations. The $\eta^4$-di-$\sigma$(CC)-di-$\sigma$(CO) structures formed by (*s*)-*trans* isomers on Pt resulted in higher values of ~210–240 kJ/mol. The $\eta^4$-di-$\sigma$(CC)-$\pi$(CO) adsorbates, in contrast, were less distorted from the gas-phase geometries (~110 kJ/mol) due to the weaker $\pi$(CO) coordination.

Also here, the increased Pauli repulsion for prenal resulted in larger molecule deformation costs related to the larger outward relaxations of metal atoms below the C=C moiety. The structural variations between the corresponding adsorption geometries of both molecules were minor for most of the bond lengths such as particularly the C1=O, C1–C2, C2=C3 bonds, and the hydrogen bond distances, showing that no significant change in the back-donation (and net electron transfer) occurred. The molecule–surface bonds from the aldehydic part agreed closely in the various adsorption geometries, and also the bond distances from the vinylic group were comparable for the corresponding adsorption modes.

Consequently, the interaction energies were very similar in the corresponding coordination types of both aldehydes. For the $\eta^1$ forms, only relatively small interaction energies ( about −31 to −47 kJ/mol) resulted because of the small relaxation costs, whereas the $\eta^2$-di-$\sigma$(CC) forms were computed to be interacting as strongly as the $\eta^3$-di-$\sigma$(CC) forms but more weakly than the $\eta^4$-di-$\sigma$(CC) ones. Therefore, the decrease of the adsorption energies from crotonaldehyde to prenal was proposed to arise primarily from larger surface and molecular distortion energies due to increased Pauli repulsions.

A closer analysis of the geometries and the vibrational frequencies of the adsorption structures revealed surprising "inconsistencies" given the weakened adsorption energies. The geometries were very similar for a given coordination to the three surfaces, which holds true in particular for the C2=C3 and C1=O bond lengths and the surface bonds that should be affected significantly by changing the surface bonding (see Refs [35, 37, 38] for structure details). Moreover, the vibrational properties of the corresponding structures were also very similar, contrary to what might be inferred from the strongly decreasing adsorption energies.

This problem can easily be seen for the $\eta^1$-top structures and the three model catalyst surfaces. In case of prenal, the most stable $\eta^1$ forms of the (*s*)-*trans* isomer show an increase of the adsorption energy from −32 kJ/mol on Pt(111) to −39 and −33 kJ/mol on $Pt_3Sn$ and $Pt_2Sn(111)$, respectively, while changing the coordination from surface Pt to Sn atoms. This increase is unique to the $\eta^1$ forms and caused by the character of the interaction from the aldehydic oxygen, which is strengthened by interaction with the electron-deficient Sn.

However, despite the increase in adsorption energies, the vibrational properties point to a weaker interaction energy: both the $\nu$(M–O) metal–molecule stretching modes, which are the primary indicators of the adsorption strength, and indirectly perturbed internal vibrations such as $\nu$(C1=O) and $\nu$(C2=C3) pointed to a decrease of the force constants (Figure 40.45)[1]. The $\nu$(M=O) frequencies decrease from 125 $cm^{-1}$ on Pt(111) to ~104–109 $cm^{-1}$ on the surface alloys and, similarly, the $\nu$(C1=O) modes shift from 1545 $cm^{-1}$ to a far less perturbed value of ~1588 $cm^{-1}$. Hence, unlike the adsorption energies imply, the surface bonding must in fact be slightly weaker than on Pt(111).

This weakening in the interaction strength with the metal atoms is indeed found in the energy decomposition analysis, which shows that the interaction terms follow the expected opposite trend to the adsorption strength. The crucial differences arise from the surface deformation energies: while the energetic cost to shape the adsorption site is almost negligible on the surface alloys, the lifting of a bonding metal atom on Pt(111) requires ~10 kJ/mol. The molecule distortion energies remain negligible for all $\eta^1$ forms, which in combination with the adsorption energies then lead to a decrease of the interaction strength by alloying: the strongest interaction is calculated on Pt(111) (−47 kJ/mol), whereas it is generally weaker on the Pt–Sn surface alloys (about −40 kJ/mol on average on $Pt_3Sn(111)$ and −37 kJ/mol on $Pt_2Sn(111)$). Thus, alloying weakens the interaction and therefore leads to the peculiar behavior of the vibrational frequencies.

1) Although the metal atoms were frozen in the vibrational calculations (i.e., both Sn and Pt essentially treated with infinite mass) and the coupling to metal phonons, ignored, the changes in the $\nu$(M-O) frequencies can still be traced back to the force constants in the harmonic picture.

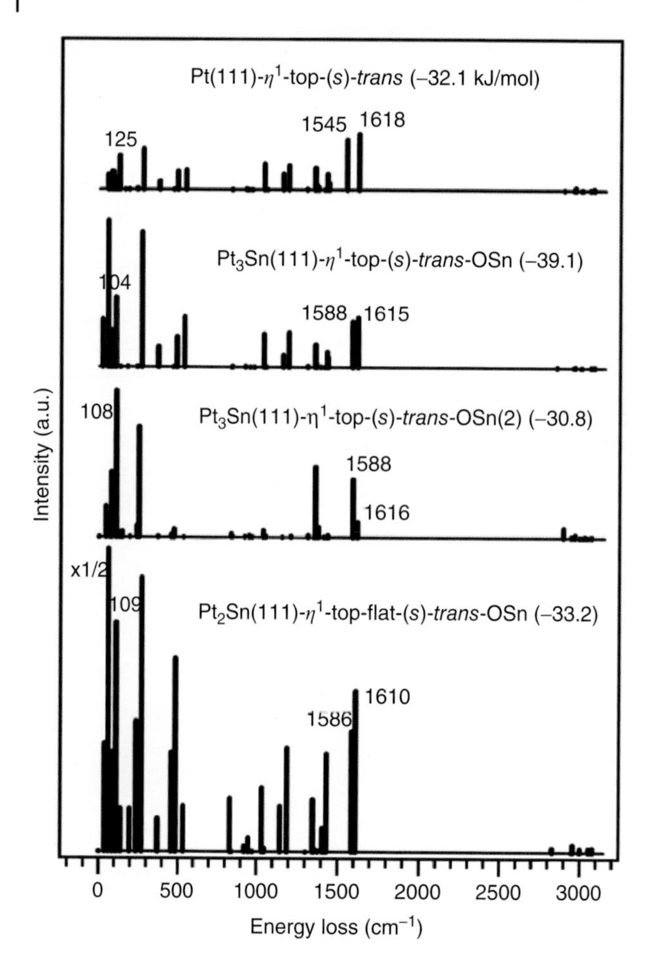


**Figure 40.45** Comparison of the computed HREEL spectra (dipole scattering limit, 4.7 eV, 60° specular) of the $\eta^1$ configurations of (*s*)-*trans* prenal on the three model catalysts. Note that the adsorption structures on the surface alloy show a coordination change from Pt to Sn atoms. Strong changes of the relative loss intensities can be observed, whereas significant frequency shifts are absent. (Reproduced from Ref. [38].)

Evidently, the adsorption energy is not always a good measure of the true interaction between the molecule and the surface – particularly for the more complex α,β-unsaturated aldehydes, which can show much larger molecular deformation penalties and relaxation requirements for the adsorption sites compared to atoms and small molecules such as, for example, CO, $H_2O$, $N_2$, or $NH_3$. For the $\eta^1$ forms, it actually masks a very different energy compromise: on Pt(111), the interaction energies are large but the surface deformations have a significant cost, whereas on Pt–Sn the interaction energy is decreased by ~15% and the deformation cost is almost completely negligible. Surprisingly, the absolute relaxations of the metal atoms are found to be even slightly larger than on Pt(111): around ~0.15 Å on $Pt_3Sn$ and even 0.25 Å on $Pt_2Sn$ versus only ~0.09 Å for the $\eta^1$ structures on the pure Pt surface, which was understood from the relaxation properties of the different clean surfaces.

The findings for the flat $\eta^2$, $\eta^3$, and $\eta^4$ structures were very similar and were exemplified for the $\eta^2$-di-σ(CC) and $\eta^3$-di-σ(CC)-σ(O) forms of crotonaldehyde. We will only review the key results for the $\eta^2$-di-σ(CC)-*E*-(*s*)-*trans* structure briefly to illustrate the role of the interaction energies and the modification of the surface relaxation costs.

The structural parameters and the HREEL spectra for the $\eta^2$-di-σ(CC) configurations of *E*-(*s*)-*trans* crotonaldehyde on the different surfaces reveal close similarities (Figure 40.46) despite the weakening of the adsorption energies from −76 kJ/mol on Pt(111) to −33 and −34 kJ/mol on $Pt_3Sn(111)$ and $Pt_2Sn(111)$,

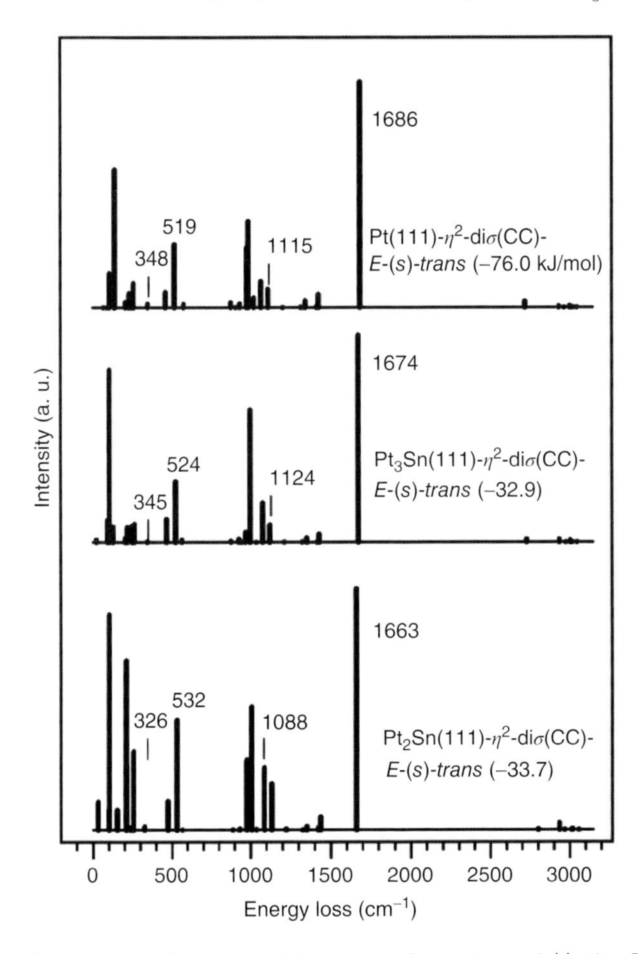


**Figure 40.46** Computed HREEL spectra of the $\eta^2$ configurations of *E*-(*s*)-*trans* crotonaldehyde on Pt(111) and the Pt–Sn surfaces. The bonding of these structures is always to two neighboring Pt atoms. Despite the alloying and the decrease of the adsorption energies, the vibrational properties remain very similar. (From Ref. [38].)

respectively. The frequencies of the indirectly perturbed $\nu$(C2=C3) stretching mode, for example, were computed at 1115, 1124, and 1081 $cm^{-1}$ on Pt, $Pt_3Sn$, and $Pt_2Sn$, respectively, suggesting rather similar interaction potentials to the surface and also a rather unchanged back-bonding component to the $\pi^*$(C=C) orbital. Also, the $\nu_{as}$(PtC2–PtC3) and $\nu_s$(PtC2–PtC3) vibrations were very similar (519, 524, and 525 $cm^{-1}$ as well as 348, 345, and 328 $cm^{-1}$).

The behavior of the structural parameters and vibrational properties of both examples are at variance with the strong decrease in adsorption energies, which has been explained by the d-band lowering [50] from Pt(111) to $Pt_3Sn$ and to $Pt_2Sn$ (from −1.93 to −2.09 and −2.12 eV, respectively [50]). The lowering is much less pronounced between both surface alloys and agrees better with the trend in the interaction energies.

The molecular deformation energies of the $\eta^2$-di-σ(CC) forms amount, on average, to ~150 kJ/mol on the surface alloys and show only small increases due to the alloying, consistent with the close agreement of the geometries to those on Pt(111). Particularly, the C=C and C=O bond length were computed to be almost unaffected by the alloying. For prenal, the substituent caused higher Pauli repulsions and, thus, ~10 kJ/mol larger molecule deformation energies.

The alloying led to considerable increase of the surface relaxations in the adsorption geometries and higher surface deformation energies. The deformations increased with Sn content and were computed to be of the same order of magnitude for corresponding types of flat structures for each molecule. The surface deformation energy costs increased with the Sn content from ~37 and 50 kJ/mol on average for the flat crotonaldehyde and prenal adsorption forms on $Pt_3Sn$ to ~51 and 61 kJ/mol, respectively, on $Pt_2Sn$. This more than doubled the energy penalty for high-hapticity structures from Pt(111) (~18 and 23 kJ/mol, respectively) and influenced the competitiveness of $\eta^3$ and $\eta^4$ modes negatively.

Significantly larger outward relaxations of the Pt atoms under the vinylic moiety occur from Pt(111) via $Pt_3Sn$(111) to $Pt_2Sn$(111) for $\eta^2$-di-σ(CC)-*E*-(*s*)-*trans* crotonaldehyde (Figure 40.47): the displacements increased from 0.20 to 0.31 and 0.38 Å for the atom bonding to vinylic C2, and from 0.18 to 0.32 and 0.51 Å for the one below C3. In addition, the Pt atoms below the C2=C3 bond shift by ~0.20 and 0.12 Å in the surface plane toward each other on $Pt_2Sn$(111), whereas these in-plane displacements are much smaller on Pt(111) and $Pt_3Sn$. The new Pt–Pt distance of 2.70 Å on $Pt_2Sn$ (bulk distance of 2.82 Å) suggests that the adsorption bonding is more favorable with a slightly contracted Pt dimer. These relaxations result in the increase of the computed surface deformation energies from 17 kJ/mol on Pt(111) to 41 and 56 kJ/mol on $Pt_3Sn$(111) and $Pt_2Sn$(111), respectively.

The apparent contradiction between the small changes in the adsorption geometries or vibrational properties and the drastic lowering of the adsorption energies can be related to the much smaller weakening of the interaction energies by the alloying in this case as well. The values obtained for the crotonaldehyde $\eta^2$-di-σ(CC) of −237, −222, and −235 kJ/mol on Pt(111), $Pt_3Sn$(111), and $Pt_2Sn$(111) are quite similar. The trend in the interaction energies was concluded to be better in line with the minor structural changes and the small vibrational shifts than the adsorption energies [38].

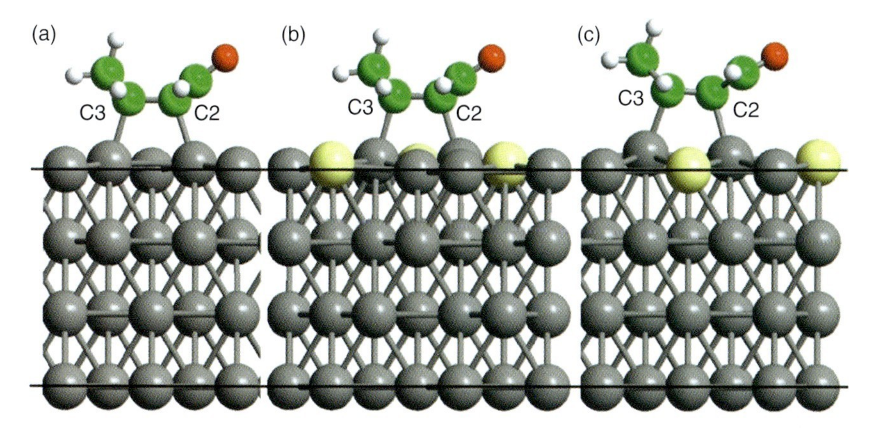


**Figure 40.47** Illustration of the outward relaxation of the Pt atoms bonding to the vinylic moiety in the $\eta^2$-di-$\sigma$(CC) configurations of *E*-(*s*)-*trans* crotonaldehyde on Pt(111) (a), $Pt_3Sn(111)$ (b), and $Pt_2Sn(111)$ (c). (From Ref. [38].)

## 40.4 Summary

In the past decades, intense studies with constantly improving experimental tools, theoretical methods, and computational resources have substantially advanced the fundamental understanding of the role of surfaces and interfaces in many areas, ranging from heterogeneous catalysis to art conservation and nature. The discovery of new details usually leads to further questions and directs the attention to aspects missed so far. We are still far from realizing a powerful, overarching model that would allow us to completely analyze and understand surface chemical or physical processes, not to mention one that would also allow prediction of new reactions or properties correctly and reliably. To a large degree, this issue stems from the enormous complexity of interface processes that present many puzzle pieces, which are often not independent of each other.

From the number of adsorption and reaction studies on olefines, dienes, and aldehydes we have discussed in this chapter, a number of general conclusions can be drawn. The tendency of hydrogen to attack the unsaturated terminal carbon atoms of the dienes, for example, or the preferential dehydrogenation of the inner carbon atoms may be used to predict the reactivity of other unsaturated molecules on metal catalysts [153, 155]. This is also shown by the TPD/TPRS studies or the spectroscopic evidence of different surface intermediates that have been detected by the authors working in this field, and good agreement with theoretical modeling.

The roles of promoters vary with the system and reaction under consideration. They generally affect the balance between adsorption strength and the activation and desorption properties of the catalyst. As evidenced by the case of Pt–Sn model catalysts, the promoter Sn modifies the preference for certain adsorption configurations and the overall bond strengths of the reactants and intermediates. The use of oxophilic promoters such as Sn, Fe, or Co with Pd or Pt hydrogenation catalysts can destablize the bonding and activation through the C=C group, which for these bifunctional unsaturated aldehydes leads to a preferential coordination of the C=O moiety to surface Sn sites. Oxygen–tin bonds are required on the surface alloys to form sufficiently stable adsorption structures. Flat structures such as the $\eta^3$ and $\eta^4$ forms were destabilized preferentially, while the $\eta^1$-OSn forms in turn became competitive.

The complete understanding of selectivity and activity of catalysts does, however, depend on more subtle balances between all involved steps as the various studies have shown. A detailed picture from the initial contact of the reactants up to the desorption of the products is necessary for a thorough argumentation.

Identification of the stable adsorption forms is a first key step for this understanding and can allow very simple qualitative arguments such as in the case of linking the transition to $\eta^1$ bonding of the α,β-unsaturated aldehydes on the Pt–Sn model catalysts to the selectivity change toward unsaturated alcohols.

Since the catalytic activation patterns are governed by the surface bonding, the characterization of the corresponding adsorption strengths of the reactant configurations can enable similarly simple rules for rationalizing the observed reactivity. A high adsorption strength as found for the di-σ bonded olefins on Pt(111) may, for example, lead to activation of the C–H bonds for decomposition to alkylidynes and subsequent site blocking, whereas a weaker bonding as in the case of the π forms favors the hydrogenation pathways.

However, besides the reactant bonding also other factors such as the actual barriers of the competing pathways, the desorption kinetics of the products, and the modification of the catalyst properties by cocatalysts or supports can modify the reactivity strongly and must be elucidated in order to obtain a full picture of the mechanisms, thermodynamics, and kinetics of the processes [218, 231, 232]. Configuration changes, for example, may occur when going from low-pressure conditions in typical studies toward elevated pressures and can lead to qualitatively different activation patterns and reaction pathway (see Chapter 39 in this volume). It appears from the various studies that the stable species of the unsaturated aldehydes and dienes at low-coverage UHV conditions may be unaltered at elevated pressures and are sufficient to rationalize the measured catalytic reactivity, but the hydrogenation of mono-olefins occurs by a reaction mechanism originating from the π forms found only at high coverages (and very low temperatures) or pressures.

Understanding the fundamental aspects of the surface bonding is, therefore, necessary for a proper understanding of the selectivity of more complex catalytic systems. The energy decomposition analysis, backed by the correlation to the geometric properties of the adsorption structures and the vibrational fingerprints, has proven to be a simple yet useful concept to analyze the bonding of polyatomic molecules on model catalysts. Despite its limitations – including the neglect of contributions from electronic changes and charge transfers of the two fragments – it provided valuable new insights: the deformation energies of the different bimetallic surface sites or the variation of the molecular relaxation costs, for example, allowed a qualitative understanding of the reasons for the configuration changes and the destabilization of flat aldehyde structures.

The trends of the adsorption energies on Pt(111), $Pt_3Sn(111)$, and $Pt_2Sn(111)$ can be linked to differences in the relaxation of the metal atom, that is, the elastic properties of the substrates, and the perturbation of the molecules. This complements the electronic picture of the bonding and leads to a deeper understanding of the differences between weakly and strongly interacting systems or between small molecules and large multifunctional adsorbates. Also, the examples of $\pi$-ethylene on the four Pt-containing metal matrices [36] or propene on the Pt–Sn surfaces [116] led to interesting new aspects, namely that pure electronic ("ligand") effects may be active to a much lesser degree than commonly suspected given such strong decreases in adsorption energies. The bonding on metal surfaces appeared to be quite local in the reviewed cases, and only the properties of the interacting metal center were decisive, and, vide infra, would probably also lead to similar electronic states just as shown in the case of ethylene.

The observation that geometry parameters, vibrational frequencies, and interactions were influenced much less by promoters than the adsorption energies suggests that it may be possible to tune the adsorption strength and the interaction strength to a certain degree independently. This raise the question of not only whether the small changes in the latter are in better agreement with the moderate lowering of the d-band center by alloying (e.g., in the case of Pt–Sn and Pt–Cu [50], where also $E_{int}$ correlated linearly with $E_d$) than the much larger changes of the adsorption energies [23] but also which of the two bonding strength indicators is a better measure of the activation. Unlike the case of simple adsorbates such as $N_2$, $NH_3$, $O_2$, or CO, where the interaction energy (bond strength) and the adsorption energy are closely related, these two properties can obviously behave differently for multifunctional molecules interacting with a mono- or bimetallic substrate in complex manners. For ammonia synthesis, the energy for dissociative adsorption $N_2$ is clearly a very good descriptor for the reactivity as seen in the Sabatier plots [233–236], but does this also hold true for larger, strongly bonding reactants with sizable relaxations?

In analogy to the Hammond–Leffler postulate, which states that "early" transition states with structural parameters closer to those of the reactants are correlated with exergonic (exothermic) reactions, the question can be posed as to what correlation exists between the perturbation of an adsorption structure or surface intermediate by bonding to the catalyst and the transition state energetics. Transition state structures still retaining close similarity to the geometries of the reactants require fewer perturbations and commonly show lower activation energies than reactions with late transitions states.

Hence, there may be an optimum perturbation of a reactant by bonding to the catalyst, which depends on the energetic balance of the rate-limiting reaction step. For an endothermic reaction, for example, the overall energetic balance might be improved if the adsorption complex on the surface would be shaped toward the products so that the geometries of transition state and adsorbed reactant structure would be shifted closer, thereby presumably lowering the related activation barrier. In this sense, if the initial adsorption step would account for part of the perturbation of the structure and the required energy cost through the relaxations and the interaction energy, also the endothermic character would be reduced for this particular step.

Since the degree of structural perturbation is linked to these relaxation costs, which are in turn linked to the interaction energy, a tuning of the adsorption strength versus the reactant activation may be possible. Too high adsorption energies may lead to surface blocking as seen for acrolein hydrogenation on Pt(111) with the product propenol being too stable on the surface [231, 232]. which opens undesired reaction pathways at higher energies such as decomposition reactions. Large interaction energies and perturbation patterns suggesting similarly sizable activation but linked to smaller adsorption energies, in contrast, such as found in the case of the prenal and crotonaldehyde on the Pt–Sn surface alloys, may shift the selectivity toward the hydrogenation channels by allowing product desorption prior to onset of other processes [35, 37, 38, 229]. Too small interaction strengths and adsorption energies, in contrast, may prevent the reaction (i.e., by leading to late transition states higher in energy) or may open the reactant desorption channels before any transformation barriers can be overcome.

Further studies are necessary to explore the relationships between the interaction strength, structural relaxation, and the energetic positions of transitions states in reactions of complex molecules on mono- and multimetallic catalyst, which might show a behavior analogous to the analysis of BEP relationships between adsorption energies and activation barriers for small molecules, but these preliminary considerations suggest that it may be possible to tune catalytic reactivity with this new strategy.

However, similar to the use of BEP relationships for the prediction of catalytic performance of new materials, also these correlations can be useful for extrapolation only if several strict assumptions hold true: most importantly, the elementary mechanisms or rate-limiting steps on different catalyst material must remain qualitatively similar and transferable. Of course, possible coordination changes or modifications of the surface reaction site ensembles need to be taken into account, which, for example, in the case of Pt–Sn alloys, strongly affect hydrogen atom bonding and, hence, hydrogenation and dehydrogenation pathways. In order to be able to take into account such modifications of reaction ensembles, more information on the adsorption and reaction behavior of new mono-, bi-, and multifunctional, multicomponent catalyst materials, on the effects of pressure (coverage) and temperature conditions, and on the the role of supports and size effects must be gathered.

Thus, although much very basic information on the case example of hydrogenation reactions on Pt-based catalysts has been compiled by the numerous studies of the past decades, many important questions and systems still remain to be investigated.

Lastly, we would like to point out that the interplay and synergy between experimental and theoretical studies in the past decade has likely produced much better insights than each approach could have achieved individually. By combining spectroscopies, microcopies, and theoretical modeling, the surface properties, bonding, and reactivity have become accessible at an unprecedented level. The analysis of highly complex spectra with theoretical means, although still a very demanding task, opens the route to the interpretation of experimental studies of much more structurally and chemically complex problems as found in real catalysis. Often, controversial topics such as promoter effects, support–catalyst interactions, or the changes correlated with different substitution patterns may be addressed in an efficient manner. Sophisticated experiments can reduce the effort for complementary theoretical studies as much as detailed modeling can be a powerful tool to guide the design of new experimental strategies. More collaborative and combined investigations will be needed to further our understanding of the complex interface chemistry and physics that form key parts of many technological, natural, or anthropogenic processes.

## Acknowledgments

One of the authors (J.H.) acknowledges support through a "Doktorandenstipendium" of the "Fonds der Chemischen Industrie (FCI)," a "Marie-Curie" scholarship of the European Union, as well as a Feodor-Lynen Rückkehrstipendium of the Alexander-von-Humboldt (AvH) Foundation. Both authors are grateful for grants from the Deutsche Forschungsgemeinschaft (DFG) and acknowledge Dr K. Gentz for his assistance in proof reading of this manuscript.

## References

1. Augustine, R.L. (1996) *Heterogenous Catalysis for the Synthetic Chemists*, Marcel Dekker, New York.
2. Molnar, A., Sarkany, A., and Varga, M. (2001) *J. Mol. Catal. A: Chem.*, **173**, 185.
3. Bartok, M. and Felföldi, K. (1985) *Stereochemistry of Heterogeneous Metal Catalysis*, John Wiley & Sons, Ltd, Chichester.
4. Christmann, K. (1988) *Surf. Sci. Rep.*, **9**, 1, and chapter of this volume...........
5. Daw, M.S. and Baskes, M.I. (1984) *Phys. Rev. B*, **29**, 6443.
6. Dong, W., Ledentu, V., Sautet, P., Eichler, A., and Hafner, J. (1998) *Surf. Sci.*, **411**, 123.
7. Hammer, B. and Norskov, J.K. (1995) *Surf. Sci.*, **343**, 211.
8. Nordlander, P., Holloway, S., and Norskov, J.K. (1984) *Surf. Sci.*, **136**, 59.
9. Ferrin, P., Kandoi, S., Nilekar, A.U., and Mavrikakis, M. (2012) *Surf. Sci.*, **606**, 679.
10. Bond, G.C. and Wells, P.B. (1964) in *Advances in Catalysis*, vol. 15 (eds D.D. Erley, H. Pines, and P.B. Weisz), Academic Press, New York, p. 91.
11. Ertl, G., Knözinger, H., Weitkamp, J. (eds) (1999) *Preparation of Solid Catalysts*, 1st edn, Wiley-VCH, Weinheim.
12. Chorkendorff, I. (2007) *Concepts of Modern Catalysis and Kinetics*, 2nd revised and enligh edn, Wiley-VCH Verlag GmbH, Weinheim.
13. Smith, G.V. and Notheisz, F. (1999) *Heterogeneous Catalysis in Organic Chemistry*, Chapter 2, Academic Press, San Diego, CA.
14. Olah, G.A. and Molnár, A. (2003) *Hydrocarbon Chemistry*, John Wiley & Sons, Inc., Hoboken, NJ.
15. Arnold, H., Döbert, F., and Gaube, J. (1997) *Handbook of Heterogeneous Catalysis*, VCH Publishers, Weinheim, p. 2165.
16. van Santen, R.A., van Leeuwen, P., Moulijn, J.A., and Averill, B.A. (1999) *Catalysis: An Integrated Approach*, 2nd edn, vol. 123, p. 3.
17. Polanyi, M. and Horiuti, J. (1934) *Trans. Faraday Soc.*, **30**, 1164.
18. Taylor, H.S. (1925) *Proc. R. Soc. London*, **108**, 105.
19. Boudart, M. (1969) *Adv. Catal.*, **20**, 153.

20. Sachtler, W.M.H. (1973) *Le Vide*, **28**, 67.
21. Nørskov, J.K., Bligaard, T., Rossmeisl, J., and Christensen, C.H. (2009) *Nat. Chem.*, **1**, 37.
22. Becker, C. and Wandelt, K. (2007) *Top. Catal.*, **46**, 151.
23. Hammer, B. and Nørskov, J.K. (2000) *Adv. Catal.*, **45**, 71.
24. Chatt, J. and Duncanson, L.A. (1953) *J. Chem. Soc.*, 2939.
25. Chatt, J., Duncanson, L.A., and Venanzi, L.M. (1955) *J. Chem. Soc. (Resumed)*, 4456.
26. Dewar, M.J. (1951) *Bull. Soc. Chim. Fr.*, **18**, C79.
27. Blyholder, G. (1964) *J. Phys. Chem.*, **68**, 2772.
28. Hoffmann, R. (1988) *Rev. Mod. Phys.*, **60**, 601.
29. Nilsson, A., Pettersson, L.G.M., Hammer, B., Bligaard, T., Christensen, C.H., and Norskov, J.K. (2005) *Catal. Lett.*, **100**, 111.
30. Norskov, J.K., Bligaard, T., Hvolbaek, B., Abild-Pedersen, F., Chorkendorff, I., and Christensen, C.H. (2008) *Chem. Soc. Rev.*, **37**, 2163.
31. Brönstedt, J.N. (1928) *Chem. Rev.*, **5**, 231.
32. Evans, M.G. and Polanyi, M. (1938) *Trans. Faraday Soc.*, **34**, 11.
33. Goda, A.M., Barteau, M.A., and Chen, J.G. (2006) *J. Phys. Chem. B*, **110**, 11823.
34. Haubrich, J., Loffreda, D., Delbecq, F., Jugnet, Y., Sautet, P., Krupski, A., Becker, C., and Wandelt, K. (2006) *Chem. Phys. Lett.*, **433**, 188.
35. Haubrich, J., Loffreda, D., Delbecq, F., Sautet, P., Jugnet, Y., Krupski, A., Becker, C., and Wandelt, K. (2008) *J. Phys. Chem. C*, **112**, 3701.
36. Haubrich, J., Becker, C., and Wandelt, K. (2009) *Surf. Sci.*, **603**, 1476.
37. Haubrich, J., Loffreda, D., Delbecq, F., Sautet, P., Jugnet, Y., Bertolini, J.C., Krupski, A., Becker, C., and Wandelt, K. (2009) *J. Phys. Chem. C*, in press.
38. Haubrich, J., Loffreda, D., Delbecq, F., Sautet, P., Jugnet, Y., Becker, C., and Wandelt, K. (2010) *J. Phys. Chem. C*, **114**, 1073.
39. Paffett, M.T. and Windham, R.G. (1989) *Surf. Sci.*, **208**, 34.
40. Batzill, M., Beck, D.E., Koel, B.E. (2000) *Surf. Sci.*, **466**, L821.
41. Teraoka, Y. (1990) *Surf. Sci.*, **235**, 249.
42. Atrei, A., Bardi, U., Wu, J.X., Zanazzi, E., and Rovida, G. (1993) *Surf. Sci.*, **290**, 286.
43. Massalsi, T.B. (1990) *Binary Alloy Phase Diagramms*, 2nd edn, ASM International, Materials Park, OH.
44. Overbury, S.H., Mullins, D.R., Paffett, M.T., and Koel, B.E. (1991) *Surf. Sci.*, **254**, 45.
45. Overbury, S.H. and Ku, Y.S. (1992) *Phys. Rev. B*, **46**, 7868.
46. Galeotti, M., Atrei, A., Bardi, U., Rovida, G., and Torrini, M. (1994) *Surf. Sci.*, **313**, 349.
47. Haubrich, J. (2007) *University of Bonn*.
48. Sachtler, W.M. (1973) *Vide-Sci. Tech. Appl.*, **28**, 67.
49. Paffett, M.T., Gebhard, S.C., Windham, R.G., and Koel, B.E. (1989) *Surf. Sci.*, **223**, 449.
50. Delbecq, F. and Sautet, P. (2003) *J. Catal.*, **220**, 115.
51. Paffett, M.T., Gebhard, S.C., Windham, R.G., and Koel, B.E. (1990) *J. Phys. Chem.*, **94**, 6831.
52. Rodriguez, J.A., Chaturvedi, S., Jirsak, T., and Hrbek, J. (1998) *J. Chem. Phys.*, **109**, 4052.
53. Sabatier, P., Senderens, J. B. (1897) *CR Hebd. Seances Acad. Sci.*, ("Action of nickel on ethylene. Ethane synthesis"), **124**, 1358.
54. Bond, G.C. (1962) *Catalysis by Metals*, Academic Press, London.
55. Ofner, H. and Zaera, F. (1997) *J. Phys. Chem. B*, **101**, 396.
56. McDougall, G. and Yates, H. (1992) *Spec. Publ. R. Soc. Chem.*, **114**, 109.
57. Zaera, F. (1996) *Langmuir*, **12**, 88.
58. Zaera, F. (1990) *J. Phys. Chem.*, **94**, 5090.
59. Kubota, J., Ichihara, S., Kondo, J.N., Domen, K., and Hirose, C. (1996) *Surf. Sci.*, **357**, 634.
60. Zaera, F. and Somorjai, G.A. (1984) *J. Am. Chem. Soc.*, **106**, 2288.
61. Rodriguez, J.A. and Goodman, D.W. (1991) *Surf. Sci. Rep.*, **14**, 1.

62. Cremer, P.S., Su, X., Chen, Y.R., and Somorjai, G.A. (1996) *J. Phys. Chem. A*, **100**, 16302.
63. Soma, Y. (1979) *J. Catal.*, **59**, 239.
64. Twigg, G.H. (1950) *Discuss. Faraday Soc.*, **8**, 152.
65. Salmeron, M. and Somorjai, G.A. (1982) *J. Phys. Chem. A*, **86**, 341.
66. Sheppard, N. (1988) *Annu. Rev. Phys. Chem.*, **39**, 589.
67. Hugenschmidt, M.B., Dolle, P., Jupille, J., and Cassuto, A. (1989) *J. Vac. Sci. Technol., A: Vac. Surf. Films*, **7**, 3312.
68. Avery, N.R. (1985) *J. Vac. Sci. Technol., A: Vac. Surf. Films*, **3**, 1459.
69. Bent, B.E. (1996) *Chem. Rev.*, **96**, 1361.
70. Zaera, F. (1995) *Chem. Rev.*, **95**, 2651.
71. Zaera, F. (2002) *J. Phys. Chem. B*, **106**, 4043.
72. Stuve, E.M. and Madix, R.J. (1985) *J. Phys. Chem.*, **89**, 105.
73. Stuve, E.M. and Madix, R.J. (1985) *J. Phys. Chem.*, **89**, 3183.
74. Baró, A.M., Ibach, H., and Bruchmann, H.D. (1981) *J. Chem. Phys.*, **74**, 4194.
75. Tsai, Y.-L. and Koel, B.E. (1997) *J. Phys. Chem. B*, **101**, 2895.
76. Essen, J.M., Haubrich, J., Becker, C., and Wandelt, K. (2007) *Surf. Sci.*, **601**, 3472.
77. Ibach, H., Hopster, H., and Sexton, B. (1977) *Appl. Surf. Sci.*, **1**, 1.
78. Ibach, H. and Lehwald, S. (1978) *J. Vac. Sci. Technol.*, **152**, 407.
79. Steininger, H., Lehwald, S., and Ibach, H. (1982) *Surf. Sci.*, **123**, 264.
80. Stohr, J., Sette, F., and Johnson, A.L. (1984) *Phys. Rev. Lett.*, **53**, 1684.
81. Demuth, J.E., Ibach, H., and Lehwald, S. (1978) *Phys. Rev. Lett.*, **40**, 1044.
82. Baro, A.M. and Ibach, H. (1980) *Surf. Sci.*, **92**, 237.
83. Kesmodel, L.L., Dubois, L.H., and Somorjai, G.A. (1978) *Chem. Phys. Lett.*, **56**, 267.
84. Redhead, P.A. (1962) *Vacuum*, **12**, 203.
85. Kesmodel, L.L., Dubois, L.H., and Somorjai, G.A. (1979) *J. Chem. Phys.*, **70**, 2180.
86. Carter, E.A. and Koel, B.E. (1990) *Surf. Sci.*, **226**, 339.
87. Berkowitz, J., Ellison, G.B., and Gutman, D. (1994) *J. Phys. Chem.*, **98**, 2744.
88. Koestner, R.J., Stohr, J., Gland, J.L., and Horsley, J.A. (1984) *Chem. Phys. Lett.*, **105**, 332.
89. Cassuto, A., Mane, M., and Jupille, J. (1991) *Surf. Sci.*, **249**, 8.
90. Cassuto, A., Kiss, J., and White, J.M. (1991) *Surf. Sci.*, **255**, 289.
91. Carter, E.A. and Goddard, W.A. (1989) *Surf. Sci.*, **209**, 243.
92. Okada, T., Kim, Y., Sainoo, Y., Komeda, T., Trenary, M., and Kawai, M. (2011) *J. Phys. Chem. Lett.*, **2**, 2263.
93. Shen, J.Y., Hill, J.M., Watwe, R.M., Spiewak, B.E., and Dumesic, J.A. (1999) *J. Phys. Chem. B*, **103**, 3923.
94. Miura, T., Kobayashi, H., and Domen, K. (2000) *J. Phys. Chem. B*, **104**, 6809.
95. Watson, G.W., Wells, R.P.K., Willock, D.J., and Hutchings, G.J. (2000) *J. Phys. Chem. B*, **104**, 6439.
96. Watson, G.W., Wells, R.P.K., Willock, D.J., and Hutchings, G.J. (2000) *Surf. Sci.*, **459**, 93.
97. Fearon, J. and Watson, G.W. (2004) 3rd Int. Conf. on Comp. Model. and Sim. of Mat., Part A, p. 589.
98. Anderson, A.B. and Choe, S.J. (1989) *J. Phys. Chem.*, **93**, 6145.
99. Silvestre, J. and Hoffmann, R. (1985) *Langmuir*, **1**, 621.
100. Baetzold, R.C. (1987) *Langmuir*, **3**, 189.
101. Sautet, P. and Paul, J.F. (1991) *Catal. Lett.*, **9**, 245.
102. Mittendorfer, F., Thomazeau, C., Raybaud, P., and Toulhoat, H. (2003) *J. Phys. Chem. B*, **107**, 12287.
103. Chen, Z.X., Aleksandrov, H.A., Basaran, D., and Rosch, N. (2010) *J. Phys. Chem. C*, **114**, 17683.
104. Neurock, M. and van Santen, R.A. (2000) *J. Phys. Chem. B*, **104**, 11127.
105. Pallassana, V., Neurock, M., Lusvardi, V.S., Lerou, J.J., Kragten, D.D., and van Santen, R.M. (2002) *J. Phys. Chem. B*, **106**, 1656.
106. Watwe, R.M., Cortright, R.D., Nørskov, J.K., and Dumesic, J.A. (2000) *J. Phys. Chem. B*, **104**, 2299.
107. Chen, Y. and Vlachos, D.G. (2010) *J. Phys. Chem. C*, **114**, 4973.
108. Zhao, Z.-J., Moskaleva, L.V., Aleksandrov, H.A., Basaran, D., and Rösch, N. (2010) *J. Phys. Chem. C*, **114**, 12190.

109. Basaran, D., Aleksandrov, H.A., Chen, Z.X., Zhao, Z.J., and Rosch, N. (2011) *J. Mol. Catal. A: Chem.*, **344**, 37.
110. Nørskov, J.K., Bligaard, T., Logadottir, A., Bahn, S., Hansen, L.B., Bollinger, M., Bengaard, H., Hammer, B., Sljivancanin, Z., Mavrikakis, M., Xu, Y., Dahl, S., and Jacobsen, C.J.H. (2002) *J. Catal.*, **209**, 275.
111. Kresse, G. and Hafner, J. (1993) *Phys. Rev. B*, **47**, 558.
112. Kresse, G. and Hafner, J. (1993) *Phys. Rev. B*, **48**, 13115.
113. Kresse, G. and Joubert, D. (1998) *Phys. Rev. B*, **59**, 1758.
114. Perdew, J.P. and Wang, Y. (1992) *Phys. Rev. B*, **45**, 13244.
115. Marsman, M., Paier, J., Stroppa, A., and Kresse, G. (2008) *J. Phys. Condens. Matter*, **20**, 064201.
116. Nykanen, L. and Honkala, K. (2011) *J. Phys. Chem. C*, **115**, 9578.
117. Grimme, S. (2006) *J. Comput. Chem.*, **27**, 1787.
118. Neugebauer, J. and Scheffler, M. (1992) *Phys. Rev. B*, **46**, 16067.
119. Hirschl, R., Delbecq, F., Sautet, P., and Hafner, J. (2003) *J. Catal.*, **217**, 354–366.
120. Ibach, H. and Mills, D.L. (1982) *Electronic Energy Loss Spectroscopy and Surface Vibrations*, Academic Press, New York.
121. Loffreda, D., Jugnet, Y., Delbecq, F., Bertolini, J.C., and Sautet, P. (2004) *J. Phys. Chem. B*, **108**, 9085.
122. Morse, P.M. (1929) *Phys. Rev.*, **34**, 57.
123. Castro, G.R., Schneider, U., Busse, H., Janssens, T., and Wandelt, K. (1992) *Surf. Sci.*, **269**, 321.
124. Schneider, U., Castro, G.R., Busse, H., Janssens, T., Wesemann, J., and Wandelt, K. (1992) *Surf. Sci.*, **269**, 316.
125. Becker, C., Haubrich, J., Wandelt, K., Delbecq, F., Loffreda, D., and Sautet, P. (2008) *J. Phys. Chem. C*, **112**, 14693.
126. Linke, R., Becker, C., Pelster, T., Tanemura, M., and Wandelt, K. (1997) *Surf. Sci.*, **377**, 655.
127. Vigné, F., Haubrich, J., Loffreda, D., Sautet, P., and Delbecq, F. (2010) *J. Catal.*, **275**, 129.
128. Becker, C., Delbecq, F., Breitbach, J., Hamm, G., Franke, D., Jager, F., and Wandelt, K. (2004) *J. Phys. Chem. B*, **108**, 18960.
129. Delbecq, F., Vigne-Maeder, F., Becker, C., Breitbach, J., and Wandelt, K. (2008) *J. Phys. Chem. C*, **112**, 555.
130. Mitoraj, M.P., Michalak, A., and Ziegler, T. (2009) *J. Chem. Theory Comput.*, **5**, 962.
131. Deubel, D.V. and Ziegler, T. (2002) *Organometallics*, **21**, 1603.
132. Kesmodel, L.L., Stair, P.C., and Somorjai, G.A. (1977) *Surf. Sci.*, **64**, 342.
133. Van Hove, M.A. and Somorjai, G.A. (1998) *J. Mol. Catal. A: Chem.*, **131**, 243.
134. Starke, U., Vanhove, M.A., and Somorjai, G.A. (1994) *Prog. Surf. Sci.*, **46**, 305.
135. Vanhove, M.A. and Somorjai, G.A. (1994) *Surf. Sci.*, **299**, 487.
136. Starke, U., Barbieri, A., Materer, N., Vanhove, M.A., and Somorjai, G.A. (1993) *Surf. Sci.*, **286**, 1.
137. Somorjai, G.A. (1991) *Langmuir*, **7**, 3176.
138. Demuth, J.E. and Ibach, H. (1979) *Surf. Sci.*, **85**, 365.
139. Demuth, J.E. (1979) *Surf. Sci.*, **84**, 315.
140. Morin, C., Simon, D., and Sautet, P. (2003) *J. Phys. Chem. B*, **107**, 2995.
141. Brizuela, G. and Castellani, N. (1999) *J. Mol. Catal. A: Chem.*, **139**, 209.
142. Brizuela, G. and Castellani, N.J. (1998) *Surf. Sci.*, **401**, 297.
143. Brizuela, G. and Hoffmann, R. (1998) *J. Phys. Chem. A*, **102**, 9618.
144. Koestner, R.J., Frost, J.C., Stair, P.C., Van Hove, M.A., and Somorjai, G.A. (1982) *Surf. Sci.*, **116**, 85.
145. Ogle, K.M., Creighton, J.R., Akhter, S., and White, J.M. (1986) *Surf. Sci.*, **169**, 246.
146. Valcarcel, A., Ricart, J.M., Clotet, A., Markovits, A., Minot, C., and Illas, F. (2002) *Surf. Sci.*, **519**, 250.
147. Avery, N.R. and Sheppard, N. (1986) *Proc. R. Soc. London, Ser. A*, **405**, 1.
148. Avery, N.R. and Sheppard, N. (1986) *Proc. R. Soc. London, Ser. A*, **405**, 27.
149. Zaera, F. and Chrysostomou, D. (2000) *Surf. Sci.*, **457**, 71.
150. Zaera, F. and Chrysostomou, D. (2000) *Surf. Sci.*, **457**, 89.

151. Valcarcel, A., Clotet, A., Ricart, J.M., Delbecq, F., and Sautet, P. (2004) *Surf. Sci.*, **549**, 121.
152. Delbecq, F. and Zaera, F. (2008) *J. Am. Chem. Soc.*, **130**, 14924.
153. Lee, I. and Zaera, F. (2005) *J. Phys. Chem. B*, **109**, 2745.
154. Lee, I. and Zaera, F. (2005) *J. Am. Chem. Soc.*, **127**, 12174.
155. Lee, I. and Zaera, F. (2007) *J. Phys. Chem. C*, **111**, 10062.
156. Lee, I., Ma, Z., Kaneko, S., and Zaera, F. (2008) *J. Am. Chem. Soc.*, **130**, 14597.
157. Tsai, Y.-L., Xu, C., and Koel, B.E. (1997) *Surf. Sci.*, **385**, 37.
158. Cassuto, A. and Tourillon, G. (1994) *Surf. Sci.*, **307**, 65.
159. Zaera, F. (1989) *J. Am. Chem. Soc.*, **111**, 8744.
160. Zaera, F. (1992) *Acc. Chem. Res.*, **25**, 260.
161. Tjandra, S. and Zaera, F. (1997) *J. Phys. Chem. B*, **101**, 1006.
162. Tjandra, S. and Zaera, F. (1999) *J. Phys. Chem. A*, **103**, 2312.
163. Chrysostomou, D., Chou, A., and Zaera, F. (2001) *J. Phys. Chem. B*, **105**, 5968.
164. Scoggins, T.B. and White, J.M. (1999) *J. Phys. Chem. B*, **103**, 9663.
165. Carter, R.N., Anton, A.B., and Apai, G. (1992) *J. Am. Chem. Soc.*, **114**, 4410.
166. Chrysostomou, D. and Zaera, F. (2001) *J. Phys. Chem. B*, **105**, 1003.
167. Jenks, C.J., Bent, B.E., and Zaera, F. (1999) *J. Phys. Chem. B*, **104**, 3017.
168. Tourillon, G., Cassuto, A., Jugnet, Y., Massardier, J., and Bertolini, J.C. (1996) *J. Chem. Soc., Faraday Trans.*, **92**, 4835.
169. Cosyns, J. (1984) *Catalyse par les Métaux*, Editions du CNRS, Paris, p. 371.
170. Saint-Lager, M.C., Jugnet, Y., Dolle, P., Piccolo, L., Baudoing-Savois, R., Bertolini, J.C., Bailly, A., Robach, O., Walker, C., and Ferrer, S. (2005) *Surf. Sci.*, **587**, 229.
171. Jugnet, Y., Sedrati, R., and Bertolini, J.C. (2005) *J. Catal.*, **229**, 252.
172. Bertolini, J.C., Cassuto, A., Jugnet, Y., Massardier, J., Tardy, B., and Tourillon, G. (1996) *Surf. Sci.*, **349**, 88.
173. Zhao, H. and Koel, B.E. (2005) *J. Catal.*, **234**, 24.
174. Boitiaux, J.P., Cosyns, J., and Robert, E. (1987) *Appl. Catal.*, **35**, 193.
175. Goetz, J., Touroude, R., and Murzin, D.Y. (1997) *Chem. Eng. Technol.*, **20**, 138.
176. Abon, M., Bertolini, J.C., and Tardy, B. (1988) *J. Chim. Phys. Chim. Biol.*, **85**, 711.
177. Furlong, B.K., Hightower, J.W., Chan, T.Y.L., Sarkany, A., and Guczi, L. (1994) *Appl. Catal., A Gen.*, **117**, 41.
178. Palczewska, W., Jablonski, A., Kaszkur, Z., Zuba, G., and Wernisch, J. (1984) *J. Mol. Catal.*, **25**, 307.
179. Massardier, J., Bertolini, J.C., Ruiz, P., and Delichere, P. (1988) *J. Catal.*, **112**, 21.
180. Ouchaib, T., Massardier, J., and Renouprez, A. (1989) *J. Catal.*, **119**, 517.
181. Bond, G.C., Webb, G., Wells, P.B., and Winterbottom, J.M. (1965) *J. Chem. Soc.*, 3218.
182. Pradier, C.M., Margot, E., Berthier, Y., and Oudar, J. (1988) *Appl. Catal.*, **43**, 177.
183. Wells, P.B. and Bates, A.J. (1968) *J. Chem. Soc. A: Inorg. Phys. Theor.*, 3064.
184. Phillips, J., Wells, P.B., and Wilson, G.R. (1969) *J. Chem. Soc. A: Inorg. Phys. Theor.*, 1351.
185. Komatsu, T., Hyodo, S., and Yashima, T. (1997) *J. Phys. Chem. B*, **101**, 5565.
186. Moyes, R.B., Wells, P.B., Grant, J., and Salman, N.Y. (2002) *Appl. Catal., A Gen.*, **229**, 251.
187. Valcarcel, A., Clotet, A., Ricart, J.M., Delbecq, F., and Sautet, P. (2005) *J. Phys. Chem. B*, **109**, 14175.
188. Santori, G.F., Casella, M.L., Siri, G.J., Aduriz, H.R., and Ferretti, O.A. (2000) *Appl. Catal., A Gen.*, **197**, 141.
189. Zhao, H. and Koel, B.E. (2004) *Surf. Sci.*, **573**, 413–425.
190. Lonergan, W.W., Xing, X.J., Zheng, R.Y., Qi, S.T., Huang, B., and Chen, J.G.G. (2011) *Catal. Today*, **160**, 61.
191. Delbecq, F. and Sautet, P. (1994) *Catal. Lett.*, **28**, 89.
192. Breinlich, C., Haubrich, J., Becker, C., Valcarcel, A., Delbecq, F., and Wandelt, K. (2007) *J. Catal.*, **251**, 123.

193. Bauer, K. and Garbe, D. (1985) *Common Fragrance and Flavor Materials*, Wiley-VCH Verlag GmbH, New York.
194. Jenck, J. and Germain, J.E. (1980) *J. Catal.*, **65**, 133.
195. Jerdev, D.I., Olivas, A., and Koel, B.E. (2002) *J. Catal.*, **205**, 278–288.
196. Ponec, V. (1997) *Appl. Catal., A Gen.*, **149**, 27.
197. Beccat, J., Bertolini, J.C., Gauthier, Y., Massardier, J., and Ruiz, P. (1990) *J. Catal.*, **126**, 451.
198. Birchem, T., Pradier, C.M., Berthier, Y., and Cordier, G. (1994) *J. Catal.*, **146**, 503.
199. Birchem, T., Pradier, C.M., Berthier, Y., and Cordier, G. (1996) *J. Catal.*, **161**, 68–77.
200. Marinelli, T.B.L.W., Naubuurs, S., and Ponec, V. (1995) *J. Catal.*, **151**, 431.
201. Marinelli, T.B.L.W. and Ponec, V. (1995) *J. Catal.*, **156**, 51–59.
202. Borgna, A., Anderson, B.G., Saib, A.M., Bluhm, H., Hävecker, M., Knop-Gericke, A., Kuiper, A.E.T., Tamminga, Y., and Niemantsverdriet, J.W. (2004) *J. Phys. Chem. B*, **108**, 14340–14347.
203. Dandekar, A., Baker, R.T.K., and Vannice, M.A. (1999) *J. Catal.*, **184**, 421.
204. Dandekar, A. and Vannice, M.A. (1999) *J. Catal.*, **183**, 344.
205. Dandekar, A., Baker, R.T.K., and Vannice, M.A. (1999) *J. Catal.*, **183**, 131.
206. Nishiyama, S., Hara, T., Tsuruya, S., and Masai, M. (1999) *J. Phys. Chem. B*, **103**, 4431.
207. Bailie, J.E., Rochester, C.H., and Hutchings, G.J. (1997) *J. Chem. Soc.-Faraday Trans.*, **93**, 4389.
208. Hutchings, G.J., King, F., Okoye, I.P., Padley, M.B., and Rochester, C.H. (1994) *J. Catal.*, **148**, 453.
209. Hutchings, G.J., King, F., Okoye, I.P., Padley, M.B., and Rochester, C.H. (1994) *J. Catal.*, **148**, 464.
210. Iwasa, N., Takizawa, M., and Arai, M. (2005) *Appl. Catal., A Gen.*, **283**, 255.
211. Reyes, P., Aguirre, M.D., Melian-Cabrera, I., Granados, M.L., and Fierro, J.L.G. (2002) *Bol. Soc. Chil. Quim.*, **47**, 547.
212. Coloma, F., Coronado, J.M., Rochester, C.H., and Anderson, J.A. (1998) *Catal. Lett.*, **51**, 155.
213. Coloma, F., Narciso-Romero, J., Sepulveda-Escribano, A., and Rodriguez-Reinoso, F. (1998) *Carbon*, **36**, 1011.
214. Goupil, D., Fouilloux, P., and Maurel, R. (1987) *React. Kinet. Catal. Lett.*, **35**, 185.
215. Makouangou, R.M., Murzin, D.Y., Dauscher, A.E., and Touroude, R.A. (1994) *Ind. Eng. Chem. Res.*, **33**, 1881.
216. Claus, P. (1998) *Top. Catal.*, **5**, 51.
217. Liberkova-Sebkova, K., Cerveny, L., and Touroude, R. (2003) *Res. Chem. Intermed.*, **29**, 609.
218. Abid, M., Ammari, F., Liberkova, K., and Touroude, R. (2003) *Science and Technology in Catalysis 2002*, vol. 145, p. 267.
219. Bournel, F., Laffon, C., Parent, P., and Tourillon, G. (1996) *Surf. Sci.*, **350**, 60.
220. Bournel, F., Laffon, C., Parent, P., and Tourillon, G. (1996) *Surf. Sci.*, **359**, 10.
221. de Jesús, J.C. and Zaera, F. (1999) *Surf. Sci.*, **430**, 99.
222. de Jesús, J.C. and Zaera, F. (1999) *J. Mol. Catal. A Chem.*, **138**, 237.
223. Kliewer, C.J., Bieri, M., and Somorjai, G.A. (2009) *J. Am. Chem. Soc.*, **131**, 9958.
224. Janin, E., Ringler, S., Weissenrieder, J., Åkermark, T., Karlsson, U.O., Göthelid, M., Nordlund, D., and Ogasawara, H. (2001) *Surf. Sci.*, **482**, 83–89.
225. Janin, E., Schenck, H., Ringler, S., Weissenrieder, J., Åkermark, T., and Göthelid, M. (2003) *J. Catal.*, **215**, 245–253.
226. Ashour, S.S., Bailie, J.E., Rochester, C.H., Thomson, J., and Hutchings, G.J. (1997) *J. Mol. Catal. A: Chem.*, **123**, 65.
227. Delbecq, F. and Sautet, P. (2002) *J. Catal.*, **211**, 398.
228. Delbecq, F. and Sautet, P. (1995) *J. Catal.*, **152**, 217.
229. Haubrich, J., Loffreda, D., Delbecq, F., Sautet, P., Jugnet, Y., Krupski, A., Becker, C., and Wandelt, K. (2011) *Phys. Chem. Chem. Phys.*, **13**, 6000.

230. Loffreda, D., Delbecq, F., and Sautet, P. (2005) *Chem. Phys. Lett.*, **405**, 434.
231. Loffreda, D., Delbecq, F., Vigne, F., and Sautet, P. (2005) *Angew. Chem.*, **44**, 5279.
232. Loffreda, D., Delbecq, F., Vigné, F., and Sautet, P. (2006) *J. Am. Chem. Soc.*, **128**, 1316.
233. Dahl, S., Logadottir, A., Jacobsen, C.J.H., and Norskov, J.K. (2001) *Appl. Catal., A Gen.*, **222**, 19.
234. Jacobsen, C.J.H., Dahl, S., Clausen, B.S., Bahn, S., Logadottir, A., and Norskov, J.K. (2001) *J. Am. Chem. Soc.*, **123**, 8404.
235. Andersson, M.P., Bligaard, T., Kustov, A., Larsen, K.E., Greeley, J., Johannessen, T., Christensen, C.H., and Norskov, J.K. (2006) *J. Catal.*, **239**, 501.
236. Greeley, J. and Mavrikakis, M. (2004) *Nat. Mater.*, **3**, 810.